W9-BMH-092

Handbook of
PATHOPHYSIOLOGY

Handbook of
PATHOPHYSIOLOGY

Springhouse Corporation
Springhouse, Pennsylvania

STAFF

Senior publisher
Donna O. Carpenter

Creative director
Jake Smith

Design director
John Hubbard

Executive editor
H. Nancy Holmes

Clinical project managers
Clare M. Brabson, RN, BSN; Joan M. Robinson, RN, MSN, CCRN

Clinical editors
Joanne Bartelmo, RN, MSN; Jill Curry, RN, BSN, CCRN; Maryann Foley, RN, BSN; Margaret Klein, RN, BSN; Lori Musolf Neri, RN, MSN, CCRN; John E. Taylor, RN, BS

Editors
Jennifer P. Kowalak (editorial project manager), Mario Cavallini, Naina D. Chohan, Audrey Selena Hughes, Peter H. Johnson, Joanne C. Poeggel

Copy editors
Richard H. Adin, Catherine Cramer, Joy Epstein, Dolores P. Matthews, Celia McCoy, Maarten Reilingh, Barbara F. Ritter

Designers
Arlene Putterman (associate design director), BJ Crim (designer), Susan Sheridan (design project manager), Joseph John Clark, Lynn Foulke, Donna S. Morris, Jeffrey Sklarow

Illustrators
Jean Gardner, Judy Newhouse

Electronic production services
Diane Paluba (manager), Joyce Rossi Biletz

Manufacturing
Deborah Meiris (director), Patricia K. Dorshaw (manager), Otto Mezei (book production manager)

Editorial and design assistants
Carol Caputo, Arlene Claffee, Tom Hasenmayer, Elfriede Young

Indexer
Barbara E. Hodgson

Cover illustration
Frontal view computed tomography scan of thorax showing cancer of the right lung/©GJLP/CNRI/Phototake

The clinical procedures described and recommended in this publication are based on research and consultation with nursing, medical, and legal authorities. To the best of our knowledge, these procedures reflect currently accepted practice; nevertheless, they can't be considered absolute and universal recommendations. For individual application, all recommendations must be considered in light of the patient's clinical condition and, before administration of new or infrequently used drugs, in light of the latest package-insert information. The authors and the publisher disclaim responsibility for any adverse effects resulting directly or indirectly from the suggested procedures, from any undetected errors, or from the reader's misunderstanding of the text.

Printed in the United States of America.

HP-D N

03 02 01 00   10 9 8 7 6 5 4 3 2 1

Library of Congress Cataloging-in-Publication Data

Handbook of Pathophysiology.
    p. ; cm.
    Includes bibliographical references and index.
    ISBN 1-58255-046-8 (flexible cover)
    1. Physiology, Pathological — Handbooks, manuals, etc. I. Springhouse Corporation.
    [DNLM: 1. Pathology — Handbooks. 2. Physiology — Handbooks. QZ 39 H2359 2000]
RB113 .H24 2000
616.07 — dc21
00-058808

Contents

Contributors

Gary J. Arnold, MD, FACS
Assistant Professor of Nursing
University of Louisiana at Lafayette
Lafayette, LA

Deborah Becker, MSN, CRNP, CS, CCRN
Lecturer, Adult Acute Care Practitioner
    Program
University of Pennsylvania
Philadelphia, PA

Marcy S. Caplin, RN, MSN
Independent Nurse Consultant
Hudson, OH

Susan B. Dickey, PhD, RN,C
Associate Professor
Temple University
Philadelphia, PA

Kay Gentieu, RN, MSN, CRNP
Coordinator Family Nurse Practitioner
    Program
Thomas Jefferson University
Philadelphia, PA

H. Dean Krimmel, RN, MSN
Independent Nurse Consultant
Collingdale, PA

Nancy LaPlante, RN, BSN
RN-Staff Nurse Emergency Department
The Chester County Hospital
West Chester, PA

Kay Luft, RN, MN, CCRN, TNCC
Assistant Professor
Saint Luke's College
Kansas City, MO

Elaine Mohn-Brown, EdD, RN
Professor, Nursing
Chemeketa Community College
Salem, OR

Roger M. Morrell, MD, PhD, FACP, FAIC
Neurologist
Lathrup Village, MI

David Toub, MD
Consultant
Lansdale, PA

Tracy S. Weintraub, RN, MSN, CNS
Instructor of Clinical Nursing
University of Southern California
Los Angeles, CA

Patricia A. Wessels, RN, MSN
Assistant Dean
Viterbo College
LaCrosse, WI

Foreword

In today's fast-paced, ever-changing health care environment, health care professionals are required to provide competent, compassionate care that integrates every aspect of prior learning. They must assess patients, relate patients' clinical symptoms to the pathophysiology associated with the disease process, interpret laboratory data, and prepare patients for the expected treatment. These actions must be completed quickly and accurately, making both the science and the art of health care more complex. Thus, clinicians need a reliable, accessible reference that incorporates all of this information, enabling them to feel confident about the quality of their care.

Modern clinical practice includes the domains of patient education and advocacy as well as the more traditional domains of providing and coordinating actual patient care. Additionally, clinical practice has moved from the traditional hospital or long-term care facility to the patient's home or to outpatient centers. Typically, there must be collaboration between physicians and other health care professionals as part of the health care team to discuss the patient's physiologic status and treatment issues. They initiate meetings with patients, families, and other members of the health care team to disseminate information about these same issues. To be confident in these aspects clinicians need a reference that enables them to obtain relevant pathophysiologic information applicable to the patient's disease status.

The *Handbook of Pathophysiology* is designed for the health care professional who enjoys the challenge of science-based practice. It is an easy-to-use reference that provides a synopsis of updated information on the major pathophysiologic disease processes. This handbook presents more than 450 diseases. The basic concepts of altered homeostasis, disease development, and disease progression are presented in an easy-to-read format. Additionally, "Pathophysiology in color" is a special section (located within chapter 9) that contains 16 full-color pages, illustrating asthma, cancer, osteoporosis, and ulcers.

The first chapter of the handbook provides an overview of the cell in health and illness. Various cell types and their normal function are discussed, including muscle and nerve cells. This provides the basis for the review of normal physiology found in each chapter. Information about pathophysiologic changes at the cellular level provides the foundation for describing alterations in the major organ systems that occur during illness.

Subsequent chapters are presented in a systems format, including a discussion of the major disorders associated with that particular body system. The pathophysiologic manifestations are described in relation to the patient's clinical presentation. Thus, the clinician can monitor physical changes and relate them directly to the disease process.

The appropriate diagnostic tests for each disease are included in each chapter. The review of expected results from these tests provides information about disease progression, remission, and resolution. This enables all members of the health care team to become active participants in the clinical decision-making process as plans are made for future care.

The usually recommended treatments are presented as well. Inclusion of this information enables the clinician to prepare for the next phase of patient care. The rationales for the treatment support the development of individualized patient education about the particular treatment.

Each chapter contains crucial age-related, cultural, or socioeconomic information related to common pathophysiologic conditions for that organ system. For example, Chapter 7, the "Respiratory System," includes a discussion of age-related triggers for asthma. There's also information about asthma triggers that patients may encounter in the workplace and the inner city. This is the type of comprehensive information this handbook includes that's applicable to most patient-care circumstances.

The appendix of the handbook includes flow charts that summarize core information for some of the less common diseases. Thus, important facts are available in a synopsis format. The clinician can readily access and refer to these flow charts when accurate information is needed very quickly.

The *Handbook of Pathophysiology* is a much needed reference for the entire health care team. For students, this handbook will complement other textual material and will be easy to use in the clinical site in conjunction with drug and diagnostic study handbooks. New clinicians will refer to this handbook to enable them to integrate patient-assessment information with the proposed plan of care. Experienced professionals will find that this reference contains information that will provide foundation knowledge to be utilized in coordinating patient care and developing patient-education information.

Joan P. Frizzell, RN, PhD
Assistant Professor
School of Nursing
LaSalle University
Philadelphia

An understanding of pathophysiology requires a review of normal physiology — how the body functions day to day, minute to minute, at the levels of cells, tissues, organs, and organisms.

HOMEOSTASIS

Every cell in the body is involved in maintaining a dynamic, steady state of internal balance, called *homeostasis*. Any change or damage at the cellular level can affect the entire body. When homeostasis is disrupted by an external stressor — such as injury, lack of nutrients, or invasion by parasites or other organisms — illness may occur. Many external stressors affect the body's internal equilibrium throughout the course of a person's lifetime. Pathophysiology can be considered as what happens when normal defenses fail.

MAINTAINING BALANCE

Three structures in the brain are responsible for maintaining homeostasis of the entire body:
▶ medulla oblongata, which is the part of the brain stem associated with vital functions such as respiration and circulation
▶ pituitary gland, which regulates the function of other glands and, thereby, a person's growth, maturation, and reproduction
▶ reticular formation, a network of nerve cells (nuclei) and fibers in the brain stem and spinal cord that help control vital reflexes such as cardiovascular function and respiration.

Homeostasis is maintained by self-regulating feedback mechanisms. These mechanisms have three components:
▶ a sensor that detects disruptions in homeostasis
▶ a control center that regulates the body's response to those disruptions
▶ an effector that acts to restore homeostasis.

An endocrine or hormone-secreting gland usually serves as the sensor. It signals the control center in the central nervous system to initiate the effector mechanism.

Feedback mechanisms exist in two varieties: positive and negative.
▶ A positive feedback mechanism moves the system away from homeostasis by enhancing a change in the system. For example, the heart pumps at increased rate and force when someone is in shock. If the shock progresses, the heart action may require more oxygen than is available. The result is heart failure.
▶ A negative feedback mechanism works to restore homeostasis by correcting a deficit in the system.

An effective negative feedback mechanism must sense a change in the body — such as a high blood glucose level — and attempt to return body functions to normal. In the case of a high blood glucose level, the effector mechanism triggers increased insulin production by the pancreas, returning blood glucose levels to normal and restoring homeostasis.

1

DISEASE AND ILLNESS

Although *disease* and *illness* are often used interchangeably, they aren't synonyms. Disease occurs when homeostasis isn't maintained. Illness occurs when a person is no longer in a state of perceived "normal" health. For example, a person may have coronary artery disease, diabetes, or asthma but not be ill all the time because his body has adapted to the disease. In such a situation, a person can perform necessary activities of daily living. Illness usually refers to subjective symptoms, that may or may not indicate the presence of disease.

The course and outcome of a disease are influenced by genetic factors (such as a tendency toward obesity), unhealthy behaviors (such as smoking), attitudes (such as being a "Type A" personality), and even the person's perception of the disease (such as acceptance or denial). Diseases are dynamic and may be manifested in a variety of ways, depending on the patient or his environment.

Cause

The cause of disease may be intrinsic or extrinsic. Inheritance, age, gender, infectious agents, or behaviors (such as inactivity, smoking, or abusing illegal drugs) can all cause disease. Diseases that have no known cause are called *idiopathic*.

Development

A disease's development is called its *pathogenesis*. Unless identified and successfully treated, most diseases progress according to a typical pattern of symptoms. Some diseases are self-limiting or resolve quickly with limited or no intervention; others are chronic and never resolve. Patients with chronic diseases may undergo periodic remissions and exacerbations.

A disease is usually detected when it causes a change in metabolism or cell division that causes signs and symptoms.

Manifestations of disease may include hypofunction (such as constipation), hyperfunction (such as increased mucus production), or increased mechanical function (such as a seizure).

How the cells respond to disease depends on the causative agent and the affected cells, tissues, and organs. The resolution of disease depends on many factors functioning over a period of time, such as extent of disease and the presence of other diseases.

Stages

Typically, diseases progress through these stages:

▶ Exposure or injury — Target tissue is exposed to a causative agent or is injured.
▶ Latency or incubation period — No signs or symptoms are evident.
▶ Prodromal period — Signs and symptoms are usually mild and nonspecific.
▶ Acute phase — The disease reaches its full intensity, possibly resulting in complications. This phase is called the subclinical acute phase if the patient can still function as though the disease wasn't present.
▶ Remission — This second latent phase occurs in some diseases and is often followed by another acute phase.
▶ Convalescence — The patient progresses toward recovery after the termination of a disease.
▶ Recovery — The patient regains health or normal functioning. No signs or symptoms of disease remain.

Stress and disease

When a stressor such as a life change occurs, a person can respond in one of two ways: by adapting successfully or by failing to adapt. A maladaptive response to stress may result in disease.

Hans Selye, a pioneer in the study of stress and disease, describes the following stages of adaptation to a stressful event: alarm, resistance, and recovery or exhaustion (See *Physical response to stress.*)

PHYSICAL RESPONSE TO STRESS

According to Hans Selye's General Adaptation Model, the body reacts to stress in the stages depicted below.

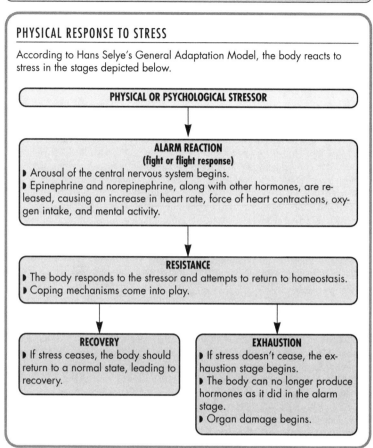

In the alarm stage, the body senses stress and arouses the central nervous system (CNS). The body releases chemicals to mobilize the fight-or-flight response. In this dual effort, the sympatho-adrenal medullary response causes the release of epinephrine and the hypothalamic pituitary adrenal axis causes the release of glucocorticoids. Both of these systems work in concert to enable the body to respond to stressors. This release is the adrenaline rush associated with panic or aggression. In the resistance stage, the body either adapts and achieves homeostasis or it fails to adapt and enters the exhaustion stage, resulting in disease.

The stress response is controlled by actions that take place in the cells of the nervous and endocrine systems. These actions try to redirect energy to the organ that is most affected by the stress, such as the heart, lungs, or brain.

Stressors may be physiologic or psychological. Physiologic stressors, such as exposure to a toxin, may elicit a harmful response leading to an identifiable illness or set of symptoms. Psychological stressors, such as the death of a loved one, may also cause a maladaptive response. Stressful events can exacerbate some chronic diseases, such as diabetes or multiple sclerosis. Effective coping strategies can prevent or reduce the harmful effects of stress.

CELL PHYSIOLOGY

The cell is the smallest living component of a living organism. Many organisms, such as bacteria, consist of one independent cell. Human beings and other large organisms consist of millions of cells. In large organisms, highly specialized cells that perform an identical function form tissue such as epithelial tissue, connective tissue, nerve tissue, and muscle tissue. Tissues, in turn, form organs (skin, skeleton, brain, and heart), which are integrated into body systems such as the CNS, cardiovascular system, and musculoskeletal system.

Cell components

Like organisms, cells are complex organizations of specialized components, each component having its own specific function. The largest components of a normal cell are the cytoplasm, the nucleus, and the cell membrane, which surrounds the internal components and holds the cell together.

Cytoplasm

The gel-like cytoplasm consists primarily of cytosol, a viscous, semitransparent fluid that is 70% to 90% water plus various proteins, salts, and sugars. Suspended in the cytosol are many tiny structures called *organelles*.

Organelles are the cell's metabolic machinery. Each performs a specific function to maintain the life of the cell. Organelles include mitochondria, ribosomes, endoplasmic reticulum, Golgi apparatus, lysosomes, peroxisomes, cytoskeletal elements, and centrosomes.

▶ *Mitochondria* are threadlike structures that produce most of the body's adenosine triphosphate (ATP). ATP contains high-energy phosphate chemical bonds that fuel many cellular activities.

▶ *Ribosomes* are the sites of protein synthesis.

▶ The *endoplasmic reticulum* is an extensive network of two varieties of membrane-enclosed tubules. The rough endoplasmic reticulum is covered with ribosomes. The smooth endoplasmic reticulum contains enzymes that synthesize lipids.

▶ The *Golgi apparatus* synthesizes carbohydrate molecules that combine with protein produced by the rough endoplasmic reticulum and lipids produced by the smooth endoplasmic reticulum to form such products as lipoproteins, glycoproteins, and enzymes.

▶ *Lysosomes* are digestive bodies that break down nutrient material as well as foreign or damaged material in cells. A membrane surrounding each lysosome separates its digestive enzymes from the rest of the cytoplasm. The enzymes digest nutrient matter brought into the cell by means of endocytosis, in which a portion of the cell membrane surrounds and engulfs matter to form a membrane-bound intracellular vesicle. The membrane of the lysosome fuses with the membrane of the vesicle surrounding the endocytosed material. The lysosomal enzymes then digest the engulfed material. Lysosomes digest the foreign matter ingested by white blood cells by a similar process called *phagocytosis*.

▶ *Peroxisomes* contain oxidases, which are enzymes that chemically reduce oxygen to hydrogen peroxide and hydrogen peroxide to water.

▶ *Cytoskeletal elements* form a network of protein structures.

▶ *Centrosomes* contain centrioles, which are short cylinders adjacent to the nucleus that take part in cell division.

▶ *Microfilaments* and *microtubules* enable the movement of intracellular vesicles (allowing axous to transport neurotransmitters) and the formation of the mitotic spin-

A LOOK AT CELL COMPONENTS

The illustration below shows the components and structures of a cell. Each part has a function in maintaining the cell's life and homeostasis.

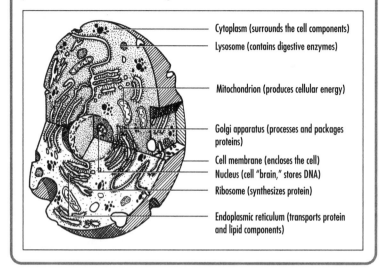

Cytoplasm (surrounds the cell components)

Lysosome (contains digestive enzymes)

Mitochondrion (produces cellular energy)

Golgi apparatus (processes and packages proteins)

Cell membrane (encloses the cell)

Nucleus (cell "brain," stores DNA)

Ribosome (synthesizes protein)

Endoplasmic reticulum (transports protein and lipid components)

dle, which connects the chromosomes during cell division.

Nucleus

The cell's control center is the nucleus, which plays a role in cell growth, metabolism, and reproduction. Within the nucleus, one or more nucleoli (dark-staining intranuclear structures) synthesize ribonucleic acid (RNA), a complex polynucleotide that controls protein synthesis. The nucleus also stores deoxyribonucleic acid (DNA), the famous double helix that carries genetic material and is responsible for cellular reproduction or division. (See *A look at cell components*.)

Cell membrane

The semipermeable cell membrane forms the cell's external boundary, separating it from other cells and from the external environment. Roughly 75Å (3/10 millionths of an inch) thick, the cell membrane con-

sists of a double layer of phospholipids with protein molecules embedded in it.

Cell division

Each cell must replicate itself for life to continue. Cells replicate by division in one of two ways: mitosis (division that results in two daughter cells with the same DNA and chromosome content as the mother cell) or meiosis (division that creates four gametocytes, each containing half the number of chromosomes of the original cell). Most cells undergo mitosis; meiosis occurs only in reproductive cells.

Mitosis

Mitosis, the type of cell division that leads to tissue growth, creates an equal division of material in the nucleus (karyokinesis) followed by division of the cell body (cytokinesis). This process yields two exact duplicates of the original cell. (See Chapter 4 for a detailed discussion of mitosis and meiosis.)

Cell functions

The basic functions of a cell are movement, conduction, absorption, secretion, excretion, respiration, and reproduction. In the human body, different cells are specialized to perform only one function; muscle cells, for example, are responsible for movement.

Movement

Some cells, such as muscle cells, working together produce movement of a specific body part or the entire organism. Muscle cells attached to bone move the extremities. When muscle cells that envelop hollow organs or cavities contract, they produce movement of contents, such as the peristaltic movement of the intestines or the ejection of blood from the heart.

Conduction

Conduction is the transmission of a stimulus, such as a nerve impulse, heat, or sound wave, from one body part to another.

Absorption

This process of absorption occurs as substances move through a cell membrane. For example, food is broken down into amino acids, fatty acids, and glucose in the digestive tract. Specialized cells in the intestine then absorb the nutrients and transport them to blood vessels, which carry them to other cells of the body. These target cells, in turn, absorb the substances, using them as energy sources or as building blocks to form or repair structural and functional cellular components.

Secretion

Some cells, such as those in the glands, release substances that are used in another part of the body. The beta cells of the islets of Langerhans of the pancreas, for example, secrete insulin, which is trans-

ported by the blood to its target cells, where the insulin facilitates the movement of glucose across cell membranes.

Excretion

Cells excrete the waste that is generated by normal metabolic processes. This waste includes such substances as carbon dioxide and certain acids and nitrogen-containing molecules.

Respiration

Cellular respiration occurs in the mitochondria, where ATP is produced. The cell absorbs oxygen; it then uses the oxygen and releases carbon dioxide during cellular metabolism. The energy stored in ATP is used in other reactions that require energy.

Reproduction

New cells are needed to replace older cells for tissue and body growth. Most cells divide and reproduce through mitosis. However, some cells, such as nerve and muscle cells, typically lose their ability to reproduce after birth.

Cell types

Each of the four types of tissue (epithelial, connective, nerve, and muscle tissue) consists of several specialized cell types, which perform specific functions.

Epithelial cell

Epithelial cells line most of the internal and external surfaces of the body, such as the epidermis of the skin, internal organs, blood vessels, body cavities, glands, and sensory organs. The functions of epithelial cells include support, protection, absorption, excretion, and secretion.

Connective tissue cell

Connective tissue cells are found in the skin, the bones and joints, the artery walls, the fascia around organs, nerves, and body fat. The types of connective tissue cells include fibroblasts (such as collagen, elastin, and reticular fibers), adipose (fat) cells, mast cells (release histamines and other substances during inflammation), and bone. The major functions of connective tissues are protection, metabolism, support, temperature maintenance, and elasticity.

Nerve cell

Two types of cells — neurons and neuroglial cells — comprise the nervous system. Neurons have a cell body, dendrites, and an axon. The dendrites carry nerve impulses to the cell body from the axons of other neurons. Axons carry impulses away from the cell body to other neurons or organs. A myelin sheath around the axon facilitates rapid conduction of impulses by keeping them within the nerve cell. Nerve cells:

▶ generate electrical impulses
▶ conduct electrical impulses
▶ influence other neurons, muscle cells, and cells of glands by transmitting those impulses.

Neuroglial cells, also called glial cells, consist of four different cell types: oligodendroglia, astrocytes, ependymal cells, and microglia. Their function is to support, nourish, and protect the neurons.

Muscle cell

Muscle cells contract to produce movement or tension. The intracellular proteins actin and myosin interact to form crossbridges that result in muscle contraction. An increase in intracellular calcium is necessary for muscle to contract.

There are three basic types of muscle cells:

▶ Skeletal (striated) muscle cells are long, cylindrical cells that extend along the entire length of the skeletal muscles. These muscles, which attach directly to the bone or are connected to the bone by tendons, are responsible for voluntary movement. By contracting and relaxing, striated muscle cells alter the length of the muscle.

▶ Smooth (nonstriated) muscle cells are present in the walls of hollow internal organs, such as the gastrointestinal (GI) and genitourinary tracts, and of blood vessels and bronchioles. Unlike striated muscle, these spindle-shaped cells contract involuntarily. By contracting and relaxing, they change the luminal diameter of the hollow structure, and thereby move substances through the organ.

▶ Cardiac muscle cells branch out across the smooth muscle of the chambers of the heart and contract involuntarily. They produce and transmit cardiac action potentials, which cause cardiac muscle cells to contract. Impulses travel from cell to cell as though no cell membrane existed.

AGE ALERT In older adults, muscle cells become smaller and many are replaced by fibrous connective tissue. The result is loss of muscle strength and mass.

PATHOPHYSIOLOGIC CHANGES

The cell faces a number of challenges through its life. Stressors, changes in the body's health, diease, and other extrinsic and intrinsic factors can change the cell's normal functioning (homeostasis).

Cell adaptation

Cells are generally able to continue functioning despite changing conditions or stressors. However, severe or prolonged stress or changes may injure or even kill cells. When cell integrity is threatened — for example, by hypoxia, anoxia, chemical injury, infection, or temperature extremes — the cell reacts in one of two ways:

▶ by drawing on its reserves to keep functioning
▶ by adaptive changes or cellular dysfunction.

If enough cellular reserve is available and the body doesn't detect abnormalities, the cell adapts. If cellular reserve is insufficient, cell death (necrosis) occurs.

ADAPTIVE CELL CHANGES

Cells adapt to changing conditions and stressors within the body in the ways shown below.

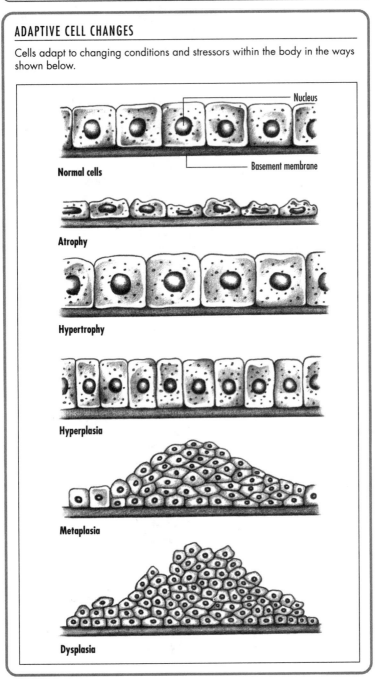

Normal cells

Nucleus

Basement membrane

Atrophy

Hypertrophy

Hyperplasia

Metaplasia

Dysplasia

Necrosis is usually localized and easily identifiable.

The cells' methods of adapting include atrophy, hypertrophy, hyperplasia, metaplasia, and dysplasia. (See *Adaptive cell changes.*)

Atrophy

Atrophy is a reduction in the size of a cell or organ that may occur when cells face reduced workload or disuse, insufficient blood flow, malnutrition, or reduced hormonal and nerve stimulation. Examples of atrophy include loss of muscle mass and tone after prolonged bed rest.

Hypertrophy

In contrast, hypertrophy is an increase in the size of a cell or organ due to an increase in workload. The three basic types of hypertrophy are *physiologic*, *compensatory*, and *pathologic*.

◗ *Physiologic hypertrophy* reflects an increase in workload that is not caused by disease — for example, the increase in muscle size caused by hard physical labor or weight training.

◗ *Compensatory hypertrophy* takes place when cell size increases to take over for nonfunctioning cells. For instance, one kidney will hypertrophy when the other is not functioning or is removed.

◗ *Pathologic hypertrophy* is a response to disease. An example is hypertrophy of the heart muscle as the muscle pumps against increasing resistance in patients with hypertension.

Hyperplasia

Hyperplasia is an increase in the number of cells caused by increased workload, hormonal stimulation, or decreased tissue density. Like hypertrophy, hyperplasia may be *physiologic*, *compensatory*, or *pathologic*.

◗ *Physiologic hyperplasia* is an adaptive response to normal changes. An example is the monthly increase in number of uterine cells that occurs in response to estrogen stimulation of the endometrium after ovulation.

◗ *Compensatory hyperplasia* occurs in some organs to replace tissue that has been removed or destroyed. For example, liver cells regenerate when part of the liver is surgically removed.

◗ *Pathologic hyperplasia* is a response to either excessive hormonal stimulation or abnormal production of hormonal growth factors. Examples include acromegaly, in which excessive growth hormone production causes bones to enlarge, and endometrial hyperplasia, in which excessive secretion of estrogen causes heavy menstrual bleeding and possibly malignant changes.

Metaplasia

Metaplasia is the replacement of one cell type with another cell type. A common cause of metaplasia is constant irritation or injury that initiates an inflammatory response. The new cell type can better endure the stress of chronic inflammation. Metaplasia may be either *physiologic* or *pathologic.*

◗ *Physiologic metaplasia* is a normal response to changing conditions and is generally transient. For example, in the body's normal response to inflammation, monocytes that migrate to inflamed tissues transform into macrophages.

◗ *Pathologic metaplasia* is a response to an extrinsic toxin or stressor and is generally irreversible. For example, after years of exposure to cigarette smoke, stratified squamous epithelial cells replace the normal ciliated columnar epithelial cells of the bronchi. Although the new cells can better withstand smoke, they don't secrete mucus nor do they have cilia to protect the airway. If exposure to cigarette smoke continues, the squamous cells can become cancerous.

Dysplasia

In dysplasia, abnormal differentiation of dividing cells results in cells that are abnormal in size, shape, and appearance. Although dysplastic cell changes aren't cancerous, they can precede cancerous changes. Common examples include dysplasia of epithelial cells of the cervix or the respiratory tract.

Cell injury

Injury to any cellular component can lead to illness as the cells lose their ability to adapt. One early indication of cell injury is a biochemical lesion that forms on the cell at the point of injury. For example, in a patient with chronic alcoholism, biochemical lesions on the cells of the immune system may increase the patient's susceptibility to infection, and cells of the pancreas and liver are affected in a way that prevents their reproduction. These cells can't return to normal functioning.

Causes of cell injury

Cell injury may result from any of several intrinsic or extrinsic causes:

▶ *Toxins.* Substances that originate in the body (endogenous factors) or outside the body (exogenous factors) may cause toxic injuries. Common endogenous toxins include products of genetically determined metabolic errors, gross malformations, and hypersensitivity reactions. Exogenous toxins include alcohol, lead, carbon monoxide, and drugs that alter cellular function. Examples of such drugs are chemotherapeutic agents used for cancer and immunosuppressants used to prevent rejection in organ transplant recipients.

▶ *Infection.* Viruses, fungi, protozoa, and bacteria can cause cell injury or death. These organisms affect cell integrity, usually by interfering with cell division, producing nonviable, mutant cells. For example, human immunodeficiency virus alters the cell when the virus is replicated in the cell's RNA.

▶ *Physical injury.* Physical injury results from a disruption in the cell or in the relationships of the intracellular organelles. Two major types of physical injury are thermal and mechanical. Causes of thermal injury include burns, radiation therapy for cancer, X-rays, and ultraviolet radiation. Causes of mechanical injury include surgery, trauma from motor vehicle accidents, and frostbite.

▶ *Deficit injury.* When a deficit of water, oxygen, or nutrients occurs, or if constant temperature and adequate waste disposal aren't maintained, normal cellular metabolism can't take place. A lack of just one of these basic requirements can cause cell disruption or death. Causes of deficit include hypoxia (inadequate oxygen), ischemia (inadequate blood supply), and malnutrition.

Irreversible cell injury occurs when there's a breakdown of organelles and cell membrane.

Cell degeneration

Degeneration is a type of nonlethal cell damage that generally occurs in the cytoplasm and that doesn't affect the nucleus. Degeneration usually affects organs with metabolically active cells, such as the liver, heart, and kidneys, and is caused by these problems:

▶ increased water in the cell or cellular swelling
▶ fatty infiltrates
▶ atrophy
▶ autophagocytosis (that is, the cell absorbs some of its own parts)
▶ pigmentation changes
▶ calcification
▶ hyaline infiltration
▶ hypertrophy
▶ dysplasia (related to chronic irritation)
▶ hyperplasia.

When changes in cells are identified, prompt health care can slow degeneration and prevent cell death. An electron microscope can help identify cellular changes, and thus diagnose a disease, before the patient complains of any symptoms. Unfortunately, many cell changes remain unidentifiable even under a microscope, making early detection of disease impossible.

Cell aging

During the normal process of aging, cells lose both structure and function. Atrophy, a decrease in size or wasting away, may indicate loss of cell structure. Hypertrophy or hyperplasia is characteristic of lost cell function. (See *Factors that affect cell aging.*)

Signs of aging occur in all body systems. Examples include diminished elasticity of blood vessels, bowel motility, muscle mass, and subcutaneous fat. Cell aging can slow down or speed up, depending on the number and extent of injuries and the amount of wear and tear on the cell.

The cell aging process limits the human life span (of course, many people die from disease before they reach the maximum life span of about 110 years). A number of theories attempt to explain the reasons behind cell aging. (See *Biological theories of aging,* pages 12 and 13.)

Cell death

Like disease, cell death may be caused by internal (intrinsic) factors that limit the cell's life span or external (extrinsic) factors that contribute to cell damage and aging. When a stressor is severe or prolonged, the cell can no longer adapt and it dies.

Cell death, or necrosis, may manifest in different ways, depending on the tissues or organs involved.

▶ *Apoptosis* is genetically programmed cell death. This accounts for the constant cell turnover in the skin's outer keratin layer and the lens of the eye.

▶ *Liquefactive necrosis* occurs when a lytic (dissolving) enzyme liquefies necrotic cells. This type of necrosis is common in the brain, which has a rich supply of lytic enzymes.

▶ In *caseous necrosis,* the necrotic cells disintegrate but the cellular pieces remain undigested for months or years. This type of necrotic tissue gets its name from its crumbly, cheeselike (caseous) appearance. It commonly occurs in lung tuberculosis.

FACTORS THAT AFFECT CELL AGING

Cell aging can be affected by the intrinsic and extrinsic factors listed below.

INTRINSIC FACTORS
▶ Congenital
▶ Degenerative
▶ Immunologic
▶ Inherited
▶ Metabolic
▶ Neoplastic
▶ Nutritional
▶ Psychogenic

EXTRINSIC FACTORS

Physical agents
▶ Chemicals
▶ Electricity
▶ Force
▶ Humidity
▶ Radiation
▶ Temperature

Infectious agents
▶ Bacteria
▶ Fungi
▶ Insects
▶ Protozoa
▶ Viruses
▶ Worms

▶ In *fat necrosis,* enzymes called *lipases* break down intracellular triglycerides into free fatty acids. These free fatty acids combine with sodium, magnesium, or calcium ions to form soaps. The tissue becomes opaque and chalky white.

▶ *Coagulative necrosis* commonly occurs when the blood supply to any organ (except the brain) is interrupted. It typically affects the kidneys, heart, and adrenal glands. Lytic (lysosomal) enzyme activity in the cells is inhibited, so that the necrotic cells maintain their shape, at least temporarily.

BIOLOGICAL THEORIES OF AGING

Various theories have been proposed to explain the process of normal aging. Biological theories attempt to explain physical aging as an involuntary process that eventually leads to cumulative changes in cells, tissues, and fluids.

THEORY
Cross-link theory
Strong chemical bonding between organic molecules in the body causes increased stiffness, chemical instability, and insolubility of connective tissue and deoxyribonucleic acid.
Free-radical theory
An increased number of unstable free radicals produces effects harmful to biologic systems, such as chromosomal changes, pigment accumulation, and collagen alteration.
Immunologic theory
An aging immune system is less able to distinguish body cells from foreign cells; as a result, it begins to attack and destroy body cells as if they were foreign. This may explain the adult onset of conditions such as diabetes mellitus, rheumatic heart disease, and arthritis. Theorists have speculated about the existence of several erratic cellular mechanisms that are capable of precipitating attacks on various tissues through autoaggression or immunodeficiencies.
Wear and tear theory
Body cells, structures, and functions wear out or are overused through exposure to internal and external stressors. Effects of the residual damage accumulate, the body can no longer resist stress, and death occurs.

▶ *Gangrenous necrosis,* a form of coagulative necrosis, typically results from a lack of blood flow and is complicated by an overgrowth and invasion of bacteria. It commonly occurs in the lower legs as a result of arteriosclerosis or in the GI tract. Gangrene can occur in one of three forms: *dry, moist (or wet),* or *gas.*

▶ *Dry gangrene* occurs when bacterial invasion is minimal. It's marked by dry, wrinkled, dark brown or blackened tissue on an extremity.

▶ *Moist (or wet) gangrene* develops with liquifactive necrosis that includes extensive lytic activity from bacteria and white blood cells to produce a liquid center in an area of tissue. It can occur in the internal organs as well as the extremities.

▶ *Gas gangrene* develops when anaerobic bacteria of the genus *Clostridium* infect

SOURCES	RETARDANTS
Lipids, proteins, carbohydrates, and nucleic acids	Restricting calories and sources of lathyrogens (antilink agents), such as chick peas
Environmental pollutants; oxidation of dietary fats, proteins, carbohydrates, and elements	Improving environmental monitoring; decreasing intake of free-radical-stimulating foods; increasing intake of vitamins A and C (mercaptans) and vitamin E
Alteration of B and T cells of the humoral and cellular systems	Immunoengineering — selective alteration and replenishment or rejuvenation of the immune system
Repeated injury or overuse; internal and external stressors (physical, psychological, social, and environmental), including trauma, chemicals, and buildup of naturally occurring wastes	Reevaluating and possibly adjusting lifestyle

tissue. It's more likely to occur with severe trauma and may be fatal. The bacteria release toxins that kill nearby cells and the gas gangrene rapidly spreads. Release of gas bubbles from affected muscle cells indicates that gas gangrene is present.

Necrotic changes
When a cell dies, enzymes inside the cell are released and start to dissolve cellular components. This triggers an acute inflammatory reaction in which white blood cells migrate to the necrotic area and begin to digest the dead cells. At this point, the dead cells — primarily the nuclei — begin to change morphologically in one of three ways:

▶ *pyknosis*, in which the nucleus shrinks, becoming a dense mass of genetic material with an irregular outline.

▶ *karyorrhexis*, in which the nucleus breaks up, strewing pieces of genetic material throughout the cell.

▶ *karyolysis*, in which hydrolytic enzymes released from intracellular structures called lysosomes simply dissolve the nucleus.

2

Cancer, also called malignant neoplasia, refers to a group of more than 100 different diseases that are characterized by DNA damage that causes abnormal cell growth and development. Malignant cells have two defining characteristics: (1) they can no longer divide and differentiate normally, and (2) they have acquired the ability to invade surrounding tissues and travel to distant sites.

In the United States, cancer accounts for more than half a million deaths each year, second only to cardiovascular disease. However, a 1999 review of the Healthy People 2010 cancer objectives by the Department of Health and Human Services had encouraging results: a reversal of a 20-year trend of increasing cancer incidence and deaths. The rates for all cancers combined and for most of the top 10 cancer sites declined between 1990 and 1996.

Worldwide, the most common malignancies include skin cancer, leukemias, lymphomas, and cancers of the breast, bone, gastrointestinal (GI) tract and associated structures, thyroid, lung, urinary tract, and reproductive tract. (See *Reviewing common cancers,* pages 43 to 52.) In the United States, the most common forms of cancer are skin, prostate, breast, lung, and colorectal. Some cancers, such as ovarian germ-cell tumors and retinoblastoma, occur predominantly in younger patients; yet, more than two-thirds of the patients who develop cancer are over age 65.

HOW DOES CANCER HAPPEN?

Most of the numerous theories about carcinogenesis suggest that it involves three steps: initiation, promotion, and progression.

Initiation

Initiation refers to the damage to or mutation of DNA that occurs when the cell is exposed to an initiating substance or event (such as chemicals, virus, or radiation) during DNA replication (transcription). Normally, enzymes detect errors in transcription and remove or repair them. But sometimes an error is missed. If regulatory proteins recognize the error and block further division, then the error may be repaired or the cell may self-destruct. If these proteins miss the error again, it becomes a permanent mutation that is passed on to future generations of cells.

Promotion

Promotion involves the exposure of the mutated cell to factors (*promoters*) that enhance its growth. This exposure may occur either shortly after initiation or years later.

Promoters may be hormones, such as estrogen; food additives, such as nitrates; or drugs, such as nicotine. Promoters can affect the mutated cell by altering:

▶ function of genes that control cell growth and duplication
▶ cell response to growth stimulators or inhibitors
▶ intercellular communication.

Progression

Some investigators believe that progression is actually a late promotion phase in which the tumor invades, metastasizes, and becomes resistant to drugs. This step is irreversible.

CAUSES

Current evidence suggests that cancer develops from a complex interaction of exposure to carcinogens and accumulated mutations in several genes. Researchers have identified approximately 100 cancer genes. Some cancer genes, called *oncogenes*, activate cell division and influence embryonic development. Other cancer genes, the *tumor-suppressor genes*, halt cell division. Normal human cells typically contain proto-oncogenes (oncogene precursors) and tumor-suppressor genes, which remain dormant unless they are transformed by genetic or acquired mutation. Common causes of acquired genetic damage are viruses, radiation, environmental and dietary carcinogens, and hormones. Other factors that interact to increase a person's likelihood of developing cancer are age, nutritional status, hormonal balance, and response to stress; these are discussed below as risk factors.

The healthy body is well equipped to defend itself against cancer. Only when the immune system and other defenses fail does cancer prevail.

Genetics

Some cancers and precancerous lesions may result from genetic predisposition either directly or indirectly. Direct causation occurs when a single gene is responsible for the cancer, as in Wilms' tumor and retinoblastoma, for example. Indirect carcinogenesis is associated with inherited conditions, such as Down syndrome or immunodeficiency diseases. Common characteristics of genetically predisposed cancer include:

▶ early onset of malignant disease

▶ increased incidence of bilateral cancer in paired organs
▶ increased incidence of multiple primary cancers in nonpaired organs
▶ abnormal chromosome complement in tumor cells.

Viruses

Viral proto-oncogenes often contain DNA that's identical to that of human oncogenes. In animal studies of viral ability to transform cells, some viruses that infect people have demonstrated the potential to cause cancer. For example, the Epstein-Barr virus, which causes infectious mononucleosis, has been linked to lymphomas.

Failure of immunosurveillance

Research suggests that cancer cells develop continually, but the immune system recognizes these cells as foreign and destroys them. This defense mechanism, termed immunosurveillance, has two major components: *cell-mediated* immune response and *humoral* immune response. Together, these two components interact to promote antibody production, cellular immunity, and immunologic memory. Researchers believe that an intact immune system is responsible for spontaneous regression of tumors. Thus, cancer development is a concern for patients who must take immunosuppressant medications.

Cell-mediated immune response

Cancer cells carry cell-surface antigens (specialized protein molecules that trigger an immune response) called tumor-associated antigens (TAAs) and tumor-specific antigens (TSAs). The cell-mediated immune response begins when T lymphocytes encounter a TAA or a TSA and become sensitized to it. After repeated contacts, the sensitized T cells release chemical factors called *lymphokines*, some of which begin to destroy the antigen. This reaction triggers the transformation of a different population of T lymphocytes into "killer T lymphocytes" targeted to cells carrying the specific antigen — in this case cancer cells.

Humoral immune response

The humoral immune response reacts to a TAA by triggering the release of antibodies from plasma cells and activating the serum-complement system to destroy the antigen-bearing cells. However, an opposing immune factor, a "blocking antibody," may enhance tumor growth by protecting malignant cells from immune destruction.

Disruption of the immune response

Immunosurveillance isn't a fail-safe system. If the immune system fails to recognize tumor cells as foreign, the immune response won't activate. The tumor will continue to grow until it's beyond the immune system's ability to destroy it. In addition to this failure of surveillance, other mechanisms may come into play.

The tumor cells may actually suppress the immune defenses. The tumor antigens may combine with humoral antibodies to form complexes that essentially hide the antigens from the normal immune defenses. These complexes could also depress further antibody production. Tumors also may change their antigenic "appearance" or produce substances that impair usual immune defenses. The tumor growth factors not only promote the growth of the tumor, but also increase the person's risk of infection. Finally, prolonged exposure to a tumor antigen may deplete the patient's lymphocytes and further impair the ability to mount an appropriate response.

The patient's population of suppressor T lymphocytes may be inadequate to defend against malignant tumors. Suppressor T lymphocytes normally assist in regulating antibody production; they also signal the immune system when an immune response is no longer needed. Certain carcinogens, such as viruses or chemicals, may weaken the immune system by destroying or damaging suppressor T cells or their precursors, and subsequently, allow for tumor growth.

Research data support the concept that cancer develops when any of several factors disrupts the immune response:

▶ *Aging cells.* As cells age, errors in copying genetic material during cell division may give rise to mutations. If the aging immune system doesn't recognize these mutations as foreign, the mutated cells may proliferate and form a tumor.

▶ *Cytotoxic drugs or steroids.* These agents decrease antibody production and destroy circulating lymphocytes.

▶ *Extreme stress or certain viral infections.* These conditions may depress the immune response, thus allowing cancer cells to proliferate.

▶ *Suppression of immune system.* Radiation, cytotoxic drug therapy, and lymphoproliferative and myeloproliferative diseases (such as lymphatic and myelocytic leukemia) depress bone marrow production and impair leukocyte function.

▶ *Acquired immunodeficiency syndrome (AIDS).* This condition weakens the cell-mediated immune response.

▶ *Cancer.* The disease itself is immunosuppressive. Advanced disease exhausts the immune system, leading to anergy (the absence of immune reactivity).

RISK FACTORS

Many cancers are related to specific environmental and lifestyle factors that predispose a person to develop cancer. Accumulating data suggest that some of these risk factors initiate carcinogenesis, other risk factors act as promoters, and some risk factors both initiate and promote the disease process.

Air pollution

Air pollution has been linked to the development of cancer, particularly lung cancer. Persons living near industries that release toxic chemicals have a documented increased risk of cancer. Many outdoor air pollutants — such as arsenic, benzene, hydrocarbons, polyvinyl chlorides, and other industrial emissions as well as vehicle exhaust — have been studied for their carcinogenic properties.

Indoor air pollution, such as from cigarette smoke and radon, also poses an increased risk of cancer. In fact, indoor air pollution is considered to be more carcinogenic than outdoor air pollution.

Tobacco

Cigarette smoking increases the risk of lung cancer more than tenfold over that of nonsmokers by late middle age. Tobacco smoke contains nitrosamines and polycyclic hydrocarbons, two carcinogens that are known to cause mutations. The risk of lung cancer from cigarette smoking correlates directly with the duration of smoking and the number of cigarettes smoked per day. Tobacco smoke is also associated with laryngeal cancer and is considered a contributing factor in cancer of the bladder, pancreas, kidney, and cervix. Research also shows that a person who stops smoking decreases his or her risk of lung cancer.

Although the risk associated with pipe and cigar smoking is similar to that of cigarette smoking, some evidence suggests that the effects are less severe. Smoke from cigars and pipes is more alkaline. This alkalinity decreases nicotine absorption in the lungs and also is more irritating to the lungs, so that the smoker doesn't inhale as readily.

Inhalation of "secondhand" smoke, or passive smoking, by nonsmokers also increases the risk of lung and other cancers. Plus use of smokeless tobacco, in which the oral tissue directly absorbs nicotine and other carcinogens, is linked to an increase in oral cancers that seldom occur in persons who don't use the product.

Alcohol

Alcohol consumption, especially in conjunction with cigarette smoking, is commonly associated with cirrhosis of the liver, a precursor to hepatocellular cancer. The risk of breast and colorectal cancers also increases with alcohol consumption. Possible mechanisms for breast cancer development include impaired removal of carcinogens by the liver, impaired immune response, and interference with cell membrane permeability of the breast tissue. Alcohol stimulates rectal cell proliferation in rats, an observation that may help explain the increased incidence of colorectal cancer in humans.

Heavy use of alcohol and cigarette smoking synergistically increases the incidence of cancers of the mouth, larynx, pharynx, and esophagus. It's likely that alcohol acts as a solvent for the carcinogenic substances found in smoke, enhancing their absorption.

Sexual and reproductive behavior

Sexual practices have been linked to specific types of cancer. The age of first sexual intercourse and the number of sexual partners are positively correlated with a woman's risk of cervical cancer. Furthermore, a woman who has had only one sexual partner is at higher risk if that partner has had multiple partners. The suspected underlying mechanism here involves virus transmission, most likely human papilloma virus (HPV). HPV types 6 and 11 are associated with genital warts. HPV is the most common cause of abnormal Papanicolaou (Pap) smears, and cervical dysplasia is a direct precursor to squamous cell carcinoma of the cervix, both of which have been linked to HPV (especially types 16 and 31).

Occupation

Certain occupations, because of exposure to specific substances, increase the risk of cancer. Persons exposed to asbestos, such as insulation installers and miners, are at risk of a specific type of lung cancer. Asbestos also may act as a promoter for other carcinogens. Workers involved in the production of dyes, rubber, paint, and beta-naphthylamine are at increased risk of bladder cancer.

Ultraviolet radiation

Exposure to ultraviolet radiation, or sunlight, causes genetic mutation in the P53 control gene. Sunlight also releases tumor necrosis factor alpha in exposed skin, possibly diminishing the immune response. Ultraviolet sunlight is a direct cause of basal and squamous cell cancers of the skin. The amount of exposure to ultraviolet radiation also correlates with the type of cancer that develops. For example, cumulative exposure to ultraviolet sunlight is associated with basal and squamous cell skin cancer, and severe episodes of burning and blistering at a young age are associated with melanoma.

Ionizing radiation

Ionizing radiation (such as X-rays) is associated with acute leukemia, thyroid, breast, lung, stomach, colon, and urinary tract cancers as well as multiple myeloma. Low doses can cause DNA mutations and chromosomal abnormalities, and large doses can inhibit cell division. This damage can directly affect carbohydrate, protein, lipid, and nucleic acids (macromolecules), or it can act on intracellular water to produce free radicals that damage the macromolecules.

Ionizing radiation also can enhance the effects of genetic abnormalities. For example, it increases the risk of cancer in persons with a genetic abnormality that affects DNA repair mechanisms. Other compounding variables include the part and percentage of the body exposed, the person's age, hormonal balance, prescribed drugs and preexisting or concurrent conditions.

Hormones

Hormones — specifically the sex steroid hormones estrogen, progesterone, and testosterone — have been implicated as promoters of breast, endometrial, ovarian, or prostate cancer.

Estrogen, which stimulates the proliferation of breast and endometrial cells, is considered a promoter for breast and endometrial cancers. Prolonged exposure to estrogen, as in women with early menarche and late menopause, increases the risk of breast cancer. Likewise, long-term use of estrogen replacement without progesterone supplementation for menopausal symptoms increases a woman's risk of endometrial cancer. Progesterone may play a protective role, counteracting estrogen's stimulatory effects.

The male sex hormones stimulate the growth of prostatic tissue. However, research fails to show an increased risk of prostatic cancer in men who take exogenous androgens.

Diet

Numerous aspects of diet are linked to an increase in cancer, including:

▶ obesity (in women only, possibly related to production of estrogen by fatty tissue), which is linked to a suspected increased risk of endometrial cancer

▶ high consumption of dietary fat (due to an increase in free radical formation), which is linked to endometrial, breast, prostatic, ovarian, and rectal cancers

▶ high consumption of smoked foods and salted fish or meats and foods containing nitrites, which may be linked to gastric cancer

▶ naturally occurring carcinogens (such as hydrazines and aflatoxin) in foods, which are linked to liver cancer

ACS GUIDELINES: DIET, NUTRITION, AND CANCER PREVENTION

Because of the numerous aspects of diet and nutrition that may contribute to the development of cancer, the American Cancer Society (ACS) has developed a list of guidelines to reduce cancer risk in persons ages 2 years and older.

▶ Choose most of the foods you eat from plant sources.
– Eat 5 or more servings of fruits and vegetables each day.
– Eat other foods from plant sources such as breads, cereals, grain products, rice, pasta, or beans several times each day.
▶ Limit your intake of high-fat foods, particularly from animal sources.
– Choose foods low in fat.
– Limit consumption of meats, especially high-fat meats.
▶ Be physically active and achieve and maintain a healthy weight.
– Be at least moderately active for 30 minutes or more on most days of the week.
– Stay within your healthy weight range.
▶ Limit your consumption of alcoholic beverages, if you drink at all.

▶ carcinogens produced by microorganisms stored in foods, which are linked to stomach cancer
▶ diet low in fiber (which slows transport through the gut), which is linked to colorectal cancer.

The American Cancer Society (ACS) has developed specific nutritional guidelines for cancer prevention. (See *ACS guidelines: Diet, nutrition, and cancer prevention.*)

PATHOPHYSIOLOGIC CHANGES

The characteristic features of cancer are rapid, uncontrollable proliferation of cells and independent spread from a primary site (site of origin) to other tissues where it establishes secondary foci (metastases). This spread occurs through circulation in the blood or lymphatic fluid, by unintentional transplantation from one site to another during surgery, and by local extension. Thus, cancer cells differ from normal cells in terms of cell size, shape, number, differentiation, and purpose or function. Plus cancer cells can travel to distant tissues and organ systems. (See *Cancer cell characteristics.*)

Cell growth

Typically, each of the billions of cells in the human body has an internal clock that tells the cell when it is time to reproduce. Mitotic reproduction occurs in a sequence called the *cell cycle.* Normal cell division occurs in direct proportion to cells lost, thus providing a mechanism for controlling growth and differentiation. These controls are absent in cancer cells, and cell production exceeds cell loss. Consequently, cancer cells enter the cell cycle more frequently and at different rates. They're most commonly found in the synthesis and mitosis phases of the cell cycle, and they spend very little time in the resting phase.

Normal cells reproduce at a rate controlled through the activity of specific control or regulator genes (called proto-oncogenes when they are functioning normally). These genes produce proteins that act as "on" and "off" switches. There is no generalized control gene; different cells respond to specific control genes. The P53 and c-myc genes are two examples of control genes: P53 can stop DNA replication if the cell's DNA has been damaged;

c-myc helps initiate DNA replication and if it senses an error in DNA replication, it can cause the cell to self-destruct.

Hormones, growth factors, and chemicals released by neighboring cells or by immune or inflammatory cells can influence control gene activity. These substances bind to specific receptors on the cell membranes and send out signals causing the control genes to stimulate or suppress cell reproduction. Examples of hormones and growth factors that affect control genes include:

▶ erythropoietin, which stimulates red blood cell proliferation

▶ epidermal growth factor, which stimulates epidermal cell proliferation

▶ insulin-like growth factor, which stimulates fat and connective tissue proliferation

▶ platelet-derived growth factor, which stimulates connective tissue cell proliferation.

Injured or infected nearby cells or those of the immune system also affect cellular reproduction. For example, interleukin, released by immune cells, stimulates cell proliferation and differentiation. Interferon, released from virus-infected and immune cells, may affect the cell's rate of reproduction.

Additionally, cells that are close to one another appear to communicate with each other through gap junctions (channels through which ions and other small molecules pass). This communication provides information to the cell about the neighboring cell types and the amount of space available. The nearby cells send out physical and chemical signals that control the rate of reproduction. For example, if the area is crowded, the nearby cells will signal the same type of cells to slow or cease reproduction, thus allowing the formation of only a single layer of cells. This feature is called *density-dependent growth inhibition*.

In cancer cells, the control genes fail to function normally. The actual control may

CANCER CELL CHARACTERISTICS

Cancer cells, which undergo uncontrolled cellular growth and development, exhibit these typical characteristics:
▶ Vary in size and shape
▶ Undergo abnormal mitosis
▶ Function abnormally
▶ Don't resemble the cell of origin
▶ Produce substances not usually associated with the original cell or tissue
▶ Aren't encapsulated
▶ Are able to spread to other sites

be lost or the gene may become damaged. An imbalance of growth factors may occur, or the cells may fail to respond to the suppressive action of the growth factors. Any of these mechanisms may lead to uncontrolled cellular reproduction.

One striking characteristic of cancer cells is that they fail to recognize the signals emitted by nearby cells about available tissue space. Instead of forming only a single layer, cancer cells continue to accumulate in a disorderly array.

The loss of control over normal growth is termed *autonomy*. This independence is further evidenced by the ability of cancer cells to break off and travel to other sites.

Differentiation

Normally, during development, cells become specialized. That is, the cells develop highly individualized characteristics that reflect their specific structure and functions in their corresponding tissue. For example, all blood cells are derived from a single stem cell that differentiates into red blood cells (RBCs), white blood cells (WBCs), platelets, monocytes, and lymphocytes. As the cells become more specialized, their reproduction and development slow down. Eventually, highly dif-

UNDERSTANDING ANAPLASIA

Anaplasia refers to the loss of differentiation, a common characteristic of cancer cells. As differentiation is lost, the cancer cells no longer demonstrate the appearance and function of the original cell.

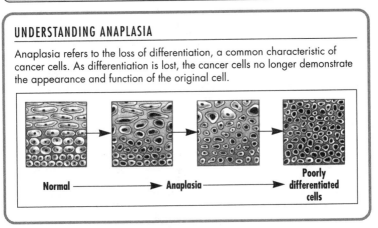

Normal ⟶ Anaplasia ⟶ Poorly differentiated cells

ferentiated cells become unable to reproduce and some, skin cells for example, are programmed to die and be replaced.

Cancer cells lose the ability to differentiate; that is, they enter a state, called *anaplasia*, in which they no longer appear or function like the original cell. (See *Understanding anaplasia.*)

Anaplasia occurs in varying degrees. The less the cells resemble the cell of origin, the more anaplastic they are said to be. As the anaplastic cells continue to reproduce, they lose the typical characteristics of the original cell.

Some anaplastic cells begin functioning as another type of cell, possibly becoming a site for hormone production. For example, oat-cell lung cancer cells often produce antidiuretic hormone (ADH), which is produced by the hypothalamus but stored in and secreted by the posterior pituitary gland.

When anaplasia occurs, cells of the same type in the same site exhibit many different shapes and sizes. Mitosis is abnormal and chromosome defects are common.

Intracellular changes

The abnormal and uncontrolled cell proliferation of cancer cells is associated with numerous changes within the cancer cell itself. These changes affect the cell membrane, cytoskeleton, and nucleus.

Cell membrane

This thin, dynamic semipermeable structure separates the cell's internal environment from its external environment. It consists of two layers of lipid molecules (called the lipid bilayer) with protein molecules attached to or embedded in each layer. The bilayer is composed of phospholipids, glycolipids, and other lipids, such as cholesterol.

The protein molecules help stabilize the structure of the membrane and participate in the transport and exchange of material between the cell and its environment. Large glycoproteins, called fibronectin, are responsible for holding the cells in place and maintaining the specific arrangement of the receptors to allow for the exchange of material.

In the cancer cell, fibronectin is defective or is broken down as it is produced, thus affecting the organization, structure, adhesion, and migration of the cells. Some of the other proteins and glycolipids are also absent or altered. These changes affect the density of the receptors on the cell membrane and the shape of the cell. Communication between the cells becomes impaired, response to growth factors is enhanced, and recognition of other cells is diminished. The result is uncontrolled growth.

Permeability of the cancer cell membrane also is altered. During its uncontrolled, rapid proliferation, the cancer cell has a much greater metabolic demand for nutrients to sustain its growth.

During normal development, cell division can occur only when the cells are anchored to nearby cells or to extracellular molecules via anchoring junctions. In cancer cells, anchoring junctions need not be present. Thus, they continue to divide and can metastasize.

Disruption or blockage of gap junctions interferes with intercellular communication. This may be the underlying mechanism by which cancer cells continue to grow and migrate, forming layers of undifferentiated cells, even in a crowded environment.

Cytoskeleton

The *cytoskeleton* is composed of protein filament networks including actin and microtubules. Normally, actin filaments exert a pull on the extracellular organic molecules that bind cells together. Microtubules control cell shape, movement, and division. In cancer cells, the functions of these components are altered. Additionally, cytoplasmic components are fewer in number and abnormally shaped. Less cellular work occurs because of a decrease in endoplasmic reticulum and mitochondria.

Nucleus

In cancer cells, nuclei are pleomorphic, meaning enlarged and of various shapes and sizes. They also are highly pigmented and have larger and more numerous nucleoli than normal. The nuclear membrane is often irregular and commonly has projections, pouches, or blebs, and fewer pores. Chromatin (uncoiled chromosomes) may clump along the outer areas of the nucleus. Breaks, deletions, translocations, and abnormal karyotypes (chromosome shape and number) are common changes in the chromosomes. The chromosome defects seem to stem from the increased mitotic rate in cancer cells. The appearance of the mitotic cancer cell under light microscopy is often described as atypical and bizarre.

Tumor development and growth

Typically, a long time passes between the initiating event and the onset of the disease. During this time, the cancer cells continue to grow, develop, and replicate, each time undergoing successive changes and further mutations.

How fast a tumor grows depends on specific characteristics of the tumor itself and the host.

Tumor growth needs

For a tumor to grow, an initiating event or events must cause a mutation that will transform the normal cell into a cancer cell. After the initial event, the tumor continues to grow only if available nutrients, oxygen, and blood supply are adequate and the immune system fails to recognize or respond to the tumor.

Effect of tumor characteristics

Two important tumor characteristics affecting growth are location of the tumor and available blood supply. The location determines the originating cell type, which, in turn, determines the cell cycle time. For example, epithelial cells have a shorter cell cycle than connective tissue cells. Thus, tumors of epithelial cells grow more rapidly than do tumors of connective tissue cells.

Tumors need an available blood supply to provide nutrients and oxygen for continued growth, and to remove wastes, but a tumor larger than 1 to 2 mm in size has typically outgrown its available blood supply. Some tumors secrete tumor angiogenesis factors, which stimulate the formation of new blood vessels, to meet the demand.

The degree of anaplasia also affects tumor growth. Remember that the more anaplastic the cells of the tumor, the less differentiated the cells and the more rapidly they divide.

Many cancer cells also produce their own growth factors. Numerous growth factor receptors are present on the cell membranes of rapidly growing cancer cells. This increase in receptors, in conjunction with the changes in the cell membranes, further enhances cancer cell proliferation.

Effect of host characteristics

Several important characteristics of the host affect tumor growth. These characteristics include age, sex, overall health status, and immune system function.

 AGE ALERT Age is an important factor affecting tumor growth. Relatively few cancers are found in children. Yet the incidence of cancer correlates directly with increasing age. This suggests that numerous or cumulative events are necessary for the initial mutation to continue, eventually forming a tumor.

Certain cancers are more prevalent in one sex than the other. For example, sex hormones influence tumor growth in breast, endometrial, cervical, and prostate cancers. Researchers believe that the hormone sensitizes the cell to the initial precipitating factor, thus promoting carcinogenesis.

Overall health status also is an important characteristic. As tumors obtain nutrients for growth from the host, they can alter normal body processes and cause cachexia. Conversely, if the person is nutritionally depleted, tumor growth may slow. Chronic tissue trauma also has been linked with tumor growth because healing involves increased cell division. And the more rapidly cells divide, the greater the likelihood of mutations.

Spread of cancer

Between the initiating event and the emergence of a detectable tumor, some or all of the mutated cells may die. The survivors, if any, reproduce until the tumor reaches a diameter of 1 to 2 mm. New blood vessels form to support continued growth and proliferation. As the cells further mutate and divide more rapidly, they become more undifferentiated. And the number of cancerous cells soon begins to exceed the number of normal cells. Eventually, the tumor mass extends, spreading into local tissues and invading the surrounding tissues. When the local tissue is blood or lymph, the tumor can gain access to the circulation. Once access is gained, tumor cells that detach or break off travel to distant sites in the body, where tumor cells can survive and form a new tumor in the new secondary site. This process is called metastasis.

Dysplasia

Not all cells that proliferate rapidly go on to become cancerous. Throughout a person's life span, various body tissues experience periods of benign rapid growth, such as during wound healing. In some cases, changes in the size, shape, and organization of the cells leads to a condition called dysplasia.

Exposure to chemicals, viruses, radiation, or chronic inflammation causes dysplastic changes that may be reversed by removing the initiating stimulus or treating its effects. However, if the stimulus is not removed, precancerous or dysplastic lesions can progress and give rise to cancer. For example, actinic keratoses, thickened patches on the skin of the face and hands of persons exposed to sunlight, are associated with the development of skin cancer. Removal of the lesions and use of sun block helps minimize the risk that the lesions will progress to skin cancer.

Knowledge about precancerous lesions and promoter events provide the rationale for early detection and screening as important preventative measures.

Localized tumor

Initially, a tumor remains localized. Recall that cancer cells communicate poorly with nearby cells. As a result, the cells continue to grow and enlarge forming a mass or clumps of cells. The mass exerts pressure on the neighboring cells, blocking their blood supply, and subsequently causing their death.

Invasive tumor

Invasion is growth of the tumor into surrounding tissues. It's actually the first step in metastasis. Five mechanisms are linked to invasion: cellular multiplication, mechanical pressure, lysis of nearby cells, reduced cell adhesion, and increased motility. Experimental data indicate that the interaction of all five mechanisms is necessary for invasion.

By their very nature, cancer cells are multiplying rapidly. As they grow, they exert pressure on surrounding cells and tissues, which eventually die because their blood supply has been cut off or blocked. Loss of mechanical resistance leads the way for the cancer cells to spread along the lines of least resistance and occupy the space once filled by the dead cells.

Vesicles on the cancer cell surface contain a rich supply of receptors for laminin, a complex glycoprotein that is a major component of the basement membrane, a thin sheet of noncellular connective tissue upon which cells rest. These receptors permit the cancer cells to attach to the basement membrane, forming a bridge-like connection. Some cancer cells produce and excrete powerful proteolytic enzymes; other cancer cells induce normal host cells to produce them. These enzymes, such as collagenases and proteases destroy the normal cells and break through their basement membrane, enabling the cancer cells to enter.

Reduced cell adhesion also is seen with cancer cells. As discussed in the section on intracellular changes, reduced cell adhesion likely results when the cell-stabilizing glycoprotein fibronectin is deficient or defective. Cancer cells also secrete a chemotactic factor that stimulates motility. Thus, the cancer cells are able to move independently into adjacent tissues, and the circulation, and then to a secondary site. Finally, cancer cells develop finger-like projections called pseudopodia that facilitate cell movement. These projections injure and kill neighboring cells and attach to vessel walls, enabling the cancer cells to enter.

Metastatic tumor

Metastatic tumors are those in which the cancer cells have traveled from the original or primary site to a second or more distant site. Most commonly, metastasis occurs through the blood vessels and lymphatic system. Tumor cells also can be transported from one body location to another by external means, such as carriage on instruments or gloves during surgery.

Hematogenous spread. Invasive tumor cells break down the basement membrane and walls of blood vessels, and the tumor sheds malignant cells into the circulation. Most of the cells die, but a few escape the host defenses and the turbulent environment of the bloodstream. From here, the surviving tumor mass of cells, called a *tumor cell embolus*, travels downstream and commonly lodges in the first capillary bed it encounters. For example, blood from most organs next enters the capillaries of the lungs, which are the most common site of metastasis.

Once lodged, the tumor cells develop a protective coat of fibrin, platelets, and clotting factors to evade detection by the immune system. Then they become attached to the epithelium, ultimately in-

HOW CANCER METASTASIZES

Cancer usually spreads through the bloodstream to other organs and tissues, as shown here.

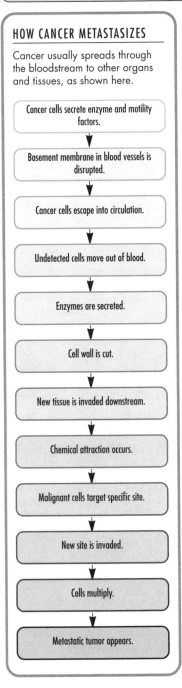

Cancer cells secrete enzyme and motility factors.

↓

Basement membrane in blood vessels is disrupted.

↓

Cancer cells escape into circulation.

↓

Undetected cells move out of blood.

↓

Enzymes are secreted.

↓

Cell wall is cut.

↓

New tissue is invaded downstream.

↓

Chemical attraction occurs.

↓

Malignant cells target specific site.

↓

New site is invaded.

↓

Cells multiply.

↓

Metastatic tumor appears.

vading the vessel wall, interstitium, and the parenchyma of the target organ. (See *How cancer metastasizes.*) To survive, the new tumor develops its own vascular network and may ultimately spread again.

Lymphatic spread. The lymphatic system is the most common route for distant metastasis. Tumor cells enter the lymphatic vessels through damaged basement membranes and are transported to regional lymph nodes. In this case, the tumor becomes trapped in the first lymph node it encounters. The consequent enlargement, possibly the first evidence of metastasis, may be due to the increased tumor growth within the node or a localized immune reaction to the tumor. The lymph node may filter out or contain some of the tumor cells, limiting the further spread. The cells that escape can enter the blood from the lymphatic circulation through plentiful connections between the venous and lymphatic systems.

Metastatic sites. Typically, the first capillary bed, whether lymphatic or vascular, encountered by the circulating tumor mass determines the location of the metastasis. For example, because the lungs receive all of the systemic venous return, they are a frequent site for metastasis. In breast cancer, the axillary lymph nodes, which are in close proximity to the breast, are a common site of metastasis. Other types of cancer seem most likely to spread to specific organs. This *organ tropism* may be a result of growth factor or hormones secreted by the target organ or chemotactic factors that attract the tumor. (See *Common sites of metastasis.*)

SIGNS AND SYMPTOMS

In most patients, the earlier the cancer is found, the more effective the treatment is likely to be and the better the prognosis. Some cancers may be diagnosed on a routine physical examination, even before the person develops any signs or symptoms.

Others may display some early warning signals. The ACS developed a mnemonic device to identify cancer-warning signs. (See *Cancer's seven warning signs*, page 28.)

Unfortunately, a person may not notice or heed the warning signs. These patients may present with some of the commoner signs and symptoms of advancing disease, such as fatigue, cachexia, pain, anemia, thrombocytopenia and leukopenia, and infection. Unfortunately, these signs and symptoms are nonspecific and can be attributed to many other disorders.

Fatigue

Patients commonly describe fatigue as feelings of weakness, being tired, and lacking energy or the ability to concentrate. The exact underlying mechanism for fatigue is not known, but it is believed to be the combined result of several pathophysiologic mechanisms.

The very existence of the tumor may contribute to fatigue. A malignant tumor needs oxygen and nutrients to grow. Thus, it depletes the surrounding tissues of adequate blood and oxygen supply. For example, a vascular tumor can cause lethargy secondary to inadequate oxygen supply to the brain. Lung cancer can interfere with gas exchange and oxygen supply to the heart and peripheral tissues. Accumulating waste products and muscle loss from the release of toxic products of metabolism or other substances from the tumor further add to the fatigue.

Other factors also play a role in fatigue. Pain can be physically and emotionally draining. Stress, anxiety, and other emotional factors further compound the problem. And if the person lacks the energy required for self-care, malnutrition, consequent lack of energy reserves, and anemia can contribute to complaints of fatigue.

Cachexia

Cachexia, a generalized wasting of fat and protein, is common in persons with can-

COMMON SITES OF METASTASIS

The chart below lists some of the more common sites of metastasis for selected cancers.

CANCER TYPE	SITES FOR METASTASIS
Breast	Axillary lymph nodes, lung, liver, bone, brain
Colorectal	Liver, lung, peritoneum
Lung	Liver, brain, bone
Ovarian	Peritoneum, diaphragm, liver, lungs
Prostate	Bone
Testicular	Lungs, liver

cer. A person with cachexia typically appears emaciated and wasted, and experiences an overall deterioration in physical status. Cachexia is characterized by anorexia (loss of appetite), alterations in taste perception, early satiety, weight loss, anemia, marked weakness, and altered metabolism of proteins, carbohydrates, and lipids.

Anorexia may accompany pain or adverse reactions to chemotherapy or radiation therapy. Diminished perception of sweet, sour, or salty sensations also contribute to anorexia. Food that once seemed seasoned and palatable now tastes bland.

Protein-calorie malnutrition may cause hypoalbuminemia, edema (the lack of serum proteins, which normally keep fluid in the blood vessels, enables fluid to escape into the tissues), muscle wasting, and immunodeficiency.

The high metabolic activity of malignant tumor cells carries with it the need for nutrients above those required for nor-

CANCER'S SEVEN WARNING SIGNS

The American Cancer Society has developed an easy way to re-member the seven warning signs of cancer. Each letter in the word *CAUTION* represents a possible warning sign that should spur an individual to see a doctor.

C hange in bowel or bladder habits

A sore that doesn't heal

U nusual bleeding or discharge

T hickening or lump in the breast or elsewhere

I ndigestion or difficulty swallowing

O bvious change in a wart or mole

N agging cough or hoarseness

mal metabolism. As cancer cells appro-priate nutrients to fuel their growth, nor-mal tissue becomes starved and depleted, and wasting begins. Under normal cir-cumstances, when starvation occurs, the body spares protein, relying on carbohy-drates and fats for energy production. However, cancer cells metabolize both protein and fatty acids to produce energy.

Patients with cancer commonly feel sat-ed after eating only a few bites of food. This feeling is believed to be the result of metabolites released from the tumor. Also, tumor necrosis factor produced by the body in response to cancer contributes to cachexia.

Pain

In cancer's early stages, pain is typically absent or mild; as cancer progresses, how-ever, the severity of the pain usually in-creases. Generally, pain is the result of one or more of the following:

- pressure

- obstruction
- invasion of sensitive tissue
- visceral surface stretching
- tissue destruction
- inflammation.

Pressure on or obstruction of nerves, blood vessels, or other tissues and organs leads to tissue hypoxia, accumulation of lactic acid, and possibly cell death. In ar-eas where space for the tumor to grow is limited, such as in the brain or bone, com-pression is a common cause of pain. Ad-ditionally, pain occurs when the viscera, which is normally hollow, is stretched by a tumor, as in GI cancer.

Cancer cells also release proteolytic en-zymes that directly injure or destroy neigh-boring cells. This injury sets up a painful inflammatory response.

Anemia

Cancer of the blood-forming cells, WBCs, or RBCs may directly cause anemia. Ane-mia in patients with metastatic cancer is commonly the result of chronic bleeding, severe malnutrition, or chemotherapy or radiation.

Leukopenia and thrombocytopenia

Typically, leukopenia and thrombocy-topenia occur when cancer invades the bone marrow. Chemotherapy and radia-tion therapy to the bones also can cause leukopenia.

Leukopenia greatly increases the pa-tient's risk of infection. The patient with thrombocytopenia is at risk of hemorrhage. Even when the platelet count is normal, platelet function may be impaired in cer-tain hematologic cancers.

Infection

Infection is common in patients with ad-vanced cancer, particularly those with myelosuppression from treatment, direct invasion of the bone marrow, development of fistulas, or immunosuppression from hormonal release in response to chronic stress. Malnutrition and anemia further increase the patient's risk of infection. Also, obstructions, effusions, and ulcerations may develop, creating a favorable environment for microbial growth.

DIAGNOSIS

A thorough history and physical examination should precede sophisticated diagnostic tests. The choice of diagnostic tests is determined by the patient's presenting signs and symptoms and the suspected body system involved. Diagnostic tests serve several purposes, including:

▶ establishing tumor presence and extent of disease

▶ determining possible sites of metastasis

▶ evaluating affected and unaffected body systems

▶ identifying the stage and grade of tumor.

Useful tests for early detection and staging of tumors include screening tests, X-rays, radioactive isotope scanning (nuclear medicine imaging), computed tomography (CT) scanning, endoscopy, ultrasonography, and magnetic resonance imaging (MRI). The single most important diagnostic tool is the biopsy for direct histologic study of the tumor tissue.

Screening tests

Screening tests are perhaps the most important diagnostic tools in the prevention and early detection of cancer. They may provide valuable information about the possibility of cancer even before the patient develops signs and symptoms. The ACS has recommended specific screening tests to aid in the early detection of cancer. (See *ACS guidelines: Early cancer detection,* page 30.)

Diagnosis by imaging

X-rays

Most commonly, X-rays are ordered to identify and evaluate changes in tissue densities. The type and location of the X-ray is determined by the patient's signs and symptoms and the suspected location of the tumor or metastases. For example, a chest X-ray may be indicated to identify lung cancer if the patient is an older smoker or to rule out lung metastasis in a patient with colorectal cancer.

Some X-rays such as those of the GI tract (barium enema) and the urinary tract (intravenous pyelogram) involve the use of contrast agents. Radiopaque substances also can be injected into the lymphatic system, and their flow can be monitored by lymphangiography. This specialized X-ray technique is helpful in evaluating tumors of the lymph nodes and metastasis. Because lymphangiography is invasive and may be difficult to interpret, CT and MRI scans have largely replaced it.

Radioactive isotope scanning

A specialized camera detects radioactive isotopes that are injected into the blood stream or ingested. The radiologist evaluates their distribution (uptake) throughout tissues, organs, and organ systems. This type of scanning provides a view of organs and regions within the organ that cannot be seen with a simple X-ray. The area of uptake is termed either a hotspot or a coldspot (an area of decreased uptake). Typically tumors are revealed as coldspots; the exception is the bone scan, in which hotspots indicate the presence of disease. Examples of organs commonly evaluated with radioactive isotope scanning include thyroid, liver, spleen, brain, and bone.

Computed tomography

Computed tomography (CT) scanning evaluates successive layers of tissue by using narrow beam X-ray to provide a cross-sectional view of the structure. It also can reveal different characteristics of tissues within a solid organ. CT scans are commonly obtained of the brain and head, body, and abdomen to evaluate for neurologic, pelvic, abdominal, and thoracic cancers.

ACS GUIDELINES: EARLY CANCER DETECTION

The following recommendations from the American Cancer Society focus on five common cancers whose survival rates can be improved if detected early.

SCREENING AREA	RECOMMENDATIONS
Generalized cancer-related checkup (including health counseling and specific examinations for malignant and nonmalignant disorders)	▶ Every three years for ages 20 to 40 ▶ Every year for ages 40 and older
Breast ▶ Mammogram ▶ Clinical breast exam ▶ Self breast exam	▶ Every year for ages 40 and older ▶ Every year for ages 40 and older ▶ Every 3 years for ages 20 to 39 ▶ Monthly for ages 20 and older
Colon and rectum (one of the examinations below) ▶ Fecal occult blood and flexible sigmoidoscopy ▶ Colonoscopy ▶ Double-contrast barium enema ▶ Digital rectal examination	*Men and women age 50 and older:* ▶ Every year ▶ Every 10 years ▶ Every 5 to 10 years ▶ At the same time as sigmoidoscopy, colonoscopy, or double-contrast barium enema
Prostate ▶ Prostate specific antigen (PSA) ▶ Digital rectal examination	*Men age 50 and older with life expectancy of at least 10 years and younger men at high risk:* ▶ Annually ▶ Annually
Cervix ▶ Pap smear ▶ Pelvic examination	*Sexually active females or females age 18 and older:* ▶ Annually, until 3 or more consecutive satisfactory examinations with normal findings; then may be performed less frequently ▶ Annually
Endometrium ▶ Tissue sampling	▶ At the onset of menopause

Ultrasonography

Ultrasonography uses high frequency sound waves to detect changes in the density of tissues that are difficult or impos- sible to observe by radiology or endoscopy. Ultrasound helps to differentiate cysts from solid tumors and is commonly used to provide information about abdominal and pelvic cancer.

Magnetic resonance imaging

MRI uses magnetic fields and radio frequencies to show a cross-sectional view of the body organs and structures. Like CT scanning, it's commonly used to evaluate neurologic, pelvic, abdominal, and thoracic cancers.

Endoscopy

Endoscopy provides a direct view of a body cavity or passageway to detect abnormalities. Common endoscopic sites include the upper and lower gastrointestinal tract, and bronchial tree of the lungs. During endoscopy, the physician can excise small tumors, aspirate fluid, or obtain tissue samples for histologic examination.

Biopsy

A biopsy, the removal of a portion of suspicious tissue, is the only definitive method to diagnose cancer. Biopsy tissue samples can be taken by curettage, fluid aspiration (pleural effusion), fine-needle aspiration (breast), dermal punch (skin or mouth), endoscopy (rectal polyps and esophageal lesions), and surgical excision (visceral tissue and nodes). The specimen then undergoes laboratory analysis for cell type and characteristics to provide information about the grade and stage of the cancer.

Tumor cell markers

Some cancer cells release substances that normally aren't present in the body or are present only in small quantities. These substances, called tumor markers or biologic markers, are produced either by the cancer cell's genetic material during growth and development or by other cells in response to the presence of cancer. Markers may be found on the cell membrane of the tumor or in the blood, cerebrospinal fluid, or urine. Tumor cell markers include hormones, enzymes, genes, antigens, and antibodies. (See *Common tumor cell markers,* pages 32 to 34.)

Tumor cell markers have many clinical uses, for example:
▶ screening people who are at high risk of cancer
▶ diagnosing a specific type of cancer in conjunction with clinical manifestations
▶ monitoring the effectiveness of therapy
▶ detecting recurrence.

Tumor cell markers provide a method for detecting and monitoring the progression of certain types of cancer. Unfortunately, several disadvantages of tumor markers may preclude their use alone. For example,
▶ By the time the tumor cell marker level is elevated, the disease may be too far advanced to treat.
▶ Most tumor cell markers are not specific enough to identify one certain type of cancer.
▶ Some nonmalignant diseases, such as pancreatitis or ulcerative colitis, also are associated with tumor cell markers.

Perhaps the worst drawback is that the absence of a tumor cell marker does not mean that a person is free of cancer. For example, mucinous ovarian cancer tumors typically do not express the ovarian cancer marker CA-125, so that a negative test doesn't eliminate the possibility of ovarian malignancy.

TUMOR CLASSIFICATION

Tumors are initially classified as benign or malignant depending on the specific features exhibited by the tumor. Typically, benign tumors are well differentiated; that is, their cells closely resemble those of the tissue of origin. Commonly encapsulated with well-defined borders, benign tumors grow slowly, often displacing but not infiltrating surrounding tissues, and therefore causing only slight damage. Benign tumors do not metastasize.

Conversely, most malignant tumors are undifferentiated to varying degrees, having cells that may differ considerably from those of the tissue of origin. They are seldom encapsulated and are often poorly

(Text continues on page 34.)

COMMON TUMOR CELL MARKERS

Tumors cell markers may be used to detect, diagnose, or treat cancer. Alone, however, they aren't sufficient for a diagnosis. Tumor cell markers may also be associated with other benign (nonmalignant) conditions. The chart below highlights some of the more commonly used tumor cell markers and their associated malignant and nonmalignant conditions.

MARKER	MALIGNANT CONDITIONS	NONMALIGNANT CONDITIONS
Alpha-fetoprotein (AFP)	▶ Endodermal sinus tumor ▶ Liver cancer ▶ Ovarian germ cell cancer ▶ Testicular germ cell cancer (specifically embryonal cell carcinoma)	▶ Ataxia-telangiectasia ▶ Cirrhosis ▶ Hepatitis ▶ Pregnancy ▶ Wiskott-Aldrich syndrome
Carcinoembryonic antigen (CEA)	▶ Bladder cancer ▶ Breast cancer ▶ Cervical cancer ▶ Colorectal cancer ▶ Kidney cancer ▶ Liver cancer ▶ Lymphoma ▶ Lung cancer ▶ Melanoma ▶ Ovarian cancer ▶ Pancreatic cancer ▶ Stomach cancer ▶ Thyroid cancer	▶ Inflammatory bowel disease ▶ Liver disease ▶ Pancreatitis ▶ Tobacco use
CA 15-3	▶ Breast cancer (usually advanced) ▶ Lung cancer ▶ Ovarian cancer ▶ Prostate cancer	▶ Benign breast disease ▶ Benign ovarian disease ▶ Endometriosis ▶ Hepatitis ▶ Lactation ▶ Pelvic inflammatory disease ▶ Pregnancy
CA 19-9	▶ Bile duct cancer ▶ Colorectal cancer ▶ Pancreatic cancer ▶ Stomach cancer	▶ Cholecystitis ▶ Cirrhosis ▶ Gallstones ▶ Pancreatitis

COMMON TUMOR CELL MARKERS *(continued)*

MARKER	MALIGNANT CONDITIONS	NONMALIGNANT CONDITIONS
CA 27-29	▶ Breast cancer ▶ Colon cancer ▶ Kidney cancer ▶ Liver cancer ▶ Lung cancer ▶ Ovarian cancer ▶ Pancreatic cancer ▶ Stomach cancer ▶ Uterine cancer	▶ Benign breast disease ▶ Endometriosis ▶ Kidney disease ▶ Liver disease ▶ Ovarian cysts ▶ Pregnancy (first trimester)
CA 125	▶ Ovarian cancer ▶ Pancreatic cancer	▶ Endometriosis ▶ Liver disease ▶ Menstruation ▶ Pancreatitis ▶ Pelvic inflammatory disease ▶ Peritonitis ▶ Pregnancy
Human chorionic gonadotropin (HCG)	▶ Choriocarcinoma ▶ Embryonal cell carcinoma ▶ Gestational trophoblastic disease ▶ Liver cancer ▶ Lung cancer ▶ Pancreatic cancer ▶ Specific dysgerminomas of the ovary ▶ Stomach cancer ▶ Testicular cancer	▶ Marijuana use ▶ Pregnancy
Lactate dehydrogenase	▶ Almost all cancers ▶ Ewing's sarcoma ▶ Leukemia ▶ Non-Hodgkin's lymphoma ▶ Testicular cancer	▶ Anemia ▶ Heart failure ▶ Hypothyroidism ▶ Lung disease ▶ Liver disease
Neuron-specific enolase (NSE)	▶ Kidney cancer ▶ Melanoma ▶ Neuroblastoma ▶ Pancreatic cancer ▶ Small cell lung cancer ▶ Testicular cancer ▶ Thyroid cancer ▶ Wilms' tumor	▶ Unknown

(continued)

COMMON TUMOR CELL MARKERS (continued)

MARKER	MALIGNANT CONDITIONS	NONMALIGNANT CONDITIONS
Prostatic acid phosphatase (PAP)	▶ Prostate cancer	▶ Benign prostatic conditions
Prostate-specific antigen (PSA)	▶ Prostate cancer	▶ Benign prostatic hypertrophy ▶ Prostatitis

delineated. They rapidly expand in all directions, causing extensive damage as they infiltrate surrounding tissues. Most malignant tumors metastasize through the blood or lymph to secondary sites.

Malignant tumors are further classified by tissue type, degree of differentiation (grading), and extent of the disease (staging). High-grade tumors are poorly differentiated and are more aggressive than low-grade tumors. Early-stage cancers carry a more favorable prognosis than later-stage cancers that have spread to nearby or distant sites.

Tissue type
Histologically, the type of tissue in which the growth originates classifies malignant tumors. Three cell layers form during the early stages of embryonic development:
▶ Ectoderm primarily forms the external embryonic covering and the structures that will come into contact with the environment.
▶ Mesoderm forms the circulatory system, muscles, supporting tissue, and most of the urinary and reproductive system.
▶ Endoderm gives rise to the internal linings of the embryo, such as the epithelial lining of the pharynx and respiratory and gastrointestinal tracts.

Carcinomas are tumors of epithelial tissue. They may originate in the endodermal tissues, which develop into internal structures, such as the stomach and intestine, or in ectodermal tissues, which develop into external structures such as the skin. Tumors arising from glandular epithelial tissue are commonly called adenocarcinomas.

Sarcomas originate in the mesodermal tissues, which develop into supporting structures, such as the bone, muscle, fat, or blood. Sarcomas may be further classified based on the specific cells involved. For example, malignant tumors arising from pigmented cells are called melanomas; from plasma cells, myelomas; and from lymphatic tissue, lymphomas.

Grading
Histologically, malignant tumors are classified by their degree of differentiation. The greater their differentiation, the greater the tumor cells' similarity to the tissue of origin. Typically, a malignant tumor is graded on a scale of 1 to 4, in order of increasing clinical severity.
▶ Grade 1: Well differentiated; cells closely resemble the tissue of origin and maintain some specialized function.
▶ Grade 2: Moderately well differentiated; cells vary somewhat in size and shape with increased mitosis.
▶ Grade 3: Poorly differentiated; cells vary widely in size and shape with little resemblance to the tissue of origin; mitosis is greatly increased.
▶ Grade 4: Undifferentiated; cells exhibit no similarity to tissue of origin.

Staging
Malignant tumors are staged (classified anatomically) by the extent of the disease.

UNDERSTANDING TNM STAGING

The TNM (tumor, node, and metastasis) system developed by the American Joint Committee on Cancer provides a consistent method for classifying malignant tumors based on the extent of the disease. It also offers a convenient structure to standardize diagnostic and treatment protocols. Some differences in classification may occur, depending on the primary cancer site.

T FOR PRIMARY TUMOR

The anatomic extent of the primary tumor depends on its size, depth of invasion, and surface spread. Tumor stages progress from TX to T4 as follows:

TX — primary tumor can't be assessed
T 0 — no evidence of primary tumor
Tis — carcinoma in situ
T1, T2, T3, T4 — increasing size or local extent (or both) of primary tumor

N FOR NODAL INVOLVEMENT

Nodal involvement reflects the tumor's spread to the lymph nodes as follows:

NX — regional lymph nodes can't be assessed
N 0 — no evidence of regional lymph node metastasis
N1, N2, N3 — increasing involvement of regional lymph nodes

M FOR DISTANT METASTASIS

Metastasis denotes the extent (or spread) of disease. Levels range from MX to M4 as follows:

MX — distant metastasis can't be assessed
M 0 — no evidence of distant metastasis
M1 — single, solitary distant metastasis
M2, M3, M4 — multiple foci or multiple organ metastasis

The most commonly used method for staging is the TNM staging system, which evaluates Tumor size, Nodal involvement, and Metastatic progress. This classification system provides an accurate tumor description that is adjustable as the disease progresses. TNM staging enables reliable comparison of treatments and survival rates among large population groups. (See *Understanding TNM staging*.)

TREATMENT

Cancer treatments include surgery, radiation therapy, chemotherapy, immunotherapy (also called biotherapy), and hormone therapy. Each may be used alone or in combination (called multimodal therapy), depending on the type, stage, localization, and responsiveness of the tumor and on limitations imposed by the patient's clinical status. Cancer treatment has four goals:

▶ cure, to eradicate the cancer and promote long-term patient survival
▶ control, to arrest tumor growth
▶ palliation, to alleviate symptoms when the disease is beyond control
▶ prophylaxis, to provide treatment when no tumor is detectable, but the patient is known to be at high risk of tumor development or recurrence.

Cancer treatment is further categorized by type according to when it is used, as follows:
▶ primary, to eradicate the disease
▶ adjuvant, in addition to primary, to eliminate microscopic disease and promote cure or improve the patient's response
▶ salvage, to manage recurrent disease.

As with any treatment regimen, complications may arise. Indeed, many complications of cancer are related to the adverse effects of treatment, such as fluid and electrolyte imbalances secondary to anorexia, vomiting, or diarrhea; bone mar-

COMMON CANCER EMERGENCIES

The following chart shows what can cause certain oncologic emergencies and the underlying malignancy.

EMERGENCIES AND CAUSE	ASSOCIATED MALIGNANCY
Cardiac tamponade	
▸ Fluid accumulation around pericardial space or pericardial thickening secondary to radiation therapy	▸ Breast cancer ▸ Leukemia ▸ Lymphoma ▸ Melanoma
Hypercalcemia	
▸ Increased bone resorption due to bone destruction or tumor-related elevation of parathyroid hormone, osteoclast-activating factor, or prostaglandin levels	▸ Breast cancer ▸ Lung cancer ▸ Multiple myeloma ▸ Renal cancer
Disseminated intravascular coagulation	
▸ Widespread clotting in arterioles and capillaries and simultaneous hemorrhage	▸ Hematologic malignancies ▸ Mucin-producing adenocarcinomas
Malignant peritoneal effusion	
▸ Seeding of tumor into the peritoneum, excess intraperitoneal fluid production or release of humoral factors by the tumor	▸ Ovarian cancer
Malignant pleural effusion	
▸ Implantation of cancer cells on pleural surface, tumor obstruction of lymphatic channels or pulmonary veins, shed of necrotic tumor cells into the pleural space or thoracic duct perforation	▸ Breast cancer ▸ GI tract cancer ▸ Leukemia ▸ Lung cancer (most common) ▸ Lymphoma ▸ Testicular cancer
Spinal cord compression	
▸ Encroachment on spinal cord or cauda equina due to metastasis or vertebral collapse and displacement of bony elements	▸ Cancer of lung, breast, kidney, gastrointestinal tract, prostate, or cervix ▸ Melanoma
Superior vena cava syndrome	
▸ Impaired venous return secondary to occlusion of vena cava	▸ Breast cancer ▸ Lymphoma ▸ Lung cancer

COMMON CANCER EMERGENCIES *(continued)*

EMERGENCIES AND CAUSE	ASSOCIATED MALIGNANCY
Syndrome of inappropriate antidiuretic hormone (SIADH)	
▸ Ectopic production by tumor; abnormal stimulation of hypothalamus-pituitary axis; mimicking or enhanced effects on kidney; may be induced by chemotherapy	▸ Bladder cancer ▸ GI tract cancer ▸ Hodgkin's disease ▸ Prostate cancer ▸ Sarcomas ▸ Small-cell lung cancer
Tumor lysis syndrome	
▸ Rapid cell destruction and turnover caused by chemotherapy or rapid tumor growth	▸ Leukemias ▸ Lymphomas

row suppression, including anemia, leukopenia, thrombocytopenia, and neutropenia; and infection. Hypercalcemia is the most common metabolic abnormality experienced by cancer patients. Pain, which accompanies all progressing cancers, can reach intolerable levels.

Certain complications are life threatening and require prompt intervention. These oncologic emergencies may result from the effects of the tumor or its byproducts, secondary involvement of other organs due to disease spread, or adverse effects of treatment. (See *Common cancer emergencies.*)

Surgery

Surgery, once the mainstay of cancer treatment, is now typically combined with other therapies. It may be performed to diagnose the disease, initiate primary treatment, or achieve palliation, and is occasionally done for prophylaxis. The surgical biopsy procedure is diagnostic surgery, and continuing surgery then removes the bulk of the tumor. When used as a primary treatment method, surgery is an attempt to remove the entire tumor (or as much as possible, by a procedure called debulking), along with surrounding tissues, including lymph nodes.

A common method of surgical removal of a small tumor mass is called wide and local excision. The tumor mass is removed along with a small or moderate amount of easily accessible surrounding tissue that is normal. A radical or modified radical excision removes the primary tumor along with lymph nodes, nearby involved structures, and surrounding structures that may be at high risk of disease spread. Often a radical excision results in some degree of disfigurement and altered functioning. Today's less-radical surgical procedures such as a lumpectomy instead of mastectomy are more acceptable to patients. The health care professional and the patient should discuss the type of surgery. Ultimately the choice belongs to the patient.

Palliative surgery is used to relieve complications, such as pain, ulceration, obstruction, hemorrhage, or pressure. Examples include a cordotomy to relieve intractable pain and bowel resection or ostomy to remove a bowel obstruction. Additionally, surgery may be performed to remove hormone-producing glands and thereby limit the growth of a hormone-sensitive tumor.

Prophylactic surgery may be done if a patient has personal or familial risk factors for a particular type of cancer. Here, nonvital tissues or organs with a high potential for developing cancer are removed. One example is prophylactic mastectomy. Much controversy exists over this type of surgery because of the possible long-term physiologic and psychological effects, although potential benefits may significantly outweigh the downside.

Radiation therapy

Radiation therapy involves the use of high-energy radiation to treat cancer. Used alone or in conjunction with other therapies, it aims to destroy dividing cancer cells while damaging normal cells as little as possible. Two types of radiation are used to treat cancer: ionizing radiation and particle bean radiation. Both target the cellular DNA. Ionizing radiation deposits energy that damages the genetic material inside the cancer cells. Normal cells are also affected but can recover. Particle bean radiation uses a special machine and fast-moving particles to treat the cancer. The particles can cause more cell damage than ionizing radiation does.

The guiding principle for radiation therapy is that the dose administered be large enough to eradicate the tumor, but small enough to minimize the adverse effects to the surrounding normal tissue. How well the treatment meets this goal is known as the *therapeutic ratio*.

Radiation interacts with oxygen in the nucleus to break strands of DNA and interacts with water in body fluids (including intracellular fluid) to form free radicals, which also damage the DNA. If this damage isn't repaired, the cells die, either immediately or when they attempt to divide. Radiation also may render tumor cells unable to enter the cell cycle. Thus, cells most vulnerable to radiation therapy are those that undergo frequent cell division, for example, cells of the bone marrow, lymph, GI epithelium, and gonads.

Therapeutic radiation may be delivered by external beam radiation or by intracavitary or interstitial implants. Use of implants requires that anyone who comes in contact with the patient while the internal radiation implants are in place must wear radiation protection.

Normal and malignant cells respond to radiation differently, depending on blood supply, oxygen saturation, previous irradiation, and immune status. Generally, normal cells recover from radiation faster than malignant cells. Success of treatment and damage to normal tissue also vary with the intensity of the radiation. Although a large, single dose of radiation has greater cellular effects than fractions of the same amount delivered sequentially, a protracted schedule allows time for normal tissue to recover between doses.

Adverse effects

Radiation may be used palliatively to relieve pain, obstruction, malignant effusions, cough, dyspnea, ulcerations, and hemorrhage. It can also promote healing of pathologic fractures after surgical stabilization and delay metastasis.

Combining radiation and surgery can minimize the need for radical surgery, prolong survival, and preserve anatomic function. For example, preoperative doses of radiation shrink a large tumor to operable size while preventing further spread of the disease during surgery. After the wound heals, postoperative doses prevent residual cancer cells from multiplying or metastasizing.

Radiation therapy has both local and systemic adverse effects, because it affects both normal and malignant cells. (See *Radiation's adverse effects*.) Systemic adverse effects, such as weakness, fatigue, anorexia, nausea, vomiting, and anemia may respond to antiemetics, steroids, frequent small meals, fluid maintenance, and rest. They are seldom severe enough to require discontinuing radiation but they may mandate a dosage adjustment.

Patients receiving radiation therapy must have frequent blood counts, particularly of WBCs and platelets if the target

RADIATION'S ADVERSE EFFECTS

Radiation therapy can cause local adverse effects depending on the area irradiated. The chart below highlights some of the more commonly seen local effects and the measures to manage them.

AREA IRRADIATED	ADVERSE EFFECT	MANAGEMENT
Head	▶ Alopecia	▶ Gentle combing and grooming ▶ Soft head cover
	▶ Mucositis	▶ Cool carbonated drinks ▶ Ice ▶ Soft, nonirritating diet ▶ Viscous lidocaine mouthwash
	▶ Monilia	▶ Medicated mouthwash
	▶ Dental caries	▶ Gingival care ▶ Prophylactic fluoride to teeth
Chest	▶ Lung tissue irritation	▶ Avoidance of persons with upper respiratory infections ▶ Humidifier, if necessary ▶ Smoking cessation ▶ Steroid therapy
	▶ Pericarditis ▶ Myocarditis	▶ Antiarrhythmic drugs
	▶ Esophagitis	▶ Analgesia ▶ Fluid maintenance ▶ Total parenteral nutrition
Kidneys	▶ Anemia ▶ Azotemia ▶ Edema ▶ Headache ▶ Hypertensive nephropathy ▶ Lassitude ▶ Nephritis	▶ Fluid and electrolyte maintenance ▶ Monitoring for signs of renal failure
Abdomen/pelvis	▶ Cramps ▶ Diarrhea	▶ Fluid and electrolyte maintenance ▶ Loperamide and diphenoxylate with atropine ▶ Low-residue diet

site involves areas of bone marrow production. Radiation also requires special skin care measures, such as covering the irradiated area with loose cotton clothing to protect it from light and avoiding deodorants, colognes, and other topical agents during treatment.

Chemotherapy

Chemotherapy includes a wide range of antineoplastic drugs, which may induce regression of a tumor and its metastasis. It's particularly useful in controlling residual disease and as an adjunct to surgery or radiation therapy. It can induce long remissions and sometimes effect cure, especially in patients with childhood leukemia, Hodgkin's disease, choriocarcinoma, or testicular cancer. As a palliative treatment, chemotherapy aims to improve the patient's quality of life by temporarily relieving pain and other symptoms.

Every dose of a chemotherapeutic agent destroys only a percentage of tumor cells. Therefore, regression of the tumor requires repeated doses of drugs. The goal is to eradicate enough of the tumor so that the immune system can destroy the remaining malignant cells.

Tumor cells that are in the active phase of cell division (called the growth fraction) are the most sensitive to chemotherapeutic agents. Nondividing cells are the least sensitive and thus are the most potentially dangerous. They must be destroyed to eradicate a malignancy completely. Therefore, repeated cycles of chemotherapy are used to destroy nondividing cells as they enter the cell cycle to begin active proliferation.

Depending on the type of cancer, one or more different categories of chemotherapeutic agents may be used. The most commonly used types of chemotherapeutic agents are:

▶ Alkylating agents and nitrosoureas inhibit cell growth and division by reacting with DNA at any phase of the cell cycle. They prevent cell replication by breaking and cross-linking DNA.

▶ Antimetabolites prevent cell growth by competing with metabolites in the production of nucleic acid, substituting themselves for purines and pyrimidines which are essential for DNA and ribonucleic acid (RNA) synthesis. They exert their effect during the S phase of the cell cycle.

▶ Antitumor antibiotics block cell growth by binding with DNA and interfering with DNA-dependent RNA synthesis. Acting in any phase of the cell cycle, they bind to DNA and generate toxic oxygen free radicals that break one or both strands of DNA.

▶ Plant (Vinca) alkaloids prevent cellular reproduction by disrupting mitosis. Acting primarily in the M phase of the cell cycle, they interfere with the formation of the mitotic spindle by binding to microtubular proteins.

▶ Hormones and hormone antagonists impair cell growth by one or both of two mechanisms. The may alter the cell environment, thereby affecting the permeability of the cell membrane, or they may inhibit the growth of hormone-susceptible tumors by changing their chemical environment. These agents include adrenocorticosteroids, androgens, gonadotropin inhibitors, and aromatase inhibitors.

Other chemotherapeutic agents include podophyllotoxins and taxanes which, like plant alkaloids, interfere with formation of the mitotic spindle, and miscellaneous agents, such as hydroxyurea and L-asparaginase, which seem to be cell-cycle specific agents but whose mode of action is unclear. (See *Chemotherapy's action in the cell cycle.*)

A combination of drugs from different categories may be used to maximize the tumor cell kill. Combination therapy typically includes drugs with different toxicities and also synergistic actions. Use of combination therapy also helps prevent the development of drug-resistant mechanisms by the tumor cells.

CHEMOTHERAPY'S ACTION IN THE CELL CYCLE

Some chemotherapeutic agents are cell-cycle specific, impairing cellular growth by causing changes in the cell during specific phases of the cell cycle. Other agents are cell-cycle nonspecific, affecting the cell at any phase during the cell cycle. The illustration below shows where the cell-cycle-specific agents work to disrupt cancer cell growth.

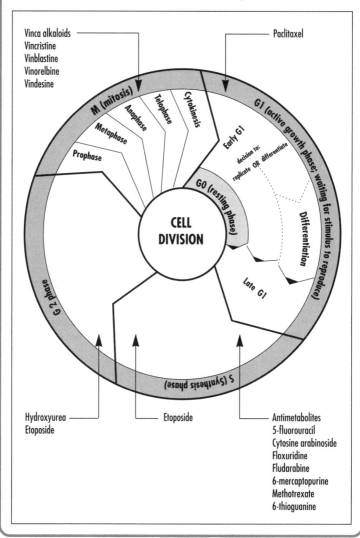

Vinca alkaloids
Vincristine
Vinblastine
Vinorelbine
Vindesine

Paclitaxel

M (mitosis)
Cytokinesis
Telophase
Anaphase
Metaphase
Prophase

G1 (active growth phase; waiting for stimulus to reproduce)
Early G1
decision to: replicate OR differentiate
Differentiation
Late G1

G0 (resting phase)

CELL DIVISION

G 2 phase

S (Synthesis phase)

Hydroxyurea
Etoposide

Etoposide

Antimetabolites
5-fluorouracil
Cytosine arabinoside
Floxuridine
Fludarabine
6-mercaptopurine
Methotrexate
6-thioguanine

Adverse effects

Chemotherapy causes numerous adverse effects that reflect the drugs' mechanism of action. Antineoplastic agents can cause transient changes in normal tissues, especially those with proliferating cells. For example, antineoplastic agents typically cause anemia, leukopenia, and thrombocytopenia because they suppress bone marrow function; vomiting because they irritate the gastrointestinal epithelial cells; and alopecia and dermatitis because they destroy hair follicles and skin cells. Many antineoplastic agents are given intravenously, and they can cause venous sclerosis and pain when administered. If extravasated, they may cause deep cutaneous necrosis, requiring debridement and skin grafting. To minimize the risk of extravasation, most drugs with the potential for direct tissue injury are now given through a central venous catheter.

The pharmacologic action of a given drug determines whether it is administered orally, subcutaneously, intramuscularly, intravenously, intracavitarily, intrathecally, or by arterial infusion. Dosages are calculated according to the patient's body surface area, with adjustments for general condition and degree of myelosuppression.

Many patients approach chemotherapy apprehensively. They need to be allowed to express their concerns and be provided with simple and truthful information. Explanations about what to expect, including possible adverse effects, can help minimize the fear and anxiety.

Hormonal therapy

Hormonal therapy is based on studies showing that certain hormones can inhibit the growth of certain cancers. For example, the luteinizing hormone-releasing hormone analogue, leuprolide, is used to treat prostate cancer. With long-term use, this hormone inhibits testosterone release and tumor growth. Tamoxifen, an antiestrogen hormonal agent, blocks estrogen receptors in breast tumor cells that require estrogen to thrive. Adrenocortical steroids are effective in treating leukemias and lymphomas because they suppress lymphocytes.

Immunotherapy

Immunotherapy, now commonly called biotherapy, is treatment with agents known as biologic response modifiers. Biologic agents are usually combined with chemotherapy or radiation therapy, and are most effective in the early stages of cancer. Many types of immunotherapy are still experimental; their availability may be restricted and their adverse effects generally unpredictable. However, several approaches appear promising.

Some biologic agents such as interferons have a direct antitumor effect, whereas other biologic agents activate or influence the immune system. Biologic agents are useful in treating hairy cell leukemia, renal cell carcinoma, and melanoma. The most widely used are the interferons and interleukin-2.

Bone marrow transplantation (BMT) is the therapy of choice for curing many malignancies that otherwise require high-dose chemotherapy or radiation, such as leukemia, lymphoma, and some breast cancers. BMT restores hematologic and immunologic function.

Two useful immunologic techniques that are not used to treat cancer directly are the use of monoclonal antibodies and colony stimulating factors (CSF). Monoclonal antibodies labeled with radioisotopes may be injected into the body to help localize tumors. The antibodies attach to surface antigens on tumor cells, where they are identified by radiologic techniques. One day soon these antibodies may be linked with toxins to destroy specific cancer cells without disturbing healthy cells. CSFs, which are naturally produced by many cells in the immune system, may be used to support the patient who has low blood counts caused by chemotherapy.

REVIEWING COMMON CANCERS

The chart below highlights the important signs and symptoms and diagnostic test results for some of the most common cancers.

TYPE AND FINDINGS	DIAGNOSTIC TEST RESULTS

Acute leukemia

▶ Sudden onset of high fever resulting from bone marrow invasion and cellular proliferation within bone marrow
▶ Thrombocytopenia and abnormal bleeding secondary to bone marrow suppression
▶ Weakness, lassitude related to anemia from bone marrow invasion
▶ Pallor and weakness related to anemia
▶ Chills and recurrent infections related to proliferation of immature nonfunctioning white blood cells
▶ Bone pain from leukemic infiltration of bone
▶ Neurologic manifestations including headache, papilledema, facial palsy, blurred vision and meningeal irritation secondary to leukemic infiltration or cerebral bleeding
▶ Liver, spleen, and lymph node enlargement related to leukemic cell infiltration

▶ Bone marrow aspiration reveals proliferation of immature white blood cells (WBCs).
▶ Complete blood count (CBC) shows thrombocytopenia and neutropenia.
▶ Differential WBC count reveals cell type.
▶ Lumbar puncture reveals leukemic infiltration to cerebrospinal fluid (CSF).

Basal cell carcinoma

▶ Noduloulcerative lesions usually on face (forehead, eyelid regions, and nasolabial folds) appearing as small, smooth, pinkish and translucent papules with telangietactic vessels crossing surface; occasionally pigmented; depressed centers with firm elevated borders with enlargement resulting from basal cell proliferation in the deepest layer of epidermis with local extension
▶ Superficial basal cell epitheliomas, commonly on chest and back, appearing as oval or irregularly shaped, lightly pigmented scaly plaques with sharply defined, threadlike elevated borders resembling psoriasis or eczema resulting from basal cell proliferation
▶ Sclerosing basal cell epitheliomas occurring on the head and neck and appearing as waxy, sclerotic yellow to white plaques without distinct borders resulting from basal cell proliferation

▶ Incisional or excisional biopsy and cytology confirm the cell type.

(continued)

REVIEWING COMMON CANCERS *(continued)*

TYPE AND FINDINGS	DIAGNOSTIC TEST RESULTS

Bladder cancer

Early stages:
▶ Commonly asymptomatic

Later:
▶ Gross painless intermittent hematuria secondary to tumor invasion
▶ Suprapubic pain after voiding from pressure exerted by the tumor or obstruction
▶ Bladder irritability and frequency related to tumor compression and invasion

▶ Cystoscopy and biopsy confirm cell type.
▶ Urinalysis reveals hematuria and malignant cytology.
▶ Excretory urography identifies large early stage tumor or infiltrating tumor.
▶ Retrograde cystography reveals changes in structure and bladder wall integrity.
▶ Pelvic arteriography confirms tumor invasion into bladder wall.
▶ Computed tomography (CT) scan reveals thickened bladder wall and enlarged retroperitoneal lymph nodes.
▶ Ultrasonography detects metastasis beyond bladder; differentiates presence of tumor from cyst.

Bone cancer

▶ Possibly asymptomatic
▶ Bone pain especially at night from tumor disruption of normal structural integrity and pressure on surrounding tissues
▶ Tender, swollen, possibly palpable mass resulting from tumor growth
▶ Pathologic fractures secondary to tumor invasion and destruction of bone causing weakening
▶ Hypercalcemia from ectopic parathyroid hormone production by the tumor or increased bone resorption
▶ Limited mobility (late in the disease) from continued tumor growth and disruption of bone strength

▶ Incisional or aspiration biopsy confirms cell type.
▶ Bone X-rays, bone scan, and CT scan reveal tumor size.
▶ Bone scan reveals increased uptake of isotope in area of tumor.
▶ Serum alkaline phosphatase is elevated.

REVIEWING COMMON CANCERS *(continued)*

TYPE AND FINDINGS	DIAGNOSTIC TEST RESULTS

Breast cancer

▶ Hard stony mass in the breast related to cellular growth
▶ Change in symmetry of breast secondary to growth of tumor on one side
▶ Skin thickening or dimpling, scaly skin around nipple or changes in nipple, edema or ulceration related to tumor cell infiltration to surrounding tissues
▶ Warm, hot, pink area from inflammation and infiltration of surrounding tissues
▶ Unusual discharge or drainage indicating tumor invasion and infiltration into the ductal system
▶ Pain related to advancement of tumor and subsequent pressure
▶ Hypercalcemia or pathologic fractures secondary to metastasis to bone

▶ Breast examination reveals lump or mass in breast.
▶ Mammography reveals presence of mass and location.
▶ Needle or surgical biopsy confirms the cell type.
▶ Ultrasonography reveals solid tumor differentiating it from a fluid filled cyst.
▶ Bone scan and CT scan reveal metastasis.
▶ Elevated alkaline phosphatase levels, liver biopsy, and liver function studies reveal liver metastasis.
▶ Hormonal receptor assay identifies tumor as hormonal dependent.

Cervical cancer

▶ Abnormal vaginal bleeding with persistent vaginal discharge and postcoital pain and bleeding related to cellular invasion and erosion of the cervical epithelium
▶ Pelvic pain secondary to pressure on surrounding tissues and nerves from cellular proliferation
▶ Vaginal leakage of urine and feces from fistulas due to erosion and necrosis of cervix
▶ Anorexia, weight loss, and anemia related to the hypermetabolic activity of cellular proliferation and increased tumor growth needs

▶ Pap smear reveals malignant cellular changes.
▶ Colposcopy identifies the presence and extent of early lesions.
▶ Biopsy confirms cell type.
▶ CT scan, nuclear imaging scan, and lymphangiography identify metastasis.

Chronic lymphocytic leukemia

▶ Slow onset of fatigue related to anemia
▶ Splenomegaly secondary to increase numbers of lysed red blood cells being filtered
▶ Hepatomegaly and lymph node enlargement from infiltration by leukemic cells
▶ Bleeding tendencies secondary to thrombocytopenia
▶ Infections related to deficient humoral immunity

▶ CBC count reveals:
– numerous abnormal lymphocytes with mild but persistently elevated WBC count
– granulocytopenia common but WBC count increasing as disease progresses
– hemoglobin levels below 11 g/dL
– neutropenia
– lymphocytosis
– thrombocytopenia.
▶ Serum globulin levels are decreased.
▶ Bone marrow aspiration and biopsy show lymphocytic invasion.

(continued)

REVIEWING COMMON CANCERS *(continued)*

TYPE AND FINDINGS	DIAGNOSTIC TEST RESULTS

Colorectal cancer

Tumor on right colon:
▶ Black tarry stools secondary to tumor erosion and necrosis of the intestinal lining
▶ Anemia secondary to increased tumor growth needs and bleeding resulting from necrosis and ulceration of mucosa
▶ Abdominal aching, pressure, or cramps secondary to pressure from tumor
▶ Weakness, fatigue, anorexia, weight loss secondary to increased tumor growth needs
▶ Vomiting as disease progresses related to possible obstruction.

Tumor on left colon:
▶ Intestinal obstruction including abdominal distention, pain, vomiting, cramps, and rectal pressure related to increasing tumor size and ulceration of mucosa
▶ Dark red or bright red blood in stools secondary to erosion and ulceration of mucosa

▶ Digital rectal examination (DRE) reveals mass.
▶ Stools for guaiac test positive.
▶ Proctosigmoidoscopy or sigmoidoscopy reveals tumor mass.
▶ Colonoscopy visualizes tumor location up to the ileocecal valve.
▶ CT scan reveals areas of possible metastasis.
▶ Barium X-ray shows location and size of lesions not manually or visually detectable.
▶ Carcinoembryonic antigen (tumor marker) may be elevated.

Esophageal cancer

▶ No early symptoms
▶ Dysphagia secondary to tumor interfering with passageway
▶ Weight loss resulting from dysphasia, tumor growth and increasing obstruction, and anorexia related to tumor growth needs
▶ Ulceration and subsequent hemorrhage from erosive effects (fungating and infiltrative) of the tumor
▶ Fistula formation and possible aspiration secondary to continued erosive tumor effects

▶ Esophageal X-ray with barium swallow and motility studies reveals structural and filling defects and reduced peristalsis.
▶ Endoscopic examination with punch and brush biopsies confirms cancer cell type.

REVIEWING COMMON CANCERS *(continued)*

TYPE AND FINDINGS	DIAGNOSTIC TEST RESULTS

Hodgkin's disease

▶ Painless swelling in one of lymph nodes (usually the cervical region) with a history of upper respiratory infection
▶ Persistent fever, night sweats, fatigue, weight loss, and malaise related to hypermetabolic state of cellular proliferation and defective immune function
▶ Back and neck pain with hyperreflexia related to epidural infiltration
▶ Extremity pain, nerve irritation, or absence of pulse due to obstruction of pressure of tumor in surrounding lymph nodes
▶ Pericardial friction rub, pericardial effusion, and neck vein engorgement secondary to direct invasion from mediastinal lymph nodes
▶ Enlargement of retroperitoneal nodes, spleen, and liver related to progression of disease and cellular infiltration

▶ Lymph node biopsy confirms presence of Reed-Sternberg cells, nodular fibrosis, and necrosis.
▶ Bone marrow, liver, mediastinal, lymph node, and spleen biopsies reveal histologic presence of cells.
▶ Chest X-ray, abdominal CT scan, lung scan, bone scan, and lymphangiography detect lymph and organ involvement.
▶ Hematologic tests show:
– mild to severe normocytic anemia
– normochromic anemia
– elevated, normal, or reduced WBC count
– differential with any combination of neutrophilia, lymphocytopenia, monocytosis, and eosinophilia.
▶ Elevated serum alkaline phosphatase indicates bone or liver involvement.

Laryngeal cancer

▶ Hoarseness persisting longer than 3 weeks related to encroachment on the true vocal cord
▶ Lump in the throat or pain or burning when drinking citrus juice or hot liquids related to tumor growth
▶ Dysphagia secondary to increasing pressure and obstruction with tumor growth
▶ Dyspnea and cough related to progressive tumor growth and metastasis
▶ Enlargement of cervical lymph nodes and pain radiating to ear related to invasion of lymphatic and subsequent pressure

▶ Laryngoscopy shows presence of tumor.
▶ Xeroradiography, biopsy, laryngeal tomography, CT scan, or laryngography identifies borders of the lesion.
▶ Chest X-ray reveals metastasis.

Liver cancer

▶ Mass in right upper quadrant with a tender nodular liver on palpation secondary to tumor cell growth
▶ Severe pain in epigastrium or right upper quadrant related to tumor size and increased pressure on surrounding tissue
▶ Weight loss, weakness, anorexia related to increased tumor growth needs
▶ Dependent edema secondary to tumor invasion and obstruction of portal veins

▶ Liver biopsy confirms cell type.
▶ Serum aspartate aminotransferase, alanine aminotransferase, alkaline phosphatase, lactic dehydrogenase, and bilirubin are elevated indicating abnormal liver function.
▶ Alpha-fetoprotein levels are elevated.
▶ Chest X-ray reveals possible metastasis.
▶ Liver scan may show filling defects.
▶ Serum electrolyte studies reveal hypernatremia and hypercalcemia; serum laboratory studies reveal hypoglycemia, leukocytosis, or hypocholesterolemia.

(continued)

REVIEWING COMMON CANCERS (continued)

TYPE AND FINDINGS	DIAGNOSTIC TEST RESULTS

Lung cancer

▶ Cough, hoarseness, wheezing, dyspnea, hemoptysis, and chest pain related to local infiltration of pulmonary membranes and vasculature
▶ Fever, weight loss, weakness, anorexia related to increased tumor growth needs from hypermetabolic state of cellular proliferation
▶ Bone and joint pain from cartilage erosion due to abnormal production of growth hormone
▶ Cushing's syndrome related to abnormal production of adrenocorticopic hormone
▶ Hypercalcemia related to abnormal production of parathyroid hormone or bone metastasis
▶ Hemoptysis, atelectasis, pneumonitis, and dyspnea from bronchial obstruction related to increasing growth
▶ Shoulder pain and unilateral paralysis of diaphragm due to phrenic nerve involvement
▶ Dysphagia related to esophageal compression
▶ Venous distention, facial, neck and chest edema secondary to obstruction of vena cava
▶ Piercing chest pain, increasing dyspnea, severe arm pain secondary to invasion of the chest wall

▶ Chest X-ray shows an advanced lesion, including size and location.
▶ Sputum cytology reveals possible cell type.
▶ CT scan of the chest delineates tumor size and relationship to surrounding structures.
▶ Bronchoscopy locates tumor; washings reveal malignant cell type.
▶ Needle lung biopsy confirms cell type.
▶ Mediastinal and supraclavicular node biopsies reveal possible metastasis.
▶ Thoracentesis shows malignant cells in pleural fluid.
▶ Bone scan, bone marrow biopsy, and CT scan of brain and abdomen reveal metastasis.

Malignant brain tumors

▶ Headache, dizziness, vertigo, nausea and vomiting, and papilledema secondary to increased intracranial pressure from tumor invasion and compression of surrounding tissues
▶ Cranial nerve dysfunction secondary to tumor invasion or compression of cranial nerves
▶ Focal deficits including motor deficits (weakness, paralysis, or gait disorders), sensory disturbances (anesthesia, paresthesia, or disturbances of vision or hearing) secondary to tumor invasion or compression of motor or sensory control areas of the brain
▶ Disturbances of higher function including defects in cognition, learning and memory

Local:
▶ Dementia, personality or behavioral changes, gait disturbances, seizures, language disorders.
▶ Sensory loss, hemianopia, cranial nerve dysfunction, ataxia, pupillary abnormalities, nystagmus, hemiparesis, and autonomic dysfunction depending on location of tumor.

▶ Stereotactic tissue biopsy confirms cell type.
▶ Neurologic assessment reveals manifestations of lesion affecting specific lobe.
▶ Skull X-ray, CT scan, MRI, and cerebral angiography identify location of mass.
▶ Brain scan reveals area of increased uptake in location of tumor.
▶ Lumbar puncture shows increased pressure and protein levels, decreased glucose levels, and, occasionally, tumor cells in CSF.

REVIEWING COMMON CANCERS *(continued)*

TYPE AND FINDINGS	DIAGNOSTIC TEST RESULTS

Melanoma

▶ Enlargement of skin lesion or nevus accompanied by changes in color, inflammation or soreness, itching, ulceration, bleeding, or textural changes secondary to malignant transformation of melanocytes in the basal layer of the epidermis or within the aggregated melanocytes of an existing nevus

Superficial spreading melanoma:
▶ Red, white, and blue color over a brown or black background with an irregular, notched margin typically on areas of chronic irritation
Nodular melanoma:
▶ Polypoidal nodule with uniformly dark discoloration appearing as a blackberry but possibly flesh colored with flecks of pigment around base
Lentigo maligna melanoma:

▶ Large flat freckle of tan, brown, black, whitish, or slate color with irregularly scattered black nodules on surface

▶ Skin biopsy with histologic examination confirms cell type and tumor thickness.

▶ Chest X-ray, CT scan of chest and abdomen, or CT of brain reveals metastasis.

▶ Bone scan reveals bone metastasis.

Multiple myeloma

▶ Severe constant back pain that increases with exercise secondary to invasion of bone
▶ Arthritic symptoms including achiness, joint swelling, and tenderness possibly from vertebral compression
▶ Pathologic fractures resulting from invasion of bone causing loss of structural integrity and strength
▶ Azotemia secondary to tumor proliferation to the kidney and pyelonephritis due to subsequent tubular damage from large amounts of Bence Jones protein, hypercalcemia, and hyperuricemia
▶ Anemia, bleeding, and infections secondary to tumor effects on bone marrow cell production
▶ Thoracic deformities and increasing vertebral complaints secondary to extension of tumor and continued vertebral compression
▶ Loss of 5 inches or more of body height due to vertebral collapse

▶ CBC shows moderate to severe anemia; differential may show 40% to 50% lymphocytes but seldom more than 3% plasma cells.
▶ Differential smear reveals Rouleaux formation from elevated erythrocyte sedimentation rate.
▶ Urine studies reveal Bence Jones protein and hypercalciuria.
▶ Bone marrow aspiration detects myelomatous cells (abnormal number of immature plasma cells).
▶ Serum electrophoresis shows elevated globulin spike that is electrophoretically and immunologically abnormal.
▶ Bone X-rays early reveal diffuse osteoporosis; in later stages, they show multiple sharply circumscribed osteolytic lesions, particularly in the skull, pelvis, and spine.

(continued)

REVIEWING COMMON CANCERS *(continued)*

TYPE AND FINDINGS	DIAGNOSTIC TEST RESULTS

Non-Hodgkin's lymphoma

▶ Lymph node swelling from cellular proliferation
▶ Dyspnea and coughing related to lymphocytic infiltration of oropharynx
▶ Enlarged tonsils and adenoids from mechanical obstruction by the tumor
▶ Abdominal pain and constipation secondary to mechanical obstruction of surrounding tissues

▶ Lymph node biopsy reveals cell type.
▶ Biopsy of tonsils, bone marrow, liver, bowel, or skin reveals malignant cells.
▶ CBC may show anemia.
▶ Uric acid level may be elevated or normal.
▶ Serum calcium levels are elevated if bone lesions are present.
▶ Serum protein levels are normal.
▶ Bone and chest X-rays, lymphangiography, liver and spleen scans, abdominal CT scan, and excretory urography show evidence of metastasis.

Ovarian cancer

▶ Vague abdominal discomfort, dyspepsia and other mild gastrointestinal complaints from increasing size of tumor exerting pressure on nearby tissues
▶ Urinary frequency, constipation from obstruction resulting from increased tumor size
▶ Pain from tumor rupture, torsion, or infection
▶ Feminizing or masculinizing effects secondary to cellular type
▶ Ascites related to invasion and infiltration of the peritoneum
▶ Pleural effusions related to pulmonary metastasis

▶ Pap smear may be normal.
▶ Abdominal ultrasound, CT, or X-ray delineates tumor presence and size.
▶ CBC may show anemia.
▶ Excretory urography reveals abnormal renal function and urinary tract abnormalities or obstruction.
▶ Chest X-ray reveals pleural effusion with distant metastasis.
▶ Barium enema shows obstruction and size of tumor.
▶ Lymphangiography reveals lymph node involvement.
▶ Mammography is normal to rule out breast cancer as the primary site.
▶ Liver functions studies are abnormal with ascites.
▶ Paracentesis fluid aspiration reveals malignant cells.
▶ Tumor markers such as carcinoembryonic antigen and human chorionic gonadotropin are positive.

Pancreatic cancer

▶ Jaundice with clay-colored stools and dark urine secondary to obstruction of bile flow from tumor in head of pancreas
▶ Recurrent thrombophlebitis from tumor cytokines acting as platelet aggregating factors
▶ Nausea and vomiting secondary to duodenal obstruction

▶ Laparotomy with biopsy confirms cell type.
▶ Ultrasound identifies location of mass.
▶ Angiography reveals vascular supply of the tumor.
▶ Endoscopic retrograde cholangiopancreatography visualizes tumor area.
▶ MRI identifies tumor location and size.

REVIEWING COMMON CANCERS (continued)

TYPE AND FINDINGS	DIAGNOSTIC TEST RESULTS

Pancreatic cancer (continued)

▶ Weight loss, anorexia, and malaise, secondary to effects of increased tumor growth needs
▶ Abdominal or back pain secondary to tumor pressure
▶ Blood in the stools from ulceration of gastrointestinal tract or ampulla of Vater

▶ Serum laboratory tests reveal increased serum bilirubin, serum amylase and serum lipase.
▶ Prothrombin time (PT) is prolonged.
▶ Elevations of aspartate aminotransferase and alanine aminotransferase indicate necrosis of liver cells.
▶ Marked elevation of alkaline phosphatase indicates biliary obstruction.
▶ Plasma insulin immunoassay shows measurable serum insulin in the presence of islet cell tumors.
▶ Hemoglobin and hematocrit may show mild anemia.
▶ Fasting blood glucose may reveal hypo- or hyperglycemia.

Prostate cancer

▶ Symptoms appearing only in late stages
▶ Difficulty initiating a urinary stream, dribbling, urine retention secondary to obstruction of urinary tract from tumor growth
▶ Hematuria (rare) from infiltration of bladder

▶ DRE reveals a small hard nodule.
▶ Prostatic surface antigen is elevated.
▶ Serum acid phosphatase levels are elevated.
▶ MRI, CT scan, and excretory urography identify tumor mass.
▶ Elevated alkaline phosphatase levels and positive bone scan indicate bone metastasis.

Renal cancer

▶ Pain resulting from tumor pressure and invasion
▶ Hematuria secondary to tumor spreading to renal pelvis
▶ Possible fever from hemorrhage or necrosis
▶ Hypertension from compression of renal artery with renal parenchymal ischemia and renin excess
▶ Polycythemia secondary to erythropoietin excess
▶ Hypercalcemia from ectopic parathyroid hormone production by the tumor or bone metastasis
▶ Urinary retention secondary to obstruction of urinary flow
▶ Pulmonary embolism secondary to renal venous obstruction

▶ CT scan, I.V. and retrograde pyelography, ultrasound, cystoscopy and nephrotomography, and renal angiography identify presence of tumor and help differentiate it from a cyst.
▶ Liver function tests show increased levels of alkaline phosphatase, bilirubin, alanine aminotransferase, aspartate aminotransferase, and prolonged PT.
▶ Urinalysis reveals gross or microscopic hematuria.
▶ CBC shows anemia, polycythemia, and increased erythrocyte sedimentation rate.
▶ Serum calcium levels are elevated.

Squamous cell carcinoma

▶ Lesions on skin of the face, ears, dorsa of hands and forearms from cell proliferation in sun-damaged areas
▶ Induration and inflammation as cell changes from nonmalignant to malignant cell
▶ Ulceration from continued cell proliferation

▶ Excisional biopsy confirms cell type.

(continued)

REVIEWING COMMON CANCERS (continued)

TYPE AND FINDINGS	DIAGNOSTIC TEST RESULTS
Stomach cancer	
▶ Chronic dyspepsia and epigastric discomfort related to tumor growth in gastric cells and destruction of mucosal barrier ▶ Weight loss, anorexia, feelings of fullness after eating, anemia, and fatigue secondary to increased tumor growth needs ▶ Blood in stools from erosion of gastric mucosa by tumor	▶ Barium X-ray with fluoroscopy shows tumor or filling defects in outline of stomach, loss of flexibility and distensibility, and abnormal mucosa with or without ulceration. ▶ Gastroscopy with fiberoptic endoscopy visualizes gastric mucosa including presence of gastric lesions for biopsy. ▶ CT scans, X-rays, liver and bone scans, and liver biopsy reveal metastasis.
Testicular cancer	
▶ Firm painless smooth testicular mass and occasional complaints of heaviness secondary to tumor growth ▶ Gynecomastia and nipple tenderness related to tumor production of chorionic gonadotropin or estrogen ▶ Urinary complaints related to ureteral obstruction ▶ Cough, hemoptysis, and shortness of breath from invasion of the pulmonary system	▶ Transillumination of testicles reveals tumor that does not transilluminate. ▶ Surgical excision and biopsy reveals cell type. ▶ Excretory urography detects ureteral deviation from para-aortic node involvement. ▶ Serum alpha-fetoprotein and beta human chorionic gonadotropin levels as tumor markers are elevated. ▶ Lymphangiography, ultrasound, and abdominal CT scan reveal mass and possible metastasis.
Thyroid cancer	
▶ Painless nodule or hard nodule in an enlarged thyroid gland or palpable lymph nodes with thyroid enlargement reflecting tumor growth ▶ Hoarseness, dysphagia, and dyspnea from increased tumor growth and pressure on surrounding structures ▶ Hyperthyroidism from excess thyroid hormone production from tumor ▶ Hypothyroidism secondary to tumor destruction of the gland	▶ Thyroid scan reveals hypofunctional nodes or cold spots. ▶ Needle biopsy confirms cell type. ▶ CT scan, ultrasound, and chest X-ray reveal medullary cancer.
Uterine (endometrial) cancer	
▶ Uterine enlargement secondary to tumor growth ▶ Persistent and unusual premenopausal bleeding or postmenopausal bleeding from erosive effects of tumor growth ▶ Pain and weight loss related to progressive infiltration and invasion of tumor cells and continued cellular proliferation	▶ Endometrial, cervical, and endocervical biopsies are positive for malignant cells, revealing cell type. ▶ Dilatation and curettage identifies malignancy in patients whose biopsies were negative. ▶ Schiller's test reveals cervix resistant to staining. ▶ Chest X-ray and CT scan reveal metastasis. ▶ Barium enema identifies possible bladder or rectal involvement.

3

The twentieth century encompassed astonishing advances in treating and preventing infection such as potent antibiotics, complex immunizations, and modern sanitation, yet infection remains the most common cause of human disease. Even in countries with advanced medical care, infectious disease remains a major cause of death. The very young and the very old are especially susceptible.

WHAT IS INFECTION?

Infection is the invasion and multiplication of microorganisms in or on body tissue that produce signs and symptoms as well as an immune response. Such reproduction injures the host by causing cell damage from microorganism-produced toxins or from intracellular multiplication, or by competing with host metabolism. Infectious diseases range from relatively mild illnesses to debilitating and lethal conditions: from the common cold through chronic hepatitis to acquired immunodeficiency syndrome (AIDS). The severity of the infection varies with the pathogenicity and number of the invading microorganisms and the strength of host defenses.

For infection to be transmitted, the following must be present: causative agent, infectious reservoir with a portal of exit, mode of transmission, a portal of entry into the host, and a susceptible host. (See *Chain of infection*, pages 54 and 55.)

RISK FACTORS

A healthy person can usually ward off infections with the body's own built-in defense mechanisms:
▶ intact skin
▶ normal flora that inhabit the skin and various organs
▶ lysozymes (enzymes that can kill microorganisms or microbes) secreted by eyes, nasal passages, glands, stomach, and genitourinary organs
▶ defensive structures such as the cilia that sweep foreign matter from the airways (See *How microbes interact with the body,* page 56.)
▶ a healthy immune system.

However, if an imbalance develops, the potential for infection increases. Risk factors for the development of infection include weakened defense mechanisms, environmental and developmental factors, and pathogen characteristics.

Weakened defense mechanisms

The body has many defense mechanisms for resisting entry and multiplication of microbes. However, a weakened immune system makes it easier for these pathogens to invade the body and launch an infectious disease. This weakened state is referred to immunodeficiency or immunocompromise.

Impaired function of white blood cells (WBCs), as well as low levels of T and B cells, characterizes immunodeficiencies. An immunodeficiency may be congenital (caused by a genetic defect and present at

CHAIN OF INFECTION

An infection can occur only if the six components depicted here are present. Removing one link in the chain prevents infection.

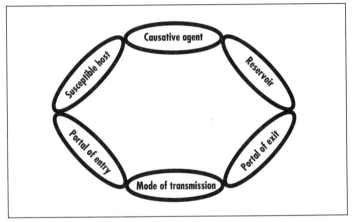

CAUSATIVE AGENT

A causative agent for infection is any microbe capable of producing disease.

RESERVOIR

The reservoir is the environment or object in or on which a microbe can survive and, in some cases, multiply. Inanimate objects, human beings, and other animals can all serve as reservoirs, providing the essential requirements for a microbe to survive at specific stages in its life cycle.

PORTAL OF EXIT

The portal of exit is the path by which an infectious agent leaves its reservoir. Usually, this portal is the site where the organism grows. Common portals of exit associated with human reservoirs include the respiratory, genitourinary, and gastrointestinal (GI) tracts; the skin and mucous membranes; and the placenta (in transplacental disease transmission from other to fetus). Blood, sputum, emesis, stool, urine, wound drainage, and genital secretions also serve as portals of exit. The portal of exit varies from one infectious agent to the next.

birth) or acquired (developed after birth). Acquired immunodeficiency may result from infection, malnutrition, chronic stress, or pregnancy. Diabetes, renal failure, and cirrhosis can suppress the immune response, as can drugs such as corticosteroids and chemotherapy.

Regardless of cause, the result of immunodeficiency is the same. The body's ability to recognize and fight pathogens is impaired. People who are immunodeficient are more susceptible to all infections, are more acutely ill when they become infected, and require a much longer time to heal.

Environmental factors

Other conditions that may weaken a person's immune defenses include poor hygiene, malnutrition, inadequate physical

MODE OF TRANSMISSION

The mode of transmission is the means by which the infectious agent passes from the portal of exit in the reservoir to the susceptible host. Infections can be transmitted through one of four modes: contact, airborne, vehicle, and vector-borne. Some organisms use more than one transmission mode to get from the reservoir to a new host. As with portals of exit, the transmission mode varies with the specific microbe.

Contact transmission is subdivided into direct contact, indirect contact, and droplet spread (contact with droplets that enter the environment).

Direct contact refers to person-to-person spread of organisms through actual physical contact.

Indirect contact occurs when a susceptible person comes in contact with a contaminated object.

Droplet transmission results from contact with contaminated respiratory secretions. It differs from airborne transmission in that the droplets don't remain suspended in the air but settle to surfaces.

Airborne transmission occurs when fine microbial particles containing pathogens remain suspended in the air for a prolonged period, and then are spread widely by air currents and inhaled.

A vehicle is a substance that maintains the life of the microbe until it's ingested or inoculated into the susceptible host. The vehicle is not harmful in itself but may harbor pathogenic microbes and thus serve as an agent of disease transmission. Examples of vehicles are water, blood, serum, plasma, medications, food, and feces.

Vector-borne transmission occurs when an intermediate carrier, or vector, such as a flea or a mosquito, transfers a microbe to another living organism. Vector-borne transmission is of most concern in tropical areas, where insects commonly transmit disease.

PORTAL OF ENTRY

Portal of entry refers to the path by which an infectious agent invades a susceptible host. Usually, this path is the same as the portal of exit.

SUSCEPTIBLE HOST

A susceptible host is also required for the transmission of infection to occur. The human body has many defense mechanisms for resisting the entry and multiplication of pathogens. When these mechanisms function normally, infection does not occur. However, in a weakened host, an infectious agent is more likely to invade the body and launch an infectious disease.

barriers, emotional and physical stressors, chronic diseases, medical and surgical treatments, and inadequate immunization.

Good hygiene promotes normal host defenses; poor hygiene increases the risk of infection. Unclean skin harbors microbes and offers an environment for them to colonize, and untended skin is more likely to allow invasion. Frequent washing removes surface microbes and maintains an intact barrier to infection, but it may damage the skin. To maintain skin integrity, lubricants and emollients may be used to prevent cracks and breaks.

The body needs a balanced diet to provide the nutrients that an effective immune system needs. Protein malnutrition inhibits the production of antibodies, without which the body can't mount an effective attack against microbe invasion. Along

HOW MICROBES INTERACT WITH THE BODY

Microbes may interact with their host in various ways.

DOUBLE BENEFIT

Some of the microorganisms of the normal human flora interact with the body in ways that mutually benefits both parties. *Escherichia coli* organisms, part of the normal intestinal flora, obtain nutrients from the human host; in return, they secrete vitamin K, which the human body needs for blood clotting.

SINGLE BENEFIT

Other microbes of the normal flora have a commensal interaction with the human body — an interaction that benefits one party (in this case, the microbes) without affecting the other.

PARASITIC INTERACTION

Some pathogenic microbes such as helminths (worms) are parasites. This means that they harm the host while they benefit from their interaction with the host.

with a balanced diet, the body needs adequate vitamins and minerals to use ingested nutrients.

Dust can facilitate transportation of pathogens. For example, dustborne spores of the fungus *aspergillus* transmit the infection. If the inhaled spores become established in the lungs, they're notoriously difficult to expel. Fortunately, persons with intact immune systems can usually resist infection with *aspergillus,* which is usually dangerous only in the presence of severe immunosuppression.

Developmental factors

The very young and very old are at higher risk for infection. The immune system doesn't fully develop until about age 6 months. An infant exposed to an infectious agent usually develops an infection. The most common type of infection in toddlers affects the respiratory tract. When young children put toys and other objects in their mouths, they increase their exposure to a variety of pathogens.

Exposure to communicable diseases continues throughout childhood, as children progress from daycare facilities to schools. Skin diseases, such as impetigo, and lice infestation commonly pass from one child to the next at this age. Accidents are common in childhood as well, and broken or abraded skin opens the way for bacterial invasion. Lack of immunization also contributes to incidence of childhood diseases.

Advancing age, on the other hand, is associated with a declining immune system, partly as a result of decreasing thymus function. Chronic diseases, such as diabetes and atherosclerosis, can weaken defenses by impairing blood flow and nutrient delivery to body systems.

PATHOGEN CHARACTERISTICS

A microbe must be present in sufficient quantities to cause a disease in a healthy human. The number needed to cause a disease varies from one microbe to the next and from host to host and may be affected by the mode of transmission. The severity of an infection depends on several factors, including the microbe's pathogenicity, that is, the likelihood that it will cause pathogenic changes or disease. Factors that affect pathogenicity include the microbe's specificity, invasiveness, quantity, virulence, toxigenicity, adhesiveness, antigenicity, and viability.

▶ *Specificity* is the range of hosts to which a microbe is attracted. Some microbes may be attracted to a wide range of both humans and animals, while others select only human or only animal hosts.

▶ *Invasiveness* (sometimes called *infectivity*) is the ability of a microbe to invade tissues. Some microbes can enter through intact skin; others can enter only if the skin or mucous membrane is broken. Some microbes produce enzymes that enhance their invasiveness.

▶ *Quantity* refers to the number of microbes that succeed in invading and reproducing in the body.

▶ *Virulence* is the severity of the disease a pathogen can produce. Virulence can vary depending on the host defenses; any infection can be life-threatening in an immunodeficient patient. Infection with a pathogen known to be particularly virulent requires early diagnosis and treatment.

▶ *Toxigenicity* is related to virulence. It describes a pathogen's potential to damage host tissues by producing and releasing toxins.

▶ *Adhesiveness* is the ability of the pathogen to attach to host tissue. Some pathogens secrete a sticky substance that helps them adhere to tissue while protecting them from the host's defense mechanisms.

▶ *Antigenicity* is the degree to which a pathogen can induce a specific immune response. Microbes that invade and localize in tissue initially stimulate a cellular response; those that disseminate quickly throughout the host's body generate an antibody response.

▶ *Viability* is the ability of a pathogen to survive outside its host. Most microbes can't live and multiply outside a reservoir, as discussed under the topic "Chain of infection."

STAGES OF INFECTION

Development of an infection usually proceeds through four stages. The first stage, *incubation*, may be almost instantaneous or last for years. During this time, the pathogen is replicating and the infected person is contagious. The *prodromal stage* (stage two) follows incubation, and the still-contagious host makes vague complaints of feeling unwell. In stage three, *acute illness*, microbes are actively de-stroying host cells and affecting specific host systems. The patient recognizes which area of the body is affected and voices complaints that are more specific. Finally, the *convalescent stage* (stage four) begins when the body's defense mechanisms have confined the microbes and healing of damaged tissue is progressing.

INFECTION-CAUSING MICROBES

Microorganisms that are responsible for infectious diseases include bacteria, viruses, fungi, parasites, mycoplasma, rickettsia, and chlamydiae.

Bacteria

Bacteria are simple one-celled microorganisms with a cell wall that protects them from many of the defense mechanisms of the human body. Although they lack a nucleus, bacteria possess all the other mechanisms they need to survive and rapidly reproduce.

Bacteria can be classified according to shape — spherical cocci, rod-shaped bacilli, and spiral-shaped spirilla. (See *Comparing bacterial shapes,* page 58.) Bacteria can also be classified according to their need for oxygen (aerobic or anaerobic), their mobility (motile or nonmotile), and their tendency to form protective capsules (encapsulated or nonencapsulated) or spores (sporulating or nonsporulating).

Bacteria damage body tissues by interfering with essential cell function or by releasing exotoxins or endotoxins, which cause cell damage. (See *How bacteria damage tissue,* page 59.) During bacterial growth, the cells release exotoxins, enzymes that damage the host cell, altering its function or killing it. Enterotoxins are a specific type of exotoxin secreted by bacteria that infect the GI tract; they affect the vomiting center of the brain and cause gastroenteritis. Exotoxins also can cause diffuse reactions in the host, such as inflammation, bleeding, clotting, and fever. Endotoxins are contained in the cell walls

COMPARING BACTERIAL SHAPES

Bacteria exist in three basic shapes: rods (bacilli), spheres (cocci), and spirals (spirilla).

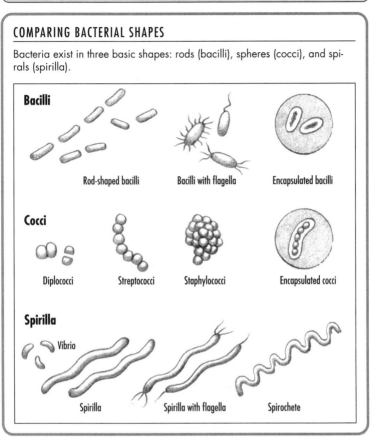

Bacilli

Rod-shaped bacilli

Bacilli with flagella

Encapsulated bacilli

Cocci

Diplococci

Streptococci

Staphylococci

Encapsulated cocci

Spirilla

Vibrio

Spirilla

Spirilla with flagella

Spirochete

of gram-negative bacteria, and they are released during lysis of the bacteria.

Examples of bacterial infection include staphylococcal wound infection, cholera, and streptococcal pneumonia. (See *Gram-positive and gram-negative bacteria,* pages 60 and 61.)

Viruses

Viruses are subcellular organisms made up only of a ribonucleic acid (RNA) nucleus or a deoxyribonucleic acid (DNA) nucleus covered with proteins. They're the smallest known organisms, so tiny that only an electron microscope can make them visible. (See *How viruses size up,* page 62.) Independent of the host cells,

viruses can't replicate. Rather, they invade a host cell and stimulate it to participate in forming additional virus particles. Some viruses destroy surrounding tissue and release toxins. (See *Viral infection of a host cell,* page 63.) Viruses lack the genes necessary for energy production. They depend on the ribosomes and nutrients of infected host cells for protein production. The estimated 400 viruses that infect humans are classified according to their size, shape, and means of transmission (respiratory, fecal, oral, sexual).

Most viruses enter the body through the respiratory, GI, and genital tracts. A few, such as human immunodeficiency virus (HIV), are transmitted through blood, bro-

HOW BACTERIA DAMAGE TISSUE

Bacteria and other infectious organisms constantly infect the human body. Some, such as the intestinal bacteria that produce vitamins, are beneficial. Others are harmful, causing illnesses ranging from the common cold to life-threatening septic shock.

To infect a host, bacteria must first enter it. They do this either by adhering to the mucosal surface and directly invading the host cell or by attaching to epithelial cells and producing toxins, which invade host cells. The result is a disruption of normal cell function or cell death (see illustration below). For example, the diphtheria toxin damages heart muscle by inhibiting protein synthesis. In addition, as some organisms multiply, they extend into deeper tissue and eventually gain access to the bloodstream.

Some toxins cause blood to clot in small blood vessels. The tissues supplied by these vessels may be deprived of blood and damaged (see illustration below).

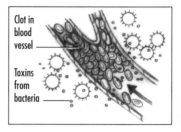

Clot in blood vessel

Toxins from bacteria

Other toxins can damage the cell walls of small blood vessels, causing leakage. This fluid loss results in decreased blood pressure, which, in turn, impairs the heart's ability to pump enough blood to vital organs (see illustration below).

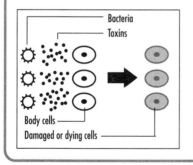

Bacteria
Toxins
Body cells
Damaged or dying cells

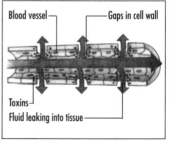

Blood vessel — Gaps in cell wall
Toxins
Fluid leaking into tissue

ken skin, and mucous membranes. Viruses can produce a wide variety of illnesses, including the common cold, herpes simplex, herpes zoster, chicken pox, infectious mononucleosis, hepatitis B and C, and rubella. Signs and symptoms depend on the status of the host cell, the specific virus, and whether the intracellular environment provides good living conditions for the virus.

Retroviruses are a unique type of virus that carry their genetic code in RNA rather than the more common carrier DNA. These RNA viruses contain the enzyme reverse transcriptase, which changes viral RNA into DNA. The host cell then incorporates the alien DNA into its own genetic material. The most notorious retrovirus today is HIV.

Fungi

Fungi have rigid walls and nuclei that are enveloped by nuclear membranes. They occur as yeast (single-cell, oval-shaped

GRAM-POSITIVE AND GRAM-NEGATIVE BACTERIA

This flowchart highlights the different types of gram positive and gram-negative bacteria.

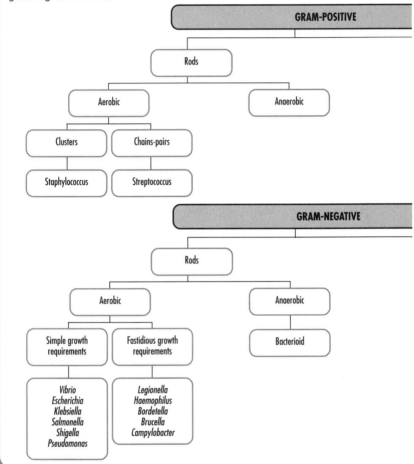

organisms) or molds (organisms with hyphae, or branching filaments). Depending on the environment, some fungi may occur in both forms. Found almost everywhere on earth, fungi live on organic matter, in water and soil, on animals and plants, and on a wide variety of unlikely materials. They can live both inside and outside their host. Superficial fungal infections cause athlete's foot and vaginal infections. *Candida albicans* is part of the body's normal flora but under certain circumstances, it can cause yeast infections of virtually any part of the body but especially the mouth, skin, vagina, and GI tract. For example, antibiotic treatment or a change in the pH of the susceptible tissues (because of a disease, such as diabetes, or use of certain drugs, such as oral contraceptives) can wipe out the normal

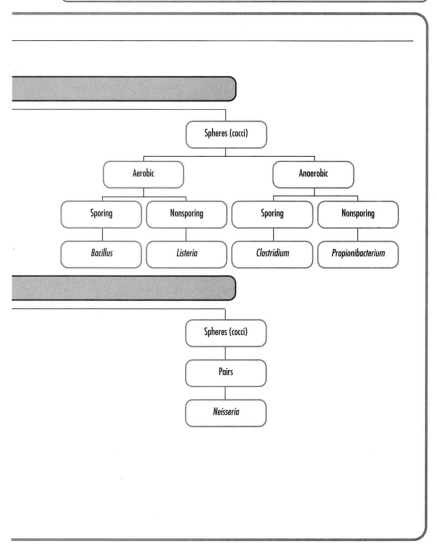

```
                          ┌─────────────────┐
                          │  Spheres (cocci) │
                          └─────────────────┘
                   ┌──────────────┴──────────────┐
            ┌─────────────┐                ┌─────────────┐
            │   Aerobic   │                │  Anaerobic  │
            └─────────────┘                └─────────────┘
          ┌───────┴───────┐              ┌───────┴───────┐
    ┌──────────┐   ┌──────────────┐  ┌──────────┐  ┌──────────────┐
    │  Sporing │   │  Nonsporing  │  │  Sporing │  │  Nonsporing  │
    └──────────┘   └──────────────┘  └──────────┘  └──────────────┘
    ┌──────────┐   ┌──────────────┐  ┌──────────┐  ┌──────────────┐
    │ Bacillus │   │   Listeria   │  │Clostridium│ │Propionibacterium│
    └──────────┘   └──────────────┘  └──────────┘  └──────────────┘
```

```
                          ┌─────────────────┐
                          │  Spheres (cocci) │
                          └─────────────────┘
                          ┌─────────────────┐
                          │      Pairs      │
                          └─────────────────┘
                          ┌─────────────────┐
                          │    Neisseria    │
                          └─────────────────┘
```

bacteria that normally keep the yeast population in check.

Parasites

Parasites are unicellular or multicellular organisms that live on or within another organism and obtain nourishment from the host. They take only the nutrients they need and usually don't kill their hosts. Examples of parasites that can produce an infection if they cause cellular damage to the host include helminths, such as pinworms and tapeworms, and arthropods, such as mites, fleas, and ticks. Helminths can infect the human gut; arthropods commonly cause skin and systemic disease.

Mycoplasmas

Mycoplasmas are bacterialike organisms, the smallest of the cellular microbes that

HOW VIRUSES SIZE UP

Viruses vary in size, appearance, and behavior. This illustration compares the sizes of selected viruses with the size of a typical bacterium, *Escherichia coli* (*E. coli*).

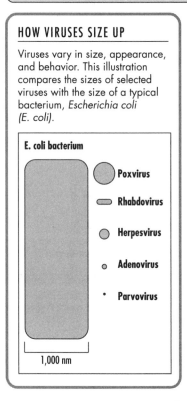

E. coli bacterium

Poxvirus

Rhabdovirus

Herpesvirus

Adenovirus

Parvovirus

1,000 nm

can live outside a host cell, although some may be parasitic. Lacking cell walls, they can assume many different shapes ranging from coccoid to filamentous. The lack of a cell wall makes them resistant to penicillin and other antibiotics that work by inhibiting cell wall synthesis. Mycoplasmas can cause primary atypical pneumonia and many secondary infections.

Rickettsia

Rickettsia are small, gram-negative, bacteria-like organisms that can cause life-threatening illness. They may be coccoid, rod-shaped, or irregularly shaped. Because they're live viruses, rickettsia require a host cell for replication. They have no cell wall, and their cell membranes are leaky; thus, they must live inside another, better protected cell. Rickettsia are transmitted by the bites of arthropod carriers, such as

lice, fleas, and ticks, and through exposure to their waste products. Rickettsial infections that occur in the United States include Rocky Mountain spotted fever, typhus, and Q fever.

Chlamydia

Chlamydiae are smaller than rickettsia and bacteria but larger than viruses. They depend on host cells for replication and are susceptible to antibiotics. Chlamydiae are transmitted by direct contact such as occurs during sexual activity. They are a common cause of infections of the urethra, bladder, fallopian tubes, and prostate gland.

PATHOPHYSIOLOGIC CHANGES

Clinical expressions of infectious disease vary, depending on the pathogen involved and the organ system affected. Most of the signs and symptoms result from host responses, which may be similar or very different from host to host. During the prodromal stage, a person will complain of some common, nonspecific signs and symptoms, such as fever, muscle aches, headache, and lethargy. In the acute stage, signs and symptoms that are more specific provide evidence of the microbe's target. However, some illnesses remain asymptomatic and are discovered only by laboratory tests.

Inflammation

The inflammatory response is a major reactive defense mechanism in the battle against infective agents. Inflammation may be the result of tissue injury, infection, or allergic reaction. Acute inflammation has two stages: vascular and cellular. In the vascular stage, arterioles at or near the site of the injury briefly constrict and then dilate, causing an increase in fluid pressure in the capillaries. The consequent movement of plasma into the interstitial space causes edema. At the same time, inflammatory cells release histamine and bradykinin, which further increase capillary per-

VIRAL INFECTION OF A HOST CELL

The virion (A) attaches to receptors on the host-cell membrane and releases enzymes (called absorption) (B) that weaken the membrane and enable the virion to penetrate the cell. The virion removes the protein coat that protects its genetic material (C), replicates (D), and matures, and then escapes from the cell by budding from the plasma membrane (E). The infection then can spread to other host cells.

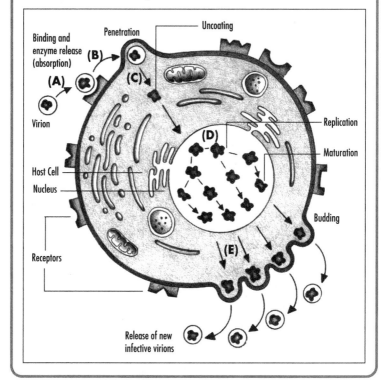

meability. Red blood cells and fluid flow into the interstitial space, contributing to edema. The extra fluid arriving in the inflamed area dilutes microbial toxins.

During the cellular stage of inflammation, white blood cells (WBCs) and platelets move toward the damaged cells. Phagocytosis of the dead cells and microorganisms begins. Platelets control any excess bleeding in the area, and mast cells arriving at the site release heparin to main-tain blood flow to the area. (See *Blocking inflammation,* page 64.)

SIGNS AND SYMPTOMS

Acute inflammation is the body's immediate response to cell injury or cell death. The cardinal signs of inflammation include redness, heat, pain, edema, and decreased function of a body part.

▶ *Redness (rubor)* results when arterioles dilate and circulation to the site increas-

BLOCKING INFLAMMATION

Several substances act to control inflammation. The flowchart below shows the progression of inflammation and the points ✳ at which drugs can reduce inflammation and pain.

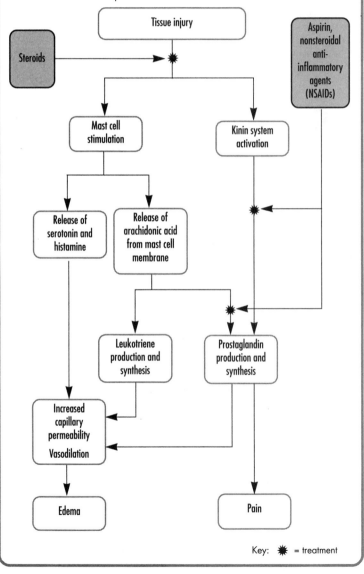

Key: ✳ = treatment

es. Filling of previously empty or partially distended capillaries causes a localized blush.

▶ *Heat (calor)* in the area results from local vasodilation, fluid leakage into the interstitial spaces, and increased blood flow to the area.

▶ *Pain (dolor)* occurs when pain receptors are stimulated by swollen tissue, local pH changes, and chemicals excreted during the inflammatory process.

▶ *Edema (tumor)* is caused by local vasodilation, leakage of fluid into interstitial spaces, and the blockage of lymphatic drainage to help wall off the inflammation.

▶ *Loss of function (functio laesa)* occurs primarily as a result of edema and pain at the site.

Fever

Fever occurs with the introduction of an infectious agent. An elevated temperature helps fight an infection because many microorganisms are unable to survive in a hot environment. When the body temperature rises too high, body cells can be damaged, particularly those of the nervous system.

Diaphoresis is the body's method to cool the body and return the temperature to "normal" for that individual. Artificial methods to reduce a slight fever can actually impair the body's defenses against infection.

Leukocytosis

The body responds to the introduction of pathogens by increasing the number and types of circulating WBCs. This process is called *leukocytosis*. In the acute or early stage, the neutrophil count increases. Bone marrow begins to release immature leukocytes, because existing neutrophils cannot meet the body's demand for defensive cells. The immature neutrophils (called "bands" in the differential WBC count) can't serve any defensive purpose.

As the acute phase comes under control and the damage is isolated, the next stage of the inflammatory process takes place. Neutrophils, monocytes, and macrophages begin the process of phagocytosis of dead tissue and bacteria. Neutrophils and monocytes identify the foreign antigen and attach to it. Then they engulf, kill, and degrade the microorganism that carries antigen on its surface. Macrophages, a mature type of monocyte, arrive at the site later and remain in the area of inflammation longer than the other cells. Besides phagocytosis, macrophages play several other key roles at the site, such as preparing the area for healing and processing antigens for a cellular immune response. An elevated monocyte count is common during resolution of any injury and in chronic infections.

Chronic inflammation

An inflammation reaction lasting longer than 2 weeks is referred to as chronic inflammation. It may or may not follow an acute process. A poorly healed wound or an unresolved infection can lead to chronic inflammation. The body may encapsulate a pathogen that it can't destroy in order to isolate it. An example of such a pathogen is mycobacteria, the cause of tuberculosis; encapsulated mycobacteria appear in X-rays as identifiable spots in the lungs.

DIAGNOSIS

Accurate assessment helps identify infectious diseases, appropriate treatment, and avoidable complications. It begins with obtaining the patient's complete medical history, performing a thorough physical examination, and performing or ordering appropriate diagnostic tests. Tests that can help identify and gauge the extent of infection include laboratory studies, radiographic tests, and scans.

Most often, the first test is a WBC count and a differential. Any elevation in overall number of WBCs is a positive result.

The differential count is the relative number of each of five types of white blood cells—neutrophils, eosinophils, basophils, lymphocytes, and monocytes. It's obtained by classifying 100 or more white cells in a stained film of peripheral blood. Multiplying the percentage value of each type by the total WBC gives the absolute number of each type of white cell. This test recognizes only that something has stimulated an immune response. Bacterial infection usually causes an elevation in the counts; viruses may cause no change or a decrease in normal WBC level.

Erythrocyte sedimentation rate (ESR) may be done as general test to reveal that an inflammatory process is occurring within the body.

The next step is to obtain a stained smear from a specific body site to determine the causative agent. Stains that may be used to visualize the microorganism include:
▶ gram stain, which identifies gram-negative or gram-positive bacteria
▶ acid-fast stain, which identifies mycobacteria and nocardia
▶ silver stain, which identifies fungi, Legionella, and Pneumocystis.

Although stains provide rapid and valuable diagnostic information, they only tentatively identify a pathogen. Confirmation requires culturing. Any body substance can be cultured, but enough growth to identify the microbe may occur as quickly as 8 hours (streptococcal) or as long as several weeks, depending how rapidly the microbe replicates. Types of cultures that may be ordered are blood, urine, sputum, throat, nasal, wound, skin, stool, and cerebrospinal fluid.

A specimen obtained for culture must not be contaminated with any other substance. For example, a urine specimen must not contain any debris from the perineum or vaginal area. If obtaining a clean urine specimen isn't possible, the patient must be catheterized to make sure that only the urine is being examined. Contaminated specimens may mislead and prolong treatment.

Additional tests that may be requested include magnetic resonance imaging to locate infection sites, chest X-rays to search the lungs for respiratory changes, and gallium scans to detect abscesses.

TREATMENTS

Treatment for infections can vary widely. Vaccines may be administered to induce a primary immune response under conditions that won't cause disease. If infection does occur, treatment is tailored to the specific causative organism. Drug therapy should be used only when it is appropriate. Supportive therapy can play an important role in fighting infections.
▶ Antibiotics work in a variety of ways, depending on the class of antibiotic. Their action is either bactericidal (killing the bacteria) or bacteriostatic (preventing the bacteria from multiplying). Antibiotics may inhibit cell wall synthesis, protein synthesis, bacterial metabolism, or nucleic acid synthesis or activity, or they may increase cell-membrane permeability. (See *Antimicrobial drugs and chemicals*.)
▶ Antifungal drugs destroy the invading microbe by increasing cell membrane permeability. The antifungal binds sterols in the cell membrane, resulting in leakage of intracellular contents, such as potassium, sodium, and nutrients.
▶ Antiviral drugs stop viral replication by interfering with DNA synthesis.

The overuse of antimicrobials has created widespread resistance to some specific drugs. Some pathogens that were once well controlled by medicines are again surfacing with increased virulence. One such is tuberculosis.

Some diseases, including most viral infections, don't respond to available drugs. Supportive care is the only recourse while the host defenses repel the invader. To help the body fight an infection, the patient should:

ANTIMICROBIAL DRUGS AND CHEMICALS

The following drugs or chemicals prevent growth of microorganisms or destroy them by a specific action.

MECHANISMS OF ACTION	AGENT
Inhibition of cell-wall synthesis	Bacitracin Carbapenems Cephalosporins Cycloserine Fosfomycin Monobactams Penicillins Vancomycin
Damage to cytoplasmic membrane	Imidazoles Polyene antifungals Polymyxins
Metabolism of nucleic acid	Nitrofurans Nitroimidazoles Quinolones Rifampin
Protein synthesis	Aminoglycosides Chloramphenicol Clindamycin Macrolides Mupirocin Spectinomycin Tetracyclines
Modification of energy metabolism	Dapsone Isoniazid Sulfonamides Trimethoprim

▶ use universal precautions to avoid spreading the infection
▶ drink plenty of fluids
▶ get plenty of rest
▶ avoid people who may have other illnesses
▶ take only over-the-counter medications appropriate for his symptoms, with and only with full knowledge about dosage, actions, and possible side effects or adverse reactions

▶ follow the doctor's orders for taking any prescription drugs and be sure to finish the entire prescription
▶ not share the prescription with others.

Infections

Infection can strike any part of the body. The accompanying chart describes a variety of infections along with their signs and symptoms and appropriate diagnostic tests. (See *Reviewing common infections,* pages 68 to 83.)

REVIEWING COMMON INFECTIONS

INFECTION AND FINDINGS	DIAGNOSIS
Bacterial infections	

Conjunctivitis

▶ hyperemia of the conjunctiva
▶ discharge
▶ tearing
▶ pain
▶ photophobia (with corneal involvement)
▶ itching and burning
Note: May also result from viral infection

▶ Culture from the conjunctiva identifies the causative organism
▶ In stained smears, predominance of monocytes indicates viral infection; of neutrophils, bacterial infection; of eosinophils an allergy—related infection.

Gonorrhea

Males:
▶ urethritis, including dysuria and purulent urethral discharge, with redness and swelling at the site of infection

Females:
▶ may be asymptomatic
▶ inflammation and a greenish yellow discharge from the cervix

Males or females:
▶ pharyngitis or tonsillitis
▶ rectal burning, itching, and bloody mucopurulent discharge

Clinical features vary according to the site involved:
▶ urethra: dysuria, urinary frequency and incontinence, purulent discharge, itching, red and edematous meatus
▶ vulva: occasional itching, burning, and pain due to exudate from an adjacent infected area
▶ vagina: engorgement, redness, swelling, and profuse purulent discharge
▶ liver: right-upper quadrant pain
▶ pelvis: severe pelvic and lower abdominal pain, muscle rigidity, tenderness, and abdominal distention; nausea, vomiting, fever, and tachycardia (may develop in patients with salpingitis or pelvic inflammatory disease (PID)

▶ Culture from the site of infection, grown on a Thayer-Martin or Transgrow medium, establishes the diagnosis by isolating N. gonorrhoeae.
▶ Gram stain shows gram-negative diplococci.
▶ Complement fixation and immunofluorescent assays of serum reveal antibody titers four times the normal rate.

REVIEWING COMMON INFECTIONS *(continued)*

INFECTION AND FINDINGS	DIAGNOSIS
Bacterial infections *(continued)*	

Lyme disease

Stage 1:

▶ red macule or papule, commonly on the site of a tick bite, which grows to over 20 inches, feels hot and itchy, and resembles a bull's eye or target; after a few days, more lesions erupt and a migratory, ringlike rash appears

▶ conjunctivitis

▶ diffuse urticaria occurs

▶ lesions are replaced by small red blotches in 3 to 4 weeks

▶ malaise and fatigue

▶ intermittent headache

▶ neck stiffness

▶ fever, chills, and achiness

▶ regional lymphadenopathy

Stage 2:

▶ neurologic abnormalities-fluctuating meningoencephalitis with peripheral and cranial neuropathy

▶ facial palsy

▶ cardiac abnormalities: brief, fluctuating atrioventricular heart block, left ventricular dysfunction, cardiomegaly

Stage 3:

▶ arthritis with marked swelling

▶ neuropsychiatric symptoms such as psychotic behavior, memory loss, dementia, and depression.

▶ encephalopathic symptoms such as headache, confusion, and difficulty concentrating

▶ ophthalmic manifestations such as iritis, keratitis, renal vasculitis, optic neuritis

▶ Blood tests, including antibody titers, enzyme-linked immunosorbent assay, and Western blot assay, may be used to identify *Borrelia burgdorferi.*

▶ Serology reveals mild anemia and elevated erythrocyte sedimentation rate (ESR), white blood cell (WBC) count, serum immunoglobulin M (IgM) level, and aspartate aminotransferase.

▶ Cerebrospinal fluid (CSF) analysis reveals presence of antibodies to *B. burgdorferi* if the disease has affected the central nervous system.

(continued)

REVIEWING COMMON INFECTIONS *(continued)*

INFECTION AND FINDINGS	DIAGNOSIS
Bacterial infections *(continued)*	

Listeriosis

▶ commonly causes asymptomatic carrier state
▶ malaise
▶ chills
▶ fever
▶ back pain

Fetuses:
▶ abortion
▶ premature delivery or stillbirth
▶ organ abscesses

Neonates:
▶ meningitis, resulting in tense fontanels
▶ irritability
▶ lethargy
▶ seizures
▶ coma

▶ *L. monocytogenes* is identified by its diagnostic tumbling motility on a wet mount of the culture.
▶ Positive culture of blood, spinal fluid, drainage from cervical or vaginal lesions, or lochia from a mother with an infected infant.

Meningitis

▶ fever
▶ chills
▶ headache
▶ nuchal rigidity
▶ vomiting
▶ photophobia
▶ lethargy
▶ coma
▶ positive Brudzinski's and Kernig's signs
▶ exaggerated and symmetrical deep tendon reflexes and opisthotonos
▶ wide pulse pressure
▶ bradycardia
▶ occasional rash
Note: May also result from viral protozoal, or fungal infection.

▶ Lumbar puncture isolates *S. pneumoniae* from CSF and shows increased CSF cell count and protein level and decreased CSF glucose level.
▶ Blood culture isolates *S. pneumoniae*.

REVIEWING COMMON INFECTIONS *(continued)*

INFECTION AND FINDINGS	DIAGNOSIS
Bacterial infections *(continued)*	

Otitis media

- ear pain
- ear drainage
- hearing loss
- fever
- lethargy
- irritability
- vertigo
- nystagmus
- tinnitus

- Otoscopy reveals obscured or distorted bony landmarks of the tympanic membrane.
- Pneumatoscopy can show decreased tympanic membrane mobility.
- Culture of the ear drainage identifies the causative organism.

Peritonitis

- sudden, severe, and diffuse abdominal pain that tends to intensify and localize in the area of the underlying disorder
- weakness and pallor
- excessive sweating
- cold skin
- decreased intestinal motility and paralytic ileus
- intestinal obstruction causes nausea, vomiting, and abdominal rigidity
- hypotension
- tachycardia
- fever
- abdominal distention

- Abdominal X-ray shows edematous and gaseous distention of the small and large bowel or in the case of visceral organ perforation, air lying under the diaphragm.
- Chest X-ray may show elevation of the diaphragm.
- Blood studies show leukocytosis.
- Paracentesis reveals bacteria, exudate, blood, pus, or urine.
- Laparotomy may be necessary to identify the underlying cause.

Pneumonia

- high temperature
- cough with purulent, yellow or bloody sputum
- dyspnea
- crackles, and decreased breath sounds
- pleuritic pain
- chills
- malaise
- tachypnea

Note: May also result from fungal or protozoal infection

- Chest X-rays confirm the diagnosis by disclosing infiltrates.
- Sputum specimen, Gram stain and culture, and sensitivity tests help differentiate the type of infection and the drugs that are effective.
- WBC count indicates leukocytosis in bacterial pneumonia, and a normal or low count in viral or mycoplasmal pneumonia.
- Blood cultures reflect bacteremia and are used to determine the causative organism.
- Arterial blood gas (ABG) levels vary, depending on severity of pneumonia and underlying lung state,
- Bronchoscopy or transtracheal aspiration allows the collection of material for culture.
- Pulse oximetry may show a reduced oxygen saturation level.

(continued)

REVIEWING COMMON INFECTIONS (continued)

INFECTION AND FINDINGS	DIAGNOSIS
Bacterial infections (continued)	

Salmonellosis

▶ fever
▶ abdominal pain, severe diarrhea with enterocolitis

Typhoidal infection:
▶ headache
▶ increasing fever
▶ constipation

▶ Blood cultures isolate the organism in typhoid fever, paratyphoid fever, and bacteremia.
▶ Stool cultures isolate the organism in typhoid fever, paratyphoid fever, and enterocolitis.
▶ Cultures of urine, bone marrow, pus, and vomitus may show the presence of *Salmonella*.

Shigellosis

Children:
▶ high fever
▶ diarrhea with tenesmus
▶ nausea, vomiting, and abdominal pain and distention
▶ irritability
▶ drowsiness
▶ stool may contain pus, mucus or blood
▶ dehydration and weight loss

Adults:
▶ sporadic, intense abdominal pain
▶ rectal irritability
▶ tenesmus
▶ headache and prostration
▶ stools may contain pus, mucus, and blood

▶ Microscopic examination of a fresh stool may reveal mucus, red blood cells, and polymorphonuclear leukocytes.
▶ Severe infection increases hemagglutinating antibodies.
▶ Sigmoidoscopy or proctoscopy may reveal typical superficial ulcerations.

REVIEWING COMMON INFECTIONS *(continued)*

INFECTION AND FINDINGS	DIAGNOSIS

Bacterial infections *(continued)*

Syphilis

Primary syphilis:
▶ chancres on the anus, fingers, lips, tongue, nipples, tonsils, or eyelids
▶ regional lymphadenopathy

Secondary syphilis:
▶ symmetrical mucocutaneous lesions
▶ general lymphadenopathy
▶ rash may be macular, papular, pustular, or nodular
▶ headache
▶ malaise
▶ anorexia, weight loss, nausea, and vomiting
▶ sore throat and slight fever
▶ alopecia
▶ brittle and pitted nails

Late syphilis:
▶ benign — gumma lesion found on any bone or organ
▶ gastric pain, tenderness, enlarged spleen
▶ anemia
▶ involvement of the upper respiratory tract; perforation of the nasal septum or palate; destruction or bones and organs
▶ fibrosis of elastic tissue of the aorta
▶ aortic insufficiency
▶ aortic aneurysm
▶ meningitis
▶ paresis
▶ personality changes
▶ arm and leg weakness

▶ Dark field examination of a lesion identifies *T. pallidum.*
▶ Fluorescent treponemal antibody absorption tests identifies antigens of *T. pallidum* in tissue, ocular fluid, CSF, tracheobronchial secretions, and exudates from lesions.

(continued)

REVIEWING COMMON INFECTIONS *(continued)*

INFECTION AND FINDINGS	DIAGNOSIS
Bacterial infections *(continued)*	

Tetanus

Localized:
▶ spasm and increased muscle tone near the wound

Systemic:
▶ marked muscle hypertonicity
▶ hyperactive deep tendon reflexes
▶ tachycardia
▶ profuse sweating
▶ low-grade fever
▶ painful, involuntary muscle contractions

▶ Diagnosis may rest on clinical features, a history of trauma, and no previous tetanus immunization.
▶ Blood cultures are positive for the organism in only a third of patients.
▶ Tetanus antibody test may return as negative.
▶ Cerebrospinal fluid pressure may rise above normal.

Toxic shock syndrome (TSS)

▶ intense myalgias
▶ fever over 104°F (40°C)
▶ vomiting and diarrhea
▶ headache
▶ decreased level of consciousness
▶ rigor
▶ conjunctival hyperemia,
▶ vaginal hyperemia and discharge
▶ deep red rash (especially on the palms and soles), desquamates (develops later)
▶ severe hypotension

▶ Isolation of *S. aureus* from vaginal discharge or lesions.
▶ Negative results on blood tests for Rocky Mountain spotted fever, leptospirosis, and measles help rule out these disorders.
▶ Diagnosis is based on clinical findings.

Tuberculosis

▶ fever and night sweats
▶ productive cough lasting longer than 3 weeks
▶ hemoptysis
▶ malaise
▶ adenopathy
▶ weight loss
▶ pleuritic chest pain
▶ symptoms of airway obstruction from lymph node involvement

▶ Chest X-ray shows nodular lesions, patchy infiltrates (mainly in upper lobes), cavity formation, scar tissue, and calcium deposits.
▶ Tuberculin skin test reveals infection at some point, but doesn't indicate active disease.
▶ Stains and cultures of sputum, CSF, urine, drainage from abscesses, or pleural fluid show heat-sensitive, nonmotile, aerobic, acid-fast bacilli.
▶ CT or MRI scans allow the evaluation of lung damage and may confirm a difficult diagnosis.
▶ Bronchoscopy shows inflammation and altered lung tissue. It also may be performed to obtain sputum if the patient can't produce an adequate sputum specimen.

REVIEWING COMMON INFECTIONS *(continued)*

INFECTION AND FINDINGS	DIAGNOSIS

Bacterial infections *(continued)*

Urinary tract infections

Cystitis:
▶ dysuria, frequency, urgency, and suprapubic pain
▶ cloudy, malodorous, and possibly bloody urine
▶ fever
▶ nausea and vomiting
▶ costovertebral angle tenderness

▶ Urine culture reveals microorganism.
▶ Urinary microscopy is positive for pyuria, hematuria, or bacteriuria.

Acute pyelonephritis:
▶ fever and shaking chills
▶ nausea, vomiting, and diarrhea
▶ symptoms of cystitis may be present
▶ tachycardia
▶ generalized muscle tenderness

Urethritis:
▶ dysuria, frequency, and pyuria

Whooping cough (Pertussis)

▶ irritating, hacking cough characteristically ending in a loud, crowing, inspiratory whoop that may expel tenacious mucus
▶ anorexia
▶ sneezing
▶ listlessness
▶ infected conjunctiva
▶ low grade fever

▶ Classic clinical findings suggest the disease.
▶ Nasopharyngeal swabs and sputum cultures show *B. pertussis.*
▶ Fluorescent antibody screening of nasopharyngeal smears is less reliable than cultures.
▶ Serology shows an elevated WBC count.

Viral infections

Chickenpox (Varicella)

▶ irritating, hacking cough characteristically ending in a loud, crowing, inspiratory whoop that may expel tenacious mucus
▶ anorexia
▶ sneezing
▶ listlessness
▶ infected conjunctiva
▶ low grade fever

▶ Characteristic clinical signs suggest the virus.
▶ Isolation of virus from vesicular fluid helps confirm the virus; Giemsa stain distinguishes varicella-zoster from vaccinia and variola viruses.
▶ Leucocyte count may be normal, low, or mildly increased.

(continued)

REVIEWING COMMON INFECTIONS *(continued)*

INFECTION AND FINDINGS	DIAGNOSIS

Viral infections *(continued)*

Cytomegalovirus infection

▶ pruritic rash of small, erythematous macules that progresses to papules and then to vesicles as a result of viremia
▶ slight fever
▶ malaise and anorexia
▶ mild, nonspecific complaints
▶ immunodeficient population: pneumonia, chorioretinitis, colitis, encephalitis, abdominal pain, diarrhea, or weight loss
▶ infants age 3 to 6 months appear asymptomatic but may develop hepatic dysfunction, hepatosplenomegaly, spider angiomas, pneumonitis, and lymphadenopathy
▶ congenital infection: jaundice, petechial rash, hepatosplenomegaly, thrombocytopenia, hemolytic anemia

▶ Virus isolated in urine, saliva, throat, cervix, WBC and biopsy specimens. Complement fixation studies, hemagglutination inhibition antibody tests, and indirect immunofluorescent test for CMV immunoglobulin M antibody (congenital infections) aid diagnosis.

Herpes simplex

Type 1:
▶ fever
▶ sore, red, swollen, throat
▶ submaxillary lymphadenopathy
▶ increased salivation, halitosis, and anorexia
▶ severe mouth pain
▶ edema of the mouth
▶ vesicles (on the tongue, gingiva, and cheeks, or anywhere in or around the mouth) on a red base that eventually rupture, leaving a painful ulcer and then yellow crusting

Type 2:
▶ tingling in the area involved
▶ malaise
▶ dysuria
▶ dyspareunia (painful intercourse)
▶ leukorrhea (white vaginal discharge containing mucus and pus cells)
▶ localized, fluid-filled vesicles that are found on the cervix, labia, perianal skin, vulva, vagina, glans penis, foreskin, and penile shaft, mouth or anus; inguinal swelling may be present

▶ Tzanck smear shows multinucleated giant cells.
▶ Herpes simplex virus culture is positive.
▶ Virus is isolated from local lesions.
▶ Tissue biopsy aids in diagnosis.
▶ Elevated antibodies and increased white blood cell count indicate primary infection.

REVIEWING COMMON INFECTIONS *(continued)*

INFECTION AND FINDINGS	DIAGNOSIS
Viral infections *(continued)*	

Herpes zoster

▶ pain within the dermatome affected
▶ fever
▶ malaise
▶ pruritus
▶ paresthesia or hyperesthesia in the trunk, arms, or legs may also occur
▶ small, red, nodular skin lesions on painful areas (nerve specific) that change to pus or fluid-filled vescicles

▶ Staining antibodies from vesicular fluid and identification under fluorescent light differentiates herpes zoster from localized herpes simplex.
▶ Examination of vesicular fluid and infected tissue shows eosinophilic intranuclear inclusions and varicella virus.
▶ Lumbar puncture shows increased pressure; CSF shows increased protein levels and possibly pleocytosis.

Human immunodeficiency virus infection (HIV)

▶ rapid weight loss
▶ dry cough
▶ recurring fever or profuse night sweats
▶ profound and unexplained fatigue
▶ swollen lymph glands in the armpits, groin, or neck
▶ diarrhea that lasts for more than a week
▶ white spots or unusual blemishes on the tongue, in the mouth, or in the throat
▶ pneumonia
▶ red, brown, pink, or purplish blotches on or under the skin or inside the mouth, nose, or eyelids
▶ memory loss, depression, and other neurologic disorders

▶ Two enzyme immunosorbent assay (EIA) tests are positive.
▶ Western blot test is positive.

Infectious mononucleosis

▶ headache
▶ malaise and fatigue
▶ sore throat
▶ cervical lymphadenopathy
▶ temperature fluctuations with an evening peak
▶ splenomegaly
▶ hepatomegaly
▶ stomatitis
▶ exudative tonsillitis, or pharyngitis
▶ maculopapular rash

▶ Monospot test is positive.
▶ WBC count is abnormally high (10,000 to 20,000/mm^2) during the second and third weeks of illness. From 50% to 70% of the total count consists of lymphocytes and monocytes, and 10% of the lymphocytes are atypical.
▶ Heterophil antibodies in serum drawn during the acute phase and at 3- to 4-week intervals increase to four times normal.
▶ Indirect immunofluorescence shows antibodies to Epstein-Barr virus and cellular antigens.

(continued)

REVIEWING COMMON INFECTIONS (continued)

INFECTION AND FINDINGS	DIAGNOSIS
Viral infections (continued)	

Mumps

- myalgia
- malaise and fever
- headache
- earache that is aggravated by chewing
- parotid gland tenderness and swelling, and pain when chewing sour or acidic liquids
- swelling of the other salivary glands

- Virus is isolated from throat washings, urine, blood or spinal fluid.
- Serologic antibody testing shows a rise in paired antibodies.
- Clinical signs and symptoms, especially parotid gland enlargement are characteristic.

Rabies

Prodromal symptoms:
- local or radiating pain or burning and a sensation of cold, pruritus, and tingling at the bite site
- malaise and fever
- headache
- nausea
- sore throat and persistent loose cough
- nervousness, anxiety, irritability, hyperesthesia, sensitivity to light and loud noises
- excessive salivation, tearing and perspiration

Excitation phase:
- intermittent hyperactivity, anxiety, apprehension
- pupillary dilation
- shallow respirations
- altered level of consciousness
- ocular palsies
- strabismus
- asymmetrical pupillary dilation or constriction
- absence of corneal reflexes
- facial muscle weakness
- forceful, painful pharyngeal muscle spasms that expel fluids from the mouth, resulting in dehydration
- swallowing problems cause frothy drooling and soon the sight, sound, or thought of water triggers uncontrollable pharyngeal muscle spasms and excessive salivation
- nuchal rigidity
- seizures
- cardiac arrhythmias

Terminal phase:
- gradual, generalized, flaccid paralysis
- peripheral vascular collapse
- coma and death

- Virus is isolated from saliva or CSF.
- Fluorescent rabies antibody (FRA) test is positive.
- WBC is elevated.

- no diagnostic tests for rabies before its onset
- histologic examination of brain tissue from human rabies victims shows perivascular inflammation of the gray matter, degeneration of neuron, and characteristic minute bodies, called Negri bodies, in the nerve cells

REVIEWING COMMON INFECTIONS *(continued)*

INFECTION AND FINDINGS	DIAGNOSIS
Viral infections *(continued)*	

Respiratory syncytial virus infection

Mild disease:
- nasal congestion
- coughing and wheezing
- malaise
- sore throat
- earache
- dyspnea
- fever

Bronchitis, bronchiolitis, pneumonia:
- nasal flaring, retraction, cyanosis, and tachypnea
- wheezes, rhonchi, and crackles
- signs such as weakness, irritability, and nuchal rigidity of central nervous system (CNS) infection may be observed

▶ Cultures of nasal and pharyngeal secretions may reveal the virus; however, this infection is so labile that cultures aren't always reliable.
▶ Serum antibody titers may be elevated.
▶ WBC may be normal to elevated.

Rubella

- maculopapular, mildly itchy rash that usually begins on the face and then spreads rapidly, often covering the trunk and extremities
- small, red macules on the soft palate
- low-grade fever
- headache
- malaise
- anorexia
- sore throat
- cough
- postauricular, suboccipital, and posterior cervical lymph node enlargement

▶ Clinical signs and symptoms are usually sufficient to make a diagnosis.
▶ Cell cultures of the throat, blood, urine, and CSF, along with convalescent serum that shows a fourfold rise in antibody titers, confirms the diagnosis.
▶ Blood tests confirm rubella-specific IgM antibody.

Rubeola

- fever
- photophobia
- malaise
- anorexia
- conjunctivitis, puffy red eyes, and rhinorrhea
- coryza
- hoarseness, hacking cough
- Koplik's spots, pruritic macular rash that becomes papular and erythematous

▶ Diagnosis rests on distinctive clinical features.
▶ Measles virus may be isolated from the blood, nasopharyngeal secretions, and urine during the febrile stage.
▶ Serum antibodies appear within 3 days.

(continued)

REVIEWING COMMON INFECTIONS *(continued)*

INFECTION AND FINDINGS	DIAGNOSIS
Fungal infections	

Chlamydial infections

Cervicitis:
- cervical erosion
- dyspareunia
- micropurulent discharge
- pelvic pain

Endometritis or salpingitis:
- pain and tenderness of the lower abdomen, cervix, uterus, and lymph nodes
- chills, fever
- breakthrough bleeding; bleeding after intercourse; and vaginal discharge
- dysuria

Urethral syndrome:
- dysuria, pyuria, and urinary frequency

Urethritis:
- dysuria, erythema, tenderness of the urethral meatus
- urinary frequency
- pruritus and urethral discharge (copious and purulent or scant and clear or mucoid.

Epididymitis:
- painful scrotal swelling
- urethral discharge

Prostatitis:
- low back pain
- urinary frequency, nocturia, and dysuria
- painful ejaculation

Proctitis:
- diarrhea
- tenesmus
- pruritus
- bloody or mucopurulent discharge
- diffuse or discrete ulceration in the rectosigmoid colon

- Swab from site of infection establishes a diagnosis of urethritis, cervicitis, salpingitis, endometritis, or proctitis.
- Culture of aspirated material establishes a diagnosis of epididymitis.
- Antigen-detection methods are the diagnostic tests of choice for identifying chlamydial infection.
- Polymerase chain reaction (PCR) test is highly sensitive and specific.

REVIEWING COMMON INFECTIONS *(continued)*

INFECTION AND FINDINGS	DIAGNOSIS

Fungal infections *(continued)*

Histoplasmosis

Primary acute histoplasmosis:
▶ may be asymptomatic or may cause symptoms of a mild respiratory illness similar to a severe cold or influenza
▶ fever
▶ malaise
▶ headache
▶ myalgia
▶ anorexia
▶ cough
▶ chest pain
▶ anemia, leukopenia, or thrombocytopenia
▶ oropharyngeal ulcers

Progressive disseminated histoplasmosis:
▶ hepatosplenomegaly
▶ general lymphadenopathy
▶ anorexia and weight loss
▶ fever and, possibly, ulceration of the tongue, palate, epiglottis, and larynx, with resulting pain, hoarseness, and dysphagia

Chronic pulmonary histoplasmosis:
▶ productive cough, dyspnea, and occasional hemoptysis
▶ weight loss
▶ extreme weakness
▶ breathlessness and cyanosis

African histoplasmosis:
▶ cutaneous nodules, papules, and ulcers
▶ lesions of the skull and long bones
▶ lymphadenopathy and visceral involvement without pulmonary lesions

▶ Culture or histology reveals the organism.
▶ Stained biopsies using Gomori's stains or periodic acid-Schiff reaction give a fast diagnosis of the disease.
▶ Positive histoplasmin skin test indicates exposure to histoplasmosis.
▶ Rising complement fixation and agglutination titers (more than 1:32) strongly suggest histoplasmosis.

(continued)

REVIEWING COMMON INFECTIONS *(continued)*

INFECTION AND FINDINGS	DIAGNOSIS
Protozoal infections	

Malaria

Benign form:
▶ chills
▶ fever
▶ headache and myalgia

Acute attacks (occur when erythrocytes rupture):
▶ chills and shaking
▶ high fever (up to 107°F/41.7°C)
▶ profuse sweating
▶ hepatosplenomegaly
▶ hemolytic anemia

Life-threatening form:
▶ persistent high fever
▶ orthostatic hypotension
▶ red blood cell sludging that leads to capillary obstruction at various sites
▶ hemiplegia
▶ seizures
▶ delirium and coma
▶ hemoptysis
▶ vomiting
▶ abdominal pain, diarrhea, and melena
▶ oliguria, anuria, uremia

▶ Peripheral blood smears of red blood cells identify the parasite.
▶ Indirect fluorescent serum antibody tests are unreliable in the acute phase.
▶ Hemoglobin levels are decreased.
▶ Leukocyte count is normal to decreased.
▶ Protein and leukocytes are present in urine sediment.

Schistosomiasis

▶ transient pruritic rash at the site of cercariae penetration
▶ fever
▶ myalgia
▶ cough

Later signs and symptoms:
▶ hepatomegaly, splenomegaly, and lymphadenopathy

S. mansoni and S. japonicum:
▶ irregular fever
▶ malaise, weakness
▶ weight loss
▶ diarrhea
▶ ascites, hepatosplenomegaly
▶ portal hypertension
▶ fistulas, intestinal stricture

S. haematobium:
▶ terminal hematuria dysuria
▶ ureteral colic

▶ Typical symptoms and a history of travel to endemic areas suggest the diagnosis.
▶ Ova in the urine or stool or a mucosal lesion biopsy confirm diagnosis.
▶ WBC count shows eosinophilia.

REVIEWING COMMON INFECTIONS (continued)

INFECTION AND FINDINGS	DIAGNOSIS
Protozoal infections (continued)	

Toxoplasmosis

Ocular toxoplasmosis:
- chorioretinitis
- yellow-white elevated cotton patches
- blurred vision
- scotoma
- pain
- photophobia

Acute toxoplasmosis:
- malaise, myalgia, headache
- fatigue
- sore throat
- fever
- cervical lymphadenopathy
- maculopapular rash

Congenital:
- hydrocephalus or microcephalus
- seizures
- jaundice
- purpura and rash

▶ Blood tests detect a specific toxoplasma antibody.
▶ Isolation of *T. gondii* in mice after their inoculation with human body fluids reveals antibodies for the disease and confirms toxoplasmosis.

Trichinosis

Stage 1 (enteric phase):
- anorexia
- nausea, vomiting, diarrhea
- abdominal pain and cramps

Stage 2 (systemic phase) and Stage 3 (muscular encystment phase):
- edema (especially of the eyelids or face)
- muscle pain
- itching and burning skin
- sweating
- skin lesions
- fever
- delirium and lethargy in severe respiratory, cardiovascular, or CNS infection

▶ Stools may contain mature worms and larvae during the invasion stage.
▶ Skeletal muscle biopsies can show encysted larvae 10 days after ingestion.
▶ Skin testing may show a positive histaminelike reactivity.
▶ Elevated acute and convalescent antibody titers confirm the diagnosis.
▶ Serology results indicate elevated aspartate aminotransferase, alanine aminotransferase, creatine kinase, and lactate dehydrogenase levels during the acute stages and an elevated eosinophil count.
▶ Lumbar puncture demonstrates CNS involvement with normal or elevated cerebrospinal fluid lymphocytes and increased protein levels.

4

Genetics is the study of heredity — the passing of physical, biochemical, and physiologic traits from biological parents to their children. In this transmission, disorders can be transmitted, and mistakes or mutations can result in disability or death.

Genetic information is carried in genes, which are strung together on the deoxyribonucleic acid (DNA) double helix to form chromosomes. Every normal human cell (except reproductive cells) has 46 chromosomes, 22 paired chromosomes called autosomes, and 2 sex chromosomes (a pair of Xs in females and an X and a Y in males). A person's individual set of chromosomes is called his karyotype. (See *Normal human karyotype.*) The human genome (structure and location of each gene on which chromosome) has been under intense study for only about 15 years. In June 2000, two teams of scientists announced the completion of the "rough draft" of the entire genome sequence. The sequence consists of more than 3.1 billion pairs of chemicals. Decoding the genome will enable people to know who is likely to get a specific inherited disease and enable researchers to eradicate or improve the treatment of many diseases. (See *The genome at a glance,* page 86.)

A word of warning at the outset. For a wide variety of reasons, not every gene that might be expressed is. Thus, the following chapter may seem to contain a great many "hedge" words — may, perhaps, some. Genetic principles are based on study of thousands of individuals. Those studies have led to generalities that are usually true, but exceptions occur. Genetics is an inexact science.

GENETIC COMPONENTS

Each of the two strands of DNA in a chromosome consists of thousands of combinations of four nucleotides — adenine (A), thymine (T), cyotosine (C), and guanine (G) — arranged in complementary triplet pairs (called codons), each of which represents a gene. The strands are loosely held together by chemical bonds between adenine and thymine or cytosine and guanine —for example, a triplet ACT on one strand is linked to the triplet TGA on the other. The looseness of the bonds allows the strands to separate easily during cell division. (See *DNA duplication: Two double helices from one,* page 87.) The genes carry a code for each trait a person inherits, from blood type to eye color to body shape and a myriad of other traits.

DNA ultimately controls the formation of essential substances throughout the life of every cell in the body. It does this through the genetic code, the precise sequence of AT and CG pairs on the DNA molecule. Genes not only control hereditary traits, transmitted from parents to offspring, but also cell reproduction and the daily functions of all cells. Genes control cell function by controlling the structures and chemicals that are synthesized within the cell. (See *How genes control cell function,* page 88.) For example, they control the formation of ribonucleic acid (RNA), which in turn controls the formation of specific proteins, most of which

NORMAL HUMAN KARYOTYPE

The illustration shows the arrangement of chromosomes (karyotype) in a normal male.

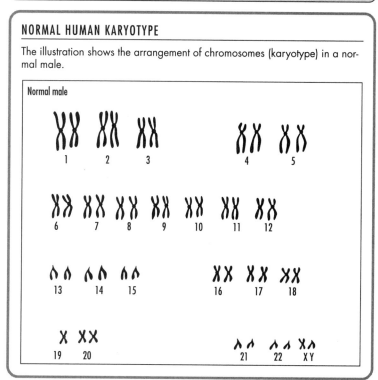

Normal male

are enzymes that catalyze chemical reactions in the cells.

TRANSMITTING TRAITS

Germ cells, or gametes (ovum and sperm), are one of two classes of cells in the body; each germ cell contains 23 chromosomes (called the *haploid* number) in its nucleus. All the other cells in the body are somatic cells, which are *diploid*, that is, they contain 23 *pairs* of chromosomes.

When ovum and sperm unite, the corresponding chromosomes pair up, so that the fertilized cell and every somatic cell of the new person has 23 pairs of chromosomes in its nucleus.

Germ cells

The body produces germ calls through a kind of cell division called meiosis. Meiosis occurs only when the body is creating haploid germ cells from their diploid precursors. Each of the 23 pairs of chromosomes in the germ cell separates, so that, when the cell then divides, each new cell (ovum or sperm) contains one set of 23 chromosomes.

Most of the genes on one chromosome are identical or almost identical to the gene on its mate. (As we discuss later, each chromosome may carry a different version of the same gene.) The location (or locus) of a gene on a chromosome is specific and doesn't vary from person to person. This allows each of the thousands of genes on a strand of DNA in an ovum to join the corresponding gene in a sperm when the chromosomes pair up at fertilization.

THE GENOME AT A GLANCE

In 1998, a gene map was released by an international consortium of radiation hybrid mapping labs containing over 30,000 distinct cDNA-based markers. The particular makeup of an individual organism is called its genome, which is made up of the alleles (or different versions of the genes) it possesses.

The genome at a glance

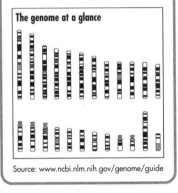

Source: www.ncbi.nlm.nih.gov/genome/guide

Determining sex

Only one pair of chromosomes in each cell — pair 23 — is involved in determining a person's sex. These are the sex chromosomes; the other 22 chromosome pairs are called autosomes. Females have two X chromosomes and males have one X and one Y chromosome.

Each gamete produced by a male contains either an X or a Y chromosome. When a sperm with an X chromosome fertilizes an ovum, the offspring is female (two X chromosomes); when a sperm with a Y chromosome fertilizes an ovum, the offspring is male (one X and one Y chromosome). Very rare errors in cell division can result in a germ cell that has no sex chromosome and/or two X chromosomes. After fertilization, the zygote may have an XO or XXY karyotype and still survive. Most other errors in sex chromosome division are incompatible with life.

Mitosis

The fertilized ovum — now called a zygote — undergoes a kind of cell division called mitosis. Before a cell divides, its chromosomes duplicate. During this process, the double helix of DNA separates into two chains; each chain serves as a template for constructing a new chain. Individual DNA nucleotides are linked into new strands with bases complementary to those in the originals. In this way, two identical double helices are formed, each containing one of the original strands and a newly formed complementary strand. These double helices are duplicates of the original DNA chain.

Mitotic cell division occurs in five phases: an inactive phase called *interphase* and four active phases: *prophase, metaphase, anaphase,* and *telophase*. (See *Five phases of mitosis,* page 89.) The result of every mitotic cell division is two new daughter cells, each genetically identical to the original and to each other. Each of the two resulting cells likewise divides, and so on, eventually forming a many-celled human embryo. Thus, each cell in a person's body (except ovum or sperm) contains an identical set of 46 chromosomes that are unique to that person.

TRAIT PREDOMINANCE

Each parent contributes one set of chromosomes (and therefore one set of genes) so that every offspring has two genes for every locus (location on the chromosome) on the autosomal chromosomes.

Some characteristics, or traits, are determined by one gene that may have many variants (alleles), such as eye color. Others, called *polygenic* traits, require the interaction of one or more genes. In addition, environmental factors may affect how a gene or genes are expressed, although the environmental factors do not affect the genetic structure.

Variations in a particular gene — such as brown, blue, or green eye color — are called alleles. A person who has identical

DNA DUPLICATION: TWO DOUBLE HELICES FROM ONE

The nucleotide, the basic structural unit of deoxyribonucleic acid (DNA), contains a phosphate group, deoxyribose, and a nitrogen base made of adenine (A), guanine (G), thymine (T), or cystosine (C). A DNA molecule's double helix forms from the twisting of nucleotide strands (shown below).

During duplication, a DNA chain separates, and new complementary chains form and link to the separated originals (parents). The result is two identical double helices — parent and daughter.

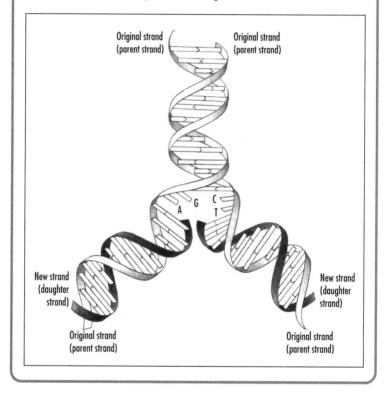

Original strand (parent strand)

Original strand (parent strand)

New strand (daughter strand)

New strand (daughter strand)

Original strand (parent strand)

Original strand (parent strand)

alleles on each chromosome is *homozygous* for that gene; if the alleles are different, they're said to be *heterozygous*.

Autosomal inheritance

For unknown reasons, on autosomal chromosomes, one allele may be more influential than the other in determining a specific trait. The more powerful, or *dominant*, gene is more likely to be expressed in the offspring than the less influential, or *recessive*, gene. Offspring will express a dominant allele when one or both chromosomes in a pair carry it. A recessive allele won't be expressed unless both chromosomes carry identical alleles. For example, a child may receive a gene for brown eyes from one parent and a gene for blue eyes from the other parent. The gene for brown eyes is dominant, and the

HOW GENES CONTROL CELL FUNCTION

This simplified diagram outlines how the genetic code directs formation of specific proteins. Some proteins are the building blocks of cell structure. Others, called enzymes, direct intracellular chemical reactions. Together, structural proteins and enzymes direct cell function.

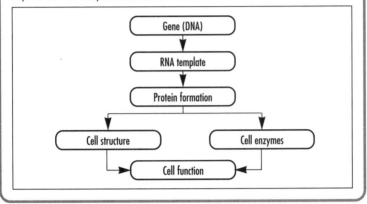

gene for blue eyes is recessive. Because the dominant gene is more likely to be expressed, the child is more likely to have brown eyes.

Sex-linked inheritance

The X and Y chromosomes are not literally a pair because the X chromosome is much larger than the Y. The male literally has less genetic material than the female, which means he has only one copy of most genes on the X chromosome. Inheritance of those genes is called *X-linked*. A man will transmit one copy of each X-linked gene to his daughters and none to his sons. A woman will transmit one copy to each child, whether male or female.

Inheritance of genes on the X chromosomes is different in another way. Some recessive genes on the X chromosomes act like dominants in females. For reasons that are not yet clear, one recessive allele will be expressed in some somatic cells and another in other somatic cells. The most common example occurs not in people but in cats. Only female cats have calico (tricolor) coat patterns. Hair color in

the cat is carried on the X chromosome. Some hair cells in females express the brown allele, others the white, and still others a third color.

Multifactorial inheritance

Environmental factors can affect the expression of some genes; this is called multifactorial inheritance. Height is a classic example of a multifactorial trait. In general, the height of offspring will be in a range between the height of the two parents. But nutritional patterns, health care, and other environmental factors also influence development. The better-nourished, healthier children of two short parents may be taller than either. Some diseases have genetic predisposition but multifactorial inheritance, that is, the gene for the disease is expressed only under certain environmental conditions.

PATHOPHYSIOLOGIC CHANGES

Autosomal disorders, sex-linked disorders, and multifactorial disorders result from damage to genes or chromosomes.

FIVE PHASES OF MITOSIS

In mitosis (used by all cells except gametes), the nuclear contents of a cell reproduce and divide, resulting in the formation of two daughter cells. The five steps, or phases, of this process are illustrated below.

INTERPHASE

During this phase, the nucleus and nuclear membrane are well defined and the nucleolus is prominent. Chromosomes replicate, each forming a double strand that remains attached at the center of each chromosome by a structure called the centromere; they appear as an indistinguishable matrix within the nucleus. Centrioles (in animal cells only, not plant cells) appear outside the nucleus.

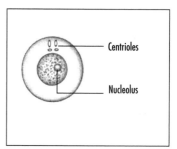

PROPHASE

In this stage, the nucleolus disappears and chromosomes become distinct. Halves of each duplicated chromosome (chromatids) remain attached by a centromere. Centrioles move to opposite sides of the cell and radiate spindle fibers.

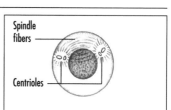

METAPHASE

Chromosomes line up randomly in the center of the cell between spindles, along the metaphase plate. The centromere of each chromosome replicates.

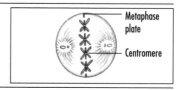

ANAPHASE

Centromeres move apart, pulling the separate chromatids (now called chromosomes) to opposite ends of the cell. In human cells, each end of the cell now contains 46 chromosomes, The number of chromosomes at each end of the cell equals the original number

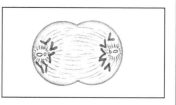

TELOPHASE

A nuclear membrane forms around each end of the cell, and spindle fibers disappear. The cytoplasm compresses and divides the cell in half. Each new cell contains the diploid number (46 in humans) of chromosomes.

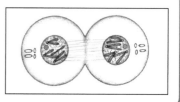

TERATOGENS AND ASSOCIATED DISORDERS

This chart lists common teratogens and their associated disorders.

INFECTIONS	ASSOCIATED DISORDERS
Toxoplasmosis Rubella Cytomegalovirus Herpes simplex Other infections (syphilis, hepatitis B, mumps, gonorrhea, parvovirus, varicella)	▶ Growth deficiency ▶ Mental retardation ▶ Hepatosplenomegaly ▶ Hearing loss ▶ Cardiac and ocular defects ▶ Active infection ▶ Carrier state

MATERNAL DISORDERS	
Diabetes mellitus	Abnormalities of ▶ spine ▶ lower extremities ▶ heart ▶ kidney ▶ external genitalia
Phenylketonuria	▶ Mental retardation ▶ Microcephaly ▶ Congenital heart defects ▶ Intrauterine growth retardation
Hyperthermia	▶ Intrauterine growth retardation ▶ CNS and neural tube defects ▶ Facial defects

DRUGS, CHEMICALS, AND PHYSICAL AGENTS	
Alcohol	▶ Fetal alcohol syndrome ▶ Learning disabilities
Anticonvulsants	▶ Intrauterine growth retardation ▶ Mental deficiency ▶ Facial abnormalities ▶ Cardiac defects ▶ Cleft lip and palate ▶ Malformed ears ▶ Genital defects
Cocaine	▶ Premature delivery ▶ Abruptio placentae ▶ Intracranial hemorrhage ▶ GI and GU abnormalities

TERATOGENS AND ASSOCIATED DISORDERS *(continued)*

DRUGS, CHEMICALS, AND PHYSICAL AGENTS	ASSOCIATED DISORDERS
Diethylstilbestrol	◗ Clear-cell adenocarcinoma of vagina ◗ Structural and functional defects of female GU tract
Lithium	◗ Congenital heart disease
Methotrexate	◗ Intrauterine growth retardation ◗ Decreased ossification of skull ◗ Prominent eyes ◗ Limb abnormalities ◗ Mild developmental delay
Radiation	◗ Microcephaly ◗ Mental retardation
Tetracycline	◗ Brown staining of decidual teeth ◗ Dental caries ◗ Enamel hypoplasia
Vitamin A derivatives	◗ Facial defects ◗ Cardiac defects ◗ CNS defects ◗ Incomplete development of thymus
Warfarin (Coumadin)	◗ Intrauterine growth retardation ◗ Mental retardation ◗ Seizures ◗ Nasal hypoplasia ◗ Abnormal calcification of axial skeleton

Adapted with permission from M., Hansen. *Pathophysiology: Foundations of Disease and Clinical Intervention.* Philadelphia: W.B. Saunders, 1998.

Some defects arise spontaneously, and others may be caused by environmental teratogens.

Environmental teratogens

Teratogens are environmental agents (infectious toxins, maternal systemic diseases, drugs, chemicals, and physical agents) that can harm the developing fetus by causing congenital structural or functional defects. Teratogens may also cause spontaneous miscarriage, complications during labor and delivery, hidden defects in later development (such as cognitive or behavioral problems), or neoplastic transformations. (See *Teratogens and associated disorders*.)

The embryonic period — the first 8 weeks after fertilization — is a vulnerable time, when specific organ systems are actively differentiating. Exposure to teratogens usually kills the embryo. During

AUTOSOMAL DOMINANT INHERITANCE

The diagram shows the inheritance pattern of an abnormal trait when one parent has recessive normal genes (aa) and the other has a dominant abnormal gene (Aa). Each child has a 50% chance of inheriting A.

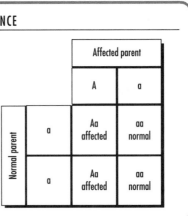

		Affected parent	
		A	a
Normal parent	a	Aa affected	aa normal
	a	Aa affected	aa normal

the fetal period, organ systems are formed and continue to mature. Exposure during this time can cause intrauterine growth retardation, cognitive abnormalities, or structural defects.

Gene errors

A permanent change in genetic material is a mutation, which may occur spontaneously or after exposure of a cell to radiation, certain chemicals, or viruses. Mutations can occur anywhere in the genome — the person's entire inventory of genes.

Every cell has built-in defenses against genetic damage. However, if a mutation isn't identified or repaired, the mutation may produce a trait different from the original trait and is transmitted to offspring during reproduction. The mutation initially causes the cell to produce some abnormal protein that makes the cell different from its ancestors. Mutations may have no effect; they may change expression of a trait, and others change the way a cell functions. Some mutations cause serious or deadly defects, such as cancer or congenital anomalies.

Autosomal disorders

In single-gene disorders, an error occurs at a single gene site on the DNA strand.

A mistake may occur in the copying and transcribing of a single codon (nucleotide triplet) through additions, deletions, or excessive repetitions.

Single-gene disorders are inherited in clearly identifiable patterns that are the same as those seen in inheritance of normal traits. Because every person has 22 pairs of autosomes and only 1 pair of sex chromosomes, most hereditary disorders are caused by autosomal defects.

Autosomal dominant transmission usually affects male and female offspring equally. If one parent is affected, each child has one chance in two of being affected. If both parents are affected, all their children will be affected. An example of this type of inheritance occurs in Marfan syndrome. (See *Autosomal dominant inheritance*.)

Autosomal recessive inheritance also usually affects male and female offspring equally. If both parents are affected, all their offspring will be affected. If both parents are unaffected but are heterozygous for the trait (carriers of the defective gene), each child has one chance in four of being affected. If only one parent is affected, none of their offspring will be affected, but all will carry the defective gene. If one parent is affected and the other is a carrier, half their children will be affect-

AUTOSOMAL RECESSIVE INHERITANCE

The diagram shows the inheritance pattern of an abnormal trait when both unaffected parents are heterozygous (Aa) for a recessive abnormal gene (a) on an autosome. As shown, each child has a one-in-four chance of being affected (aa), a one-in-four chance of having two normal genes (AA) and no chance of transmittal, and a 50% chance of being a carrier (Aa) who can transmit the gene.

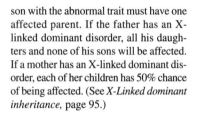

		Heterozygous parent Aa	
		A	a
Heterozygous parent Aa	A	AA Normal	Aa Carrier
	a	Aa Carrier	aa Affected

ed. (See *Autosomal recessive inheritance*.) Autosomal recessive disorders may occur when there is no family history of the disease.

Sex-linked disorders

Genetic disorders caused by genes located on the sex chromosomes are termed sex-linked disorders. Most sex-linked disorders are passed on the X chromosome, usually as recessive traits. Because males have only one X chromosome, a single X-linked recessive gene can cause disease in a male. Females receive two X chromosomes, so they can be homozygous for a disease allele, homozygous for a normal allele, or heterozygous.

Most people who express X-linked recessive traits are males with unaffected parents. In rare cases, the father is affected and the mother is a carrier. All daughters of an affected male will be carriers. Sons of an affected male will be unaffected, and the unaffected sons aren't carriers. Unaffected male children of a female carrier don't transmit the disorder. Hemophilia is an example of an X-linked inheritance disorder. (See *X-Linked recessive inheritance*, page 94.)

Characteristics of X-linked dominant inheritance include evidence of the inherited trait in the family history. A person with the abnormal trait must have one affected parent. If the father has an X-linked dominant disorder, all his daughters and none of his sons will be affected. If a mother has an X-linked dominant disorder, each of her children has 50% chance of being affected. (See *X-Linked dominant inheritance,* page 95.)

Multifactorial disorders

Most multifactorial disorders result from the effects of several different genes and an environmental component. In polygenic inheritance, each gene has a small additive effect, and the effect of a combination of genetic errors in a person is unpredictable. Multifactorial disorders can result from a less-than-optimum expression of many different genes, not from a specific error.

Some multifactorial disorders are apparent at birth, such as cleft lip, cleft palate, congenital heart disease, anencephaly, clubfoot, and myelomeningocele. Others don't become apparent until later, such as type II diabetes mellitus, hypertension, hyperlipidemia, most autoimmune diseases, and many cancers. Multifactorial disorders that develop during adulthood are often believed to be strongly related to environmental factors, not only in in-

X-LINKED RECESSIVE INHERITANCE

The diagram shows the children of a normal parent and a parent with a recessive gene on the X chromosome (shown by an open dot). All daughters of an affected male will be carriers. The son of a female carrier may inherit a recessive gene on the X chromosome and be affected by the disease. Unaffected sons can't transmit the disorder.

		Normal mother	
		X	X
Affected father	X̶	X̶X Carrier daughter	XX̶ Carrier daughter
	Y	XY Normal son	XY Normal son

		Carrier mother	
		X̶	X
Normal father	X	X̶X Carrier daughter	XX Normal daughter
	Y	X̶Y Affected son	XY Normal son

cidence but also in the degree of expression.

Environmental factors of maternal or paternal origin include the use of chemicals (such as drugs, alcohol, or hormones), exposure to radiation, general health, and age. Maternal factors include infections during pregnancy, existing diseases, nutritional factors, exposure to high altitude, maternal–fetal blood incompatibility, and poor prenatal care.

Chromosome defects

Aberrations in chromosome structure or number cause a class of disorders called congenital anomalies, or birth defects. The aberration may be loss, addition, or re-arrangement genetic material. If the remaining genetic material is sufficient to maintain life, an endless variety of clinical manifestations may occur. Most clinically significant chromosome aberrations arise during meiosis. Meiosis is an incredibly complex process that can go wrong in many ways. Potential contribut-

ing factors include maternal age, radiation, and use of some therapeutic or recreational drugs.

Translocation, the shifting or moving of a chromosome, occurs when chromosomes split apart and rejoin in an abnormal arrangement. The cells still have a normal amount of genetic material, so often there are no visible abnormalities. However, the children of parents with translocated chromosomes may have serious genetic defects, such as monosomies or trisomies. Parental age doesn't seem to be a factor in translocation.

Errors in chromosome number

During both meiosis and mitosis, chromosomes normally separate in a process called *disjunction*. Failure to separate, called nondisjunction, causes an unequal distribution of chromosomes between the two resulting cells. If nondisjunction occurs during mitosis soon after fertilization, it may affect all the resulting cells.

X-LINKED DOMINANT INHERITANCE

The diagram shows the children of a normal parent and a parent with an abnormal, X-linked dominant gene on the X chromosome (shown by the dot on the X). When the father is affected, only his daughters have the abnormal gene. When the mother is affected, both sons and daughters may be affected.

		Normal mother	
		X	X
Affected father	X̣	X̣X Affected daughter	X̣X Affected daughter
	Y	XY Normal son	XY Normal son

		Affected mother	
		X̣	X
Normal father	X	X̣X Affected daughter	XX Normal daughter
	Y	X̣Y Affected son	XY Normal son

Gain or loss of chromosomes is usually caused by nondisjunction of autosomes or sex chromosomes during meiosis. The incidence of nondisjunction increases with parental age. (See *Chromosomal disjunction and nondisjunction,* page 96.)

The presence of one chromosome less than the normal number is called *monosomy*; an autosomal monosomy is nonviable. The presence of an extra chromosome is called a trisomy. A mixture of both trisomic and normal cells results in mosaicism, which is the presence of two or more cell lines in the same person. The effect of mosaicism depends on the proportion and anatomic location of abnormal cells.

DISORDERS

This section discusses disorders in the context of their pattern of inheritance as well as environmental factors. The alphabetically listed disorders have the following patterns of inheritance:

◗ Autosomal recessive: cystic fibrosis, phenylketonuria, sickle cell anemia, Tay-Sachs disease
◗ Autosomal dominant: Marfan syndrome
◗ X-linked recessive: hemophilia
◗ Polygenic multifactorial: cleft lip/cleft palate, spina bifida
◗ Chromosome number: Down syndrome.

Cleft lip and cleft palate

Cleft lip and cleft palate may occur separately or in combination. They originate in the second month of pregnancy if the front and sides of the face and the palatine shelves fuse imperfectly. Cleft lip with or without cleft palate occurs twice as often in males than females. Cleft palate without cleft lip is more common in females.

Cleft lip deformities can occur unilaterally, bilaterally, or, rarely, in the midline. Only the lip may be involved, or the defect may extend into the upper jaw or nasal cavity. (See *Types of cleft deformities,* page 97.) Incidence is highest in children with a family history of cleft defects.

CHROMOSOMAL DISJUNCTION AND NONDISJUNCTION

The illustration shows normal disjunction and nondisjunction of an ovum. When disjunction proceeds normally, fertilization with a normal sperm results in a zygote with the correct number of chromosomes. In nondisjunction, the sister chromatids fail to separate; the result is one trisomic cell and one monosomic cell.

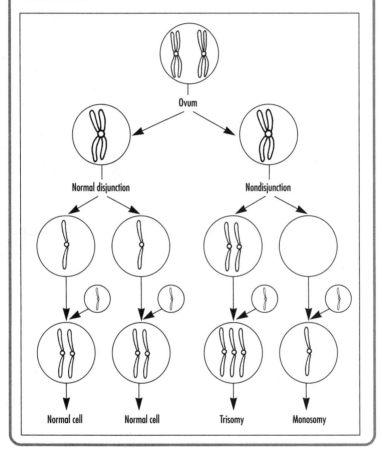

Cleft lip with or without cleft palate occurs in about 1 in 600 to 1 in 1,250 births among whites; the incidence is lower in blacks and greater in Japanese populations.

Causes
Possible causes include:
▶ chromosomal abnormality (trisomy 13)
▶ exposure to teratogens during fetal development
▶ combined genetic and environmental factors.

TYPES OF CLEFT DEFORMITIES

The following illustrations show variations of cleft lip and cleft palate.

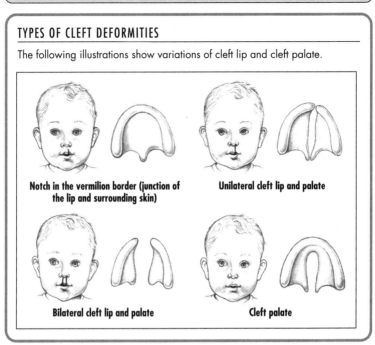

Notch in the vermilion border (junction of the lip and surrounding skin)

Unilateral cleft lip and palate

Bilateral cleft lip and palate

Cleft palate

Pathophysiology

During the second month of pregnancy, the front and sides of the face and the palatine shelves develop. Because of a chromosomal abnormality, exposure to teratogens, genetic abnormality, or environmental factors, the lip or palate fuses imperfectly.

The deformity may range from a simple notch to a complete cleft. A cleft palate may be partial or complete. A complete cleft includes the soft palate, the bones of the maxilla, and the alveolus on one or both sides of the premaxilla.

A double cleft is the most severe of the deformities. The cleft runs from the soft palate forward to either side of the nose. A double cleft separates the maxilla and premaxilla into freely moving segments. The tongue and other muscles can displace the segments, enlarging the cleft.

Signs and symptoms
▶ Obvious cleft lip or cleft palate

▶ Feeding difficulties due to incomplete fusion of the palate.

Complications
Complications may include:
▶ malnutrition, because the abnormal lip and palate affect nutritional intake
▶ hearing impairment, often due to middle-ear damage or recurrent infections
▶ permanent speech impediment, even after surgical repair.

Diagnosis
▶ Clinical presentation, obvious at birth
▶ Prenatal targeted ultrasound.

Treatment
Correcting cleft lip or palate may involve:
▶ surgical correction of cleft lip in the first few days of life to permit sucking
▶ orthodontic prosthesis to improve sucking
▶ surgical correction of cleft lip at 8 weeks to 8 months to allow maternal bonding

and rule out associated congenital anomalies

▶ surgical correction of cleft palate at 12 to 18 months, after the infant gains weight and is infection-free

▶ speech therapy to correct speech patterns

▶ use of a contoured speech bulb attached to the posterior of a denture to occlude the nasopharynx when a wide horseshoe defect makes surgery impossible (to help the child develop intelligible speech)

▶ adequate nutrition for normal growth and development

▶ use of a large soft nipple with large holes, such as a lamb's nipple, to improve feeding patterns and promote nutrition.

 AGE ALERT Daily use of folic acid before conception decreases the risk for isolated (not associated with another genetic or congenital malformation) cleft lip or palate by up to 25%. Women of childbearing age should be encouraged to take a daily multivitamin containing folic acid until menopause or until they're no longer fertile.

Cystic fibrosis

In cystic fibrosis, dysfunction of the exocrine glands affects multiple organ systems. The disease affects males as well as females and is the most common fatal genetic disease in white children.

Cystic fibrosis is accompanied by many complications and now carries an average life expectancy of 28 years. The disorder is characterized by chronic airway infection leading to bronchiectasis, bronchiolectasis, exocrine pancreatic insufficiency, intestinal dysfunction, abnormal sweat gland function, and reproductive dysfunction.

CULTURAL DIVERSITY The incidence of cystic fibrosis varies with ethnic origin. It occurs in 1 of 3,000 births in whites of North America and northern European descent, 1 in 17,000 births in blacks, and 1 in 90,000 births in the Asian population in Hawaii.

Causes

The responsible gene is on chromosome 7q; it encodes a membrane-associated protein called the cystic fibrosis transmembrane regulator (CFTR). The exact function of CFTR remains unknown, but it appears to help regulate chloride and sodium transport across epithelial membranes.

Causes of CF include:

▶ coding found on as many as 350 alleles

▶ autosomal recessive inheritance.

Pathophysiology

Most cases arise from the mutation that affects the genetic coding for a single amino acid, resulting in a protein (the CFTR) that doesn't function properly. The CFTR resembles other transmembrane transport proteins, but it lacks the phenylalanine in the protein produced by normal genes. This regulator interferes with cAMP-regulated chloride channels and other ions by preventing adenosine triphosphate from binding to the protein or by interfering with activation by protein kinase.

The mutation affects volume-absorbing epithelia (in the airways and intestines), salt-absorbing epithelia (in sweat ducts), and volume-secretory epithelia (in the pancreas). Lack of phenylalanine leads to dehydration, increasing the viscosity of mucus-gland secretions, leading to obstruction of glandular ducts. CF has a varying effect on electrolyte and water transport.

Signs and symptoms

Signs and symptoms may include:

▶ thick secretions and dehydration due to ionic imbalance

▶ chronic airway infections by *Staphylococcus aureus, Pseudomonas aeruginosa*, and *Pseudomonas cepacea,* possibly due to abnormal airway surface fluids and failure of lung defenses

▶ dyspnea due to accumulation of thick secretions in bronchioles and alveoli

▶ paroxysmal cough due to stimulation of the secretion-removal reflex

▶ barrel chest, cyanosis, and clubbing of fingers and toes from chronic hypoxia

▶ crackles on auscultation due to thick, airway-occluding secretions

▶ wheezes heard on auscultation due to constricted airways

▶ retention of bicarbonate and water due to the absence of the CFTR chloride channel in the pancreatic ductile epithelia; limits membrane function and leads to retention of pancreatic enzymes, chronic cholecystitis and cholelithiasis, and ultimate destruction of the pancreas

▶ obstruction of the small and large intestine due to inhibited secretion of chloride and water and excessive absorption of liquid

▶ biliary cirrhosis due to retention of biliary secretions

▶ fatal shock and arrhythmias due to hyponatremia and hypochloremia from sodium lost in sweat

▶ failure to thrive: poor weight gain, poor growth, distended abdomen, thin extremities, and sallow skin with poor turgor due to malabsorption

▶ clotting problems, retarded bone growth, and delayed sexual development due to deficiency of fat-soluble vitamins

▶ rectal prolapse in infants and children due to malnutrition and wasting of perirectal supporting tissues

▶ esophageal varices due to cirrhosis and portal hypertension.

Complications

Complications may include:

▶ obstructed glandular ducts (leading to peribronchial thickening) due to increased viscosity of bronchial, pancreatic, and other mucus gland secretions

▶ atelectasis or emphysema due to respiratory effects

▶ diabetes, pancreatitis, and hepatic failure due to effects on the intestines, pancreas, and liver

▶ malnutrition and malabsorption of fat-soluble vitamins (A, D, E, and K) due to deficiencies of trypsin, amylase, and lipase (from obstructed pancreatic ducts, preventing the conversion and absorption of fat and protein in the intestinal tract)

▶ lack of sperm in the semen (azoospermia)

▶ secondary amenorrhea and increased mucus in the reproductive tracts, blocking the passage of ova.

Diagnosis

The following tests help diagnose cystic fibrosis:

▶ test to detect elevated sodium chloride levels in sweat of 60 mEq/L or greater

AGE ALERT The sweat test may be inaccurate in very young infants because they may not produce enough sweat for a valid test. The test may need to be repeated.

▶ chest X-ray to reveal early signs of obstructive lung disease

▶ sputum culture to detect organisms that chronically colonize

▶ electrolyte status to detect dehydration

▶ stool analysis showing absence of trypsin (suggesting pancreatic insufficiency)

▶ DNA testing to detect abnormal gene and determine carrier status and for prenatal diagnosis in families with an affected child.

Treatment

Possible treatments include:

▶ hypertonic radiocontrast materials delivered by enema to treat acute obstructions due to meconium ileus

▶ breathing exercises and chest percussion to clear pulmonary secretions

▶ antibiotics to treat lung infection, guided by sputum culture results

▶ drugs to increase mucus clearance

▶ inhaled beta-adrenergic agonists to control airway constriction

▶ pancreatic enzyme replacement to maintain adequate nutrition

▶ sodium-channel blocker to decrease sodium reabsorption from secretions and improve viscosity

▶ uridine triphosphate to stimulate chloride secretion by a non-CFTR

▶ salt supplements to replace electrolytes lost through sweat

KARYOTYPE OF DOWN SYNDROME

Normally, each autosome is one of a pair. A patient with Down syndrome, or trisomy 21, has an extra chromosome 21.

Male with Down syndrome

▶ recombinant human DNase (Pulmozyme), a DNA-splitting enzyme, to help liquefy mucus
▶ recombinant alpha-antitrypsin to counteract excessive proteolytic activity produced during airway inflammation
▶ gene therapy to introduce normal CFTR into affected epithelial cells
▶ transplantation of heart or lungs in severe organ failure.

Down syndrome

Down syndrome, or trisomy 21, is a spontaneous chromosome abnormality that causes characteristic facial features, other distinctive physical abnormalities, and mental retardation; 60% of affected persons have cardiac defects. It occurs in 1 of 650 to 700 live births. Improved treatment for heart defects, respiratory and other infections, and acute leukemia has sig-

nificantly increased life expectancy. Fetal and neonatal mortality rates remain high, usually resulting from complications of associated heart defects.

Causes

Causes of Down syndrome include:
▶ Advanced age (when the mother is over 35 at delivery or the father is over 42)
▶ Cumulative effects of environmental factors, such as radiation and viruses.

Pathophysiology

Nearly all cases of Down syndrome result from trisomy 21 (3 copies of chromosome 21). The result is a karyotype of 47 chromosomes instead of the usual 46. (See *Karyotype of Down syndrome*.) In 4% of the patients, Down syndrome results from an unbalanced translocation or chromosomal rearrangement in which the long arm of chromosome 21 breaks and attaches to another chromosome.

Some affected persons and some asymptomatic parents may have chromosomal mosaicism, a mixture of two cell types, some with the normal 46 and some with an extra chromosome 21.

Signs and symptoms

AGE ALERT The physical signs of Down syndrome are apparent at birth. The infant is lethargic and has distinctive craniofacial features.

Other signs and symptoms include:
▶ distinctive facial features (low nasal bridge, epicanthic folds, protruding tongue, and low-set ears); small open mouth and large tongue
▶ single transverse crease on the palm (Simian crease)
▶ small white spots on the iris (Brushfield's spots)
▶ mental retardation (estimated IQ of 20 to 50)
▶ developmental delay due to hypotonia and decreased cognitive processing
▶ congenital heart disease, mainly septal defects and especially of the endocardial cushion
▶ impaired reflexes due to decreased muscle tone in limbs.

Complications

Possible complications include:
▶ early death due to cardiac complications
▶ increased susceptibility to leukemia
▶ premature senile dementia, usually in the fourth decade if the patient survives
▶ increased susceptibility to acute and chronic infections
▶ strabismus and cataracts as the child grows
▶ poorly developed genitalia and delayed puberty (females may menstruate and be fertile; males may be infertile, with low serum testosterone levels and often with undescended testes).

Diagnosis

Diagnostic tests include:
▶ definitive karyotype
▶ amniocentesis for prenatal diagnosis, recommended for pregnant women over 34, even with a negative family history
▶ prenatal targeted ultrasonography for duodenal obstruction or an atrioventricular canal defect (suggestive of Down syndrome)
▶ blood tests for reduced alpha-fetoprotein levels (suggestive of Down syndrome).

Treatment

▶ Surgery to correct heart defects and other related congenital abnormalities
▶ Antibiotics for recurrent infections
▶ Plastic surgery to correct characteristic facial traits (especially protruding tongue; possibly improving speech, reducing susceptibility to dental caries, and resulting in fewer orthodontic problems)
▶ Early intervention programs and supportive therapies to maximize mental and physical capabilities
▶ Thyroid hormone replacement for hypothyroidism.

Hemophilia

Hemophilia is an X-linked recessive bleeding disorder; the severity and prognosis of bleeding vary with the degree of deficiency, or nonfunction, and the site of bleeding. Hemophilia occurs in 20 of 100,000 male births and results from a deficiency of specific clotting factors.

Hemophilia A, or classic hemophilia, is a deficiency of clotting factor VIII; it is more common than type B, affecting more than 80% of all hemophiliacs. Hemophilia B, or Christmas disease, affects 15% of all hemophiliacs and results from a deficiency of factor IX. There's no relationship between factor VIII and factor IX inherited defects.

Causes

▶ Defect in a specific gene on the X chromosome that codes for factor VIII synthesis (hemophilia A)

▶ More than 300 different base-pair substitutions involving the factor IX gene on the X chromosome (hemophilia B).

Pathophysiology

Hemophilia is an X-linked recessive genetic disease causing abnormal bleeding because of specific clotting factor malfunction. Factors VIII and IX are components of the intrinsic clotting pathway; factor IX is an essential factor and factor VIII is a critical cofactor. Factor VIII accelerates the activation of factor X by several thousandfold. Excessive bleeding occurs when these clotting factors are reduced by more than 75%. A deficiency or nonfunction of factor VIII causes hemophilia A, and a deficiency or nonfunction of factor IX causes hemophilia B.

Hemophilia may be severe, moderate, or mild, depending on the degree of activation of clotting factors. Patients with severe disease have no detectable factor VIII or factor IX activity. Moderately afflicted patients have 1% to 4% of normal clotting activity, and mildly afflicted patients have 5% to 25% of normal clotting activity.

A person with hemophilia forms a platelet plug at a bleeding site, but clotting factor deficiency impairs the ability to form a stable fibrin clot. Delayed bleeding is more common than immediate hemorrhage.

Signs and symptoms

Signs and symptoms may include:
▶ spontaneous bleeding in severe hemophilia (prolonged or excessive bleeding after circumcision is often the first sign)
▶ excessive or continued bleeding or bruising after minor trauma or surgery
▶ large subcutaneous and deep intramuscular hematomas due to mild trauma
▶ prolonged bleeding in mild hemophilia after major trauma or surgery, but no spontaneous bleeding after minor trauma
▶ pain, swelling, and tenderness due to bleeding into joints (especially weight-bearing joints)

▶ internal bleeding, often manifested as abdominal, chest, or flank pain
▶ hematuria from bleeding into kidney
▶ hematemesis or tarry stools from bleeding into the GI tract.

Complications

Complications may include:
▶ peripheral neuropathy, pain, paresthesia, and muscle atrophy due to bleeding near peripheral nerves
▶ ischemia and gangrene due to impaired blood flow through a major vessel distal to bleed
▶ decreased tissue perfusion and hypovolemic shock (shown as restlessness, anxiety, confusion, pallor, cool and clammy skin, chest pain, decreased urine output, hypotension, and tachycardia).

Diagnosis

▶ Specific coagulation factor assays to diagnose the type and severity of hemophilia
▶ Factor VIII assay of 0% to 30% of normal and prolonged activated partial thromboplastin time (hemophilia A)
▶ Deficient factor IX and normal factor VIII levels (hemophilia B)
▶ Normal platelet count and function, bleeding time, and prothrombin time (hemophilia A and B).

Treatment

Treatment of hemophilia includes:
▶ cryoprecipitated or lyophilized antihemophilic factor to increase clotting factor levels (to permit normal hemostasis in hemophilia A)
▶ factor IX concentrate during bleeding episodes (hemophilia B)
▶ cold compresses or ice bags and elevation of bleeding site to slow or stop flow
▶ analgesics to control pain
▶ aminocaproic acid (Amicar) for oral bleeding (inhibits plasminogen activator substances)
▶ prophylactic desmopressin (DDAVP) before dental procedures or minor surgery to release stored von Willebrand's factor and factor VIII (to reduce bleeding)

❱ no IM injections, such as analgesics, due to possible hematoma at the site

❱ no aspirin or aspirin-containing medications due to decreased platelet adherence and possible increased bleeding.

AGE ALERT To help prevent injury, young children should wear clothing with padded patches on the knees and elbows. Older children should avoid contact sports.

Marfan syndrome

Marfan syndrome is a rare degenerative, generalized disease of the connective tissue. It results from elastin and collagen defects and causes ocular, skeletal, and cardiovascular anomalies. Death occurs from cardiovascular complications from early infancy to adulthood. The syndrome occurs in 1 of 20,000 individuals, affecting males and females equally.

Cause
❱ Autosomal dominant mutation.

Pathophysiology
The syndrome is caused by a mutation in a single allele of a gene located on chromosome 15; the gene codes for fibrillin, a glycoprotein component of connective tissue. These small fibers are abundant in large blood vessels and the suspensory ligaments of the ocular lenses. The effect on connective tissue is varied and includes excessive bone growth, ocular disorders, and cardiac defects.

Signs and symptoms
Signs and symptoms may include:
❱ increased height, long extremities, and arachnodactyly (long spider-like fingers) due to effects on long bones and joints and excessive bone growth
❱ defects of sternum (funnel chest or pigeon breast, for example), chest asymmetry, scoliosis, and kyphosis
❱ hypermobile joints due to effects on connective tissue
❱ nearsightedness due to elongated ocular globe

❱ lens displacement due to altered connective tissue
❱ valvular abnormalities (redundancy of leaflets, stretching of chordae tendineae, and dilation of valvulae annulus)
❱ mitral valve prolapse due to weakened connective tissue
❱ aortic regurgitation due to dilation of aortic root and ascending aorta.

Complications
Possible complications include:
❱ weak joints and ligaments, predisposing to injury
❱ cataracts due to lens displacement
❱ retinal detachments and retinal tears
❱ severe mitral valve regurgitation due to mitral valve prolapse
❱ spontaneous pneumothorax due to chest wall instability
❱ inguinal and incisional hernias
❱ dilation of the dural sac (portion of the dura mater beyond caudal end of the spinal cord).

Diagnosis
❱ Positive family history in one parent (85% of patients) or negative family history (15%; suggesting a mutation, possibly from advanced paternal age)
❱ Clinical presentation and history of the disease in close relatives
❱ Presence of lens displacement and aneurysm of the ascending aorta without other symptoms or familial tendency
❱ Detection of fibrillin defects in cultured skin
❱ X-rays confirming abnormalities
❱ Echocardiogram showing dilation of the aortic root
❱ DNA analysis of the gene.

Treatment
Treatment for Marfan syndrome may involve:
❱ surgical repair of aneurysms to prevent rupture
❱ surgical correction of ocular deformities to improve vision

CHARACTERISTICS OF SICKLED CELLS

Normal red blood cells and sickled cells vary in shape, life span, oxygen-carrying capacity, and the rate at which they're destroyed. The illustration shows normal and sickled cells and lists the major differences.

NORMAL RED BLOOD CELLS

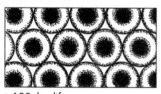

- 120-day life span
- Hemoglobin has normal oxygen-carrying capacity
- 12 to 14 g/ml of hemoglobin
- Red blood cells destroyed at normal rate

SICKLED CELLS

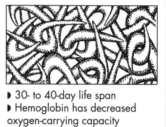

- 30- to 40-day life span
- Hemoglobin has decreased oxygen-carrying capacity
- 6 to 9 g/ml of hemoglobin
- Red blood cells destroyed at accelerated rate

- steroid and sex hormone therapy to induce early epiphyseal closure and limit adult height
- beta-adrenergic blockers to delay or prevent aortic dilation
- surgical replacement of aortic valve and mitral valve
- mechanical bracing and physical therapy for mild scoliosis if curvature > 20 degrees
- surgery for scoliosis if curvature > 45 degrees.

Sickle cell anemia

Sickle cell anemia is a congenital hemolytic anemia resulting from defective hemoglobin molecules. Half the patients with sickle cell anemia die by their early twenties, and few live to middle age.

 CULTURAL DIVERSITY Sickle cell anemia occurs primarily in persons of African and Mediterranean descent, but it also affects other populations. Although most common in tropical Africans and people of African descent, it also occurs in Puerto Rico, Turkey, India, the Middle East, and the Mediterranean.

Cause
- Mutation of hemoglobin S gene (heterozygous inheritance results in sickle cell trait, usually an asymptomatic condition).

Pathophysiology
Sickle cell anemia results from substitution of the amino acid valine for glutamic acid in the hemoglobin S gene encoding the beta chain of hemoglobin. Abnormal hemoglobin S, found in the red blood cells of patients, becomes insoluble during hypoxia. As a result, these cells become rigid, rough, and elongated, forming a crescent or sickle shape. (See *Characteristics of sickled cells*.) The sickling produces hemolysis. The altered cells also pile up in the capillaries and smaller blood vessels, making the blood more viscous. Normal circulation is impaired, causing pain, tissue infarctions, and swelling.

Each patient with sickle cell anemia has a different hypoxic threshold and different factors that trigger a sickle cell crisis.

SICKLE CELL CRISIS

Infection, exposure to cold, high altitudes, overexertion, or other situations that cause cellular oxygen deprivation may trigger a sickle cell crisis. The deoxygenated, sickle-shaped red blood cells stick to the capillary wall and each other, blocking blood flow and causing cellular hypoxia. The crisis worsens as tissue hypoxia and acidic waste products cause more sickling and cell damage. With each new crisis, organs and tissues are slowly destroyed, especially the spleen and kidneys.

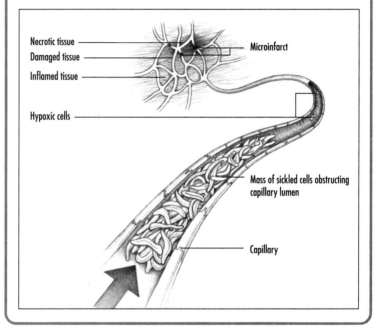

Necrotic tissue

Damaged tissue

Inflamed tissue

Hypoxic cells

Microinfarct

Mass of sickled cells obstructing capillary lumen

Capillary

Illness, exposure to cold, stress, acidotic states, or a pathophysiologic process that pulls water out of the sickle cells precipitates a crisis in most patients. (See *Sickle cell crisis.*) The blockages then cause anoxic changes that lead to further sickling and obstruction.

Signs and symptoms

AGE ALERT Symptoms of sickle cell anemia don't develop until after the age of 6 months because fetal hemoglobin protects infants for the first few months after birth.

Signs and symptoms may include:

▶ severe pain in the abdomen, thorax, muscle, or bones (characterizes painful crisis)
▶ jaundice, dark urine, and low-grade fever due to blood vessel obstruction by rigid, tangled, sickle cells (leading to tissue anoxia and possibly necrosis)
▶ *Streptococcus pneumoniae* sepsis due to autosplenectomy (splenic damage and scarring in patients with long-term disease)
▶ aplastic crisis due to bone marrow depression (associated with infection, usually viral)
▶ megaloblastic crisis due to bone marrow depression, decreased bone marrow

activity, and hemolysis (characterized by pallor, lethargy, sleepiness, and dyspnea, with possible coma)
▶ acute sequestration crisis (rare; affects infants aged 8 months to 2 years; may cause lethargy, pallor, and hypovolemic shock) due to the sudden massive entrapment of cells in spleen and liver
▶ hemolytic crisis (rare; usually affects patients who also have glucose-6-phosphate dehydrogenase deficiency; degenerative changes cause liver congestion and enlargement and chronic jaundice worsens).

Complications
Complications may include:
▶ retinopathy, nephropathy, and cerebral vessel occlusion due to organ infarction
▶ hypovolemic shock and death due to massive entrapment of cells
▶ necrosis
▶ infection and gangrene.

Diagnosis
▶ Positive family history and typical clinical features
▶ Hemoglobin electrophoresis, showing hemoglobin S
▶ Electrophoresis of umbilical cord blood to provide screening for all newborns at risk
▶ Stained blood smear showing sickle cells
▶ Low red blood cell counts, elevated white blood cell and platelet counts, decreased erythrocyte sedimentation rate, increased serum iron levels, decreased red blood cell survival, and reticulocytosis (hemoglobin levels may be low or normal)
▶ Lateral chest X-ray showing "Lincoln log" deformity in the vertebrae of many adults and some adolescents.

Treatment
Possible treatments include:
▶ packed red blood cell transfusion to correct hypovolemia (if hemoglobin levels decrease)
▶ sedation and analgesics, such as meperidine (Demerol) or morphine sulfate, for pain

▶ oxygen administration to correct hypoxia
▶ large amounts of oral or I.V. fluids to correct hypovolemia and prevent dehydration and vessel occlusion
▶ prophylactic penicillin before the age of 4 months to prevent infection
▶ warm compresses to painful areas to promote venous drainage
▶ iron and folic acid supplements to prevent anemia.

AGE ALERT Vaccines to prevent illness and anti-infectives, such as low-dose penicillin, should be considered to prevent complications in patients with sickle cell anemia.

Spina bifida
Spina bifida is the incomplete fusion of one or more vertebrae, resulting in dimpling of the area (spina bifida occulta) or protrusion of the spinal tissue (spina bifida cystica). Spina bifida occulta occurs in as many as 25% of births, and spina bifida cystica occurs in 1 in 1,000 births in the United States. The incidence varies greatly with countries and regions.

CULTURAL DIVERSITY The incidence of spina bifida is significantly greater in the British Isles and low in southern China and Japan. In the United States, North and South Carolina have twice the incidence as most other parts of the country.

Causes
Causes of spina bifida include:
▶ teratogenic insults before the 26th day of gestation
▶ isolated birth defect or part of a multiple chromosomal malformation (trisomy 18 or 13)
▶ environmental (such as lack of folic acid in the mother's diet) and genetic factors.

Pathophysiology
Neural tube closure normally occurs at 24 days' gestation in the cranial region and continues distally, with closure of the lumbar regions by 26 days. As the nervous system develops and differentiates, it's

vulnerable to teratogenic effects. Teratogenic insults during this critical time can cause structural defects, intrauterine growth retardation, and cognitive abnormalities.

The specific cause of spinal bifida, incomplete fusion of the vertebrae in the developing nervous system of the embryo, is unknown, but neural tube defects have been associated with such noninfectious maternal disorders as folic acid deficiency.

Spina bifida occulta rarely affects the structure or function of the cord and peripheral nerve roots. Spina bifida cystica can occur as a meningocele, in which the meninges protrude in a cerebrospinal fluid-filled sac, or as a myelomeningocele, in which peripheral nerves, root segments, or the spinal cord also protrude. Varying degrees of sensory and motor dysfunction below the level of the lesion are present.

Signs and symptoms

Signs and symptoms depend on the type and severity of neural tube defect:
▶ weak feet or bowel and bladder disturbances due to rapid growth phases and abnormal adherence of the spinal cord to other tissues (occasionally with spina bifida occulta)
▶ dimple, tuft of hair, soft fatty deposits, port wine nevi, or combination over spine (spina bifida occulta)
▶ saclike protrusion over the spine due to meningocele or myelomeningocele (spina bifida cystica)
▶ possible permanent neurologic dysfunction due to meningocele or myelomeningocele (spina bifida cystica).

Complications

Complications may include:
▶ paralysis below the level of the defect
▶ infection, such as meningitis.

Diagnosis

▶ Amniocentesis to detect elevated alpha-fetoprotein levels (indicates presence of open neural tube defect)
▶ Acetylcholinesterase levels (can confirm diagnosis)
▶ Four-marker screen (maternal serum alpha-fetoprotein, human chorionic gonadotropin [HCG], free alpha-subunit-HCG, and unconjugated estriol) of women not scheduled for amniocentesis; elevated levels suggest a defect
▶ Physical examination to detect meningocele and myelomeningocele
▶ X-ray to show bone defect or palpation to detect defect (spina bifida occulta)
▶ Myelography to confirm spina bifida occulta from other spinal abnormalities.

Treatment

▶ Usually no treatment (spina bifida occulta)
▶ Neurosurgic closure (treatment of meningocele)
▶ Repair of the sac and supportive measures to promote independence and prevent further complications (treatment of myelomeningocele).

Tay-Sachs disease

Tay-Sachs disease, also known as GM_2 gangliosidosis, is the most common lipid-storage disease.

AGE ALERT Progressive mental and motor deterioration often causes death before the age of 5 years. Tay-Sachs disease appears in fewer than 100 infants born each year in the United States.

 CULTURAL DIVERSITY Tay-Sachs affects persons of Eastern European Jewish (Ashkenazi) ancestry about 100 times more often than the general population, occurring in about 1 in 3,600 live births in this ethnic group.

Cause

▶ Congenital deficiency of the enzyme hexosaminidase A.

CULTURAL DIVERSITY About 1 in 30 Ashkenazi Jews, French Canadians, and American Cajuns are heterozygous carriers. If two such carriers have children, each of their offspring has a 25% chance of having Tay-Sachs disease.

Pathophysiology

Tay-Sachs disease is an autosomal recessive disorder in which the enzyme hexosaminidase A is absent or deficient. This enzyme is necessary to metabolize gangliosides, water-soluble glycolipids found primarily in the central nervous system (CNS). Without hexosaminidase A, lipid pigments accumulate and progressively destroy and demyelinate the CNS cells.

Signs and symptoms

Signs and symptoms of Tay-Sachs disease may include:
▶ exaggerated Moro reflex (also called startle reflex) at birth and apathy (response only to loud sounds) by age 3 to 6 months due to demyelination of CNS cells
▶ inability to sit up, lift the head, or grasp objects; difficulty turning over; progressive vision loss due to CNS involvement
▶ deafness, blindness, seizure activity, paralysis, spasticity, and continued neurologic deterioration (by 18 months of age)
▶ recurrent bronchopneumonia due to diminished protective reflexes.

Complications

Complications may include:
▶ blindness
▶ generalized paralysis
▶ recurrent bronchopneumonia, usually fatal by 5 years of age.

Diagnosis

▶ Clinical features
▶ Serum analysis showing deficient hexosaminidase A.

 CULTURAL DIVERSITY Diagnostic screening is recommended for all couples of Ashkenazi Jewish ancestry and for others with a familial history of the disease. A blood test can detect carriers.

Treatment

Treatment for Tay-Sachs disease includes the following:
▶ tube feedings to provide nutritional supplements
▶ suctioning and postural drainage to maintain a patent airway
▶ skin care to prevent pressure ulcers in bedridden children
▶ laxatives to relieve neurogenic constipation.

5

The body is mostly liquid — various electrolytes dissolved in water. Electrolytes are ions (electrically charged versions) of essential elements — predominantly sodium (Na^+), chloride (Cl^-), oxygen (O_2), hydrogen (H^+), bicarbonate (HCO_3^-), calcium (Ca^{2+}), potassium (K^+), sulfate (SO_4^{2-}), and phosphate (PO_4^{3-}). Only ionic forms of elements can dissolve or combine with other elements. Electrolyte balance must remain in a narrow range for the body to function. The kidneys maintain chemical balance throughout the body by producing and eliminating urine. They regulate the volume, electrolyte concentration, and acid-base balance of body fluids; detoxify and eliminate wastes; and regulate blood pressure by regulating fluid volume. The skin and lungs also play a role in fluid and electrolyte balance. Sweating results in loss of sodium and water; every breath contains water vapor.

FLUID BALANCE

The kidneys maintain fluid balance in the body by regulating the amount and components of fluid inside and around the cells.

Intracellular fluid

The fluid inside each cell is called the intracellular fluid (ICF). Each cell has its own mixture of components in the intracellular fluid, but the amounts of these substances are similar in every cell. Intracellular fluid contains large amounts of potassium, magnesium, and phosphate ions.

Extracellular fluid

The fluid in the spaces outside the cells, called extracellular fluid (ECF), is constantly moving. Normally, ECF includes blood plasma and interstitial fluid (the fluid between cells in tissues); in some pathologic states it accumulates in a so-called third space, the space around organs in the chest or abdomen.

ECF is rapidly transported through the body by circulating blood and between blood and tissue fluids by fluid and electrolyte exchange across the capillary walls. ECF contains large amounts of sodium, chloride, and bicarbonate ions, plus such cell nutrients as oxygen, glucose, fatty acids, and amino acids. It also contains CO_2, transported from the cells to the lungs for excretion, and other cellular products, transported from the cells to the kidneys for excretion.

The kidneys maintain the volume and composition of ECF and, to a lesser extent, intracellular fluid by continually exchanging water and ionic solutes, such as hydrogen, sodium, potassium, chloride, bicarbonate, sulfate, and phosphate ions, across the cell membranes of the renal tubules.

Fluid exchange

Two sets of forces determine the exchange of fluid between blood plasma and interstitial fluid. All four forces act to equalize concentrations of fluids, electrolytes, and proteins on both sides of the capillary wall.

Forces that tend to move fluid from the vessels to the interstitial fluid are:

▶ hydrostatic pressure of blood (the outward pressure of plasma against the walls of capillaries)

▶ osmotic pressure of tissue fluid (the tendency of ions to move across a semipermeable membrane — the capillary wall — from an area of greater concentration to one of lower concentration)

Forces that tend to move fluid into vessels are:

▶ oncotic pressure of plasma proteins (similar to osmosis, but because proteins can't cross the vessel wall, they attract fluid into the area of greater concentration)

▶ hydrostatic pressure of interstitial fluid (inward pressure against the capillary walls).

Hydrostatic pressure at the arteriolar end of the capillary bed is greater than at the venular end. Oncotic pressure of plasma increases slightly at the venular end as fluid escapes. When the endothelial barrier (capillary wall) is normal and intact, fluid escapes at the arteriolar end of the capillary bed and is returned at the venular end. The small amount of fluid lost from the capillaries into the interstitial tissue spaces is drained off through the lymphatic system and returned to the bloodstream.

ACID–BASE BALANCE

Regulation of the ECF environment involves the ratio of acid to base, measured clinically as pH. In physiology, all positively charged ions are acids and all negatively charged ions are bases. To regulate acid–base balance, the kidneys secrete hydrogen ions (acid), reabsorb sodium (acid) and bicarbonate ions (base), acidify phosphate salts, and produce ammonium ions (acid). This keeps the blood at its normal pH of 7.37 to 7.43. The following are important pH boundaries:

▶ <6.8   incompatible with life
▶ <7.2   cell function seriously impaired
▶ <7.35   acidosis
▶ 7.37 to 7.43   normal
▶ >7.45   alkalosis
▶ >7.55   cell function seriously impaired
▶ >7.8   incompatible with life.

PATHOPHYSIOLOGIC MANIFESTATIONS OF ELECTROLYTE IMBALANCE

The regulation of intracellular and extracellular electrolyte concentrations depends on:

▶ balance between the intake of substances containing electrolytes and the output of electrolytes in urine, feces, and sweat

▶ transport of fluid and electrolytes between extracellular and intracellular fluid.

Fluid imbalance occurs when regulatory mechanisms can't compensate for abnormal intake and output at any level from the cell to the organism. Fluid and electrolyte imbalances include edema, isotonic alterations, hypertonic alterations, hypotonic alterations, and electrolyte imbalances. Disorders of fluid volume or osmolarity (concentration of electrolytes in the fluid) result. Many conditions also affect capillary exchange, resulting in fluid shifts.

Edema

Despite almost constant interchange through the endothelial barrier, the body maintains a steady state of extracellular water balance between the plasma and interstitial fluid. Increased fluid volume in the interstitial spaces is called edema. It's classified as localized or systemic. Obstruction of the veins or lymphatic system or increased vascular permeability usually causes localized edema in the affected area, such as the swelling around an injury. Systemic, or generalized edema, may be due to heart failure or renal disease. Massive systemic edema is called anasarca.

Edema results from abnormal expansion of the interstitial fluid or the accumulation of fluid in a third space, such as

CAUSES OF EDEMA

Edema results when excess fluid accumulates in the interstitial spaces. The chart shows the causes and effects of this fluid accumulation.

CAUSE	UNDERLYING CONDITION
Increased hydrostatic pressure	Heart failure Constrictive pericarditis Venous thrombosis Cirrhosis
Hypoproteinemia	Cirrhosis Malnutrition Nephrotic syndrome Gastroenteropathy
Lymphatic obstruction	Cancer Inflammatory scarring Radiation
Sodium retention	Excessive salt intake Increased tubular reabsorption of sodium Reduced renal perfusion
Increased endothelial permeability	Inflammation Burns Trauma Allergic or immunologic reactions

the peritoneum (ascites), pleural cavity (hydrothorax), or pericardial sac (pericardial effusion). (See *Causes of edema*.)

Tonicity

Many fluid and electrolyte disorders are classified according to how they affect osmotic pressure, or tonicity. Tonicity describes the relative concentrations of electrolytes (osmotic pressure) on both sides of a semipermeable membrane (the cell wall or the capillary wall). The word *normal* in this context refers to the usual electrolyte concentration of physiologic fluids. Normal saline has a sodium chloride concentration of 0.9%.

▶ Isotonic solutions have the same electrolyte concentration and therefore the same osmotic pressure.

▶ Hypertonic solutions have a greater than normal concentration of some essential electrolyte, usually sodium.

▶ Hypotonic solutions have a lower than normal concentration of some essential electrolyte, also usually sodium.

Isotonic alterations

Isotonic alterations or disorders don't make the cells swell or shrink because osmosis doesn't occur. They occur when intracellular and extracellular fluids have equal osmotic pressure, but there's a dramatic change in total-body fluid volume. Examples include blood loss from penetrating trauma or expansion of fluid volume if a patient receives too much normal saline.

Hypertonic alterations

Hypertonic alterations occur when the ECF is more concentrated than the ICF. Water flows out of the cell through the semipermeable cell membrane, causing cell shrinkage. This can occur when a patient is given hypertonic (>0.9%) saline, when severe dehydration causes hypernatremia (high Na^+ concentration in blood), or when renal disease causes sodium retention.

Hypotonic alterations

When the ECF becomes hypotonic, osmotic pressure forces some ECF into the cells, causing them to swell. Overhydration is the most common cause; as water dilutes the ECF, it becomes hypotonic with respect to the ICF. Water moves into the cells until balance is restored. In extreme hypotonicity, cells may swell until they burst and die.

Alterations in electrolyte balance

The major electrolytes are the cations (positively charged ions) sodium, potassium, calcium, and magnesium and the anions (negatively charged ions) chloride, phosphate, and bicarbonate. The body continuously attempts to maintain intracellular and extracellular equilibrium of electrolytes. Too much or too little of any electrolyte will affect most body systems.

Sodium and potassium

Sodium is the major cation in ECF, and potassium is the major cation in ICF. Especially in nerves and muscles, communication within and between cells involves changes (repolarization and depolarization) in surface charge on the cell membrane. During repolarization, an active transport mechanism in the cell membrane, called the sodium–potassium pump, continually shifts sodium into and potassium out of cells; during depolarization, the process is reversed.

Physiologic roles of sodium cations include:

▶ maintaining tonicity of ECF
▶ regulating acid–base balance by renal reabsorption of sodium ion (base) and excretion of hydrogen ion (acid)
▶ facilitating nerve conduction and neuromuscular function
▶ facilitating glandular secretion
▶ maintaining water balance.

Physiologic roles of potassium include:

▶ maintain cell electrical neutrality
▶ facilitate cardiac muscle contraction and electrical conductivity
▶ facilitate neuromuscular transmission of nerve impulses
▶ maintain acid–base balance.

Chloride

Chloride is mainly an extracellular anion; it accounts for two-thirds of all serum anions. Secreted by the stomach mucosa as hydrochloric acid, it provides an acid medium for digestion and enzyme activation. Chloride also:

▶ helps maintain acid–base and water balances
▶ influences the tonicity of ECF
▶ facilitates exchange of oxygen and CO_2 in red blood cells
▶ helps activate salivary amylase, which triggers the digestive process.

Calcium

Calcium is indispensable in cell permeability, bone and teeth formation, blood coagulation, nerve impulse transmission, and normal muscle contraction. Hypocalcemia can cause tetany and seizures; hypercalcemia can cause cardiac arrhythmias and coma.

Magnesium

Magnesium is present in smaller quantity, but physiologically it is as significant as the other major electrolytes. The major function of magnesium is to enhance neuromuscular communication. Other functions include:

▶ stimulating parathyroid hormone secretion, which regulates intracellular calcium

FLUID AND ELECTROLYTE IMPLICATIONS OF BLOOD PRESSURE FINDINGS

Blood pressure reflects changes in fluid and electrolyte status.

BLOOD PRESSURE	FLUID AND ELECTROLYTE STATUS
Normal	▶ Hemodynamic stability ▶ Initial hemodynamic instability
Hypotension	▶ Fluid volume deficit ▶ Potassium imbalance ▶ Calcium imbalance ▶ Magnesium imbalance ▶ Acidosis
Hypertension	▶ Fluid volume excess ▶ Hypernatremia

▶ activating many enzymes in carbohydrate and protein metabolism
▶ facilitating cell metabolism
▶ facilitating sodium, potassium, and calcium transport across cell membranes
▶ facilitating protein transport.

Phosphate

The phosphate anion is involved in cellular metabolism as well as neuromuscular regulation and hematologic function. Phosphate reabsorption in the renal tubules is inversely related to calcium levels, which means that an increase in urinary phosphorous triggers calcium reabsorption and vice versa.

Effects of electrolyte imbalance

Electrolyte imbalances can affect all body systems. Too much or too little potassium or too little calcium or magnesium can increase the excitability of the cardiac muscle, causing arrhythmias. Multiple neurologic symptoms may result from electrolyte imbalance, ranging from disorientation or confusion to a completely depressed central nervous system (CNS). Too much or too little sodium or too much potassium can cause oliguria. Blood pressure may be increased or decreased. (See *Fluid and electrolyte implications of blood pressure findings*.) The GI tract is particularly susceptible to electrolyte imbalance:
▶ too much potassium—abdominal cramps, nausea, and diarrhea
▶ too little potassium—paralytic ileus
▶ too much magnesium—nausea, vomiting, and diarrhea
▶ too much calcium—nausea, vomiting, and constipation.

DISORDERS OF ELECTROLYTE BALANCE

Fluid and electrolyte balance is essential for health. Many factors, such as illness, injury, surgery, and treatments, can disrupt a patient's fluid and electrolyte balance. Even a patient with a minor illness is at risk for fluid and electrolyte imbalance. (See *Electrolyte imbalances*, pages 114 to 119.)

Hypovolemia

Water content of the human body progressively decreases from birth to old age, as follows:

(Text continues on page 118.)

ELECTROLYTE IMBALANCES

Signs and symptoms of a fluid and electrolyte imbalance are often subtle. Blood chemistry tests help diagnose and evaluate electrolyte imbalances.

ELECTROLYTE IMBALANCE	SIGNS AND SYMPTOMS
Hyponatremia	▸ Muscle twitching and weakness due to osmotic swelling of cells ▸ Lethargy, confusion, seizures, and coma due to altered neurotransmission ▸ Hypotension and tachycardia due to decreased extracellular circulating volume ▸ Nausea, vomiting, and abdominal cramps due to edema affecting receptors in the brain or vomiting center of the brain stem ▸ Oliguria or anuria due to renal dysfunction
Hypernatremia	▸ Agitation, restlessness, fever, and decreased level of consciousness due to altered cellular metabolism ▸ Hypertension, tachycardia, pitting edema, and excessive weight gain due to water shift from intracellular to extracellular fluid ▸ Thirst, increased viscosity of saliva, rough tongue due to fluid shift ▸ Dyspnea, respiratory arrest, and death from dramatic increase in osmotic pressure
Hypokalemia	▸ Dizziness, hypotension, arrhythmias, electrocardiogram (ECG) changes, and cardiac arrest due to changes in membrane excitability ▸ Nausea, vomiting, anorexia, diarrhea, decreased peristalsis, and abdominal distention due to decreased bowel motility ▸ Muscle weakness, fatigue, and leg cramps due to decreased neuromuscular excitability
Hyperkalemia	▸ Tachycardia changing to bradycardia, ECG changes, and cardiac arrest due to hypopolarization and alterations in repolarization ▸ Nausea, diarrhea, and abdominal cramps due to decreased gastric motility ▸ Muscle weakness and flaccid paralysis due to inactivation of membrane sodium channels
Hypochloremia	▸ Muscle hypertonicity and tetany ▸ Shallow, depressed breathing ▸ Usually associated with hyponatremia and its characteristic symptoms, such as muscle weakness and twitching

DIAGNOSTIC TEST RESULTS

- Serum sodium < 135 mEq/L
- Decreased urine specific gravity
- Decreased serum osmolality
- Urine sodium > 100 mEq/24 hours
- Increased red blood cell count

- Serum sodium > 145 mEq/L
- Urine sodium < 40 mEq/24 hours
- High serum osmolality

- Serum potassium < 3.5 mEq/L
- Coexisting low serum calcium and magnesium levels not responsive to treatment for hypokalemia usually suggest hypomagnesemia
- Metabolic alkalosis
- ECG changes include flattened T waves, elevated U waves, depressed ST segment

- Serum potassium > 5 mEq/L
- Metabolic acidosis
- ECG changes include tented and elevated T waves, widened QRS complex, prolonged PR interval, flattened or absent P waves, depressed ST segment

- Serum chloride < 98 mEq/L
- Serum pH > 7.45 (supportive value)
- Serum CO_2 > 32 mEq/L (supportive value)

(continued)

ELECTROLYTE IMBALANCES *(continued)*

ELECTROLYTE IMBALANCE	SIGNS AND SYMPTOMS
Hyperchloremia	▸ Deep, rapid breathing ▸ Weakness ▸ Diminished cognitive ability, possibly leading to coma
Hypocalcemia	▸ Anxiety, irritability, twitching around the mouth, laryngospasm, seizures, Chvostek's and Trousseau's signs due to enhanced neuromuscular irritability ▸ Hypotension and arrhythmias due to decreased calcium influx
Hypercalcemia	▸ Drowsiness, lethargy, headaches, irritability, confusion, depression, or apathy due to decreased neuromuscular irritability (increased threshold) ▸ Weakness and muscle flaccidity due to depressed neuromuscular irritability and release of acetylcholine at the myoneural junction ▸ Bone pain and pathological fractures due to calcium loss from bones ▸ Heart block due to decreased neuromuscular irritability ▸ Anorexia, nausea, vomiting, constipation, and dehydration due to hyperosmolarity ▸ Flank pain due to kidney stone formation
Hypomagnesemia	▸ Nearly always coexists with hypokalemia and hypocalcemia ▸ Hyperirritability, tetany, leg and foot cramps, positive Chvostek's and Trousseau's signs, confusion, delusions, and seizures due to alteration in neuromuscular transmission ▸ Arrhythmias, vasodilation, and hypotension due to enhanced inward sodium current or concurrent effects of calcium and potassium imbalance

DIAGNOSTIC TEST RESULTS

▶ Serum chloride > 108 mEq/L
▶ Serum pH < 7.35, serum CO_2 < 22 mEq/L (supportive values)

▶ Serum calcium < 8.5 mg/dl
▶ Low platelet count
▶ ECG shows lengthened QT interval, prolonged ST segment, arrhythmias
▶ Possible changes in serum protein because half of serum calcium is bound to albumin

▶ Serum calcium > 10.5 mg/dl
▶ ECG shows signs of heart block and shortened QT interval
▶ Azotemia
▶ Decreased parathyroid hormone level
▶ Sulkowitch urine test shows increased calcium precipitation

▶ Serum magnesium < 1.5 mEq/L
▶ Coexisting low serum potassium and calcium levels

(continued)

ELECTROLYTE IMBALANCES (continued)

ELECTROLYTE IMBALANCE	SIGNS AND SYMPTOMS
Hypermagnesemia	▶ Hypermagnesemia is uncommon, caused by decreased renal excretion (renal failure) or increased intake of magnesium ▶ Diminished reflexes, muscle weakness to flaccid paralysis due to suppression of acetylcholine release at the myoneural junction, blocking neuromuscular transmission and reducing cell excitablity ▶ Respiratory distress secondary to respiratory muscle paralysis ▶ Heart block, bradycardia due to decreased inward sodium current ▶ Hypotension due to relaxation of vascular smooth muscle and reduction of vascular resistance by displacing calcium from the vascular wall surface
Hypophosphatemia	▶ Muscle weakness, tremor, and paresthesia due to deficiency of adenosine triphosphate ▶ Peripheral hypoxia due to 2,3-diphosphoglycerate deficiency
Hyperphosphatemia	▶ Usually asymptomatic unless leading to hypocalcemia, with tetany and seizures

▶ in the newborn, as much as 75% of body weight
▶ in adults, about 60% of body weight
▶ in the elderly, about 55%
Most of the decrease occurs in the first 10 years of life. Hypovolemia, or ECF volume deficit, is the isotonic loss of body fluids, that is, relatively equal losses of sodium and water.

 AGE ALERT Infants are at risk for hypovolemia because their bodies need to have a higher proportion of water to total body weight.

Causes

Excessive fluid loss, reduced fluid intake, third-space fluid shift, or a combination of these factors can cause ECF volume loss.

Causes of fluid loss include:
▶ hemorrhage
▶ excessive perspiration
▶ renal failure with polyuria
▶ abdominal surgery
▶ vomiting or diarrhea
▶ nasogastric drainage
▶ diabetes mellitus with polyuria or diabetes insipidus
▶ fistulas
▶ excessive use of laxatives
▶ excessive diuretic therapy
▶ fever.
    Possible causes of reduced fluid intake include:
▶ dysphagia
▶ coma
▶ environmental conditions preventing fluid intake

DIAGNOSTIC TEST RESULTS

▶ Serum magnesium > 2.5 mEq/L
▶ Coexisting elevated potassium and calcium levels

▶ Serum phosphate < 2.5 mg/dl
▶ Urine phosphate > 1.3 g/24 hours

▶ Serum phosphate > 4.5 mg/dl
▶ Serum calcium < 9 mg/dl
▶ Urine phosphorus < 0.9 g/24 hours

▶ psychiatric illness.
Fluid shift may be related to:
▶ burns (during the initial phase)
▶ acute intestinal obstruction
▶ acute peritonitis
▶ pancreatitis
▶ crushing injury
▶ pleural effusion
▶ hip fracture (1.5 to 2 L of blood may accumulate in tissues around the fracture).

Pathophysiology

Hypovolemia is an isotonic disorder. Fluid volume deficit decreases capillary hydrostatic pressure and fluid transport. Cells are deprived of normal nutrients that serve as substrates for energy production, metabolism, and other cellular functions. Decreased renal blood flow triggers the renin–angiotensin system to increase sodium and water reabsorption. The cardiovascular system compensates by increasing heart rate, cardiac contractility, venous constriction, and systemic vascular resistance, thus increasing cardiac output and mean arterial pressure. Hypovolemia also triggers the thirst response, releasing more antidiuretic hormone and producing more aldosterone.

When compensation fails, hypovolemic shock occurs in the following sequence:
▶ decreased intravascular fluid volume
▶ diminished venous return, which reduces preload and decreases stroke volume
▶ reduced cardiac output
▶ decreased mean arterial pressure
▶ impaired tissue perfusion

ESTIMATING FLUID LOSS

The following assessment parameters indicate the severity of fluid loss.

MINIMAL FLUID LOSS

Intravascular volume loss of 10% to 15% is regarded as minimal. Signs and symptoms include:
▶ slight tachycardia
▶ normal supine blood pressure
▶ positive postural vital signs, including a decrease in systolic blood pressure more than 10 mm Hg or an increase in pulse rate more than 20 beats/minute
▶ increased capillary refill time (> 3 seconds)
▶ urine output greater than 30 ml/hour
▶ cool, pale skin on arms and legs
▶ anxiety.

MODERATE FLUID LOSS

Intravascular volume loss of about 25% is regarded as moderate. Signs and symptoms include:
▶ rapid, thready pulse
▶ supine hypotension
▶ cool truncal skin
▶ urine output of 10 to 30 ml/hour
▶ severe thirst
▶ restlessness, confusion, or irritability.

SEVERE FLUID LOSS

Intravascular volume loss of 40% or more is regarded as severe. Signs and symptoms include:
▶ marked tachycardia
▶ marked hypotension
▶ weak or absent peripheral pulses
▶ cold, mottled, or cyanotic skin
▶ urine output less than 10 ml/hour
▶ unconsciousness.

▶ decreased oxygen and nutrient delivery to cells
▶ multisystem organ failure.

Signs and symptoms

Signs and symptoms depend on the amount of fluid loss. (See *Estimating fluid loss.*) These may include:
▶ orthostatic hypotension due to increased systemic vascular resistance and decreased cardiac output
▶ tachycardia induced by the sympathetic nervous system to increase cardiac output and mean arterial pressure
▶ thirst to prompt ingestion of fluid (increased ECF osmolality stimulates the thirst center in the hypothalamus)
▶ flattened neck veins due to decreased circulating blood volume
▶ sunken eyeballs due to decreased volume of total-body fluid and consequent dehydration of connective tissue and aqueous humor
▶ dry mucous membranes due to decreased body fluid volume (glands that produce fluids to moisten and protect the vascular mucous membranes fail, so they dry rapidly)
▶ diminished skin turgor due to decreased fluid in the dermal layer (making skin less pliant)
▶ rapid weight loss due to acute loss of body fluid

 AGE ALERT In hypovolemic infants younger than 4 months, the posterior and anterior fontanels are sunken when palpated. Between 4 and 18 months, the posterior fontanel is normally closed, but the anterior fontanel is sunken in hypovolemic infants.

▶ decreased urine output due to decreased renal perfusion from renal vasoconstriction
▶ prolonged capillary refill time due to increased systemic vascular resistance.

Complications

Possible complications of hypovolemia include:
▶ shock

◗ acute renal failure
◗ death.

Diagnosis

No single diagnostic finding confirms hypovolemia, but the following test results are suggestive:

◗ increased blood urea nitrogen (BUN) level (early sign)
◗ elevated serum creatinine level (late sign)
◗ increased serum protein, hemoglobin, and hematocrit (unless caused by hemorrhage, when loss of blood elements causes subnormal values)
◗ rising blood glucose
◗ elevated serum osmolality; except in hyponatremia, where serum osmolality is low
◗ serum electrolyte and arterial blood gas (ABG) analysis may reflect associated clinical problems due to underlying cause of hypovolemia or treatment regimen.

If the patient has no underlying renal disorder, typical urinalysis findings include:

◗ urine specific gravity > 1.030
◗ increased urine osmolality
◗ urine sodium level < 50 mEq/L.

Treatment

Possible treatments for hypovolemia include:

◗ oral fluids (may be adequate in mild hypovolemia if the patient is alert enough to swallow and can tolerate it)
◗ parenteral fluids to supplement or replace oral therapy (moderate to severe hypovolemia; choice of parenteral fluid depends on type of fluids lost, severity of hypovolemia, and patient's cardiovascular, electrolyte, and acid–base status)
◗ fluid resuscitation by rapid I.V. administration (severe volume depletion; depending on patient's condition, 100 to 500 ml of fluid over 15 minutes to 1 hour; fluid bolus may be given more quickly if needed)
◗ blood or blood products (with hemorrhage)
◗ antidiarrheals as needed

◗ antiemetics as needed
◗ I.V. dopamine (Intropin) or norepinephrine (Levophed) to increase cardiac contractility and renal perfusion (if patient remains symptomatic after fluid replacement)
◗ autotransfusion (for some patients with hypovolemia caused by trauma).

CULTURAL DIVERSITY Indian patients should be questioned about the use of Ayurvedic cleansing practices, such as the use of laxatives, diuretics, and emetics, which can affect drug absorption and metabolic balance.

Hypervolemia

The expansion of ECF volume, called hypervolemia, may involve the interstitial or intravascular space. Hypervolemia develops when excess sodium and water are retained in about the same proportions. It is always secondary to an increase in total-body sodium content, which causes water retention. Usually the body can compensate and restore fluid balance.

Causes

Conditions that increase the risk for sodium and water retention include:

◗ heart failure
◗ cirrhosis of the liver
◗ nephrotic syndrome
◗ corticosteroid therapy
◗ low dietary protein intake
◗ renal failure.

Sources of excessive sodium and water intake include:

◗ parenteral fluid replacement with normal saline or lactated Ringer's solution
◗ blood or plasma replacement
◗ dietary intake of water, sodium chloride, or other salts.

Fluid shift to the ECF compartment may follow:

◗ remobilization of fluid after burn treatment
◗ hypertonic fluids, such as mannitol (Osmitrol) or hypertonic saline solution
◗ colloid oncotic fluids such as albumin.

Pathophysiology

Increased ECF volume causes the following sequence of events:

▶ circulatory overload
▶ increased cardiac contractility and mean arterial pressure
▶ increased capillary hydrostatic pressure
▶ shift of fluid to the interstitial space
▶ edema.

Elevated mean arterial pressure inhibits secretion of antidiuretic hormone and aldosterone and consequent increased urinary elimination of water and sodium. These compensatory mechanisms usually restore normal intravascular volume. If hypervolemia is severe or prolonged or the patient has a history of cardiovascular dysfunction, compensatory mechanisms may fail, and heart failure and pulmonary edema may ensue.

Signs and symptoms

Possible signs and symptoms of hypervolemia include:

▶ rapid breathing due to fewer red blood cells per milliliter of blood (dilution causes a compensatory increase in respiratory rate to increase oxygenation)
▶ dyspnea (labored breathing) due to increased fluid volume in pleural spaces
▶ crackles (gurgling or bubbling sounds on auscultation) due to elevated hydrostatic pressure in pulmonary capillaries
▶ rapid, bounding pulse due to increased cardiac contractility (from circulatory overload)
▶ hypertension (unless heart is failing) due to circulatory overload (causes increased mean arterial pressure)
▶ distended neck veins due to increased blood volume and increased preload
▶ moist skin (compensatory to increase water excretion through perspiration)
▶ acute weight gain due to increased volume of total-body fluid from circulatory overload (best indicator of ECF volume excess)
▶ edema (increased mean arterial pressure leads to increased capillary hydrostatic pressure, causing fluid shift from plasma to interstitial spaces)
▶ S_3 gallop (abnormal heart sound due to rapid filling and volume overload of the ventricles during diastole).

Complications

Possible complications of hypervolemia include:

▶ skin breakdown
▶ acute pulmonary edema with hypoxemia.

Diagnosis

No single diagnostic test confirms the disorder, but the following findings indicate hypervolemia:

▶ decreased serum potassium and blood urea nitrogen (BUN) due to hemodilution (increased serum potassium and BUN usually indicate renal failure or impaired renal perfusion)
▶ decreased hematocrit due to hemodilution
▶ normal serum sodium (unless associated sodium imbalance is present)
▶ low urine sodium excretion (usually < 10 mEq/day because edematous patient is retaining sodium)
▶ increased hemodynamic values (including pulmonary artery, pulmonary artery wedge, and central venous pressures).

Treatment

Possible treatments for hypervolemia include:

▶ restricted sodium and water intake
▶ preload reduction agents, such as morphine, furosemide (Lasix), and nitroglycerin (Nitro-Bid), and afterload reduction agents, such as hydralazine (Apresoline) and captopril (Capoten) for pulmonary edema.

 AGE ALERT Carefully monitor I.V. fluid administration rate and patient response, especially in elderly patients or those with impaired cardiac or renal function, who are particularly vulnerable to acute pulmonary edema.

For severe hypervolemia or renal failure, the patient may undergo renal replacement therapy, including:
▶ hemodialysis or peritoneal dialysis
▶ continuous arteriovenous hemofiltration (allows removal of excess fluid from critically ill patients who may not need dialysis; the patient's arterial pressure serves as a natural pump, driving blood through the arterial line)
▶ continuous venovenous hemofiltration (similar to arteriovenous hemofiltration, but a mechanical pump is used when mean arterial pressure is < 60 mm Hg).

Supportive measures include:
▶ oxygen administration
▶ use of thromboembolic disease support hose to help mobilize edematous fluid
▶ bed rest
▶ treatment of underlying condition that caused or contributed to hypervolemia.

PATHOPHYSIOLOGIC MANIFESTATIONS OF ACID-BASE IMBALANCE

Acid–base balance is essential to life. Concepts related to imbalance include acidemia, acidosis, alkalemia, alkalosis, and compensation.

Acidemia

Acidemia is an arterial pH of less than 7.35, which reflects a relative excess of acid in the blood. The hydrogen ion content in ECF increases, and the hydrogen ions move to the ICF. To keep the intracellular fluid electrically neutral, an equal amount of potassium leaves the cell, creating a relative hyperkalemia.

Acidosis

Acidosis is a systemic increase in hydrogen ion concentration. If the lungs fail to eliminate CO_2 or if volatile (carbonic acid) or nonvolatile (lactic) acid products of metabolism accumulate, hydrogen ion concentration rises. Acidosis can also occur if persistent diarrhea causes loss of basic bicarbonate anions or the kidneys fail to reabsorb bicarbonate or secrete hydrogen ions.

Alkalemia

Alkalemia is arterial blood pH > 7.45, which reflects a relative excess of base in the blood. In alkalemia, an excess of hydrogen ions in the ICF forces them into the ECF. To keep the ICF electrically neutral, potassium moves from the ECF to the ICF, creating a relative hypokalemia.

Alkalosis

Alkalosis is a bodywide decrease in hydrogen ion concentration. An excessive loss of CO_2 during hyperventilation, loss of nonvolatile acids during vomiting, or excessive ingestion of base may decrease hydrogen ion concentration.

Compensation

The lungs and kidneys, along with a number of chemical buffer systems in the intracellular and extracellular compartments, work together to maintain plasma pH in the range of 7.35 to 7.45. For a description of acid–base values and compensatory mechanisms, see *Interpreting ABGs,* page 124.

Buffer systems

A buffer system consists of a weak acid (that doesn't readily release free hydrogen ions) and a corresponding base, such as sodium bicarbonate. These buffers resist or minimize a change in pH when an acid or base is added to the buffered solution. Buffers work in seconds.

The four major buffers or buffer systems are:
▶ carbonic acid — bicarbonate system (the most important, works in lungs)
▶ hemoglobin–oxyhemoglobin system (works in red blood cells) hemoglobin binds free H+, blood flows through lungs, H+ combines with CO_2.
▶ other protein buffers (in ECF and ICF)
▶ phosphate system (primarily in ICF).

When primary disease processes alter either the acid or base component of the

INTERPRETING ABGs

This chart compares abnormal arterial blood gas (ABG) values and their significance for patient care.

DISORDER	pH	Paco$_2$ (mm Hg)
Normal	*7.35 to 7.45*	*35 to 45*
Respiratory acidosis	< 7.35	> 45
Respiratory alkalosis	> 7.45	< 35
Metabolic acidosis	< 7.35	< 35
Metabolic alkalosis	> 7.45	> 45

ratio, the lungs or kidneys (whichever is not affected by the disease process) act to restore the ratio and normalize pH. Because the body's mechanisms that regulate pH occur in stepwise fashion over time, the body tolerates gradual changes in pH better than abrupt ones.

Compensation by the kidneys

If a respiratory disorder causes acidosis or alkalosis, the kidneys respond by altering their handling of hydrogen and bicarbonate ions to return the pH to normal. Renal compensation begins hours to days after a respiratory alteration in pH. Despite this delay, renal compensation is powerful.

▶ Acidemia: the kidneys excrete excess hydrogen ions, which may combine with phosphate or ammonia to form titratable acids in the urine. The net effect is to *raise* the concentration of bicarbonate ions in the ECF and so restore acid–base balance.

▶ Alkalemia: the kidneys excrete excess bicarbonate ions, usually with sodium ions.

The net effect is to *reduce* the concentration of bicarbonate ions in the ECF and restore acid–base balance.

Compensation by the lungs

If acidosis or alkalosis results from a metabolic or renal disorder, the respiratory system regulates the respiratory rate to return the pH to normal. The partial pressure of CO_2 in arterial blood (Paco$_2$) reflects CO_2 levels proportionate to blood pH. As the concentration of the gas increases, so does its partial pressure. Within minutes after the slightest change in Paco$_2$, central chemoreceptors in the medulla that regulate the rate and depth of ventilation detect the change.

▶ Acidemia increases respiratory rate and depth to eliminate CO_2.

▶ Alkalemia decreases respiratory rate and depth to retain CO_2.

HCO$_3^-$ (mEq/L)	COMPENSATION
22 to 26	
▶ Acute: may be normal ▶ Chronic: > 26	▶ Renal: increased secretion and excretion of acid; compensation takes 24 hours to begin ▶ Respiratory: rate increases to expel CO_2
▶ Acute: normal ▶ Chronic: < 22	▶ Renal: decreased H$^+$ secretion and active secretion of HCO$_3^-$ into urine
< 22	▶ Respiratory: lungs expel more CO_2 by increasing rate and depth of respirations
> 26	▶ Respiratory: hypoventilation is immediate but limited because of ensuing hypoxemia ▶ Renal: more effective but slow to excrete less acid and more base

DISORDERS OF ACID–BASE BALANCE

Acid–base disturbances can cause respiratory acidosis or alkalosis or metabolic acidosis or alkalosis.

Respiratory acidosis

Respiratory acidosis is an acid–base disturbance characterized by reduced alveolar ventilation. The patient's pulmonary system can't clear enough CO_2 from the body. This leads to hypercapnia (PaCO_2 > 45 mm Hg) and acidosis (pH < 7.35). Respiratory acidosis can be acute (due to a sudden failure in ventilation) or chronic (in long-term pulmonary disease). Any compromise in the essential components of breathing — ventilation, perfusion, and diffusion — may cause respiratory acidosis.

Prognosis depends on the severity of the underlying disturbance as well as the patient's general clinical condition. The prognosis is least optimistic for a patient with a debilitating disorder.

Causes

Factors leading to respiratory acidosis include:

▶ drugs (narcotics, general anesthetics, hypnotics, alcohol, and sedatives decrease the sensitivity of the respiratory center)
▶ Central nervous system (CNS) trauma (injury to the medulla may impair ventilatory drive)
▶ cardiac arrest (acute)
▶ sleep apnea
▶ chronic metabolic alkalosis as respiratory compensatory mechanisms try to normalize pH
▶ ventilation therapy (use of high-flow oxygen in patients with chronic respiratory disorders suppresses the patient's hypoxic drive to breathe; high positive end-expiratory pressure in the presence of reduced cardiac output may cause hypercapnia due to large increases in alveolar dead space)
▶ neuromuscular diseases, such as myasthenia gravis, Guillain-Barré syndrome, and poliomyelitis (respiratory muscles

cannot respond properly to respiratory drive)
▶ airway obstruction or parenchymal lung disease (interferes with alveolar ventilation)
▶ chronic obstructive pulmonary disease (COPD) or asthma
▶ severe adult respiratory distress syndrome (reduced pulmonary blood flow and poor exchange of CO_2 and oxygen between the lungs and blood)
▶ chronic bronchitis
▶ large pneumothorax
▶ extensive pneumonia
▶ pulmonary edema.

Pathophysiology

When pulmonary ventilation decreases, $Paco_2$ is increased, and the CO_2 level rises in all tissues and fluids, including the medulla and cerebrospinal fluid. Retained CO_2 combines with water (H_2O) to form carbonic acid (H_2CO_3). The carbonic acid dissociates to release free hydrogen (H^+) and bicarbonate (HCO_3^-) ions. Increased $Paco_2$ and free hydrogen ions stimulate the medulla to increase respiratory drive and expel CO_2.

As pH falls, 2,3-diphosphoglycerate (2,3-DPG) accumulates in red blood cells, where it alters hemoglobin so it releases oxygen. This reduced hemoglobin, which is strongly alkaline, picks up hydrogen ions and CO_2 and removes them from the serum.

As respiratory mechanisms fail, rising $Paco_2$ stimulates the kidneys to retain bicarbonate and sodium ions and excrete hydrogen ions As a result, more sodium bicarbonate ($NaHCO_3$) is available to buffer free hydrogen ions. Some hydrogen is excreted in the form of ammonium ion ($NH4^+$), neutralizing ammonia, which is an important CNS toxin.

As the hydrogen ion concentration overwhelms compensatory mechanisms, hydrogen ions move into the cells and potassium ions move out. Without enough oxygen, anaerobic metabolism produces lactic acid. Electrolyte imbalances and acidosis critically depress neurologic and cardiac functions.

Signs and symptoms

Clinical features vary according to the severity and duration of respiratory acidosis, the underlying disease, and the presence of hypoxemia. Carbon dioxide and hydrogen ions dilate cerebral blood vessels and increase blood flow to the brain, causing cerebral edema and depressing CNS activity.

Possible signs and symptoms include:
▶ restlessness
▶ confusion
▶ apprehension
▶ somnolence
▶ fine or flapping tremor (asterixis)
▶ coma
▶ headaches
▶ dyspnea and tachypnea
▶ papilledema
▶ depressed reflexes
▶ hypoxemia, unless the patient is receiving oxygen.

Respiratory acidosis may also cause cardiovascular abnormalities, including:
▶ tachycardia
▶ hypertension
▶ atrial and ventricular arrhythmias
▶ hypotension with vasodilation (bounding pulses and warm periphery, in severe acidosis).

Complications

Possible complications include:
▶ profound CNS and cardiovascular deterioration due to dangerously low blood pH (< 7.15)
▶ myocardial depression (leading to shock and cardiac arrest)
▶ elevated $Paco_2$ despite optimal treatment (in chronic lung disease).

Diagnosis

The following tests help diagnose respiratory acidosis:
▶ arterial blood gas (ABG) analysis, showing $Paco_2 > 45$ mm Hg; pH < 7.35 to 7.45; and normal HCO_3^- in the acute stage and

elevated HCO_3^- in the chronic stage (confirms the diagnosis)
▶ chest X-ray (often shows such causes as heart failure, pneumonia, COPD, and pneumothorax)
▶ potassium > 5 mEq/L
▶ low serum chloride
▶ acidic urine pH (as the kidneys excrete hydrogen ions to return blood pH to normal)
▶ drug screening (may confirm suspected drug overdose).

Treatment

Effective treatment of respiratory acidosis requires correction of the underlying source of alveolar hypoventilation. Treatment of pulmonary causes of respiratory acidosis includes:
▶ removal of a foreign body from the airway
▶ artificial airway through endotracheal intubation or tracheotomy and mechanical ventilation (if the patient can't breathe spontaneously)
▶ increasing the partial pressure of arterial oxygen (PaO_2) to at least 60 mm Hg and pH to > 7.2 to avoid cardiac arrhythmias
▶ aerosolized or I.V. bronchodilators to open constricted airways
▶ antibiotics to treat pneumonia
▶ chest tubes to correct pneumothorax
▶ positive end-expiratory pressure to prevent alveolar collapse
▶ thrombolytic or anticoagulant therapy for massive pulmonary emboli
▶ bronchoscopy to remove excessive retained secretions.
    Treatment for patients with COPD includes:
▶ bronchodilators
▶ oxygen at low flow rates (more oxygen than the person's normal removes the hypoxic drive, further reducing alveolar ventilation)
▶ corticosteroids
▶ gradual reduction in $PaCO_2$ to baseline to provide sufficient chloride and potassium ions to enhance renal excretion of

bicarbonate (in chronic respiratory acidosis).
    Other treatments include:
▶ drug therapy for such conditions as myasthenia gravis
▶ dialysis or charcoal to remove toxic drugs
▶ correction of metabolic alkalosis
▶ careful administration of I.V. sodium bicarbonate.

Respiratory alkalosis

Respiratory alkalosis is an acid–base disturbance characterized by a $PaCO_2$ < 35 mm Hg and blood pH > 7.45; alveolar hyperventilation is the cause. Hypocapnia (below normal $PaCO_2$) occurs when the lungs eliminate more CO_2 than the cells produce.
    Respiratory alkalosis is the most common acid–base disturbance in critically ill patients and, when severe, has a poor prognosis.

Causes

Conditions that may cause or contribute to respiratory alkalosis include:
▶ acute hypoxemia, pneumonia, interstitial lung disease, pulmonary vascular disease, and acute asthma (may stimulate the respiratory control center, causing the patient to breathe faster and deeper)
▶ anxiety (may cause deep rapid breathing)
▶ hypermetabolic states such as fever and sepsis (especially gram-negative sepsis)
▶ excessive mechanical ventilation
▶ injury to the CNS respiratory control center
▶ salicylate (aspirin) intoxication
▶ salicylate toxicity
▶ metabolic acidosis
▶ hepatic failure
▶ pregnancy (progesterone increases ventilation and reduces $PaCO_2$ by as much as 5 to 10 mm Hg).

Pathophysiology

When pulmonary ventilation increases more than needed to maintain normal CO_2

levels, excessive amounts of CO_2 are exhaled. The consequent hypocapnia leads to a chemical reduction of carbonic acid, excretion of hydrogen and bicarbonate ions, and a rising pH.

In defense against the increasing serum pH, the hydrogen–potassium buffer system pulls hydrogen ions out of the cells and into the blood in exchange for potassium ions. The hydrogen ions entering the blood combine with available bicarbonate ions to form carbonic acid, and the pH falls.

Hypocapnia stimulates the carotid and aortic bodies as well as the medulla, increasing the heart rate (which hypokalemia can further aggravate) but not the blood pressure. At the same time, hypocapnia causes cerebral vasoconstriction and decreased cerebral blood flow. It also overexcites the medulla, pons, and other parts of the autonomic nervous system. When hypocapnia lasts more than 6 hours, the kidneys secrete more bicarbonate and less hydrogen. Full renal adaptation to respiratory alkalosis requires normal volume status and renal function, and it may take several days.

Continued low $Paco_2$ and the vasoconstriction it causes increases cerebral and peripheral hypoxia. Severe alkalosis inhibits calcium ionization; as calcium ions become unavailable, nerves and muscles become progressively more excitable. Eventually, alkalosis overwhelms the CNS and heart.

Signs and symptoms

Possible signs and symptoms of respiratory alkalosis include:
▶ deep, rapid breathing (possibly more than 40 breaths/minute and much like the Kussmaul's respirations that characterize diabetic acidosis) usually causing CNS and neuromuscular disturbances (cardinal sign of respiratory alkalosis)
▶ light-headedness or dizziness due to decreased cerebral blood flow
▶ agitation
▶ circumoral and peripheral paresthesias

▶ carpopedal spasms, twitching (possibly progressing to tetany), and muscle weakness.

Complications

Possible complications of severe respiratory alkalosis include:
▶ cardiac arrhythmias that may not respond to conventional treatment as the hemoglobin–oxygen buffer system becomes overwhelmed
▶ hypocalcemic tetany, seizures
▶ periods of apnea if pH remains high and $Paco_2$ remains low.

Diagnosis

The following test results indicate respiratory alkalosis:
▶ arterial blood gas (ABG) analysis showing $Paco_2$ <35 mm Hg; elevated pH in proportion to decrease in $Paco_2$ in the acute stage but decreasing toward normal in the chronic stage; normal HCO_3^- in acute stage but less than normal in the chronic stage (confirms respiratory alkalosis, rules out respiratory compensation for metabolic acidosis)
▶ serum electrolyte studies (detect metabolic disorders causing compensatory respiratory alkalosis)
▶ ECG findings (may indicate cardiac arrhythmias)
▶ low chloride (in severe respiratory alkalosis)
▶ toxicology screening (for salicylate poisoning)
▶ basic urine pH as kidneys excrete HCO_3^- to raise blood pH.

Treatment

Possible treatments to correct the underlying condition include:
▶ removal of ingested toxins
▶ treatment of fever or sepsis
▶ oxygen for acute hypoxemia
▶ treatment of CNS disease
▶ having patient breathe into a paper bag to increase CO_2 and help relieve anxiety (for hyperventilation caused by severe anxiety)

▶ adjustment of tidal volume and minute ventilation in patients on mechanical ventilation to prevent hyperventilation (by monitoring ABG analysis results).

Metabolic acidosis

Metabolic acidosis is an acid–base disorder characterized by excess acid and deficient HCO_3^- caused by an underlying nonrespiratory disorder. A primary decrease in plasma HCO_3^- causes pH to fall. It can occur with increased production of a nonvolatile acid (such as lactic acid), decreased renal clearance of a nonvolatile acid (as in renal failure), or loss of HCO_3^- (as in chronic diarrhea). Symptoms result from action of compensatory mechanisms in the lungs, kidneys, and cells.

 AGE ALERT Metabolic acidosis is more prevalent among children, who are vulnerable to acid–base imbalance because their metabolic rates are rapid and ratios of water to total-body weight are low.

Severe or untreated metabolic acidosis can be fatal. The prognosis improves with prompt treatment of the underlying cause and rapid reversal of the acidotic state.

Causes

Metabolic acidosis usually results from excessive fat metabolism in the absence of usable carbohydrates (produces more ketoacids than the metabolic process can handle). Possible causes are:
▶ excessive acid accumulation
▶ deficient HCO_3^- stores
▶ decreased acid excretion by the kidneys
▶ a combination of these factors.

Conditions that may cause or contribute to metabolic acidosis include:
▶ diabetic ketoacidosis
▶ chronic alcoholism
▶ malnutrition or a low-carbohydrate, high-fat diet
▶ anaerobic carbohydrate metabolism (decreased tissue oxygenation or perfusion— as in cardiac pump failure after myocardial infarction, pulmonary or hepatic disease, shock, or anemia — forces a shift from aerobic to anaerobic metabolism, causing a corresponding increase in lactic acid level)
▶ underexcretion of metabolized acids or inability to conserve base due to renal insufficiency and failure (renal acidosis)
▶ diarrhea, intestinal malabsorption, or loss of sodium bicarbonate from the intestines, causing bicarbonate buffer system to shift to the acidic side (e.g., ureteroenterostomy and Crohn's disease can also induce metabolic acidosis)
▶ salicylate intoxication (overuse of aspirin), exogenous poisoning, or less frequently, Addison's disease (increased excretion of sodium and chloride and retention of potassium)
▶ inhibited secretion of acid due to hypoaldosteronism or the use of potassium-sparing diuretics.

Pathophysiology

As acid (H^+) starts to accumulate in the body, chemical buffers (plasma HCO_3^- and proteins) in the cells and ECF bind the excess hydrogen ions.

Excess hydrogen ions that the buffers can't bind decrease blood pH and stimulate chemoreceptors in the medulla to increase respiration. The consequent fall of $Paco_2$ frees hydrogen ions to bind with HCO_3^-. Respiratory compensation occurs in minutes but isn't sufficient to correct the acidosis.

Healthy kidneys try to compensate by secreting excess hydrogen ions into the renal tubules. These ions are buffered by either phosphate or ammonia and excreted into the urine in the form of weak acid. For each hydrogen ion secreted into the renal tubules, the tubules reabsorb and return to the blood one Na^+ and one HCO_3^-.

The excess hydrogen ions in ECF passively diffuse into cells. To maintain the balance of charge across the membranes, the cells release potassium ions. Excess hydrogen ions change the normal balance of potassium, sodium, and calcium ions and thereby impair neural excitability.

Signs and symptoms

In mild acidosis, symptoms of the under-lying disease may hide the direct clinical evidence. Signs and symptoms include:

‣ headache and lethargy progressing to drowsiness, CNS depression, Kussmaul's respirations (as the lungs attempt to com-pensate by blowing off CO_2), hypoten-sion, stupor, and (if condition is severe and untreated) coma and death

‣ associated GI distress leading to anorex-ia, nausea, vomiting, diarrhea, and possi-bly dehydration

‣ warm, flushed skin due to a pH-sensi-tive decrease in vascular response to sym-pathetic stimuli

‣ fruity-smelling breath from fat catabo-lism and excretion of accumulated ace-tone through the lungs due to underlying diabetes mellitus.

Complications

Metabolic acidosis depresses the CNS and, if untreated, may lead to:

‣ weakness, flaccid paralysis
‣ coma
‣ ventricular arrhythmias, possibly car-diac arrest.

In the metabolic acidosis of chronic re-nal failure, HCO_3^- is drawn from bone to buffer hydrogen ions; the results include:

‣ growth retardation in children
‣ bone disorders such as renal osteodys-trophy.

Diagnosis

The following test results confirm the di-agnosis of metabolic acidosis:

‣ arterial pH < 7.35 (as low as 7.10 in severe acidosis); $Paco_2$ normal or < 34 mm Hg as respiratory compensatory mecha-nisms take hold; HCO_3^- may be < 22 mEq/L).

The following test results support the diagnosis of metabolic acidosis:

‣ urine pH < 4.5 in the absence of renal disease (as the kidneys excrete acid to raise blood pH)

‣ serum potassium > 5.5 mEq/L from chemical buffering

‣ glucose > 150 mg/dl
‣ serum ketone bodies in diabetes
‣ elevated plasma lactic acid in lactic aci-dosis

‣ anion gap > 14 mEq/L in high-anion gap metabolic acidosis, lactic acidosis, ke-toacidosis, aspirin overdose, alcohol poi-soning, renal failure, or other conditions characterized by accumulation of organ-ic acids, sulfates or phosphates

‣ anion gap 12 mEq/L or less in normal anion gap metabolic acidosis from HCO_3^- loss, GI or renal loss, increased acid load (hyperalimentation fluids), rapid I.V. saline administration, or other conditions char-acterized by loss of bicarbonate.

Treatment

Treatment aims to correct the acidosis as quickly as possible by addressing both the symptoms and the underlying cause. Mea-sures may include:

‣ sodium bicarbonate I.V. for severe high anion gap to neutralize blood acidity in patients with pH < 7.20 and HCO_3^- loss; monitor plasma electrolytes, especially potassium, during sodium bicarbonate therapy (potassium level may fall as pH rises)

‣ I.V. lactated Ringer's solution to correct normal anion gap metabolic acidosis and ECF volume deficit

‣ evaluation and correction of electrolyte imbalances

‣ correction of the underlying cause (e.g., in diabetic ketoacidosis, continuous low-dose I.V. insulin infusion)

‣ mechanical ventilation to maintain res-piratory compensation, if needed

‣ antibiotic therapy to treat infection
‣ dialysis for patients with renal failure or certain drug toxicities

‣ antidiarrheal agents for diarrhea-induced HCO_3^- loss

‣ monitor for secondary changes due to hypovolemia, such as falling blood pres-sure (in diabetic acidosis)

‣ position patient to prevent aspiration (metabolic acidosis commonly causes vomiting)

‣ seizure precautions.

Metabolic alkalosis

Metabolic alkalosis occurs when low levels of acid or high HCO_3^- cause metabolic, respiratory, and renal responses, producing characteristic symptoms (most notably, hypoventilation). This condition is always secondary to an underlying cause. With early diagnosis and prompt treatment, prognosis is good, but untreated metabolic alkalosis may lead to coma and death.

Causes

Metabolic alkalosis results from loss of acid, retention of base, or renal mechanisms associated with low serum levels of potassium and chloride.

Causes of critical acid loss include:
‣ chronic vomiting
‣ nasogastric tube drainage or lavage without adequate electrolyte replacement
‣ fistulas
‣ use of steroids and certain diuretics (furosemide [Lasix], thiazides, and ethacrynic acid [Edecrin])

 CULTURAL DIVERSITY Various Chinese herbal cures contain benzodiazepines, steroids, and nonsteroidal anti-inflammatory drugs that may produce unexpected adverse effects, such as muscle weakness and cushingoid features.

‣ massive blood transfusions
‣ Cushing's disease, primary hyperaldosteronism, and Bartter's syndrome (lead to sodium and chloride retention and urinary loss of potassium and hydrogen).

Excessive HCO_3^- retention causing chronic hypercapnia can result from:
‣ excessive intake of bicarbonate of soda or other antacids (usually for treatment of gastritis or peptic ulcer)
‣ excessive intake of absorbable alkali (as in milk alkali syndrome, often seen in patients with peptic ulcers)

‣ excessive amounts of I.V. fluids, high concentrations of bicarbonate or lactate
‣ respiratory insufficiency.

Alterations in extracellular electrolyte levels that can cause metabolic alkalosis include:
‣ low chloride (as chloride diffuses out of the cell, hydrogen diffuses into the cell)
‣ low plasma potassium causing increased hydrogen ion excretion by the kidneys.

Pathophysiology

Chemical buffers in the ECF and ICF bind HCO_3^- that accumulates in the body. Excess unbound HCO_3^- raises blood pH, which depresses chemoreceptors in the medulla, inhibiting respiration and raising $Paco_2$. Carbon dioxide combines with water to form carbonic acid. Low oxygen levels limit respiratory compensation.

When the blood HCO_3^- rises to 28 mEq/L or more, the amount filtered by the renal glomeruli exceeds the reabsorptive capacity of the renal tubules. Excess HCO_3^- is excreted in the urine, and hydrogen ions are retained. To maintain electrochemical balance, sodium ions and water are excreted with the bicarbonate ions.

When hydrogen ion levels in ECF are low, hydrogen ions diffuse passively out of the cells and, to maintain the balance of charge across the cell membrane, extracellular potassium ions move into the cells. As intracellular hydrogen ion levels fall, calcium ionization decreases, and nerve cells become more permeable to sodium ions. As sodium ions move into the cells, they trigger neural impulses, first in the peripheral nervous system and then in the CNS.

Signs and symptoms

Clinical features of metabolic alkalosis result from the body's attempt to correct the acid-base imbalance, primarily through hypoventilation. Signs and symptoms include:
‣ irritability, picking at bedclothes (carphology), twitching, and confusion due to decreased cerebral perfusion

▶ nausea, vomiting, and diarrhea (which aggravate alkalosis)

▶ cardiovascular abnormalities due to hypokalemia

▶ respiratory disturbances (such as cyanosis and apnea) and slow, shallow respirations

▶ diminished peripheral blood flow during repeated blood pressure checks may provoke carpopedal spasm in the hand (Trousseau's sign, a possible sign of impending tetany).

Complications

Uncorrected metabolic alkalosis may progress to:

▶ seizures

▶ coma.

Diagnosis

Findings indicating metabolic alkalosis include:

▶ blood pH > 7.45 and HCO_3^- > 29 mEq/L (confirm diagnosis)

▶ $Paco_2$ > 45 mm Hg (indicates attempts at respiratory compensation)

▶ low potassium (< 3.5 mEq/L), calcium (< 8.9 mg/dl), and chloride (< 98 mEq/L)

▶ urine pH about 7

▶ alkaline urine after the renal compensatory mechanism begins to excrete bicarbonate

▶ ECG may show low T wave, merging with a P wave, and atrial or sinus tachycardia.

Treatment

The goal of treatment is to correct the underlying cause of metabolic alkalosis. Possible treatments include:

▶ *cautious* use of ammonium chloride I.V. (rarely) or HCl to restore ECF hydrogen and chloride levels;

▶ KCl and normal saline solution (except in heart failure); usually sufficient to replace losses from gastric drainage

▶ discontinuation of diuretics and supplementary KCl (metabolic alkalosis from potent diuretic therapy)

▶ oral or I.V. acetazolamide (Diamox; enhances renal bicarbonate excretion) to correct metabolic alkalosis without rapid volume expansion (acetazolamide also enhances potassium excretion, so potassium may be given before drug).

6

The cardiovascular system begins its activity when the fetus is barely 4 weeks old and is the last system to cease activity at the end of life. This body system is so vital that its activity helps define the presence of life.

The heart, arteries, veins, and lymphatics form the cardiovascular network that serves the body's transport system. This system brings life-supporting oxygen and nutrients to cells, removes metabolic waste products, and carries hormones from one part of the body to another.

The cardiovascular system, often called the circulatory system, may be divided into two branches: pulmonary and systemic circulations. In *pulmonary circulation,* blood picks up oxygen and liberates the waste product carbon dioxide. In *systemic circulation* (which includes coronary circulation), blood carries oxygen and nutrients to all active cells and transports waste products to the kidneys, liver, and skin for excretion.

Circulation requires normal heart function, which propels blood through the system by continuous rhythmic contractions. Blood circulates through three types of vessels: arteries, veins, and capillaries. The sturdy, pliable walls of the arteries adjust to the volume of blood leaving the heart. The aorta is the major artery arching out of the left ventricle; its segments and sub-branches ultimately divide into minute, thin-walled (one cell thick) capillaries. Capillaries pass the blood to the veins, which return it to the heart. In the veins, valves prevent blood backflow.

PATHOPHYSIOLOGIC MANIFESTATIONS

Pathophysiologic manifestations of cardiovascular disease may stem from aneurysm, cardiac shunts, embolus, release of cardiac enzymes, stenosis, thrombus, and valve incompetence.

Aneurysm

An aneurysm is a localized outpouching or dilation of a weakened arterial wall. This weakness can be the result of either atherosclerotic plaque formation that erodes the vessel wall, or the loss of elastin and collagen in the vessel wall. Congenital abnormalities in the media of the arterial wall, trauma, and infections such as syphilis may lead to aneurysm formation. A ruptured aneurysm may cause massive hemorrhage and death.

Several types of aneurysms can occur:
▶ A *saccular aneurysm* occurs when increased pressure in the artery pushes out a pouch on one side of the artery, creating a bulge. (See *Types of aortic aneurysms,* page 134.)
▶ A *fusiform aneurysm* develops when the arterial wall weakens around its circumference, creating a spindle-shaped aneurysm.
▶ A *dissecting aneurysm* occurs when blood is forced between the layers of the arterial wall, causing them to separate and creating a false lumen.
▶ A *false aneurysm* develops when there is a break in all layers of the arterial wall and blood leaks out but is contained by

133

TYPES OF AORTIC ANEURYSMS

 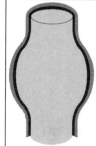 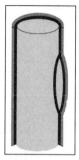 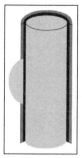

SACCULAR ANEURYSM
Unilateral pouch-like bulge with a narrow neck

FUSIFORM ANEURYSM
A spindle-shaped bulge encompass-ing the entire diam-eter of the vessel

DISSECTING ANEURYSM
A hemorrhagic separation of the medial lay-ers of the vessel wall, which cre-ate a false lu-men

FALSE ANEURYSM
A pulsating hematoma re-sulting from trauma and of-ten mistaken for an abdominal aneurysm

surrounding structures, creating a pulsatile hematoma.

Cardiac shunts

A cardiac shunt provides communication between the pulmonary and systemic circulations. Before birth, shunts between the right and left sides of the heart and between the aorta and pulmonary artery are a normal part of fetal circulation. Following birth, however, the mixing of pulmonary and systemic blood or the movement of blood between the left and right sides of the heart is abnormal. Blood flows through a shunt from an area of high pressure to an area of low pressure or from an area of high resistance to an area of low resistance.

Left-to-right shunts

In a left-to-right shunt, blood flows from the left side of the heart to the right side through an atrial or ventricular defect, or from the aorta to the pulmonary circula-

tion through a patent ductus arteriosus. Because the blood in the left side of the heart is rich in oxygen, a left-to-right shunt delivers oxygenated blood back to the right side of the heart or to the lungs. Consequently, a left-to-right shunt that occurs as a result of a congenital heart defect is called an *acyanotic defect*.

In a left-to-right shunt, pulmonary blood flow increases as blood is continually recirculated to the lungs, leading to hypertrophy of the pulmonary vessels. The increased amounts of blood circulated from the left side of the heart to the right side can result in right-sided heart failure. Eventually, left-sided heart failure may also occur.

Right-to-left shunts

A right-to-left shunt occurs when blood flows from the right side of the heart to the left side such as occurs in tetralogy of Fallot, or from the pulmonary artery directly into the systemic circulation through a patent ductus arteriosus. Because blood returning to the right side of the heart and the pulmonary artery is low in oxygen, a right-to-left shunt adds deoxygenated blood to the systemic circulation, causing hypoxia and cyanosis. Congenital defects that involve right-to-left shunts are therefore called *cyanotic defects.* Common manifestations of a right-to-left shunt related to poor tissue and organ perfusion include fatigue, increased respiratory rate, and clubbing of the fingers.

Embolus

An embolus is a substance that circulates from one location in the body to another, through the bloodstream. Although most emboli are blood clots from a thrombus, they may also consist of pieces of tissue, an air bubble, amniotic fluid, fat, bacteria, tumor cells, or a foreign substance.

Emboli that originate in the venous circulation, such as from deep vein thrombosis, travel to the right side of the heart to the pulmonary circulation and eventually lodge in a capillary, causing pulmonary infarction and even death. Most emboli in the arterial system originate from the left side of the heart from conditions such as arrhythmias, valvular heart disease, myocardial infarction, heart failure, or endocarditis. Arterial emboli may lodge in organs, such as the brain, kidneys, or extremities, causing ischemia or infarction.

Release of cardiac enzymes and proteins

When the heart muscle is damaged, the integrity of the cell membrane is impaired, and intracellular contents — including cardiac enzymes and proteins — are released and can be measured in the bloodstream. The release follows a characteristic rising and falling of values. The released enzymes include creatine kinase, lactate dehydrogenase, and aspartate aminotransferase; the proteins released include troponin T, troponin I, and myoglobin. (See *Release of cardiac enzymes and proteins,* page 136.)

Stenosis

Stenosis is the narrowing of any tubular structure such as a blood vessel or heart valve. When an artery is stenosed, the tissues and organs perfused by that blood vessel may become ischemic, function abnormally, or die. An occluded vein may result in venous congestion and chronic venous insufficiency.

When a heart valve is stenosed, blood flow through that valve is reduced, causing blood to accumulate in the chamber behind the valve. Pressure in that chamber rises in order to pump against the resistance of the stenosed valve. Consequently, the heart has to work harder, resulting in hypertrophy. Hypertrophy and an increase in workload raise the oxygen demands of the heart. A heart with diseased coronary arteries may not be able to sufficiently increase oxygen supply to meet the increased demand.

When stenosis occurs in a valve on the left side of the heart, the increased pressure leads to greater pulmonary venous pressure and pulmonary congestion. As pulmonary vascular resistance rises, right-sided heart failure may occur. Stenosis in a valve on the right side of the heart causes an increase in pressures on the right side of the heart, leading to systemic venous congestion.

Thrombus

A thrombus is a blood clot, consisting of platelets, fibrin, and red and white blood cells, that forms anywhere within the vascular system, such as the arteries, veins, heart chambers, or heart valves.

Three conditions, known as Virchow's triad, promote thrombus formation: endothelial injury, sluggish blood flow, and increased coagulability. When a blood vessel wall is injured, the endothelial lining attracts platelets and other inflammatory mediators, which may stimulate clot formation. Sluggish or abnormal blood flow

RELEASE OF CARDIAC ENZYMES AND PROTEINS

Because they're released by damaged tissue, serum enzymes and isoenzymes — catalytic proteins that vary in concentration in specific organs — can help identify the compromised organ and assess the extent of damage. After acute myocardial infarction (MI), cardiac enzymes and proteins rise and fall in a characteristic pattern, as shown in the graph below.

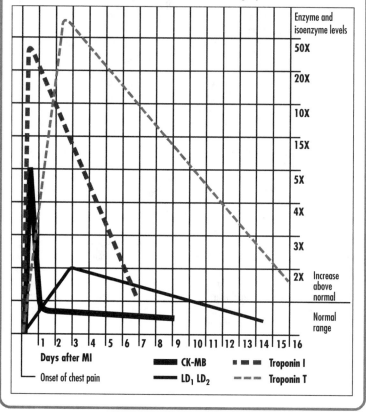

also promotes thrombus formation by allowing platelets and clotting factors to accumulate and adhere to the blood vessel walls. Conditions that increase the coagulability of blood also promote clot formation.

The consequences of thrombus formation include occlusion of the blood vessel or the formation of an embolus (if a portion of a thrombus breaks loose and trav-

els through the circulatory system until it lodges in a smaller vessel).

Valve incompetence

Valve incompetence, also called insufficiency or regurgitation, occurs when valve leaflets do not completely close. Incompetence may affect valves of the veins or the heart.

In the veins, valves keep the blood flowing in one direction, toward the heart. When valve leaflets close improperly, blood flows backward and pools above, causing that valve to weaken and become incompetent. Eventually, the veins become distended, which may result in varicose veins, chronic venous insufficiency, and venous ulcers. Blood clots may form as blood flow becomes sluggish.

In the heart, incompetent valves allow blood to flow in both directions through the valve, increasing the volume of blood that must be pumped (as well as the heart's workload) and resulting in hypertrophy. As blood volume in the heart increases, the involved heart chambers dilate to accommodate the increased volume. Although incompetence may occur in any of the valves of the heart, it's more common in the mitral and aortic valves.

DISORDERS

Atrial septal defect

In this acyanotic congenital heart defect, an opening between the left and right atria allows the blood to flow from left to right, resulting in ineffective pumping of the heart, thus increasing the risk of heart failure.

The three types of atrial septal defects (ASDs) are:

▶ an *ostium secundum defect,* the most common type, which occurs in the region of the fossa ovalis and, occasionally, extends inferiorly, close to the vena cava

▶ a *sinus venosus defect* that occurs in the superior-posterior portion of the atrial septum, sometimes extending into the vena cava, and is almost always associated with abnormal drainage of pulmonary veins into the right atrium

▶ an *ostium primum defect* that occurs in the inferior portion of the septum primum and is usually associated with atrioventricular valve abnormalities (cleft mitral valve) and conduction defects.

ASD accounts for about 10% of congenital heart defects and appears almost twice as often in females as in males, with a strong familial tendency. Although an ASD is usually a benign defect during infancy and childhood, delayed development of symptoms and complications makes it one of the most common congenital heart defects diagnosed in adults.

The prognosis is excellent in asymptomatic patients and in those with uncomplicated surgical repair, but poor in patients with cyanosis caused by large, untreated defects.

Causes

The cause of an ASD is unknown. Ostium primum defects commonly occur in patients with Down syndrome.

Pathophysiology

In an ASD, blood shunts from the left atrium to the right atrium because the left atrial pressure is normally slightly higher than the right atrial pressure. This shunt results in right heart volume overload, affecting the right atrium, right ventricle, and pulmonary arteries. Eventually, the right atrium enlarges, and the right ventricle dilates to accommodate the increased blood volume. If pulmonary artery hypertension develops, increased pulmonary vascular resistance and right ventricular hypertrophy follow. In some adults, irreversible pulmonary artery hypertension causes reversal of the shunt direction, which results in unoxygenated blood entering the systemic circulation, causing cyanosis.

Signs and symptoms

The following are signs and symptoms of an ASD:

▶ fatigue after exertion due to decreased cardiac output from the left ventricle

▶ early to midsystolic murmur at the second or third left intercostal space, caused by extra blood passing through the pulmonic valve

▶ low-pitched diastolic murmur at the lower left sternal border, more pronounced on inspiration, resulting from increased tricuspid valve flow in patients with large shunts

▶ fixed, widely split S_2 due to delayed closure of the pulmonic valve, resulting from an increased volume of blood

▶ systolic click or late systolic murmur at the apex, resulting from mitral valve prolapse in older children with ASD

▶ clubbing and cyanosis, if right-to-left shunt develops.

 AGE ALERT An infant may be cyanotic because he has a cardiac or pulmonary disorder. Cyanosis that worsens with crying is most likely associated with cardiac causes because crying increases pulmonary resistance to blood flow, resulting in an increased right-to-left shunt. Cyanosis that improves with crying is most likely due to pulmonary causes as deep breathing improves tidal volume.

Complications

Complications of an ASD may include:
▶ physical underdevelopment
▶ respiratory infections
▶ heart failure
▶ atrial arrhythmias
▶ mitral valve prolapse.

Diagnosis

The following tests help diagnose atrial septal defect:

▶ Chest X-rays show an enlarged right atrium and right ventricle, a prominent pulmonary artery, and increased pulmonary vascular markings.

▶ Electrocardiography results may be normal but often show right axis deviation, prolonged PR interval, varying degrees of right bundle branch block, right ventricular hypertrophy, atrial fibrillation (particularly in severe cases after age 30) and, in ostium primum defect, left axis deviation.

▶ Echocardiography measures right ventricular enlargement, may locate the defect, and shows volume overload in the right side of the heart. It may reveal right ventricular and pulmonary artery dilation.

▶ Cardiac catheterization may confirm an ASD. Right atrial blood is more oxygenated than superior vena caval blood, indicating a left-to-right shunt, and determines the degree of shunting and pulmonary vascular disease. Dye injection shows the defect's size and location, the location of pulmonary venous drainage, and the competence of the atrioventricular valves.

Treatment

Correcting an ASD typically involves:
▶ surgery to repair the defect by age 3 to 6, using a patch of pericardium or prosthetic material. A small defect may be sutured closed. Monitor for arrhythmias postoperatively because edema of the atria may interfere with sinoatrial node function.

▶ valve repair if heart valves are involved
▶ antibiotic prophylaxis to prevent infective endocarditis
▶ antiarrhythmic medication to treat arrhythmias.

Cardiac arrhythmias

In arrhythmias, abnormal electrical conduction or automaticity changes heart rate and rhythm. Arrhythmias vary in severity, from those that are mild, asymptomatic, and require no treatment (such as sinus arrhythmia, in which heart rate increases and decreases with respiration) to catastrophic ventricular fibrillation, which requires immediate resuscitation. Arrhythmias are generally classified according to their origin (ventricular or supraventricular). Their effect on cardiac output and blood pressure, partially influenced by the site of origin, determines their clinical significance.

Causes

Common causes of arrhythmias include:
▶ congenital defects
▶ myocardial ischemia or infarction
▶ organic heart disease

- drug toxicity
- degeneration of the conductive tissue
- connective tissue disorders
- electrolyte imbalances
- cellular hypoxia
- hypertrophy of the heart muscle
- acid-base imbalances
- emotional stress.

However, each arrhythmia may have its own specific causes. (See *Types of cardiac arrhythmias,* pages 140 to 147.)

Pathophysiology
Arrhythmias may result from enhanced automaticity, reentry, escape beats, or abnormal electrical conduction. (See *Comparing normal and abnormal conduction,* pages 148 and 149.)

Signs and symptoms
Signs and symptoms of arrhythmias result from reduced cardiac output and altered perfusion to the organs, and may include:
- dyspnea
- hypotension
- dizziness, syncope, and weakness
- chest pain
- cool, clammy skin
- altered level of consciousness
- reduced urinary output.

Complications
Possible complications of arrhythmias include:
- sudden cardiac death
- myocardial infarction
- heart failure
- thromboembolism.

Diagnosis
- Electrocardiography detects arrhythmias as well as ischemia and infarction that may result in arrhythmias.
- Laboratory testing may reveal electrolyte abnormalities, acid-base abnormalities, or drug toxicities that may cause arrhythmias.

- Holter monitoring detects arrhythmias and effectiveness of drug therapy during a patient's daily activities.
- Exercise testing may detect exercise-induced arrhythmias.
- Electrophysiologic testing identifies the mechanism of an arrhythmia and the location of accessory pathways; it also assesses the effectiveness of antiarrhythmic drugs.

Treatment
Follow the specific treatment guidelines for each arrhythmia. (See *Types of cardiac arrhythmias*, pages 140 to 147.)

Cardiac tamponade
Cardiac tamponade is a rapid, unchecked rise in pressure in the pericardial sac that compresses the heart, impairs diastolic filling, and reduces cardiac output. The rise in pressure usually results from blood or fluid accumulation in the pericardial sac. Even a small amount of fluid (50 to 100 ml) can cause a serious tamponade if it accumulates rapidly.

Prognosis depends on the rate of fluid accumulation. If fluid accumulates rapidly, cardiac tamponade requires emergency lifesaving measures to prevent death. A slow accumulation and rise in pressure may not produce immediate symptoms because the fibrous wall of the pericardial sac can gradually stretch to accommodate as much as 1 to 2 L of fluid.

Causes
Cause of cardiac tamponade may include:
- idiopathic causes (e.g., Dressler's syndrome)
- effusion (from cancer, bacterial infections, tuberculosis and, rarely, acute rheumatic fever)
- hemorrhage from trauma (such as gunshot or stab wounds of the chest)
- hemorrhage from nontraumatic causes (such as anticoagulant therapy in patients with pericarditis or rupture of the heart or great vessels)
- viral or postirradiation pericarditis

TYPES OF CARDIAC ARRHYTHMIAS

This chart reviews many common cardiac arrhythmias and outlines their features, causes, and treatments. Use a normal electrocardiogram strip, if available, to compare normal cardiac rhythm configurations with the rhythm strips below. Characteristics of normal sinus rhythm include:

▶ ventricular and atrial rates of 60 to 100 beats/minute
▶ regular and uniform QRS complexes and P waves
▶ PR interval of 0.12 to 0.20 second
▶ QRS duration < 0.12 second
▶ identical atrial and ventricular rates, with constant PR intervals.

ARRHYTHMIA AND FEATURES

Sinus tachycardia

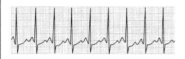

▶ Atrial and ventricular rates regular
▶ Rate > 100 beats/minute; rarely, > 160 beats/minute
▶ Normal P wave preceding each QRS complex

Sinus bradycardia

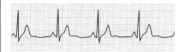

▶ Regular atrial and ventricular rates
▶ Rate < 60 beats/minute
▶ Normal P waves preceding each QRS complex

Paroxysmal supraventricular tachycardia (PSVT)

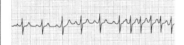

▶ Atrial and ventricular rates regular
▶ Heart rate > 160 beats/minute; rarely exceeds 250 beats/minute
▶ P waves regular but aberrant; difficult to differentiate from preceding T wave
▶ P wave preceding each QRS complex
▶ Sudden onset and termination of arrhythmia

CAUSES	TREATMENT
▶ Normal physiologic response to fever, exercise, anxiety, pain, dehydration; may also accompany shock, left ventricular failure, cardiac tamponade, hyperthyroidism, anemia, hypovolemia, pulmonary embolism, and anterior wall myocardial infarction (MI) ▶ May also occur with atropine, epinephrine, isoproterenol, quinidine, caffeine, alcohol, and nicotine use	▶ Correction of underlying cause ▶ Propranolol for symptomatic patients
▶ Normal, in well-conditioned heart, as in an athlete ▶ Increased intracranial pressure; increased vagal tone due to straining during defecation, vomiting, intubation, or mechanical ventilation; sick sinus syndrome; hypothyroidism; and inferior wall MI ▶ May also occur with anticholinesterase, beta blocker, digoxin, and morphine use	▶ For low cardiac output, dizziness, weakness, altered level of consciousness, or low blood pressure; advanced cardiac life support (ACLS) protocol for administration of atropine ▶ Temporary pacemaker or isoproterenol if atropine fails; may need permanent pacemaker
▶ Intrinsic abnormality of atrioventricular (AV) conduction system ▶ Physical or psychological stress, hypoxia, hypokalemia, cardiomyopathy, congenital heart disease, MI, valvular disease, Wolff-Parkinson-White syndrome, cor pulmonale, hyperthyroidism, and systemic hypertension ▶ Digoxin toxicity; use of caffeine, marijuana, or central nervous system stimulants	▶ If patient is unstable, prepare for immediate cardioversion ▶ If patient is stable, apply vagal stimulation, Valsalva's maneuver, carotid sinus massage ▶ Adenosine by rapid intravenous (I.V.) bolus injection to rapidly convert arrhythmia ▶ If patient is stable, determine QRS complex width. For wide complex width, follow ACLS protocol for lidocaine and procainamide. For narrow complex width and normal or elevated blood pressure, follow ACLS protocol for verapamil and consider digoxin, beta blockers, and diltiazem. For narrow complex width with low or unstable blood pressure (and for ineffective drug response for others), use synchronized cardioversion.

(continued)

TYPES OF CARDIAC ARRHYTHMIAS *(continued)*

ARRHYTHMIA AND FEATURES

Atrial flutter

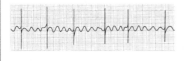

- ▶ Atrial rhythm at regular rate; 250 to 400 beats/minute
- ▶ Ventricular rate variable, depending on degree of atrioventricular (AV) block (usually 60 to 100 beats/minute)
- ▶ Sawtooth P-wave configuration possible (F waves)
- ▶ QRS complexes uniform in shape, but often irregular in rate

Atrial fibrillation

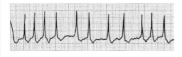

- ▶ Atrial rhythm grossly irregular; rate > 400 beats/minute
- ▶ Ventricular rate grossly irregular
- ▶ QRS complexes of uniform configuration and duration
- ▶ PR interval indiscernible
- ▶ No P waves, or P waves that appear as erratic, irregular, baseline fibrillatory waves

Junctional rhythm

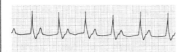

- ▶ Atrial and ventricular rates regular; atrial rate 40 to 60 beats/minute; ventricular rate usually 40 to 60 beats/minute (60 to 100 beats/minute is accelerated junctional rhythm)
- ▶ P waves preceding, hidden within (absent), or after QRS complex; inverted if visible
- ▶ PR interval (when present) < 0.12 second
- ▶ QRS complex configuration and duration normal, except in aberrant conduction

First-degree AV block

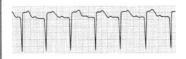

- ▶ Atrial and ventricular rates regular
- ▶ PR interval > 0.20 second
- ▶ P wave precedes QRS complex
- ▶ QRS complex normal

Second-degree AV block
Mobitz I (Wenckebach)

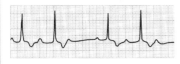

- ▶ Atrial rhythm regular
- ▶ Ventricular rhythm irregular
- ▶ Atrial rate exceeds ventricular rate
- ▶ PR interval progressively, but only slightly, longer with each cycle until QRS complex disappears (dropped beat); PR interval shorter after dropped beat

CAUSES	TREATMENT
▶ Heart failure, tricuspid or mitral valve disease, pulmonary embolism, cor pulmonale, inferior wall MI, and pericarditis ▶ Digoxin toxicity	▶ If patient is unstable with a ventricular rate > 150 beats/minute, prepare for immediate cardioversion ▶ If patient is stable, drug therapy may include diltiazem, beta blockers, verapamil, digoxin, procainamide, or quinidine
▶ Heart failure, chronic obstructive pulmonary disease, thyrotoxicosis, constrictive pericarditis, ischemic heart disease, sepsis, pulmonary embolus, rheumatic heart disease, hypertension, mitral stenosis, atrial irritation, or complication of coronary bypass or valve replacement surgery ▶ Nifedipine and digoxin use	▶ If patient is unstable with a ventricular rate > 150 beats/minute, prepare for immediate cardioversion ▶ If patient is stable, drug therapy may include diltiazem, beta blockers, verapamil, digoxin, procainamide, or ibutilide, given I.V.
▶ Inferior wall MI or ischemia, hypoxia, vagal stimulation, and sick sinus syndrome ▶ Acute rheumatic fever ▶ Valve surgery ▶ Digoxin toxicity	▶ Atropine for symptomatic slow rate ▶ Pacemaker insertion if patient doesn't respond to drugs ▶ Discontinuation of digoxin if appropriate
▶ May be seen in healthy persons ▶ Inferior wall MI or ischemia, hypothyroidism, hypokalemia, and hyperkalemia ▶ Digoxin toxicity; use of quinidine, procainamide, or propranolol	▶ Cautious use of digoxin ▶ Correction of underlying cause ▶ Possibly atropine if PR interval > 0.26 second or bradycardia develops
▶ Inferior wall MI, cardiac surgery, acute rheumatic fever, and vagal stimulation ▶ Digoxin toxicity; use of propranolol, quinidine, or procainamide	▶ Treatment of underlying cause ▶ Atropine or temporary pacemaker for symptomatic bradycardia ▶ Discontinuation of digoxin if appropriate

(continued)

TYPES OF CARDIAC ARRHYTHMIAS *(continued)*

ARRHYTHMIA AND FEATURES

Second-degree AV block
Mobitz II

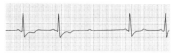

▶ Atrial rate regular
▶ Ventricular rhythm regular or irregular, with varying degree of block
▶ P-P interval constant
▶ QRS complexes periodically absent

Third-degree AV block
(complete heart block)

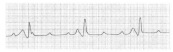

▶ Atrial rate regular
▶ Ventricular rate slow and regular
▶ No relation between P waves and QRS complexes
▶ No constant PR interval
▶ QRS interval normal (nodal pacemaker) or wide and bizarre (ventricular pacemaker)

Premature ventricular contraction (PVC)

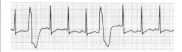

▶ Atrial rate regular
▶ Ventricular rate irregular
▶ QRS complex premature, usually followed by a compensatory pause
▶ QRS complex wide and distorted, usually > 0.14 second
▶ Premature QRS complexes occurring singly, in pairs, or in threes, alternating with normal beats; focus from one or more sites
▶ Ominous when clustered, multifocal, with R wave on T pattern

---

Ventricular tachycardia

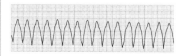

▶ Ventricular rate 140 to 220 beats/minute, regular or irregular
▶ QRS complexes wide, bizarre, and independent of P waves
▶ P waves not discernible
▶ May start and stop suddenly

CAUSES	TREATMENT
▶ Severe coronary artery disease, anterior wall MI, and acute myocarditis ▶ Digoxin toxicity	▶ Atropine or isoproterenol for symptomatic bradycardia ▶ Temporary or permanent pacemaker ▶ Discontinuation of digoxin if appropriate
▶ Inferior or anterior wall MI, congenital abnormality, rheumatic fever, hypoxia, postoperative complication of mitral valve replacement, Lev's disease (fibrosis and calcification that spreads from cardiac structures to the conductive tissue), and Lenègre's disease (conductive tissue fibrosis) ▶ Digoxin toxicity	▶ Atropine or isoproterenol for symptomatic bradycardia ▶ Temporary or permanent pacemaker
▶ Heart failure; old or acute MI, ischemia, or contusion; myocardial irritation by ventricular catheter or a pacemaker; hypercapnia; hypokalemia; and hypocalcemia ▶ Drug toxicity (digoxin, aminophylline, tricyclic antidepressants, beta-blockers, isoproterenol, or dopamine) ▶ Caffeine, tobacco, or alcohol use ▶ Psychological stress, anxiety, pain, or exercise	▶ If warranted, lidocaine, procainamide, or bretylium I.V. ▶ Treatment of underlying cause ▶ Discontinuation of drug causing toxicity ▶ Potassium chloride I.V. if PVC induced by hypokalemia
▶ Myocardial ischemia, MI, or aneurysm; coronary artery disease; rheumatic heart disease; mitral valve prolapse; heart failure; cardiomyopathy; ventricular catheters; hypokalemia; hypercalcemia; and pulmonary embolism ▶ Digoxin, procainamide, epinephrine, or quinidine toxicity ▶ Anxiety	▶ With pulse: If hemodynamically stable with ventricular rate < 150 beats/minute, follow ACLS protocol for administration of lidocaine, procainamide, or bretylium; if drugs are ineffective, initiate synchronized cardioversion ▶ If ventricular rate > 150 beats/minute, follow ACLS protocol for immediate synchronized cardioversion, followed by antiarrhythmic agents ▶ Pulseless: Initiate cardiopulmonary resuscitation (CPR); follow ACLS protocol for defibrillation, endotracheal (ET) intubation, and administration of epinephrine, lidocaine, bretylium, magnesium sulfate, or procainamide

(continued)

TYPES OF CARDIAC ARRYTHMIAS *(continued)*

ARRHYTHMIA AND FEATURES

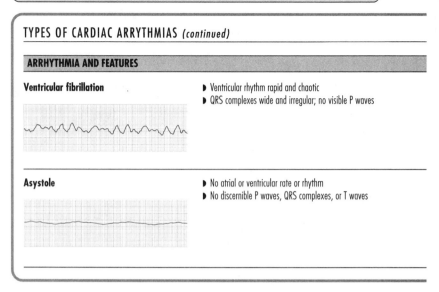

Ventricular fibrillation

▶ Ventricular rhythm rapid and chaotic
▶ QRS complexes wide and irregular; no visible P waves

Asystole

▶ No atrial or ventricular rate or rhythm
▶ No discernible P waves, QRS complexes, or T waves

▶ chronic renal failure requiring dialysis
▶ drug reaction from procainamide, hydralazine, minoxidil, isoniazid, penicillin, methysergide maleate, or daunorubicin
▶ connective tissue disorders (such as rheumatoid arthritis, systemic lupus erythematosus, rheumatic fever, vasculitis, and scleroderma)
▶ acute myocardial infarction.

Pathophysiology

In cardiac tamponade, the progressive accumulation of fluid in the pericardial sac causes compression of the heart chambers. This compression obstructs blood flow into the ventricles and reduces the amount of blood that can be pumped out of the heart with each contraction. (See *Understanding cardiac tamponade,* page 150.)

Each time the ventricles contract, more fluid accumulates in the pericardial sac. This further limits the amount of blood that can fill the ventricular chambers, especially the left ventricle, during the next cardiac cycle.

The amount of fluid necessary to cause cardiac tamponade varies greatly; it may be as little as 200 ml when the fluid ac-

cumulates rapidly or more than 2,000 ml if the fluid accumulates slowly and the pericardium stretches to adapt.

Signs and symptoms

The following signs and symptoms may occur:
▶ elevated central venous pressure (CVP) with neck vein distention due to increased jugular venous pressure
▶ muffled heart sounds caused by fluid in the pericardial sac
▶ pulsus paradoxus (an inspiratory drop in systemic blood pressure greater than 15 mm Hg) due to impaired diastolic filling
▶ diaphoresis and cool clammy skin caused by a drop in cardiac output
▶ anxiety, restlessness, and syncope due to a drop in cardiac output
▶ cyanosis due to reduced oxygenation of the tissues
▶ weak, rapid pulse in response to a drop in cardiac output
▶ cough, dyspnea, orthopnea, and tachypnea because the lungs are compressed by an expanding pericardial sac.

CAUSES	TREATMENT
▶ Myocardial ischemia, MI, untreated ventricular tachycardia, R-on-T phenomenon, hypokalemia, hyperkalemia, hypercalcemia, alkalosis, electric shock, and hypothermia ▶ Digoxin, epinephrine, or quinidine toxicity	▶ Initiate CPR; follow ACLS protocol for defibrillation, ET intubation, and administration of epinephrine, lidocaine, bretylium, magnesium sulfate, or procainamide
▶ Myocardial ischemia, MI, aortic valve disease, heart failure, hypoxia, hypokalemia, severe acidosis, electric shock, ventricular arrhythmia, AV block, pulmonary embolism, heart rupture, cardiac tamponade, hyperkalemia, and electromechanical dissociation ▶ Cocaine overdose	▶ Continue CPR, follow ACLS protocol for ET intubation, administration of epinephrine and atropine, and possible transcutaneous pacing

Complications
Reduced cardiac output may be fatal without prompt treatment.

Diagnosis
▶ Chest X-rays show slightly widened mediastinum and possible cardiomegaly. The cardiac silhouette may have a goblet-shaped appearance.

▶ Electrocardiography (ECG) may show low-amplitude QRS complex and electrical alternans, an alternating beat-to-beat change in amplitude of the P wave, QRS complex, and T wave. Generalized ST-segment elevation is noted in all leads. An ECG is used to rule out other cardiac disorders; it may reveal changes produced by acute pericarditis.

▶ Pulmonary artery catheterization detects increased right atrial pressure, right ventricular diastolic pressure, and CVP.

▶ Echocardiography may reveal pericardial effusion with signs of right ventricular and atrial compression.

Treatment
Correcting cardiac tamponade typically involves:

▶ supplemental oxygen to improve oxygenation

▶ continuous ECG and hemodynamic monitoring in an intensive care unit to detect complications and monitor effects of therapy

▶ pericardiocentesis (needle aspiration of the pericardial cavity) to reduce fluid in the pericardial sac and improve systemic arterial pressure and cardiac output. A catheter may be left in the pericardial space attached to a drainage bag to allow for continuous drainage of fluid

▶ pericardectomy — the surgical creation of an opening to remove accumulated fluid from the pericardial sac

▶ resection of a portion or all of the pericardium to allow full communication with the pleura, if repeated pericardiocentesis fails to prevent recurrence

▶ trial volume loading with crystalloids such as intravenous 0.9% normal saline to maintain systolic blood pressure

▶ inotropic drugs, such as isoproterenol or dopamine, to improve myocardial contractility until fluid in the pericardial sac can be removed

COMPARING NORMAL AND ABNORMAL CONDUCTION

NORMAL CARDIAC CONDUCTION

The conduction system of the heart, shown below, begins at the heart's pacemaker, the sinoatrial (SA) node. When an impulse leaves the SA node, it travels through the atria along Bach-mann's bundle and the internodal pathways to the atrioventricular (AV) node and then down the bundle of His, along the bundle branches and, finally, down the Purkinje fibers to the ventricles.

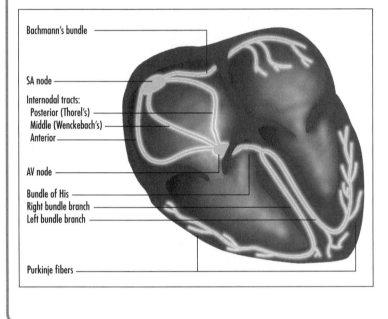

Bachmann's bundle

SA node

Internodal tracts:
Posterior (Thorel's)
Middle (Wenckebach's)
Anterior

AV node

Bundle of His
Right bundle branch
Left bundle branch

Purkinje fibers

❯ in traumatic injury, a blood transfusion or a thoracotomy to drain reaccumulating fluid or to repair bleeding sites may be necessary

❯ heparin-induced tamponade requires administration of heparin antagonist protamine sulfate to stop bleeding

❯ warfarin-induced tamponade may necessitate use of vitamin K to stop bleeding.

Cardiomyopathy

Cardiomyopathy generally applies to disease of the heart muscle fibers, and it occurs in three main forms: dilated, hypertrophic, and restrictive (extremely rare).

Cardiomyopathy is the second most common direct cause of sudden death; coronary artery disease is first. Approximately 5 to 8 per 100,000 Americans have *dilated congestive cardiomyopathy,* the most common type. At greatest risk of cardiomyopathy are males and blacks; other risk factors include hypertension, pregnancy, viral infections, and alcohol use. Because dilated cardiomyopathy is usually not diagnosed until its advanced stages, the prognosis is generally poor. The course of *hypertrophic cardiomyopathy* is variable. Some patients progressively deteriorate, whereas others remain stable for years. It is estimated that almost 50% of all sud-

ABNORMAL CARDIAC CONDUCTION

Altered automaticity, reentry, or conduction disturbances may cause cardiac arrhythmias.

Altered automaticity

Enhanced automaticity is the result of partial depolarization, which may increase the intrinsic rate of the SA node or latent pacemakers, or may induce ectopic pacemakers to reach threshold and depolarize.

Automaticity may be enhanced by drugs such as epinephrine, atropine, and digoxin and conditions such as acidosis, alkalosis, hypoxia, myocardial infarction, hypokalemia, and hypocalcemia. Examples of arrhythmias caused by enhanced automaticity include atrial fibrillation and flutter; supraventricular tachycardia; premature atrial, junctional, and ventricular complexes; ventricular tachycardia and fibrillation; and accelerated idioventricular and junctional rhythms.

Reentry

Ischemia or deformation causes an abnormal circuit to develop within conductive fibers. Although current flow is blocked in one direction within the circuit, the descending impulse can travel in the other direction. By the time the impulse completes the circuit, the previously depolarized tissue within the circuit is no longer refractory to stimulation.

Conditions that increase the likelihood of reentry include hyperkalemia, myocardial ischemia, and the use of certain antiarrhythmic drugs. Reentry may be responsible for dysrhythmias such as paroxysmal supraventricular tachycardia; premature atrial, junctional, and ventricular complexes; and ventricular tachycardia.

An alternative reentry mechanism depends on the presence of a congenital accessory pathway linking the atria and the ventricles outside the AV junction, for example, Wolff-Parkinson-White syndrome.

Conduction disturbances

Conduction disturbances occur when impulses are conducted too quickly or too slowly. Possible causes include trauma, drug toxicity, myocardial ischemia, myocardial infarction, and electrolyte abnormalities. The atrioventricular blocks occur as a result of conduction disturbances.

den deaths in competitive athletes age 35 or younger are due to hypertrophic cardiomyopathy. If severe, *restrictive cardiomyopathy* is irreversible.

Causes

Most patients with cardiomyopathy have idiopathic, or primary, disease, but some are secondary to identifiable causes. (See *Comparing the cardiomyopathies,* pages 152 and 153.) Hypertrophic cardiomyopathy is almost always inherited as a non–sex-linked autosomal dominant trait.

Pathophysiology

Dilated cardiomyopathy results from extensively damaged myocardial muscle fibers. Consequently, there is reduced contractility in the left ventricle. As systolic function declines, stroke volume, ejection fraction, and cardiac output fall. As end-diastolic volumes rise, pulmonary congestion may occur. The elevated end-diastolic volume is a compensatory response to preserve stroke volume despite a reduced ejection fraction. The sympathetic nervous system is also stimulated to increase heart rate and contractility. The kidneys are stimulated to retain sodium and water to maintain cardiac output, and vaso-

UNDERSTANDING CARDIAC TAMPONADE

NORMAL HEART AND PERICARDIUM

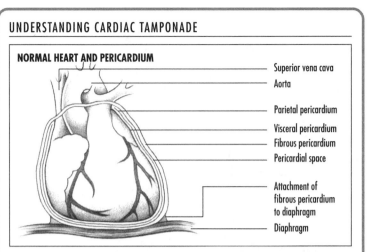

- Superior vena cava
- Aorta
- Parietal pericardium
- Visceral pericardium
- Fibrous pericardium
- Pericardial space
- Attachment of fibrous pericardium to diaphragm
- Diaphragm

The pericardial sac, which surrounds and protects the heart, is composed of several layers. The fibrous pericardium is the tough outermost membrane; the inner membrane, called the serous membrane, consists of the visceral and parietal layers. The visceral layer clings to the heart and is also known as the epicardial layer of the heart. The parietal layer lies between the visceral layer and the fibrous pericardium. The pericardial space — between the visceral and parietal layers — contains 10 to 30 ml of pericardial fluid. This fluid lubricates the layers and minimizes friction when the heart contracts.

CARDIAC TAMPONADE

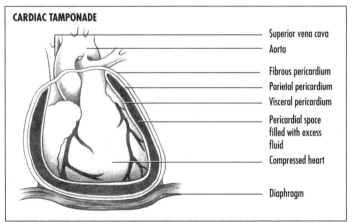

- Superior vena cava
- Aorta
- Fibrous pericardium
- Parietal pericardium
- Visceral pericardium
- Pericardial space filled with excess fluid
- Compressed heart
- Diaphragm

In cardiac tamponade, blood or fluid fills the pericardial space, compressing the heart chambers, increasing intracardiac pressure, and obstructing venous return. As blood flow into the ventricles falls, so does cardiac output. Without prompt treatment, low cardiac output can be fatal.

constriction also occurs as the renin-angiotensin system is stimulated. When these compensatory mechanisms can no longer maintain cardiac output, the heart begins to fail. Left ventricular dilation occurs as venous return and systemic vascular resistance rise. Eventually, the atria also dilate as more work is required to pump blood into the full ventricles. Cardiomegaly occurs as a consequence of dilation of the atria and ventricles. Blood pooling in the ventricles increases the risk of emboli.

 AGE ALERT Barth syndrome is a rare genetic disorder that can cause dilated cardiomyopathy in boys. This syndrome may be associated with skeletal muscle changes, short stature, neutropenia, and increased susceptibility to bacterial infections. Evidence of dilated cardiomyopathy may appear as early as the first few days or months of life.

Unlike dilated cardiomyopathy, which affects systolic function, *hypertrophic cardiomyopathy* primarily affects diastolic function. The features of hypertrophic cardiomyopathy include asymmetrical left ventricular hypertrophy; hypertrophy of the intraventricular septum; rapid, forceful contractions of the left ventricle; impaired relaxation; and obstruction to left ventricular outflow. The hypertrophied ventricle becomes stiff, noncompliant, and unable to relax during ventricular filling. Consequently, ventricular filling is reduced and left ventricular filling pressure rises, causing a rise in left atrial and pulmonary venous pressures and leading to venous congestion and dyspnea. Ventricular filling time is further reduced as a compensatory response to tachycardia. Reduced ventricular filling during diastole and obstruction to ventricular outflow lead to low cardiac output. If papillary muscles become hypertrophied and do not close completely during contraction, mitral regurgitation occurs. Moreover, intramural coronary arteries are abnormally small and may not be sufficient to supply the hypertrophied muscle with enough blood and

oxygen to meet the increased needs of the hyperdynamic muscle.

Restrictive cardiomyopathy is characterized by stiffness of the ventricle caused by left ventricular hypertrophy and endocardial fibrosis and thickening, thus reducing the ability of the ventricle to relax and fill during diastole. Moreover, the rigid myocardium fails to contract completely during systole. As a result, cardiac output falls.

Signs and symptoms

Clinical manifestations of *dilated cardiomyopathy* may include:
◗ shortness of breath, orthopnea, dyspnea on exertion, paroxysmal nocturnal dyspnea, fatigue, and a dry cough at night due to left-sided heart failure
◗ peripheral edema, hepatomegaly, jugular venous distention, and weight gain caused by right-sided heart failure
◗ peripheral cyanosis associated with a low cardiac output
◗ tachycardia as a compensatory response to low cardiac output
◗ pansystolic murmur associated with mitral and tricuspid insufficiency secondary to cardiomegaly and weak papillary muscles
◗ S_3 and S_4 gallop rhythms associated with heart failure
◗ irregular pulse if atrial fibrillation exists.

Clinical manifestations of *hypertrophic cardiomyopathy* may include:
◗ angina caused by the inability of the intramural coronary arteries to supply enough blood to meet the increased oxygen demands of the hypertrophied heart
◗ syncope resulting from arrhythmias or reduced ventricular filling leading to a reduced cardiac output
◗ dyspnea due to elevated left ventricular filling pressure
◗ fatigue associated with a reduced cardiac output
◗ systolic ejection murmur along the left sternal border and at the apex caused by mitral regurgitation

COMPARING THE CARDIOMYOPATHIES

Cardiomyopathies include a variety of structural or functional abnormalities of the ventricles. They are grouped into three main pathophysiologic types — dilated, hypertrophic, and restrictive. These conditions may lead to heart failure by impairing myocardial structure and function.

NORMAL HEART	DILATED CARDIOMYOPATHY

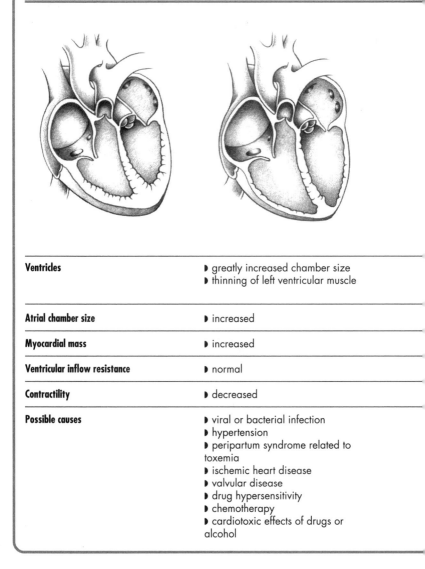

Ventricles	▶ greatly increased chamber size ▶ thinning of left ventricular muscle
Atrial chamber size	▶ increased
Myocardial mass	▶ increased
Ventricular inflow resistance	▶ normal
Contractility	▶ decreased
Possible causes	▶ viral or bacterial infection ▶ hypertension ▶ peripartum syndrome related to toxemia ▶ ischemic heart disease ▶ valvular disease ▶ drug hypersensitivity ▶ chemotherapy ▶ cardiotoxic effects of drugs or alcohol

HYPERTROPHIC CARDIOMYOPATHY	RESTRICTIVE CARDIOMYOPATHY

▶ normal or decreased chamber size ▶ left ventricular hypertrophy ▶ thickened interventricular septum	▶ decreased ventricular chamber size ▶ left ventricular hypertrophy
▶ increased	▶ increased
▶ increased	▶ normal
▶ increased	▶ increased
▶ increased or decreased	▶ normal or decreased
▶ autosomal dominant trait ▶ hypertension ▶ obstructive valvular disease ▶ thyroid disease	▶ amyloidosis ▶ sarcoidosis ▶ hemochromatosis ▶ infiltrative neoplastic disease

COMPARING DIAGNOSTIC TESTS IN CARDIOMYOPATHY

TEST	DILATED CARDIOMYOPATHY
Electrocardiography	Biventricular hypertrophy, sinus tachycardia, atrial enlargement, atrial and ventricular arrhythmias, bundle branch block, and ST-segment and T-wave abnormalities
Echocardiography	Left ventricular thrombi, global hypokinesia, enlarged atria, left ventricular dilation and, possibly, valvular abnormalities
Chest X-ray	Cardiomegaly, pulmonary congestion, pulmonary venous hypertension, and pleural or pericardial effusions
Cardiac catheterization	Elevated left atrial and left ventricular end-diastolic pressures, left ventricular enlargement, and mitral and tricuspid incompetence; may identify coronary artery disease as a cause
Radionuclide studies	Left ventricular dilation and hypokinesis, reduced ejection fraction

▶ peripheral pulse with a characteristic double impulse (pulsus biferiens) caused by powerful left ventricular contractions and rapid ejection of blood during systole
▶ abrupt arterial pulse secondary to vigorous left ventricular contractions
▶ irregular pulse if an enlarged atrium causes atrial fibrillation.

Clinical manifestations of *restrictive cardiomyopathy* may include:
▶ fatigue, dyspnea, orthopnea, chest pain, edema, liver engorgement, peripheral cyanosis, pallor, and S_3 or S_4 gallop rhythms due to heart failure

▶ systolic murmurs caused by mitral and tricuspid insufficiency.

Complications
Possible complications of cardiomyopathy include:
▶ heart failure
▶ arrhythmias
▶ systemic or pulmonary embolization
▶ sudden death.

Diagnosis
The following tests help diagnose cardiomyopathy:

HYPERTROPHIC CARDIOMYOPATHY	RESTRICTIVE CARDIOMYOPATHY
Left ventricular hypertrophy, ST-segment and T-wave abnormalities, left anterior hemiblock, Q waves in precordial and inferior leads, ventricular arrhythmias and, possibly, atrial fibrillation	Low voltage, hypertrophy, atrioventricular conduction defects, and arrhythmias
Asymmetrical thickening of the left ventricular wall, increased thickness of the intraventricular septum and abnormal motion of the anterior mitral leaflet during systole, and occluding left ventricular outflow in obstructive disease	Increased left ventricular muscle mass, normal or reduced left ventricular cavity size, and normal systolic function; rules out constrictive pericarditis
Cardiomegaly	Cardiomegaly, pericardial effusion, and pulmonary congestion Increased left ventricular end-diastolic pressure; rules out constrictive pericarditis
Elevated ventricular end-diastolic pressure and, possibly, mitral insufficiency, hyperdynamic systolic function, left ventricular outflow obstruction	Normal or reduced systolic function and myocardial infiltration
Reduced left ventricular volume, increased muscle mass, and ischemia	Left ventricular hypertrophy with restricted ventricular filling

▶ Echocardiography confirms dilated cardiomyopathy.

▶ Chest X-ray may reveal cardiomegaly associated with any of the cardiomyopathies.

▶ Cardiac catheterization with possible heart biopsy can be definitive with hypertrophic cardiomyopathy.

▶ Diagnosis requires elimination of other possible causes of heart failure and arrhythmias. (See *Comparing diagnostic tests in cardiomyopathy*.)

Treatment

Correction of *dilated cardiomyopathy* may involve:

▶ treatment of the underlying cause, if identifiable

▶ angiotensin-converting enzyme (ACE) inhibitors, as first-line therapy, to reduce afterload through vasodilation

▶ diuretics, taken with ACE inhibitors, to reduce fluid retention

▶ digoxin, for patients not responding to ACE inhibitor and diuretic therapy, to improve myocardial contractility

CLASSIFYING HEART FAILURE

The New York Heart Association (NYHA) classification is a universal gauge of heart failure severity based on physical limitations.

CLASS I: MINIMAL
▶ No limitations
▶ Ordinary physical activity doesn't cause undue fatigue, dyspnea, palpitations, or angina

CLASS II: MILD
▶ Slightly limited physical activity
▶ Comfortable at rest
▶ Ordinary physical activity results in fatigue, palpitations, dyspnea, or angina

CLASS III: MODERATE
▶ Markedly limited physical activity
▶ Comfortable at rest
▶ Less than ordinary activity produces symptoms

CLASS IV: SEVERE
▶ Patient unable to perform any physical activity without discomfort
▶ Angina or symptoms of cardiac inefficiency may develop at rest

▶ hydralazine and isosorbide dinitrate, in combination, to produce vasodilation
▶ beta-adrenergic blockers for patients with New York Heart Association class II or III heart failure. (See *Classifying heart failure*.)
▶ antiarrhythmics such as amiodarone, used cautiously, to control arrhythmias
▶ cardioversion to convert atrial fibrillation to sinus rhythm
▶ pacemaker insertion to correct arrhythmias
▶ anticoagulants (controversial) to reduce the risk of emboli

▶ revascularization, such as coronary artery bypass graft surgery, if dilated cardiomyopathy is due to ischemia
▶ valvular repair or replacement, if dilated cardiomyopathy is due to valve dysfunction
▶ heart transplantation in patients refractory to medical therapy
▶ lifestyle modifications, such as smoking cessation; low-fat, low-sodium diet; physical activity; and abstinence from alcohol.

Correction of *hypertrophic cardiomyopathy* may involve:
▶ beta-adrenergic blockers to slow the heart rate, reduce myocardial oxygen demands, and increase ventricular filling by relaxing the obstructing muscle, thereby increasing cardiac output
▶ antiarrhythmic drugs, such as amiodarone, to reduce arrhythmias
▶ cardioversion to treat atrial fibrillation
▶ anticoagulation to reduce risk of systemic embolism with atrial fibrillation
▶ verapamil and diltiazem to reduce ventricular stiffness and elevated diastolic pressures
▶ ablation of the atrioventricular node and implantation of a dual-chamber pacemaker (controversial), in patients with obstructive hypertrophic cardiomyopathy and ventricular tachycardias, to reduce the outflow gradient by altering the pattern of ventricular contraction
▶ implantable cardioverter-defibrillator to treat ventricular arrhythmias
▶ ventricular myotomy or myectomy (resection of the hypertrophied septum) to ease outflow tract obstruction and relieve symptoms
▶ mitral valve replacement to treat mitral regurgitation
▶ cardiac transplantation for intractable symptoms.

Correction of *restrictive cardiomyopathy* may involve:
▶ treatment of the underlying cause, such as administering deferoxamine to bind iron in restrictive cardiomyopathy due to hemochromatosis

◗ although no therapy exists for restricted ventricular filling, digoxin, diuretics, and a restricted sodium diet may ease the symptoms of heart failure

◗ oral vasodilators may control intractable heart failure.

Coarctation of the aorta

Coarctation is a narrowing of the aorta, usually just below the left subclavian artery, near the site where the ligamentum arteriosum (the remnant of the ductus arteriosus, a fetal blood vessel) joins the pulmonary artery to the aorta. Coarctation may occur with aortic valve stenosis (usually of a bicuspid aortic valve) and with severe cases of hypoplasia of the aortic arch, patent ductus arteriosus (PDA), and ventricular septal defect (VSD). The obstruction to blood flow results in ineffective pumping of the heart and increases the risk for heart failure.

This acyanotic condition accounts for about 7% of all congenital heart defects in children and is twice as common in males as in females. When coarctation of the aorta occurs in females, it's often associated with Turner's syndrome, a chromosomal disorder that causes ovarian dysgenesis.

The prognosis depends on the severity of associated cardiac anomalies. If corrective surgery is performed before isolated coarctation induces severe systemic hypertension or degenerative changes in the aorta, the prognosis is good.

Causes

Although the cause of this defect is unknown, it may be associated with Turner's syndrome.

Pathophysiology

Coarctation of the aorta may develop as a result of spasm and constriction of the smooth muscle in the ductus arteriosus as it closes. Possibly, this contractile tissue extends into the aortic wall, causing narrowing. The obstructive process causes hypertension in the aortic branches above

the constriction (arteries that supply the arms, neck, and head) and diminished pressure in the vessel below the constriction.

Restricted blood flow through the narrowed aorta increases the pressure load on the left ventricle and causes dilation of the proximal aorta and ventricular hypertrophy.

As oxygenated blood leaves the left ventricle, a portion travels through the arteries that branch off the aorta proximal to the coarctation. If PDA is present, the rest of the blood travels through the coarctation, mixes with deoxygenated blood from the PDA, and travels to the legs. If the PDA is closed, the legs and lower portion of the body must rely solely on the blood that gets through the coarctation.

Untreated, this condition may lead to left-sided heart failure and, rarely, to cerebral hemorrhage and aortic rupture. If VSD accompanies coarctation, blood shunts from left to right, straining the right side of the heart. This leads to pulmonary hypertension and, eventually, right-sided heart hypertrophy and failure.

If coarctation is asymptomatic in infancy, it usually remains so throughout adolescence as collateral circulation develops to bypass the narrowed segment.

Signs and symptoms

The following signs and symptoms may occur:

◗ during the first year of life, an infant may display tachypnea, dyspnea, pulmonary edema, pallor, tachycardia, failure to thrive, cardiomegaly, and hepatomegaly due to heart failure

◗ claudication due to reduced blood flow to the legs

◗ hypertension in the upper body due to increased pressure in the arteries proximal to the coarctation

◗ headache, vertigo, and epistaxis secondary to hypertension

◗ upper extremity blood pressure greater than lower extremity blood pressure because blood flow through the coarctation is greater to the upper body than to the lower body

▶ pink upper extremities and cyanotic lower extremities due to reduced oxygenated blood reaching the legs

▶ absent or diminished femoral pulses due to restricted blood flow to the lower extremities through the constricted aorta

▶ continuous midsystolic murmur due to left-to-right shunting of the blood; the murmur is best heard at the base of the heart

▶ chest and arms may be more developed than the legs because circulation to legs is restricted.

Complications
Possible complications of this defect include:

▶ heart failure

▶ severe hypertension

▶ cerebral aneurysms and hemorrhage

▶ rupture of the aorta

▶ aortic aneurysm

▶ infective endocarditis.

Diagnosis
The following tests help diagnose coarctation of the aorta:

▶ Physical examination reveals the cardinal signs — resting systolic hypertension in the upper body, absent or diminished femoral pulses, and a wide pulse pressure.

▶ Chest X-rays may demonstrate left ventricular hypertrophy, heart failure, a wide ascending and descending aorta, and notching of the undersurfaces of the ribs due to erosion by collateral circulation.

▶ Electrocardiography may reveal left ventricular hypertrophy.

▶ Echocardiography may show increased left ventricular muscle thickness, coexisting aortic valve abnormalities, and the coarctation site.

▶ Cardiac catheterization evaluates collateral circulation and measures pressure in the right and left ventricles and in the ascending and descending aortas (on both sides of the obstruction). Aortography locates the site and extent of coarctation.

Treatment
Correction of coarctation of the aorta may involve:

▶ digoxin, diuretics, oxygen, and sedatives in infants with heart failure

▶ prostaglandin infusion to keep the ductus open

▶ antibiotic prophylaxis against infective endocarditis before and after surgery

▶ antihypertensive therapy for children with previous undetected coarctation until surgery is performed

▶ preparation of the infant with heart failure or hypertension for early surgery, or else surgery is delayed until the preschool years. A flap of the left subclavian artery may be used to reconstruct the aorta. Balloon angioplasty or resection with end-to-end anastomosis or use of a tubular graft may also be performed.

Coronary artery disease
Coronary artery disease (CAD) results from the narrowing of the coronary arteries over time due to atherosclerosis. The primary effect of CAD is the loss of oxygen and nutrients to myocardial tissue because of diminished coronary blood flow. As the population ages, the prevalence of CAD is rising. Approximately 11 million Americans have CAD, and it occurs more often in males, whites, and in the middle-aged and elderly. With proper care, the prognosis for CAD is favorable.

Causes
CAD is commonly caused by atherosclerosis. Less common causes of reduced coronary artery blood flow include:

▶ dissecting aneurysm

▶ infectious vasculitis

▶ syphilis

▶ congenital defects.

Pathophysiology
Fatty, fibrous plaques progressively narrow the coronary artery lumina, reducing the volume of blood that can flow through them and leading to myocardial ischemia. (See Atherosclerotic plaque development.)

ATHEROSCLEROTIC PLAQUE DEVELOPMENT

The coronary arteries are made of three layers: intima (the innermost layer, media (the middle layer), and adventitia (the outermost layer).

Damaged by risk factors, a fatty streak begins to build up on the intimal layer.

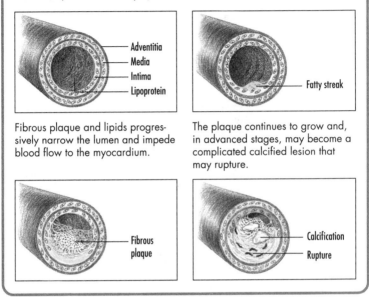

Adventitia
Media
Intima
Lipoprotein

Fatty streak

Fibrous plaque and lipids progressively narrow the lumen and impede blood flow to the myocardium.

The plaque continues to grow and, in advanced stages, may become a complicated calcified lesion that may rupture.

Fibrous plaque

Calcification
Rupture

As atherosclerosis progresses, luminal narrowing is accompanied by vascular changes that impair the ability of the diseased vessel to dilate. This causes a precarious balance between myocardial oxygen supply and demand, threatening the myocardium beyond the lesion. When oxygen demand exceeds what the diseased vessel can supply, localized myocardial ischemia results.

Myocardial cells become ischemic within 10 seconds of a coronary artery occlusion. Transient ischemia causes reversible changes at the cellular and tissue levels, depressing myocardial function. Untreated, this can lead to tissue injury or necrosis. Within several minutes, oxygen deprivation forces the myocardium to shift from aerobic to anaerobic metabolism,

leading to accumulation of lactic acid and reduction of cellular pH.

The combination of hypoxia, reduced energy availability, and acidosis rapidly impairs left ventricular function. The strength of contractions in the affected myocardial region is reduced as the fibers shorten inadequately, resulting in less force and velocity. Moreover, wall motion is abnormal in the ischemic area, resulting in less blood being ejected from the heart with each contraction. Restoring blood flow through the coronary arteries restores aerobic metabolism and contractility. However, if blood flow is not restored, myocardial infarction results.

Signs and symptoms

The following signs and symptoms may occur:

TYPES OF ANGINA

There are four types of angina:
▶ *Stable angina:* pain is predictable in frequency and duration and is relieved by rest and nitroglycerin.
▶ *Unstable angina:* pain increases in frequency and duration and is more easily induced; it indicates a worsening of coronary artery disease that may progress to myocardial infarction.
▶ *Prinzmetal's* or *variant angina:* pain is caused by spasm of the coronary arteries; it may occur spontaneously and may not be related to physical exercise or emotional stress.
▶ *Microvascular angina:* impairment of vasodilator reserve causes angina-like chest pain in a person with normal coronary arteries.

▶ angina, the classic sign of CAD, results from a reduced supply of oxygen to the myocardium. It may be described as burning, squeezing, or tightness in the chest that may radiate to the left arm, neck, jaw, or shoulder blade (See *Types of angina.*)
▶ nausea and vomiting as a result of reflex stimulation of the vomiting centers by pain
▶ cool extremities and pallor caused by sympathetic stimulation
▶ diaphoresis due to sympathetic stimulation
▶ xanthelasma (fat deposits on the eyelids) occurs secondary to hyperlipidemia and atherosclerosis.

 AGE ALERT CAD may be asymptomatic in the older adult because of a decrease in sympathetic response. Dyspnea and fatigue are two key signals of ischemia in an active, older adult.

Complications
Complications of CAD include:

▶ arrhythmias
▶ myocardial infarction.

Diagnosis
The following tests help diagnose coronary artery disease:
▶ Electrocardiography may be normal between anginal episodes. During angina, it may show ischemic changes, such as T-wave inversion, ST-segment depression and, possibly, arrhythmias. ST-segment elevation suggests Prinzmetal's angina.
▶ Exercise testing may be performed to detect ST-segment changes during exercise, indicating ischemia, and to determine a safe exercise prescription.
▶ Coronary angiography reveals location and degree of coronary artery stenosis or obstruction, collateral circulation, and the condition of the artery beyond the narrowing.
▶ Myocardial perfusion imaging with thallium-201 may be performed during treadmill exercise to detect ischemic areas of the myocardium; they appear as "cold spots."
▶ Stress echocardiography may show abnormal wall motion.

Treatment
Treatment of CAD may involve:
▶ nitrates, such as nitroglycerin (given sublingually, orally, transdermally, or topically in ointment form) or isosorbide dinitrate (given sublingually or orally) to reduce myocardial oxygen consumption
▶ beta-adrenergic blockers to reduce the workload and oxygen demands of the heart by reducing heart rate and peripheral resistance to blood flow
▶ calcium channel blockers to prevent coronary artery spasm
▶ antiplatelet drugs to minimize platelet aggregation and the risk of coronary occlusion
▶ antilipemic drugs to reduce serum cholesterol or triglyceride levels
▶ antihypertensive drugs to control hypertension

▶ estrogen replacement therapy to reduce the risk for CAD in postmenopausal women

▶ coronary artery bypass graft (CABG) surgery to restore blood flow by bypassing an occluded artery using another vessel

▶ "key hole" or minimally invasive surgery, an alternative to traditional CABG using fiber-optic cameras inserted through small cuts in the chest, to correct blockages in one or two accessible arteries

▶ angioplasty, to relieve occlusion in patients without calcification and partial occlusion

▶ laser angioplasty to correct occlusion by vaporizing fatty deposits

▶ rotational atherectomy to remove arterial plaque with a high-speed burr

▶ stent placement in a reopened artery to hold the artery open

▶ lifestyle modifications to reduce further progression of CAD; these include smoking cessation, regular exercise, maintaining an ideal body weight, and following a low-fat, low-sodium diet.

Heart failure

A syndrome rather than a disease, heart failure occurs when the heart can't pump enough blood to meet the metabolic needs of the body. Heart failure results in intravascular and interstitial volume overload and poor tissue perfusion. An individual with heart failure experiences reduced exercise tolerance, a reduced quality of life, and a shortened life span.

Although the most common cause of heart failure is coronary artery disease, it also occurs in infants, children, and adults with congenital and acquired heart defects. The incidence of heart failure rises with age. Approximately 1% of people older than age 50 experience heart failure; it occurs in 10% of people older than age 80. About 700,000 Americans die of heart failure each year. Mortality from heart failure is greater for males, blacks, and the elderly.

Although advances in diagnostic and therapeutic techniques have greatly improved the outlook for patients with heart failure, the prognosis still depends on the underlying cause and its response to treatment.

Causes
Causes of heart failure may be divided into four general categories. (See *Causes of heart failure,* page 162.)

Pathophysiology
Heart failure may be classified according to the side of the heart affected (left- or right-sided heart failure) or by the cardiac cycle involved (systolic or diastolic dysfunction).

Left-sided heart failure. This type of heart failure occurs as a result of ineffective left ventricular contractile function. As the pumping ability of the left ventricle fails, cardiac output falls. Blood is no longer effectively pumped out into the body; it backs up into the left atrium and then into the lungs, causing pulmonary congestion, dyspnea, and activity intolerance. If the condition persists, pulmonary edema and right-sided heart failure may result. Common causes include left ventricular infarction, hypertension, and aortic and mitral valve stenosis.

Right-sided heart failure. Right-sided heart failure results from ineffective right ventricular contractile function. Consequently, blood is not pumped effectively through the right ventricle to the lungs, causing blood to back up into the right atrium and into the peripheral circulation. The patient gains weight and develops peripheral edema and engorgement of the kidney and other organs. It may be due to an acute right ventricular infarction or a pulmonary embolus. However, the most common cause is profound backward flow due to left-sided heart failure.

CAUSES OF HEART FAILURE

CAUSE	EXAMPLES
Abnormal cardiac muscle function	▶ Myocardial infarction ▶ Cardiomyopathy
Abnormal left ventricular volume	▶ Valvular insufficiency ▶ High-output states: chronic anemia arteriovenous fistula thyrotoxicosis pregnancy septicemia beriberi infusion of large volume of intravenous fluids in a short time period
Abnormal left ventricular pressure	▶ Hypertension ▶ Pulmonary hypertension ▶ Chronic obstructive pulmonary disease ▶ Aortic or pulmonic valve stenosis
Abnormal left ventricular filling	▶ Mitral valve stenosis ▶ Tricuspid valve stenosis ▶ Atrial myxoma ▶ Constrictive pericarditis ▶ Atrial fibrillation ▶ Impaired ventricular relaxation: hypertension myocardial hibernation myocardial stunning

Systolic dysfunction. Systolic dysfunction occurs when the left ventricle can't pump enough blood out to the systemic circulation during systole and the ejection fraction falls. Consequently, blood backs up into the pulmonary circulation and pressure rises in the pulmonary venous system. Cardiac output falls; weakness, fatigue, and shortness of breath may occur. Causes of systolic dysfunction include myocardial infarction and dilated cardiomyopathy.

Diastolic dysfunction. Diastolic dysfunction occurs when the ability of the left ventricle to relax and fill during diastole is reduced and the stroke volume falls. Therefore, higher volumes are needed in the ventricles to maintain cardiac output. Consequently, pulmonary congestion and peripheral edema develop. Diastolic dysfunction may occur as a result of left ventricular hypertrophy, hypertension, or restrictive cardiomyopathy. This type of heart failure is less common than systolic dysfunction, and its treatment is not as clear.

All causes of heart failure eventually lead to reduced cardiac output, which triggers compensatory mechanisms such as increased sympathetic activity, activation of the renin-angiotensin-aldosterone system, ventricular dilation, and hypertrophy. These mechanisms improve cardiac output at the expense of increased ventricular work.

Increased sympathetic activity — a response to decreased cardiac output and blood pressure— enhances peripheral vascular resistance, contractility, heart rate, and venous return. Signs of increased sympathetic activity, such as cool extremities and clamminess, may indicate impending heart failure.

Increased sympathetic activity also restricts blood flow to the kidneys, causing them to secrete renin which, in turn, converts angiotensinogen to angiotensin I, which then becomes angiotensin II — a potent vasoconstrictor. Angiotensin causes the adrenal cortex to release aldosterone, leading to sodium and water retention and an increase in circulating blood volume. This renal mechanism is helpful; however, if it persists unchecked, it can aggravate heart failure as the heart struggles to pump against the increased volume.

In ventricular dilation, an increase in end-diastolic ventricular volume (preload) causes increased stroke work and stroke volume during contraction, stretching cardiac muscle fibers so that the ventricle can accept the increased intravascular volume. Eventually, the muscle becomes stretched beyond optimum limits and contractility declines.

In ventricular hypertrophy, an increase in ventricular muscle mass allows the heart to pump against increased resistance to the outflow of blood, improving cardiac output. However, this increased muscle mass also increases the myocardial oxygen requirements. An increase in the ventricular diastolic pressure necessary to fill the enlarged ventricle may compromise diastolic coronary blood flow, limiting the oxygen supply to the ventricle, and causing ischemia and impaired muscle contractility.

In heart failure, counterregulatory substances — prostaglandins and atrial natriuretic factor — are produced in an attempt to reduce the negative effects of volume overload and vasoconstriction caused by the compensatory mechanisms.

The kidneys release the prostaglandins, prostacyclin and prostaglandin E_2, which are potent vasodilators. These vasodilators also act to reduce volume overload produced by the renin-angiotensin-aldosterone system by inhibiting sodium and water reabsorption by the kidneys.

Atrial natriuretic factor is a hormone that is secreted mainly by the atria in response to stimulation of the stretch receptors in the atria caused by excess fluid volume. Atrial natriuretic factor works to counteract the negative effects of sympathetic nervous system stimulation and the renin-angiotensin-aldosterone system by producing vasodilation and diuresis.

Signs and symptoms

Early clinical manifestations of *left-sided heart failure* include:
▶ dyspnea caused by pulmonary congestion
▶ orthopnea as blood is redistributed from the legs to the central circulation when the patient lies down at night
▶ paroxysmal nocturnal dyspnea due to the reabsorption of interstitial fluid when lying down and reduced sympathetic stimulation while sleeping
▶ fatigue associated with reduced oxygenation and a lack of activity
▶ nonproductive cough associated with pulmonary congestion.

Later clinical manifestations of left-sided heart failure may include:
▶ crackles due to pulmonary congestion
▶ hemoptysis resulting from bleeding veins in the bronchial system caused by venous distention
▶ point of maximal impulse displaced toward the left anterior axillary line caused by left ventricular hypertrophy

▶ tachycardia due to sympathetic stimulation

▶ S₃ heart sound caused by rapid ventricular filling

▶ S₄ heart sound resulting from atrial contraction against a noncompliant ventricle

▶ cool, pale skin resulting from peripheral vasoconstriction

▶ restlessness and confusion due to reduced cardiac output.

Clinical manifestations of *right-sided heart failure* include:

▶ elevated jugular venous distention due to venous congestion

▶ positive hepatojugular reflux and hepatomegaly secondary to venous congestion

▶ right upper quadrant pain caused by liver engorgement

▶ anorexia, fullness, and nausea may be caused by congestion of the liver and intestines

▶ nocturia as fluid is redistributed at night and reabsorbed

▶ weight gain due to the retention of sodium and water

▶ edema associated with fluid volume excess

▶ ascites or anasarca caused by fluid retention.

 CULTURAL DIVERSITY In the Chinese culture, disagreement or discomfort isn't typically displayed openly. Direct questioning and vigilant assessment skills are necessary to ensure that a patient's quiet nature doesn't mask signs and symptoms that may be life-threatening.

Complications

Acute complications of heart failure include:

▶ pulmonary edema
▶ acute renal failure
▶ arrhythmias.

Chronic complications include:

▶ activity intolerance
▶ renal impairment
▶ cardiac cachexia
▶ metabolic impairment

▶ thromboembolism.

Diagnosis

The following tests help diagnose heart failure:

▶ Chest X-rays show increased pulmonary vascular markings, interstitial edema, or pleural effusion and cardiomegaly.

▶ Electrocardiography may indicate hypertrophy, ischemic changes, or infarction, and may also reveal tachycardia and extrasystoles.

▶ Laboratory testing may reveal abnormal liver function tests and elevated blood urea nitrogen and creatinine levels.

▶ Echocardiography may reveal left ventricular hypertrophy, dilation, and abnormal contractility.

▶ Pulmonary artery monitoring typically demonstrates elevated pulmonary artery and pulmonary artery wedge pressures, left ventricular end-diastolic pressure in left-sided heart failure, and elevated right atrial pressure or central venous pressure in right-sided heart failure.

▶ Radionuclide ventriculography may reveal an ejection fraction less than 40%; in diastolic dysfunction, the ejection fraction may be normal.

Treatment

Correction of heart failure may involve:

▶ treatment of the underlying cause, if known

▶ angiotensin-converting enzyme (ACE) inhibitors to patients with left ventricle dysfunction to reduce production of angiotensin II, resulting in preload and afterload reduction

AGE ALERT Older adults may require lower doses of ACE inhibitors because of impaired renal clearance. Monitor for severe hypotension, signifying a toxic effect.

▶ digoxin for patients with heart failure due to left ventricular systolic dysfunction to increase myocardial contractility, improve cardiac output, reduce the volume of the ventricle, and decrease ventricular stretch

▶ diuretics to reduce fluid volume overload and venous return

▶ beta-adrenergic blockers in patients with New York Heart Association class II or class III heart failure caused by left ventricular systolic dysfunction to prevent remodeling (See *Classifying heart failure,* page 156.)

▶ diuretics, nitrates, morphine, and oxygen to treat pulmonary edema

▶ lifestyle modifications (to reduce symptoms of heart failure) such as weight loss (if obese); limited sodium (to 3 g/day) and alcohol intake; reduced fat intake; smoking cessation; reduced stress; and development of an exercise program. Heart failure is no longer a contraindication to exercise and cardiac rehabilitation.

 CULTURAL DIVERSITY Asian Americans consume large amounts of sodium. Encourage an Asian patient to substitute fresh vegetables, herbs, and spices for canned foods, monosodium glutamate, and soy sauce.

▶ coronary artery bypass surgery or angioplasty for heart failure due to coronary artery disease

▶ cardiac transplantation in patients receiving aggressive medical treatment but still experiencing limitations or repeated hospitalizations

▶ other surgery or invasive procedures may be recommended in patients with severe limitations or repeated hospitalizations, despite maximal medical therapy. Some are controversial and may include cardiomyoplasty, insertion of an intra-aortic balloon pump, partial left ventriculectomy, use of a mechanical ventricular assist device, and implanting an internal cardioverter-defibrillator.

 AGE ALERT Heart failure in children occurs mainly as a result of congenital heart defects. Therefore, treatment guidelines are directed toward the specific cause.

Hypertension

Hypertension, an elevation in diastolic or systolic blood pressure, occurs as two major types: essential (primary) hypertension, the most common, and secondary hypertension, which results from renal disease or another identifiable cause. Malignant hypertension is a severe, fulminant form of hypertension common to both types. Hypertension is a major cause of cerebrovascular accident, cardiac disease, and renal failure.

Hypertension affects 15% to 20% of adults in the United States. The risk of hypertension increases with age and is higher for blacks than whites and in those with less education and lower income. Men have a higher incidence of hypertension in young and early middle adulthood; thereafter, women have a higher incidence.

Essential hypertension usually begins insidiously as a benign disease, slowly progressing to a malignant state. If untreated, even mild cases can cause major complications and death. Carefully managed treatment, which may include lifestyle modifications and drug therapy, improves prognosis. Untreated, it carries a high mortality rate. Severely elevated blood pressure (hypertensive crisis) may be fatal.

Causes

Risk factors for *primary hypertension* include:

▶ family history

▶ advancing age

AGE ALERT Older adults may have isolated systolic hypertension (ISH), in which just the systolic blood pressure is elevated, as atherosclerosis causes a loss of elasticity in large arteries. Previously, it was believed that ISH was a normal part of the aging process and should not be treated. Results of the Systolic Hypertension in the Elderly Program (SHEP), however, found that treating ISH with antihypertensive drugs lowered the incidence of stroke, coronary artery disease (CAD), and left ventricular heart failure.

▶ race (most common in blacks)

CULTURAL DIVERSITY Blacks are at increased risk for primary hypertension when predisposition to low plasma renin levels diminishes ability to excrete excess sodium. Hypertension develops at an earlier age and, at any age, it is more severe than in whites.

▶ obesity
▶ tobacco use
▶ high intake of sodium
▶ high intake of saturated fat
▶ excessive alcohol consumption
▶ sedentary lifestyle
▶ stress
▶ excess renin
▶ mineral deficiencies (calcium, potassium, and magnesium)
▶ diabetes mellitus.

Causes of *secondary hypertension* include:

▶ coarctation of the aorta
▶ renal artery stenosis and parenchymal disease
▶ brain tumor, quadriplegia, and head injury
▶ pheochromocytoma, Cushing's syndrome, hyperaldosteronism, and thyroid, pituitary, or parathyroid dysfunction
▶ oral contraceptives, cocaine, epoetin alfa, sympathetic stimulants, monoamine oxidase inhibitors taken with tyramine, estrogen replacement therapy, and nonsteroidal anti-inflammatory drugs
▶ pregnancy-induced hypertension
▶ excessive alcohol consumption.

Pathophysiology

Arterial blood pressure is a product of total peripheral resistance and cardiac output. Cardiac output is increased by conditions that increase heart rate or stroke volume, or both. Peripheral resistance is increased by factors that increase blood viscosity or reduce the lumen size of vessels, especially the arterioles.

Several theories help to explain the development of hypertension, including:

▶ changes in the arteriolar bed causing increased peripheral vascular resistance

▶ abnormally increased tone in the sympathetic nervous system that originates in the vasomotor system centers, causing increased peripheral vascular resistance
▶ increased blood volume resulting from renal or hormonal dysfunction
▶ an increase in arteriolar thickening caused by genetic factors, leading to increased peripheral vascular resistance
▶ abnormal renin release, resulting in the formation of angiotensin II, which constricts the arteriole and increases blood volume. (See *Understanding blood pressure regulation.*)

Prolonged hypertension increases the workload of the heart as resistance to left ventricular ejection increases. To increase contractile force, the left ventricle hypertrophies, raising the oxygen demands and workload of the heart. Cardiac dilation and failure may occur when hypertrophy can no longer maintain sufficient cardiac output. Because hypertension promotes coronary atherosclerosis, the heart may be further compromised by reduced blood flow to the myocardium, resulting in angina or myocardial infarction (MI). Hypertension also causes vascular damage, leading to accelerated atherosclerosis and target organ damage, such as retinal injury, renal failure, stroke, and aortic aneurysm and dissection.

The pathophysiology of secondary hypertension is related to the underlying disease. For example:

▶ The most common cause of secondary hypertension is chronic renal disease. Insult to the kidney from chronic glomerulonephritis or renal artery stenosis interferes with sodium excretion, the renin-angiotensin-aldosterone system, or renal perfusion, causing blood pressure to rise.
▶ In Cushing's syndrome, increased cortisol levels raise blood pressure by increasing renal sodium retention, angiotensin II levels, and vascular response to norepinephrine.
▶ In primary aldosteronism, increased intravascular volume, altered sodium concentrations in vessel walls, or very high

UNDERSTANDING BLOOD PRESSURE REGULATION

Hypertension may result from a disturbance in one of the following intrinsic mechanisms.

RENIN-ANGIOTENSIN SYSTEM

The renin-angiotensin system acts to increase blood pressure through the following mechanisms:
▶ sodium depletion, reduced blood pressure, and dehydration stimulate renin release
▶ renin reacts with angiotensin, a liver enzyme, and converts it to angiotensin I, which increases preload and afterload
▶ angiotensin I converts to angiotensin II in the lungs; angiotensin II is a potent vasoconstrictor that targets the arterioles
▶ circulating angiotensin II works to increase preload and afterload by stimulating the adrenal cortex to secrete aldosterone; this increases blood volume by conserving sodium and water.

AUTOREGULATION

Several intrinsic mechanisms work to change an artery's diameter to maintain tissue and organ perfusion despite fluctuations in systemic blood pressure. These mechanisms include stress relaxation and capillary fluid shifts:
▶ in stress relaxation, blood vessels gradually dilate when blood pressure rises to reduce peripheral resistance
▶ in capillary fluid shift, plasma moves between vessels and extravascular spaces to maintain intravascular volume.

SYMPATHETIC NERVOUS SYSTEM

When blood pressure drops, baroreceptors in the aortic arch and carotid sinuses decrease their inhibition of the medulla's vasomotor center. The consequent increases in sympathetic stimulation of the heart by norepinephrine increases cardiac output by strengthening the contractile force, raising the heart rate, and augmenting peripheral resistance by vasoconstriction. Stress can also stimulate the sympathetic nervous system to increase cardiac output and peripheral vascular resistance.

ANTIDIURETIC HORMONE

The release of antidiuretic hormone can regulate hypotension by increasing reabsorption of water by the kidney. With reabsorption, blood plasma volume increases, thus raising blood pressure.

aldosterone levels cause vasoconstriction and increased resistance.
▶ Pheochromocytoma is a chromaffin cell tumor of the adrenal medulla that secretes epinephrine and norepinephrine. Epinephrine increases cardiac contractility and rate, whereas norepinephrine increases peripheral vascular resistance.

Signs and symptoms

Although hypertension is frequently asymptomatic, the following signs and symptoms may occur:
▶ elevated blood pressure readings on at least two consecutive occasions after initial screening

WHAT HAPPENS IN HYPERTENSIVE CRISIS

Hypertensive crisis is a severe rise in arterial blood pressure caused by a disturbance in one or more of the regulating mechanisms. If untreated, hypertensive crisis may result in renal, cardiac, or cerebral complications and, possibly, death.

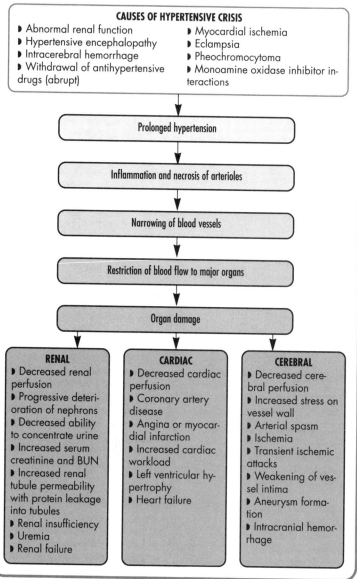

CAUSES OF HYPERTENSIVE CRISIS

- Abnormal renal function
- Hypertensive encephalopathy
- Intracerebral hemorrhage
- Withdrawal of antihypertensive drugs (abrupt)
- Myocardial ischemia
- Eclampsia
- Pheochromocytoma
- Monoamine oxidase inhibitor interactions

Prolonged hypertension

↓

Inflammation and necrosis of arterioles

↓

Narrowing of blood vessels

↓

Restriction of blood flow to major organs

↓

Organ damage

RENAL
- Decreased renal perfusion
- Progressive deterioration of nephrons
- Decreased ability to concentrate urine
- Increased serum creatinine and BUN
- Increased renal tubule permeability with protein leakage into tubules
- Renal insufficiency
- Uremia
- Renal failure

CARDIAC
- Decreased cardiac perfusion
- Coronary artery disease
- Angina or myocardial infarction
- Increased cardiac workload
- Left ventricular hypertrophy
- Heart failure

CEREBRAL
- Decreased cerebral perfusion
- Increased stress on vessel wall
- Arterial spasm
- Ischemia
- Transient ischemic attacks
- Weakening of vessel intima
- Aneurysm formation
- Intracranial hemorrhage

AGE ALERT Because many older adults have a wide auscultatory gap — the hiatus between the first Korotkoff sound and the next sound — failure to pump the blood pressure cuff up high enough can lead to missing the first beat and underestimating the systolic blood pressure. To avoid missing the first Korotkoff sound, palpate the radial artery and inflate the cuff to a point approximately 20 mm beyond which the pulse beat disappears.

▶ occipital headache (may worsen on rising in the morning as a result of increased intracranial pressure); nausea and vomiting may also occur
▶ epistaxis possibly due to vascular involvement
▶ bruits (which may be heard over the abdominal aorta or carotid, renal, and femoral arteries) caused by stenosis or aneurysm
▶ dizziness, confusion, and fatigue caused by decreased tissue perfusion due to vasoconstriction of blood vessels
▶ blurry vision as a result of damage to the retina
▶ nocturia caused by an increase in blood flow to the kidneys and an increase in glomerular filtration
▶ edema caused by increased capillary pressure.

If secondary hypertension exists, other signs and symptoms may be related to the cause. For example, Cushing's syndrome may cause truncal obesity and purple striae, whereas patients with pheochromocytoma may develop headache, nausea, vomiting, palpitations, pallor, and profuse perspiration.

Complications

Complications of hypertension include:
▶ hypertensive crisis, peripheral arterial disease, dissecting aortic aneurysm, CAD, angina, MI, heart failure, arrhythmias, and sudden death (See *What happens in hypertensive crisis.*)
▶ transient ischemic attacks, cerebrovascular accident, retinopathy, and hypertensive encephalopathy

▶ renal failure.

Diagnosis

The following tests help diagnose hypertension:
▶ Serial blood pressure measurements may be useful.
▶ Urinalysis may show protein, casts, red blood cells, or white blood cells, suggesting renal disease; presence of catecholamines associated with pheochromocytoma; or glucose, suggesting diabetes.
▶ Laboratory testing may reveal elevated blood urea nitrogen and serum creatinine levels suggestive of renal disease, or hypokalemia indicating adrenal dysfunction (primary hyperaldosteronism).
▶ Complete blood count may reveal other causes of hypertension, such as polycythemia or anemia.
▶ Excretory urography may reveal renal atrophy, indicating chronic renal disease. One kidney smaller than the other suggests unilateral renal disease.
▶ Electrocardiography may show left ventricular hypertrophy or ischemia.
▶ Chest X-rays may show cardiomegaly.
▶ Echocardiography may reveal left ventricular hypertrophy.

Treatment

Hypertension may be treated by following the 1997 revised guidelines of the Sixth Report of the Joint National Committee on Prevention, Detection, Evaluation, and Treatment of High Blood Pressure to determine the approach to treatment according to the patient's blood pressure, risk factors, and target organ damage. (See *Risk stratification and treatment,* page 170.)
▶ diuretics (thiazide diuretics such as hydrochlorothiazide, loop diuretics such as furosemide, and combination diuretics such as hydrochlorothiazide-spironolactone) to reduce excess fluid volume

RISK STRATIFICATION AND TREATMENT

BLOOD PRESSURE STAGES	RISK GROUP A (No major risk factors No TOD/CCD*)	RISK GROUP B (At least 1 risk factor, not including diabetes; no TOD/CCD)	RISK GROUP C (TOD/CCD and/or diabetes, with or without other risk factors)
High normal (130–139/85–89)	Lifestyle modification	Lifestyle modification	Drug therapy for those with heart failure, renal insufficiency, or diabetes Lifestyle modification
Stage 1 (140–159/90–99)	Lifestyle modification (up to 12 months)	Lifestyle modification † (up to 6 months)	Drug therapy Lifestyle modification
Stages 2 and 3 (>160/>100)	Drug therapy	Drug therapy	Drug therapy Lifestyle modification

Notes:
* TOD/CCD indicates target organ disease/clinical cardiovascular disease.
† For patients with multiple risk factors, clinicians should consider drugs as initial therapy plus lifestyle modifications.

 CULTURAL DIVERSITY According to the treatment guidelines issued by the National Institutes of Health in 1997, drug therapy for blacks should consist of calcium channel blockers and diuretics.

▶ beta blockers (such as metoprolol) to reduce heart rate and contractility, and to dilate the blood vessels

 CULTURAL DIVERSITY Asians are twice as sensitive as whites to propranolol and are able to metabolize and clear this drug more rapidly. Hypertensive whites are more responsive to beta blockers than are hypertensive blacks.

▶ calcium channel blockers such as diltiazem to reduce heart rate and contractility; these agents are also effective against vasospasm

▶ angiotensin-converting enzyme (ACE) inhibitors such as captopril or angiotensin

II receptor blockers such as valsartan to produce vasodilation

▶ alpha-receptor blockers such as doxazosin to produce vasodilation

▶ alpha-receptor agonists such as clonidine to lower peripheral vascular resistance

 AGE ALERT Older adults are at an increased risk for adverse effects of antihypertensives, especially orthostatic hypotension. Lower doses may be needed.

▶ treatment of underlying cause of secondary hypertension and controlling hypertensive effects

▶ treatment of hypertensive emergencies with a parenteral vasodilator such as nitroprusside or an adrenergic inhibitor, or oral administration of a selected drug, such as nifedipine, captopril, clonidine, or labetalol, to rapidly reduce blood pressure

▶ lifestyle modifications, including weight control; limited alcohol, saturated fat, and sodium (2.4 g/day) intake; regular exercise; and smoking cessation

▶ inclusion of adequate amounts of calcium, magnesium, and potassium in the diet.

Myocardial infarction

In myocardial infarction (MI) — also known as a heart attack — reduced blood flow through one of the coronary arteries results in myocardial ischemia and necrosis. In cardiovascular disease, the leading cause of death in the United States and Western Europe, death usually results from cardiac damage or complications of MI. Each year, approximately 900,000 people in the United States experience MI. Mortality is high when treatment is delayed, and almost half of sudden deaths due to MI occur before hospitalization, within 1 hour of the onset of symptoms. The prognosis improves if vigorous treatment begins immediately.

Causes

Predisposing risk factors include:

▶ positive family history

▶ gender (men and postmenopausal women are more susceptible to MI than premenopausal women, although the incidence is rising among women, especially those who smoke and take oral contraceptives)

▶ hypertension

▶ smoking

▶ elevated serum triglyceride, total cholesterol, and low-density lipoprotein levels

▶ obesity

▶ excessive intake of saturated fats

▶ sedentary lifestyle

▶ aging

▶ stress or type A personality

▶ drug use, especially cocaine and amphetamines.

Pathophysiology

MI results from occlusion of one or more of the coronary arteries. Occlusion can stem from atherosclerosis, thrombosis, platelet aggregation, or coronary artery stenosis or spasm. If coronary occlusion causes prolonged ischemia, lasting longer than 30 to 45 minutes, irreversible myocardial cell damage and muscle death occur. All MIs have a central area of necrosis or infarction surrounded by an area of potentially viable hypoxic injury. This zone may be salvaged if circulation is restored, or it may progress to necrosis. The zone of injury, in turn, is surrounded by an area of viable ischemic tissue. (See *Zones of myocardial infarction,* page 172.) Although ischemia begins immediately, the size of the infarct can be limited if circulation is restored within 6 hours.

Several changes occur after MI. Cardiac enzymes and proteins are released by the infarcted myocardial cells, which are used in the diagnosis of an MI. (See *Release of cardiac enzymes and proteins,* page 136.) Within 24 hours, the infarcted muscle becomes edematous and cyanotic. During the next several days, leukocytes infiltrate the necrotic area and begin to remove necrotic cells, thinning the ventricular wall. Scar formation begins by the

ZONES OF MYOCARDIAL INFARCTION

Myocardial infarction has a central area of necrosis surrounded by a zone of injury that may recover if revascularization occurs. This zone of injury is surrounded by an outer ring of reversible ischemia. Characteristic electrocardiographic changes are associated with each zone.

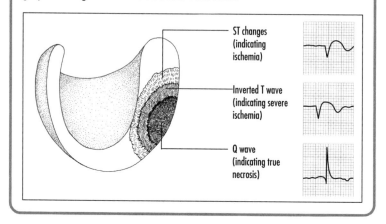

ST changes (indicating ischemia)

Inverted T wave (indicating severe ischemia)

Q wave (indicating true necrosis)

third week after MI, and by the sixth week, scar tissue is well established.

The scar tissue that forms on the necrotic area inhibits contractility. When this occurs, the compensatory mechanisms (vascular constriction, increased heart rate, and renal retention of sodium and water) try to maintain cardiac output. Ventricular dilation may also occur in a process called remodeling. Functionally, an MI may cause reduced contractility with abnormal wall motion, altered left ventricular compliance, reduced stroke volume, reduced ejection fraction, and elevated left ventricular end-diastolic pressure.

Signs and symptoms

The following signs and symptoms may occur:

▶ persistent, crushing substernal chest pain that may radiate to the left arm, jaw, neck, or shoulder blades caused by reduced oxygen supply to the myocardial cells; it may be described as heavy, squeezing, or crushing

AGE ALERT Many older adults do not have chest pain with MI, but experience atypical symptoms such as fatigue, dyspnea, falls, tingling of the extremities, nausea, vomiting, weakness, syncope, and confusion.

▶ cool extremities, perspiration, anxiety, and restlessness due to the release of catecholamines

▶ blood pressure and pulse initially elevated as a result of sympathetic nervous system activation. If cardiac output is reduced, blood pressure may fall. Bradycardia may be associated with conduction disturbances

▶ fatigue and weakness caused by reduced perfusion to skeletal muscles

▶ nausea and vomiting as a result of reflex stimulation of vomiting centers by pain fibers or from vasovagal reflexes

▶ shortness of breath and crackles reflecting heart failure

▶ low-grade temperature in the days following acute MI due to the inflammatory response

PINPOINTING MYOCARDIAL INFARCTION

Depending on location, ischemia or infarction causes changes in the following electrocardiographic leads.

TYPE OF MYOCARDIAL INFARCTION	LEADS
Inferior	II, III, aV_F
Anterior	V_3, V_4
Septal	V_1, V_2
Lateral	I, aV_L, V_5, V_6
Anterolateral	I, aV_L, V_3-V_6
Posterior	V_1 or V_2
Right ventricular	II, III, aV_F, V_{1R} – V_{4R}

▶ jugular venous distention reflecting right ventricular dysfunction and pulmonary congestion
▶ S_3 and S_4 heart sounds reflecting ventricular dysfunction
▶ loud holosystolic murmur in apex possibly caused by papillary muscle rupture
▶ reduced urine output secondary to reduced renal perfusion and increased aldosterone and antidiuretic hormone.

Complications
Complications of MI include:
▶ arrhythmias
▶ cardiogenic shock
▶ heart failure causing pulmonary edema
▶ pericarditis
▶ rupture of the atrial or ventricular septum, ventricular wall, or valves
▶ mural thrombi causing cerebral or pulmonary emboli
▶ ventricular aneurysms
▶ myocardial rupture
▶ extensions of the original infarction.

Diagnosis
The following tests help diagnose MI:

▶ Serial 12-lead electrocardiography (ECG) may reveal characteristic changes, such as serial ST-segment depression in non–Q-wave MI (subendocardial MI that affects the innermost myocardial layer) and ST-segment elevation in Q-wave MI (transmural MI with damage extending through all myocardial layers). An ECG can also identify the location of MI, arrhythmias, hypertrophy, and pericarditis. (See *Pinpointing myocardial infarction*.)
▶ Serial cardiac enzymes and proteins may show a characteristic rise and fall of cardiac enzymes, specifically CK-MB, and the proteins troponin T and I, and myoglobin to confirm the diagnosis of MI. (See *Release of cardiac enzymes and proteins*, page 136.)
▶ Laboratory testing may reveal elevated white blood cell count and erythrocyte sedimentation rate due to inflammation, and increased glucose levels following the release of catecholamines.
▶ Echocardiography may show ventricular wall motion abnormalities and may detect septal or papillary muscle rupture.
▶ Chest X-rays may show left-sided heart failure or cardiomegaly.

▶ Nuclear imaging scanning using thallium-201 and technetium 99m can be used to identify areas of infarction and areas of viable muscle cells.

▶ Cardiac catheterization may be used to identify the involved coronary artery as well as to provide information on ventricular function and pressures and volumes within the heart.

Treatment

Treatment of an MI typically involves following the treatment guidelines recommended by the American College of Cardiology/American Heart Association (ACC/AHA) Task Force on Practice Guidelines. These include:

▶ assessment of patients with chest pain in the Emergency Department within 10 minutes of an MI because at least 50% of deaths take place within 1 hour of the onset of symptoms. Moreover, thrombolytic therapy is most effective when started within the first 6 hours after the onset of symptoms

▶ oxygen by nasal cannula for 2 to 3 hours to increase oxygenation of the blood (See *Blocking myocardial infarction,* pages 176 and 177.)

▶ nitroglycerin sublingually to relieve chest pain, unless systolic blood pressure is less than 90 mm Hg or heart rate is less than 50 or greater than 100 beats per minute

▶ morphine or meperidine (Demerol) for analgesia because pain stimulates the sympathetic nervous system, leading to an increase in heart rate and vasoconstriction

▶ aspirin 160 to 325 mg/day indefinitely, to inhibit platelet aggregation

▶ continuous cardiac monitoring to detect arrhythmias and ischemia

▶ intravenous thrombolytic therapy to patients with chest pain of at least 30 minutes' duration who reach the hospital within 12 hours of the onset of symptoms (unless contraindications exist) and whose ECG shows new left bundle branch block or ST-segment elevation of at least 1 to 2 mm in two or more ECG leads. The greatest benefit of reperfusion therapy, however, occurs when reperfusion takes place within 6 hours of the onset of chest pain

▶ intravenous heparin for patients who have received tissue plasminogen activator (tPA) to increase the chances of patency in the affected coronary artery. Limited evidence exists that intravenous or subcutaneous heparin is beneficial in patients with acute MI treated with nonspecific thrombolytic drugs, such as streptokinase or anistreplase

▶ percutaneous transluminal coronary angioplasty (PTCA) may be an alternative to thrombolytic therapy if it can be performed in a timely manner in an institution with personnel skilled in the procedure

▶ limitation of physical activity for the first 12 hours to reduce cardiac workload, thereby limiting the area of necrosis

▶ keeping atropine, lidocaine, transcutaneous pacing patches or a transvenous pacemaker, a defibrillator, and epinephrine readily available to treat arrhythmias. The ACC/AHA doesn't recommend the prophylactic use of antiarrhythmic drugs during the first 24 hours

▶ intravenous nitroglycerin for 24 to 48 hours in patients without hypotension, bradycardia, or excessive tachycardia to reduce afterload and preload and relieve chest pain

▶ early intravenous beta blockers to patients with an evolving acute MI followed by oral therapy, as long as there are no contraindications, to reduce heart rate and myocardial contractile force, thereby reducing myocardial oxygen requirements

▶ angiotensin-converting enzyme inhibitors in patients with an evolving MI with ST-segment elevation or left bundle branch block, but without hypotension or other contraindications, to reduce afterload and preload and prevent remodeling

▶ if needed, magnesium sulfate for 24 hours to correct hypomagnesemia

▶ angiography and possible percutaneous or surgical revascularization for patients with spontaneous or provoked myocardial ischemia following an acute MI

exercise testing before discharge to determine adequacy of medical therapy and to obtain baseline information for an appropriate exercise prescription; it can also determine functional capacity and stratify the patient's risk of a subsequent cardiac event

cardiac risk modification program of weight control; a low-fat, low-cholesterol diet; smoking cessation; and regular exercise to reduce cardiac risk.

Myocarditis

Myocarditis is focal or diffuse inflammation of the cardiac muscle (myocardium). It may be acute or chronic and can occur at any age. In many cases, myocarditis fails to produce specific cardiovascular symptoms or electrocardiogram (ECG) abnormalities, and recovery is usually spontaneous without residual defects. Occasionally, myocarditis is complicated by heart failure; in rare cases, it leads to cardiomyopathy.

Causes

Common causes of myocarditis include:

viral infections (most common cause in the United States and western Europe), such as Coxsackie virus A and B strains and, possibly, poliomyelitis, influenza, Epstein-Barr virus, human immunodeficiency virus, cytomegalovirus, measles, mumps, rubeola, rubella, and adenoviruses and echoviruses

bacterial infections, such as diphtheria, tuberculosis, typhoid fever, tetanus, and staphylococcal, pneumococcal, and gonococcal infections

hypersensitive immune reactions, including acute rheumatic fever and postcardiotomy syndrome

radiation therapy — large doses of radiation to the chest in treating lung or breast cancer

toxins such as lead, chemicals, cocaine, and chronic alcoholism

parasitic infections, especially South American trypanosomiasis (Chagas' disease) in infants and immunosuppressed adults; also, toxoplasmosis

fungal infections, including candidiasis and aspergillosis

helminthic infections such as trichinosis.

Pathophysiology

Damage to the myocardium occurs when an infectious organism triggers an autoimmune, cellular, and humoral reaction. The resulting inflammation may lead to hypertrophy, fibrosis, and inflammatory changes of the myocardium and conduction system. The heart muscle weakens and contractility is reduced. The heart muscle becomes flabby and dilated and pinpoint hemorrhages may develop.

Signs and symptoms

The following signs and symptoms may occur:

nonspecific symptoms such as fatigue, dyspnea, palpitations, and fever caused by systemic infection

mild, continuous pressure or soreness in the chest (occasionally) related to inflammation

tachycardia due to a compensatory sympathetic response

S_3 and S_4 gallops as a result of heart failure

murmur of mitral insufficiency may be heard, if papillary muscles involved

pericardial friction rub, if pericarditis exists

if myofibril degeneration occurs, it may lead to right-sided and left-sided heart failure, with cardiomegaly, neck vein distention, dyspnea, edema, pulmonary congestion, persistent fever with resting or exertional tachycardia disproportionate to the degree of fever, and supraventricular and ventricular arrhythmias.

Complications

Complications of myocarditis include:

recurrence of myocarditis

chronic valvulitis (when it results from rheumatic fever)

dilated cardiomyopathy

BLOCKING MYOCARDIAL INFARCTION

This chart shows how treatments can be applied to myocardial infarction at various stages of its development.

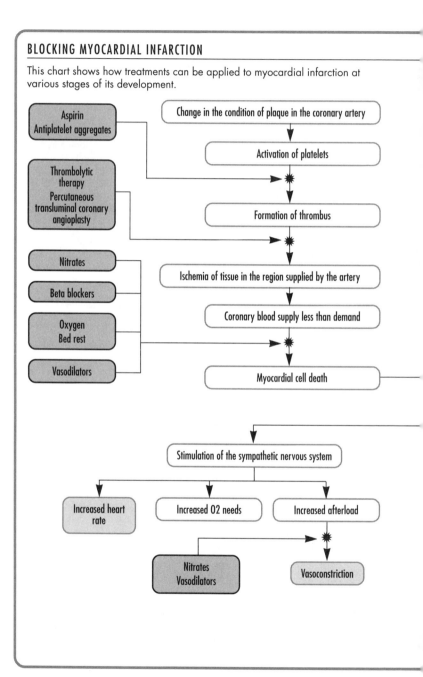

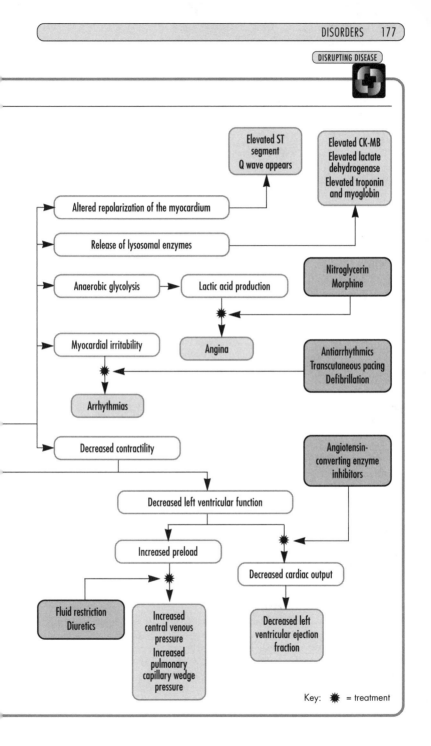

Key: ✳ = treatment

▶ arrhythmias and sudden death
▶ heart failure
▶ pericarditis
▶ ruptured myocardial aneurysm
▶ thromboembolism.

Diagnosis
▶ History reveals recent febrile upper respiratory infection.
▶ Laboratory testing may reveal elevated levels of creatine kinase (CK), CK-MB, aspartate aminotransferase, and lactate dehydrogenase. Also, inflammation and infection can cause elevated white blood cell count and erythrocyte sedimentation rate.
▶ Antibody titers may be elevated, such as antistreptolysin-O titer in rheumatic fever.
▶ Electrocardiography may reveal diffuse ST-segment and T-wave abnormalities, conduction defects (prolonged PR interval, bundle branch block, or complete heart block), supraventricular arrhythmias, and ventricular extrasystoles.
▶ Chest X-rays may show an enlarged heart and pulmonary vascular congestion.
▶ Echocardiography may demonstrate some degree of left ventricular dysfunction.
▶ Radionuclide scanning may identify inflammatory and necrotic changes characteristic of myocarditis.
▶ Laboratory cultures of stool, throat, and other body fluids may identify bacterial or viral causes of infection.
▶ Endomyocardial biopsy may confirm diagnosis. A negative biopsy does not exclude the diagnosis.

Treatment
Correction of myocarditis may involve:
▶ antibiotics to treat bacterial infections
▶ antipyretics to reduce fever and decrease stress on the heart
▶ bed rest to reduce oxygen demands and the workload on the heart
▶ restricted activity to minimize myocardial oxygen consumption; supplemental oxygen therapy; sodium restriction and diuretics to decrease fluid retention; an-

giotensin-converting enzyme inhibitors; and digoxin to increase myocardial contractility for patients with heart failure. Administer digoxin cautiously because some patients may show a paradoxical sensitivity even to small doses
▶ antiarrhythmic drugs, such as quinidine or procainamide, to treat arrhythmias; use cautiously because these drugs may depress myocardial contractility. A temporary pacemaker may be inserted if complete atrioventricular block occurs
▶ anticoagulation to prevent thromboembolism
▶ corticosteroids and immunosuppressants, although controversial, may be used to combat life-threatening complications such as intractable heart failure
▶ nonsteroidal anti-inflammatory drugs are contraindicated during the acute phase (first 2 weeks) because they increase myocardial damage
▶ cardiac assist devices or transplantation as a last resort in severe cases resistant to treatment.

Patent ductus arteriosus
The ductus arteriosus is a fetal blood vessel that connects the pulmonary artery to the descending aorta, just distal to the left subclavian artery. Normally, the ductus closes within days to weeks after birth. In patent ductus arteriosus (PDA), the lumen of the ductus remains open after birth. This creates a left-to-right shunt of blood from the aorta to the pulmonary artery and results in recirculation of arterial blood through the lungs. Initially, PDA may produce no clinical effects, but over time it can precipitate pulmonary vascular disease, causing symptoms to appear by age 40. PDA affects twice as many females as males and is the most common acyanotic congenital heart defect found in adults.

The prognosis is good if the shunt is small or surgical repair is effective. Otherwise, PDA may advance to intractable heart failure, which may be fatal.

Causes

PDA is associated with:

▶ premature birth, probably as a result of abnormalities in oxygenation or the relaxant action of prostaglandin E, which prevents ductal spasm and contracture necessary for closure
▶ rubella syndrome
▶ coarctation of the aorta
▶ ventricular septal defect
▶ pulmonary and aortic stenosis
▶ living at high altitudes.

Pathophysiology

The ductus arteriosus normally closes as prostaglandin levels from the placenta fall and oxygen levels rise. This process should begin as soon as the newborn takes its first breath, but may take as long as 3 months in some children.

In PDA, relative resistances in pulmonary and systemic vasculature and the size of the ductus determine the quantity of blood that is shunted from left to right. Because of increased aortic pressure, oxygenated blood is shunted from the aorta through the ductus arteriosus to the pulmonary artery. The blood returns to the left side of the heart and is pumped out to the aorta once more.

The left atrium and left ventricle must accommodate the increased pulmonary venous return, in turn increasing filling pressure and workload on the left side of the heart and causing left ventricular hypertrophy and possibly heart failure. In the final stages of untreated PDA, the left-to-right shunt leads to chronic pulmonary artery hypertension that becomes fixed and unreactive. This causes the shunt to reverse so that unoxygenated blood enters systemic circulation, causing cyanosis.

Signs and symptoms

The following signs and symptoms may occur:

▶ respiratory distress with signs of heart failure in infants, especially those who are premature, due to the tremendous volume of blood shunted to the lungs through a patent ductus and the increased workload on the left side of the heart
▶ classic machinery murmur (Gibson murmur), a continuous murmur heard throughout systole and diastole in older children and adults due to shunting of blood from the aorta to the pulmonary artery throughout systole and diastole. It is best heard at the base of the heart, at the second left intercostal space under the left clavicle. The murmur may obscure S_2. However, in a right-to-left shunt, this murmur may be absent
▶ thrill palpated at the left sternal border caused by the shunting of blood from the aorta to the pulmonary artery
▶ prominent left ventricular impulse due to left ventricular hypertrophy
▶ bounding peripheral pulses (Corrigan's pulse) due to the high-flow state
▶ widened pulse pressure because of an elevated systolic blood pressure and, primarily, a drop in diastolic blood pressure as blood is shunted through the PDA, thus reducing peripheral resistance
▶ slow motor development caused by heart failure
▶ failure to thrive as a result of heart failure
▶ fatigue and dyspnea on exertion may develop in adults with undetected PDA.

Complications

Possible complications of PDA may include:

▶ infective endocarditis
▶ heart failure
▶ recurrent pneumonia.

Diagnosis

The following tests help diagnose patent ductus arteriosus:

▶ Chest X-rays may show increased pulmonary vascular markings, prominent pulmonary arteries, and enlargement of the left ventricle and aorta.
▶ Electrocardiography may be normal or may indicate left atrial or ventricular hypertrophy and, in pulmonary vascular disease, biventricular hypertrophy.

• Echocardiography detects and estimates the size of a PDA. It also reveals an enlarged left atrium and left ventricle, or right ventricular hypertrophy from pulmonary vascular disease.

• Cardiac catheterization shows higher pulmonary arterial oxygen content than right ventricular content because of the influx of aortic blood. Increased pulmonary artery pressure indicates a large shunt or, if it exceeds systemic arterial pressure, severe pulmonary vascular disease. Cardiac catheterization allows for the calculation of blood volume crossing the ductus, and can rule out associated cardiac defects. Injection of contrast agent can conclusively demonstrate PDA.

Treatment
Correction of PDA may involve the following:

• surgery to ligate the ductus if medical management can't control heart failure. Asymptomatic infants with PDA don't require immediate treatment. If symptoms are mild, surgical ligation of the PDA is usually delayed until age 1

• indomethacin (a prostaglandin inhibitor) to induce ductus spasm and closure in premature infants

• prophylactic antibiotics to protect against infective endocarditis

• treatment of heart failure with fluid restriction, diuretics, and digoxin

• other therapy, including cardiac catheterization, to deposit a plug or umbrella in the ductus to stop shunting.

Pericarditis
Pericarditis is an inflammation of the pericardium — the fibroserous sac that envelops, supports, and protects the heart. It occurs in both acute and chronic forms. Acute pericarditis can be fibrinous or effusive, with purulent, serous, or hemorrhagic exudate. Chronic constrictive pericarditis is characterized by dense fibrous pericardial thickening. The prognosis depends on the underlying cause but is generally good in acute pericarditis, unless constriction occurs.

Causes
Common causes of pericarditis include:

• bacterial, fungal, or viral infection (infectious pericarditis)

• neoplasms (primary, or metastases from lungs, breasts, or other organs)

• high-dose radiation to the chest

• uremia

• hypersensitivity or autoimmune disease, such as acute rheumatic fever (most common cause of pericarditis in children), systemic lupus erythematosus, and rheumatoid arthritis

• previous cardiac injury, such as myocardial infarction (Dressler's syndrome), trauma, or surgery (post-cardiotomy syndrome), that leaves the pericardium intact but causes blood to leak into the pericardial cavity

• drugs such as hydralazine or procainamide

• idiopathic factors (most common in acute pericarditis)

• aortic aneurysm with pericardial leakage (less common)

• myxedema with cholesterol deposits in the pericardium (less common).

Pathophysiology
Pericardial tissue damaged by bacteria or other substances results in the release of chemical mediators of inflammation (prostaglandins, histamines, bradykinins, and serotonin) into the surrounding tissue, thereby initiating the inflammatory process. Friction occurs as the inflamed pericardial layers rub against each other. Histamines and other chemical mediators dilate vessels and increase vessel permeability. Vessel walls then leak fluids and protein (including fibrinogen) into tissues, causing extracellular edema. Macrophages already present in the tissue begin to phagocytize the invading bacteria and are joined by neutrophils and monocytes. After several days, the area fills with an exudate composed of necrotic tissue and

dead and dying bacteria, neutrophils, and macrophages. Eventually, the contents of the cavity autolyze and are gradually reabsorbed into healthy tissue.

A pericardial effusion develops if fluid accumulates in the pericardial cavity. Cardiac tamponade results when there is a rapid accumulation of fluid in the pericardial space, compressing the heart and preventing it from filling during diastole, and resulting in a drop in cardiac output. (See "Cardiac tamponade.")

Chronic constrictive pericarditis develops if the pericardium becomes thick and stiff from chronic or recurrent pericarditis, encasing the heart in a stiff shell and preventing the heart from properly filling during diastole. This causes an increase in both left- and right-sided filling pressures, leading to a drop in stroke volume and cardiac output.

Signs and symptoms

The following signs and symptoms of pericarditis may occur:

▶ pericardial friction rub caused by the roughened pericardial membranes rubbing against one another; although rub may be heard intermittently, it's best heard when the patient leans forward and exhales

▶ sharp and often sudden pain, usually starting over the sternum and radiating to the neck (especially the left trapezius ridge), shoulders, back, and arms due to inflammation and irritation of the pericardial membranes. The pain is often pleuritic, increasing with deep inspiration and decreasing when the patient sits up and leans forward, pulling the heart away from the diaphragmatic pleurae of the lungs.

▶ shallow, rapid respirations to reduce pleuritic pain

▶ mild fever caused by the inflammatory process

▶ dyspnea, orthopnea, and tachycardia as well as other signs of heart failure may occur as fluid builds up in the pericardial space causing pericardial effusion, a major complication of acute pericarditis

▶ muffled and distant heart sounds due to the buildup of fluid

▶ pallor, clammy skin, hypotension, pulsus paradoxus, neck vein distention and, eventually, cardiovascular collapse may occur with the rapid fluid accumulation of cardiac tamponade

▶ fluid retention, ascites, hepatomegaly, jugular venous distention, and other signs of chronic right-sided heart failure may occur with chronic constrictive pericarditis as the systemic venous pressure gradually rises

▶ pericardial knock in early diastole along the left sternal border produced by restricted ventricular filling

▶ Kussmaul's sign, increased jugular venous distention on inspiration, occurs due to restricted right-sided filling.

Diagnosis

The following tests help diagnose pericarditis:

▶ Electrocardiography may reveal diffuse ST-segment elevation in the limb leads and most precordial leads that reflects the inflammatory process. Upright T waves are present in most leads. QRS segments may be diminished when pericardial effusion exists. Arrhythmias, such as atrial fibrillation and sinus arrhythmias, may occur. In chronic constrictive pericarditis, there may be low-voltage QRS complexes, T-wave inversion or flattening, and P mitrale (wide P waves) in leads I, II, and V_6.

▶ Laboratory testing may reveal an elevated erythrocyte sedimentation rate as a result of the inflammatory process or a normal or elevated white blood cell count, especially in infectious pericarditis; blood urea nitrogen may detect uremia as a cause of pericarditis.

▶ Blood cultures may identify an infectious cause.

▶ Antistreptolysin-O titers may be positive if pericarditis is due to rheumatic fever.

▶ Purified protein derivative skin test may be positive if pericarditis is due to tuberculosis.

● Echocardiography may show an echo-free space between the ventricular wall and the pericardium, and reduced pumping action of the heart.
● Chest X-rays may be normal with acute pericarditis. The cardiac silhouette may be enlarged with a water bottle shape caused by fluid accumulation, if pleural effusion is present.

Treatment
Correcting pericarditis typically involves:
● bed rest as long as fever and pain persist, to reduce metabolic needs
● treatment of the underlying cause, if it can be identified
● nonsteroidal anti-inflammatory drugs, such as aspirin and indomethacin, to relieve pain and reduce inflammation
● corticosteroids if nonsteroidal anti-inflammatory drugs are ineffective and no infection exists; corticosteroids must be administered cautiously because episodes may recur when therapy is discontinued
● antibacterial, antifungal, or antiviral therapy if an infectious cause is suspected
● partial pericardectomy, for recurrent pericarditis, to create a window that allows fluid to drain into the pleural space
● total pericardectomy may be necessary in constrictive pericarditis to permit adequate filling and contraction of the heart
● pericardiocentesis to remove excess fluid from the pericardial space
● idiopathic pericarditis may be benign and self-limiting.

Raynaud's disease
Raynaud's disease is one of several primary arteriospastic disorders characterized by episodic vasospasm in the small peripheral arteries and arterioles, precipitated by exposure to cold or stress. This condition occurs bilaterally and usually affects the hands or, less often, the feet. Raynaud's disease is most prevalent in females, particularly between puberty and age 40. It is a benign condition, requiring no specific treatment and causing no serious sequelae.

Raynaud's phenomenon, however, a condition often associated with several connective disorders — such as scleroderma, systemic lupus erythematosus, or polymyositis — has a progressive course, leading to ischemia, gangrene, and amputation. Distinguishing between the two disorders is difficult because some patients who experience mild symptoms of Raynaud's disease for several years may later develop overt connective tissue disease, especially scleroderma.

Causes
Although family history is a risk factor, the cause of this disorder is unknown.
    Raynaud's phenomenon may develop secondary to:
● connective tissue disorders, such as scleroderma, rheumatoid arthritis, systemic lupus erythematosus, or polymyositis
● pulmonary hypertension
● thoracic outlet syndrome
● arteriocclusive disease
● myxedema
● trauma
● serum sickness
● exposure to heavy metals
● previous damage from cold exposure
● long-term exposure to cold, vibrating machinery (such as operating a jackhammer), or pressure to the fingertips (such as occurs in typists and pianists).

Pathophysiology
Although the cause is unknown, several theories account for the reduced digital blood flow, including:
● intrinsic vascular wall hyperactivity to cold
● increased vasomotor tone due to sympathetic stimulation
● antigen-antibody immune response (the most likely theory because abnormal immunologic test results accompany Raynaud's phenomenon).

Signs and symptoms
The following signs and symptoms may occur:

▶ blanching of the fingers bilaterally after exposure to cold or stress as vasoconstriction or vasospasm reduces blood flow. This is followed by cyanosis due to increased oxygen extraction resulting from sluggish blood flow. As the spasm resolves, the fingers turn red as blood rushes back into the arterioles

▶ cold and numbness may occur during the vasoconstrictive phase due to ischemia

▶ throbbing, aching pain, swelling, and tingling may occur during the hyperemic phase

▶ trophic changes, such as sclerodactyly, ulcerations, or chronic paronychia may occur as a result of ischemia in long-standing disease.

Complications
Cutaneous gangrene may occur as a result of prolonged ischemia, necessitating amputation of one or more digits (although extremely rare).

Diagnosis
The following tests help diagnose Raynaud's disease:

▶ *Clinical criteria* include skin color changes induced by cold or stress; bilateral involvement; absence of gangrene or, if present, minimal cutaneous gangrene; normal arterial pulses; and patient history of symptoms for at least 2 years.

▶ *Antinuclear antibody (ANA) titer* to identify autoimmune disease as an underlying cause of Raynaud's phenomenon; further tests must be performed if ANA titer is positive.

▶ *Arteriography* to rule out arterial occlusive disease.

▶ *Doppler ultrasonography* may show reduced blood flow if symptoms result from arterial occlusive disease.

Treatment
Treatment of this disorder typically involves:

▶ teaching the patient to avoid triggers such as cold, mechanical, or chemical injury

▶ encouraging the patient to cease smoking and avoid decongestants and caffeine to reduce vasoconstriction

▶ keeping fingers and toes warm to reduce vasoconstriction

▶ calcium channel blockers, such as nifedipine, diltiazem, and nicardipine, to produce vasodilation and prevent vasospasm

▶ adrenergic blockers, such as phenoxybenzamine or reserpine, which may improve blood flow to fingers or toes

▶ biofeedback and relaxation exercises to reduce stress and improve circulation

▶ sympathectomy to prevent ischemic ulcers by promoting vasodilation (necessary in less than 25% of patients)

▶ amputation if ischemia causes ulceration and gangrene.

Rheumatic fever and rheumatic heart disease
A systemic inflammatory disease of childhood, acute rheumatic fever develops after infection of the upper respiratory tract with group A beta-hemolytic streptococci. It mainly involves the heart, joints, central nervous system, skin, and subcutaneous tissues, and often recurs. Rheumatic heart disease refers to the cardiac manifestations of rheumatic fever and includes pancarditis (myocarditis, pericarditis, and endocarditis) during the early acute phase and chronic valvular disease later. Cardiac involvement develops in up to 50% of patients.

Worldwide, 15 to 20 million new cases of rheumatic fever are reported each year. The disease strikes most often during cool, damp weather in the winter and early spring. In the United States, it is most common in the North.

Rheumatic fever tends to run in families, lending support to the existence of genetic predisposition. Environmental factors also seem to be significant in the development of the disorder. For example, in lower socioeconomic groups, the incidence is highest in children between ages 5 and 15, probably due to malnutrition and crowded living conditions.

Patients without carditis or with mild carditis have a good long-term prognosis. Severe pancarditis occasionally produces fatal heart failure during the acute phase. Of patients who survive this complication, about 20% die within 10 years. Antibiotic therapy has greatly reduced the mortality of rheumatic heart disease. In 1950, approximately 15,000 people in the United States died of the disease compared with an estimated 5,000 deaths in 1996.

Causes

Rheumatic fever is caused by group A beta-hemolytic streptococcal pharyngitis.

Pathophysiology

Rheumatic fever appears to be a hypersensitivity reaction to a group A beta-hemolytic streptococcal infection. Because very few persons (3%) with streptococcal infections contract rheumatic fever, altered host resistance must be involved in its development or recurrence. The antigens of group A streptococci bind to receptors in the heart, muscle, brain, and synovial joints, causing an autoimmune response. Because of a similarity between the antigens of the streptococcus bacteria and the antigens of the body's own cells, antibodies may attack healthy body cells by mistake.

Carditis may affect the endocardium, myocardium, or pericardium during the early acute phase. Later, the heart valves may be damaged, causing chronic valvular disease.

Pericarditis produces a serofibrinous effusion. Myocarditis produces characteristic lesions called Aschoff's bodies (fibrin deposits surrounded by necrosis) in the interstitial tissue of the heart, as well as cellular swelling and fragmentation of interstitial collagen. These lesions lead to formation of progressively fibrotic nodules and interstitial scars.

Endocarditis causes valve leaflet swelling, erosion along the lines of leaflet closure, and blood, platelet, and fibrin deposits, which form bead-like vegetation. Eventually, the valve leaflets become scarred, lose their elasticity, and begin to adhere to each other. Endocarditis strikes the mitral valve most often in females and the aortic valve in males. In both sexes, it occasionally affects the tricuspid valve and, rarely, the pulmonic valve.

Signs and symptoms

The classic symptoms of rheumatic fever and rheumatic heart disease include:
▶ polyarthritis or migratory joint pain, caused by inflammation, occurs in most patients. Swelling, redness, and signs of effusion usually accompany such pain, which most often affects the knees, ankles, elbows, and hips
▶ erythema marginatum, a nonpruritic, macular, transient rash on the trunk or inner aspects of the upper arms or thighs, that gives rise to red lesions with blanched centers
▶ subcutaneous nodules — firm, movable, and nontender, about 3 mm to 2 cm in diameter, usually near tendons or bony prominences of joints, especially the elbows, knuckles, wrists, and knees. They often accompany carditis and may last a few days to several weeks
▶ chorea — rapid jerky movements — may develop up to 6 months after the original streptococcal infection. Mild chorea may produce hyperirritability, a deterioration in handwriting, or inability to concentrate. Severe chorea causes purposeless, nonrepetitive, involuntary muscle spasms; poor muscle coordination; and weakness.

Other signs and symptoms include:
▶ report of a streptococcal infection a few days to 6 weeks earlier; it occurs in 95% of those with rheumatic fever
▶ temperature of at least 100.4° F (38° C) due to infection and inflammation
▶ a new mitral or aortic heart murmur or a worsening murmur in a person with a preexisting murmur

JONES CRITERIA FOR DIAGNOSING RHEUMATIC FEVER

The Jones criteria are used to standardize the diagnosis of rheumatic fever. Diagnosis requires that the patient have either two major criteria, or one major criterion and two minor criteria, plus evidence of a previous streptococcal infection.

MAJOR CRITERIA	MINOR CRITERIA
▶ Carditis	▶ Fever
▶ Migratory polyarthritis	▶ Arthralgia
▶ Sydenham's chorea	▶ Elevated acute phase reactants
▶ Subcutaneous nodules	▶ Prolonged PR interval
▶ Erythema marginatum	

▶ pericardial friction rub caused by inflamed pericardial membranes rubbing against one another, if pericarditis exists
▶ chest pain, often pleuritic, due to inflammation and irritation of the pericardial membranes. Pain may increase with deep inspiration and decrease when the patient sits up and leans forward, pulling the heart away from the diaphragmatic pleurae of the lungs
▶ dyspnea, tachypnea, nonproductive cough, bibasilar crackles, and edema due to heart failure in severe rheumatic carditis.

Complications
Possible complications of rheumatic fever and rheumatic heart disease include:
▶ destruction of the mitral and aortic valves
▶ pancarditis (pericarditis, myocarditis, and endocarditis)
▶ heart failure.

Diagnosis
The following tests help diagnose rheumatic fever:
▶ Jones criteria revealing either two major criteria, or one major criterion and two minor criteria, plus evidence of a previous group A streptococcal infection, are necessary for diagnosis. (See *Jones Criteria for diagnosing rheumatic fever.*)
▶ Laboratory testing may reveal an elevated white blood cell count and erythro-

cyte sedimentation rate during the acute phase.
▶ Hemoglobin and hematocrit may show slight anemia due to suppressed erythropoiesis during inflammation.
▶ C-reactive protein may be positive, especially during the acute phase.
▶ Cardiac enzyme levels may be increased in severe carditis.
▶ Antistreptolysin-O titer may be elevated in 95% of patients within 2 months of onset.
▶ Throat cultures may continue to show the presence of group A beta-hemolytic streptococci; however, they usually occur in small numbers.
▶ Electrocardiography may show changes that are not diagnostic, but PR interval is prolonged in 20% of patients.
▶ Chest X-rays may show normal heart size or cardiomegaly, pericardial effusion, or heart failure.
▶ Echocardiography can detect valvular damage and pericardial effusion, and can measure chamber size and provide information on ventricular function.
▶ Cardiac catheterization provides information on valvular damage and left ventricular function.

Treatment
Typically, treatment of these disorders involves:

▶ prompt treatment of all group A beta-hemolytic streptococcal pharyngitis with oral penicillin V or intramuscular benzathine penicillin G; erythromycin is given for patients with penicillin hypersensitivity

▶ salicylates to relieve fever and pain and minimize joint swelling

▶ corticosteroids if the patient has carditis or if salicylates fail to relieve pain and inflammation

▶ strict bed rest for about 5 weeks for the patient with active carditis to reduce cardiac demands

▶ bed rest, sodium restriction, angiotensin-converting enzyme inhibitors, digoxin, and diuretics to treat heart failure

▶ corrective surgery, such as commissurotomy (separation of adherent, thickened valve leaflets of the mitral valve), valvuloplasty (inflation of a balloon within a valve), or valve replacement (with a prosthetic valve) for severe mitral or aortic valvular dysfunction that causes persistent heart failure

▶ secondary prevention of rheumatic fever, which begins after the acute phase subsides with monthly intramuscular injections of penicillin G benzathine or daily doses of oral penicillin V or sulfadiazine; treatment usually continues for at least 5 years or until age 21, whichever is longer

▶ prophylactic antibiotics for dental work and other invasive or surgical procedures to prevent endocarditis.

Shock

Shock is not a disease but rather a clinical syndrome leading to reduced perfusion of tissues and organs and, eventually, organ dysfunction and failure. Shock can be classified into three major categories based on the precipitating factors: distributive (neurogenic, septic, and anaphylactic); cardiogenic; and hypovolemic shock. Even with treatment, shock has a high mortality rate once the body's compensatory mechanisms fail. (See *Types of shock.*)

Causes

Causes of *neurogenic shock* may include:
▶ spinal cord injury
▶ spinal anesthesia
▶ vasomotor center depression
▶ severe pain
▶ medications
▶ hypoglycemia.

Causes of *septic shock* may include:
▶ gram-negative bacteria (most common cause)
▶ gram-positive bacteria
▶ viruses, fungi, *Rickettsiae*, parasites, yeast, protozoa, or mycobacteria.

AGE ALERT The immature immune system of newborns and infants and the weakened immune system of older adults, often accompanied by chronic illness, make these populations more susceptible to septic shock.

Causes of *anaphylactic shock* may include:
▶ medications
▶ vaccines
▶ venom
▶ foods
▶ contrast media
▶ ABO-incompatible blood.

Causes of *cardiogenic shock* may include:
▶ myocardial infarction (most common cause)
▶ heart failure
▶ cardiomyopathy
▶ arrhythmias
▶ obstruction
▶ pericardial tamponade
▶ tension pneumothorax
▶ pulmonary embolism.

Causes of *hypovolemic shock* may include:
▶ blood loss (most common cause)
▶ gastrointestinal fluid loss
▶ burns
▶ renal loss (diabetic ketoacidosis, diabetes insipidus, adrenal insufficiency)
▶ fluid shifts
▶ ascites

TYPES OF SHOCK

DISTRIBUTIVE SHOCK

In this type of shock, vasodilation causes a state of hypovolemia.

▶ *Neurogenic shock.* A loss of sympathetic vasoconstrictor tone in the vascular smooth muscle and reduced autonomic function lead to widespread arterial and venous vasodilation. Venous return is reduced as blood pools in the venous system, leading to a drop in cardiac output and hypotension.

▶ *Septic shock.* An immune response is triggered when bacteria release endotoxins. In response, macrophages secrete tumor necrosis factor (TNF) and interleukins. These mediators, in turn, are responsible for increase release of platelet-activating factor (PAF), prostaglandins, leukotrienes, thromboxane A_2, kinins, and complement. The consequences are vasodilation and vasoconstriction, increased capillary permeability, reduced systemic vascular resistance, microemboli, and an elevated cardiac output. Endotoxins also stimulate the release of histamine, further increasing capillary permeability. Moreover, myocardial depressant factor, TNF, PAF, and other factors depress myocardial function. Cardiac output falls, resulting in multisystem organ failure.

▶ *Anaphylactic shock.* Triggered by an allergic reaction, anaphylactic shock occurs when a person is exposed to an antigen to which he has already been sensitized. Exposure results in the production of specific immunoglobulin E (IgE) antibodies by plasma cells that bind to membrane receptors on mast cells and basophils. On reexposure, the antigen binds to IgE antibodies or crosslinked IgE receptors, triggering the release of powerful chemical mediators from mast cells. IgG or IgM enters into the reaction and activates the release of complement factors. At the same time, the chemical mediators bradykinin and leukotrienes induce vascular collapse by stimulating contraction of certain groups of smooth muscles and by increasing vascular permeability, leading to decreased peripheral resistance and plasma leakage into the extravascular tissues, thereby reducing blood volume and causing hypotension, hypovolemic shock, and cardiac dysfunction. Bronchospasm and laryngeal edema also occur.

CARDIOGENIC SHOCK

In cardiogenic shock, the left ventricle can't maintain an adequate cardiac output. Compensatory mechanisms increase heart rate, strengthen myocardial contractions, promote sodium and water retention, and cause selective vasoconstriction. However, these mechanisms increase myocardial workload and oxygen consumption, which reduces the heart's ability to pump blood, especially if the patient has myocardial ischemia. Consequently, blood backs up, resulting in pulmonary edema. Eventually, cardiac output falls and multisystem organ failure develops as the compensatory mechanisms fail to maintain perfusion.

HYPOVOLEMIC SHOCK

When fluid is lost from the intravascular space through external losses or the shift of fluid from the vessels to the interstitial or intracellular spaces, venous return to the heart is reduced. This reduction in preload decreases ventricular filling, leading to a drop in stroke volume. Then, cardiac output falls, causing reduced perfusion of the tissues and organs.

▶ peritonitis
▶ hemothorax.

Pathophysiology

There are three basic stages common to each type of shock: the compensatory, progressive, and irreversible or refractory stages.

Compensatory stage. When arterial pressure and tissue perfusion are reduced, compensatory mechanisms are activated to maintain perfusion to the heart and brain. As the baroreceptors in the carotid sinus and aortic arch sense a drop in blood pressure, epinephrine and norepinephrine are secreted to increase peripheral resistance, blood pressure, and myocardial contractility. Reduced blood flow to the kidney activates the renin-angiotensin-aldosterone system, causing vasoconstriction and sodium and water retention, leading to increased blood volume and venous return. As a result of these compensatory mechanisms, cardiac output and tissue perfusion are maintained.

Progressive stage. This stage of shock begins as compensatory mechanisms fail to maintain cardiac output. Tissues become hypoxic because of poor perfusion. As cells switch to anaerobic metabolism, lactic acid builds up, producing metabolic acidosis. This acidotic state depresses myocardial function. Tissue hypoxia also promotes the release of endothelial mediators, which produce vasodilation and endothelial abnormalities, leading to venous pooling and increased capillary permeability. Sluggish blood flow increases the risk of disseminated intravascular coagulation.

Irreversible (refractory) stage. As the shock syndrome progresses, permanent organ damage occurs as compensatory mechanisms can no longer maintain cardiac output. Reduced perfusion damages cell membranes, lysosomal enzymes are released, and energy stores are depleted,

possibly leading to cell death. As cells use anaerobic metabolism, lactic acid accumulates, increasing capillary permeability and the movement of fluid out of the vascular space. This loss of intravascular fluid further contributes to hypotension. Perfusion to the coronary arteries is reduced, causing myocardial depression and a further reduction in cardiac output. Eventually, circulatory and respiratory failure occur. Death is inevitable.

Signs and symptoms

In the *compensatory stage* of shock, signs and symptoms may include:
▶ tachycardia and bounding pulse due to sympathetic stimulation
▶ restlessness and irritability related to cerebral hypoxia
▶ tachypnea to compensate for hypoxia
▶ reduced urinary output secondary to vasoconstriction
▶ cool, pale skin associated with vasoconstriction; warm, dry skin in septic shock due to vasodilation.

In the *progressive stage* of shock, signs and symptoms may include:
▶ hypotension as compensatory mechanisms begin to fail
▶ narrowed pulse pressure associated with reduced stroke volume
▶ weak, rapid, thready pulse caused by decreased cardiac output
▶ shallow respirations as the patient weakens
▶ reduced urinary output as poor renal perfusion continues
▶ cold, clammy skin caused by vasoconstriction
▶ cyanosis related to hypoxia.

AGE ALERT Hypotension, altered level of consciousness, and hyperventilation may be the only signs of septic shock in infants and the elderly.

In the *irreversible stage,* clinical findings may include:
▶ unconsciousness and absent reflexes caused by reduced cerebral perfusion, acid-

base imbalance, or electrolyte abnormalities

▶ rapidly falling blood pressure as decompensation occurs

▶ weak pulse caused by reduced cardiac output

▶ slow, shallow or Cheyne-Stokes respirations secondary to respiratory center depression

▶ anuria related to renal failure.

Complications

Possible complications of shock include:

▶ acute respiratory distress syndrome

▶ acute tubular necrosis

▶ disseminated intravascular coagulation (DIC)

▶ cerebral hypoxia

▶ death.

Diagnosis

The following tests help diagnose shock:

▶ Hematocrit may be reduced in hemorrhage or elevated in other types of shock due to hypovolemia.

▶ Blood, urine, and sputum cultures may identify the organism responsible for septic shock.

▶ Coagulation studies may detect coagulopathy from DIC.

▶ Laboratory testing may reveal increased white blood cell count and erythrocyte sedimentation rate due to injury and inflammation; elevated blood urea nitrogen and creatinine levels due to reduced renal perfusion; serum lactate may be increased secondary to anaerobic metabolism; and serum glucose may be elevated in early stages of shock as liver releases glycogen stores in response to sympathetic stimulation.

▶ Cardiac enzymes and proteins may be elevated, indicating myocardial infarction as a cause of cardiogenic shock.

▶ Arterial blood gas analysis may reveal respiratory alkalosis in early shock associated with tachypnea, respiratory acidosis in later stages associated with respiratory depression, and metabolic acidosis in later stages secondary to anaerobic metabolism.

▶ Urine specific gravity may be high in response to effects of antidiuretic hormone.

▶ Chest X-rays may be normal in early stages; pulmonary congestion may be seen in later stages.

▶ Hemodynamic monitoring may reveal characteristic patterns of intracardiac pressures and cardiac output, which are used to guide fluid and drug management. (See *Putting hemodynamic monitoring to use,* page 190.)

▶ Electrocardiography determines the heart rate and detects arrhythmias, ischemic changes, and myocardial infarction.

▶ Echocardiography determines left ventricular function and reveals valvular abnormalities.

Treatment

Correction of shock typically involves the following measures:

▶ identification and treatment of the underlying cause, if possible

▶ maintaining a patent airway; preparing for intubation and mechanical ventilation if the patient develops respiratory distress

▶ supplemental oxygen to increase oxygenation

▶ continuous cardiac monitoring to detect changes in heart rate and rhythm; administration of antiarrhythmics, as necessary

▶ initiating and maintaining at least two intravenous lines with large-gauge needles for fluid and drug administration

▶ intravenous fluids, crystalloids, colloids, or blood products, as necessary, to maintain intravascular volume.

Additional therapy for *hypovolemic shock* may include:

▶ pneumatic antishock garment, although controversial, which may be applied to control both internal and external hemorrhage by direct pressure

▶ fluids such as normal saline or lactated Ringer's solution, initially, to restore filling pressures

PUTTING HEMODYNAMIC MONITORING TO USE

Hemodynamic monitoring provides information on intracardiac pressures and cardiac output. To understand intracardiac pressures, picture the cardiovascular system as a continuous loop with constantly changing pressure gradients that keep the blood moving.

RIGHT ATRIAL PRESSURE (RAP), OR CENTRAL VENOUS PRESSURE (CVP)

The RAP reflects right atrial, or right heart, function and end-diastolic pressure.

▶ **Normal:** 1 to 6 mm Hg (1.34 to 8 cm H$_2$O). (To convert mm Hg to cm H$_2$O, multiply mm Hg by 1.34)

▶ **Elevated value suggests:** right ventricular (RV) failure, volume overload, tricuspid valve stenosis or regurgitation, constrictive pericarditis, pulmonary hypertension, cardiac tamponade, or RV infarction.

▶ **Low value suggests:** reduced circulating blood volume.

RIGHT VENTRICULAR PRESSURE (RVP)

RV systolic pressure normally equals pulmonary artery systolic pressure; RV end-diastolic pressure, which equals right atrial pressure, reflects RV function.

▶ **Normal**: systolic, 15 to 25 mm Hg; diastolic, 0 to 8 mm Hg.

▶ **Elevated value suggests:** mitral stenosis or insufficiency, pulmonary disease, hypoxemia, constrictive pericarditis, chronic heart failure, atrial and ventricular septal defects, and patent ductus arteriosus.

PULMONARY ARTERY PRESSURE

Pulmonary artery systolic pressure reflects right ventricular function and pulmonary circulation pressures. Pulmonary artery diastolic pressure (PADP) reflects left ventricular (LV) pressures, specifically left ventricular end-diastolic pressure.

▶ **Normal:** Systolic, 15 to 25 mm Hg; diastolic, 8 to 15 mm Hg; mean, 10 to 20 mm Hg.

▶ **Elevated value suggests:** LV failure, increased pulmonary blood flow (left or right shunting, as in atrial or ventricular septal defects), and in any condition causing increased pulmonary arteriolar resistance.

PULMONARY CAPILLARY WEDGE PRESSURE (PCWP)

PCWP reflects left atrial and LV pressures unless the patient has mitral stenosis. Changes in PCWP reflect changes in LV filling pressure. The heart momentarily relaxes during diastole as it fills with blood from the pulmonary veins; this permits the pulmonary vasculature, left atrium, and left ventricle to act as a single chamber.

▶ **Normal**: mean pressure, 6 to 12 mm Hg.

▶ **Elevated value suggests:** LV failure, mitral stenosis or insufficiency, and pericardial tamponade.

▶ **Low value suggests:** hypovolemia.

LEFT ATRIAL PRESSURE

This value reflects left ventricular end-diastolic pressure in patients without mitral valve disease.

▶ **Normal:** 6 to 12 mm Hg.

CARDIAC OUTPUT

Cardiac output is the amount of blood ejected by the heart each minute.

▶ **Normal:** 4 to 8 liters; varies with a patient's weight, height, and body surface area. Adjusting the cardiac output to the patient's size yields a measurement called the cardiac index.

packed red blood cells in hemorrhagic shock to restore blood loss and improve oxygen-carrying capacity of the blood.

Additional measures for *cardiogenic shock* may include:

▶ inotropic drugs such as dopamine, dobutamine, amrinone, and epinephrine, to increase contractility of the heart and increase cardiac output

▶ vasodilators, such as nitroglycerin or nitroprusside, given with a vasopressor to reduce the workload of the left ventricle

▶ diuretics to reduce preload, if patient has fluid volume overload

▶ intra-aortic balloon pump therapy to reduce the work of the left ventricle by decreasing systemic vascular resistance. Diastolic pressure is increased, resulting in improved coronary artery perfusion

▶ thrombolytic therapy or coronary artery revascularization to restore coronary artery blood flow, if cardiogenic shock is due to acute myocardial infarction

▶ emergency surgery to repair papillary muscle rupture or ventricular septal defect, if either is the cause of cardiogenic shock

▶ ventricular assist device to assist the pumping action of the heart when intraaortic balloon pump and drug therapy fail

▶ cardiac transplantation, which may be considered when other medical and surgical therapeutic measures fail.

Correction of *septic shock* may also include:

▶ antibiotic therapy to eradicate the causative organism

▶ inotropic and vasopressor drugs, such as dopamine, dobutamine, and norepinephrine, to improve perfusion and maintain blood pressure

▶ although still investigational, monoclonal antibodies to tumor necrosis factor, endotoxin, and interleukin-1, to counteract mediators of septic shock.

Additional therapy for *neurogenic shock* may include:

▶ vasopressor drugs to raise blood pressure by vasoconstriction

▶ fluid replacement to maintain blood pressure and cardiac output.

Tetralogy of Fallot

Tetralogy of Fallot is a combination of four cardiac defects: ventricular septal defect (VSD), right ventricular outflow tract obstruction (pulmonary stenosis), right ventricular hypertrophy, and dextroposition of the aorta, with overriding of the VSD. Blood shunts from right to left through the VSD, allowing unoxygenated blood to mix with oxygenated blood and resulting in cyanosis. This cyanotic heart defect sometimes coexists with other congenital acyanotic heart defects, such as patent ductus arteriosus or atrial septal defect. It accounts for about 10% of all congenital defects and occurs equally in males and females. Before surgical advances made correction possible, about one-third of these children died in infancy.

Causes

The cause of tetralogy of Fallot is unknown. It may be associated with:

▶ fetal alcohol syndrome

▶ thalidomide use during pregnancy.

Pathophysiology

In tetralogy of Fallot, unoxygenated venous blood returning to the right side of the heart may pass through the VSD to the left ventricle, bypassing the lungs, or it may enter the pulmonary artery, depending on the extent of the pulmonic stenosis. Rather than originating from the left ventricle, the aorta overrides both ventricles.

The VSD usually lies in the outflow tract of the right ventricle and is generally large enough to permit equalization of right and left ventricular pressures. However, the ratio of systemic vascular resistance to pulmonary stenosis affects the direction and magnitude of shunt flow across the VSD. Severe obstruction of right ventricular outflow produces a right-to-left shunt, causing decreased systemic arterial oxygen saturation, cyanosis, reduced pulmonary blood flow, and hypoplasia of the entire pulmonary vasculature. Right ventricular hypertrophy develops in response to the extra force needed to push blood into the stenotic pulmonary artery. Milder forms of pulmonary stenosis result in a left-to-right shunt or no shunt at all.

Signs and symptoms

The following signs and symptoms may occur:

▶ cyanosis, the hallmark of tetralogy of Fallot, is caused by a right-to-left shunt

▶ cyanotic or "blue" spells (Tet spells), characterized by dyspnea; deep, sighing respirations; bradycardia; fainting; seizures; and loss of consciousness following exercise, crying, straining, infection, or fever. It may result from reduced oxygen to the brain because of increased right-to-left shunting, possibly caused by spasm of the right ventricular outflow tract, increased systemic venous return, or decreased systemic arterial resistance

▶ clubbing, diminished exercise tolerance, increasing dyspnea on exertion, growth retardation, and eating difficulties in older children due to poor oxygenation

▶ squatting with shortness of breath to reduce venous return of unoxygenated blood from the legs and to increase systemic arterial resistance

▶ loud systolic murmur best heard along the left sternal border, which may diminish or obscure the pulmonic component of S_2

▶ continuous murmur of the ductus in a patient with a large patent ductus, which may obscure systolic murmur

▶ thrill at the left sternal border caused by abnormal blood flow through the heart

▶ obvious right ventricular impulse and prominent inferior sternum associated with right ventricular hypertrophy.

Complications

Possible complications of tetralogy of Fallot include:

▶ pulmonary thrombosis
▶ venous thrombosis
▶ cerebral embolism
▶ infective endocarditis
▶ risk of spontaneous abortion, premature birth, and low-birth-weight infants born to women with tetralogy of Fallot.

Diagnosis

The following tests help diagnose tetralogy of Fallot:

▶ Chest X-rays may demonstrate decreased pulmonary vascular marking (depending on the severity of the pulmonary obstruction), an enlarged right ventricle, and a boot-shaped cardiac silhouette.

▶ Electrocardiography shows right ventricular hypertrophy, right axis deviation and, possibly, right atrial hypertrophy.

▶ Echocardiography identifies septal overriding of the aorta, the VSD, and pulmonary stenosis, and detects the hypertrophied walls of the right ventricle.

▶ Laboratory testing reveals diminished oxygen saturation and polycythemia (hematocrit may be more than 60%) if the cyanosis is severe and longstanding, predisposing the patient to thrombosis.

▶ Cardiac catheterization confirms the diagnosis by providing visualization of pulmonary stenosis, the VSD, and the overriding aorta and ruling out other cyanotic heart defects. This test also measures the degree of oxygen saturation in aortic blood.

Treatment

Tetralogy of Fallot may be managed by:

▶ a knee-chest position, and administration of oxygen and morphine to improve oxygenation

▶ palliative surgery with a Blalock-Taussig procedure, which joins the subclavian artery to the pulmonary artery to enhance blood flow to the lungs to reduce hypoxia

▶ prophylactic antibiotics to prevent infective endocarditis or cerebral abscesses

▶ phlebotomy to reduce polycythemia

corrective surgery to relieve pulmonary stenosis and close the VSD, directing left ventricular outflow to the aorta.

Transposition of the great arteries

In this cyanotic congenital heart defect, the great arteries are reversed such that the aorta arises from the right ventricle and the pulmonary artery from the left ventricle, producing two noncommunicating circulatory systems (pulmonic and systemic). The right-to-left shunting of blood leads to an increased risk of heart failure and anoxia. Transposition accounts for about 5% of all congenital heart defects and often coexists with other congenital heart defects, such as ventricular septal defect (VSD), VSD with pulmonary stenosis, atrial septal defect (ASD), and patent ductus arteriosus (PDA). It affects two to three times more males than females.

Causes
The cause of this disorder is unknown.

Pathophysiology
Transposition of the great arteries results from faulty embryonic development. Oxygenated blood returning to the left side of the heart is carried back to the lungs by a transposed pulmonary artery. Unoxygenated blood returning to the right side of the heart is carried to the systemic circulation by a transposed aorta.

Communication between the pulmonary and systemic circulations is necessary for survival. In infants with isolated transposition, blood mixes only at the patent foramen ovale and at the patent ductus arteriosus, resulting in slight mixing of unoxygenated systemic blood and oxygenated pulmonary blood. In infants with concurrent cardiac defects, greater mixing of blood occurs.

Signs and symptoms
The following signs and symptoms may occur:

cyanosis and tachypnea that worsens with crying within the first few hours after birth, when no other heart defects exist that allow mixing of systemic and pulmonary blood. Cyanosis may be minimized with associated defects such as ASD, VSD, or PDA

gallop rhythm, tachycardia, dyspnea, hepatomegaly, and cardiomegaly within days to weeks due to heart failure

loud S_2 because the anteriorly transposed aorta is directly behind the sternum

murmurs of ASD, VSD, or PDA

diminished exercise tolerance, fatigability, and clubbing due to reduced oxygenation.

Complications
Transposition of the great arteries may be complicated by:

heart failure

infective endocarditis.

Diagnosis
The following tests help diagnose transposition of the great arteries:

Chest X-rays are normal in the first days after birth. Within days to weeks, right atrial and right ventricular enlargement characteristically cause the heart to appear oblong. X-ray may also show increased pulmonary vascular markings, except when pulmonary stenosis exists.

Electrocardiography typically reveals right axis deviation and right ventricular hypertrophy but may be normal in a neonate.

Echocardiography demonstrates the reversed position of the aorta and pulmonary artery, and records echoes from both semilunar valves simultaneously, due to aortic valve displacement. It also detects other cardiac defects.

Cardiac catheterization reveals decreased oxygen saturation in left ventricular blood and aortic blood; increased right atrial, right ventricular, and pulmonary artery oxygen saturation; and right ventricular systolic pressure equal to systemic pressure. Dye injection reveals the transposed vessels and the presence of any other cardiac defects.

▶ Arterial blood gas analysis indicates hypoxia and secondary metabolic acidosis.

Treatment
Treatment of this disorder may involve:

▶ prostaglandin infusion to keep the ductus arteriosus patent until surgical correction

▶ atrial balloon septostomy (Rashkind procedure) during cardiac catheterization if needed as a palliative measure until surgery can be performed; enlarges the patent foramen ovale and thereby improves oxygenation and alleviates hypoxia by allowing greater mixing of blood from the pulmonary and systemic circulations

▶ digoxin and diuretics after atrial balloon septostomy to lessen heart failure until the infant is ready to withstand corrective surgery (usually between birth and age 1)

▶ surgery to correct transposition, although the procedure depends on the physiology of the defect.

Valvular heart disease
In valvular heart disease, three types of mechanical disruption can occur: stenosis, or narrowing, of the valve opening; incomplete closure of the valve; or prolapse of the valve. Valvular disorders in children and adolescents most commonly occur as a result of congenital heart defects, whereas in adults, rheumatic heart disease is a common cause.

Causes
The causes of valvular heart disease are varied and are different for each type of valve disorder. (See *Types of valvular heart disease,* pages 196 and 197.)

Pathophysiology
Pathophysiology of valvular heart disease varies according to the valve and the disorder.

Mitral regurgitation. Any abnormality of the mitral leaflets, mitral annulus, chordae tendineae, papillary muscles, left atrium, or left ventricle can lead to mitral regurgitation. Blood from the left ventricle flows back into the left atrium during systole, causing the atrium to enlarge to accommodate the backflow. As a result, the left ventricle also dilates to accommodate the increased volume of blood from the atrium and to compensate for diminishing cardiac output. Ventricular hypertrophy and increased end-diastolic pressure result in increased pulmonary artery pressure, eventually leading to left-sided and right-sided heart failure.

Mitral stenosis. Narrowing of the valve by valvular abnormalities, fibrosis, or calcification obstructs blood flow from the left atrium to the left ventricle. Consequently, left atrial volume and pressure rise and the chamber dilates. Greater resistance to blood flow causes pulmonary hypertension, right ventricular hypertrophy, and right-sided heart failure. Also, inadequate filling of the left ventricle produces low cardiac output.

Aortic regurgitation. Blood flows back into the left ventricle during diastole, causing fluid overload in the ventricle, which dilates and hypertrophies. The excess volume causes fluid overload in the left atrium and, finally, the pulmonary system. Left-sided heart failure and pulmonary edema eventually result.

Aortic stenosis. Increased left ventricular pressure tries to overcome the resistance of the narrowed valvular opening. The added workload increases the demand for oxygen, and diminished cardiac output causes poor coronary artery perfusion, ischemia of the left ventricle, and left-sided heart failure.

Pulmonic stenosis. Obstructed right ventricular outflow causes right ventricular hypertrophy, eventually resulting in right-sided heart failure.

Signs and symptoms

The clinical manifestations vary according to the type of valvular defects. (See *Types of valvular heart disease,* pages 196 and 197, for specific clinical features of each valve disorder.)

Complications

Possible complications of valvular heart disease include:
▶ heart failure
▶ pulmonary edema
▶ thromboembolism
▶ endocarditis.

Diagnosis

The diagnosis of valvular heart disease can be made through cardiac catheterization, chest X-rays, echocardiography, or electrocardiography. (See *Types of valvular heart disease,* pages 196 and 197.)

Treatment

Correcting this disorder typically involves:
▶ digoxin, a low-sodium diet, diuretics, vasodilators, and especially angiotensin-converting enzyme inhibitors to treat left ventricular failure
▶ oxygen in acute situations, to increase oxygenation
▶ anticoagulants to prevent thrombus formation around diseased or replaced valves
▶ prophylactic antibiotics before and after surgery or dental care to prevent endocarditis
▶ nitroglycerin to relieve angina in conditions such as aortic stenosis
▶ beta-adrenergic blockers or digoxin to slow the ventricular rate in atrial fibrillation or atrial flutter
▶ cardioversion to convert atrial fibrillation to sinus rhythm
▶ open or closed commissurotomy to separate thick or adherent mitral valve leaflets
▶ balloon valvuloplasty to enlarge the orifice of a stenotic mitral, aortic, or pulmonic valve
▶ annuloplasty or valvuloplasty to reconstruct or repair the valve in mitral regurgitation
▶ valve replacement with a prosthetic valve for mitral and aortic valve disease.

Varicose veins

Varicose veins are dilated, tortuous veins, engorged with blood and resulting from improper venous valve function. They can be primary, originating in the superficial veins, or secondary, occurring in the deep veins.

Primary varicose veins tend to be familial and to affect both legs; they are twice as common in females as in males. They account for approximately 90% of varicose veins; about 10% to 20% of Americans have primary varicose veins. Usually, secondary varicose veins occur in one leg. Both types are more common in middle adulthood.

Without treatment, varicose veins continue to enlarge. Although there is no cure, certain measures such as walking and use of compression stockings can reduce symptoms. Surgery may remove varicose veins but the condition can occur in other veins.

Causes

Primary varicose veins can result from:
▶ congenital weakness of the valves or venous wall
▶ conditions that produce prolonged venous stasis or increased intra-abdominal pressure such as pregnancy, obesity, constipation, or wearing tight clothes
▶ occupations that necessitate standing for an extended period of time
▶ family history of varicose veins.
Secondary varicose veins can result from:
▶ deep vein thrombosis
▶ venous malformation
▶ arteriovenous fistulas
▶ trauma to the venous system
▶ occlusion.

Pathophysiology

Veins are thin-walled, distensible vessels with valves that keep blood flowing in one direction. Any condition that weakens, de-

TYPES OF VALVULAR HEART DISEASE

CAUSES AND INCIDENCE	CLINICAL FEATURES
Mitral stenosis ▶ Results from rheumatic fever (most common cause) ▶ Most common in females ▶ May be associated with other congenital anomalies	▶ Dyspnea on exertion, paroxysmal nocturnal dyspnea, orthopnea, weakness, fatigue, and palpitations ▶ Peripheral edema, jugular vein distention, ascites, and hepatomegaly (right ventricular failure) ▶ Crackles, atrial fibrillation, and signs of systemic emboli ▶ Auscultation reveals a loud S_1 or opening snap and a diastolic murmur at the apex
Mitral insufficiency ▶ Results from rheumatic fever, hypertrophic cardiomyopathy, mitral valve prolapse, myocardial infarction, severe left ventricular failure, or ruptured chordae tendineae ▶ Associated with other congenital anomalies such as transposition of the great arteries ▶ Rare in children without other congenital anomalies	▶ Orthopnea, dyspnea, fatigue, angina, and palpitations ▶ Peripheral edema, jugular vein distention, and hepatomegaly (right ventricular failure) ▶ Tachycardia, crackles, and pulmonary edema ▶ Auscultation reveals a holosystolic murmur at apex, a possible split S_2, and an S_3
Aortic insufficiency ▶ Results from rheumatic fever, syphilis, hypertension, or endocarditis, or may be idiopathic ▶ Associated with Marfan syndrome ▶ Most common in males ▶ Associated with ventricular septal defect, even after surgical closure	▶ Dyspnea, cough, fatigue, palpitations, angina, and syncope ▶ Pulmonary congestion, left ventricular failure, and "pulsating" nail beds (Quincke's sign) ▶ Rapidly rising and collapsing pulses (pulsus biferiens), cardiac arrhythmias, and widened pulse pressure ▶ Auscultation reveals an S_3 and a diastolic blowing murmur at left sternal border ▶ Palpation and visualization of apical impulse in chronic disease
Aortic stenosis ▶ Results from congenital aortic bicuspid valve (associated with coarctation of the aorta), congenital stenosis of valve cusps, rheumatic fever, or atherosclerosis in the elderly ▶ Most common in males	▶ Dyspnea on exertion, paroxysmal nocturnal dyspnea, fatigue, syncope, angina, and palpitations ▶ Pulmonary congestion, and left ventricular failure ▶ Diminished carotid pulses, decreased cardiac output, and cardiac arrhythmias; may have pulsus alternans ▶ Auscultation reveals systolic murmur heard at base or in carotids and, possibly, an S_4
Pulmonic stenosis ▶ Results from congenital stenosis of valve cusp or rheumatic heart disease (infrequent) ▶ Associated with tetralogy of Fallot	▶ Asymptomatic or symptomatic with dyspnea on exertion, fatigue, chest pain, and syncope ▶ May cause jugular distention/right ventricular failure ▶ Auscultation reveals a systolic murmur at the left sternal border and a split S_2 with a delayed or absent pulmonic component

DIAGNOSTIC MEASURES

▶ *Cardiac catheterization:* diastolic pressure gradient across valve; elevated left atrial and pulmonary capillary wedge pressures (PCWP) > 15 mm Hg with severe pulmonary hypertension; elevated right-sided heart pressure with decreased cardiac output (CO); and abnormal contraction of the left ventricle
▶ *Chest X-rays:* left atrial and ventricular enlargement, enlarged pulmonary arteries, and mitral valve calcification
▶ *Echocardiography:* thickened mitral valve leaflets and left atrial enlargement
▶ *Electrocardiography:* left atrial hypertrophy, atrial fibrillation, right ventricular hypertrophy, and right axis deviation

▶ *Cardiac catheterization:* mitral regurgitation with increased left ventricular end-diastolic volume and pressure, increased atrial pressure and PCWP, and decreased CO
▶ *Chest X-rays:* left atrial and ventricular enlargement and pulmonary venous congestion
▶ *Echocardiography:* abnormal valve leaflet motion, and left atrial enlargement
▶ *Electrocardiography:* may show left atrial and ventricular hypertrophy, sinus tachycardia, and atrial fibrillation

▶ *Cardiac catheterization:* reduction in arterial diastolic pressures, aortic regurgitation, other valvular abnormalities, and increased left ventricular end-diastolic pressure
▶ *Chest X-rays:* left ventricular enlargement and pulmonary venous congestion
▶ *Echocardiography:* left ventricular enlargement, alterations in mitral valve movement (indirect indication of aortic valve disease), and mitral thickening
▶ *Electrocardiography:* sinus tachycardia, left ventricular hypertrophy, and left atrial hypertrophy in severe disease

▶ *Cardiac catheterization:* pressure gradient across valve (indicating obstruction), and increased left ventricular end-diastolic pressures
▶ *Chest X-rays:* valvular calcification, left ventricular enlargement, and pulmonary vein congestion
▶ *Echocardiography:* thickened aortic valve and left ventricular wall, possibly coexistent with mitral valve stenosis
▶ *Electrocardiography:* left ventricular hypertrophy

▶ *Cardiac catheterization:* increased right ventricular pressure, decreased pulmonary artery pressure, and abnormal valve orifice
▶ *Electrocardiography:* may show right ventricular hypertrophy, right axis deviation, right atrial hypertrophy, and atrial fibrillation

stroys, or distends these valves allows the backflow of blood to the previous valve. If a valve cannot hold the pooling blood, it can become incompetent, allowing even more blood to flow backward. As the volume of venous blood builds, pressure in the vein increases and the vein becomes distended. As the veins are stretched, their walls weaken and they lose their elasticity. As the veins enlarge, they become lumpy and tortuous. As hydrostatic pressure increases, plasma is forced out of the veins and into the surrounding tissues, resulting in edema.

People who stand for prolonged periods of time may also develop venous pooling because there is no muscular contraction in the legs forcing blood back up to the heart. If the valves in the veins are too weak to hold the pooling blood, they begin to leak, allowing blood to flow backward.

Signs and symptoms
The following signs and symptoms may occur:
▶ dilated, tortuous, purplish, ropelike veins, particularly in the calves, due to venous pooling
▶ edema of the calves and ankles due to deep vein incompetence
▶ leg heaviness that worsens in the evening and in warm weather; caused by venous pooling
▶ dull aching in the legs after prolonged standing or walking, which may be due to tissue breakdown
▶ aching during menses as a result of increased fluid retention.

Complications
Possible complications of varicose veins include:
▶ blood clots secondary to venous stasis
▶ venous stasis ulcers
▶ chronic venous insufficiency.

AGE ALERT As a person ages, veins dilate and stretch, increasing susceptibility to varicose veins and chronic venous insufficiency. Because the skin is very friable and can easily break down, ulcers in the elderly caused by chronic venous insufficiency may take longer to heal.

Diagnosis
The following tests help diagnose varicose veins:
▶ Manual compression test detects a palpable impulse when the vein is firmly occluded at least 8" above the point of palpation, indicating incompetent valves in the vein.
▶ Trendelenburg's test (retrograde filling test) detects incompetent deep and superficial vein valves.
▶ Photoplethysmography characterizes venous blood flow by noting changes in the skin's circulation.
▶ Doppler ultrasonography detects the presence or absence of venous backflow in deep or superficial veins.
▶ Venous outflow and reflux plethysmography detects deep venous occlusion; this test is invasive and not routinely used.
▶ Ascending and descending venography demonstrates venous occlusion and patterns of collateral flow.

Treatment
Correction of this disorder typically involves:
▶ if possible, treatment of the underlying cause, such as an abdominal tumor or obesity
▶ antiembolism stockings or elastic bandages to counteract swelling by supporting the veins and improving circulation
▶ a regular exercise program to promote muscular contraction to force blood through the veins and reduce venous pooling
▶ injection of a sclerosing agent into small to medium-sized varicosities

▶ surgical stripping and ligation of severe varicose veins

▶ phlebectomy, removing the varicose vein through small incisions in the skin, may be performed in an outpatient setting.

Additional treatment measures include the following:

▶ Discourage the patient from wearing constrictive clothing that interferes with venous return.

▶ Encourage the obese patient to lose weight, to reduce increased intra-abdominal pressure.

▶ Elevate the legs above the heart whenever possible to promote venous return.

▶ Instruct the patient to avoid prolonged standing or sitting because these actions enhance venous pooling.

Ventricular septal defect

In a ventricular septal defect (VSD), the most common acyanotic congenital heart disorder, an opening in the septum between the ventricles allows blood to shunt between the left and right ventricles. This results in ineffective pumping of the heart and increases the risk for heart failure.

VSDs account for up to 30% of all congenital heart defects. The prognosis is good for defects that close spontaneously or are correctable surgically, but poor for untreated defects, which are sometimes fatal in children by age 1, usually from secondary complications.

Causes

A VSD may be associated with the following conditions:

▶ fetal alcohol syndrome

▶ Down syndrome and other autosomal trisomies

▶ renal anomalies

▶ patent ductus arteriosus and coarctation of the aorta

▶ prematurity.

Pathophysiology

In infants with VSD, the ventricular septum fails to close completely by the eighth week of gestation. VSDs are located in the membranous or muscular portion of the ventricular septum and vary in size. Some defects close spontaneously; in other defects, the septum is entirely absent, creating a single ventricle. Small VSDs are likely to close spontaneously. Large VSDs should be surgically repaired before pulmonary vascular disease occurs or while it is still reversible.

VSD isn't readily apparent at birth because right and left pressures are approximately equal and pulmonary artery resistance is elevated. Alveoli are not yet completely opened, so blood doesn't shunt through the defect. As the pulmonary vasculature gradually relaxes, between 4 and 8 weeks after birth, right ventricular pressure decreases, allowing blood to shunt from the left to the right ventricle. Initially, large VSD shunts cause left atrial and left ventricular hypertrophy. Later, an uncorrected VSD causes right ventricular hypertrophy due to increasing pulmonary resistance. Eventually, biventricular heart failure and cyanosis (from reversal of the shunt direction) occur. Fixed pulmonary hypertension may occur much later in life with right-to-left shunting (Eisenmenger's syndrome), causing cyanosis and clubbing of the nail beds.

Signs and symptoms

Signs and symptoms of a VSD may include:

▶ thin, small infants who gain weight slowly when a large VSD is present secondary to heart failure

▶ loud, harsh systolic murmur heard best along the left sternal border at the third or fourth intercostal space, caused by abnormal blood flow through the VSD; murmur is widely transmitted

▶ palpable thrill caused by turbulent blood flow between the ventricles through a small VSD

▶ loud, widely split pulmonic component of S_2 caused by increased pressure gradient across the VSD

▶ displacement of point of maximal impulse to the left due to hypertrophy of the heart

 AGE ALERT Typically, in infants the apical impulse is palpated over the fourth intercostal space, just to the left of the midclavicular line. In children older than age 7, it's palpated over the fifth intercostal space. When the heart is enlarged, the apical beat is displaced to the left or downward.

▶ prominent anterior chest secondary to cardiac hypertrophy

▶ liver, heart, and spleen enlargement because of systemic congestion

▶ feeding difficulties associated with heart failure

▶ diaphoresis, tachycardia, and rapid, grunting respirations secondary to heart failure

▶ cyanosis and clubbing if right-to-left shunting occurs later in life secondary to pulmonary hypertension.

Complications

Complications of a VSD may include:
▶ pulmonary hypertension
▶ infective endocarditis
▶ pneumonia
▶ heart failure
▶ Eisenmenger's syndrome
▶ aortic regurgitation (if the aortic valve is involved).

Diagnosis

The following tests help diagnose ventricular septal defect:

▶ Chest X-rays appear normal in small defects. In large VSDs, the X-ray may show cardiomegaly, left atrial and left ventricular enlargement, and prominent vascular markings.

▶ Electrocardiography may be normal with small VSDs, whereas in large VSDs it may show left and right ventricular hypertrophy, suggestive of pulmonary hypertension.

▶ Echocardiography can detect VSD in the septum, estimate the size of the left-to-right shunt, suggest pulmonary hyperten-

sion, and identify associated lesions and complications.

▶ Cardiac catheterization determines the size and exact location of the VSD and the extent of pulmonary hypertension; it also detects associated defects. It calculates the degree of shunting by comparing the blood oxygen saturation in each ventricle. The oxygen saturation of the right ventricle is greater than normal because oxygenated blood is shunted from the left ventricle to the right.

Treatment

Typically, correction of a VSD may involve:

▶ early surgical correction for a large VSD, usually performed using a patch graft, before heart failure and irreversible pulmonary vascular disease develop

▶ placement of a permanent pacemaker, which may be necessary after VSD repair if complete heart block develops from interference with the bundle of His during surgery

▶ surgical closure of small defects using sutures. They may not be surgically repaired if the patient has normal pulmonary artery pressure and a small shunt

▶ pulmonary artery banding to normalize pressures and flow distal to the band and to prevent pulmonary vascular disease if the child has other defects and will benefit from delaying surgery

▶ digoxin, sodium restriction, and diuretics before surgery to prevent heart failure

▶ prophylactic antibiotics before and after surgery to prevent infective endocarditis.

7

The respiratory system's major function is gas exchange, in which air enters the body on inhalation (inspiration), travels throughout the respiratory passages, exchanging oxygen for carbon dioxide at the tissue level, and carbon dioxide is expelled on exhalation (expiration).

The upper airway — composed of the nose, mouth, pharynx, and larynx — allows airflow into the lungs. This area is responsible for warming, humidifying, and filtering the air, thereby protecting the lower airway from foreign matter.

The lower airway consists of the trachea, mainstem bronchi, secondary bronchi, bronchioles, and terminal bronchioles. These structures are anatomic dead spaces and function only as passageways for moving air into and out of the lung. Distal to each terminal bronchiole is the acinus, which consists of respiratory bronchioles, alveolar ducts, and alveolar sacs. The bronchioles and ducts function as conduits, and the alveoli are the chief units of gas exchange. These final subdivisions of the bronchial tree make up the lobules — the functional units of the lungs. (See *Structure of the lobule,* page 202.)

In addition to warming, humidifying, and filtering inspired air, the lower airway protects the lungs with several defense mechanisms. *Clearance mechanisms* include the cough reflex and mucociliary system. The mucociliary system produces mucus, trapping foreign particles. Foreign matter is then swept to the upper airway for expectoration by specialized fingerlike projections called cilia. A breakdown in the epithelium of the lungs or the mu-

cociliary system can cause the defense mechanisms to malfunction, and pollutants and irritants then enter and inflame the lungs. The lower airway also provides *immunologic protection* and initiates *pulmonary injury responses.*

The external component of respiration (ventilation or breathing) delivers inspired air to the lower respiratory tract and alveoli. Contraction and relaxation of the respiratory muscles moves air into and out of the lungs.

Normal expiration is passive; the inspiratory muscles cease to contract, and the elastic recoil of the lungs and the chest wall causes them to contract again. These actions raise the pressure within the lungs to above atmospheric pressure, moving air from the lungs to the atmosphere.

An adult lung contains an estimated 300 million alveoli; each alveolus is supplied by many capillaries. To reach the capillary lumen, oxygen must cross the alveolar capillary membrane.

The pulmonary alveoli promote gas exchange by diffusion — the passage of gas molecules through respiratory membranes. In diffusion, oxygen passes to the blood, and carbon dioxide, a byproduct of cellular metabolism, passes out of the blood and is channeled away.

Circulating blood delivers oxygen to the cells of the body for metabolism and transports metabolic wastes and carbon dioxide from the tissues back to the lungs. When oxygenated arterial blood reaches tissue capillaries, the oxygen diffuses from the blood into the cells because of an oxygen tension gradient. The amount of oxy-

201

STRUCTURE OF THE LOBULE

Each lobule contains terminal bronchioles and the acinus. The acinus consists of respiratory bronchioles and the alveolar sacs.

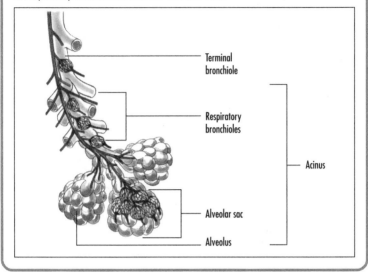

Terminal bronchiole

Respiratory bronchioles

Acinus

Alveolar sac

Alveolus

gen available to cells depends on the concentration of hemoglobin (the principal carrier of oxygen) in the blood; the regional blood flow; the arterial oxygen content; and cardiac output.

Because circulation is continuous, carbon dioxide does not normally accumulate in tissues. Carbon dioxide produced during cellular respiration diffuses from tissues to regional capillaries and is transported by the systemic venous circulation. When carbon dioxide reaches the alveolar capillaries, it diffuses into the alveoli, where the partial pressure of carbon dioxide is lower. Carbon dioxide is removed from the alveoli during exhalation.

For effective gas exchange, ventilation and perfusion at the alveolar level must match closely. (See *Understanding ventilation and perfusion*.) The ratio of ventilation to perfusion is called the $\dot{V}/\dot{Q}$ ratio. A $\dot{V}/\dot{Q}$ mismatch can result from ventilation-perfusion dysfunction or altered lung mechanics.

The amount of air reaching the lungs carrying oxygen depends on lung volume and capacity, compliance, and resistance to airflow. Changes in compliance can occur in either the lung or the chest wall. Destruction of the lung's elastic fibers, which occurs in adult respiratory distress syndrome, decreases lung compliance. The lungs become stiff, making breathing difficult. The alveolar capillary membrane may also be affected, causing hypoxia. Chest wall compliance is affected by disorders causing thoracic deformity, muscle spasm, and abdominal distention.

Respiration is also controlled neurologically by the lateral medulla oblongata of the brain stem. Impulses travel down the phrenic nerves to the diaphragm and then down the intercostal nerves to the intercostal muscles between the ribs. The rate and depth of respiration are controlled similarly.

Apneustic and pneumotaxic centers in the pons of the midbrain influence the pat-

UNDERSTANDING VENTILATION AND PERFUSION

Effective gas exchange depends on the relationship between ventilation and perfusion, expressed as the $\dot{V}/\dot{Q}$ ratio. The diagrams below show what happens when the $\dot{V}/\dot{Q}$ ratio is normal and abnormal.

NORMAL VENTILATION AND PERFUSION

When the $\dot{V}/\dot{Q}$ ratio is matched, un-oxygenated blood from the venous system returns to the right ventricle through the pulmonary artery to the lungs, carrying carbon dioxide. The arteries branch into the alveolar capillaries, where gas exchange occurs.

INADEQUATE PERFUSION (DEAD-SPACE VENTILATION)

When the $\dot{V}/\dot{Q}$ ratio is high, ventilation is normal but alveolar perfusion is reduced or absent (illustrated by the perfusion blockage). This results from a perfusion defect, such as pulmonary embolism or a disorder that decreases cardiac output.

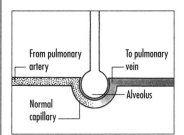

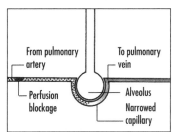

INADEQUATE VENTILATION (SHUNT)

When the $\dot{V}/\dot{Q}$ ratio is low, pulmonary circulation is adequate, but oxygen is inadequate for normal diffusion (illustrated by the ventilation blockage). A portion of the blood flowing through the pulmonary vessels does not become oxygenated.

INADEQUATE VENTILATION AND PERFUSION (SILENT UNIT)

The silent unit indicates an absence of ventilation and perfusion to the lung area (illustrated by blockages in both perfusion and ventilation). The silent unit may try to compensate for this $\dot{V}/\dot{Q}$ imbalance by delivering blood flow to better ventilated lung areas.

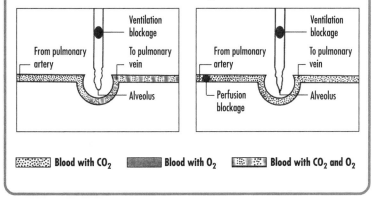

tern of breathing. Stimulation of the lower pontine apneustic center (by trauma, tumor, or cerebrovascular accident) produces forceful inspiratory gasps alternating with weak expiration. This pattern does not occur if the vagi are intact. The apneustic center continually excites the medullary inspiratory center and thus facilitates inspiration. Signals from the pneumotaxic center and afferent impulses from the vagus nerve inhibit the apneustic center and "turn off" inspiration.

In addition, chemoreceptors respond to the hydrogen ion concentration of arterial blood (pH), the partial pressure of arterial carbon dioxide ($Paco_2$), and the partial pressure of arterial oxygen (Pao_2). Central chemoreceptors respond indirectly to arterial blood by sensing changes in the pH of the cerebrospinal fluid (CSF). $Paco_2$ also helps regulate ventilation by impacting the pH of CSF. If $Paco_2$ is high, the respiratory rate increases; if $Paco_2$ is low, the respiratory rate decreases. Information from peripheral chemoreceptors in the carotid and aortic bodies also responds to decreased Pao_2 and decreased pH. Either of these changes results in increased respiratory drive within minutes.

PATHOPHYSIOLOGIC MANIFESTATIONS

Pathophysiologic manifestations of respiratory disease may stem from atelectasis, bronchiectasis, cyanosis, and hypoxemia.

Atelectasis

Atelectasis occurs when the alveolar sacs or entire lung segments expand incompletely, producing a partial or complete lung collapse. This phenomenon removes certain regions of the lung from gas exchange, allowing unoxygenated blood to pass unchanged through these regions and resulting in hypoxia. Atelectasis may be chronic or acute, and often occurs in patients undergoing upper abdominal or thoracic surgery. There are two major caus-

es of collapse due to atelectasis: absorptional atelectasis, secondary to bronchial or bronchiolar obstruction, and compression atelectasis.

Absorption atelectasis

Bronchial occlusion, which prevents air from entering the alveoli distal to the obstruction, can cause absorption atelectasis — the air present in the alveoli is absorbed gradually into the bloodstream, and eventually the alveoli collapse. This may result from intrinsic or extrinsic bronchial obstruction. The most frequent intrinsic cause is retained secretions or exudate forming mucous plugs. Disorders such as cystic fibrosis, chronic bronchitis, or pneumonia increase the risk of absorption atelectasis. Extrinsic bronchial atelectasis usually results from occlusion caused by foreign bodies, bronchogenic carcinoma, and scar tissue.

Impaired production of surfactant can also cause absorption atelectasis. Increasing surface tension of the alveolus due to reduced surfactant leads to collapse.

Compression atelectasis

Compression atelectasis results from external compression, which drives the air out and causes the lung to collapse. This may result from upper abdominal surgical incisions, rib fractures, pleuritic chest pain, tight chest dressings, and obesity (which elevates the diaphragm and reduces tidal volume). These situations inhibit full lung expansion or make deep breathing painful, thus resulting in this disorder.

Bronchiectasis

Bronchiectasis is marked by chronic abnormal dilation of the bronchi and destruction of the bronchial walls, and can occur throughout the tracheobronchial tree. It may also be confined to a single segment or lobe. This disorder is usually bilateral in nature and involves the basilar segments of the lower lobes.

There are three forms of bronchiectasis: cylindrical, fusiform (varicose), and saccular (cystic). (See *Forms of bronchiectasis.*) It results from conditions associated with repeated damage to bronchial walls with abnormal mucociliary clearance, which causes a breakdown of supporting tissue adjacent to the airways. (See *Causes of bronchiectasis,* page 206.)

In patients with bronchiectasis, sputum stagnates in the dilated bronchi and leads to secondary infection, characterized by inflammation and leukocytic accumulations. Additional debris collects within and occludes the bronchi. Increasing pressure from the retained secretions induces mucosal injury.

Cyanosis

Cyanosis is a bluish discoloration of the skin and mucous membranes. In most populations, it is readily detectable by a visible blue tinge on the nail beds and the lips. Central cyanosis indicates a decreased oxygen saturation of the hemoglobin in arterial blood, which is best observed in the buccal mucous membranes and the lips. Peripheral cyanosis is a slowed blood circulation of the fingers and toes that is best visualized by examining the nail bed area.

 CULTURAL DIVERSITY In patients with black or dark complexions, cyanosis may not be evident in the lip area or the nail beds. A better indicator in these individuals is to assess the membranes of the oral mucosa (buccal mucous membranes) and of the conjunctivae of the eyes.

Cyanosis is caused by desaturation with oxygen or reduced hemoglobin amounts. It develops when 5 g of hemoglobin is desaturated, even if hemoglobin counts are adequate or reduced. Conditions that result in cyanosis include decreased arterial oxygenation (indicated by low Pao_2), pulmonary or cardiac right-to-left shunts, decreased cardiac output, anxiety, and a cold environment.

An individual who is not cyanotic does not necessarily have adequate oxygenation. Inadequate oxygenation of the tis-

---

FORMS OF BRONCHIECTASIS

The three types of bronchiectasis are cylindrical, fusiform (or varicose), and saccular. In cylindrical bronchiectasis, bronchioles are usually symmetrically dilated, whereas in fusiform bronchiectasis, bronchioles are deformed. In saccular bronchiectasis, large bronchi become enlarged and balloonlike.

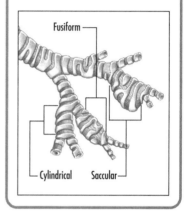

---

sues occurs in severe anemia, resulting in inadequate hemoglobin concentration. It also occurs in carbon monoxide poisoning, in which hemoglobin binds to carbon monoxide instead of to oxygen. Although assessment of these patients does not reveal cyanosis, oxygenation is inadequate.

Others may appear cyanotic even though oxygenation is adequate — as in polycythemia, an abnormal increase in red blood cell count. Because the hemoglobin count is increased and oxygenation occurs at a normal rate, the patient may still present with cyanosis.

Cyanosis as a presenting condition must be interpreted in relation to the patient's underlying pathophysiology. Diagnosis of inadequate oxygenation may be confirmed by analyzing arterial blood gases and obtaining Pao_2 measurements.

CAUSES OF BRONCHIECTASIS

Bronchiectasis results from conditions associated with repeated damage to bronchial walls and with abnormal mucociliary clearance, leading to a breakdown in the supporting tissue adjacent to the airways. Such conditions include:

▶ cystic fibrosis
▶ immune disorders (agammaglobulinemia)
▶ recurrent bacterial respiratory tract infections that were inadequately treated (tuberculosis)
▶ complications of measles, pneumonia, pertussis, or influenza
▶ obstruction (from a foreign body, tumor, or stenosis) with recurrent infection
▶ inhalation of corrosive gas or repeated aspiration of gastric juices
▶ congenital anomalies, such as bronchomalacia, congenital bronchiectasis, and Kartagener's syndrome (bronchiectasis, sinusitis, and dextrocardia)
▶ rare disorders such as immotile cilia syndrome.

Hypoxemia

Hypoxemia is reduced oxygenation of the arterial blood, evidenced by reduced Pao_2 of arterial blood gases. It is caused by respiratory alterations, whereas hypoxia is a diminished oxygenation of tissues at the cellular level that may be caused by conditions affecting other body systems that are unrelated to alterations of pulmonary functioning. Low cardiac output or cyanide poisoning can result in hypoxia, in addition to alterations of respiration. Hypoxia can occur anywhere in the body. If hypoxia occurs in the blood, it is termed hypoxemia. Hypoxemia can lead to tissue hypoxia.

Hypoxemia can be caused by decreased oxygen content (Po_2) of inspired gas, hypoventilation, diffusion abnormalities, abnormal $\dot{V}/\dot{Q}$ ratios, and pulmonary right-to-left shunts. The physiologic mechanism for each cause of hypoxemia is variable. (See *Major causes of hypoxemia*.)

DISORDERS

Respiratory disorders can be acute or chronic. The following disorders include examples from each.

Adult respiratory distress syndrome

Adult respiratory distress syndrome (ARDS) is a form of pulmonary edema that can quickly lead to acute respiratory failure. Also known as shock lung, stiff lung, white lung, wet lung, or Da Nang lung, ARDS may follow direct or indirect injury to the lung. However, its diagnosis is difficult, and death can occur within 48 hours of onset if not promptly diagnosed and treated. A differential diagnosis needs to rule out cardiogenic pulmonary edema, pulmonary vasculitis, and diffuse pulmonary hemorrhage.

Causes

Common causes of ARDS include:
▶ injury to the lung from trauma (most common cause) such as airway contusion
▶ trauma-related factors, such as fat emboli, sepsis, shock, pulmonary contusions, and multiple transfusions, which increase the likelihood that microemboli will develop
▶ anaphylaxis
▶ aspiration of gastric contents
▶ diffuse pneumonia, especially viral pneumonia
▶ drug overdose, such as heroin, aspirin, or ethchlorvynol
▶ idiosyncratic drug reaction to ampicillin or hydrochlorothiazide
▶ inhalation of noxious gases, such as nitrous oxide, ammonia, or chlorine
▶ near drowning

MAJOR CAUSES OF HYPOXEMIA

The chart below lists the major causes of hypoxemia and contributing factors.

MAJOR CAUSE	CONTRIBUTING FACTORS
Decrease in inspired oxygen	High altitudes, inhaling poorly oxygenated gases, or breathing in an enclosed space
Hypoventilation	Respiratory center inappropriately stimulated (such as by oversedation, overdosage, or neurologic damage), chronic obstructive pulmonary disease
Alveolar capillary diffusion abnormality	Emphysema, conditions resulting in fibrosis, or pulmonary edema
Ventilation-perfusion ($\dot{V}/\dot{Q}$) mismatch	Asthma, chronic bronchitis, or pneumonia
Shunting	Adult respiratory distress syndrome, idiopathic respiratory distress syndrome of the newborn, or atelectasis

▶ oxygen toxicity
▶ sepsis
▶ coronary artery bypass grafting
▶ hemodialysis
▶ leukemia
▶ acute miliary tuberculosis
▶ pancreatitis
▶ thrombotic thrombocytopenic purpura
▶ uremia
▶ venous air embolism.

Pathophysiology

Injury in ARDS involves both the alveolar epithelium and the pulmonary capillary epithelium. A cascade of cellular and biochemical changes is triggered by the specific causative agent. Once initiated, this injury triggers neutrophils, macrophages, monocytes, and lymphocytes to produce various cytokines. The cytokines promote cellular activation, chemotaxis, and adhesion. The activated cells produce inflammatory mediators, including oxidants, proteases, kinins, growth factors, and neuropeptides, which initiate the complement cascade, intravascular coagulation, and fibrinolysis.

These cellular triggers result in increased vascular permeability to proteins, affecting the hydrostatic pressure gradient of the capillary. Elevated capillary pressure, such as results from insults of fluid overload or cardiac dysfunction in sepsis, greatly increases interstitial and alveolar edema, which is evident in dependent lung areas and can be visualized as whitened areas on chest X-rays. Alveolar closing pressure then exceeds pulmonary pressures, and alveolar closure and collapse begin.

In ARDS, fluid accumulation in the lung interstitium, the alveolar spaces, and the small airways causes the lungs to stiffen, thus impairing ventilation and reducing oxygenation of the pulmonary capillary blood. The resulting injury reduces normal blood flow to the lungs. Damage can occur directly —by aspiration of gastric contents and inhalation of noxious gases —or indirectly —from chemical mediators released in response to systemic disease.

Platelets begin to aggregate and release substances, such as serotonin, bradykinin, and histamine, which attract and activate neutrophils. These substances inflame and damage the alveolar membrane and later increase capillary permeability. In the early stages of ARDS, signs and symptoms may be undetectable.

Additional chemotactic factors released include endotoxins (such as those present in septic states), tumor necrosis factor, and interleukin-1 (IL-1). The activated neutrophils release several inflammatory mediators and platelet aggravating factors that damage the alveolar capillary membrane and increase capillary permeability.

Histamines and other inflammatory substances increase capillary permeability, allowing fluids to move into the interstitial space. Consequently, the patient experiences tachypnea, dyspnea, and tachycardia. As capillary permeability increases, proteins, blood cells, and more fluid leak out, increasing interstitial osmotic pressure and causing pulmonary edema. Tachycardia, dyspnea, and cyanosis may occur. Hypoxia (usually unresponsive to increasing fraction of inspired oxygen [FIO_2]), decreased pulmonary compliance, crackles, and rhonchi develop. The resulting pulmonary edema and hemorrhage significantly reduce lung compliance and impair alveolar ventilation.

The fluid in the alveoli and decreased blood flow damage surfactant in the alveoli. This reduces the ability of alveolar cells to produce more surfactant. Without surfactant, alveoli and bronchioles fill with fluid or collapse, gas exchange is impaired, and the lungs are much less compliant. Ventilation of the alveoli is further decreased. The burden of ventilation and gas exchange shifts to uninvolved areas of the lung, and pulmonary blood flow is shunted from right to left. The work of breathing is increased, and the patient may develop thick frothy sputum and marked hypoxemia with increasing respiratory distress.

Mediators released by neutrophils and macrophages also cause varying degrees of pulmonary vasoconstriction, resulting in pulmonary hypertension. The result of these changes is a $\dot{V}/\dot{Q}$ mismatch. Although the patient responds with an increased respiratory rate, sufficient oxygen cannot cross the alveolar capillary membrane. Carbon dioxide continues to cross easily and is lost with every exhalation. As both oxygen and carbon dioxide levels in the blood decrease, the patient develops increasing tachypnea, hypoxemia, and hypocapnia (low $PaCO_2$).

Pulmonary edema worsens and hyaline membranes form. Inflammation leads to fibrosis, which further impedes gas exchange. Fibrosis progressively obliterates alveoli, respiratory bronchioles, and the interstitium. Functional residual capacity decreases and shunting becomes more serious. Hypoxemia leads to metabolic acidosis. At this stage, the patient develops increasing $PaCO_2$, decreasing pH and PaO_2, decreasing bicarbonate levels, and mental confusion. (See *Looking at adult respiratory distress syndrome.*)

The end result is respiratory failure. Systemically, neutrophils and inflammatory mediators cause generalized endothelial damage and increased capillary permeability throughout the body. Multisystem organ dysfunction syndrome (MODS) occurs as the cascade of mediators affects each system. Death may occur from the influence of both ARDS and MODS.

LOOKING AT ADULT RESPIRATORY DISTRESS SYNDROME

These diagrams show the process and progress of ARDS.

In phase 1 of this syndrome, injury reduces normal blood flow to the lungs. Platelets aggregate and release histamine (H), serotonin (S), and bradykinin (B).

In phase 4, decreased blood flow and fluids in the alveoli damage surfactant and impair the cell's ability to produce more. The alveoli then collapse, thus impairing gas exchange.

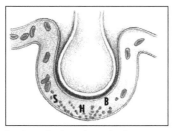

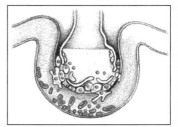

In phase 2, the released substances inflame and damage the alveolar capillary membrane, increasing capillary permeability. Fluids then shift into the interstitial space.

In phase 5, oxygenation is impaired, but carbon dioxide easily crosses the alveolar capillary membrane and is expired. Blood oxygen and carbon dioxide levels are low.

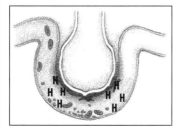

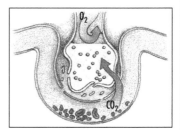

In phase 3, capillary permeability increases and proteins and fluids leak out, increasing interstitial osmotic pressure and causing pulmonary edema.

In phase 6, pulmonary edema worsens and inflammation leads to fibrosis. Gas exchange is further impeded.

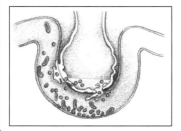

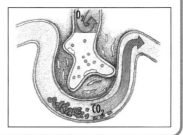

Signs and symptoms

The following signs and symptoms may occur:

▶ rapid, shallow breathing and dyspnea, which occur hours to days after the initial injury in response to decreasing oxygen levels in the blood

▶ increased rate of ventilation due to hypoxemia and its effects on the pneumotaxic center

▶ intercostal and suprasternal retractions due to the increased effort required to expand the stiff lung

▶ crackles and rhonchi, which are audible and result from fluid accumulation in the lungs

▶ restlessness, apprehension, and mental sluggishness, which occur as the result of hypoxic brain cells

▶ motor dysfunction, which occurs as hypoxia progresses

▶ tachycardia, which signals the heart's effort to deliver more oxygen to the cells and vital organs

▶ respiratory acidosis, which occurs as carbon dioxide accumulates in the blood and oxygen levels decrease

▶ metabolic acidosis, which eventually results from failure of compensatory mechanisms.

Complications

Possible complications of ARDS include:

▶ hypotension
▶ decreased urine output
▶ metabolic acidosis
▶ respiratory acidosis
▶ MODS
▶ ventricular fibrillation
▶ ventricular standstill.

Diagnosis

The following tests help diagnose ARDS:

▶ Arterial blood gas (ABG) analysis with the patient breathing room air initially reveals a reduced Pao_2 (less than 60 mm Hg) and a decreased $Paco_2$ (less than 35 mm Hg). Hypoxemia, despite increased supplemental oxygen, is the hallmark of ARDS; the resulting blood pH reflects respiratory alkalosis. As ARDS worsens, ABG values show respiratory acidosis evident by an increasing $Paco_2$ (over 45 mm Hg), metabolic acidosis evident by a decreasing bicarbonate (HCO_3 less than 22 mEq/L), and a declining Pao_2 despite oxygen therapy.

▶ Pulmonary artery catheterization identifies the cause of edema by measuring pulmonary capillary wedge pressure (PCWP of 12 mm Hg or less in ARDS).

▶ Pulmonary artery mixed venous blood indicates hypoxemia.

▶ Serial chest X-rays in early stages show bilateral infiltrates; in later stages, lung fields with a ground-glass appearance and "whiteouts" of both lung fields (with irreversible hypoxemia) may be observed.

▶ Sputum analysis, including Gram stain and culture and sensitivity, identifies causative organisms.

▶ Blood cultures identify infectious organisms.

▶ Toxicology testing screens for drug ingestion.

▶ Serum amylase rules out pancreatitis.

Treatment

Therapy is focused on correcting the causes of ARDS and preventing progression of hypoxemia and respiratory acidosis; it may involve:

▶ mechanical ventilation and intubation to increase lung volume, open airways, and improve oxygenation

▶ positive end-expiratory pressure (may be added to increase lung volume and open alveoli)

▶ pressure-controlled inverse ratio ventilation to reverse the conventional inspiration-to-expiration ratio and minimize the risk of barotrauma (mechanical breaths are pressure-limited to prevent increased damage to the alveoli)

▶ permissive hypercapnia to limit peak inspiratory pressure (although carbon dioxide removal is compromised, treatment is not given for subsequent changes in blood hydrogen and oxygen concentration)

▶ sedatives, narcotics, or neuromuscular blockers such as vecuronium bromide, which may be given during mechanical ventilation to minimize restlessness, oxygen consumption, and carbon dioxide production and to facilitate ventilation
▶ high-dose corticosteroids (may be given when ARDS is due to fatty emboli, to optimize cellular membranes)
▶ sodium bicarbonate, which may reverse severe metabolic acidosis
▶ intravenous fluid administration to maintain blood pressure by treating hypovolemia
▶ vasopressors to maintain blood pressure
▶ antimicrobial drugs to treat nonviral infections
▶ diuretics to reduce interstitial and pulmonary edema
▶ correction of electrolyte and acid-base imbalances to maintain cellular integrity, particularly the sodium-potassium pump
▶ fluid restriction to prevent increase of interstitial and alveolar edema.

Asbestosis

Considered a form of pneumoconiosis, asbestosis is characterized by diffuse interstitial pulmonary fibrosis. Prolonged exposure to airborne particles causes pleural plaques and tumors of the pleura and peritoneum. Asbestosis may develop 15 to 20 years after regular exposure to asbestos has ended. It is a potent co-carcinogen and increases the smoker's risk for lung cancer. An asbestos worker who smokes is 90 times more likely to develop lung cancer than a smoker who has never worked with asbestos.

Causes

Common causes of this disorder include:
▶ prolonged inhalation of asbestos fibers; people at high risk include workers in the mining, milling, construction, fireproofing, and textile industries
▶ asbestos used in paints, plastics, and brake and clutch linings

▶ family members of asbestos workers who may be exposed to stray fibers from the worker's clothing
▶ exposure to fibrous asbestos dust in deteriorating buildings or in waste piles from asbestos plants.

Pathophysiology

Asbestosis occurs when lung spaces become filled with asbestos fibers. The inhaled asbestos fibers (about 50 μ $\times$ 0.5 μ in size) travel down the airway and penetrate respiratory bronchioles and alveolar walls. Coughing attempts to expel the foreign matter. Mucus production and goblet cells are stimulated to protect the airway from the debris and aid in expectoration. Fibers then become encased in a brown, iron-rich proteinlike sheath in sputum or lung tissue, called asbestosis bodies. Chronic irritation by the fibers continues to affect the lower bronchioles and alveoli. The foreign material and inflammation swell airways, and fibrosis develops in response to the chronic irritation. Interstitial fibrosis may develop in lower lung zones, affecting lung parenchyma and the pleurae. Raised hyaline plaques may form in the parietal pleura, the diaphragm, and the pleura adjacent to the pericardium. Hypoxia develops as more alveoli and lower airways are affected.

Signs and symptoms

The following signs and symptoms may occur:
▶ dyspnea on exertion
▶ dyspnea at rest with extensive fibrosis
▶ severe, nonproductive cough in nonsmokers
▶ productive cough in smokers
▶ clubbed fingers due to chronic hypoxia
▶ chest pain (often pleuritic) due to pleural irritation
▶ recurrent respiratory tract infections as pulmonary defense mechanisms begin to fail
▶ pleural friction rub due to fibrosis
▶ crackles on auscultation attributed moving through thickened sputu

▶ decreased lung inflation due to lung stiffness
▶ recurrent pleural effusions due to fibrosis
▶ decreased forced expiratory volume due to diminished alveoli
▶ decreased vital capacity due to fibrotic changes.

Complications

Possible complications of asbestosis include:
▶ pulmonary fibrosis due to progression of asbestosis
▶ respiratory failure
▶ pulmonary hypertension
▶ cor pulmonale.

Diagnosis

▶ Chest X-rays may show fine, irregular, linear, and diffuse infiltrates. Extensive fibrosis is revealed by a honeycomb or ground-glass appearance. Chest X-rays may also show pleural thickening and calcification, bilateral obliteration of the costophrenic angles and, in later stages, an enlarged heart with a classic "shaggy" border.
▶ Pulmonary function studies may identify decreased vital capacity, forced vital capacity (FVC), and total lung capacity; decreased or normal forced expiratory volume in 1 second (FEV_1); a normal ratio, or FEV_1 to FVC; and reduced diffusing capacity for carbon monoxide when fibrosis destroys alveolar walls and thickens the alveolar capillary membrane.
▶ Arterial blood gas analysis may reveal decreased Pao_2 and $Paco_2$ from hyperventilation.

Treatment

Asbestosis can't be cured. The goal of treatment is to relieve symptoms and control complications; it may involve:
▶ chest physiotherapy (such as controlled coughing and postural drainage with chest percussion and vibration) to help relieve respiratory signs and symptoms and manage hypoxia and cor pulmonale

▶ aerosol therapy to liquefy mucus
▶ inhaled mucolytics to liquefy and mobilize secretions
▶ increased fluid intake to 3 L daily
▶ antibiotics to treat respiratory tract infections
▶ oxygen administration to relieve hypoxia
▶ diuretics to decrease inflammation and edema
▶ digoxin to enhance cardiac output
▶ salt restriction to prevent fluid retention and thickened secretions.

Asthma

Asthma is a chronic reactive airway disorder causing episodic airway obstruction that results from bronchospasms, increased mucus secretion, and mucosal edema. It is a type of chronic obstructive pulmonary disease (COPD), a long-term pulmonary disease characterized by increased airflow resistance; other types of COPD include chronic bronchitis and emphysema.

Although asthma strikes at any age, about 50% of patients are younger than age 10; twice as many boys as girls are affected in this age group. One-third of patients develops asthma between ages 10 and 30, and the incidence is the same in both sexes in this age group. Moreover, approximately one-third of all patients share the disease with at least one immediate family member.

Asthma may result from sensitivity to extrinsic or intrinsic allergens. Extrinsic, or atopic, asthma begins in childhood; typically, patients are sensitive to specific external allergens.

AGE ALERT Extrinsic asthma is commonly accompanied by other hereditary allergies, such as eczema and allergic rhinitis, in childhood populations.

Intrinsic, or nonatopic, asthmatics react to internal, nonallergenic factors; external substances cannot be implicated in patients with intrinsic asthma. Most episodes occur after a severe respiratory tract infection, especially in adults. However, many asthmatics, especially children, have both intrinsic and extrinsic asthma.

CULTURAL DIVERSITY A significant number of adults acquire an allergic form of asthma or exacerbation of existing asthma from exposure to agents in the workplace. Irritants such as chemicals in flour, acid anhydrides, toluene di-isocyanates, screw flies, river flies, and excreta of dust mites in carpet have been identified as agents that trigger asthma.

Causes
Extrinsic allergens include:
▶ pollen
▶ animal dander
▶ house dust or mold
▶ kapok or feather pillows
▶ food additives containing sulfites
▶ other sensitizing substances.
  Intrinsic allergens include:
▶ irritants
▶ emotional stress
▶ fatigue
▶ endocrine changes
▶ temperature variations
▶ humidity variations
▶ exposure to noxious fumes
▶ anxiety
▶ coughing or laughing
▶ genetic factors (see below).

Pathophysiology
There are two genetic influences identified with asthma, namely the ability of an individual to develop asthma (atopy) and the tendency to develop hyperresponsiveness of the airways independent of atopy. A locus of chromosome 11 associated with atopy contains an abnormal gene that encodes a part of the immunoglobulin E (IgE) receptor. Environmental factors interact with inherited factors to cause asthmatic reactions with associated bronchospasms.

In asthma, bronchial linings overreact to various stimuli, causing episodic smooth muscle spasms that severely constrict the airways. (See *Pathophysiology of asthma,* page 214.) IgE antibodies, attached to histamine-containing mast cells and receptors on cell membranes, initiate intrinsic asthma attacks. When exposed to an antigen such as pollen, the IgE antibody combines with the antigen.

On subsequent exposure to the antigen, mast cells degranulate and release mediators. Mast cells in the lung interstitium are stimulated to release both histamine and the slow-reacting substance of anaphylaxis. Histamine attaches to receptor sites in the larger bronchi, where it causes swelling in smooth muscles. Mucous membranes become inflamed, irritated, and swollen. The patient may experience dyspnea, prolonged expiration, and an increased respiratory rate.

The slow-reacting substance of anaphylaxis (SRS-A) attaches to receptor sites in the smaller bronchi and causes local swelling of the smooth muscle. SRS-A also causes prostaglandins to travel via the bloodstream to the lungs, where they enhance the effect of histamine. A wheeze may be audible during coughing; the higher the pitch, the narrower is the bronchial lumen. Histamine stimulates the mucous membranes to secrete excessive mucus, further narrowing the bronchial lumen. Goblet cells secrete viscous mucus that is difficult to cough up, resulting in coughing, rhonchi, increased-pitch wheezing, and increased respiratory distress. Mucosal edema and thickened secretions further block the airways. (See *Looking at a bronchiole in asthma,* page 215.)

On inhalation, the narrowed bronchial lumen can still expand slightly, allowing air to reach the alveoli. On exhalation, increased intrathoracic pressure closes the bronchial lumen completely. Air enters but cannot escape. The patient develops a barrel chest and hyperresonance to percussion.

Mucus fills the lung bases, inhibiting alveolar ventilation. Blood is shunted to alveoli in other lung parts, but still can't compensate for diminished ventilation.

PATHOPHYSIOLOGY OF ASTHMA

In asthma, hyperresponsiveness of the airways and bronchospasms occur. These illustrations show the progression of an asthma attack.

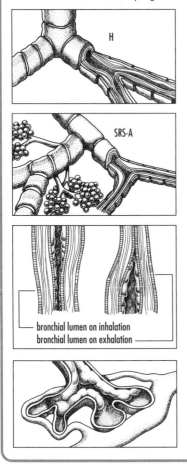

bronchial lumen on inhalation
bronchial lumen on exhalation

▶ Histamine (H) attaches to receptor sites in larger bronchi, causing swelling of the smooth muscles.

▶ Slow-reacting substance of anaphylaxis (SRS-A) attaches to receptor sites in the smaller bronchi and causes swelling of smooth muscle there. SRS-A also causes prostaglandins to travel via the bloodstream to the lungs, where they enhance histamine's effects.

▶ Histamine stimulates the mucous membranes to secrete excessive mucus, further narrowing the bronchial lumen. On inhalation, the narrowed bronchial lumen can still expand slightly; however, on exhalation, the increased intrathoracic pressure closes the bronchial lumen completely.

▶ Mucus fills lung bases, inhibiting alveolar ventilation. Blood is shunted to alveoli in other parts of the lungs, but it still can't compensate for diminished ventilation.

Hyperventilation is triggered by lung receptors to increase lung volume because of trapped air and obstructions. Intrapleural and alveolar gas pressures rise, causing a decreased perfusion of alveoli. Increased alveolar gas pressure, decreased ventilation, and decreased perfusion result in uneven $\dot{V}/\dot{Q}$ ratios and mismatching within different lung segments.

Hypoxia triggers hyperventilation by stimulation of the respiratory center, which in turn decreases $Paco_2$ and increases pH, resulting in a respiratory alkalosis. As the obstruction to the airways increases in severity, more alveoli are affected. Ventilation and perfusion remain inadequate, and CO_2 retention develops. Respiratory

LOOKING AT A BRONCHIOLE IN ASTHMA

Asthma is characterized by bronchospasms, increased mucus secretion, and mucosal edema, which contribute to airway narrowing and obstruction. Shown below is a normal bronchiole in cross section and an obstructed bronchiole, as it occurs in asthma.

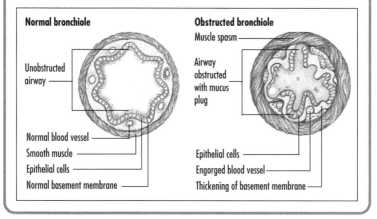

Normal bronchiole

Unobstructed airway

Normal blood vessel
Smooth muscle
Epithelial cells
Normal basement membrane

Obstructed bronchiole

Muscle spasm

Airway obstructed with mucus plug

Epithelial cells
Engorged blood vessel
Thickening of basement membrane

acidosis results and respiratory failure occurs.

If status asthmaticus occurs, hypoxia worsens and expiratory flows and volumes decrease even further. If treatment is not initiated, the patient begins to tire out. (See *Averting an asthma attack,* page 216.) Acidosis develops as arterial carbon dioxide increases. The situation becomes life-threatening as no air becomes audible upon auscultation (a silent chest) and $Paco_2$ rises to over 70 mm Hg.

Signs and symptoms

Patients with *mild asthma* have adequate air exchange and are asymptomatic between attacks; they may have the following signs and symptoms:

▶ wheezing due to edema of the airways
▶ coughing due to stimulation of the cough reflex to eliminate the lungs of excess mucus and irritants
▶ histamine-induced production of thick, clear, or yellow mucus
▶ dyspnea on exertion due to narrowing of airways and inability to take in the increased oxygen that is required for exercise.

Patients with *moderate asthma* have normal or below normal air exchange and may exhibit:

▶ respiratory distress at rest due to narrowed airways and decreased oxygenation to the tissues
▶ hyperpnea (abnormal increase in the depth and rate of respiration) due to the body's attempt to take in more oxygen
▶ barrel chest due to air trapping and retention
▶ diminished breath sounds due to air trapping.

Patients with *severe asthma* have continuous signs and symptoms that include:

▶ marked respiratory distress due to failure of compensatory mechanisms and decreased oxygenation levels
▶ marked wheezing due to increased edema and increased mucus in the lower airways
▶ absent breath sounds due to severe bronchoconstriction and edema

AVERTING AN ASTHMA ATTACK

The following flow chart shows pathophysiologic changes that occur with asthma. Treatments and interventions show where the physiologic cascade would be altered to stop an asthma attack.

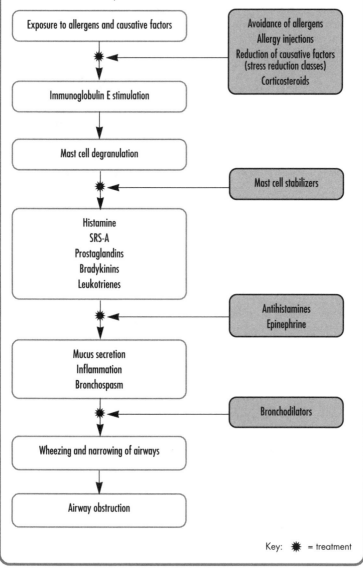

Exposure to allergens and causative factors

Avoidance of allergens
Allergy injections
Reduction of causative factors
(stress reduction classes)
Corticosteroids

Immunoglobulin E stimulation

Mast cell degranulation

Mast cell stabilizers

Histamine
SRS-A
Prostaglandins
Bradykinins
Leukotrienes

Antihistamines
Epinephrine

Mucus secretion
Inflammation
Bronchospasm

Bronchodilators

Wheezing and narrowing of airways

Airway obstruction

Key: ✳ = treatment

▶ pulsus paradoxus greater than 10 mm Hg
▶ chest wall contractions due to use of accessory muscles.

Complications
Possible complications include:
▶ status asthmaticus
▶ respiratory failure.

Diagnosis
The following tests help diagnose asthma:
▶ Pulmonary function studies reveal signs of airway obstructive disease, low-normal or decreased vital capacity, and increased total lung and residual capacities. Pulmonary function may be normal between attacks. Pao_2 and $Paco_2$ usually are decreased, except in severe asthma, when $Paco_2$ may be normal or increased, indicating severe bronchial obstruction.
▶ Serum IgE levels may increase from an allergic reaction.
▶ Sputum analysis may indicate presence of Curschmann's spirals (casts of airways), Charcot-Leyden crystals, and eosinophils.
▶ Complete blood count with differential reveals increased eosinophil count.
▶ Chest X-rays can be used to diagnose or monitor the progress of asthma and may show hyperinflation with areas of atelectasis.
▶ Arterial blood gas analysis detects hypoxemia (decreased Pao_2; decreased, normal, or increasing $Paco_2$) and guides treatment.
▶ Skin testing may identify specific allergens; results read in 1 or 2 days detect an early reaction, and after 4 or 5 days reveal a late reaction.
▶ Bronchial challenge testing evaluates the clinical significance of allergens identified by skin testing.
▶ Electrocardiography shows sinus tachycardia during an attack; severe attack may show signs of cor pulmonale (right axis deviation, peaked P wave) that resolve after the attack.

Treatment
Correcting asthma typically involves:
▶ prevention, by identifying and avoiding precipitating factors such as environmental allergens or irritants, which is the best treatment
▶ desensitization to specific antigens — helpful if the stimuli can't be removed entirely — which decreases the severity of attacks of asthma with future exposure
▶ bronchodilators (such as theophylline, aminophylline, epinephrine, albuterol, metaproterenol, and terbutaline) to decrease bronchoconstriction, reduce bronchial airway edema, and increase pulmonary ventilation
▶ corticosteroids (such as hydrocortisone and methylprednisolone) to decrease bronchoconstriction, reduce bronchial airway edema, and increase pulmonary ventilation
▶ subcutaneous epinephrine to counteract the effects of mediators of an asthma attack
▶ mast cell stabilizers (cromolyn sodium and nedocromil sodium), effective in patients with atopic asthma who have seasonal disease. When given prophylactically, they block the acute obstructive effects of antigen exposure by inhibiting the degranulation of mast cells, thereby preventing the release of chemical mediators responsible for anaphylaxis
▶ low-flow humidified oxygen, which may be needed to treat dyspnea, cyanosis, and hypoxemia. However, the amount delivered should maintain Pao_2 between 65 and 85 mm Hg, as determined by arterial blood gas analysis
▶ mechanical ventilation — necessary if the patient doesn't respond to initial ventilatory support and drugs, or develops respiratory failure
▶ relaxation exercises such as yoga to help increase circulation and to help a patient recover from an asthma attack.

Chronic bronchitis
Chronic bronchitis is inflammation of the bronchi caused by irritants or infection. A

CHANGES IN CHRONIC BRONCHITIS

In chronic bronchitis, irritants inflame the tracheobronchial tree over time, leading to increased mucus production and a narrowed or blocked airway. As the inflammation continues, goblet and epithelial cells hypertrophy. Because the natural defense mechanisms are blocked, the airways accumulate debris in the respiratory tract. Shown below is a cross section of these changes.

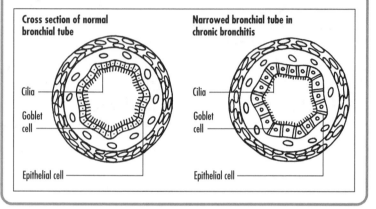

Cross section of normal bronchial tube

Cilia

Goblet cell

Epithelial cell

Narrowed bronchial tube in chronic bronchitis

Cilia

Goblet cell

Epithelial cell

form of chronic obstructive pulmonary disease (COPD), bronchitis may be classified as acute or chronic. In chronic bronchitis, hypersecretion of mucus and chronic productive cough last for 3 months of the year and occur for at least 2 consecutive years. The distinguishing characteristic of bronchitis is obstruction of airflow.

CULTURAL DIVERSITY COPD is more prevalent in an urban versus rural environment, and is also related to occupational factors (mineral or organic dusts).

AGE ALERT Children of parents who smoke are at higher risk for respiratory tract infection that can lead to chronic bronchitis.

Causes

Common causes of chronic bronchitis include:

▶ exposure to irritants
▶ cigarette smoking
▶ genetic predisposition
▶ exposure to organic or inorganic dusts
▶ exposure to noxious gases
▶ respiratory tract infection.

Pathophysiology

Chronic bronchitis occurs when irritants are inhaled for a prolonged time. The irritants inflame the tracheobronchial tree, leading to increased mucus production and a narrowed or blocked airway. As the inflammation continues, changes in the cells lining the respiratory tract result in resistance of the small airways and severe $\dot{V}/\dot{Q}$ imbalance, which decreases arterial oxygenation.

Chronic bronchitis results in hypertrophy, hyperplasia of the mucous glands, increased goblet cells, ciliary damage, squamous metaplasia of the columnar epithelium, and chronic leukocytic and lymphocytic infiltration of bronchial walls. (See *Changes in chronic bronchitis.*) Hypersecretion of the goblet cells blocks the free movement of the cilia, which normally sweep dust, irritants, and mucus away from the airways. With mucus and debris accumulating in the airway, the defenses are altered, and the individual is prone to respiratory tract infections.

Additional effects include widespread inflammation, airway narrowing, and mucus within the airways. Bronchial walls become inflamed and thickened from edema and accumulation of inflammatory cells, and the effects of smooth muscle bronchospasm further narrow the lumen. Initially, only large bronchi are involved but eventually, all airways are affected. Airways become obstructed and closure occurs, especially on expiration. The gas is then trapped in the distal portion of the lung. Hypoventilation occurs, leading to a $\dot{V}/\dot{Q}$ mismatch and resultant hypoxemia.

Hypoxemia and hypercapnia occur secondary to hypoventilation. Pulmonary vascular resistance (PVR) increases as inflammatory and compensatory vasoconstriction in hypoventilated areas narrows the pulmonary arteries. Increased PVR leads to increased afterload of the right ventricle. With repeated inflammatory episodes, scarring of the airways occurs and permanent structural changes develop. Respiratory infections can trigger acute exacerbations, and respiratory failure can occur.

Patients with chronic bronchitis have a diminished respiratory drive. The resulting chronic hypoxia causes the kidneys to produce erythropoietin, which stimulates excessive red blood cell production and leads to polycythemia. Although hemoglobin levels are high, the amount of reduced (not fully oxygenated) hemoglobin that is in contact with oxygen is low; therefore, cyanosis occurs.

Signs and symptoms
The following signs and symptoms may occur:
▶ copious gray, white, or yellow sputum due to hypersecretion of goblet cells
▶ productive cough to expectorate mucus that is produced by the lungs
▶ dyspnea due to obstruction of airflow to the lower tracheobronchial tree

▶ cyanosis related to diminished oxygenation and cellular hypoxia; reduced oxygen is supplied to the tissues
▶ use of accessory muscles for breathing due to compensated attempts to supply the cells with increased oxygen
▶ tachypnea due to hypoxia
▶ pedal edema due to right-sided heart failure
▶ neck vein distention due to right-sided heart failure
▶ weight gain due to edema
▶ wheezing due to air moving through narrowed respiratory passages
▶ prolonged expiratory time due to the body's attempt to keep airways patent
▶ rhonchi due to air moving through narrow, mucus-filled passages
▶ pulmonary hypertension caused by involvement of small pulmonary arteries, due to inflammation in the bronchial walls and spasms of pulmonary blood vessels from hypoxia.

Complications
Possible complications of this disorder include:
▶ cor pulmonale (right ventricular hypertrophy with right-sided heart failure) due to increased right ventricular end-diastolic pressure
▶ pulmonary hypertension
▶ heart failure, resulting in increased venous pressure, liver engorgement, and dependent edema
▶ acute respiratory failure.

Diagnosis
The following tests help diagnose chronic bronchitis:
▶ Chest X-rays may show hyperinflation and increased bronchovascular markings.
▶ Pulmonary function studies indicate increased residual volume, decreased vital capacity and forced expiratory flow, and normal static compliance and diffusing capacity.
▶ Arterial blood gas analysis reveals decreased Pao_2 and normal or increased $Paco_2$.

▶ Sputum analysis may reveal many microorganisms and neutrophils.

▶ Electrocardiography may show atrial arrhythmias; peaked P waves in leads II, III, and aV_F; and occasionally, right ventricular hypertrophy.

Treatment

Correcting chronic bronchitis typically involves:

▶ avoidance of air pollutants (most effective)

▶ smoking cessation

▶ antibiotics to treat recurring infections

▶ bronchodilators to relieve bronchospasms and facilitate mucociliary clearance

▶ adequate hydration to liquefy secretions

▶ chest physiotherapy to mobilize secretions

▶ ultrasonic or mechanical nebulizers to loosen and mobilize secretions

▶ corticosteroids to combat inflammation

▶ diuretics to reduce edema

▶ oxygen to treat hypoxia.

Chronic obstructive pulmonary disease

Chronic obstructive pulmonary disease (COPD), also called chronic obstructive lung disease (COLD), results from emphysema, chronic bronchitis, asthma, or a combination of these disorders. Usually, more than one of these underlying conditions coexist; bronchitis and emphysema often occur together. (See "Asthma," "Chronic bronchitis," and "Emphysema" for a review of these conditions.)

COPD is the most common lung disease and affects an estimated 17 million Americans; the incidence is rising. The disease is not always symptomatic and may cause only minimal disability. However, COPD worsens with time.

Causes

Common causes of COPD may include:

▶ cigarette smoking

▶ recurrent or chronic respiratory tract infections

▶ air pollution

▶ allergies

▶ familial and hereditary factors such as deficiency of alpha$_1$-antitrypsin.

Pathophysiology

Smoking, one of the major causes of COPD, impairs ciliary action and macrophage function and causes inflammation in the airways, increased mucus production, destruction of alveolar septa, and peribronchiolar fibrosis. Early inflammatory changes may reverse if the patient stops smoking before lung disease becomes extensive.

The mucus plugs and narrowed airways cause air trapping, as in chronic bronchitis and emphysema. Hyperinflation occurs to the alveoli on expiration. On inspiration, airways enlarge, allowing air to pass beyond the obstruction; on expiration, airways narrow and gas flow is prevented. Air trapping (also called ball valving) occurs commonly in asthma and chronic bronchitis. (See *Air trapping in chronic obstructive pulmonary disease.*)

Signs and symptoms

The following signs and symptoms may occur:

▶ reduced ability to perform exercises or do strenuous work due to diminished pulmonary reserve

▶ productive cough due to stimulation of the reflex by mucus

▶ dyspnea on minimal exertion

▶ frequent respiratory tract infections

▶ intermittent or continuous hypoxemia

▶ grossly abnormal pulmonary function studies

▶ thoracic deformities.

Complications

Possible complications of COPD include:

▶ overwhelming disability

▶ cor pulmonale

▶ severe respiratory failure

▶ death.

AIR TRAPPING IN CHRONIC OBSTRUCTIVE PULMONARY DISEASE

In chronic obstructive pulmonary disease, mucus plugs and narrowed airways trap air (also called ball valving). During inspiration, the airways enlarge and gas enters; on expiration, the airways narrow and air can't escape. This commonly occurs in asthma and chronic bronchitis.

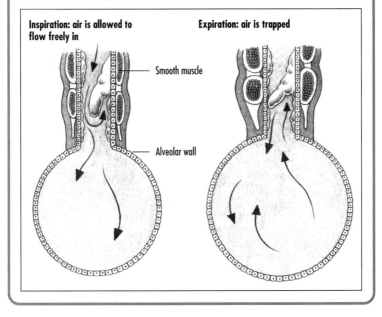

Inspiration: air is allowed to flow freely in

Expiration: air is trapped

Smooth muscle

Alveolar wall

Diagnosis

The following tests help diagnose COPD:
▶ Arterial blood gas analysis determines oxygen need by indicating degree of hypoxia and helps avoid carbon dioxide narcosis.
▶ Chest X-rays support underlying diagnosis.
▶ Pulmonary function studies support diagnosis of underlying condition.
▶ Electrocardiography may show arrhythmias consistent with hypoxemia.

Treatment

Managing COPD typically involves:
▶ bronchodilators to alleviate bronchospasms and enhance mucociliary clearance of secretions
▶ effective coughing to remove secretions
▶ postural drainage to help mobilize secretions
▶ chest physiotherapy to mobilize secretions
▶ low oxygen concentrations as needed (high flow rates of O_2 can lead to narcosis)
▶ antibiotics to allow treatment of respiratory tract infections
▶ pneumococcal vaccination and annual influenza vaccinations as important preventive measures
▶ smoking cessation
▶ installation in the home of an air conditioner with an air filter and avoidance of allergens, which may be helpful
▶ increased fluid intake to thin mucus
▶ use of a humidifier to thin secretions.

Cor pulmonale

Cor pulmonale (also called right ventricular failure) is a condition in which hypertrophy and dilation of the right ventricle develop secondary to disease affecting the structure or function of the lungs or their vasculature. It can occur at the end stage of various chronic disorders of the lungs, pulmonary vessels, chest wall, and respiratory control center. Cor pulmonale doesn't occur with disorders stemming from congenital heart disease or with those affecting the left side of the heart.

 CULTURAL DIVERSITY Cor pulmonale is more prevalent in countries where the incidence of obstructive lung disease is high, such as in the United Kingdom.

About 85% of patients with cor pulmonale also have chronic obstructive pulmonary disease (COPD), and about 25% of patients with bronchial COPD eventually develop cor pulmonale. The disorder is most common in smokers and in middle-aged and elderly males; however, its incidence in females is rising. Because cor pulmonale occurs late in the course of the individual's underlying condition and with other irreversible diseases, the prognosis is poor.

 AGE ALERT In children, cor pulmonale may be a complication of cystic fibrosis, hemosiderosis, upper airway obstruction, scleroderma, extensive bronchiectasis, neuromuscular diseases that affect respiratory muscles, or abnormalities of the respiratory control area.

Causes

Common causes of cor pulmonale include:
▶ disorders that affect the pulmonary parenchyma
▶ COPD
▶ bronchial asthma
▶ primary pulmonary hypertension
▶ vasculitis
▶ pulmonary emboli
▶ external vascular obstruction resulting from a tumor or aneurysm

▶ kyphoscoliosis
▶ pectus excavatum (funnel chest)
▶ muscular dystrophy
▶ poliomyelitis
▶ obesity
▶ high altitude.

Pathophysiology

In cor pulmonale, pulmonary hypertension increases the heart's workload. To compensate, the right ventricle hypertrophies to force blood through the lungs. As long as the heart can compensate for the increased pulmonary vascular resistance, signs and symptoms reflect only the underlying disorder.

Severity of right ventricular enlargement in cor pulmonale is due to increased afterload. An occluded vessel impairs the heart's ability to generate enough pressure. Pulmonary hypertension results from the increased blood flow needed to oxygenate the tissues.

In response to hypoxia, the bone marrow produces more red blood cells, causing polycythemia. The blood's viscosity increases, which further aggravates pulmonary hypertension. This increases the right ventricle's workload, causing heart failure. (See *Cor pulmonale: An overview.*)

In chronic obstructive disease, increased airway obstruction makes airflow worse. The resulting hypoxia and hypercarbia can have vasodilatory effects on systemic arterioles. However, hypoxia increases pulmonary vasoconstriction. The liver becomes palpable and tender because it is engorged and displaced downward by the low diaphragm. Hepatojugular reflux may occur.

Compensatory mechanisms begin to fail and larger amounts of blood remain in the right ventricle at the end of diastole, causing ventricular dilation. Increasing intrathoracic pressures impede venous return and raise jugular venous pressure. Peripheral edema can occur and right ventricular hypertrophy increases progressively. The main pulmonary arteries en-

COR PULMONALE: AN OVERVIEW

Although pulmonary restrictive disorders (such as fibrosis or obesity), obstructive disorders (such as bronchitis), or primary vascular disorders (such as recurrent pulmonary emboli) may cause cor pulmonale, these disorders share the following common pathway.

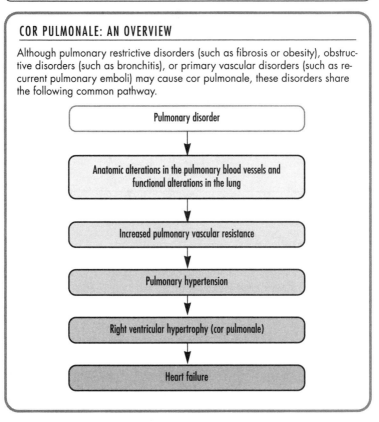

large, pulmonary hypertension increases, and heart failure occurs.

Signs and symptoms

Patients in *early stages* of cor pulmonale may present with:

▶ chronic productive cough to clear secretions from the lungs
▶ exertional dyspnea due to hypoxia
▶ wheezing respirations as airways narrow
▶ fatigue and weakness due to hypoxemia.
  Patients with *progressive cor pulmonale* may present with:
▶ dyspnea at rest due to hypoxemia
▶ tachypnea due to response to decreased oxygenation to the tissues
▶ orthopnea due to pulmonary edema

▶ dependent edema due to right-sided heart failure
▶ distended neck veins due to pulmonary hypertension
▶ enlarged, tender liver related to polycythemia and decreased cardiac output
▶ hepatojugular reflux (distention of the jugular vein induced by pressing over the liver) due to right-sided heart failure
▶ right upper quadrant discomfort due to liver involvement
▶ tachycardia due to decreased cardiac output and increasing hypoxia
▶ weakened pulses due to decreased cardiac output
▶ decreased cardiac output

▶ pansystolic murmur at the lower left sternal border with tricuspid insufficiency, which increases in intensity when the patient inhales.

Complications

Possible complications of cor pulmonale include:

▶ biventricular failure as the heart hypertrophies in an attempt to circulate the blood
▶ hepatomegaly
▶ edema
▶ ascites
▶ pleural effusions
▶ thromboembolism due to polycythemia.

Diagnosis

The following tests help diagnose cor pulmonale:

▶ Pulmonary artery catheterization shows increased right ventricular and pulmonary artery pressures, resulting from increased pulmonary vascular resistance. Both right ventricular systolic and pulmonary artery systolic pressures are over 30 mm Hg, and pulmonary artery diastolic pressure is higher than 15 mm Hg.

▶ Echocardiography demonstrates right ventricular enlargement.

▶ Angiography shows right ventricular enlargement.

▶ Chest X-rays reveal large central pulmonary arteries and right ventricular enlargement.

▶ Arterial blood gas analysis detects decreased Pao_2 (usually less than 70 mm Hg and rarely more than 90 mm Hg).

▶ Electrocardiography shows arrhythmias, such as premature atrial and ventricular contractions and atrial fibrillation during severe hypoxia, and also right bundle branch block, right axis deviation, prominent P waves, and an inverted T wave in right precordial leads.

▶ Pulmonary function studies reflect underlying pulmonary disease.

▶ Magnetic resonance imaging measures right ventricular mass, wall thickness, and ejection fraction.

▶ Cardiac catheterization measures pulmonary vascular pressures.

▶ Laboratory testing may reveal hematocrit typically over 50%; serum hepatic tests may show an elevated level of aspartate aminotransferase levels with hepatic congestion and decreased liver function, and serum bilirubin levels may be elevated if liver dysfunction and hepatomegaly exist.

Treatment

Therapy of cor pulmonale has three aims: reducing hypoxemia and pulmonary vasoconstriction, increasing exercise tolerance, and correcting the underlying condition when possible. Treatment may involve:

▶ bed rest to reduce myocardial oxygen demands

▶ digoxin to increase the strength of contraction of the myocardium

▶ antibiotics to treat an underlying respiratory tract infection

▶ a potent pulmonary artery vasodilator, such as diazoxide, nitroprusside, or hydralazine, to reduce primary pulmonary hypertension

▶ continuous administration of low concentrations of oxygen to decrease pulmonary hypertension, polycythemia, and tachypnea

▶ mechanical ventilation to reduce the workload of breathing in the acute disease

▶ a low-sodium diet with restricted fluid to reduce edema

▶ phlebotomy to decrease excess red blood cell mass that occurs with polycythemia

▶ small doses of heparin to decrease the risk of thromboembolism

▶ tracheotomy, which may be required if the patient has an upper airway obstruction

▶ corticosteroids to treat vasculitis or an underlying autoimmune disorder.

Emphysema

Emphysema, a form of chronic obstructive pulmonary disease, is the abnormal, permanent enlargement of the acini ac companied by destruction of alveolar walls. Obstruction results from tissue changes rather than mucus production, which occurs with asthma and chronic bronchitis. The distinguishing characteristic of emphysema is airflow limitation caused by lack of elastic recoil in the lungs.

Emphysema appears to be more prevalent in males than females; about 65% of patients with well-defined emphysema are men and 35% are women.

 AGE ALERT Aging is a risk factor for emphysema. Senile emphysema results from degenerative changes; stretching occurs without destruction in the smooth muscle. Connective tissue is not usually affected.

Causes
Emphysema is usually caused by:
◗ deficiency of alpha$_1$-antitrypsin
◗ cigarette smoking.

Pathophysiology
Primary emphysema has been linked to an inherited deficiency of the enzyme alpha$_1$-antitrypsin, a major component of alpha$_1$-globulin. Alpha$_1$-antitrypsin inhibits the activation of several proteolytic enzymes; deficiency of this enzyme is an autosomal recessive trait that predisposes an individual to develop emphysema because proteolysis in lung tissues is not inhibited. Homozygous individuals have up to an 80% chance of developing lung disease; if the individual smokes, he has a greater chance of developing emphysema. Patients who develop emphysema before or during their early forties and those who are nonsmokers are believed to have a deficiency of alpha$_1$-antitrypsin.

In emphysema, recurrent inflammation is associated with the release of proteolytic enzymes from lung cells. This causes irreversible enlargement of the air spaces distal to the terminal bronchioles. Enlargement of air spaces destroys the alveolar walls, which results in a breakdown of elasticity and loss of fibrous and mus-

cle tissue, thus making the lungs less compliant.

In normal breathing, the air moves into and out of the lungs to meet metabolic needs. A change in airway size compromises the ability of the lungs to circulate sufficient air. In patients with emphysema, recurrent pulmonary inflammation damages and eventually destroys the alveolar walls, creating large air spaces. (See *A look at abnormal alveoli,* page 226.) The alveolar septa are initially destroyed, eliminating a portion of the capillary bed and increasing air volume in the acinus. This breakdown leaves the alveoli unable to recoil normally after expanding and results in bronchiolar collapse on expiration. The damaged or destroyed alveolar walls cannot support the airways to keep them open. (See *Air trapping in emphysema,* page 227.) The amount of air that can be expired passively is diminished, thus trapping air in the lungs and leading to overdistention. Hyperinflation of the alveoli produces bullae (air spaces) and air spaces adjacent to the pleura (blebs). Septal destruction also decreases airway calibration. Part of each inspiration is trapped due to increased residual volume and decreased calibration. Septal destruction may affect only the respiratory bronchioles and alveolar ducts, leaving alveolar sacs intact (centriacinar emphysema), or it can involve the entire acinus (panacinar emphysema), with damage more random and involving the lower lobes of the lungs.

 AGE ALERT Panacinar emphysema tends to occur in the elderly with alpha$_1$-antitrypsin deficiency, whereas centriacinar emphysema occurs in smokers with chronic bronchitis.

Associated pulmonary capillary destruction usually allows a patient with severe emphysema to match ventilation to perfusion. This process prevents the development of cyanosis. The lungs are usually enlarged; therefore, the total lung capacity and residual volume increase.

A LOOK AT ABNORMAL ALVEOLI

In the patient with emphysema, recurrent pulmonary inflammation damages and eventually destroys the alveolar walls, creating large air spaces. The damaged alveoli can't recoil normally after expanding, and so bronchioles collapse on expiration, trapping air in the lungs and causing overdistention. As the alveolar walls are destroyed, the lungs become enlarged, and the total lung capacity and residual volume then increase. Shown below are changes that occur during emphysema.

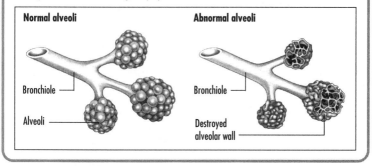

Normal alveoli

Bronchiole

Alveoli

Abnormal alveoli

Bronchiole

Destroyed alveolar wall

Signs and symptoms

The following signs and symptoms may occur:

▶ tachypnea related to decreased oxygenation

▶ dyspnea on exertion, which is often the initial symptom

▶ barrel-shaped chest due to the lungs overdistending and overinflating

▶ prolonged expiration and grunting, which occur because the accessory muscles are used for inspiration and abdominal muscles are used for expiration

▶ decreased breath sounds due to air-trapping in the alveoli and destruction of alveoli

▶ clubbed fingers and toes related to chronic hypoxic changes

▶ decreased tactile fremitus on palpation as air moves through poorly functional alveoli

▶ decreased chest expansion due to hypoventilation

▶ hyperresonance on chest percussion due to overinflated air spaces

▶ crackles and wheezing on inspiration as bronchioles collapse.

Complications

Possible complications of emphysema include:

▶ right ventricular hypertrophy (cor pulmonale)

▶ respiratory failure

▶ recurrent respiratory tract infections.

Diagnosis

▶ Chest X-rays in advanced disease may show a flattened diaphragm, reduced vascular markings at the lung periphery, over-aeration of the lungs, a vertical heart, enlarged anteroposterior chest diameter, and large retrosternal air space.

▶ Pulmonary function studies indicate increased residual volume and total lung capacity, reduced diffusing capacity, and increased inspiratory flow.

▶ Arterial blood gas analysis usually reveals reduced Pao_2 and a normal $Paco_2$ until late in the disease process.

▶ Electrocardiography may show tall, symmetrical P waves in leads II, III, and aV_F; vertical QRS axis and signs of right ventricular hypertrophy are seen late in the disease.

AIR TRAPPING IN EMPHYSEMA

Once alveolar walls are damaged or destroyed, they can't support and keep the airways open. The alveolar walls then lose their capability of elastic recoil. Collapse then occurs on expiration, as shown below.

Normal expiration: note normal recoil and the open bronchiole

Impaired expiration: note decreased elastic recoil and narrowed bronchiol

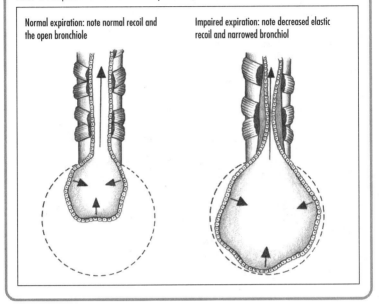

▶ Complete blood count usually reveals an increased hemoglobin level late in the disease when the patient has persistent severe hypoxia.

Treatment

Correcting this disorder typically involves:
▶ avoiding smoking to preserve remaining alveoli
▶ avoiding air pollution to preserve remaining alveoli
▶ bronchodilators, such as beta-adrenergic blockers and albuterol and ipratropium bromide, to reverse bronchospasms and promote mucociliary clearance
▶ antibiotics to treat respiratory tract infections
▶ pneumovax to prevent pneumococcal pneumonia
▶ adequate hydration to liquefy and mobilize secretions

▶ chest physiotherapy to mobilize secretions
▶ oxygen therapy at low settings to correct hypoxia
▶ flu vaccine to prevent influenza
▶ mucolytics to thin secretions and aid in expectoration of mucus
▶ aerosolized or systemic corticosteroids
▶ transtracheal catheterization to enable the patient to receive oxygen therapy at home.

Idiopathic respiratory distress syndrome of the newborn

Also known as respiratory distress syndrome (RDS) and hyaline membrane disease, this disorder is the most common cause of neonatal death; in the United States alone, it kills about 40,000 newborns every year.

AGE ALERT The syndrome occurs most exclusively in infants born before 37 weeks' gestation; it occurs in about 60% of those born before gestational week 28.

Idiopathic respiratory distress syndrome (IRDS) of the newborn is marked by widespread alveolar collapse. Occurring mainly in premature infants and in sudden infant death syndrome, it strikes apparently healthy infants. It is most common in infants of diabetic mothers and in those delivered by cesarean section, or it may occur suddenly after antepartum hemorrhage.

In IRDS of the newborn, the premature infant develops a widespread alveolar collapse due to surfactant deficiency. If untreated, the syndrome causes death within 72 hours of birth in up to 14% of infants weighing under 5.5 lbs (2,500 g). Aggressive management and mechanical ventilation can improve the prognosis, although some surviving infants are left with bronchopulmonary dysplasia. Mild cases of the syndrome subside after about 3 days.

Causes
Common causes of this syndrome include:
▶ lack of surfactant
▶ premature birth.

Pathophysiology
Surfactant, a lipoprotein present in alveoli and respiratory bronchioles, helps to lower surface tension, maintain alveolar patency, and prevent alveolar collapse, particularly at the end of expiration.

Although the neonatal airways are developed by gestational week 27, the intercostal muscles are weak, and the alveoli and capillary blood supply are immature. Surfactant deficiency causes a higher surface tension. The alveoli are not allowed to maintain patency and begin to collapse.

With alveolar collapse, ventilation is decreased and hypoxia develops. The resulting pulmonary injury and inflammatory reaction lead to edema and swelling of the interstitial space, thus impeding gas exchange between the capillaries and the functional alveoli. The inflammation also stimulates production of hyaline membranes composed of white fibrin accumulation in the alveoli. These deposits further reduce gas exchange in the lung and decrease lung compliance, resulting in increased work of breathing.

Decreased alveolar ventilation results in decreased $\dot{V}/\dot{Q}$ ratio and pulmonary arteriolar vasoconstriction. The pulmonary vasoconstriction can result in increased right cardiac volume and pressure, causing blood to be shunted from the right atrium through a patent foramen ovale to the left atrium. Increased pulmonary resistance also results in deoxygenated blood passing through the ductus arteriosus, totally bypassing the lungs, and causing a right-to-left shunt. The shunt further increases hypoxia.

Because of immature lungs and an already increased metabolic rate, the infant must expend more energy to ventilate collapsed alveoli. This further increases oxygen demand and contributes to cyanosis. The infant attempts to compensate with rapid shallow breathing, causing an initial respiratory alkalosis as carbon dioxide is expelled. The increased effort at lung expansion causes respirations to slow and respiratory acidosis to occur, leading to respiratory failure.

Signs and symptoms
The following signs and symptoms may occur:

AGE ALERT Suspect idiopathic respiratory distress syndrome of the newborn in a patient with a history that includes preterm birth (before gestational week 28), cesarean delivery, maternal history of diabetes, or antepartal hemorrhage.
▶ rapid, shallow respirations due to hypoxia
▶ intercostal, subcostal, or sternal retractions due to hypoxia
▶ nasal flaring due to hypoxia
▶ audible expiratory grunting; the grunting is a natural compensatory mechanism

that produces positive end-expiratory pressure (PEEP) to prevent further alveolar collapse
▶ hypotension due to cardiac failure
▶ peripheral edema due to cardiac failure
▶ oliguria due to vasoconstriction of kidneys.

In severe cases, the following may also occur:
▶ apnea due to respiratory failure
▶ bradycardia due to cardiac failure
▶ cyanosis from hypoxemia, right-to-left shunting through the foramen ovale, or right-to-left shunting through the atelectatic lung areas
▶ pallor due to decreased circulation
▶ frothy sputum due to pulmonary edema and atelectasis
▶ low body temperature, resulting from an immature nervous system and inadequate subcutaneous fat
▶ diminished air entry and crackles on auscultation due to atelectasis.

Complications
Possible complications include:
▶ respiratory failure
▶ cardiac failure
▶ bronchopulmonary dysplasia.

Diagnosis
The following tests help diagnose IRDS:
▶ Chest X-rays may be normal for the first 6 to 12 hours in 50% of patients, although later films show a fine reticulonodular pattern and dark streaks, indicating air-filled, dilated bronchioles.
▶ Arterial blood gas analysis reveals a diminished Pao_2 level; a normal, decreased, or increased $Paco_2$ level; and a reduced pH, indicating a combination of respiratory and metabolic acidosis.
▶ Lecithin-sphingomyelin ratio helps to assess prenatal lung development and infants at risk for this syndrome; this test is usually ordered if a cesarean section will be performed before gestational week 36.

Treatment
Correcting this syndrome typically involves:
▶ warm, humidified, oxygen-enriched gases administered by oxygen hood or, if such treatment fails, by mechanical ventilation to promote adequate oxygenation and reverse hypoxia
▶ mechanical ventilation with PEEP or continuous positive airway pressure (CPAP) administered by a tight-fitting face mask or endotracheal tube; this forces the alveoli to remain open on expiration and promotes increased surface area for exchange of oxygen and carbon dioxide
▶ high-frequency oscillation ventilation if the neonate can't maintain adequate gas exchange; this provides satisfactory minute volume (the total air breathed in 1 minute) with lower airway pressures
▶ radiant warmer or an Isolette to help maintain thermoregulation and reduce metabolic demands
▶ intravenous fluids to promote adequate hydration and maintain circulation with capillary refill of less than 2 seconds; fluid and electrolyte balance is also maintained
▶ sodium bicarbonate to control acidosis
▶ tube feedings or total parenteral nutrition to maintain adequate nutrition if the neonate is too weak to eat
▶ drug therapy with pancuronium bromide, a paralytic agent, preventing spontaneous respiration during mechanical ventilation
▶ prophylactic antibiotics for underlying infections
▶ diuretics to reduce pulmonary edema
▶ synthetic surfactant to prevent atelectasis and maintain alveolar integrity
▶ vitamin E to prevent complications associated with oxygen therapy

AGE ALERT Corticosteroids may be administered to the mother to stimulate surfactant production in a fetus at high risk for preterm birth.
▶ delayed delivery of an infant (if premature labor) to possibly prevent idiopathic respiratory distress syndrome.

Pneumothorax

Pneumothorax is an accumulation of air in the pleural cavity that leads to partial or complete lung collapse. When the air between the visceral and parietal pleurae collects and accumulates, increasing tension in the pleural cavity can cause the lung to progressively collapse. Air is trapped in the intrapleural space and determines the degree of lung collapse. Venous return to the heart may be impeded to cause a life-threatening condition called tension pneumothorax.

The most common types of pneumothorax are open, closed, and tension.

Causes

Common causes of *open pneumothorax* include:

▶ penetrating chest injury (gunshot or stab wound)
▶ insertion of a central venous catheter
▶ chest surgery
▶ transbronchial biopsy
▶ thoracentesis or closed pleural biopsy.

Causes of *closed pneumothorax* include:

▶ blunt chest trauma
▶ air leakage from ruptured blebs
▶ rupture resulting from barotrauma caused by high intrathoracic pressures during mechanical ventilation
▶ tubercular or cancerous lesions that erode into the pleural space
▶ interstitial lung disease, such as eosinophilic granuloma.

Tension pneumothorax may be caused by:

▶ penetrating chest wound treated with an air-tight dressing
▶ fractured ribs
▶ mechanical ventilation
▶ high-level positive end-expiratory pressure that causes alveolar blebs to rupture
▶ chest tube occlusion or malfunction.

Pathophysiology

A rupture in the visceral or parietal pleura and chest wall causes air to accumulate and separate the visceral and parietal pleu-

rae. Negative pressure is destroyed and the elastic recoil forces are affected. The lung recoils by collapsing toward the hilus.

Open pneumothorax (also called sucking chest wound or communicating pneumothorax) results when atmospheric air (positive pressure) flows directly into the pleural cavity (negative pressure). As the air pressure in the pleural cavity becomes positive, the lung collapses on the affected side, resulting in decreased total lung capacity, vital capacity, and lung compliance. $\dot{V}/\dot{Q}$ imbalances lead to hypoxia.

Closed pneumothorax occurs when air enters the pleural space from within the lung, causing increased pleural pressure, which prevents lung expansion during normal inspiration. Spontaneous pneumothorax is another type of closed pneumothorax.

 AGE ALERT Spontaneous pneumothorax is common in older patients with chronic pulmonary disease, but it may also occur in healthy, tall, young adults.

Both types of closed pneumothorax can result in a collapsed lung with hypoxia and decreased total lung capacity, vital capacity, and lung compliance. The range of lung collapse is between 5% and 95%.

Tension pneumothorax results when air in the pleural space is under higher pressure than air in the adjacent lung. The air enters the pleural space from the site of pleural rupture, which acts as a one-way valve. Air is allowed to enter into the pleural space on inspiration but cannot escape as the rupture site closes on expiration. More air enters on inspiration and air pressure begins to exceed barometric pressure. Increasing air pressure pushes against the recoiled lung, causing compression atelectasis. Air also presses against the mediastinum, compressing and displacing the heart and great vessels. The air cannot escape, and the accumulating pressure causes the lung to collapse. As air continues to accumulate and intrapleural pressures rise, the mediastinum shifts away from the affected side and decreases ve-

UNDERSTANDING TENSION PNEUMOTHORAX

In tension pneumothorax, air accumulates intrapleurally and cannot escape. As intrapleural pressure rises, the ipsilateral lung is affected and also collapses.

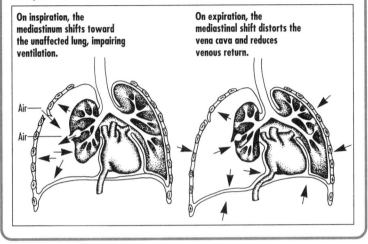

On inspiration, the mediastinum shifts toward the unaffected lung, impairing ventilation.

On expiration, the mediastinal shift distorts the vena cava and reduces venous return.

Air

Air

nous return. This forces the heart, trachea, esophagus, and great vessels to the unaffected side, compressing the heart and the contralateral lung. Without immediate treatment, this emergency can rapidly become fatal. (See *Understanding tension pneumothorax.*)

Signs and symptoms
The following signs and symptoms may occur:

▶ sudden, sharp pleuritic pain exacerbated by chest movement, breathing, and coughing
▶ asymmetrical chest wall movement due to collapse of the lung
▶ shortness of breath due to hypoxia
▶ cyanosis due to hypoxia
▶ respiratory distress
▶ decreased vocal fremitus related to collapse of the lung
▶ absent breath sounds on the affected side due to collapse of the lung
▶ chest rigidity on the affected side due to decreased expansion

▶ tachycardia due to hypoxia
▶ crackling beneath the skin on palpation (subcutaneous emphysema), which is due to air leaking into the tissues.

Tension pneumothorax produces the most severe respiratory symptoms, including:
▶ decreased cardiac output
▶ hypotension due to decreased cardiac output
▶ compensatory tachycardia
▶ tachypnea due to hypoxia
▶ lung collapse due to air or blood in the intrapleural space
▶ mediastinal shift due to increasing tension
▶ tracheal deviation to the opposite side
▶ distended neck veins due to intrapleural pressure, mediastinal shift, and increased cardiovascular pressure
▶ pallor related to decreased cardiac output
▶ anxiety related to hypoxia
▶ weak and rapid pulse due to decreased cardiac output.

Complications

Possible complications include:
▶ decreased cardiac output
▶ hypoxemia
▶ cardiac arrest.

Diagnosis

The following tests help diagnose pneumothorax:
▶ Chest X-rays confirm the diagnosis by revealing air in the pleural space and, possibly, a mediastinal shift.
▶ Arterial blood gas analysis may reveal hypoxemia, possibly with respiratory acidosis and hypercapnia. Pao_2 levels may decrease at first, but typically return to normal within 24 hours.

Treatment

Treatment depends on the type of pneumothorax.

Spontaneous pneumothorax with less than 30% of lung collapse, no signs of increased pleural pressure, and no dyspnea or indications of physiologic compromise, may be corrected with:
▶ bed rest to conserve energy and reduce oxygenation demands
▶ monitoring of blood pressure and pulse for early detection of physiologic compromise
▶ monitoring of respiratory rate to detect early signs of respiratory compromise
▶ oxygen administration to enhance oxygenation and improve hypoxia
▶ aspiration of air with a large-bore needle attached to a syringe to restore negative pressure within the pleural space.

Correction of pneumothorax with more than 30% of lung collapse may include:
▶ thoracostomy tube placed in the second or third intercostal space in the midclavicular line to try to re-expand the lung by restoring negative intrapleural pressure
▶ connection of the thoracostomy tube to underwater seal or to low-pressure suction to re-expand the lung
▶ if recurrent spontaneous pneumothorax, thoracotomy and pleurectomy may be per-

formed, which causes the lung to adhere to the parietal pleura.

Open (traumatic) pneumothorax may be corrected with:
▶ chest tube drainage to re-expand the lung
▶ surgical repair of the lung.

Correction of tension pneumothorax typically involves:
▶ immediate treatment with large-bore needle insertion into the pleural space through the second intercostal space to re-expand the lung
▶ insertion of a thoracostomy tube if large amounts of air escape through the needle after insertion
▶ analgesics to promote comfort and encourage deep breathing and coughing.

Pulmonary edema

Pulmonary edema is an accumulation of fluid in the extravascular spaces of the lungs. It is a common complication of cardiac disorders and may occur as a chronic condition or may develop quickly and rapidly become fatal.

Causes

Pulmonary edema is caused by left-sided heart failure due to:
▶ arteriosclerosis
▶ cardiomyopathy
▶ hypertension
▶ valvular heart disease.

Factors that predispose the patient to pulmonary edema include:
▶ barbiturate or opiate poisoning
▶ cardiac failure
▶ infusion of excessive volume of intravenous fluids or overly rapid infusion
▶ impaired pulmonary lymphatic drainage (from Hodgkin's disease or obliterative lymphangitis after radiation)
▶ inhalation of irritating gases
▶ mitral stenosis and left atrial myxoma (which impairs left atrial emptying)
▶ pneumonia
▶ pulmonary venoocclusive disease.

Pathophysiology

Normally, pulmonary capillary hydrostatic pressure, capillary oncotic pressure, capillary permeability, and lymphatic drainage are in balance. When this balance changes, or the lymphatic drainage system is obstructed, fluid infiltrates into the lung and pulmonary edema results. If pulmonary capillary hydrostatic pressure increases, the compromised left ventricle requires increased filling pressures to maintain adequate cardiac output. These pressures are transmitted to the left atrium, pulmonary veins, and pulmonary capillary bed, forcing fluids and solutes from the intravascular compartment into the interstitium of the lungs. As the interstitium overloads with fluid, fluid floods the peripheral alveoli and impairs gas exchange.

If colloid osmotic pressure decreases, the hydrostatic force that regulates intravascular fluids (the natural pulling force) is lost because there is no opposition. Fluid flows freely into the interstitium and alveoli, impairing gas exchange and leading to pulmonary edema. (See *Understanding pulmonary edema,* page 234.)

A blockage of the lymph vessels can result from compression by edema or tumor fibrotic tissue, and by increased systemic venous pressure. Hydrostatic pressure in the large pulmonary veins rises, the pulmonary lymphatic system cannot drain correctly into the pulmonary veins, and excess fluid moves into the interstitial space. Pulmonary edema then results from the accumulation of fluid.

Capillary injury, such as occurs in adult respiratory distress syndrome or with inhalation of toxic gases, increases capillary permeability. The injury causes plasma proteins and water to leak out of the capillary and move into the interstitium, increasing the interstitial oncotic pressure, which is normally low. As interstitial oncotic pressure begins to equal capillary oncotic pressure, the water begins to move out of the capillary and into the lungs, resulting in pulmonary edema.

Signs and symptoms

Early signs and symptoms may include:
▶ dyspnea on exertion due to hypoxia
▶ paroxysmal nocturnal dyspnea due to decreased expansion of the lungs
▶ orthopnea due to decreased ability of the diaphragm to expand
▶ cough due to stimulation of cough reflex by excessive fluid
▶ mild tachypnea due to hypoxia
▶ increased blood pressure due to increased pulmonary pressures and decreased oxygenation
▶ dependent crackles as air moves through fluid in the lungs
▶ neck vein distention due to decreased cardiac output and increased pulmonary vascular resistance
▶ tachycardia due to hypoxia.

Later stages of pulmonary edema may include the following signs and symptoms:
▶ labored, rapid respiration due to hypoxia
▶ more diffuse crackles as air moves through fluid in the lungs
▶ cough, producing frothy, bloody sputum
▶ increased tachycardia due to hypoxemia
▶ arrhythmias due to hypoxic myocardium
▶ cold, clammy skin due to peripheral vasoconstriction
▶ diaphoresis due to decreased cardiac output and shock
▶ cyanosis due to hypoxia
▶ decreased blood pressure due to decreased cardiac output and shock
▶ thready pulse due to decreased cardiac output and shock.

Complications

Possible complications include:
▶ respiratory failure
▶ respiratory acidosis
▶ cardiac arrest.

Diagnosis

The following tests help diagnose pulmonary edema:
▶ Arterial blood gas analysis usually reveals hypoxia with variable $Paco_2$, de-

UNDERSTANDING PULMONARY EDEMA

In pulmonary edema, diminished function of the left ventricle causes blood to back up into pulmonary veins and capillaries. The increasing capillary hydrostatic pressure pushes fluid into the interstitial spaces and alveoli. The following illustrations show a normal alveolus and an alveolus affected by pulmonary edema.

Normal alveolus

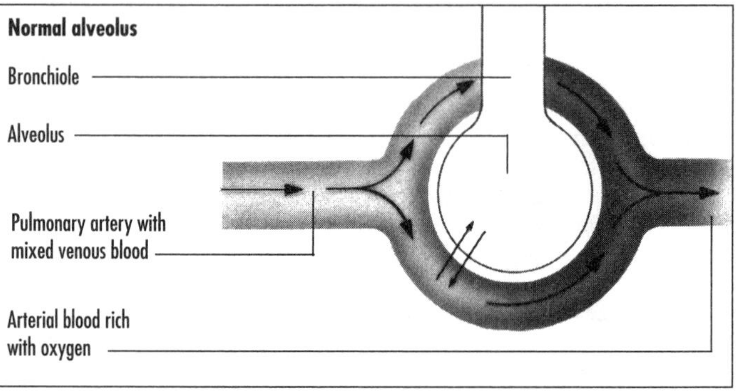

- Bronchiole
- Alveolus
- Pulmonary artery with mixed venous blood
- Arterial blood rich with oxygen

Alveolus in pulmonary edema

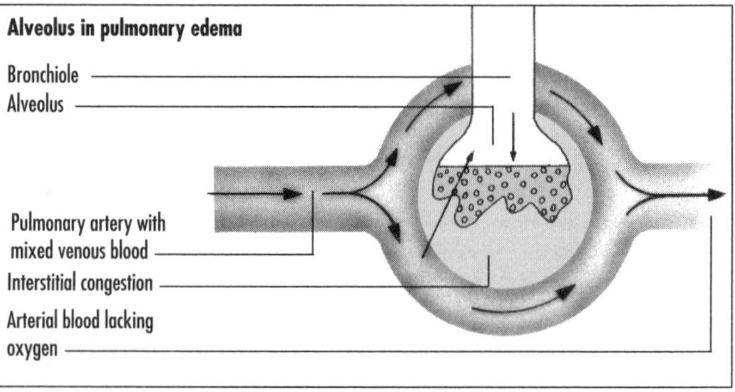

- Bronchiole
- Alveolus
- Pulmonary artery with mixed venous blood
- Interstitial congestion
- Arterial blood lacking oxygen

pending on the patient's degree of fatigue. Respiratory acidosis may occur.

▶ Chest X-rays show diffuse haziness of the lung fields and, usually, cardiomegaly and pleural effusion.

▶ Pulse oximetry may reveal decreasing Sao_2 levels.

▶ Pulmonary artery catheterization identifies left-sided heart failure and helps rule out adult respiratory distress syndrome.

▶ Electrocardiography may show previous or current myocardial infarction.

Treatment

Correcting this disorder typically involves:

▶ high concentrations of oxygen administered by nasal cannula to enhance gas exchange and improve oxygenation

▶ assisted ventilation to improve oxygen delivery to the tissues and promote acid-base balance

▶ diuretics, such as furosemide, ethacrynic acid, and bumetanide, to increase urination, which helps mobilize extravascular fluid

- positive inotropic agents, such as digoxin and amrinone, to enhance contractility in myocardial dysfunction
- pressor agents to enhance contractility and promote vasoconstriction in peripheral vessels
- antiarrhythmics for arrhythmias related to decreased cardiac output
- arterial vasodilators such as nitroprusside to decrease peripheral vascular resistance, preload, and afterload
- morphine to reduce anxiety and dyspnea, and to dilate the systemic venous bed, promoting blood flow from pulmonary circulation to the periphery.

Pulmonary hypertension

Pulmonary hypertension is indicated by a resting systolic pulmonary artery pressure (PAP) above 30 mm Hg and a mean PAP above 18 mm Hg. Primary or idiopathic pulmonary hypertension is characterized by increased PAP and increased pulmonary vascular resistance. This form is most common in women ages 20 to 40 and is usually fatal within 3 to 4 years.

 AGE ALERT Mortality is highest in pregnant women.

Secondary pulmonary hypertension results from existing cardiac or pulmonary disease, or both. The prognosis in secondary pulmonary hypertension depends on the severity of the underlying disorder.

The patient may have no signs or symptoms of the disorder until lung damage becomes severe. In fact, it may not be diagnosed until an autopsy is performed.

Causes

Causes of *primary pulmonary hypertension* are unknown, but may include:
- hereditary factors
- altered immune mechanisms.

Secondary pulmonary hypertension results from hypoxemia as the result of the following.

Conditions causing alveolar hypoventilation:
- chronic obstructive pulmonary disease

- sarcoidosis
- diffuse interstitial pneumonia
- malignant metastases
- scleroderma
- obesity
- kyphoscoliosis.

Conditions causing vascular obstruction:
- pulmonary embolism
- vasculitis
- left atrial myxoma
- idiopathic venoocclusive disease
- fibrosing mediastinitis
- mediastinal neoplasm.

Conditions causing primary cardiac disease:
- patent ductus arteriosus
- atrial septal defect
- ventricular septal defect.

Conditions causing acquired cardiac disease:
- rheumatic valvular disease
- mitral stenosis.

Pathophysiology

In primary pulmonary hypertension, the smooth muscle in the pulmonary artery wall hypertrophies for no reason, narrowing the small pulmonary artery (arterioles) or obliterating it completely. Fibrous lesions also form around the vessels, impairing distensibility and increasing vascular resistance. Pressures in the left ventricle, which receives blood from the lungs, remain normal. However, the increased pressures generated in the lungs are transmitted to the right ventricle, which supplies the pulmonary artery. Eventually, the right ventricle fails (cor pulmonale). Although oxygenation is not severely affected initially, hypoxia and cyanosis eventually occur. Death results from cor pulmonale.

Alveolar hypoventilation can result from diseases caused by alveolar destruction or from disorders that prevent the chest wall from expanding sufficiently to allow air into the alveoli. The resulting decreased ventilation increases pulmonary vascular resistance. Hypoxemia resulting from this $\dot{V}/\dot{Q}$ mismatch also causes vasoconstriction, further increasing vascular resistance and resulting in pulmonary hypertension.

Coronary artery disease or mitral valvular disease causing increased left ventricular filling pressures may cause secondary pulmonary hypertension. Ventricular septal defect and patent ductus arteriosus cause secondary pulmonary hypertension by increasing blood flow through the pulmonary circulation via left-to-right shunting. Pulmonary emboli and chronic destruction of alveolar walls, as in emphysema, cause secondary pulmonary hypertension by obliterating or obstructing the pulmonary vascular bed. Secondary pulmonary hypertension can also occur by vasoconstriction of the vascular bed, such as through hypoxemia, acidosis, or both. Conditions resulting in vascular obstruction can also cause pulmonary hypertension because blood is not allowed to flow appropriately through the vessels.

Secondary pulmonary hypertension can be reversed if the disorder is resolved. If hypertension persists, hypertrophy occurs in the medial smooth muscle layer of the arterioles. The larger arteries stiffen and hypertension progresses. Pulmonary pressures begin to equal systemic blood pressure, causing right ventricular hypertrophy and eventually cor pulmonale.

Primary cardiac diseases may be congenital or acquired. Congenital defects cause a left-to-right shunt, re-routing blood through the lungs twice and causing pulmonary hypertension. Acquired cardiac diseases, such as rheumatic valvular disease and mitral stenosis, result in left ventricular failure that diminishes the flow of oxygenated blood from the lungs. This increases pulmonary vascular resistance and right ventricular pressure.

Signs and symptoms
The following signs and symptoms may occur:

▶ increasing dyspnea on exertion from left ventricular failure

▶ fatigue and weakness from diminished oxygenation to the tissues

▶ syncope due to diminished oxygenation to brain cells

▶ difficulty breathing due to left ventricular failure

▶ shortness of breath due to left ventricular failure

▶ pain with breathing due to lactic acidosis buildup in the tissues

▶ ascites due to right ventricular failure

▶ neck vein distention due to right ventricular failure

▶ restlessness and agitation due to hypoxia

▶ decreased level of consciousness, confusion, and memory loss due to hypoxia

▶ decreased diaphragmatic excursion and respiration due to hypoventilation

▶ possible displacement of point of maximal impulse beyond the midclavicular line due to fluid accumulation

▶ peripheral edema due to right ventricular failure

▶ easily palpable right ventricular lift due to altered cardiac output and pulmonary hypertension

▶ reduced carotid pulse

▶ palpable and tender liver due to pulmonary hypertension

▶ tachycardia due to hypoxia

▶ systolic ejection murmur due to pulmonary hypertension and altered cardiac output

▶ split S_2, S_3, and S_4 sounds due to pulmonary hypertension and altered cardiac output

▶ decreased breath sounds due to fluid accumulation in the lungs

▶ loud, tubular breath sounds due to fluid accumulation in the lungs.

Complications
Possible complications of pulmonary hypertension include:

▶ cor pulmonale

▶ cardiac failure

▶ cardiac arrest.

Diagnosis

The following tests help diagnose pulmonary hypertension:

▶ Arterial blood gas analysis reveals hypoxemia.

▶ Electrocardiography in right ventricular hypertrophy shows right axis deviation and tall or peaked P waves in inferior leads.

▶ Cardiac catheterization reveals increased PAP, with systolic pressure above 30 mm Hg. It may also show an increased pulmonary capillary wedge pressure (PCWP) if the underlying cause is left atrial myxoma, mitral stenosis, or left ventricular failure; otherwise, PCWP is normal.

▶ Pulmonary angiography detects filling defects in pulmonary vasculature, such as those that develop with pulmonary emboli.

▶ Pulmonary function studies may show decreased flow rates and increased residual volume in underlying obstructive disease; in underlying restrictive disease, they may show reduced total lung capacity.

▶ Radionuclide imaging detects abnormalities in right and left ventricular functioning.

▶ Open lung biopsy may determine the type of disorder.

▶ Echocardiography allows the assessment of ventricular wall motion and possible valvular dysfunction. It can also demonstrate right ventricular enlargement, abnormal septal configuration consistent with right ventricular pressure overload, and reduction in left ventricular cavity size.

▶ Perfusion lung scanning may produce normal or abnormal results, with multiple patchy and diffuse filling defects that don't suggest pulmonary embolism.

Treatment

Managing this disorder typically involves:
▶ oxygen therapy to correct hypoxemia and resulting increased pulmonary vascular resistance

▶ fluid restriction in right ventricular failure to decrease workload of the heart

▶ digoxin to increase cardiac output

▶ diuretics to decrease intravascular volume and extravascular fluid accumulation

▶ vasodilators to reduce myocardial workload and oxygen consumption

▶ calcium channel blockers to reduce myocardial workload and oxygen consumption

▶ bronchodilators to relax smooth muscles and increase airway patency

▶ beta-adrenergic blockers to improve oxygenation

▶ treatment of the underlying cause to correct pulmonary edema

▶ heart-lung transplant in severe cases.

Respiratory failure

When the lungs can't adequately maintain arterial oxygenation or eliminate carbon dioxide, acute respiratory failure results, which can lead to tissue hypoxia. In patients with normal lung tissue, respiratory failure is indicated by a $Paco_2$ above 50 mm Hg and a Pao_2 below 50 mm Hg. These levels do not apply to patients with chronic obstructive pulmonary disease (COPD), who have a consistently high $Paco_2$ (hypercapnia) and a low Pao_2 (hypoxemia). Acute deterioration in arterial blood gas values for these patients and corresponding clinical deterioration signify acute respiratory failure.

Causes

Conditions that can result in alveolar hypoventilation, $\dot{V}/\dot{Q}$ mismatch, or right-to-left shunting can lead to respiratory failure; these include:
▶ COPD
▶ bronchitis
▶ pneumonia
▶ bronchospasm
▶ ventilatory failure
▶ pneumothorax
▶ atelectasis
▶ cor pulmonale
▶ pulmonary edema
▶ pulmonary emboli
▶ central nervous system disease
▶ central nervous system trauma

▶ central nervous system depressant drugs, such as anesthetics, sedation, and hypnotics

▶ neuromuscular diseases, such as poliomyelitis or amyotrophic lateral sclerosis.

Pathophysiology

Respiratory failure results from impaired gas exchange. Any condition associated with alveolar hypoventilation, $\dot{V}/\dot{Q}$ mismatch, and intrapulmonary (right-to-left) shunting can cause acute respiratory failure if left untreated.

Decreased oxygen saturation may result from alveolar hypoventilation, in which chronic airway obstruction reduces alveolar minute ventilation. Pao_2 levels fall and $Paco_2$ levels rise, resulting in hypoxemia.

Hypoventilation can occur from a decrease in the rate or duration of inspiratory signal from the respiratory center, such as with central nervous system conditions or trauma or central nervous system depressant drugs. Neuromuscular diseases, such as poliomyelitis or amytrophic lateral sclerosis, can result in alveolar hypoventilation if the condition affects normal contraction of the respiratory muscles. The most common cause of alveolar hypoventilation is airway obstruction, often seen with COPD (emphysema or bronchitis).

The most common cause of hypoxemia — V/Q imbalance — occurs when conditions such as pulmonary embolism or adult respiratory distress syndrome interrupt normal gas exchange in a specific lung region. Too little ventilation with normal blood flow or too little blood flow with normal ventilation may cause the imbalance, resulting in decreased Pao_2 levels and, thus, hypoxemia.

Decreased Fio_2 is also a cause of respiratory failure, although it is uncommon. Hypoxemia results from inspired air that does not contain adequate oxygen to establish an adequate gradient for diffusion into the blood — for example, at high altitudes or in confined, enclosed spaces.

The hypoxemia and hypercapnia characteristic of respiratory failure stimulate strong compensatory responses by all of the body systems, including the respiratory, cardiovascular, and central nervous systems. In response to hypoxemia, for example, the sympathetic nervous system triggers vasoconstriction, increases peripheral resistance, and increases the heart rate. Untreated $\dot{V}/\dot{Q}$ imbalances can lead to right-to-left shunting in which blood passes from the heart's right side to its left without being oxygenated.

Tissue hypoxemia occurs, resulting in anaerobic metabolism and lactic acidosis. Respiratory acidosis occurs from hypercapnia. Heart rate increases, stroke volume increases, and heart failure may occur. Cyanosis occurs due to increased amounts of unoxygenated blood. Hypoxia of the kidneys results in release of erythropoietin from renal cells, which causes the bone marrow to increase production of red blood cells — an attempt by the body to increase the blood's oxygen-carrying capacity.

The body responds to hypercapnia with cerebral depression, hypotension, circulatory failure, and an increased heart rate and cardiac output. Hypoxemia or hypercapnia (or both) causes the brain's respiratory control center first to increase respiratory depth (tidal volume) and then to increase the respiratory rate. As respiratory failure worsens, intercostal, supraclavicular, and suprasternal retractions may also occur.

Signs and symptoms

The following signs and symptoms may occur:

▶ cyanosis of oral mucosa, lips, and nail beds due to hypoxemia

▶ nasal flaring due to hypoxia

▶ ashen skin due to vasoconstriction

▶ use of accessory muscles for respiration

▶ restlessness, anxiety, agitation, and confusion due to hypoxic brain cells and alteration in level of consciousness
▶ tachypnea due to hypoxia
▶ cold, clammy skin due to vasoconstriction
▶ dull or flat sound on percussion if the patient has atelectasis or pneumonia
▶ diminished breath sounds over areas of hypoventilation.

Complications
Possible complications include:
▶ tissue hypoxia
▶ metabolic acidosis
▶ cardiac arrest.

Diagnosis
▶ Arterial blood gas analysis indicates respiratory failure by deteriorating values and a pH below 7.35. Patients with COPD may have a lower than normal pH compared with their previous levels.
▶ Chest X-rays identify pulmonary diseases or conditions, such as emphysema, atelectasis, lesions, pneumothorax, infiltrates, and effusions.
▶ Electrocardiography can demonstrate arrhythmias; these are commonly found with cor pulmonale and myocardial hypoxia.
▶ Pulse oximetry reveals a decreasing Sao_2.
▶ White blood cell count detects underlying infection.
▶ Abnormally low hemoglobin and hematocrit levels signal blood loss, which indicates decreased oxygen-carrying capacity.
▶ Hypokalemia may result from compensatory hyperventilation, the body's attempt to correct acidosis.
▶ Hypochloremia usually occurs in metabolic alkalosis.
▶ Blood cultures may identify pathogens.
▶ Pulmonary artery catheterization helps to distinguish pulmonary and cardiovascular causes of acute respiratory failure and monitors hemodynamic pressures.

Treatment
Correcting this disorder typically involves:
▶ oxygen therapy to promote oxygenation and raise Pao_2
▶ mechanical ventilation with an endotracheal or a tracheostomy tube if needed to provide adequate oxygenation and to reverse acidosis
▶ high-frequency ventilation, if patient doesn't respond to treatment, to force the airways open, promoting oxygenation and preventing collapse of alveoli
▶ antibiotics to treat infection
▶ bronchodilators to maintain patency of the airways
▶ corticosteroids to decrease inflammation
▶ fluid restrictions in cor pulmonale to reduce volume and cardiac workload
▶ positive inotropic agents to increase cardiac output
▶ vasopressors to maintain blood pressure
▶ diuretics to reduce edema and fluid overload
▶ deep breathing with pursed lips if patient is not intubated and mechanically ventilated to help keep airway patent
▶ incentive spirometry to increase lung volume.

Sudden infant death syndrome
Sudden infant death syndrome (SIDS) is the leading cause of death among apparently healthy infants, ages 1 month to 1 year. Also called crib death, it occurs at a rate of 2 in every 1,000 live births; about 7,000 infants die of SIDS in the United States each year. The peak incidence occurs between ages 2 and 4 months.

CULTURAL DIVERSITY The incidence of SIDS is slightly higher in preterm infants, Inuit infants, disadvantaged black infants, infants of mothers younger than age 20, and infants of multiple births. The incidence is also 10 times higher in SIDS siblings and in infants of mothers who are drug addicts.

Causes

Common causes of SIDS include:
▶ hypoxemia, possibly due to apnea or immature respiratory system
▶ re-breathing of carbon dioxide, as occurs when infant is face down on the mattress.

Pathophysiology

It has been suggested that the infant with SIDS may have damage to the respiratory control center in the brain from chronic hypoxemia. The infant may also have periods of sleep apnea and eventually dies during an episode.

Normally, increased carbon dioxide levels stimulate the respiratory center to initiate breathing until very high levels actually depress the ventilatory effort. In infants who experience SIDS or near-miss episodes of SIDS, the child may not respond to increasing carbon dioxide levels, showing only depressed ventilation. In these infants, an episode of apnea may occur and carbon dioxide levels increase; however, the child is not stimulated to breathe. Apnea continues until very high levels of carbon dioxide completely suppress the ventilatory effort and the child ceases to breathe.

Signs and symptoms

The following signs and symptoms may occur:
▶ history indicating that the infant was found not breathing
▶ mottling of the skin due to cyanosis
▶ apnea and absence of pulse due to severe hypoxemia.

Complications

Possible complications include:
▶ death
▶ brain damage from near-miss episodes.

Diagnosis

Autopsy is performed to rule out other causes of death.

Treatment

Treatment measures associated with SIDS typically involve:
▶ resuscitation to restore circulation and oxygenation (usually futile)
▶ emotional support to the family
▶ prevention of SIDS in high-risk infants or those who have had near-miss episodes.

 AGE ALERT Infants at risk for SIDS should be monitored at home on an apnea monitor until the age of vulnerability has passed. Also, parents should be educated in prompt emergency treatment of detected apnea.

The nervous system coordinates and organizes the functions of all body systems. This intricate network of interlocking receptors and transmitters is a dynamic system that controls and regulates every mental and physical function. It has three main divisions:

▶ Central nervous system (CNS): the brain and spinal cord (See *Reviewing the central nervous system*, pages 242 and 243.)

▶ Peripheral nervous system: the motor and sensory nerves, which carry messages between the CNS and remote parts of the body (See *Reviewing the peripheral nervous system*, pages 244 and 245.)

▶ Autonomic nervous system: actually part of the peripheral nervous system, regulates involuntary functions of the internal organs.

The fundamental unit that participates in all nervous system activity is the neuron, a highly specialized cell that receives and transmits electrochemical nerve impulses through delicate, threadlike fibers that extend from the central cell body. Axons carry impulses away from the cell body; dendrites carry impulses to it. Most neurons have several dendrites but only one axon.

▶ *Sensory* (or *afferent*) neurons transmit impulses from receptors to the spinal cord or the brain.

▶ *Motor* (or *efferent*) neurons transmit impulses from the CNS to regulate activity of muscles or glands.

▶ *Interneurons*, also known as *connecting* or *association* neurons, carry signals through complex pathways between sensory and motor neurons. Interneurons ac-

count for 99% of all the neurons in the nervous system.

From birth to death, the nervous system efficiently organizes and controls the smallest action, thought, or feeling; monitors communication and instinct for survival; and allows introspection, wonder, abstract thought, and self-awareness. Together, the CNS and peripheral nervous system keep a person alert, awake, oriented, and able to move about freely without discomfort and with all body systems working to maintain homeostasis.

Thus, any disorder affecting the nervous system can cause signs and symptoms in any and all body systems. Patients with nervous system disorders commonly have signs and symptoms that are elusive, subtle, and sometimes latent.

PATHOPHYSIOLOGIC CONCEPTS

Typically, disorders of the nervous system involve some alteration in arousal, cognition, movement, muscle tone, homeostatic mechanisms, or pain. Most disorders cause more than one alteration, and the close intercommunication between the CNS and peripheral nervous system means that one alteration may lead to another.

Arousal

Arousal refers to the level of consciousness, or state of awareness. A person who is aware of himself and the environment and can respond to the environment in specific ways is said to be fully conscious. Full consciousness requires that the

REVIEWING THE CENTRAL NERVOUS SYSTEM

The central nervous system (CNS) includes the brain and spinal cord. The brain consists of the cerebrum, cerebellum, brain stem, and primitive structures that lie below the cerebrum: the diencephalon, limbic system, and reticular activating system (RAS). The spinal cord is the primary pathway for messages between peripheral areas of the body and the brain. It also mediates reflexes.

CEREBRUM

The left and right cerebral hemispheres are joined by the corpus callosum, a mass of nerve fibers that allows communication between corresponding centers in the right and left hemispheres. Each hemisphere is divided into four lobes, based on anatomic landmarks and functional differences. The lobes are named for the cranial bones that lie over them (frontal, temporal, parietal, and occipital).

▶ frontal lobe — influences personality, judgment, abstract reasoning, social behavior, language expression, and movement (in the motor portion)
▶ temporal lobe — controls hearing, language comprehension, and storage and recall of memories (although memories are stored throughout the brain)
▶ parietal lobe — interprets and integrates sensations, including pain, temperature, and touch; also interprets size, shape, distance, and texture (The parietal lobe of the nondominant hemisphere, usually the right, is especially important for awareness of body schema [shape].)
▶ occipital lobe — functions primarily in interpreting visual stimuli.

The cerebral cortex, the thin surface layer of the cerebrum, is composed of gray matter (unmyelinated cell bodies). The surface of the cerebrum has convolutions (gyri) and creases or fissures (sulci).

CEREBELLUM

The cerebellum, which also has two hemispheres, maintains muscle tone, coordinates muscle movement, and controls balance.

BRAIN STEM

Composed of the pons, midbrain, and medulla oblongata, the brain stem relays messages between upper and lower levels of the nervous system. The cranial nerves originate from the midbrain, pons, and medulla.

▶ pon — connects the cerebellum with the cerebrum and the midbrain to the medulla oblongata, and contains one of the respiratory centers
▶ midbrain — mediates the auditory and visual reflexes
▶ medulla oblongata — regulates respiratory, vasomotor, and cardiac function.

PRIMITIVE STRUCTURES

The diencephalon contains the thalamus and hypothalamus, which lie beneath the cerebral hemispheres. The thalamus relays all sensory stimuli (except olfactory) as they ascend to the cerebral cortex. Thalamic functions include primitive awareness of pain, screening of incoming stimuli, and focusing of attention. The hypothalamus controls or affects body temperature, appetite, water balance, pituitary secretions, emotions, and autonomic functions, including sleep and wake cycles.

The limbic system lies deep within the temporal lobe. It initiates primitive drives (hunger, aggression, and sexual and emotional arousal) and screens all sensory messages traveling to the cerebral cortex.

REVIEWING THE CENTRAL NERVOUS SYSTEM *(continued)*

RETICULAR ACTIVATING SYSTEM

The RAS, a diffuse network of hyperexcitable neurons fanning out from the brain stem through the cerebral cortex, screens all incoming sensory information and channels it to appropriate areas of the brain for interpretation. RAS activity also stimulates wakefulness.

SPINAL CORD

The spinal cord joins the brain stem at the level of the foramen magnum and terminates near the second lumbar vertebra.

A cross section of the spinal cord reveals a central H-shaped mass of gray matter divided into dorsal (posterior) and ventral (anterior) horns. Gray matter in the dorsal horns relays sensory (afferent) impulses; in the ventral horns, motor (efferent) impulses. White matter (myelinated axons of sensory and motor nerves) surrounds these horns and forms the ascending and descending tracts.

---

reticular activating system, higher systems in the cerebral cortex, and thalamic connections are intact and functioning properly. Several mechanisms can alter arousal:

▶ direct destruction of the reticular activating system and its pathways

▶ destruction of the entire brainstem, either directly by invasion or indirectly by impairment of its blood supply

▶ compression of the reticular activating system by a disease process, either from direct pressure or compression as structures expand or herniate.

Those mechanisms may result from structural, metabolic, and psychogenic disturbances:

▶ Structural changes include infections, vascular problems, neoplasms, trauma, and developmental and degenerative conditions. They usually are identified by their location relative to the tentorial plate, the double fold of dura that supports the temporal and occipital lobes and separates the cerebral hemispheres from the brain stem and cerebellum. Those above the tentorial plate are called *supratentorial,* while those below are called *infratentorial.*

▶ Metabolic changes that affect the nervous system include hypoxia, electrolyte disturbances, hypoglycemia, drugs, and toxins, both endogenous and exogenous.

Essentially any systemic disease can affect the nervous system.

▶ Psychogenic changes are commonly associated with mental and psychiatric illnesses. Ongoing research has linked neuroanatomy and neurophysiology of the CNS and supporting structures, including neurotransmitters, with certain psychiatric illnesses. For example, dysfunction of the limbic system has been associated with schizophrenia, depression, and anxiety disorders.

Decreased arousal may be a result of diffuse or localized dysfunction in supratentorial areas:

▶ Diffuse dysfunction reflects damage to the cerebral cortex or underlying subcortical white matter. Disease is the most common cause of diffuse dysfunction; other causes include neoplasms, closed head trauma with subsequent bleeding, and pus accumulation.

▶ Localized dysfunction reflects mechanical forces on the thalamus or hypothalamus. Masses (such as bleeding, infarction, emboli, and tumors) may directly impinge on the deep diencephalic structures or herniation may compress them.

Stages of altered arousal

An alteration in arousal usually begins with some interruption or disruption in the

REVIEWING THE PERIPHERAL NERVOUS SYSTEM

The peripheral nervous system consists of the cranial nerves (CN), the spinal nerves, and the autonomic nervous system.

CRANIAL NERVES

The 12 pairs of cranial nerves transmit motor or sensory messages, or both, primarily between the brain or brain stem and the head and neck. All cranial nerves, except for the olfactory and optic nerves, originate from the midbrain, pons, or medulla oblongata. The cranial nerves are sensory, motor, or mixed (both sensory and motor) as follows:

▶ olfactory (CN I) — Sensory: smell
▶ optic (CN II) — Sensory: vision
▶ oculomotor (CN III) — Motor: extraocular eye movement (superior, medial, and inferior lateral), pupillary constriction, and upper eyelid elevation
▶ trochlear (CN IV) — Motor: extraocular eye movement (inferior medial)
▶ trigeminal (CN V) — Sensory: transmitting stimuli from face and head, corneal reflex; Motor: chewing, biting, and lateral jaw movements
▶ abducens (CN VI) — Motor: extraocular eye movement (lateral)
▶ facial (CN VII) — Sensory: taste receptors (anterior two-thirds of tongue); Motor: Facial muscle movement, including muscles of expression (those in the forehead and around the eyes and mouth)
▶ acoustic (CN VIII) — Sensory: hearing, sense of balance
▶ glossopharyngeal (CN IX) — Motor: swallowing movements; Sensory: sensations of throat; taste receptors (posterior one-third of tongue)
▶ vagus (CN X). Motor — movement of palate, swallowing, gag reflex; activity of the thoracic and abdominal viscera, such as heart rate and peristalsis; Sensory: sensations of throat, larynx, and thoracic and abdominal viscera (heart, lungs, bronchi, and GI tract)
▶ spinal accessory (CN XI) — Motor: shoulder movement, head rotation
▶ hypoglossal (CN XII) — Motor: tongue movement.

SPINAL NERVES

The 31 pairs of spinal nerves are named according to the vertebra immediately below their exit point from the spinal cord. Each spinal nerve contains of afferent (sensory) and efferent (motor) neurons, which carry messages to and from particular body regions, called dermatomes.

AUTONOMIC NERVOUS SYSTEM

The autonomic nervous system (ANS) innervates all internal organs. Sometimes known as the visceral efferent nerves, autonomic nerves carry messages to the viscera from the brain stem and neuroendocrine system. The ANS has two major divisions: the sympathetic (thoracolumbar) nervous system and the parasympathetic (craniosacral) nervous system.

Sympathetic nervous system

Sympathetic nerves exit the spinal cord between the levels of the 1st thoracic and 2nd lumbar vertebrae; hence the name thoracolumbar. These preganglionic neurons enter small relay stations (ganglia) near the cord. The ganglia form a chain that disseminates the impulse to postganglionic neurons, which reach many organs and glands, and can produce widespread, generalized responses.

The physiologic effects of sympathetic activity include:
▶ vasoconstriction
▶ elevated blood pressure
▶ enhanced blood flow to skeletal muscles

REVIEWING THE PERIPHERAL NERVOUS SYSTEM *(continued)*

▶ increased heart rate and contractility
▶ heightened respiratory rate
▶ smooth muscle relaxation of the bronchioles, GI tract, and urinary tract
▶ sphincter contraction
▶ pupillary dilation and ciliary muscle relaxation
▶ increased sweat gland secretion
▶ reduced pancreatic secretion.

Parasympathetic nervous system

The fibers of the parasympathetic, or craniosacral, nervous system leave the CNS by way of the cranial nerves from the midbrain and medulla and with the spinal nerves between the 2nd and 4th sacral vertebrae (S2 to S4).

After leaving the CNS, the long preganglionic fiber of each parasympathetic nerve travels to a ganglion near a particular organ or gland, and the short postganglionic fiber enters the organ or gland. Parasympathetic nerves have a specific response involving only one organ or gland.

The physiologic effects of parasympathetic system activity include:
▶ reduced heart rate, contractility, and conduction velocity
▶ bronchial smooth muscle constriction
▶ increased GI tract tone and peristalsis with sphincter relaxation
▶ urinary system sphincter relaxation and increased bladder tone
▶ vasodilation of external genitalia, causing erection
▶ pupillary constriction
▶ increased pancreatic, salivary, and lacrimal secretions.

The parasympathetic system has little effect on mental or metabolic activity.

---

diencephalon. When this occurs, the patient shows evidence of dullness, confusion, lethargy, and stupor. Continued decreases in arousal result from midbrain dysfunction and are evidenced by a deepening of the stupor. Eventually, if the medulla and pons are affected, coma results.

A patient may move back and forth between stages or levels of arousal, depending on the cause of the altered arousal state, initiation of treatment, and response to the treatment. Typically, if the underlying problem is not or cannot be corrected, then the patient will progress through the various stages of decreased consciousness, termed *rostral-caudal* progression.

Six levels of altered arousal or consciousness have been identified. (See *Stages of altered arousal*, page 246.) Typically, five areas of neurologic function are evaluated to help identify the cause of altered arousal:
▶ level of consciousness (includes awareness and cognitive functioning, which reflect cerebral status)
▶ pattern of breathing (helps localize cause to cerebral hemisphere or brain stem)
▶ pupillary changes (reflects level of brainstem function; the brainstem areas that control arousal are anatomically next to the areas that control the pupils)
▶ eye movement and reflex responses (help identify the level of brainstem dysfunction and its mechanism, such as destruction or compression)
▶ motor responses (help identify the level, side, and severity of brain dysfunction).

Cognition

Cognition is the ability to be aware and to perceive, reason, judge, remember, and to use intuition. It reflects higher function-

STAGES OF ALTERED AROUSAL

This chart highlights the six levels or stages of altered arousal and their manifestations.

STAGE	MANIFESTATIONS
Confusion	▶ Loss of ability to think rapidly and clearly ▶ Impaired judgment and decision making
Disorientation	▶ Beginning loss of consciousness ▶ Disorientation to time progresses to include disorientation to place ▶ Impaired memory ▶ Lack of recognition of self (last to go)
Lethargy	▶ Limited spontaneous movement or speech ▶ Easy to arouse by normal speech or touch ▶ Possible disorientation to time, place, or person
Obtundation	▶ Mild to moderate reduction in arousal ▶ Limited responsiveness to environment ▶ Ability to fall asleep easily without verbal or tactile stimulation from others ▶ Ability to answer questions with minimum response
Stupor	▶ State of deep sleep or unresponsiveness ▶ Arousable (motor or verbal response only to vigorous and repeated stimulation) ▶ Withdrawal or grabbing response to stimulation
Coma	▶ Lack of motor or verbal response to external environment or any stimuli ▶ No response to noxious stimuli, such as deep pain ▶ Unable to be aroused to any stimulus

ing of the cerebral cortex, including the frontal, parietal, and temporal lobes, and portions of the brainstem. Typically, an alteration in cognition results from direct destruction by ischemia and hypoxia, or from indirect destruction by compression or the effects of toxins and chemicals.

Altered cognition may manifest as agnosia, aphasia, or dysphasia:

▶ *Agnosia* is a defect in the ability to recognize the form or nature of objects. Usually, agnosia involves only one sense — hearing, vision, or touch.

▶ *Aphasia* is loss of the ability to comprehend or produce language.

▶ *Dysphasia* is impairment to the ability to comprehend or use symbols in either verbal or written language, or to produce language.

Dysphasia typically arises from the left cerebral hemisphere, usually the frontotemporal region. However, different types of dysphasia occur, depending on the specific area of the brain involved. For example, a dysfunction in the posterioinferior frontal lobe (Broca's area) causes a motor dysphasia in which the patient cannot find the words to speak and has difficulty writing and repeating words. Dysfunction in the pathways connecting the primary auditory area to the auditory association areas in the middle third of the left superior temporal gyrus causes a form of dysphasia called word deafness: the patient has fluent speech, but comprehension of the spoken word and ability to repeat speech are impaired. Rather than words, the patient hears only noise that has no meaning, yet reading comprehension and writing ability are intact.

Dementia

Dementia is loss of more than one intellectual or cognitive function, which interferes with ability to function in daily life. The patient may experience a problem with orientation, general knowledge and information, vigilance (attentiveness, alertness, and watchfulness), recent memory, remote memory, concept formulation, abstraction (the ability to generalize about nonconcrete thoughts and ideas), reasoning, or language use.

The underlying mechanism is a defect in the neuronal circuitry of the brain. The extent of dysfunction reflects the total quantity of neurons lost and the area where this loss occurred. Processes that have been associated with dementia include:

▶ degeneration
▶ cerebrovascular disorders
▶ compression
▶ effects of toxins
▶ metabolic conditions
▶ biochemical imbalances
▶ demyelinization
▶ infection.

Three major types of dementia have been identified: amnestic, intentional, and cognitive. Each type affects a specific area of the brain, resulting in characteristic impairments:

▶ *Amnestic dementia* typically results from defective neuronal circuitry in the temporal lobe. Characteristically, the patient exhibits difficulty in naming things, loss of recent memory, and loss of language comprehension.

▶ *Intentional dementia* results from a defect in the frontal lobe. The patient is easily distracted and, although able to follow simple commands, can't carry out such sequential executive functions as planning, initiating, and regulating behavior or achieving specific goals. The patient may exhibit personality changes and a flat affect. Possibly appearing accident prone, he may lose motor function, as evidenced by a wide shuffling gait, small steps, muscle rigidity, abnormal reflexes, incontinence of bowel and bladder, and, possibly, total immobility.

▶ *Cognitive dementia* reflects dysfunctional neuronal circuitry in the cerebral cortex. Typically, the patient loses remote memory, language comprehension, and mathematical skills, and has difficulty with visual-spatial relationships.

Movement

Movement involves a complex array of activities controlled by the cerebral cortex, the pyramidal system, the extrapyramidal system, and the motor units (the axon of the lower motor neuron from the anterior horn cell of the spinal cord and the muscles innervated by it). A problem in any one of these areas can affect movement. (See *Reviewing motor impulse transmission,* page 248.)

For movement to occur, the muscles must change their state from one of contraction to relaxation or vice versa. A change in muscle innervation anywhere along the motor pathway will affect movement. Certain neurotransmitters, such as dopamine, play a role in altered movement.

Alterations in movement typically include excessive movement (*hyperkinesia*) or decreased movement (*hypokinesia*). Hyperkinesia is a broad category that includes many different types of abnormal movements. Each type of hyperkinesia is associated with a specific underlying pathophysiologic mechanism affecting the

REVIEWING MOTOR IMPULSE TRANSMISSION

Motor impulses that originate in the motor cortex of the frontal lobe travel through upper motor neurons of the pyramidal or extrapyramidal tract to the lower motor neurons of the peripheral nervous system.

In the pyramidal tract, most impulses from the motor cortex travel through the internal capsule to the medulla, where they cross (decussate) to the opposite side and continue down the spinal cord as the lateral corticospinal tract, ending in the anterior horn of the gray matter at a specific spinal cord level. Some fibers do not cross in the medulla but continue down the anterior corticospinal tract and cross near the level of termination in the anterior horn. The fibers of the pyramidal tract are considered upper motor neurons. In the anterior horn of the spinal cord, upper motor neurons relay impulses to the lower motor neurons, which carry them via the spinal and peripheral nerves to the muscles, producing a motor response.

Motor impulses that regulate involuntary muscle tone and muscle control travel along the extrapyramidal tract from the premotor area of the frontal lobe to the pons of the brain stem, where they cross to the opposite side. The impulses then travel down the spinal cord to the anterior horn, where they are relayed to lower motor neurons for ultimate delivery to the muscles.

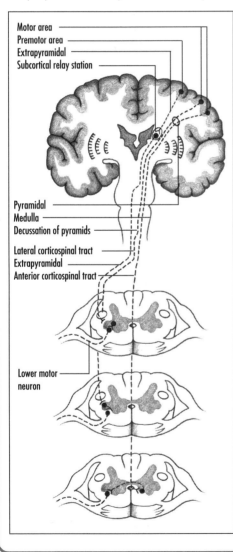

Motor area
Premotor area
Extrapyramidal
Subcortical relay station

Pyramidal
Medulla
Decussation of pyramids

Lateral corticospinal tract
Extrapyramidal
Anterior corticospinal tract

Lower motor neuron

brain or motor pathway. (See *Types of hyperkinesia,* pages 250 and 251.) Hypokinesia usually involves loss of voluntary control, even though peripheral nerve and muscle functions are intact. The types of hypokinesia include paresis, akinesia, bradykinesia, and loss of associated movement.

Paresis

Paresis is a partial loss of motor function (paralysis) and of muscle power, which the patient will often describe as weakness. Paresis can result from dysfunction of any of the following:

▶ the upper motor neurons in the cerebral cortex, subcortical white matter, internal capsule, brain stem, or spinal cord

▶ the lower motor neurons in the brainstem motor nuclei and anterior horn of the spinal cord, or problems with their axons as they travel to the skeletal muscle

▶ the motor units affecting the muscle fibers or the neuromuscular junction.

Upper motor neurons. Upper motor neuron dysfunction reflects an interruption in the pyramidal tract and consequent decreased activation of the lower motor neurons innervating one or more areas of the body. Upper motor neuron dysfunction usually affects more than one muscle group, and generally affects distal muscle groups more severely than proximal groups. Onset of spastic muscle tone over several days to weeks commonly accompanies upper motor neuron paresis, unless the dysfunction is acute. In acute dysfunction, flaccid tone and loss of deep tendon reflexes indicates spinal shock, caused by a severe, acute lesion below the foramen magnum. Incoordination associated with upper motor neuron paresis manifests as slow coarse movement with abnormal rhythm.

Lower motor neurons. Lower motor neurons are of two basic types: large (alpha) and small (gamma). Dysfunction of the large motor neurons of the anterior

horn of the spinal cord, the motor nuclei of the brainstem, and their axons causes impairment of voluntary and involuntary movement. The extent of paresis is directly correlated to the number of large lower motor neurons affected. If only a small portion of the large motor neurons are involved, paresis occurs; if all motor units are affected, the result is paralysis.

The small motor neurons play two necessary roles in movement: maintaining muscle tone and protecting the muscle from injury. Usually when the large motor neurons are affected, dysfunction of the small motor neurons causes reduced or absent muscle tone, flaccid paresis, and paralysis.

Motor units. The muscles innervated by motor neurons in the anterior horn of the spinal cord may also be affected. Paresis results from a decrease in the number or force of activated muscle fibers in the motor unit. The action potential of each motor unit decreases so that additional motor units are needed more quickly to produce the power necessary to move the muscle. Dysfunction of the neuromuscular junction causes paresis in a similar fashion; however, the functional capability of the motor units to function is lost, not the actual number of units.

Akinesia

Akinesia is a partial or complete loss of voluntary and associated movements, as well as a disturbance in the time needed to perform a movement. Often caused by dysfunction of the extrapyramidal tract, akinesia is associated with dopamine deficiency at the synapse or a defect in the postsynaptic receptors for dopamine.

Bradykinesia

Bradykinesia refers to slow voluntary movements that are labored, deliberate, and hard to initiate. The patient has difficulty performing movements consecutively and at the same time. Like akine-

(Text continues on page 252.)

TYPES OF HYPERKINESIA

This chart summarizes some of the most common types of hyperkinesias, their manifestations, and the underlying pathophysiologic mechanisms involved in their development.

TYPE	MANIFESTATIONS
Akathisia	▶ Ranges from mildly compulsive movement (usually legs) to severely frenzied motion ▶ Partly voluntary, with ability to suppress for short periods ▶ Relief obtained by performing motion
Asterixis	▶ Irregular flapping-hand movement ▶ More prominent when arms outstretched
Athetosis	▶ Slow, sinuous, irregular movements in the distal extremities ▶ Characteristic hand posture ▶ Slow, fluctuating grimaces
Ballism	▶ Severe, wild, flinging, stereotypical limb movements ▶ Present when awake or asleep ▶ Usually on one side of the body
Chorea	▶ Random, irregular, involuntary, rapid contractions of muscle groups ▶ Nonrepetitive ▶ Diminishes with rest, disappears during sleep ▶ Increases during emotional stress or attempts at voluntary movement
Hyperactivity	▶ Prolonged, generalized, increased activity ▶ Mainly involuntary but possibly subject to voluntary control ▶ Continual changes in body posture or excessive performance of a simple activity at inappropriate times
Intentional cerebellar tremor	▶ Tremor secondary to movement ▶ Most severe when nearing end of the movement
Myoclonus	▶ Shock-like contractions ▶ Throwing limb movements ▶ Random occurrence ▶ Triggered by startle ▶ Present even during sleep
Parkinsonian tremor	▶ Regular, rhythmic, slow flexion and extension contraction ▶ Primarily affects metacarpophalangeal and wrist joints ▶ Disappears with voluntary movement
Wandering	▶ Moving about without attention to environment

MECHANISMS

Possible association with impaired dopaminergic transmission

Believed to result from build up of toxins not broken down by the liver (i.e., ammonia)

Believed to result from injury to the putamen of the basal ganglion

Injury to subthalamus nucleus, causing inhibition of the nucleus

Excess concentration or heightened sensitivity to dopamine in the basal ganglia

Possibly due to injury to frontal lobe and reticular activating system

Errors in the feedback from the periphery and goal-directed movement due to disease of dentate nucleus and superior cerebellar peduncle

Irritability of nervous system and spontaneous discharge of neurons in the cerebral cortex, cerebellum, reticular activating system and spinal cord

Loss of inhibitory effects of dopamine in basal ganglia

Possibly due to bilateral injury to globus pallidus or putamen

sia, bradykinesia involves a disturbance in the time needed to perform a movement.

Loss of associated neurons

Movement involves not only the innervation of specific muscles to *accomplish* an action, but also the work of other innervated muscles that *enhance* the action. Loss of associated neurons involves alterations in movement that accompany the usual habitual voluntary movements for skill, grace, and balance. For example, when a person expresses emotion, the muscles of the face change and the posture relaxes. Loss of associated neurons involving emotional expression would result in a flat, blank expression and a stiff posture. Loss of associated neurons necessary for locomotion would result in a decrease in arm and shoulder movement, hip swinging, and rotation of the cervical spine.

Muscle tone

Like movement, muscle tone involves complex activities controlled by the cerebral cortex, pyramidal system, extrapyramidal system, and motor units. Normal muscle tone is the slight resistance that occurs in response to passive movement. When one muscle contracts, reciprocal muscles relax to permit movement with only minimal resistance. For example, when the elbow is flexed, the biceps muscle contracts and feels firm and the triceps muscle is somewhat relaxed and soft; with continued flexion, the biceps relax and the triceps contract. Thus, when a joint is moved through range of motion, the resistance is normally smooth, even, and constant.

The two major types of altered muscle tone are hypotonia (decreased muscle tone) and hypertonia (increased muscle tone).

Hypotonia

Hypotonia (also referred to as muscle flaccidity) typically reflects cerebellar damage, but rarely it may result from pure pyramidal tract damage.

Hypotonia is thought to involve a decrease in muscle spindle activity as a result of a decrease in neuron excitability. Flaccidity generally occurs with loss of nerve impulses from the motor unit responsible for maintaining muscle tone.

It may be localized to a limb or muscle group, or it may be generalized, affecting the entire body. Flaccid muscles can be moved rapidly with little or no resistance; eventually they become limp and atrophy.

Hypertonia

Hypertonia is increased resistance to passive movement. There are four types of hypertonia:

❯ Spasticity is hyperexcitability of stretch reflexes caused by damage to the motor, premotor, and supplementary motor areas and lateral corticospinal tract. (See *How spasticity develops*.)

❯ Paratonia (gegenhalten) is variance in resistance to passive movement in direct proportion to the force applied; the cause is frontal lobe injury.

❯ Dystonia is sustained, involuntary twisting movements resulting from slow muscle contraction; the cause is lack of appropriate inhibition of reciprocal muscles.

❯ Rigidity, or constant, involuntary muscle contraction, is resistance in both flexion and extension; causes are damage to basal ganglion (cog-wheel or lead-pipe rigidity) or loss of cerebral cortex inhibition or cerebellar control (gamma and alpha rigidity).

Hypertonia usually leads to atrophy of unused muscles. However, in some cases, if the motor reflex arc remains functional but is not inhibited by the higher centers, the overstimulated muscles may hypertrophy.

Homeostatic mechanisms

For proper function, the brain must maintain and regulate pressure inside the skull (intracranial pressure) as it also maintains the flow of oxygen and nutrients to its tissues. Both of these are accomplished by

HOW SPASTICITY DEVELOPS

Motor activity is controlled by pyramidal and extrapyramidal tracts that originate in the motor cortex, basal ganglia, brain stem, and spinal cord. Nerve fibers from the various tracts converge and synapse at the anterior horn of the spinal cord. Together they maintain segmental muscle tone by modulating the stretch reflex arc. This arc, shown in a simplified version below, is basically a negative feedback loop in which muscle stretch (stimulation) causes reflexive contraction (inhibition), thus maintaining muscle length and tone.

Damage to certain tracts results in loss of inhibition and disruption of the stretch reflex arc. Uninhibited muscle stretch produces exaggerated, uncontrolled muscle activity, accentuating the reflex arc and eventually resulting in spasticity.

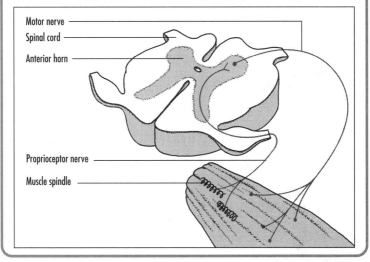

Motor nerve
Spinal cord
Anterior horn
Proprioceptor nerve
Muscle spindle

balancing changes in blood flow and cerebrospinal fluid (CSF) volume.

Constriction and dilation of the cerebral blood vessels help to regulate intracranial pressure and delivery of nutrients to the brain. These vessels respond to changes in concentrations of carbon dioxide, oxygen, and hydrogen ions. For example, if the CO_2 concentration in blood increases, the gas combines with body fluids to form carbonic acid, which eventually releases hydrogen. An increase in hydrogen ion concentration causes the cerebral vessels to dilate, increasing blood flow to the brain and cerebral perfusion and, subsequently, causing a drop in hydrogen ion concentration. A decrease in oxygen concentration also stimulates cerebral vasodilation, increasing blood flow and oxygen delivery to the brain.

Should these normal autoregulatory mechanisms fail, the abnormal blood chemistry stimulates the sympathetic nervous system to cause vasoconstriction of the large and medium-sized cerebral arteries. This helps to prevent increases in blood pressure from reaching the smaller cerebral vessels.

CSF volume remains relatively constant. However, should intracranial pressure rise, even as little as 5 mm Hg, the arachnoid villi open and excess CSF drains into the venous system.

The blood brain barrier also helps to maintain homeostasis in the brain. This barrier is composed of tight junctions between the endothelial cells of the cerebral vessels and neuroglial cells and is relatively impermeable to most substances. However, some substances required for metabolism pass through the blood brain barrier, depending on their size, solubility, and electrical charge. This barrier also regulates water flow from the blood, thereby helping to maintain the volume within the skull.

Increased intracranial pressure

Intracranial pressure (ICP) is the pressure exerted by the brain tissue, CSF, and cerebral blood (intracranial components) against the skull. Since the skull is a rigid structure, a change in the volume of the intracranial contents triggers a reciprocal change in one or more of the intracranial components to maintain consistent pressure. Any condition that alters the normal balance of the intracranial components—including increased brain volume, increased blood volume, or increased CSF volume—can increase ICP.

Initially the body uses its compensatory mechanisms (described above) to attempt to maintain homeostasis and lower ICP. But if these mechanisms become overwhelmed and are no longer effective, ICP continues to rise. Cerebral blood flow diminishes and perfusion pressure falls. Ischemia leads to cellular hypoxia, which initiates vasodilation of cerebral blood vessels in an attempt to increase cerebral blood flow. Unfortunately, this only causes the ICP to increase further. As the pressure continues to rise, compression of brain tissue and cerebral vessels further impairs cerebral blood flow.

If ICP continues to rise, the brain begins to shift under the extreme pressure and may herniate to an area of lesser pressure. When the herniating brain tissue's blood supply is compromised, cerebral ischemia and hypoxia worsen. The herniation increases pressure in the area where the pressure was lower, thus impairing its blood supply. As ICP approaches systemic blood pressure, cerebral perfusion slows even more, ceasing when ICP equals systemic blood pressure. (See *What happens when ICP rises.*)

Cerebral edema

Cerebral edema is an increase in the fluid content of brain tissue that leads to an increase in the intracellular or extracellular fluid volume. Cerebral edema may result from an initial injury to the brain tissue or it may develop in response to cerebral ischemia, hypoxia, and hypercapnia.

Cerebral edema is classified in four types — vasogenic, cytotoxic, ischemic, or interstitial — depending on the underlying mechanism responsible for the increased fluid content:

▶ *Vasogenic:* Injury to the vasculature increases capillary permeability and disruption of blood brain barrier; leakage of plasma proteins into the extracellular spaces pulls water into the brain parenchyma.

▶ *Cytotoxic (metabolic):* Toxins cause failure of the active transport mechanisms. Loss of intracellular potassium and influx of sodium (and water) cause cells in the brain to swell.

▶ *Ischemic:* Due to cerebral infarction and initially confined to intracellular compartment; after several days, released lysosomes from necrosed cells disrupt blood brain barrier.

▶ *Interstitial:* Movement of CSF from ventricles to extracellular spaces increases brain volume.

Regardless of the type of cerebral edema, blood vessels become distorted and brain tissue is displaced, ultimately leading to herniation.

Pain

Pain is the result of a complex series of steps from a site of injury to the brain,

WHAT HAPPENS WHEN ICP RISES

Intracranial pressure (ICP) is the pressure exerted within the intact skull by the intracranial volume — about 10% blood, 10% CSF, and 80% brain tissue. The rigid skull has little space for expansion of these substances.

The brain compensates for increases in ICP by regulating the volume of the three substances in the following ways:

▶ limiting blood flow to the head

▶ displacing CSF into the spinal canal

▶ increasing absorption or decreasing production of CSF — withdrawing water from brain tissue and excreting it through the kidneys.

When compensatory mechanisms become overworked, small changes in volume lead to large changes in pressure. The following chart will help you to understand increased ICP's pathophysiology.

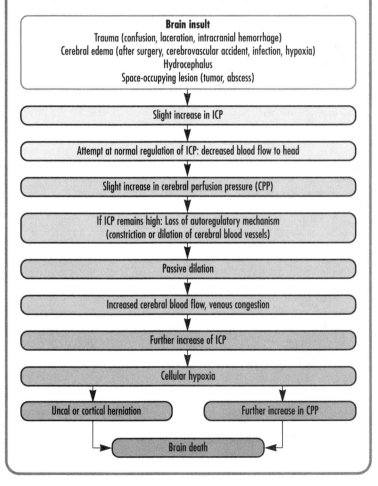

Brain insult
Trauma (confusion, laceration, intracranial hemorrhage)
Cerebral edema (after surgery, cerebrovascular accident, infection, hypoxia)
Hydrocephalus
Space-occupying lesion (tumor, abscess)

↓

Slight increase in ICP

↓

Attempt at normal regulation of ICP: decreased blood flow to head

↓

Slight increase in cerebral perfusion pressure (CPP)

↓

If ICP remains high: Loss of autoregulatory mechanism (constriction or dilation of cerebral blood vessels)

↓

Passive dilation

↓

Increased cerebral blood flow, venous congestion

↓

Further increase of ICP

↓

Cellular hypoxia

↓   ↓

Uncal or cortical herniation   Further increase in CPP

↓   ↓

Brain death

which interprets the stimuli as pain. Pain that originates outside the nervous system is termed *nociceptive* pain; pain in the nervous system is *neurogenic* or *neuropathic* pain.

Nociception

Nociception begins when noxious stimuli reach pain fibers. Sensory receptors called nociceptors — which are free nerve endings in the tissues — are stimulated by various agents, such as chemicals, temperature, or mechanical pressure. If a stimulus is sufficiently strong, impulses travel via the afferent nerve fibers along sensory pathways to the spinal cord, where they initiate autonomic and motor reflexes. The information also continues to travel to the brain, which perceives it as pain. Several theories have been developed in an attempt to explain pain. (See *Theories of pain,* pages 258 to 261.) Nociception consists of four steps: transduction, transmission, modulation, and perception.

Transduction. Transduction is the conversion of noxious stimuli into electrical impulses and subsequent depolarization of the nerve membrane. These electrical impulses are created by *algesic* substances, which sensitize the nociceptors and are released at the site of injury or inflammation. Examples include potassium and hydrogen ions, serotonin, histamine, prostaglandins, bradykinin, and substance P.

Transmission. A-delta fibers and C fibers transmit pain sensations from the tissues to the central nervous system.

A-delta fibers are small diameter, lightly myelinated fibers. Mechanical or thermal stimuli elicit a rapid or fast response. These fibers transmit well-localized, sharp, stinging, or pin-pricking type pain sensations. A-delta fibers connect with secondary neuron groupings on the dorsal horn of the spinal cord.

C fibers are smaller and unmyelinated. They connect with second order neurons in lamina I and II (the latter includes the substantia gelatinosa, an area in which pain is modulated). C fibers respond to chemical stimuli, rather than heat or pressure, triggering a slow pain response, usually within 1 second. This dull ache or burning sensation is not well localized and leads to two responses: an acute response transmitted immediately through fast pain pathways, which prompts the person to evade the stimulus, and lingering pain transmitted through slow pathways, which persists or worsens.

The A-delta and C fibers carry the pain signal from the peripheral tissues to the dorsal horn of the spinal cord. Excitatory and inhibitory interneurons and projection cells (neurons that connect pathways in the cerebral cortex of the CNS and peripheral nervous system) carry the signal to the brain by way of crossed and uncrossed pathways. An example of a crossed pathway is the spinothalamic tract, which enters the brain stem and ends in the thalamus. Sensory impulses travel from the medial and lateral lemniscus (tract) to the thalamus and brainstem. From the thalamus, other neurons carry the information to the sensory cortex, where pain is perceived and understood.

Another example of a crossed pathway is the ascending spinoreticulothalamic tract, which is responsible for the psychological components of pain and arousal. At this site, neurons synapse with interneurons before they cross to the opposite side of the cord and make their way to the medulla, and, eventually, the reticular activating system, mesencephalon, and thalamus. Impulses then are transmitted to the cerebral cortex, limbic system, and basal ganglia.

Once stimuli are delivered, responses from the brain must be relayed back to the original site. Several pathways carry the information in the dorsolateral white columns to the dorsal horn of the spinal cord. Some corticospinal tract neurons end in the dorsal horn and allow the brain to pay selective attention to certain stimuli while ignoring others. It allows transmission of the primary signal while suppressing the tendency for signals to spread to adjacent neurons.

Modulation. Modulation refers to modifications in pain transmission. Some neurons from the cerebral cortex and brainstem activate inhibitory processes, thus modifying the transmission. Substances — such as serotonin from the mesencephalon, norepinephrine from the pons, and endorphins from the brain and spinal cord — inhibit pain transmission by decreasing the release of nociceptive neurotransmitters. Spinal reflexes involving motor neurons may initiate a protective action, such as withdrawal from a pinprick, or may enhance the pain, as when trauma causes a muscle spasm in the injured area.

Perception. Perception is the end result of pain transduction, transmission, and modulation. It encompasses the emotional, sensory, and subjective aspects of the pain experience. Pain perception is thought to occur in the cortical structures of the somatosensory cortex and limbic system. Alertness, arousal, and motivation are believed to result from the action of the reticular activating system and limbic system. Cardiovascular responses and typical fight or flight responses are thought to involve the medulla and hypothalamus.

The following three variables contribute to the wide variety of individual pain experiences:

◗ *Pain threshold:* level of intensity at which a stimulus is perceived as pain
◗ *Perceptual dominance:* existence of pain at another location that is given more attention
◗ *Pain tolerance:* duration or intensity of pain to be endured before a response is initiated.

Neurogenic pain
Neurogenic pain is associated with neural injury. Pain results from spontaneous discharges from the damaged nerves, spontaneous dorsal root activity, or degenera-

tion of modulating mechanisms. Neurogenic pain does not activate nociceptors, and there is no typical pathway for transmission.

DISORDERS

Alzheimer's disease
Alzheimer's disease is a degenerative disorder of the cerebral cortex, especially the frontal lobe, which accounts for more than half of all cases of dementia. About 5% of people over the age of 65 have a severe form of this disease, and 12% have a mild to moderate form. Alzheimer's disease is estimated to affect approximately 4 million Americans; by 2040, that figure may exceed 6 million.

AGE ALERT In the elderly, Alzheimer's disease accounts for over 50% of all dementias, and the highest prevalence is among those over 85. It is the fourth leading cause of death among the elderly, after heart disease, cancer, and stroke.

This primary progressive disease has a poor prognosis. Typically, the duration of illness is 8 years, and patients die 2 to 5 years after onset of debilitating brain symptoms.

Causes
The exact cause of Alzheimer's disease is unknown. Factors that have been associated with its development include:

◗ *Neurochemical :* deficiencies in the neurotransmitters acetylcholine, somatostatin, substance P, and norepinephrine
◗ *Environmental:* repeated head trauma; exposure to aluminum or manganese
◗ *Viral:* slow CNS viruses
◗ *Genetic immunologic:* abnormalities on chromosomes 14 or 21; depositions of beta amyloid protein.

(Text continues on page 260.)

THEORIES OF PAIN

Over the years numerous theories have attempted to explain the sensation of pain and describe how it occurs. No single theory alone provides a complete explanation. This chart highlights some of the major theories about pain.

THEORY	MAJOR ASSUMPTIONS
Specificity	▶ Four types of cutaneous sensation (touch, warmth, cold, pain); each results from stimulation of specific skin receptor sites and neural pathways dedicated to one of the four sensations. ▶ Specific pain neurons (nociceptors) transmit pain sensation along specific pain fibers. ▶ At synapses in the *substantia gelatinosa*, pain impulses cross to the opposite side of the cord and ascend the specific pain pathways of spinothalamic tract to the thalamus and the pain receptor areas of the cerebral cortex.
Intensity	▶ Pain results from excessive stimulation of sensory receptors. Disorders or processes causing pain create an intense summation of non-noxious stimuli.
Pattern	▶ Nonspecific receptors transmit specific patterns (characterized by the length of the pain sensation, the amount of involved tissue, and the summation of impulses) from the skin to the spinal cord, leading to pain perception.
Neuromatrix	▶ A pattern theory. ▶ Sensations imprinted in the brain. Sensory inputs may trigger a pattern of sensation from the neuromatrix (a proposed network of neurons looping between the thalamus and the cortex, and the cortex and the limbic system). ▶ Sensation pattern is possible without the sensory trigger.
Gate Control	▶ Pain is transmitted from skin via the small diameter A-delta and C fibers to cells of the substantia gelatinosa in the dorsal horn, where interconnections between other sensory pathways exist. ▶ Stimulation of the large-diameter fast, myelinated A-beta and A-alpha fibers closes gate, which restricts transmission of the impulse to the CNS and diminishes perception of pain. ▶ Large fiber stimulation possible through massage, scratching or rubbing the skin, or through electrical stimulation. Concurrent firing of pain and touch paths reduces transmission and perception of the pain impulses but not of touch impulses. ▶ An increase in small-fiber activity inhibits the substantia gelatinosa cells, "opening the gate" and increasing pain transmission and perception. ▶ Substantia gelatinosa acts as a gate-control system to modulate (inhibit) the flow of nerve impulses from peripheral fibers to the central nervous system. ▶ Central trigger cells (T cells) act as a central nervous system control to stimulate selective brain processes that influence the gate-control system. Inhibition of T cells closes the gate, pain impulses are not transmitted to the brain. ▶ T cell activation of neural mechanisms in the brain is responsible for pain perception and response; transmitters partly regulate the release of substance P, the peptide that conveys pain information. Pain modulation is also partly controlled by the neurotransmitters, enkephalin and serotonin. ▶ Persistent pain initiates a gradual decline in the fraction of impulses that pass through the various gates. ▶ Descending efferent impulses from the brain may be responsible for closing, partially opening, or completely opening the gate.

COMMENT

▶ Focuses on the direct relationship between the pain stimulus and perception; does not account for adaptation to pain and the psychosocial factors modulating it.

▶ Does not explain existence of intense stimuli not perceived as pain.

▶ Includes some components of the intensity theory; pain possibly a response to intense stimulation of the sensory receptors regardless of receptor type or pathway.

▶ Explains existence of phantom pain.

▶ Provides the basis for use of massage and electrical stimulation in pain management; being used to develop additional theories and models.

(continued)

THEORIES OF PAIN (continued)

THEORY	MAJOR ASSUMPTIONS
Melzack-Casey Conceptual Model of Pain	▶ Three major psychological dimensions of pain: – sensory-discriminative from thalamus and somatosensory cortex – motivational-affective from the reticular formation – cognitive-evaluative. ▶ Interactions among the three produce descending inhibitory influences that alter pain input to the dorsal horn and ultimately modify the sensory pain experience and motivational-affective dimensions. ▶ Pain is localized and identified by its characteristics, evaluated by past experiences and undergoes further cognitive processing. The complex sensory, motivational, and cognitive interactions determine motor activities and behaviors associated with the pain experience.

Pathophysiology

The brain tissue of patients with Alzheimer's disease exhibits three distinct and characteristic features:

▶ neurofibrillatory tangles (fibrous proteins)
▶ neuritic plaques (composed of degenerating axons and dendrites)
▶ granulovascular changes.

Additional structural changes include cortical atrophy, ventricular dilation, deposition of amyloid (a glycoprotein) around the cortical blood vessels, and reduced brain volume. Selective loss of cholinergic neurons in the pathways to the frontal lobes and hippocampus, areas that are important for memory and cognitive functions, also are found. Examination of the brain after death commonly reveals an atrophic brain, often weighing less than 1000 g (normal, 1380 g).

Signs and symptoms

The typical signs and symptoms reflect neurologic abnormalities associated with the disease:

▶ gradual loss of recent and remote memory, loss of sense of smell, and flattening of affect and personality
▶ difficulty with learning new information
▶ deterioration in personal hygiene
▶ inability to concentrate
▶ increasing difficulty with abstraction and judgment
▶ impaired communication
▶ severe deterioration in memory, language, and motor function
▶ loss of coordination
▶ inability to write or speak
▶ personality changes, wanderings
▶ nocturnal awakenings
▶ loss of eye contact and fearful look
▶ signs of anxiety, such as wringing of hands
▶ acute confusion, agitation, compulsiveness or fearfulness when overwhelmed with anxiety
▶ disorientation and emotional lability
▶ progressive deterioration of physical and intellectual ability.

Complications

The most common complications include:
▶ injury secondary to violent behavior or wandering
▶ pneumonia and other infections
▶ malnutrition
▶ dehydration
▶ aspiration
▶ death.

Diagnosis

Alzheimer's disease is diagnosed by exclusion; that is, by ruling out other disorders as the cause for the patient's signs and symptoms. The only true way to confirm Alzheimer's disease is by finding pathological changes in the brain at autopsy. However, the following diagnostic tests may be useful:

▶ Positron emission tomography shows changes in the metabolism of the cerebral cortex.

▶ Computed tomography shows evidence of early brain atrophy in excess of that which occurs in normal aging.

▶ Magnetic resonance imaging shows no lesion as the cause of the dementia.

▶ Electroencephalogram shows evidence of slowed brain waves in the later stages of the disease.

▶ Cerebral blood flow studies shows abnormalities in blood flow.

Treatment

No cure or definitive treatment exists for Alzheimer's disease. Therapy may include the following:

▶ cerebral vasodilators such as ergoloid mesylates, isoxsuprine, and cyclandelate to enhance cerebral circulation

▶ hyperbaric oxygen to increase oxygenation to the brain

▶ psychostimulants, such as methylphenidate, to enhance the patient's mood

▶ antidepressants if depression appears to exacerbate the dementia

▶ tacrine, an anticholinesterase agent, to help improve memory deficits

▶ choline salts, lecithin, physostigmine, or an experimental agent such as deanol, enkephalins, or naloxone to possibly slow disease process

▶ reduction in use of antacids, aluminum cooking utensils, and deodorants that contain aluminum, to possibly control or reduce exposure to aluminum (a possible risk factor).

Amyotrophic lateral sclerosis

Commonly called *Lou Gehrig's disease,* after the New York Yankees first baseman who died of this disorder, amyotrophic lateral sclerosis (ALS) is the most common of the motor neuron diseases causing muscular atrophy. Other motor neuron diseases include progressive muscular atrophy and progressive bulbar palsy. Onset usually occurs between age 40 and age 70. A chronic, progressively debilitating disease, ALS may be fatal in less than 1 year or continue for 10 years or more, depending on the muscles affected. More than 30,000 Americans have ALS; about 5,000 new cases are diagnosed each year; and the disease affects three times as many men as women.

Causes

The exact cause of ALS is unknown, but about 5% to 10% of cases have a genetic component—an autosomal dominant trait that affects men and women equally.

Several mechanisms have been postulated, including:
▶ a slow-acting virus
▶ nutritional deficiency related to a disturbance in enzyme metabolism
▶ metabolic interference in nucleic acid production by the nerve fibers
▶ autoimmune disorders that affect immune complexes in the renal glomerulus and basement membrane.

Precipitating factors for acute deterioration include trauma, viral infections, and physical exhaustion.

Pathophysiology

ALS progressively destroys the upper and lower motor neurons. It does not affect cranial nerves III, IV, and VI, and therefore some facial movements, such as blinking, persist. Intellectual and sensory functions are not affected.

Some believe that glutamate — the primary excitatory neurotransmitter of the CNS — accumulates to toxic levels at the synapses. The affected motor units are no longer innervated and progressive degeneration of axons causes loss of myelin. Some nearby motor nerves may sprout axons in an attempt to maintain function, but, ultimately, nonfunctional scar tissue replaces normal neuronal tissue.

Signs and symptoms

Typical signs and symptoms of ALS include:
▶ fasciculations accompanied by spasticity, atrophy, and weakness, due to degeneration of the upper and lower motor neurons, and loss of functioning motor units, especially in the muscles of the forearms and the hands
▶ impaired speech, difficulty chewing and swallowing, choking, and excessive drooling from degeneration of cranial nerves V, IX, X, and XII
▶ difficulty breathing, especially if the brainstem is affected
▶ muscle atrophy due to loss of innervation.

Mental deterioration doesn't usually occur, but patients may become depressed as a reaction to the disease. Progressive bulbar palsy may cause crying spells or inappropriate laughter.

Complications

The most common complications include:
▶ respiratory infections
▶ respiratory failure
▶ aspiration.

Diagnosis

Although no diagnostic tests are specific to ALS, the following may aid in the diagnosis:
▶ Electromyography shows abnormalities of electrical activity in involved muscles.
▶ Muscle biopsy shows atrophic fibers interspersed between normal fibers.
▶ Nerve conduction studies show normal results.
▶ Computed tomography and electroencephalogram (EEG) show normal results and thus rule out multiple sclerosis, spinal cord neoplasm, polyarteritis, syringomyelia, myasthenia gravis, and progressive muscular dystrophy.

Treatment

ALS has no cure. Treatment is supportive and may include:
▶ diazepam, dantrolene, or baclofen for decreasing spasticity
▶ quinidine to relieve painful muscle cramps
▶ thyrotropin-releasing hormone (I.V. or intrathecally) to temporarily improve mo-

tor function (successful only in some patients)

▶ riluzole to modulate glutamate activity and slow disease progression

▶ respiratory, speech, and physical therapy to maintain function as much as possible

▶ psychological support to assist with coping with this progressive, fatal illness.

Arteriovenous malformations

Arteriovenous malformations (AVMs) are tangled masses of thin-walled, dilated blood vessels between arteries and veins that do not connect by capillaries. AVMs are common in the brain, primarily in the posterior portion of the cerebral hemispheres. Abnormal channels between the arterial and venous system mix oxygenated and unoxygenated blood, and thereby prevent adequate perfusion of brain tissue.

AVMs range in size from a few millimeters to large malformations extending from the cerebral cortex to the ventricles. Usually more than one AVM is present. Males and females are affected equally, and some evidence exists that AVMs occur in families. Most AVMs are present at birth; however, symptoms typically do not occur until the person is 10 to 20 years of age.

Causes

Causes of AVMs may include:

▶ congenital, due to a hereditary defect

▶ acquired from penetrating injuries, such as trauma.

Pathophysiology

AVMs lack the typical structural characteristics of the blood vessels. The vessels of an AVM are very thin; one or more arteries feed into the AVM, causing it to appear dilated and torturous. The typically high-pressured arterial flow moves into the venous system through the connecting channels to increase venous pressure, engorging and dilating the venous structures. An aneurysm may develop. If the AVM is large enough, the shunting can deprive the surrounding tissue of adequate blood flow. Additionally, the thin-walled vessels may ooze small amounts of blood or actually rupture, causing hemorrhage into the brain or subarachnoid space.

Signs and symptoms

Typically the patient experiences few, if any, signs and symptoms unless the AVM is large, leaks, or ruptures. Possible signs and symptoms include:

▶ chronic mild headache and confusion from AVM dilation, vessel engorgement, and increased pressure

▶ seizures secondary to compression of the surrounding tissues by the engorged vessels

▶ systolic bruit over carotid artery, mastoid process, or orbit, indicating turbulent blood flow

▶ focal neurologic deficits (depending on the location of the AVM) resulting from compression and diminished perfusion

▶ symptoms of intracranial (intracerebral, subarachnoid, or subdural) hemorrhage, including sudden severe headache, seizures, confusion, lethargy, and meningeal irritation from bleeding into the brain tissue or subarachnoid space

▶ hydrocephalus from AVM extension into the ventricular lining.

Complications

Complications depend on the severity (location and size) of the AVM. This includes:

▶ aneurysm development and subsequent rupture

▶ hemorrhage (intracerebral, subarachnoid, or subdural, depending on the location of the AVM)

▶ hydrocephalus.

Diagnosis

A definitive diagnosis depends on these diagnostic tests:

▶ Cerebral arteriogram confirms the presence of AVMs and evaluates blood flow.

▶ Doppler ultrasonography of cerebrovascular system indicates abnormal, turbulent blood flow.

Treatment
Treatment can be supportive, corrective, or both, including:
▶ support measures, including aneurysm precautions to prevent possible rupture
▶ surgery — block dissection, laser, or ligation — to repair the communicating channels and remove the feeding vessels
▶ embolization or radiation therapy if surgery is not possible, to close the communicating channels and feeder vessels and thus reduce the blood flow to the AVM.

Cerebral palsy
The most common cause of crippling in children, cerebral palsy (CP) is a group of neuromuscular disorders caused by prenatal, perinatal, or postnatal damage to the upper motor neurons. Although nonprogressive, these disorders may become more obvious as an affected infant grows.

The three major types of cerebral palsy — spastic, athetoid, and ataxic — may occur alone or in combination. Motor impairment may be minimal (sometimes apparent only during physical activities such as running) or severely disabling. Common associated defects are seizures, speech disorders, and mental retardation.

Cerebral palsy occurs in an estimated 1.5 to 5 per 1,000 live births per year. Incidence is highest in premature infants (anoxia plays the greatest role in contributing to cerebral palsy) and in those who are small for gestational age. Almost half of the children with CP are mentally retarded, approximately one-fourth have seizure disorders, and more than three-fourths have impaired speech. Additionally, children with CP often have dental abnormalities, vision and hearing defects, and reading disabilities.

Cerebral palsy is slightly more common in males than in females and is more common in whites than in other ethnic groups. The prognosis varies. Treatment may make a near-normal life possible for children with mild impairment. Those with severe impairment require special services and schooling.

Causes
The exact of CP is unknown; however, conditions resulting in cerebral anoxia, hemorrhage, or other CNS damage are probably responsible. Potential causes vary with time of damage.

Prenatal causes include:
▶ maternal infection (especially rubella)
▶ exposure to radiation
▶ anoxia
▶ toxemia
▶ maternal diabetes
▶ abnormal placental attachment
▶ malnutrition
▶ isoimmunization.

Perinatal and birth factors may include:
▶ forceps delivery
▶ breech presentation
▶ placenta previa
▶ abruptio placentae
▶ depressed maternal vital signs from general or spinal anesthesia
▶ prolapsed cord with delay in blood delivery to the head
▶ premature birth
▶ prolonged or unusually rapid labor
▶ multiple births (especially infants born last)
▶ infection or trauma during infancy.

Postnatal causes include:
▶ kernicterus resulting from erythroblastosis fetalis
▶ brain infection or tumor
▶ head trauma
▶ prolonged anoxia
▶ cerebral circulatory anomalies causing blood vessel rupture
▶ systemic disease resulting in cerebral thrombosis or embolus.

Pathophysiology
In the early stages of brain development, a lesion or abnormality causes structural and functional defects that in turn cause impaired motor function or cognition.

Even though the defects are present at birth, problems may not be apparent until months later, when the axons have become myelinated and the basal ganglia are mature.

Signs and symptoms

Shortly after birth, the infant with CP may exhibit some typical signs and symptoms, including:
▶ excessive lethargy or irritability
▶ high-pitched cry
▶ poor head control
▶ weak sucking reflex.

Additional physical findings that may suggest CP include:
▶ delayed motor development (inability to meet major developmental milestones)
▶ abnormal head circumference, typically smaller than normal for age (because the head grows as the brain grows)
▶ abnormal postures, such as straightening legs when on back, toes down; holding head higher than normal when prone due to arching of back
▶ abnormal reflexes (neonatal reflexes lasting longer than expected, extreme reflexes, or clonus)
▶ abnormal muscle tone and performance (scooting on back to crawl, toe-first walking).

Each type of cerebral palsy typically produces a distinctive set of clinical features, although some children display a mixed form of the disease. (See *Assessing signs of CP,* page 266.)

Complications

Complications depend on the type of CP and the severity of the involvement. Possible complications include:
▶ contractures
▶ skin breakdown and ulcer formation
▶ muscle atrophy
▶ malnutrition
▶ seizure disorders
▶ speech, hearing, and vision problems
▶ language and perceptual deficits
▶ mental retardation
▶ dental problems

▶ respiratory difficulties, including aspiration from poor gag and swallowing reflexes.

Diagnosis

No diagnostic tests are specific to CP. However, neurologic screening will exclude other possible conditions, such as infection, spina bifida, or muscular dystrophy. Diagnostic tests that may be performed include:
▶ Developmental screening reveals delay in achieving milestones.
▶ Vision and hearing screening demonstrates degree of impairment.
▶ Electroencephalogram identifies the source of seizure activity.

Treatment

Cerebral palsy can't be cured, but proper treatment can help affected children reach their full potential within the limits set by this disorder. Such treatment requires a comprehensive and cooperative effort, involving doctors, nurses, teachers, psychologists, the child's family, and occupational, physical, and speech therapists. Home care is often possible. Treatment usually includes:
▶ braces, casts, or splints and special appliances, such as adapted eating utensils and a low toilet seat with arms, to help these children perform activities of daily living independently
▶ an artificial urinary sphincter for the incontinent child who can use the hand controls
▶ range-of-motion exercises to minimize contractures
▶ anticonvulsant to control seizures
▶ muscle relaxants (sometimes) to reduce spasticity
▶ surgery to decrease spasticity or correct contractures
▶ muscle transfer or tendon lengthening surgery to improve function of joints
▶ rehabilitation including occupational, physical, and speech therapy to maintain or improve functional abilities.

ASSESSING SIGNS OF CP

Each type of cerebral palsy (CP) is manifested by specific signs. This chart highlights the major signs and symptoms associated with each type of CP. The manifestations reflect impaired upper motor neuron function and disruption of the normal stretch reflex.

TYPE OF CP	SIGNS AND SYMPTOMS
Spastic CP (due to impairment of the pyramidal tract [most common type])	▶ Hyperactive deep tendon reflexes ▶ Increased stretch reflexes ▶ Rapid alternating muscle contraction and relaxation ▶ Muscle weakness ▶ Underdevelopment of affected limbs ▶ Muscle contraction in response to manipulation ▶ Tendency toward contractures ▶ Typical walking on toes with a scissors gait, crossing one foot in front of the other

Athetoid CP (due to impairment of the extrapyramidal tract)	▶ Involuntary movements usually affecting arms more severely than legs, including:
	  – grimacing
	  – wormlike writhing
	  – dystonia
	  – sharp jerks
	▶ Difficulty with speech due to involuntary facial movements
	▶ Increasing severity of movements during stress; decreased with relaxation and disappearing entirely during sleep

Ataxic CP (due to impairment of the extrapyramidal tract)	▶ Disturbed balance ▶ Incoordination (especially of the arms) ▶ Hypoactive reflexes ▶ Nystagmus ▶ Muscle weakness ▶ Tremor ▶ Lack of leg movement during infancy ▶ Wide gait as the child begins to walk ▶ Sudden or fine movements impossible (due to ataxia)
Mixed CP	▶ Spasticity and athetoid movements ▶ Ataxic and athetoid movements (resulting in severe impairment)

Cerebrovascular accident

A cerebrovascular accident (CVA), also known as a stroke or brain attack, is a sudden impairment of cerebral circulation in one or more blood vessels. A CVA interrupts or diminishes oxygen supply, and often causes serious damage or necrosis in the brain tissues. The sooner the circulation returns to normal after the CVA, the better chances are for complete recovery. However, about half of patients who survive a CVA remain permanently disabled and experience a recurrence within weeks, months, or years.

CVA is the third most common cause of death in the United States and the most common cause of neurologic disability. It strikes over 500,000 persons per year and is fatal in approximately half of these persons.

 AGE ALERT Although stroke may occur in younger persons, most patients experiencing stroke are over the age of 65 years. In fact, the risk of CVA doubles with each passing decade after the age of 55.

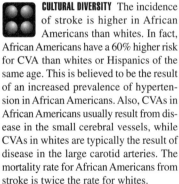 **CULTURAL DIVERSITY** The incidence of stroke is higher in African Americans than whites. In fact, African Americans have a 60% higher risk for CVA than whites or Hispanics of the same age. This is believed to be the result of an increased prevalence of hypertension in African Americans. Also, CVAs in African Americans usually result from disease in the small cerebral vessels, while CVAs in whites are typically the result of disease in the large carotid arteries. The mortality rate for African Americans from stroke is twice the rate for whites.

Causes

CVA typically results from one of three causes:

▶ thrombosis of the cerebral arteries supplying the brain, or of the intracranial vessels occluding blood flow. (See *Types of CVA*, page 268.)
▶ embolism from thrombus outside the brain, such as in the heart, aorta, or common carotid artery.
▶ hemorrhage from an intracranial artery or vein, such as from hypertension, ruptured aneurysm, AVM, trauma, hemorrhagic disorder, or septic embolism.

Risk factors that have been identified as predisposing a patient to CVA include:
▶ hypertension

▶ family history of CVA
▶ history of transient ischemic attacks (TIAs) (See *Understanding TIAs*, page 269.)
▶ cardiac disease, including arrhythmias, coronary artery disease, acute myocardial infarction, dilated cardiomyopathy, and valvular disease
▶ diabetes
▶ familial hyperlipidemia
▶ cigarette smoking
▶ increased alcohol intake
▶ obesity, sedentary lifestyle
▶ use of oral contraceptives.

Pathophysiology

Regardless of the cause, the underlying event is deprivation of oxygen and nutrients. Normally, if the arteries become blocked, autoregulatory mechanisms help maintain cerebral circulation until collateral circulation develops to deliver blood to the affected area. If the compensatory mechanisms become overworked, or if cerebral blood flow remains impaired for more than a few minutes, oxygen deprivation leads to infarction of brain tissue. The brain cells cease to function because they can neither store glucose or glycogen for use nor engage in anaerobic metabolism.

A thrombotic or embolic stroke causes ischemia. Some of the neurons served by the occluded vessel die from lack of oxygen and nutrients. This results in cerebral infarction, in which tissue injury triggers an inflammatory response that in turn increases intracranial pressure. Injury to surrounding cells disrupts metabolism and leads to changes in ionic transport, localized acidosis, and free radical formation. Calcium, sodium, and water accumulate in the injured cells, and excitatory neurotransmitters are released. Consequent continued cellular injury and swelling set up a vicious cycle of further damage.

When hemorrhage is the cause, impaired cerebral perfusion causes infarction, and the blood itself acts as a space-occupying mass, exerting pressure on the

TYPES OF CVA

Cerebrovascular accidents (CVAs) are typically classified as ischemic or hemorrhagic depending on the underlying cause. This chart describes the major types of CVAs.

TYPE OF CVA	DESCRIPTION
Ischemic: Thrombotic	▶ Most common cause of CVA ▶ Frequently the result of atherosclerosis; also associated with hypertension, smoking, diabetes ▶ Thrombus in extracranial or intracranial vessel blocks blood flow to the cerebral cortex ▶ Carotid artery most commonly affected extracranial vessel ▶ Common intracranial sites include bifurcation of carotid arteries, distal intracranial portion of vertebral arteries, and proximal basilar arteries ▶ May occur during sleep or shortly after awakening; during surgery; or after a myocardial infarction
Ischemic: Embolic	▶ Second most common type of CVA ▶ Embolus from heart or extracranial arteries floats into cerebral bloodstream and lodges in middle cerebral artery or branches ▶ Embolus commonly originates during atrial fibrillation ▶ Typically occurs during activity ▶ Develops rapidly
Ischemic: Lacunar	▶ Subtype of thrombotic CVA ▶ Hypertension creates cavities deep in white matter of the brain, affecting the internal capsule, basal ganglia, thalamus, and pons ▶ Lipid coating lining of the small penetrating arteries thickens and weakens wall, causing microaneurysms and dissections
Hemorrhagic	▶ Third most common type of CVA ▶ Typically caused by hypertension or rupture of aneurysm ▶ Diminished blood supply to area supplied by ruptured arteriy and compression by accumulated blood

brain tissues. The brain's regulatory mechanisms attempt to maintain equilibrium by increasing blood pressure to maintain cerebral perfusion pressure. The increased intracranial pressure forces CSF out, thus restoring the balance. If the hemorrhage is small, this may be enough to keep the patient alive with only minimal neurologic deficits. But if the bleeding is heavy, intracranial pressure increases rapidly and perfusion stops. Even if the pressure returns to normal, many brain cells die.

Initially, the ruptured cerebral blood vessels may constrict to limit the blood loss. This vasospasm further compromises blood flow, leading to more ischemia and cellular damage. If a clot forms in the vessel, decreased blood flow also promotes ischemia. If the blood enters the subarachnoid space, meningeal irritation occurs. The blood cells that pass through the vessel wall into the surrounding tissue also may break down and block the arachnoid villi, causing hydrocephalus.

Signs and symptoms

The clinical features of CVA vary according to the affected artery and the region of the brain it supplies, the severity of the damage, and the extent of collateral circulation developed. A CVA in one hemisphere causes signs and symptoms on the opposite side of the body; a CVA that damages cranial nerves affects structures on the same side as the infarction. General symptoms of a CVA include:
- unilateral limb weakness
- speech difficulties
- numbness on one side
- headache
- visual disturbances (diplopia, hemianopsia, ptosis)
- dizziness
- anxiety
- altered level of consciousness.

Additionally, symptoms are usually classified by the artery affected. Signs and symptoms associated with middle cerebral artery involvement include:
- aphasia
- dysphasia
- visual field deficits
- hemiparesis of affected side (more severe in face and arm than leg).

Symptoms associated with carotid artery involvement include:
- weakness
- paralysis
- numbness
- sensory changes
- visual disturbances on the affected side
- altered level of consciousness
- bruits
- headaches
- aphasia

> ## UNDERSTANDING TIAs
>
> A transient ischemic attack (TIA) is an episode of neurologic deficit resulting from cerebral ischemia. The recurrent attacks may last from seconds to hours and clear within 12 to 24 hours. It is usually considered a warning sign for cerebrovascular accident (CVA). In fact, TIAs have been reported in over one-half of the patients who have developed a CVA, usually within 2 to 5 years.
>
> In a TIA, microemboli released from a thrombus may temporarily interrupt blood flow, especially in the small distal branches of the brain's arterial tree. Small spasms in those arterioles may impair blood flow and also precede a TIA.
>
> The most distinctive features of TIAs are transient focal deficits with complete return of function. The deficits usually involve some degree of motor or sensory dysfunction. They may range to loss of consciousness and loss of motor or sensory function, only for a brief time. Commonly the patient experiences weakness in the lower part of the face and arms, hands, fingers, and legs on the side opposite the affected region. Other manifestations may include transient dysphagia, numbness or tingling of the face and lips, double vision, slurred speech, and dizziness.

- ptosis.

Symptoms associated with vertebrobasilary artery involvement include:
- weakness on the affected side
- numbness around lips and mouth
- visual field deficits
- diplopia
- poor coordination
- dysphagia
- slurred speech

- dizziness
- nystagmus
- amnesia
- ataxia.

Signs and symptoms associated with anterior cerebral artery involvement include:
- confusion
- weakness
- numbness, especially in the legs on the affected side
- incontinence
- loss of coordination
- impaired motor and sensory functions
- personality changes.

Signs and symptoms associated with posterior cerebral artery involvement include:
- visual field deficits (homonymous hemianopsia)
- sensory impairment
- dyslexia
- preservation (abnormally persistent replies to questions)
- coma
- cortical blindness
- absence of paralysis (usually).

Complications
Complications vary with the severity and type of CVA, but may include:
- unstable blood pressure (from loss of vasomotor control)
- cerebral edema
- fluid imbalances
- sensory impairment
- infections, such as pneumonia
- altered level of consciousness
- aspiration
- contractures
- pulmonary embolism
- death.

Diagnosis
- Computed tomography identifies ischemic stroke within first 72 hours of symptom onset, and evidence of hemorrhagic stroke (lesions larger than 1 cm) immediately.

- Magnetic resonance imaging assists in identifying areas of ischemia or infarction and cerebral swelling.
- Cerebral angiography reveals disruption or displacement of the cerebral circulation by occlusion, such as stenosis or acute thrombus, or hemorrhage.
- Digital subtraction angiography shows evidence of occlusion of cerebral vessels, lesions, or vascular abnormalities.
- Carotid duplex scan identifies stenosis greater than 60%.
- Brain scan shows ischemic areas but may not be conclusive for up to 2 weeks after CVA.
- Single photon emission computed tomography (SPECT) and positron emission tomography (PET) identifies areas of altered metabolism surrounding lesions not yet able to be detected by other diagnostic tests.
- Transesophageal echocardiogram reveals cardiac disorders, such as atrial thrombi, atrial septal defect or patent foramen ovale, as causes of thrombotic CVA.
- Lumbar puncture reveals bloody CSF when CVA is hemorrhagic.
- Ophthalmoscopy may identify signs of hypertension and atherosclerotic changes in retinal arteries.
- Electroencephalogram helps identify damaged areas of the brain.

Treatment
Treatment is supportive to minimize and prevent further cerebral damage. Measures include:
- ICP management with monitoring, hyperventilation (to decrease partial pressure of arterial CO_2 to lower ICP), osmotic diuretics (mannitol, to reduce cerebral edema), and corticosteroids (dexamethasone, to reduce inflammation and cerebral edema)
- stool softeners to prevent straining, which increases ICP
- anticonvulsants to treat or prevent seizures

surgery for large cerebellar infarction to remove infarcted tissue and decompress remaining live tissue

aneurysm repair to prevent further hemorrhage

percutaneous transluminal angioplasty or stent insertion to open occluded vessels.

For ischemic CVA:

thrombolytic therapy (tPA, alteplase [Activase]) within the first 3 hours after onset of symptoms to dissolve the clot, remove occlusion, and restore blood flow, thus minimizing cerebral damage (See *Treating ischemic CVA,* pages 272 and 273.)

anticoagulant therapy (heparin, warfarin) to maintain vessel patency and prevent further clot formation.

For TIAs:

antiplatelet agents (aspirin, ticlopidine) to reduce the risk of platelet aggregation and subsequent clot formation (for patients with TIAs)

carotid endarterectomy (for TIA) to open partially occluded carotid arteries.

For hemorrhagic CVAs:

analgesics such as acetaminophen to relieve headache associated with hemorrhagic CVA.

Guillain-Barré syndrome

Also known as infectious polyneuritis, Landry-Guillain-Barré syndrome, or acute idiopathic polyneuritis, Guillain-Barré syndrome is an acute, rapidly progressive, and potentially fatal form of polyneuritis that causes muscle weakness and mild distal sensory loss.

This syndrome can occur at any age but is most common between ages 30 and 50. It affects both sexes equally. Recovery is spontaneous and complete in about 95% of patients, although mild motor or reflex deficits may persist in the feet and legs. The prognosis is best when symptoms clear before 15 to 20 days after onset.

This syndrome occurs in three phases:

Acute phase begins with onset of the first definitive symptom and ends 1 to 3 weeks

later. Further deterioration does not occur after the acute phase.

Plateau phase lasts several days to 2 weeks.

Recovery phase is believed to coincide with remyelination and regrowth of axonal processes. It extends over 4 to 6 months, but may last up to 2 to 3 years if the disease was severe.

Causes

The precise cause of Guillain-Barré syndrome is unknown, but it may be a cell-mediated immune response to a virus.

About 50% of patients with Guillain-Barré syndrome have a recent history of minor febrile illness, usually an upper respiratory tract infection or, less often, gastroenteritis. When infection precedes the onset of Guillain-Barré syndrome, signs of infection subside before neurologic features appear.

Other possible precipitating factors include:

surgery

rabies or swine influenza vaccination

Hodgkin's or other malignant disease

systemic lupus erythematosus.

Pathophysiology

The major pathologic manifestation is segmental demyelination of the peripheral nerves. This prevents normal transmission of electrical impulses along the sensorimotor nerve roots. Because this syndrome causes inflammation and degenerative changes in both the posterior (sensory) and the anterior (motor) nerve roots, signs of sensory and motor losses occur simultaneously. (See *Understanding sensorimotor nerve degeneration,* page 274.) Additionally, autonomic nerve transmission may be impaired.

Signs and symptoms

Symptoms are progressive and include:

symmetrical muscle weakness (major neurologic sign) appearing in the legs first

(Text continues on page 274.)

TREATING ISCHEMIC CVA

In an ischemic cerebrovascular accident (CVA), a thrombus occludes a cerebral vessel or one of its branches and blocks blood flow to the brain. The thrombus may either have formed in that vessel or have lodged there after traveling through the circulation from another site, such as the heart. Prompt treatment with thrombolytic agents or anticoagulants helps to min-

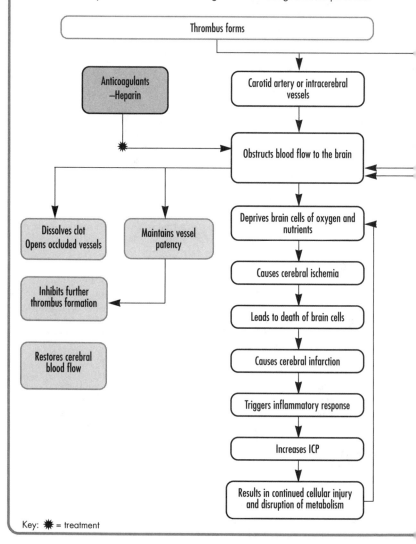

Key: ✳ = treatment

imize the effects of the occlusion. This flowchart shows how these drugs disrupt an ischemic CVA, thus minimizing the effects of cerebral ischemia and infarction. Keep in mind that thrombolytic agents should be used only within 3 hours after onset of the patient's symptoms.

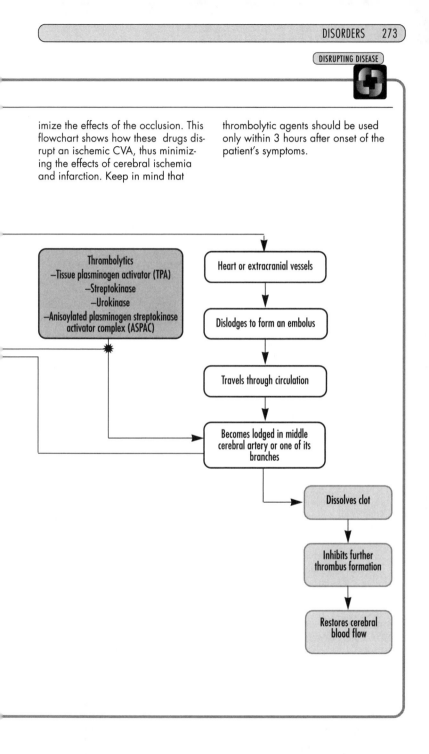

Thrombolytics
–Tissue plasminogen activator (TPA)
–Streptokinase
–Urokinase
–Anisoylated plasminogen streptokinase activator complex (ASPAC)

Heart or extracranial vessels

Dislodges to form an embolus

Travels through circulation

Becomes lodged in middle cerebral artery or one of its branches

Dissolves clot

Inhibits further thrombus formation

Restores cerebral blood flow

UNDERSTANDING SENSORIMOTOR NERVE DEGENERATION

Guillain-Barré syndrome attacks the peripheral nerves so that they can't transmit messages to the brain correctly. Here's what goes wrong.

The myelin sheath degenerates for unknown reasons. This sheath covers the nerve axons and conducts electrical impulses along the nerve pathways. Degeneration brings inflammation, swelling, and patchy demyelination. As this disorder destroys myelin, the nodes of Ranvier (at the junction of the myelin sheaths) widen. This delays and impairs impulse transmission along both the dorsal and anterior nerve roots.

Because the dorsal nerve roots handle sensory function, the patient may experience tingling and numbness. Similarly, because the anterior nerve roots are responsible for motor function, impairment causes varying weakness, immobility, and paralysis.

(ascending type) and then extending to the arms and facial nerves within 24 to 72 hours, from impaired anterior nerve root transmission

▶ muscle weakness developing in the arms first (descending type), or in the arms and legs simultaneously, from impaired anterior nerve root transmission

▶ muscle weakness absent or affecting only the cranial nerves (in mild forms)

▶ paresthesia, sometimes preceding muscle weakness but vanishing quickly, from impairment of the dorsal nerve root transmission

▶ diplegia, possibly with ophthalmoplegia (ocular paralysis), from impaired motor nerve root transmission and involvement of cranial nerves III, IV, and VI

▶ dysphagia or dysarthria and, less often, weakness of the muscles supplied by cranial nerve XI (spinal accessory nerve)

▶ hypotonia and areflexia from interruption of the reflex arc.

Complications
Common complications include:
▶ thrombophlebitis
▶ pressure ulcers
▶ muscle wasting
▶ sepsis
▶ joint contractures
▶ aspiration
▶ respiratory tract infections
▶ mechanical respiratory failure

▶ sinus tachycardia or bradycardia
▶ hypertension and postural hypotension
▶ loss of bladder and bowel sphincter control.

Diagnosis
▶ Cerebrospinal fluid analysis by lumbar puncture reveals elevated protein levels, peaking in 4 to 6 weeks, probably a result of widespread inflammation of the nerve roots; the CSF white blood cell count remains normal, but in severe disease, CSF pressure may rise above normal.

▶ Complete blood count shows leukocytosis with immature forms early in the illness, then quickly returns to normal.

▶ Electromyography possibly shows repeated firing of the same motor unit, instead of widespread sectional stimulation.

▶ Nerve conduction velocities show slowing soon after paralysis develops.

▶ Serum immunoglobulin levels reveal elevated levels from inflammatory response.

Treatment
▶ Primarily supportive, treatments include endotracheal intubation or tracheotomy if respiratory muscle involvement causes difficulty in clearing secretions.

▶ Trial dose (7 days) of prednisone is given to reduce inflammatory response if the disease is relentlessly progressive; if prednisone produces no noticeable improvement, the drug is discontinued.

▶ Plasmapheresis is useful during the initial phase but of no benefit if begun 2 weeks after onset.

▶ Continuous electrocardiogram monitoring alerts for possible arrhythmias from autonomic dysfunction; propranolol treats tachycardia and hypertension, or atropine given for bradycardia; volume replacement for severe hypotension.

Head trauma

Head trauma describes any traumatic insult to the brain that results in physical, intellectual, emotional, social, or vocational changes. Young children 6 months to 2 years of age, persons 15 to 24 years of age, and the elderly are at highest risk for head trauma. Risk in men is double the risk in women.

CULTURAL DIVERSITY African Americans and persons of any ethnicity living in poor socioeconomic groups appear to be at greatest risk for head trauma.

Head trauma is generally categorized as closed or open trauma. Closed trauma, or blunt trauma as it is sometimes called, is more common. It typically occurs when the head strikes a hard surface or a rapidly moving object strikes the head. The dura is intact, and no brain tissue is exposed to the external environment. In open trauma, as the name suggests, an opening in the scalp, skull, meninges, or brain tissue, including the dura, exposes the cranial contents to the environment, and the risk of infection is high.

Mortality from head trauma has declined with advances in preventative measures such as seat belts and airbags, quicker response and transport times, and improved treatment, including the development of regional trauma centers. Advances in technology have increased the effectiveness of rehabilitative services, even for patients with severe head injuries.

Causes

▶ Transportation/motor vehicle crashes (number one cause)

▶ Falls
▶ Sports-related accidents
▶ Crime and assaults.

Pathophysiology

The brain is shielded by the cranial vault (hair, skin, bone, meninges, and CSF), which intercepts the force of a physical blow. Below a certain level of force (the absorption capacity), the cranial vault prevents energy from affecting the brain. The degree of traumatic head injury usually is proportional to the amount of force reaching the cranial tissues. Furthermore, unless ruled out, neck injuries should be presumed present in patients with traumatic head injury.

Closed trauma is typically a sudden acceleration-deceleration or coup/contrecoup injury. In coup/contrecoup, the head hits a relatively stationary object, injuring cranial tissues near the point of impact (coup); then the remaining force pushes the brain against the opposite side of the skull, causing a second impact and injury (contrecoup). Contusions and lacerations may also occur during contrecoup as the brain's soft tissues slide over the rough bone of the cranial cavity. Also, the cerebrum may endure rotational shear, damaging the upper midbrain and areas of the frontal, temporal, and occipital lobes.

Open trauma may penetrate the scalp, skull, meninges, or brain. Open head injuries are usually associated with skull fractures, and bone fragments often cause hematomas and meningeal tears with consequent loss of CSF.

Signs and symptoms

Types of head trauma include concussion, contusion, epidural hematoma, subdural hematoma, intracerebral hematoma, and skull fractures. Each is associated with specific signs and symptoms. (See *Types of head trauma*, pages 276 to 279.)

Complications

▶ Increased ICP
▶ Infection (open trauma)

(Text continues on page 278.)

TYPES OF HEAD TRAUMA

This chart summarizes the signs and symptoms and diagnostic test findings
for the different types of head trauma.

TYPE	DESCRIPTION
Concussion (closed head injury)	▶ A blow to the head hard enough to make the brain hit the skull but not hard enough to cause a cerebral contusion causes temporary neural dysfunction. ▶ Recovery is usually complete within 24 to 48 hours. ▶ Repeated injuries exact a cumulative toll on the brain.
Contusion (bruising of brain tissue; more serious than concussion)	▶ Most common in 20 to 40 year olds. ▶ Most result from arterial bleeding. ▶ Blood commonly accumulates between skull and dura. Injury to middle meningeal artery in parietotemporal area is most common and is frequently accompanied by linear skull fractures in temporal region over middle meningeal artery. ▶ Less commonly arises from dural venous sinuses.
Epidural hematoma	▶ Acceleration-deceleration or coup-contrecoup injuries disrupt normal nerve functions in bruised area. ▶ Injury is directly beneath the site of impact when the brain rebounds against the skull from the force of a blow (a beating with a blunt instrument, for example), when the force of the blow drives the brain against the opposite side of the skull, or when the head is hurled forward and stopped abruptly (as in an automobile crash when a driver's head strikes the windshield). ▶ Brain continues moving and slaps against the skull (acceleration), then rebounds (deceleration). Brain may strike bony prominences inside the skull (especially the sphenoidal ridges), causing intracranial hemorrhage or hematoma that may result in tentorial herniation.

SIGNS AND SYMPTOMS

▶ Short-term loss of consciousness secondary to disruption of RAS, possibly due to abrupt pressure changes in the areas responsible for consciousness, changes in polarity of the neurons, ischemia, or structural distortion of neurons

▶ Vomiting from localized injury and compression

▶ Anterograde and retrograde amnesia (patient can't recall events immediately after the injury or events that led up to the traumatic incident) correlating with severity of injury; all related to disruption of RAS

▶ Irritability or lethargy from localized injury and compression

▶ Behavior out of character due to focal injury

▶ Complaints of dizziness, nausea, or severe headache due to focal injury and compression

▶ Severe scalp wounds from direct injury

▶ Labored respiration and loss of consciousness secondary to increased pressure from bruising

▶ Drowsiness, confusion, disorientation, agitation, or violence from increased ICP associated with trauma

▶ Hemiparesis related to interrupted blood flow to the site of injury

▶ Decorticate or decerebrate posturing from cortical damage or hemispheric dysfunction

▶ Unequal pupillary response from brain stem involvement.

▶ Brief period of unconsciousness after injury reflects the concussive effects of head trauma, followed by a lucid interval varying from 10-15 minutes to hours or, rarely, days.

▶ Severe headache

▶ Progressive loss of consciousness and deterioration in neurologic signs results from expanding lesion and extrusion of medial portion of temporal lobe through tentorial opening.

▶ Compression of brainstem by temporal lobe causes clinical manifestations of intracranial hypertension.

▶ Deterioration in level of consciousness results from compression of brainstem reticular formation as temporal lobe herniates on its upper portion.

▶ Respirations, initially deep and labored, become shallow and irregular as brainstem is impacted.

▶ Contralateral motor deficits reflect compression of corticospinal tracts that pass through the brainstem.

DIAGNOSTIC TEST FINDINGS

▶ Computed tomography(CT) reveals no sign of fracture, bleeding or other nervous system lesion.

▶ CT shows changes in tissue density, possible displacement of the surrounding structures, and evidence of ischemic tissue, hematomas, and fractures.

▶ Lumbar puncture with CSF analysis reveals increased pressure and blood (not performed if hemorrhage is suspected).

▶ EEG recordings directly over area of contusion reveal progressive abnormalities by appearance of high-amplitude theta and delta waves.

▶ CT or magnetic resonance imaging (MRI) identifies abnormal masses or structural shifts within the cranium

(continued)

TYPES OF HEAD TRAUMA *(continued)*

TYPE	DESCRIPTION
Epidural hematoma *(continued)*	
Subdural hematoma	▶ Meningeal hemorrhages, resulting from accumulation of blood in subdural space (between dura mater and arachnoid) are most common. ▶ May be acute, subacute, and chronic: unilateral or bilateral ▶ Usually associated with torn connecting veins in cerebral cortex; rarely from arteries. ▶ Acute hematomas are a surgical emergency.
Intracerebral hematoma	▶ Subacute hematomas have better prognosis because venous bleeding tends to be slower. ▶ Traumatic or spontaneous disruption of cerebral vessels in brain parenchyma cause neurologic deficits, depending on site and amount of bleeding. ▶ Shear forces from brain movement frequently cause vessel laceration and hemorrhage into the parenchyma. ▶ Frontal and temporal lobes are common sites. Trauma is associated with few intracerebral hematomas; most caused by result of hypertension.
Skull fractures	▶ 4 types: linear, comminuted, depressed, basilar ▶ Fractures of anterior and middle fossae are associated with severe head trauma and are more common than those of posterior fossa. ▶ Blow to the head causes one or more of the types. May not be problematic unless brain is exposed or bone fragments are driven into neural tissue.

▶ Respiratory depression and failure
▶ Brain herniation.

Diagnosis

Each type of head trauma is associated with specific diagnostic findings. (See *Types of head trauma*, pages 276 to 279.)

SIGNS AND SYMPTOMS	DIAGNOSTIC TEST FINDINGS
▶ Ipsilateral (same-side) pupillary dilation due to compression of third cranial nerve ▶ Seizures possible from high ICP ▶ Continued bleeding leads to progressive neurologic degeneration, evidenced by bilateral pupillary dilation, bilateral decerebrate response, increased systemic blood pressure, decreased pulse, and profound coma with irregular respiratory patterns.	
▶ Similar to epidural hematoma but significantly slower in onset because bleeding is typically of venous origin	▶ CT, x-rays, and arteriography reveal mass and altered blood flow in the area, confirming hematoma. ▶ CT or MRI reveals evidence of masses and tissue shifting. ▶ CSF is yellow and has relatively low protein (chronic subdural hematoma).
▶ Unresponsive immediately or experiencing a lucid period before lapsing into a coma from increasing ICP and mass effect of hemorrhage ▶ Possible motor deficits and decorticate or decerebrate responses from compression of corticospinal tracts and brain stem	▶ CT or cerebral arteriography identifies bleeding site. CSF pressure elevated pressure; fluid may appear bloody or xanthochromic (yellow or straw-colored) from hemoglobin breakdown.
▶ Possibly asymptomatic, depending on underlying brain trauma. ▶ Discontinuity and displacement of bone structure with severe fracture ▶ Motor sensory and cranial nerve dysfunction with associated facial fractures ▶ Persons with anterior fossa basilar skull fractures may have periorbital ecchymosis (raccoon eyes), anosmia (loss of smell due to first cranial nerve involvement) and pupil abnormalities (second and third cranial nerve involvement). ▶ CSF rhinorrhea (leakage through nose), CSF otorrhea (leakage from the ear), hemotympanium (blood accumulation at the tympanic membrane), ecchymosis over the mastoid bone (Battle's sign), and facial paralysis (seventh cranial nerve injury) accompany middle fossa basilar skull fractures. ▶ Signs of medullary dysfunction such as cardiovascular and respiratory failure accompany posterior fossa basilar skull fracture.	▶ CT and MRI reveal intracranial hemorrhage from ruptured blood vessels and swelling. ▶ Skull x-ray may reveal fracture. ▶ Lumbar puncture contraindicated by expanding lesions.

Treatment

Surgical treatment includes:
▶ evacuation of the hematoma or a craniotomy to elevate or remove fragments that have been driven into the brain, and to extract foreign bodies and necrotic tissue, thereby reducing the risk of infection and further brain damage from fractures.

Supportive treatment includes:

close observation to detect changes in neurologic status suggesting further damage or expanding hematoma

▶ cleaning and debridement of any wounds associated with skull fractures

▶ diuretics, such as mannitol, and corticosteroids, such as dexamethasone, are given to reduce cerebral edema

▶ analgesics, such as acetaminophen, are given to relieve complaints of headache

▶ anticonvulsants, such as phenytoin, to prevent and treat seizures

▶ respiratory support, including mechanical ventilation and endotracheal intubation, is given as indicated for respiratory failure from brainstem involvement

▶ prophylactic antibiotics are given to prevent the onset of meningitis from CSF leakage associated with skull fractures.

Herniated intervertebral disk

Also called a ruptured or slipped disk or a herniated nucleus pulposus, a herniated disk occurs when all or part of the nucleus pulposus — the soft, gelatinous, central portion of an intervertebral disk — is forced through the disk's weakened or torn outer ring (anulus fibrosus).

Herniated disks usually occur in adults (mostly men) under age 45. About 90% of herniated disks occur in the lumbar and lumbosacral regions, 8% occur in the cervical area, and 1% to 2% occur in the thoracic area. Patients with a congenitally small lumbar spinal canal or with osteophyte formation along the vertebrae may be more susceptible to nerve root compression and more likely to have neurologic symptoms.

Causes

The two major causes of herniated intervertebral disk are:

▶ severe trauma or strain

▶ intervertebral joint degeneration.

 AGE ALERT In older patients whose disks have begun to degenerate, minor trauma may cause herniation.

Pathophysiology

An intervertebral disk has two parts: the soft center called the nucleus pulposus and the tough, fibrous surrounding ring called the anulus fibrosus. The nucleus pulposus acts as a shock absorber, distributing the mechanical stress applied to the spine when the body moves. Physical stress, usually a twisting motion, can tear or rupture the anulus fibrosus so that the nucleus pulposus herniates into the spinal canal. The vertebrae move closer together and in turn exert pressure on the nerve roots as they exit between the vertebrae. Pain and possibly sensory and motor loss follow. A herniated disk also can occur with intervertebral joint degeneration. If the disk has begun to degenerate, minor trauma may cause herniation.

Herniation occurs in three steps:

▶ *protrusion:* nucleus pulposus presses against the anulus fibrosus

▶ *extrusion:* nucleus pulposus bulges forcibly though the anulus fibrosus, pushing against the nerve root

▶ *sequestration:* annulus gives way as the disk's core bursts presses against the nerve root.

Signs and symptoms

Signs and symptoms include:

▶ severe low back pain to the buttocks, legs, and feet, usually unilaterally, from compression of nerve roots supplying these areas

▶ sudden pain after trauma, subsiding in a few days, and then recurring at shorter intervals and with progressive intensity

▶ sciatic pain following trauma, beginning as a dull pain in the buttocks; Valsalva's maneuver, coughing, sneezing, and bending intensify the pain, which is often accompanied by muscle spasms from pressure and irritation of the sciatic nerve root

▶ sensory and motor loss in the area innervated by the compressed spinal nerve root and, in later stages, weakness and atrophy of leg muscles.

Complications

Complications are dependent on the severity and the specific site of herniation. Common complications include:

▶ neurologic deficits
▶ bowel and bladder problems.

Diagnosis

▶ Straight leg raising test is positive only if the patient has posterior leg (sciatic) pain, not back pain.
▶ Lasègue's test reveals resistance and pain as well as loss of ankle or knee-jerk reflex, indicating spinal root compression.
▶ Spinal X-rays rule out other abnormalities but may not diagnose a herniated disk because a marked disk prolapse may not be apparent on a normal X-ray.
▶ Myelogram, computed tomography, and magnetic resonance imaging show spinal canal compression by herniated disk material.

Treatment

Treatment may include:

▶ heat applications to aid in pain relief
▶ exercise program to strengthen associated muscles and prevent further deterioration
▶ anti-inflammatory agents, such as aspirin and NSAIDs, to reduce inflammation and edema at the site of injury; rarely, corticosteroids such as dexamethasone for the same purpose; muscle relaxants, such as diazepam, methocarbamol, or cyclobenzdiazaprin, to minimize muscle spasm from nerve root irritation
▶ surgery, including laminectomy to remove the protruding disk, spinal fusion to overcome segmental instability; or both together to stabilize the spine
▶ chemonucleolysis (injection of the enzyme chymopapain into the herniated disk) to dissolve the nucleus pulposus; microdiskectomy to remove fragments of the nucleus pulposus.

Huntington's disease

Also called Huntington's chorea, hereditary chorea, chronic progressive chorea, and adult chorea, Huntington's disease is a hereditary disorder in which degeneration of the cerebral cortex and basal ganglia causes chronic progressive chorea (involuntary and irregular movements) and cognitive deterioration, ending in dementia.

Huntington's disease usually strikes people between ages 25 and 55 (the average age is 35), affecting men and women equally. However, 2% of cases occur in children, and 5% occur as late as age 60. Death usually results 10 to 15 years after onset from suicide, heart failure, or pneumonia.

Causes

The actual cause of this disorder is unknown. However, it is transmitted as an autosomal dominant trait, which either sex can transmit and inherit. Each child of an affected parent has a 50% chance of inheriting it; the child who doesn't inherit it can't transmit it. Because of hereditary transmission and delayed expression, Huntington's disease is prevalent in areas where affected families have lived for several generations. Genetic testing is now available to families with a known history of the disease.

Pathophysiology

Huntington's disease involves a disturbance in neurotransmitter substances, primarily gamma aminobutyric acid (GABA) and dopamine. In the basal ganglia, frontal cortex, and cerebellum, GABA neurons are destroyed and replaced by glial cells. The consequent deficiency of GABA (an inhibitory neurotransmitter) results in a relative excess of dopamine and abnormal neurotransmission along the affected pathways.

Signs and symptoms

The onset of this disease is insidious. The patient eventually becomes totally dependent — emotionally and physically — through loss of musculoskeletal control.

Neurologic manifestations include:

▶ Progressively severe choreic movements are due to the relative excess of dopamine. Such movements are rapid, often violent, and purposeless.

▶ Choreic movements are initially unilateral and more prominent in the face and arms than in the legs. They progress from mild fidgeting to grimacing, tongue smacking, dysarthria (indistinct speech), emotion-related athetoid (slow, twisting, snakelike) movements (especially of the hands) from injury to the basal ganglion, and torticollis due to shortening of neck muscles.

▶ Bradykinesia (slow movement) is often accompanied by rigidity.

▶ Impairment of both voluntary and involuntary movement is due to the combination of chorea, bradykinesia, and normal muscle strength.

▶ Dysphagia occurs in most patients in the advanced stages.

▶ Dysarthria may be complicated by perseveration (persistent repetition of a reply), oral apraxia (difficulty coordinating movement of the mouth), and aprosody (inability to accurately reproduce or interpret the tone of language).

Cognitive signs and symptoms may include:

▶ dementia, an early indication of the disease, from dysfunction of the subcortex without significant impairment of immediate memory

▶ problems with recent memory due to retrieval rather than encoding problems

▶ deficits of executive function (planning, organizing, regulating, and programming) from frontal lobe involvement

▶ impaired impulse control.

The patient also may exhibit psychiatric symptoms, often before movement problems occur. Psychiatric symptoms may include:

▶ depression and possible mania (earliest symptom) related to altered levels of dopamine and GABA

▶ personality changes including irritability, lability, impulsiveness, and aggressive behavior.

Complications

Common complications of Huntington's disease include:

▶ choking
▶ aspiration
▶ pneumonia
▶ heart failure
▶ infections.

Diagnosis

▶ Genetic testing reveals autosomal dominant trait.

▶ Positron emission tomography confirms disorder.

▶ Pneumoencephalogram reveals characteristic butterfly dilation of brain's lateral ventricles.

▶ Computed tomography and magnetic resonance imaging show brain atrophy.

Treatment

No known cure exists for Huntington's disease. Treatment is symptom-based, supportive, and protective. It may include:

▶ haloperidol or diazepam to modify choreic movements and control behavioral manifestations and depression

▶ psychotherapy to decrease anxiety and stress and manage psychiatric symptoms

▶ institutionalization to manage progressive mental deterioration and self-care deficits.

Hydrocephalus

An excessive accumulation of CSF within the ventricular spaces of the brain, hydrocephalus occurs most often in neonates. It can also occur in adults as a result of injury or disease. In infants, hydrocephalus enlarges the head, and in both infants and adults, the resulting compression can damage brain tissue.

With early detection and surgical intervention, the prognosis improves but remains guarded. Even after surgery, complications may persist, such as developmental delay, impaired motor function, and vision loss. Without surgery, the prognosis is poor. Mortality may result from increased intracranial pressure (ICP) in people of all ages; infants may die of infection and malnutrition.

Causes

Hydrocephalus may result from:
▶ obstruction in CSF flow (noncommunicating hydrocephalus)
▶ faulty absorption of CSF (communicating hydrocephalus).

Risk factors associated with the development of hydrocephalus in infants may include:
▶ intrauterine infection
▶ intracranial hemorrhage from birth trauma or prematurity.

In older children and adults, risk factors may include:
▶ meningitis
▶ mastoiditis
▶ chronic otitis media
▶ brain tumors or intracranial hemorrhage.

Pathophysiology

In noncommunicating hydrocephalus, the obstruction occurs most frequently between the third and fourth ventricles, at the aqueduct of Sylvius, but it can also occur at the outlets of the fourth ventricle (foramina of Luschka and Magendie) or, rarely, at the foramen of Monro. This obstruction may result from faulty fetal development, infection (syphilis, granulomatous diseases, meningitis), a tumor, a cerebral aneurysm, or a blood clot (after intracranial hemorrhage).

In communicating hydrocephalus, faulty absorption of CSF may result from surgery to repair a myelomeningocele, adhesions between meninges at the base of the brain, or meningeal hemorrhage. Rarely, a tumor in the choroid plexus causes overproduction of CSF and consequent hydrocephalus.

In either type, both CSF pressure and volume increase. Obstruction in the ventricles causes dilation, stretching, and disruption of the lining. Underlying white matter atrophies. Compression of brain tissue and cerebral blood vessels leads to ischemia and, eventually, cell death.

Signs and symptoms

In infants, the signs and symptoms typically include:
▶ enlargement of the head clearly disproportionate to the infant's growth (most characteristic sign) from the increased CSF volume
▶ distended scalp veins from increased CSF pressure
▶ thin, shiny, fragile-looking scalp skin from the increase in CSF pressure
▶ underdeveloped neck muscles from increased weight of the head
▶ depressed orbital roof with downward displacement of the eyes and prominent sclerae from increased pressure
▶ high-pitched, shrill cry, irritability, and abnormal muscle tone in the legs from neurologic compression
▶ projectile vomiting from increased ICP
▶ skull widening to accommodate increased pressure.

In adults and older children, indicators of hydrocephalus include:
▶ decreased level of consciousness (LOC) from increasing ICP
▶ ataxia from compression of the motor areas
▶ incontinence
▶ impaired intellect.

Complications

Complications may include:
▶ mental retardation
▶ impaired motor function
▶ vision loss
▶ brain herniation
▶ death from increased ICP
▶ infection
▶ malnutrition
▶ shunt infection (following surgery)
▶ septicemia (following shunt insertion)
▶ paralytic ileus, adhesions, peritonitis, and intestinal perforation (following shunt insertion).

Diagnosis

▶ Skull X-rays show thinning of the skull with separation of sutures and widening of the fontanelles.

MOST COMMON SITES OF CEREBRAL ANEURYSM

Cerebral aneurysms usually arise at the arterial bifurcation in the Circle of Willis and its branches. This illustration shows the most common sites around this circle.

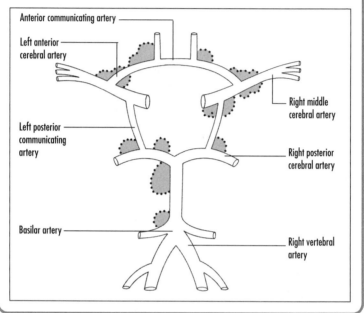

▶ Angiography shows vessel abnormalities caused by stretching.

▶ Computed tomography and magnetic resonance imaging reveal variations in tissue density and fluid in the ventricular system.

▶ Lumbar puncture reveals increased fluid pressure from communicating hydrocephalus.

▶ Ventriculography shows ventricular dilation with excess fluid.

Treatment

The only treatment for hydrocephalus is surgical correction, by insertion of:

▶ ventriculoperitoneal shunt, which transports excess fluid from the lateral ventricle into the peritoneal cavity.

▶ ventriculoatrial shunt (less common), which drains fluid from the brain's lateral ventricle into the right atrium of the heart, where the fluid makes its way into the venous circulation.

Supportive care is also warranted.

Intracranial aneurysm

In an intracranial, or cerebral, aneurysm a weakness in the wall of a cerebral artery causes localized dilation. Its most common form is the berry aneurysm, a sac-like outpouching in a cerebral artery. Cerebral aneurysms usually arise at an arterial junction in the circle of Willis, the circular anastomosis forming the major cerebral arteries at the base of the brain. (See *Most common sites of cerebral aneurysm.*) Cerebral aneurysms often rupture and cause subarachnoid hemorrhage.

Incidence is slightly higher in women than in men, especially those in their late 40s or early to mid-50s, but a cerebral aneurysm may occur at any age in either sex. The prognosis is guarded. About half of all patients who suffer a subarachnoid hemorrhage die immediately. Of those who survive untreated, 40% die from the effects of hemorrhage and another 20% die later from recurring hemorrhage. New treatments are improving the prognosis.

Causes
Causes may include:
▶ congenital defect
▶ degenerative process
▶ combination of both
▶ trauma.

Pathophysiology
Blood flow exerts pressure against a congenitally weak arterial wall, stretching it like an overblown balloon and making it likely to rupture. Such a rupture is followed by a subarachnoid hemorrhage, in which blood spills into the space normally occupied by CSF. Sometimes, blood also spills into brain tissue, where a clot can cause potentially fatal increased ICP and brain tissue damage.

Signs and symptoms
The patient may exhibit premonitory symptoms resulting from oozing of blood into the subarachnoid space. The symptoms, which may persist for several days, include:
▶ headache, intermittent nausea
▶ nuchal rigidity
▶ stiff back and legs.

Usually, however, the rupture occurs abruptly and without warning, causing:
▶ sudden severe headache caused by increased pressure from bleeding into a closed space.
▶ nausea and projectile vomiting related to increased pressure.
▶ altered level of consciousness, including deep coma, depending on the severity and location of bleeding, from increased pressure caused by increased cerebral blood volume.
▶ meningeal irritation, resulting in nuchal rigidity, back and leg pain, fever, restlessness, irritability, occasional seizures,

---

DETERMINING SEVERITY OF AN INTRACRANIAL ANEURYSM RUPTURE

The severity of symptoms varies from patient to patient, depending on the site and amount of bleeding. Five grades characterize ruptured cerebral aneurysm:
▶ *Grade I: minimal bleeding* — The patient is alert with no neurologic deficit; he may have a slight headache and nuchal rigidity.
▶ *Grade II: mild bleeding* — The patient is alert, with a mild to severe headache and nuchal rigidity; he may have third-nerve palsy.
▶ *Grade III: moderate bleeding* — The patient is confused or drowsy, with nuchal rigidity and, possibly, a mild focal deficit.
▶ *Grade IV: severe bleeding* — The patient is stuporous, with nuchal rigidity and, possibly, mild to severe hemiparesis.
▶ *Grade V: moribund (often fatal)* — If the rupture is nonfatal, the patient is in a deep coma or decerebrate.

---

photophobia, and blurred vision, secondary to bleeding into the meninges.
▶ hemiparesis, hemisensory defects, dysphagia, and visual defects from bleeding into the brain tissues.
▶ diplopia, ptosis, dilated pupil, and inability to rotate the eye from compression on the oculomotor nerve if aneurysm is near the internal carotid artery.

Typically, the severity of a ruptured intracranial aneurysm is graded according to the patient's signs and symptoms. (See *Determining severity of an intracranial aneurysm rupture.*)

Complications
The major complications associated with cerebral aneurysm include:
▶ death from increased ICP and brain herniation

▶ rebleeding
▶ vasospasm.

Diagnosis
▶ Cerebral angiography (the test of choice) reveals altered cerebral blood flow, vessel lumen dilation, and differences in arterial filling.
▶ Computed tomography reveals subarachnoid or ventricular bleeding with blood in subarachnoid space and displaced midline structures.
▶ Magnetic resonance imaging shows a cerebral blood flow void.
▶ Skull X-rays may reveal calcified wall of the aneurysm and areas of bone erosion.

Treatment
Treatment may include:
▶ bedrest in a quiet, darkened room with minimal stimulation, to reduce risk of rupture if it hasn't occurred
▶ surgical repair by clipping, ligation, or wrapping
▶ avoidance of coffee, other stimulants, and aspirin to reduce risk of rupture and elevation of blood pressure, which increases risk of rupture
▶ codeine or another analgesic as needed to maintain rest and minimize risk of pressure changes leading to rupture
▶ hydralazine or another antihypertensive agent, if the patient is hypertensive, to reduce risk of rupture
▶ calcium channel blockers to decrease spasm and subsequent rebleeding
▶ corticosteroids to reduce cerebral edema
▶ phenytoin or another anticonvulsant to prevent or treat seizures secondary to pressure and tissue irritation from bleeding
▶ phenobarbital or another sedative to prevent agitation leading to hypertension and reduce risk of rupture
▶ aminocaproic acid, an inhibitor of fibrinolysis, to minimize the risk of rebleeding by delaying blood clot lysis (drug's effectiveness under dispute).

Meningitis
In meningitis, the brain and the spinal cord meninges become inflamed, usually as a result of bacterial infection. Such inflammation may involve all three meningeal membranes — the dura mater, arachnoid, and pia mater.

In most patients, respiratory symptoms precede onset of meningitis. Approximately half of patients develop meningitis over 1 to 7 days; about 20% develop the disease in 1 to 3 weeks after onset of respiratory symptoms; and about 25% develop severe meningitis suddenly without respiratory symptoms.

If the disease is recognized early and the infecting organism responds to treatment, the prognosis is good and complications are rare. However, mortality in untreated meningitis is 70% to 100%. The prognosis is poorer for infants and elderly.

Causes
Meningitis is almost always a complication of bacteremia, especially from the following:
▶ pneumonia
▶ empyema
▶ osteomyelitis
▶ endocarditis.

Other infections associated with the development of meningitis include:
▶ sinusitis
▶ otitis media
▶ encephalitis
▶ myelitis
▶ brain abscess, usually caused by *Neisseria meningitidis, Haemophilus influenzae, Streptococcus pneumoniae,* and *Escherichia coli.*

Meningitis may follow trauma or invasive procedures, including:
▶ skull fracture
▶ penetrating head wound
▶ lumbar puncture
▶ ventricular shunting.

Aseptic meningitis may result from a virus or other organism. Sometimes no causative organism can be found.

Pathophysiology

Meningitis often begins as an inflammation of the pia-arachnoid, which may progress to congestion of adjacent tissues and destroy some nerve cells.

The microorganism typically enters the CNS by one of four routes:
◗ the blood (most common)
◗ a direct opening between the CSF and the environment as a result of trauma
◗ along the cranial and peripheral nerves
◗ through the mouth or nose.

Microorganisms can be transmitted to an infant via the intrauterine environment.

The invading organism triggers an inflammatory response in the meninges. In an attempt to ward off the invasion, neutrophils gather in the area and produce an exudate in the subarachnoid space, causing the CSF to thicken. The thickened CSG flows less readily around the brain and spinal cord, and it can block the arachnoid villi, obstructing flow of CSF and causing hydrocephalus.

The exudate also:
◗ exacerbates the inflammatory response, increasing the pressure in the brain.
◗ can extend to the cranial and peripheral nerves, triggering additional inflammation.
◗ irritates the meninges, disrupting their cell membranes and causing edema.

The consequences are elevated ICP, engorged blood vessels, disrupted cerebral blood supply, possible thrombosis or rupture, and, if ICP is not reduced, cerebral infarction. Encephalitis also may ensue as a secondary infection of the brain tissue.

In aseptic meningitis, lymphocytes infiltrate the pia-arachnoid layers, but usually not as severely as in bacterial meningitis, and no exudate is formed. Thus, this type of meningitis is self-limiting.

Signs and symptoms

The signs of meningitis typically include:
◗ fever, chills, and malaise resulting from infection and inflammation
◗ headache, vomiting and, rarely, papilledema (inflammation and edema of the optic nerve) from increased ICP.

Signs of meningeal irritation include:
◗ nuchal rigidity
◗ positive Brudzinski's and Kernig's signs
◗ exaggerated and symmetrical deep tendon reflexes
◗ opisthotonos (a spasm in which the back and extremities arch backward so that the body rests on the head and heels).

Other features of meningitis may include:
◗ sinus arrhythmias from irritation of the nerves of the autonomic nervous system
◗ irritability from increasing ICP
◗ photophobia, diplopia, and other visual problems from cranial nerve irritation
◗ delirium, deep stupor, and coma from increased ICP and cerebral edema.

An infant may show signs of infection, but most are simply fretful and refuse to eat. In an infant, vomiting can lead to dehydration, which prevents formation of a bulging fontanelle, an important sign of increased ICP.

As the illness progresses, twitching, seizures (in 30% of infants), or coma may develop. Most older children have the same symptoms as adults. In subacute meningitis, onset may be insidious.

Complications

Complications may include:
◗ increased ICP
◗ hydrocephalus
◗ cerebral infarction
◗ cranial nerve deficits including optic neuritis and deafness
◗ encephalitis
◗ paresis or paralysis
◗ endocarditis
◗ brain abscess
◗ syndrome of inappropriate antidiuretic hormone (SIADH)
◗ seizures
◗ coma.

In children, complications may include:
◗ mental retardation
◗ epilepsy

▶ unilateral or bilateral sensory hearing loss

▶ subdural effusions.

Diagnosis

▶ Lumbar puncture shows elevated CSF pressure (from obstructed CSF outflow at the arachnoid villi), cloudy or milky-white CSF, high protein level, positive Gram stain and culture (unless a virus is responsible), and decreased glucose concentration.

▶ Positive Brudzinski's and Kernig's signs indicate meningeal irritation.

▶ Cultures of blood, urine, and nose and throat secretions reveal the offending organism.

▶ Chest X-ray may reveal pneumonitis or lung abscess, tubercular lesions, or granulomas secondary to a fungal infection

▶ Sinus and skull X-rays may identify cranial osteomyelitis or paranasal sinusitis as the underlying infectious process, or skull fracture as the mechanism for entrance of microorganism.

▶ White blood cell count reveals leukocytosis.

▶ Computed tomography may reveal hydrocephalus or rule out cerebral hematoma, hemorrhage, or tumor as the underlying cause.

Treatment

Treatment may include:

▶ usually, I.V. antibiotics for at least 2 weeks, followed by oral antibiotics selected by culture and sensitivity testing

▶ digoxin, to control arrhythmias

▶ mannitol to decrease cerebral edema

▶ anticonvulsant (usually given I.V.) or a sedative to reduce restlessness and prevent or control seizure activity

▶ aspirin or acetaminophen to relieve headache and fever.

Supportive measures include:

▶ bed rest to prevent increases in ICP

▶ fever reduction to prevent hyperthermia and increased metabolic demands that may increase ICP

▶ fluid therapy (given cautiously if cerebral edema and increased ICP present) to prevent dehydration

▶ appropriate therapy for any coexisting conditions, such as endocarditis or pneumonia

▶ possible prophylactic antibiotics after ventricular shunting procedures, skull fracture, or penetrating head wounds, to prevent infection (use is controversial).

Staff should take droplet precautions (in addition to standard precautions) for meningitis caused by *H. influenzae* and *N. meningitidis,* until 24 hours after the start of effective therapy.

Multiple sclerosis

Multiple sclerosis (MS) causes demyelination of the white matter of the brain and spinal cord and damage to nerve fibers and their targets. Characterized by exacerbations and remissions, MS is a major cause of chronic disability in young adults. It usually becomes symptomatic between the ages of 20 and 40 (the average age of onset is 27). MS affects three women for every two men and five whites for every nonwhite. Incidence is generally higher among urban populations and upper socioeconomic groups. A family history of MS and living in a cold, damp climate increase the risk.

The prognosis varies. MS may progress rapidly, disabling the patient by early adulthood or causing death within months of onset. However, 70% of patients lead active, productive lives with prolonged remissions.

Several types of MS have been identified. Terms to describe MS types include:

▶ *Elapsing-remitting* — clear relapses (or acute attacks or exacerbations) with full recovery or partial recovery and lasting disability. The disease does not worsen between the attacks.

▶ *Primary progressive* — steady progression from the onset with minor recovery or plateaus. This form is uncommon and may involve different brain and spinal cord damage than other forms.

▶ *Secondary progressive* — begins as a pattern of clear-cut relapses and recovery. This form becomes steadily progressive and worsens between acute attacks

▶ *Progressive relapsing* — steadily progressive from the onset, but also has clear acute attacks. This form is rare.

Causes

The exact cause of MS is unknown, but current theories suggest that a slow-acting or latent viral infection triggers an autoimmune response. Other theories suggest that environmental and genetic factors may also be linked to MS.

Certain conditions appear to precede onset or exacerbation, including:
▶ emotional stress
▶ fatigue (physical or emotional)
▶ pregnancy
▶ acute respiratory infections.

Pathophysiology

In multiple sclerosis, sporadic patches of axon demyelination and nerve fiber loss occur throughout the central nervous system, inducing widely disseminated and varied neurologic dysfunction. (See *How myelin breaks down*, page 290.)

New evidence of nerve fiber loss may provide an explanation for the invisible neurologic deficits experienced by many patients with MS. The axons determine the presence or absence of function; loss of myelin does not correlate with loss of function.

Signs and symptoms

Signs and symptoms depend on the extent and site of myelin destruction, the extent of remyelination, and the adequacy of subsequent restored synaptic transmission. Flares may be transient, or they may last for hours or weeks, possibly waxing and waning with no predictable pattern, varying from day to day, and being bizarre and difficult for the patient to describe. Clinical effects may be so mild that the patient is unaware of them or so intense that they are debilitating. Typical first signs and symptoms related to conduction deficits and impaired impulse transmission along the nerve fiber include:
▶ visual problems
▶ sensory impairment, such as burning, pins and needles, and electrical sensations
▶ fatigue.

Other characteristic changes include:
▶ ocular disturbances — optic neuritis, diplopia, ophthalmoplegia, blurred vision, and nystagmus from impaired cranial nerve dysfunction and conduction deficits to the optic nerve
▶ muscle dysfunction — weakness, paralysis ranging from monoplegia to quadriplegia, spasticity, hyperreflexia, intention tremor, and gait ataxia from impaired motor reflex
▶ urinary disturbances — incontinence, frequency, urgency, and frequent infections from impaired transmission involving sphincter innervation
▶ bowel disturbances — involuntary evacuation or constipation from altered impulse transmission to internal sphincter
▶ fatigue — often the most debilitating symptom
▶ speech problems — poorly articulated or scanning speech and dysphagia from impaired transmission to the cranial nerves and sensory cortex.

Complications

Complications may include:
▶ injuries from falls
▶ urinary tract infection
▶ constipation
▶ joint contractures
▶ pressure ulcers
▶ rectal distention
▶ pneumonia
▶ depression.

Diagnosis

Because early symptoms may be mild, years may elapse between onset and diagnosis. Diagnosis of this disorder requires evidence of two or more neurologic attacks. Periodic testing and close observation are necessary, perhaps for years, de-

HOW MYELIN BREAKS DOWN

Myelin speeds electrical impulses to the brain for interpretation. This lipoprotein complex formed of glial cells or oligodendrocytes protects the neuron's axon much like the insulation on an electrical wire. Its high electrical resistance and low capacitance allow the myelin to conduct nerve impulses from one node of Ranvier to the next.

Myelin is susceptible to injury; for example, by hypoxemia, toxic chemicals, vascular insufficiencies, or autoimmune responses. The sheath becomes inflamed, and the membrane layers break down into smaller components that become well-circumscribed plaques (filled with microglial elements, macroglia, and lymphocytes). This process is called demyelination.

The damaged myelin sheath cannot conduct normally. The partial loss or dispersion of the action potential causes neurologic dysfunction.

Abnormal neuron

Axon

Node of Ranvier

Plaque

Myelin sheath

pending on the course of the disease. Spinal cord compression, foramen magnum tumor (which may mimic the exacerbations and remissions of MS), multiple small strokes, syphilis or another infection, thyroid disease, and chronic fatigue syndrome must be ruled out.

The following tests may be useful:

▶ Magnetic resonance imaging reveals multifocal white matter lesions.

▶ Electroencephalogram (EEG) reveals abnormalities in brain waves in one-third of patients.

▶ Lumbar puncture shows normal total CSF protein but elevated immunoglobulin G (gamma globulin, or IgG); IgG reflects hyperactivity of the immune system due to chronic demyelination. An elevated CSF IgG is significant only when serum IgG is normal. CSF white blood cell count may be elevated.

▶ CSF electrophoresis detects bands of IgG in most patients, even when the percentage of IgG in CSF is normal. Presence of kappa light chains provide additional support to the diagnosis.

▶ Evoked potential studies (visual, brain stem, auditory, and somatosensory) reveal slowed conduction of nerve impulses in most patients.

Treatment

The aim of treatment is threefold: Treat the acute exacerbation, treat the disease process, and treat the related signs and symptoms.

▶ I.V. methylprednisolone followed by oral therapy reduces edema of the myelin sheath, for speeding recovery for acute attacks. Other drugs, such as azathioprine (Imuran) or methotrexate and cytoxin, may be used.

▶ Interferon and glatiramen (a combination of 4 amino acids) possibly may reduce frequency and severity of relapses, and slow central nervous system damage.

▶ Stretching and range-of-motion exercises, coupled with correct positioning, may relieve the spasticity resulting from opposing muscle groups relaxing and contracting at the same time; helpful in relaxing muscles and maintaining function.

▶ Baclofen and tizanidine may be used to treat spasticity. For severe spasticity, botulinum toxin injections, intrathecal injections, nerve blocks, and surgery may be necessary.

▶ Frequent rest periods, aerobic exercise, and cooling techniques (air conditioning, breezes, water sprays) may minimize fatigue. Fatigue is characterized by an overwhelming feeling of exhaustion that can occur at any time of the day without warning. The cause is unknown. Changes in environmental conditions, such as heat and humidity, can aggravate fatigue.

▶ Amantadine (Symmetrel), pemoline (Cylert), and methylphenidate (Ritalin) have proven beneficial, as have antidepressants to manage fatigue.

▶ Bladder problems (failure to store urine, failure to empty the bladder or, more commonly, both) are managed by such strategies as drinking cranberry juice, or insertion of an indwelling catheter and suprapubic tubes. Intermittent self-catheterization and postvoid catheterization programs are helpful, as are anticholinergic medications in some patients.

▶ Bowel problems (constipation and involuntary evacuation) are managed by such measures as increasing fiber, using bulking agents, and bowel-training strategies, such as daily suppositories and rectal stimulation.

▶ Low-dose tricyclic antidepressants, phenytoin, or carbamazepine may manage sensory symptoms such as pain, numbness, burning, and tingling sensations.

▶ Adaptive devices and physical therapy assist with motor dysfunction, such as problems with balance, strength, and muscle coordination.

▶ Beta blockers, sedatives, or diuretics may be used to alleviate tremors.

▶ Speech therapy may manage dysarthria.

▶ Antihistamines, vision therapy, or exercises may minimize vertigo.

▶ Vision therapy or adaptive lenses may manage visual problems.

Myasthenia gravis

Myasthenia gravis causes sporadic but progressive weakness and abnormal fatigability of striated (skeletal) muscles; symptoms are exacerbated by exercise and repeated movement and relieved by anticholinesterase drugs. Usually, this disorder affects muscles innervated by the cranial nerves (face, lips, tongue, neck, and throat), but it can affect any muscle group.

Myasthenia gravis follows an unpredictable course of periodic exacerbations and remissions. There is no known cure. Drug treatment has improved the prognosis and allows patients to lead relatively normal lives, except during exacerbations. When the disease involves the respiratory system, it may be life-threatening.

Myasthenia gravis affects 1 in 25,000 people at any age, but incidence peaks between the ages of 20 and 40. It's three times more common in women than in men in this age-group, but after age 40, the incidence is similar.

About 20% of infants born to myasthenic mothers have transient (or occasionally persistent) myasthenia. This disease may coexist with immune and thyroid disorders; about 15% of myasthenic patients have thymomas. Remissions occur in about 25% of patients.

IMPAIRED TRANSMISSION IN MYASTHENIA GRAVIS

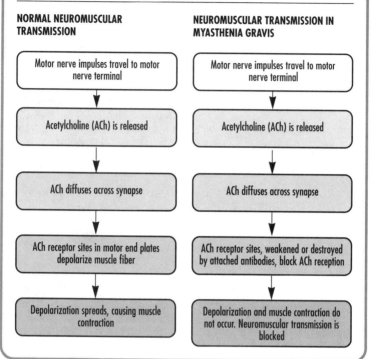

NORMAL NEUROMUSCULAR TRANSMISSION

Motor nerve impulses travel to motor nerve terminal

↓

Acetylcholine (ACh) is released

↓

ACh diffuses across synapse

↓

ACh receptor sites in motor end plates depolarize muscle fiber

↓

Depolarization spreads, causing muscle contraction

NEUROMUSCULAR TRANSMISSION IN MYASTHENIA GRAVIS

Motor nerve impulses travel to motor nerve terminal

↓

Acetylcholine (ACh) is released

↓

ACh diffuses across synapse

↓

ACh receptor sites, weakened or destroyed by attached antibodies, block ACh reception

↓

Depolarization and muscle contraction do not occur. Neuromuscular transmission is blocked

Causes

The exact cause of myasthenia gravis is unknown. However, it is believed to be the result of:

▶ autoimmune response
▶ ineffective acetylcholine release
▶ inadequate muscle fiber response to acetylcholine.

Pathophysiology

Myasthenia gravis causes a failure in transmission of nerve impulses at the neuromuscular junction. The site of action is the postsynaptic membrane. Theoretically, antireceptor antibodies block, weaken or reduce the number of acetylcholine receptors available at each neuromuscular junction and thereby impair muscle depolarization necessary for movement. (See

Impaired transmission in myasthenia gravis.)

Signs and symptoms

Myasthenia gravis may occur gradually or suddenly. Its signs and symptoms include the following:

▶ weak eye closure, ptosis, and diplopia from impaired neuromuscular transmission to the cranial nerves supplying the eye muscles
▶ skeletal muscle weakness and fatigue, increasing through the day but decreasing with rest (In the early stages, easy fatigability of certain muscles may appear with no other findings. Later, it may be severe enough to cause paralysis.)
▶ progressive muscle weakness and accompanying loss of function depending on muscle group affected; becoming more

intense during menses and after emotional stress, prolonged exposure to sunlight or cold, or infections

❱ blank and expressionless facial appearance and nasal vocal tones secondary to impaired transmission of cranial nerves innervating the facial muscles

❱ frequent nasal regurgitation of fluids, and difficulty chewing and swallowing from cranial nerve involvement

❱ drooping eyelids from weakness of facial and extraocular muscles

❱ weakened neck muscles with head tilting back to see (Neck muscles may become too weak to support the head without bobbing.)

❱ weakened respiratory muscles, decreased tidal volume and vital capacity from impaired transmission to the diaphragm making breathing difficult and predisposing to pneumonia and other respiratory tract infections

❱ respiratory muscle weakness (myasthenic crisis) possibly severe enough to require an emergency airway and mechanical ventilation.

Complications
Complications may include:
❱ respiratory distress
❱ pneumonia
❱ aspiration
❱ myasthenic crisis.

Diagnosis
❱ Tensilon test confirms diagnosis of myasthenia gravis, revealing temporarily improved muscle function within 30 to 60 seconds after I.V. injection of edrophonium or neostigmine and lasting up to 30 minutes.

❱ Electromyography with repeated neural stimulation shows progressive decrease in muscle fiber contraction.

❱ Serum antiacetylcholine antibody titer may be elevated.

❱ Chest X-ray reveals thymoma (in approximately 15% of patients).

Treatment
Treatment may include:

❱ anticholinesterase drugs, such as neostigmine and pyridostigmine to counteract fatigue and muscle weakness, and allow about 80% of normal muscle function (drugs less effective as disease worsens)

❱ immunosuppressant therapy with corticosteroids, azathioprine, cyclosporine, and cyclophosphamide used in a progressive fashion (when the previous drug response is poor, the next one is used) to decrease the immune response toward acetylcholine receptors at the neuromuscular junction

❱ IgG during acute relapses or plasmapheresis in severe exacerbations to suppress the immune system

❱ thymectomy to remove thymomas and possibly induce remission in some cases of adult-onset myasthenia

❱ tracheotomy, positive-pressure ventilation, and vigorous suctioning to remove secretions for treatment of acute exacerbations that cause severe respiratory distress

❱ discontinuation of anticholinesterase drugs in myasthenic crisis, until respiratory function improves — Myasthenic crisis requires immediate hospitalization and vigorous respiratory support.

Parkinson's disease
Named for James Parkinson, the English doctor who wrote the first accurate description of the disease in 1817, Parkinson's disease (also known as shaking palsy) characteristically produces progressive muscle rigidity, akinesia, and involuntary tremor. Deterioration is a progressive process. Death may result from complications, such as aspiration pneumonia or some other infection.

Parkinson's disease is one of the most common crippling diseases in the United States. It strikes 1 in every 100 people over age 60 and affects men more often than women. Roughly 60,000 new cases are diagnosed annually in the United States alone, and incidence is predicted to increase as the population ages.

Causes

The cause of Parkinson's disease is unknown. However, study of the extrapyramidal brain nuclei (corpus striatum, globus pallidus, substantia nigra) has established the following:

▶ Dopamine deficiency prevents affected brain cells from performing their normal inhibitory function in the CNS.

▶ Some cases are caused by exposure to toxins, such as manganese dust or carbon monoxide.

Pathophysiology

Parkinson's disease is a degenerative process involving the dopaminergic neurons in the substantia nigra (the area of the basal ganglia that produces and stores the neurotransmitter dopamine). This area plays an important role in the extrapyramidal system, which controls posture and coordination of voluntary motor movements.

Normally, stimulation of the basal ganglia results in refined motor movement because acetylcholine (excitatory) and dopamine (inhibitory) release are balanced. Degeneration of the dopaminergic neurons and loss of available dopamine leads to an excess of excitatory acetylcholine at the synapse, and consequent rigidity, tremors, and bradykinesia.

Other nondopaminergic neurons may be affected, possibly contributing to depression and the other non-motor symptoms associated with this disease. Also, the basal ganglia is interconnected to the hypothalamus, potentially affecting autonomic and endocrine function as well.

Current research on the pathogenesis of Parkinson's disease focuses on damage to the substantia nigra from oxidative stress. Oxidative stress is believed to diminish brain iron content, impair mitochondrial function, inhibit antioxidant and protective systems, reduce glutathione secretion, and damage lipids, proteins, and DNA. Brain cells are less capable of repairing oxidative damage than are other tissues.

Signs and symptoms

Signs and symptoms may include:

▶ muscle rigidity, akinesia, and an insidious tremor beginning in the fingers (unilateral pill-roll tremor) that increases during stress or anxiety and decreases with purposeful movement and sleep; secondary to loss of inhibitory dopamine activity at the synapse

▶ muscle rigidity with resistance to passive muscle stretching, which may be uniform (lead-pipe rigidity) or jerky (cogwheel rigidity) secondary to depletion of dopamine

▶ akinesia causing difficulty walking (gait lacks normal parallel motion and may be retropulsive or propulsive) from impaired dopamine action

▶ high-pitched, monotone voice from dopamine depletion

▶ drooling secondary to impaired regulation of motor function

▶ mask-like facial expression from depletion of dopamine

▶ loss of posture control (the patient walks with body bent forward) from loss of motor control due to dopamine depletion

▶ dysarthria, dysphagia, or both

▶ oculogyric crises (eyes are fixed upward, with involuntary tonic movements) or blepharospasm (eyelids are completely closed)

▶ excessive sweating from impaired autonomic dysfunction

▶ decreased motility of gastrointestinal and genitourinary smooth muscle from impaired autonomic transmission

▶ orthostatic hypotension from impaired vascular smooth muscle response

▶ oily skin secondary to inappropriate androgen production controlled by the hypothalamus pituitary axis.

Complications

Complications may include:

▶ injury from falls
▶ aspiration
▶ urinary tract infections
▶ pressure ulcers.

Diagnosis

Generally, diagnostic tests are of little value in identifying Parkinson's disease. Diagnosis is based on the patient's age and history, and on the characteristic clinical picture. However, urinalysis may support the diagnosis by revealing decreased dopamine levels.

A conclusive diagnosis is possible only after ruling out other causes of tremor, involutional depression, cerebral arteriosclerosis, and, in patients under age 30, intracranial tumors, Wilson's disease, or phenothiazine or other drug toxicity.

Treatment

The aim of treatment is to relieve symptoms and keep the patient functional as long as possible. Treatment includes:

◗ levodopa, a dopamine replacement most effective during early stages and given in increasing doses until symptoms are relieved or side effects appear. (Because side effects can be serious, levodopa is frequently given in combination with carbidopa to halt peripheral dopamine synthesis.)

◗ alternative drug therapy, including anticholinergics such as trihexyphenidyl; antihistamines such as diphenhydramine; and amantadine, an antiviral agent, or Selegiline, an enzyme-inhibiting agent, when levodopa is ineffective to conserve dopamine and enhance the therapeutic effect of levodopa.

◗ stereotactic neurosurgery when drug therapy fails to destroy the ventrolateral nucleus of the thalamus to prevent involuntary movement. (This is most effective in young, otherwise healthy persons with unilateral tremor or muscle rigidity. Neurosurgery can only *relieve* symptoms, not cure the disease.)

◗ physical therapy, including active and passive range of motion exercises, routine daily activities, walking, and baths and massage to help relax muscles; complement drug treatment and neurosurgery attempting to maintain normal muscle tone and function.

Seizure disorder

Seizure disorder, or epilepsy, is a condition of the brain characterized by susceptibility to recurrent seizures (paroxysmal events associated with abnormal electrical discharges of neurons in the brain). Primary seizure disorder or epilepsy is idiopathic without apparent structural changes in the brain. Secondary epilepsy, characterized by structural changes or metabolic alterations of the neuronal membranes, causes increased automaticity.

Epilepsy is believed to affect 1% to 2% of the population; approximately 2 million people have been diagnosed with epilepsy. The incidence is highest in childhood and old age. The prognosis is good if the patient adheres strictly to prescribed treatment.

Causes

About half of all epilepsy cases are idiopathic; possible causes of other cases include:

◗ birth trauma (inadequate oxygen supply to the brain, blood incompatibility, or hemorrhage)
◗ perinatal infection
◗ anoxia
◗ infectious diseases (meningitis, encephalitis, or brain abscess)
◗ ingestion of toxins (mercury, lead, or carbon monoxide)
◗ brain tumors
◗ inherited disorders or degenerative disease, such as phenylketonuria or tuberous sclerosis
◗ head injury or trauma
◗ metabolic disorders, such as hypoglycemia and hypoparathyroidism
◗ cerebrovascular accident (hemorrhage, thrombosis, or embolism).

Pathophysiology

Some neurons in the brain may depolarize easily or be hyperexcitable; this *epileptogenic focus* fires more readily than normal when stimulated. In these neurons, the membrane potential at rest is less negative or inhibitory connections are miss-

SEIZURE TYPES

The various types of seizures — partial, generalized, status epilepticus, or unclassified — have distinct signs and symptoms.

PARTIAL SEIZURES

Arising from a localized area of the brain, partial seizures cause focal symptoms. These seizures are classified by their effect on consciousness and whether they spread throughout the motor pathway, causing a generalized seizure.

▶ A *simple partial seizure* begins locally and generally does not cause an alteration in consciousness. It may present with sensory symptoms (lights flashing, smells, hearing hallucinations), autonomic symptoms (sweating, flushing, pupil dilation), and psychic symptoms (dream states, anger, fear). The seizure lasts for a few seconds and occurs without preceding or provoking events. This type can be motor or sensory.

▶ A *complex partial seizure* alters consciousness. Amnesia for events that occur during and immediately after the seizure is a differentiating characteristic. During the seizure, the patient may follow simple commands. This seizure generally lasts for 1 to 3 minutes.

GENERALIZED SEIZURES

As the term suggests, generalized seizures cause a generalized electrical abnormality within the brain. They can be convulsive or nonconvulsive, and include several types:

▶ *Absence seizures* occur most often in children, although they may affect adults. They usually begin with a brief change in level of consciousness, indicated by blinking or rolling of the eyes, a blank stare, and slight mouth movements. The patient retains his posture and continues pre-seizure activity without difficulty. Typ-

ically, each seizure lasts from 1 to 10 seconds. If not properly treated, seizures can recur as often as 100 times a day. An absence seizure is a nonconvulsive seizure, but it may progress to a generalized tonic-clonic seizure.

▶ *Myoclonic seizures* (bilateral massive epileptic myoclonus) are brief, involuntary muscular jerks of the body or extremities, which may be rhythmic. Consciousness is not usually affected.

▶ *Generalized tonic-clonic seizures* typically begin with a loud cry, precipitated by air rushing from the lungs through the vocal cords. The patient then loses consciousness and falls to the ground. The body stiffens (tonic phase) and then alternates between episodes of muscle spasm and relaxation (clonic phase). Tongue biting, incontinence, labored breathing, apnea, and subsequent cyanosis may occur. The seizure stops in 2 to 5 minutes, when abnormal electrical conduction ceases. When the patient regains consciousness, he is confused and may have difficulty talking. If he can talk, he may complain of drowsiness, fatigue, headache, muscle soreness, and arm or leg weakness. He may fall into a deep sleep after the seizure.

▶ *Atonic seizures* are characterized by a general loss of postural tone and a temporary loss of consciousness. They occur in young children and are sometimes called "drop attacks" because they cause the child to fall.

STATUS EPILEPTICUS

Status epilepticus is a continuous seizure state that can occur in all seizure types. The most life-threatening example is generalized tonic-clonic status epilepticus, a con-

SEIZURE TYPES *(continued)*

tinuous generalized tonic-clonic seizure. Status epilepticus is accompanied by respiratory distress leading to hypoxia or anoxia. It can result from abrupt withdrawal of anticonvulsant medications, hypoxic encephalopathy, acute head trauma, metabolic encephalopathy, or septicemia secondary to encephalitis or meningitis.

UNCLASSIFIED SEIZURES
This category is reserved for seizures that do not fit the characteristics of partial or generalized seizures or status epilepticus. Included as unclassified are events that lack the data to make a more definitive diagnosis.

ing, possibly as a result of decreased gamma-amino butyric acid (GABA) activity or localized shifts in electrolytes.

On stimulation, the epileptogenic focus fires and spreads electrical current to surrounding cells. These cells fire in turn and the impulse cascades to one side of the brain (a partial seizure), both sides of the brain (a generalized seizure), or cortical, subcortical, and brain stem areas.

The brain's metabolic demand for oxygen increases dramatically during a seizure. If this demand isn't met, hypoxia and brain damage ensue. Firing of inhibitory neurons causes the excitatory neurons to slow their firing and eventually stop. If this inhibitory action doesn't occur, the result is status epilepticus: one seizure occurring right after another and another; without treatment the anoxia is fatal.

Signs and symptoms
The hallmark of epilepsy is recurring seizures, which can be classified as partial, generalized, status epilepticus, or unclassified (some patients may be affected by more than one type). (See *Seizure types*.)

Complications
Complications may include:
▶ hypoxia or anoxia from airway occlusion
▶ traumatic injury
▶ brain damage

▶ depression and anxiety.

Diagnosis
Clinically, the diagnosis of epilepsy is based on the occurrence of one or more seizures, and proof or the assumption that the condition that caused them is still present. Diagnostic tests that help support the findings include:
▶ Computed tomography or magnetic resonance imaging reveal abnormalities.
▶ Electroencephalogram (EEG) reveals paroxysmal abnormalities to confirm the diagnosis and provide evidence of the continuing tendency to have seizures. In tonic-clonic seizures, high, fast voltage spikes are present in all leads; in absence seizures, rounded spike wave complexes are diagnostic. A negative EEG doesn't rule out epilepsy, because the abnormalities occur intermittently.
▶ Skull X-ray may show evidence of fractures or shifting of the pineal gland, bony erosion, or separated sutures.
▶ Serum chemistry blood studies may reveal hypoglycemia, electrolyte imbalances, elevated liver enzymes, and elevated alcohol levels, providing clues to underlying conditions that increase the risk of seizure activity.

Treatment
Treatment may include:
▶ drug therapy specific to the type of seizure, including phenytoin, carbamazepine, phenobarbital, and primidone

for generalized tonic-clonic seizures and complex partial seizures — I.V. fosphenytoin (Cerebyx) is an alternative to phenytoin (Dilantin) that is just as effective, with a long half-life and minimal CNS depression (stable for 120 days at room temperature and compatible with many frequently used I.V. solutions; can be administered rapidly without the adverse cardiovascular effects that occur with phenytoin)

▶ valproic acid, clonazepam, and ethosuximide for absence seizures

▶ gabapentin (Neurontin) and felbamate as other anticonvulsant drugs

▶ surgical removal of a demonstrated focal lesion, if drug therapy is ineffective

▶ surgery to remove the underlying cause, such as a tumor, abscess, or vascular problem

▶ vagus nerve stimulator implant may help reduce the incidence of focal seizure

▶ I.V. diazepam, lorazepam, phenytoin, or phenobarbital for status epilepticus

▶ administration of dextrose (when seizures are secondary to hypoglycemia) or thiamine (in chronic alcoholism or withdrawal).

Spinal cord trauma

Spinal injuries include fractures, contusions, and compressions of the vertebral column, usually as the result of trauma to the head or neck. The real danger lies in spinal cord damage — cutting, pulling, twisting, or compression. Damage may involve the entire cord or be restricted to one half, and it can occur at any level. Fractures of the 5th, 6th, or 7th cervical, 12th thoracic, and 1st lumbar vertebrae are most common.

Causes

The most serious spinal cord trauma typically results from:

▶ motor vehicle accidents

▶ falls

▶ sports injuries

▶ diving into shallow water

▶ gunshot or stab wounds.

Less serious injuries commonly occur from:

▶ lifting heavy objects

▶ minor falls.

Spinal dysfunction may also result from:

▶ hyperparathyroidism

▶ neoplastic lesions.

Pathophysiology

Like head trauma, spinal cord trauma results from acceleration, deceleration or other deforming forces usually applied from a distance. Mechanisms involved with spinal cord trauma include:

▶ hyperextension from acceleration-deceleration forces and sudden reduction in the anteroposterior diameter of the spinal cord

▶ hyperflexion from sudden and excessive force, propelling the neck forward or causing an exaggerated movement to one side

▶ vertical compression from force being applied from the top of the cranium along the vertical axis through the vertebra

▶ rotational forces from twisting, which adds shearing forces.

Injury causes microscopic hemorrhages in the gray matter and pia-arachnoid. The hemorrhages gradually increase in size until all of the gray matter is filled with blood, which causes necrosis. From the gray matter, the blood enters the white matter, where it impedes the circulation within the spinal cord. Ensuing edema causes compression and decreases the blood supply. Thus, the spinal cord loses perfusion and becomes ischemic. The edema and hemorrhage are greatest at and approximately two segments above and below the injury. The edema temporarily adds to the patient's dysfunction by increasing pressure and compressing the nerves. Edema near the 3rd to 5th cervical vertebrae may interfere with phrenic nerve impulse transmission to the diaphragm and inhibit respiratory function.

In the white matter, circulation usually returns to normal in approximately 24 hours. However, in the gray matter, an in-

COMPLICATIONS OF SPINAL CORD INJURY

Of the following three sets of complications, only autonomic dysreflexia requires emergency attention.

AUTONOMIC DYSREFLEXIA

Also known as autonomic hyperreflexia, autonomic dysreflexia is a serious medical condition that occurs after resolution of spinal shock. Emergency recognition and management is a must.

Autonomic dysreflexia should be suspected in the patient with:
▶ spinal cord trauma at or above level T6
▶ bradycardia
▶ hypertension and a severe pounding headache
▶ cold or goose-fleshed skin below the lesion.

Dysreflexia is caused by noxious stimuli, most commonly a distended bladder or skin lesion.

Treatment focuses on eliminating the stimulus; rapid identification and removal may avoid the need for pharmacologic control of the headache and hypertension.

SPINAL SHOCK

Spinal shock is the loss of autonomic, reflex, motor, and sensory activity below the level of the cord lesion. It occurs secondary to damage of the spinal cord.

Signs of spinal shock include:
▶ flaccid paralysis
▶ loss of deep tendon and perianal reflexes
▶ loss of motor and sensory function.

Until spinal shock has resolved (usually 1 to 6 weeks after injury), the extent of actual cord damage cannot be assessed. The earliest indicator of resolution is the return of reflex activity.

NEUROGENIC SHOCK

This abnormal vasomotor response occurs secondary to disruption of sympathetic impulses from the brain stem to the thoracolumbar area, and is seen most frequently in patients with cervical cord injury. This temporary loss of autonomic function below the level of injury causes cardiovascular changes.

Signs of neurogenic shock include:
▶ orthostatic hypotension
▶ bradycardia
▶ loss of the ability to sweat below the level of the lesion.

Treatment is symptomatic. Symptoms resolve when spinal cord edema resolves.

---

flammatory reaction prevents restoration of circulation. Phagocytes appear at the site within 36 to 48 hours after the injury, macrophages engulf degenerating axons, and collagen replaces the normal tissue. Scarring and meningeal thickening leaves the nerves in the area blocked or tangled.

Signs and symptoms

▶ Muscle spasm and back pain that worsens with movement. In cervical fractures, pain may cause point tenderness; in dorsal and lumbar fractures, it may radiate to other body areas such as the legs.

▶ Mild paresthesia to quadriplegia and shock, if the injury damages the spinal cord. In milder injury, such symptoms may be delayed several days or weeks.

Specific signs and symptoms depend on injury type and degree. (See *Types of spinal cord injury*, pages 300 and 301.)

Complications

▶ autonomic dysreflexia
▶ spinal shock
▶ neurogenic shock. (See *Complications of spinal cord injury*.)

TYPES OF SPINAL CORD INJURY

Injury to the spinal cord can be classified as complete or incomplete. An incomplete spinal injury may be an anterior cord syndrome, central cord syndrome or Brown-Sequard syndrome, depending on the area of the cord affected. This chart highlights the characteristic signs and symptoms of each.

TYPE	DESCRIPTION
Complete transection	▶ All tracts of the spinal cord completely disrupted ▶ All functions involving the spinal cord below the level of transection lost ▶ Complete and permanent loss
Incomplete transection: Central cord syndrome	▶ Center portion of cord affected ▶ Typically from hyperextension injury
Incomplete transection: Anterior cord syndrome	▶ Occlusion of anterior spinal artery ▶ Occlusion from pressure of bone fragments
Incomplete transection: Brown-Sequard Syndrome	▶ Hemisection of cord affected ▶ Most common in stabbing and gunshot wounds ▶ Damage to cord on only one side

Diagnosis

▶ Spinal X-rays, the most important diagnostic measure, detect the fracture.

▶ Thorough neurologic evaluation locates the level of injury and detects cord damage.

▶ Lumbar puncture may show increased CSF pressure from a lesion or trauma in spinal compression.

▶ Computed tomography or magnetic resonance imaging reveals spinal cord edema and compression and may reveal a spinal mass.

Treatment

Treatment may include:

▶ immediate immobilization to stabilize the spine and prevent cord damage (primary treatment); use of sandbags on both sides of the patient's head, a hard cervical collar, or skeletal traction with skull tongs or a halo device for cervical spine injuries

▶ high doses of methylprednisolone to reduce inflammation with evidence of cord injury

▶ bed rest on firm support (such as a bed board), analgesics, and muscle relaxants for treatment of stable lumbar and dorsal fractures for several days until the fracture stabilizes.

SIGNS AND SYMPTOMS

▶ Loss of motor function (quadriplegia) with cervical cord transection; paraplegia with thoracic cord transection
▶ Muscle flaccidity
▶ Loss of all reflexes and sensory function below level of injury
▶ Bladder and bowel atony
▶ Paralytic ileus
▶ Loss of vasomotor tone in lower body parts with low and unstable blood pressure
▶ Loss of perspiration below level of injury
▶ Dry pale skin
▶ Respiratory impairment

▶ Motor deficits greater in upper than lower extremities
▶ Variable degree of bladder dysfunction

▶ Loss of motor function below level of injury
▶ Loss of pain and temperature sensations below level of injury
▶ Intact touch, pressure, position and vibration senses

▶ Ipsilateral paralysis or paresis below the level of the injury
▶ Ipsilateral loss of touch, pressure, vibration, and position sense below level of injury
▶ Contralateral loss of pain and temperature sensations below level of injury

▶ plaster cast or a turning frame to treat unstable dorsal or lumbar fracture
▶ laminectomy and spinal fusion for severe lumbar fractures
▶ neurosurgery to relieve the pressure when the damage results in compression of the spinal column — If the cause of compression is a metastatic lesion, chemotherapy and radiation may relieve it.
▶ treatment of surface wounds accompanying the spinal injury; tetanus prophylaxis unless the patient has had recent immunization
▶ exercises to strengthen the back muscles and a back brace or corset to provide support while walking

▶ rehabilitation to maintain or improve functional level.

9

The gastrointestinal (GI) system has the critical task of supplying essential nutrients to fuel all the physiologic and pathophysiologic activities of the body. Its functioning profoundly affects the quality of life through its impact on overall health. The GI system has two major components: the alimentary canal, or GI tract, and the accessory organs. A malfunction anywhere in the system can produce far-reaching metabolic effects, eventually threatening life itself.

The alimentary canal is a hollow muscular tube that begins in the mouth and ends at the anus. It includes the oral cavity, pharynx, esophagus, stomach, small intestine, and large intestine. Peristalsis propels the ingested material along the tract; sphincters prevent its reflux. Accessory glands and organs include the salivary glands, liver, biliary duct system (gallbladder and bile ducts), and pancreas.

Together, the GI tract and accessory organs serve two major functions: digestion (breaking down food and fluids into simple chemicals that can be absorbed into the bloodstream and transported throughout the body) and elimination of waste products from the body through defecation.

PATHOPHYSIOLOGIC CONCEPTS

Disorders of the GI system often manifest as vague, nonspecific complaints or problems that reflect disruption in one or more of the system's functions. For example, movement through the GI tract can be slowed, accelerated, or blocked, and secretion, absorption, or motility can be altered. As a result, one patient may present with several problems, the most common being anorexia, constipation, diarrhea, dysphagia, jaundice, nausea, and vomiting.

Anorexia

Anorexia is a loss of appetite or a lack of desire for food. Nausea, abdominal pain, and diarrhea may accompany it. Anorexia can result from dysfunction, such as cancer, heart disease, or renal disease, in the gastrointestinal system or other systems.

Normally, a physiologic stimulus is responsible for the sensation of hunger. Falling blood glucose levels stimulate the hunger center in the hypothalamus; rising blood fat and amino acid levels promote satiety. Hunger is also stimulated by contraction of an empty stomach and suppressed when the GI tract becomes distended, possibly as a result of stimulation of the vagus nerve. Sight, touch, and smell play subtle roles in controlling the appetite center.

In anorexia, the physiologic stimuli are present but the person has no appetite or desire to eat. Slow gastric emptying or gastric stasis can cause anorexia. High levels of neurotransmitters such as serotonin (may contribute satiety) and excess cortisol levels (may suppress hypothalamic control of hunger) have been implicated.

Constipation

Constipation is hard stools and difficult or infrequent defecation, as defined by a decrease in the number of stools per week. It is defined individually, because normal bowel habits range from 2 to 3 episodes of stool passage per day to one per week. Causes of constipation include dehydration, consumption of a low bulk diet, a sedentary lifestyle, lack of regular exercise, and frequent repression of the urge to defecate.

When a person is dehydrated or delays defecation, more fluid is absorbed from the intestine, the stool becomes harder, and constipation ensues. High fiber diets cause water to be drawn into the stool by osmosis, thereby keeping stool soft and encouraging movement through the intestine. High fiber diets also causes intestinal dilation, which stimulates peristalsis. Conversely, a low fiber diet would contribute to constipation.

AGE ALERT The elderly typically experience a decrease in intestinal motility in addition to a slowing and dulling of neural impulses in the GI tract. Many older persons restrict fluid intake to prevent waking at night to use the bathroom or because of a fear of incontinence. This places them at risk for dehydration and constipation.

A sedentary lifestyle or lack of exercise can cause constipation because exercise stimulates the gastrointestinal tract and promotes defecation. Antacids, opiates, and other drugs that inhibit bowel motility also lead to constipation.

Stress stimulates the sympathetic nervous system, and GI motility slows. Absence or degeneration in the neural pathways of the large intestine also contributes to constipation. And other conditions, such as spinal cord trauma, multiple sclerosis, intestinal neoplasms, and hypothyroidism, can cause constipation.

Diarrhea

Diarrhea is an increase in the fluidity or volume of feces and the frequency of defecation. Factors that affect stool volume and consistency include water content of the colon and the presence of unabsorbed food, unabsorbable material, and intestinal secretions. Large-volume diarrhea is usually the result of an excessive amount of water, secretions, or both in the intestines. Small-volume diarrhea is usually caused by excessive intestinal motility. Diarrhea may also be caused by a parasympathetic stimulation of the gut initiated by psychological factors such as fear or stress.

The three major mechanisms of diarrhea are osmosis, secretion, and motility:

▶ *Osmotic diarrhea:* The presence of nonabsorbable substance, such as synthetic sugar, or increased numbers of osmotic particles in the intestine, increases osmotic pressure and draws excess water into the intestine, thereby increasing the weight and volume of the stool.

▶ *Secretory diarrhea:* A pathogen or tumor irritates the muscle and mucosal layers of the intestine. The consequent increase in motility and secretions (water, electrolytes, and mucus) results in diarrhea.

▶ *Motility diarrhea:* Inflammation, neuropathy, or obstruction causes a reflex increase in intestinal motility that may expel the irritant or clear the obstruction.

Dysphagia

Dysphagia — difficulty swallowing — can be caused by a mechanical obstruction of the esophagus or by impaired esophageal motility secondary to another disorder. Mechanical obstruction is characterized as intrinsic or extrinsic.

Intrinsic obstructions originate in the esophagus itself. Causes of intrinsic tumors include tumors, strictures, and diverticular herniations. Extrinsic obstructions originate outside of the esophagus and narrow the lumen by exerting pressure on the esophageal wall. Most extrinsic obstruction results from a tumor.

Distention and spasm at the site of the obstruction during swallowing may cause pain. Upper esophageal obstruction causes pain 2 to 4 seconds after swallowing; lower esophageal obstructions, 10 to 15 seconds after swallowing. If a tumor is

WHAT HAPPENS IN SWALLOWING

Before peristalsis can begin, the neural pattern to initiate swallowing, illustrated here, must occur:

▶ Food reaching the back of the mouth stimulates swallowing receptors that surround the pharyngeal opening.

▶ The receptors transmit impulses to the brain by way of the sensory por-tions of the trigeminal (V) and glos-sopharyngeal (IX) nerves.

▶ The brain's swallowing center re-lays motor impulses to the esopha-gus by way of the trigeminal (V), glossopharyngeal (IX), vagus (X), and hypoglossal (XII) nerves.

▶ Swallowing occurs.

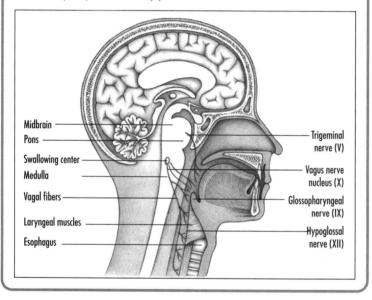

present, dysphagia begins with difficulty swallowing solids and eventually pro-gresses to difficulty swallowing semi-solids and liquids. Impaired motor func-tion makes both liquids and solids diffi-cult to swallow.

Neural or muscular disorders can also interfere with voluntary swallowing or peristalsis. This is known as functional dysphagia. Causes of functional dyspha-gia include dermatomyositis, cerebrovas-cular accident, Parkinson's disease, or achalasia. (See *What happens in swallow-ing*.) Malfunction of the upper esophageal striated muscles interferes with the vol-untary phase of swallowing.

In achalasia, the esophageal ganglion-ic cells are thought to have degenerated, and the cardiac sphincter of the stomach cannot relax. The lower end of the esoph-agus loses neuromuscular coordination and muscle tone, and food accumulates, causing hypertrophy and dilation. Even-tually, accumulated food raises the hy-drostatic pressure and forces the sphinc-ter open, and small amounts of food slow-ly move into the stomach.

Jaundice

Jaundice — yellow pigmentation of the skin and sclera — is caused by an excess accumulation of bilirubin in the blood. Bilirubin, a product of red blood cell breakdown, accumulates when production exceeds metabolism and excretion. This imbalance can result from excessive release of bilirubin precursors into the bloodstream or from impairment of its hepatic uptake, metabolism, or excretion. (See *Jaundice: Impaired bilirubin metabolism*, page 306.) Jaundice occurs when bilirubin levels exceed 34 to 43 mmol/L (2.0 to 2.5 mg/dl), which is about twice the upper limit of the normal range. Lower levels of bilirubin may cause detectable jaundice in patients with fair skin, and jaundice may be difficult to detect in patients with dark skin.

CULTURAL DIVERSITY Jaundice in dark-skinned persons may appear as yellow staining in the sclera, hard palate, and palmar or plantar surfaces.

The three main types of jaundice are hemolytic jaundice, hepatocellular jaundice, or obstructive jaundice:

▶ *Hemolytic jaundice:* When red blood cell lysis exceeds the liver's capacity to conjugate bilirubin (binding bilirubin to a polar group makes it water soluble and able to be excreted by the kidneys), hemolytic jaundice occurs. Causes include transfusion reactions, sickle cell anemia, thalassemia, and autoimmune disease.

▶ *Hepatocellular jaundice:* Hepatocyte dysfunction limits uptake and conjugation of bilirubin. Liver dysfunction can occur in hepatitis, cancer, cirrhosis, or congenital disorders, and some drugs can cause it.

▶ *Obstructive jaundice:* When the flow of bile out of the liver (through the hepatic duct) or through the bile duct is blocked, the liver can conjugate bilirubin, but the bilirubin can't reach the small intestine. Blockage of the hepatic duct by stones or a tumor is considered an intrahepatic cause of obstructive jaundice. A blocked bile duct is an extrahepatic cause. Gallstones or a tumor may obstruct the bile duct.

Nausea

Nausea is feeling the desire to vomit. It may occur independently of vomiting, or it may precede or accompany it. Specific neural pathways have not been identified, but increased salivation, diminished functional activities of the stomach, and altered small intestinal motility have been associated with nausea. Nausea may also be stimulated by high brain centers.

Vomiting

Vomiting is the forceful oral expulsion of gastric contents. The gastric musculature provides the ejection force. The gastric fundus and gastroesophageal sphincter relax, and forceful contractions of the diaphragm and abdominal wall muscles increase intraabdominal pressure. This, combined with the annular contraction of the gastric pylorus, forces gastric contents into the esophagus. Increased intrathoracic pressure then moves the gastric content from the esophagus to the mouth.

Vomiting is controlled by two centers in the medulla: the vomiting center and the chemoreceptor trigger zone. The vomiting center initiates the actual act of vomiting. It is stimulated by the gastrointestinal tract, from higher brainstem and cortical centers, and from the chemoreceptor trigger zone. The chemoreceptor trigger zone can't induce vomiting by itself. Various stimuli or drugs activate the zone, such as apomorphine, levodopa, digitalis, bacterial toxins, radiation, and metabolic abnormalities. The activated zone sends impulses to the medullary vomiting center, and the following sequence begins:

▶ The abdominal muscles and diaphragm contract.

▶ Reverse peristalsis begins, causing intestinal material to flow back into the stomach, distending it.

▶ The stomach pushes the diaphragm into the thoracic cavity, raising the intrathoracic pressure.

▶ The pressure forces the upper esophageal sphincter open, the glottis closes, and the soft palate blocks the nasopharynx.

JAUNDICE: IMPAIRED BILIRUBIN METABOLISM

Jaundice occurs in three forms: prehepatic, hepatic, and posthepatic. In all three, bilirubin levels in the blood increase.

PREHEPATIC JAUNDICE

Certain conditions and disorders, such as transfusion reactions and sickle cell anemia, cause massive hemolysis.

▶ Red blood cells rupture faster than the liver can conjugate bilirubin.

▶ Large amounts of unconjugated bilirubin pass into the blood.

▶ Intestinal enzymes convert bilirubin to water-soluble urobilinogen for excretion in urine and stools. (Unconjugated bilirubin is insoluble in water, so it can't be directly secreted in urine.)

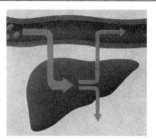

HEPATIC JAUNDICE

The liver becomes unable to conjugate or excrete bilirubin, leading to increased blood levels of conjugated and unconjugated bilirubin. This occurs in such disorders as hepatitis, cirrhosis, and metastatic cancer, and during prolonged use of drugs metabolized by the liver.

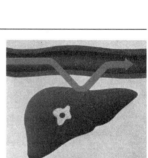

POSTHEPATIC JAUNDICE

In biliary and pancreatic disorders, bilirubin forms at its normal rate.

▶ Inflammation, scar tissue, tumor, or gallstones block the flow of bile into the intestines.

▶ Water-soluble conjugated bilirubin accumulates in the blood.

▶ The bilirubin is excreted in the urine.

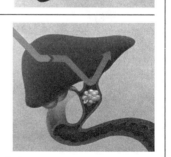

▶ The pressure also forces the material up through the sphincter and out through the mouth.

Both nausea and vomiting are manifestations of other disorders, such as acute abdominal emergencies, infections of the intestinal tract, central nervous system disorders, myocardial infarction, congestive heart failure, metabolic and endocrinologic disorders, or as the side effect of many drugs. Vomiting may also be psychogenic, resulting from emotional or psychological disturbance.

DISORDERS

Appendicitis
The most common major surgical disease, appendicitis is inflammation and obstruction of the vermiform appendix. Appendicitis may occur at any age and affects both sexes equally; however, between puberty and age 25, it's more prevalent in men. Since the advent of antibiotics, the incidence and the death rate of appendicitis have declined; if untreated, this disease is invariably fatal.

Causes
Causes may include:
▶ mucosal ulceration
▶ fecal mass
▶ stricture
▶ barium ingestion
▶ viral infection.

Pathophysiology
Mucosal ulceration triggers inflammation, which temporarily obstructs the appendix. The obstruction blocks mucus outflow. Pressure in the now distended appendix increases, and the appendix contracts. Bacteria multiply, and inflammation and pressure continue to increase, restricting blood flow to the organ and causing severe abdominal pain.

Signs and symptoms
Signs and symptoms may include:
▶ abdominal pain caused by inflammation of the appendix and bowel obstruction and distention
▶ anorexia after the onset of pain
▶ nausea or vomiting caused by the inflammation
▶ low-grade temperature from systemic manifestation of inflammation and leukocytosis
▶ tenderness from inflammation.

Complications
Complications may include:
▶ wound infection
▶ intraabdominal abscess
▶ fecal fistula
▶ intestinal obstruction
▶ incisional hernia
▶ peritonitis
▶ death.

Diagnosis
▶ White blood cell count is moderately high with an increased number of immature cells.
▶ X-ray with radiographic contrast agent reveals failure of the appendix to fill with contrast.

Treatment
Treatment may include:
▶ maintenance of NPO status until surgery
▶ high Fowler's position to aid in pain relief
▶ GI intubation for decompression
▶ appendectomy
▶ antibiotics to treat infection if peritonitis occurs
▶ parental replacement of fluid and electrolytes to reverse possible dehydration resulting from surgery or nausea and vomiting.

Cholecystitis
Cholecystitis — acute or chronic inflammation causing painful distention of the gallbladder — is usually associated with a gallstone impacted in the cystic duct. Cholecystitis accounts for 10% to 25% of all patients requiring gallbladder surgery. The acute form is most common among middle-aged women; the chronic form, among the elderly. The prognosis is good with treatment.

Causes
▶ Gallstones (the most common cause)
▶ Poor or absent blood flow to the gallbladder
▶ Abnormal metabolism of cholesterol and bile salts.

Pathophysiology

In acute cholecystitis, inflammation of the gallbladder wall usually develops after a gallstone lodges in the cystic duct. (See *Understanding gallstone formation.*) When bile flow is blocked, the gallbladder becomes inflamed and distended. Bacterial growth, usually *Escherichia coli*, may contribute to the inflammation. Edema of the gallbladder (and sometimes the cystic duct) obstructs bile flow, which chemically irritates the gallbladder. Cells in the gallbladder wall may become oxygen starved and die as the distended organ presses on vessels and impairs blood flow. The dead cells slough off, and an exudate covers ulcerated areas, causing the gallbladder to adhere to surrounding structures.

Signs and symptoms

▶ Acute abdominal pain in the right upper quadrant that may radiate to the back, between the shoulders, or to the front of the chest secondary to inflammation and irritation of nerve fibers
▶ Colic due to the passage of gallstones along the bile duct
▶ Nausea and vomiting triggered by to the inflammatory response
▶ Chills related to fever
▶ Low-grade fever secondary to inflammation
▶ Jaundice from obstruction of the common bile duct by stones.

Complications

▶ Perforation and abscess formation
▶ Fistula formation
▶ Gangrene
▶ Empyema
▶ Cholangitis
▶ Hepatitis
▶ Pancreatitis
▶ Gallstone ileus
▶ Carcinoma.

Diagnosis

▶ X-ray reveals gallstones if they contain enough calcium to be radiopaque; also helps disclose porcelain gallbladder (hard, brittle gall bladder due to calcium deposited in wall), limy bile, and gallstone ileus.
▶ Ultrasonography detects gallstones as small as 2 mm and distinguishes between obstructive and nonobstructive jaundice.
▶ Technetium-labeled scan indicates reveals cystic duct obstruction and acute or chronic cholecystitis if ultrasound doesn't visualize the gallbladder.
▶ Percutaneous transhepatic cholangiography or cholesystoscopy supports the diagnosis of obstructive jaundice and reveals calculi in the ducts.
▶ Levels of serum alkaline phosphate, lactate dehydrogenase, aspartate aminotransferase, and total bilirubin are high; serum amylase slightly elevated; and icteric index elevated.
▶ White blood cell counts are slightly elevated during cholecystitis attack.

Treatment

▶ Cholecystectomy to surgically remove the inflamed gallbladder
▶ Choledochostomy to surgically create an opening into the common bile duct for drainage
▶ Percutaneous transhepatic cholecytostomy
▶ Endoscopic retrograde cholangiopancreatography for removal of gallstones
▶ Lithotripsy to break up gallstones and relieve obstruction
▶ Oral chenodeoxycholic acid or ursodeoxycholic acid to dissolve stones
▶ Low fat diet to prevent attacks
▶ Vitamin K to relieve itching, jaundice, and bleeding tendencies due to vitamin K deficiencies
▶ Antibiotics for use during acute attack for treatment of infection
▶ Nasogastric tube insertion during acute attack for abdominal decompression.

Cirrhosis

Cirrhosis is a chronic disease characterized by diffuse destruction and fibrotic regeneration of hepatic cells. As necrotic

UNDERSTANDING GALLSTONE FORMATION

Abnormal metabolism of cholesterol and bile salts plays an important role in gallstone formation. The liver makes bile continuously. The gall bladder concentrates and stores it until the duodenum signals it needs it to help digest fat. Changes in the composition of bile may allow gallstones to form. Changes to the absorptive ability of the gallbladder lining may also contribute to gallstone formation.

TOO MUCH CHOLESTEROL

Certain conditions, such as age, obesity, and estrogen imbalance, cause the liver to secrete bile that's abnormally high in cholesterol or lacking the proper concentration of bile salts. The coronary arteries are made of three layers: intima (the innermost layer, media (the middle layer), and adventitia (the outermost layer).

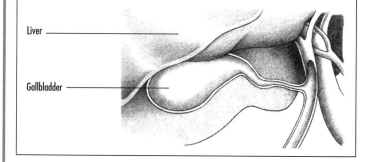

Liver

Gallbladder

INSIDE THE GALLBLADDER

When the gallbladder concentrates this bile, inflammation may occur. Excessive reabsorption of water and bile salts makes the bile less soluble. Cholesterol, calcium, and bilirubin precipitate into gallstones.

Fat entering the duodenum causes the intestinal mucosa to secrete the hormone cholecystokinin, which stimulates the gallbladder to contract and empty. If a stone lodges in the cystic duct, the gallbladder contracts but can't empty.

Contracting gallbladder

Obstructing gallstone

(continued)

UNDERSTANDING GALLSTONE FORMATION *(continued)*

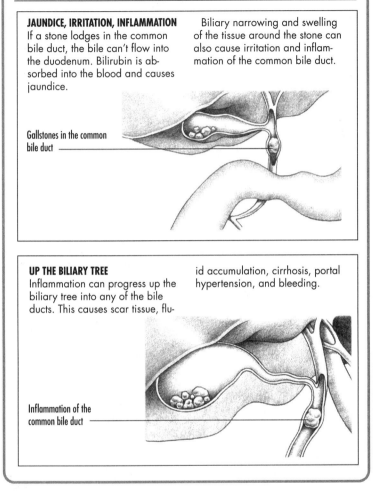

JAUNDICE, IRRITATION, INFLAMMATION
If a stone lodges in the common bile duct, the bile can't flow into the duodenum. Bilirubin is absorbed into the blood and causes jaundice.

Biliary narrowing and swelling of the tissue around the stone can also cause irritation and inflammation of the common bile duct.

Gallstones in the common bile duct

UP THE BILIARY TREE
Inflammation can progress up the biliary tree into any of the bile ducts. This causes scar tissue, flu-

id accumulation, cirrhosis, portal hypertension, and bleeding.

Inflammation of the common bile duct

---

tissue yields to fibrosis, this disease damages liver tissue and normal vasculature, impairs blood and lymph flow, and ultimately causes hepatic insufficiency. It's twice as common in men as in women, and is especially prevalent among malnourished persons over the age of 50 with chronic alcoholism. Mortality is high; many patients die within 5 years of onset.

Causes
Cirrhosis may be a result of a wide range of diseases. The following clinical types of cirrhosis reflect its diverse etiology.

Hepatocellular disease. This group includes the following disorders:
◗ Postnecrotic cirrhosis accounts for 10% to 30% of patients and stems from vari-

UNDERSTANDING ASTHMA

WHAT IS ASTHMA?
Asthma is a chronic lung disease characterized by airway inflammation and narrowing. Hypersensitivity to various stimuli or allergens causes muscles around the bronchial tubes to tighten, producing bronchospasm and constricted breathing. Asthma is either *intrinsic* or *extrinsic*. For both types, an asthma attack follows the same process. (See *An asthma attack*, pages C2 and C3, and *Extrinsic asthma*, page C4.)

INTRINSIC ASTHMA
This type of asthma, which has no obvious external cause, is more common in children under age 3 and adults over age 30. It's usually preceded by severe respiratory infection that affects nerves or cells near the surface of the bronchial tubes, as shown below. Triggers — such as irritants; exercise; fatigue; stress; and emotional, endocrine, temperature, and humidity changes — may provoke bronchoconstriction and an asthma attack.

Bronchial damage in intrinsic asthma

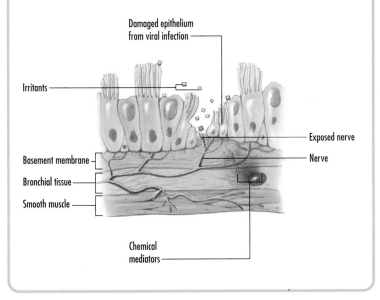

Damaged epithelium from viral infection

Irritants

Exposed nerve

Nerve

Basement membrane

Bronchial tissue

Smooth muscle

Chemical mediators

AN ASTHMA ATTACK

A person having an asthma attack has difficulty breathing and increases the use of chest muscles to assist with breathing. During an attack, muscles surrounding the bronchial tubes tighten (bronchospasm), narrowing the air passage and interrupting the normal flow of air into and out of the lungs. Air flow

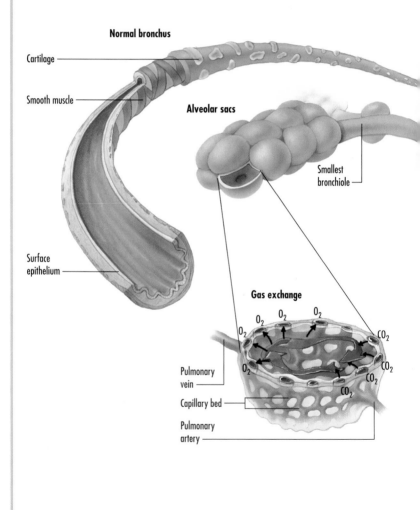

Normal bronchus

Cartilage

Smooth muscle

Alveolar sacs

Smallest bronchiole

Surface epithelium

Gas exchange

O_2 O_2 O_2 O_2 O_2 CO_2 CO_2 O_2 CO_2 CO_2

Pulmonary vein

Capillary bed

Pulmonary artery

is further interrupted by an increase in mucous secretion, formation of mucous plugs, and swelling of the bronchial tubes. If an attack continues, bronchospasm and mucous build-up traps air in the alveolar sacs, which hinders air exchange.

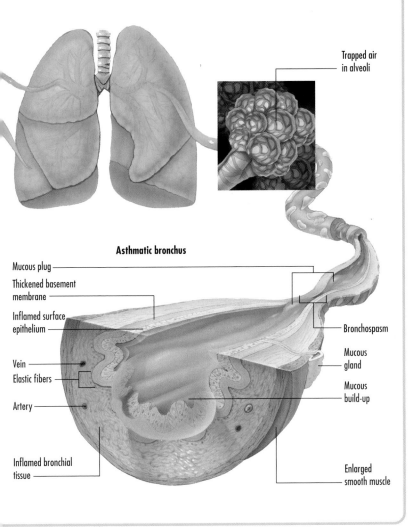

Trapped air in alveoli

Asthmatic bronchus

Mucous plug

Thickened basement membrane

Inflamed surface epithelium

Vein

Elastic fibers

Artery

Inflamed bronchial tissue

Bronchospasm

Mucous gland

Mucous build-up

Enlarged smooth muscle

EXTRINSIC ASTHMA

Persons with extrinsic asthma are hyper-responsive to environmental allergens, such as pollen, ragweed, animal dander, dust, mold, aerosols, food additives, tobacco smoke, and air pollution. Most common in children and adolescents, this type fades with age and avoidance of specific allergens.

First exposure

Allergens entering the body for the first time (1) are inhaled and absorbed into lung tissues (2). There, they trigger immune cells to produce abnormal amounts of allergen-specific defense proteins called immunoglobulin E (IgE) antibodies (3). The antibodies attach to mast cells (4), which contain chemical mediators that are responsible for allergic signs and symptoms. These cells accumulate in tissues exposed to the environment, such as mucous membranes of the respiratory tract.

Second exposure

On subsequent exposure, IgE antibodies recognize the allergens (5) and trigger the mast cells to release histamine and other chemical mediators (6). An allergic reaction (7) follows in the lungs: a hyperreactive response, producing bronchoconstriction and bronchospasm; and an inflammatory response (8), producing bronchial edema, secretions, and thick sticky mucus.

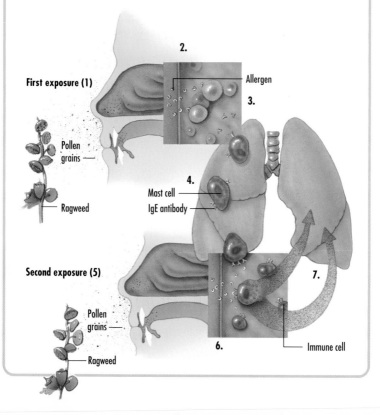

First exposure (1)

2.

Allergen

3.

Pollen grains

Ragweed

4.

Mast cell

IgE antibody

Second exposure (5)

7.

Pollen grains

Ragweed

6.

Immune cell

UNDERSTANDING CANCER

WHAT IS CANCER?

Cancer results from a destructive (malignant) transformation (carcinogenesis) of normal cells, causing them to enlarge and divide more rapidly than normal.

Cancer cells tend to be very aggressive and out of control. In contrast, a benign tumor is a localized mass of slowly multiplying cells resembling its original tissue. It's seldom life-threatening.

Normal cells

Precancerous cells

Cancer cells

HOW DOES CANCER SPREAD?

Cancer cells may invade nearby tissues or spread (metastasize) to other organs, traveling through the circulatory system or the lymphatic system or by seeding into an organ or a body cavity. The pattern and extent of spread determine the cancer's stage.

1. Circulatory system
Cancer cells may travel through the bloodstream, with the liver and the lungs as the most common destinations.

2. Lymphatic system
Cancer cells may move through this series of channels from the tissues to lymph nodes and eventually into the circulatory system.

3. Seeding
Cancer may penetrate an organ or move into a body cavity (chest or abdomen) and spread throughout that area.

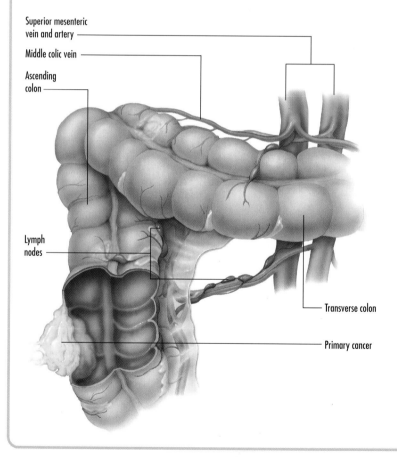

Superior mesenteric vein and artery

Middle colic vein

Ascending colon

Lymph nodes

Transverse colon

Primary cancer

STAGING CANCER

Measuring the extent that cells have spread is called staging. Depending on the type of cancer, a variety of imaging techniques, such as computer tomography (CT) scan or magnetic resonance imaging (MRI), may be used to stage a tumor.

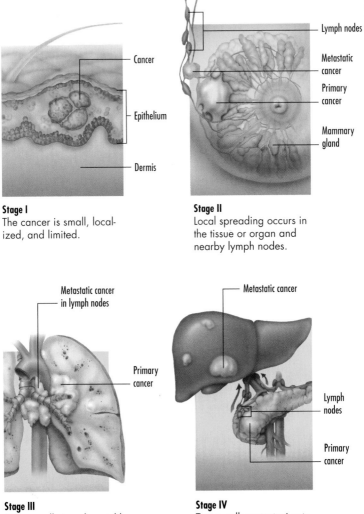

Lymph nodes

Metastatic cancer

Primary cancer

Mammary gland

Cancer

Epithelium

Dermis

Stage I
The cancer is small, localized, and limited.

Stage II
Local spreading occurs in the tissue or organ and nearby lymph nodes.

Metastatic cancer in lymph nodes

Primary cancer

Metastatic cancer

Lymph nodes

Primary cancer

Stage III
Cancer cells invade neighboring tissues and lymph nodes.

Stage IV
Tumor cells metastasize to other tissues and organs.

TREATING CANCER

Due to the variety of cancers, the ideal treatment can range from observation to complicated surgical removal with aggressive therapy. Surgery, radiation therapy, and chemotherapy can be used in combination or as individual treatments. In surgery, the diseased part of the body is removed. Neighboring healthy lymph nodes and tissues may also be removed to help control the cancer's spread.

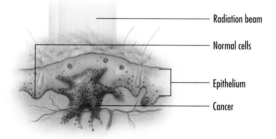

Radiation beam

Normal cells

Epithelium

Cancer

Radiation therapy

High energy rays, focused in a beam, are used to damage the cancer cells and stop their reproduction. This local therapy is used to shrink a cancer's size either before surgical removal or after, to kill any remaining cancer cells.

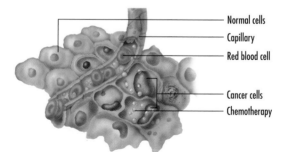

Normal cells

Capillary

Red blood cell

Cancer cells

Chemotherapy

Chemotherapy

Drugs given at specified intervals to enhance effectiveness and limit adverse effects disrupt the ability of cancer cells to divide. These drugs travel through the blood to act on dividing cells — both normal *and* cancerous — throughout the entire body.

UNDERSTANDING OSTEOPOROSIS

WHAT IS OSTEOPOROSIS?

Osteoporosis is a metabolic disease of the skeleton that reduces the amount of bone tissue. Bones weaken as local cells resorb, or take up, bone tissue. Trabecular bone at the core becomes less dense, and cortical bone on the perimeter loses thickness. (See *Controlling mineral balance*, pages C10 and C11, and *Bone formation and resorption*, page C12.)

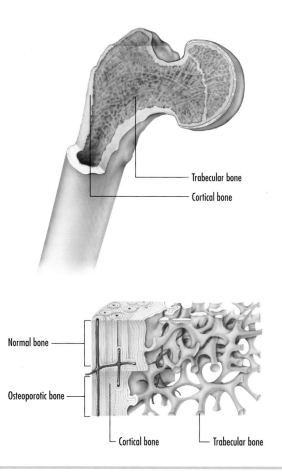

Trabecular bone

Cortical bone

Normal bone

Osteoporotic bone

Cortical bone

Trabecular bone

CONTROLLING MINERAL BALANCE

Normally, the blood absorbs calcium from the digestive system and deposits it into the bones. In osteoporosis, blood levels of calcium are reduced because of dietary calcium deficiency, inability of the intestines to absorb calcium, or postmenopausal estrogen deficiency. To maintain the blood calcium level as close to normal as possible, resorption from the bones increases, causing osteoporosis.

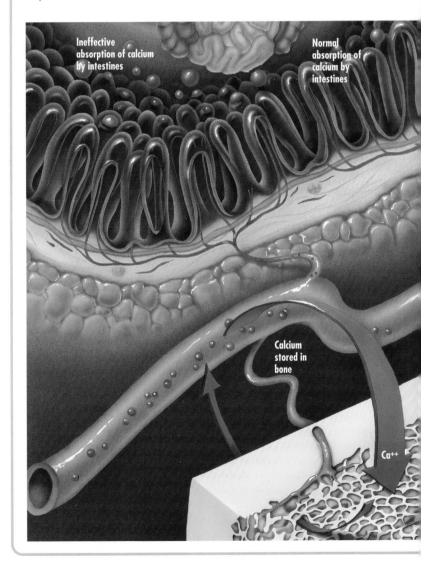

Ineffective absorption of calcium by intestines

Normal absorption of calcium by intestines

Calcium stored in bone

Ca++

In addition to enhancing bone resorption, low calcium enhances the effects of two other factors: Parathyroid hormone (PTH) and vitamin D. PTH is produced by the parathyroid glands, which are buried in the thyroid gland. Vitamin D is supplied by the diet, produced in the skin as a reaction to sunlight, and processed into a very potent form in the liver and kidneys. Both substances stimulate calcium absorption from the intestine and increase resorption from the bone. This results in an increased sacrifice of bone calcium to maintain normal levels of calcium in the blood.

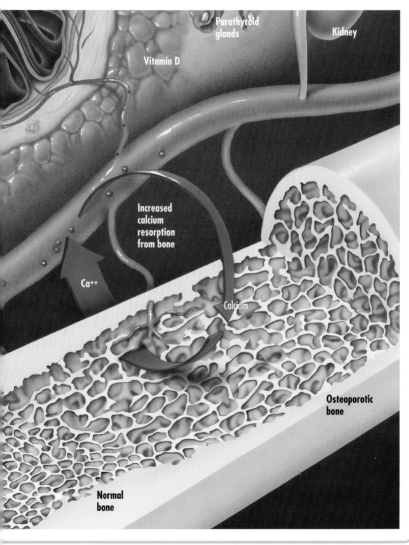

BONE FORMATION AND RESORPTION

Bone consists of 30% organic and 70% mineral substances. The organic portion, called *osteoid*, acts as the matrix or framework for the mineral portion.

Bone cells called *osteoblasts* produce the osteoid matrix. The mineral portion, which consists of calcium and other minerals, hardens the osteoid matrix.

Osteoclasts are large bone cells that reshape mature bone by resorbing the mineral and organic components. Bone formation and resorption are normal, continuous processes. However, in osteoporosis, osteoblasts continue to produce bone, but resorption by osteoclasts exceeds bone formation.

Calcium minerals

Osteoid matrix

Osteoblasts

Osteoclasts

Trabecular bone

UNDERSTANDING ULCERS

WHAT CAUSES ULCERS?

Ulcers result from three main causes: infection with *Helicobacter pylori*, use of nonsteroidal anti-inflammatory drugs (NSAIDS), and pathologic hypersecretion of gastric acids.

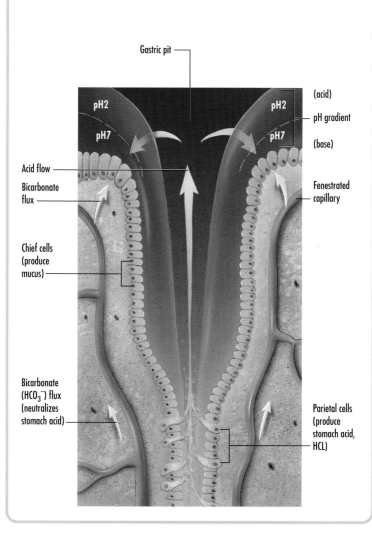

Gastric pit

(acid)

pH2 — pH2

pH gradient

pH7 — pH7

(base)

Acid flow

Bicarbonate flux

Fenestrated capillary

Chief cells (produce mucus)

Bicarbonate (HCO$_3^-$) flux (neutralizes stomach acid)

Parietal cells (produce stomach acid, HCL)

TYPES OF ULCERS

Ulcers may occur anywhere in the esophagus, stomach, pylorus, duodenum, or jejunum, with the crater of the ulcer penetrating through one or more layers of tissue. If this damage is recurring or if healing doesn't take place, the crater may penetrate the wall and extend to adjacent tissues and organs, such as the pancreas.

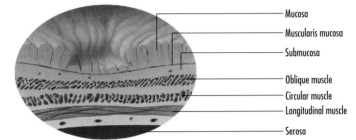

Mucosa
Muscularis mucosa
Submucosa
Oblique muscle
Circular muscle
Longitudinal muscle
Serosa

Erosion
Penetration of only the superficial layer

Acute ulcer
Penetration into the muscular layer

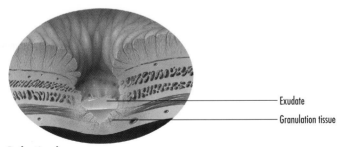

Exudate
Granulation tissue

Perforating ulcer
Penetration of the wall, creating a passage for gastric acids, other fluids, and air to enter adjacent spaces

A CLOSER LOOK AT *HELICOBACTER PYLORI*

The bacteria *Helicobacter pylori* are a contributing factor in chronic gastritis (chronic inflammation of stomach mucosa) and in ulcer formation. They're typically seen within the muscular layers and between cells that line the gastric pits. These bacteria cause tissue inflammation, which can lead to ulcers.

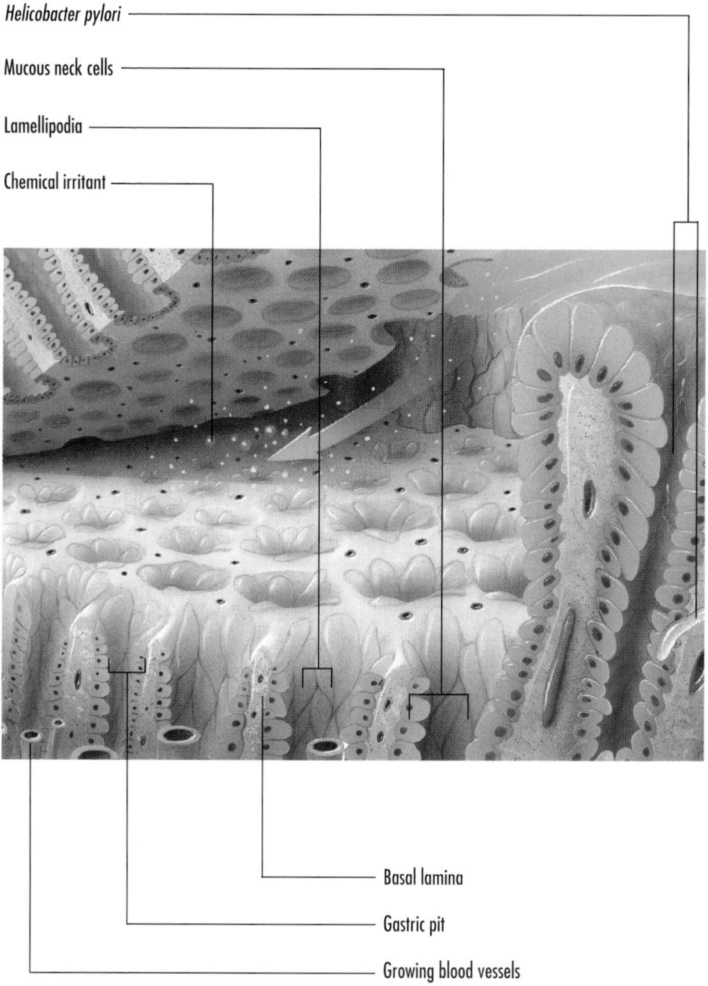

Helicobacter pylori

Mucous neck cells

Lamellipodia

Chemical irritant

Basal lamina

Gastric pit

Growing blood vessels

HOW ULCERS HEAL

In the absence of an effective mucous barrier, an irritant can cause the basal lamina to exfoliate or slough off. A process called rapid reepithelialization, shown below, replenishes the damaged epithelium and repairs any defects.

Lamellipodia

Mucous-secreting surface cells

 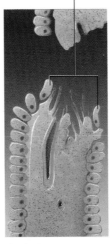 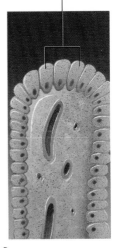

Exfoliation

In the process of exfoliation, irritants such as NSAIDS or excess gastric acid damage the epithelial cells that line the stomach. This exposes the underlying basal lamina to the irritants, causing it to detach and be sloughed off.

Restitution

Rapid reepithelialization occurs constantly and normally, repairing stomach tissues. In this process, mucous neck cells that line the gastric pits divide at a rapid rate and send out lamellipodia ("thin feet"), which move up to cover the regenerating basal lamina.

Recovery

The mucous neck cells transform back into more typical cuboidal surface cells, restoring the normal architecture of the basal lamina.

ous types of hepatitis (such as Types A, B, C, D viral hepatitis) or toxic exposures.

▶ Laënnec's cirrhosis, also called portal, nutritional, or alcoholic cirrhosis, is the most common type and is primarily caused by hepatitis C. Liver damage results from malnutrition (especially dietary protein) and chronic alcohol ingestion. Fibrous tissue forms in portal areas and around central veins.

▶ Autoimmune disease such as sarcoidosis or chronic inflammatory bowel disease may cause cirrhosis.

Cholestatic diseases. This group includes diseases of the biliary tree (biliary cirrhosis resulting from bile duct diseases suppressing bile flow) and sclerosing cholangitis.

Metabolic diseases. This group includes disorders such as Wilson's disease, alpha$_1$-antitrypsin, and hemochromatosis (pigment cirrhosis).

Other types of cirrhosis. Other types of cirrhosis include Budd-Chiari syndrome (epigastric pain, liver enlargement, and ascites due to hepatic vein obstruction), cardiac cirrhosis, and cryptogenic cirrhosis. Cardiac cirrhosis is rare; the liver damage results from right heart failure. Cryptogenic refers to cirrhosis of unknown etiology.

Pathophysiology

Cirrhosis begins with hepatic scarring or fibrosis. The scar begins as an increase in extracellular matrix components — fibril-forming collagens, proteoglycans, fibronectin, and hyaluronic acid. The site of collagen deposition varies with the cause. Hepatocyte function is eventually impaired as the matrix changes. Fat-storing cells are believed to be the source of the new matrix components. Contraction of these cells may also contribute to disruption of the lobular architecture and obstruction of the flow of blood or bile. Cel-

lular changes producing bands of scar tissue also disrupt the lobular structure.

Signs and symptoms

The following are signs and symptoms of the early stages:

▶ anorexia from distaste for certain foods

▶ nausea and vomiting from inflammatory response and systemic effects of liver inflammation

▶ diarrhea from malabsorption

▶ dull abdominal ache from liver inflammation.

The following are signs and symptoms of the late stages:

▶ respiratory — pleural effusion, limited thoracic expansion due to abdominal ascites; interferes with efficient gas exchange and causes hypoxia

▶ central nervous system — progressive signs or symptoms of hepatic encephalopathy, including lethargy, mental changes, slurred speech, asterixis, peripheral neuritis, paranoia, hallucinations, extreme obtundation, and coma — secondary to loss of ammonia to urea conversion and consequent delivery of toxic ammonia to the brain

▶ hematologic — bleeding tendencies (nosebleeds, easy bruising, bleeding gums), anemia resulting from thrombocytopenia (secondary to splenomegaly and decreased vitamin K absorption), splenomegaly, and portal hypertension

▶ endocrine — testicular atrophy, menstrual irregularities, gynecomastia, and loss of chest and axillary hair from decreased hormone metabolism

▶ skin — severe pruritus secondary to jaundice from bilirubinemia; extreme dryness and poor tissue turgor related to malnutrition; abnormal pigmentation, spider angiomas, palmar erythema, and jaundice related to impaired hepatic function

▶ hepatic — jaundice from decreased bilirubin metabolism; hepatomegaly secondary to liver scarring and portal hypertension; ascites and edema of the legs from portal hypertension and decreased plasma proteins; hepatic encephalopathy from ammonia toxicity; and hepatorenal syndrome from advanced liver disease and subsequent renal failure

▶ miscellaneous — musty breath secondary to ammonia build up; enlarged superficial abdominal veins due to portal hypertension; pain in the right upper abdominal quadrant that worsens when patient sits up or leans forward, due to inflammation and irritation of area nerve fibers; palpable liver or spleen due to organomegaly; temperature of 101° to 103° F (38° to 39° C) due to inflammatory response; pain and increased temperature from liver inflammation

▶ hemorrhage from esophageal varices resulting from portal hypertension. (See *What happens in portal hypertension.*)

Complications

Complications may include:
▶ ascites
▶ portal hypertension
▶ jaundice
▶ coagulopathy
▶ hepatic encephalopathy
▶ bleeding esophageal varices; acute GI bleeding
▶ liver failure
▶ renal failure.

Diagnosis

▶ Liver biopsy reveals tissue destruction and fibrosis.
▶ Abdominal X-ray shows enlarged liver, cysts, or gas within the biliary tract or liver, liver calcification, and massive fluid accumulation (ascites).
▶ Computed tomography and liver scans show liver size, abnormal masses, and hepatic blood flow and obstruction.
▶ Esophagogastroduodenoscopy reveals bleeding esophageal varices, stomach irritation or ulceration, or duodenal bleeding and irritation.
▶ Blood studies reveal elevated liver enzymes, total serum bilirubin, and indirect bilirubin; decreased total serum albumin and protein; prolonged prothrombin time; decreased hemoglobin, hematocrit, and

serum electrolytes; and deficiency of vitamins A, C, and K.
▶ Urine studies show increased bilirubin and urobilirubinogen.
▶ Fecal studies show decreased fecal urobilirubinogen.

Treatment

▶ Vitamins and nutritional supplements to help heal damaged liver cells and improve nutritional status
▶ Antacids to reduce gastric distress and decrease the potential for gastrointestinal bleeding
▶ Potassium-sparing diuretics to reduce fluid accumulation
▶ Vasopressin to treat esophageal varices
▶ Esophagogastric intubation with multilumen tubes to control bleeding from esophageal varices or other hemorrhage sites by using balloons to exert pressure on bleeding site.
▶ Gastric lavage until the contents are clear; with antacids and histamine antagonists if bleeding is secondary to a gastric ulcer
▶ Esophageal balloon tamponade to compress bleeding vessels and stop blood loss from esophageal varices
▶ Paracentesis to relieve abdominal pressure and remove ascitic fluid
▶ Surgical shunt placement to divert ascites into venous circulation, leading to weight loss, increased abdominal girth, increased sodium excretion from the kidneys, and improved urine output
▶ Sclerosing agents injected into oozing vessels to cause clotting and sclerosis
▶ Insertion of portosystemic shunts to control bleeding from esophageal varices and decrease portal hypertension (diverts a portion of the portal vein blood flow away from the liver; seldom performed).

Crohn's disease

Crohn's disease, also known as regional enteritis or granulomatous colitis, is inflammation of any part of the gastrointestinal tract (usually the proximal portion of the colon and less commonly the ter-

WHAT HAPPENS IN PORTAL HYPERTENSION

Portal hypertension (elevated pressure in the portal vein) occurs when blood flow meets increased resistance. This common result of cirrhosis may also stem from mechanical obstruction and occlusion of the hepatic veins (Budd-Chiari syndrome).

As the pressure in the portal vein rises, blood backs up into the spleen and flows through collateral channels to the venous system, bypassing the liver. Thus, portal hypertension causes:

▶ splenomegaly with thrombocytopenia

▶ dilated collateral veins (esophageal varices, hemorrhoids, or prominent abdominal veins)
▶ ascites.

In many patients, the first sign of portal hypertension is bleeding esophageal varices (dilated tortuous veins in the submucosa of the lower esophagus). Esophageal varices commonly cause massive hematemesis, requiring emergency care to control hemorrhage and prevent hypovolemic shock.

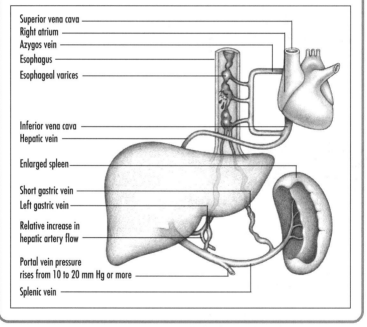

Superior vena cava
Right atrium
Azygos vein
Esophagus
Esophageal varices

Inferior vena cava
Hepatic vein

Enlarged spleen

Short gastric vein
Left gastric vein

Relative increase in hepatic artery flow

Portal vein pressure rises from 10 to 20 mm Hg or more

Splenic vein

minal ileum), extending through all layers of the intestinal wall. It may also involve regional lymph nodes and the mesentery. Crohn's disease is most prevalent in adults ages 20 to 40.

CULTURAL DIVERSITY Crohn's disease is two to three times more common in Ashkenazic Jews and least common in African Americans. Up to 20% of patients have a positive family history for the disease.

BOWEL CHANGES IN CROHN'S DISEASE

As Crohn's disease progresses, fibrosis thickens the bowel wall and narrows the lumen. Narrowing — or stenosis — can occur in any part of the intestine and cause varying degrees of intestinal obstruction. At first, the mucosa may appear normal, but as the disease progresses it takes on a "cobblestone" appearance as shown.

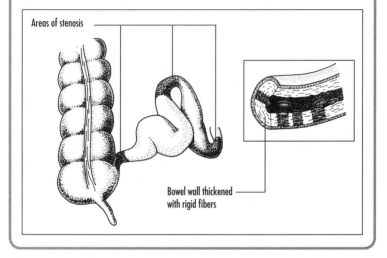

Areas of stenosis

Bowel wall thickened with rigid fibers

Causes

The exact cause is unknown but the following conditions may contribute:

▶ lymphatic obstruction
▶ allergies
▶ immune disorders
▶ infection
▶ genetic predisposition.

Pathophysiology

Whatever the cause of Crohn's disease, inflammation spreads slowly and progressively. Enlarged lymph nodes block lymph flow in the submucosa. Lymphatic obstruction leads to edema, mucosal ulceration and fissures, abscesses, and sometimes granulomas. Mucosal ulcerations are called "skipping lesions" because they are not continuous, as in ulcerative colitis.

Oval, elevated patches of closely packed lymph follicles — called Peyer's patches — develop in the lining of the small intestine. Subsequent fibrosis thickens the bowel wall and causes stenosis, or narrowing of the lumen. (See *Bowel changes in Crohn's disease.*) The serous membrane becomes inflamed (serositis), inflamed bowel loops adhere to other diseased or normal loops, and diseased bowel segments become interspersed with healthy ones. Finally, diseased parts of the bowel become thicker, narrower, and shorter.

Signs and symptoms

Signs and symptoms include:

▶ steady, colicky pain in the right lower quadrant due to acute inflammation and nerve fiber irritation
▶ cramping due to acute inflammation
▶ tenderness due to acute inflammation
▶ weight loss secondary to diarrhea and malabsorption

diarrhea due to bile salt malabsorption, loss of healthy intestinal surface area, and bacterial growth

steatorrhea secondary to fat malabsorption

bloody stools secondary to bleeding from inflammation and ulceration.

Complications

Complications may include:

anal fistula

perineal abscess

fistulas to the bladder or vagina or to the skin in an old scar area

intestinal obstruction

nutrient deficiencies from poor digestion and malabsorption of bile salts and vitamin B_{12}

fluid imbalances.

Diagnosis

Fecal occult test reveals minute amounts of blood in stools.

Small bowel X-ray shows irregular mucosa, ulceration, and stiffening.

Barium enema reveals the string sign (segments of stricture separated by normal bowel) and possibly fissures and narrowing of the bowel.

Sigmoidoscopy and colonoscopy reveal patchy areas of inflammation (helps to rule out ulcerative colitis), with cobblestone-like mucosal surface. With colon involvement, ulcers may be seen.

Biopsy reveals granulomas in up to half of all specimens.

Blood tests reveal increased white blood cell count and erythrocyte sedimentation rate, and decreased potassium, calcium, magnesium, and hemoglobin levels.

Treatment

Corticosteroids to reduce inflammation and, subsequently, diarrhea, pain, and bleeding

Immunosuppressants to suppress the response to antigens

Sulfasalazine to reduce inflammation

Metronidazole to treat perianal complications

Antidiarrheals to combat diarrhea (not used with patients with significant bowel obstruction)

Narcotic analgesics to control pain and diarrhea

Stress reduction and reduced physical activity to rest the bowel and allow it to heal

Vitamin supplements to replace and compensate for the bowel's inability to absorb vitamins

Dietary changes (elimination of fruits, vegetables, high fiber foods, dairy products, spicy and fatty foods, foods that irritate the mucosa, carbonated or caffeinated beverages, and other foods or liquids that stimulate excessive intestinal activity) to decrease bowel activity while still providing adequate nutrition

Surgery, if necessary, to repair bowel perforation and correct massive hemorrhage, fistulas or acute intestinal obstruction; colectomy with ileostomy in patients with extensive disease of the large intestine and rectum.

Diverticular disease

In diverticular disease, bulging pouches (diverticula) in the gastrointestinal wall push the mucosal lining through the surrounding muscle. Although the most common site for diverticula is in the sigmoid colon, they may develop anywhere, from the proximal end of the pharynx to the anus. Other typical sites include the duodenum, near the pancreatic border or the ampulla of Vater, and the jejunum.

CULTURAL DIVERSITY Diverticular disease is common in Western countries, suggesting that a low-fiber diet reduces stool bulk and leads to excessive colonic motility. This consequent increased intraluminal pressure causes herniation of the mucosa.

Diverticular disease of the stomach is rare and is usually a precursor of peptic or neoplastic disease. Diverticular disease of the ileum (Meckel's diverticulum) is the most common congenital anomaly of the gastrointestinal tract.

PATHOGENESIS OF DIVERTICULAR DISEASE

The etiology of diverticular disease hasn't been determined. It's thought to arise from a disordered colonic motility pattern. These diagrams compare normal and abnormal patterns.

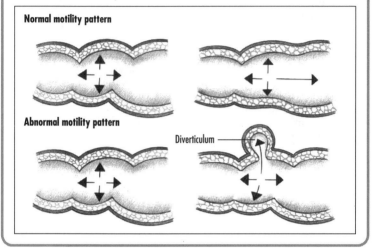

Normal motility pattern

Abnormal motility pattern

Diverticulum

Diverticular disease has two clinical forms:

▶ diverticulosis, in which diverticula are present but don't cause symptoms

▶ diverticulitis, in which diverticula are inflamed and may cause potentially fatal obstruction, infection, or hemorrhage.

 AGE ALERT Diverticular disease is most prevalent in men over age 40 and persons who eat a low fiber diet. More than half of patients older than 50 years have colonic diverticula.

Causes

▶ Diminished colonic motility and increased intraluminal pressure

▶ Low fiber diet

▶ Defects in colon wall strength.

Pathophysiology

Diverticula probably result from high intraluminal pressure on an area of weakness in the gastrointestinal wall, where blood vessels enter. (See *Pathogenesis of diverticular disease*.) Diet may be a contributing factor, because insufficient fiber reduces fecal residue, narrows the bowel lumen, and leads to high intra-abdominal pressure during defecation.

In diverticulitis, retained undigested food and bacteria accumulates in the diverticular sac. This hard mass cuts off the blood supply to the thin walls of the sac, making them more susceptible to attack by colonic bacteria. Inflammation follows and may leading to perforation, abscess, peritonitis, obstruction, or hemorrhage. Occasionally, the inflamed colon segment may adhere to the bladder or other organs and cause a fistula.

Signs and symptoms

Typically the patient with diverticulosis is asymptomatic and will remain so unless diverticulitis develops.

Mild diverticulitis. In mild diverticulitis, signs and symptoms include:

▶ moderate left lower abdominal pain secondary to inflammation of diverticula

▶ low-grade fever from trapping of bacteria-rich stool in the diverticula

▶ leukocytosis from infection secondary to trapping of bacteria-rich stool in the diverticula.

Severe diverticulitis. In severe diverticulitis, signs and symptoms include:

▶ abdominal rigidity from rupture of the diverticula, abscesses, and peritonitis

▶ left lower quadrant pain secondary to rupture of the diverticula and subsequent inflammation and infection

▶ high fever, chills, hypotension from sepsis, and shock from the release of fecal material from the rupture site

▶ microscopic or massive hemorrhage from rupture of diverticulum near a vessel.

Chronic diverticulitis. In chronic diverticulitis, signs and symptoms include:

▶ constipation, ribbon-like stools, intermittent diarrhea, and abdominal distention resulting from intestinal obstruction (possible when fibrosis and adhesions narrow the bowel's lumen)

▶ abdominal rigidity and pain, diminishing or absent bowel sounds, nausea, and vomiting secondary to intestinal obstruction.

Diagnosis

▶ Upper GI series confirms or rules out diverticulosis of the esophagus and upper bowel.

▶ Barium enema reveals filling of diverticula, which confirms diagnosis.

▶ Biopsy reveals evidence of benign disease, ruling out cancer.

▶ Blood studies show an elevated erythrocyte sedimentation rate in diverticulitis.

Treatment

▶ Liquid or bland diet, stool softeners, and occasional doses of mineral oil for symptomatic diverticulosis to relieve symptoms,

minimize irritation, and lessen the risk of progression to diverticulitis

▶ High-residue diet for treatment of diverticulosis after pain has subsided to help decrease intra-abdominal pressure during defecation

▶ Exercise to increase the rate of stool passage

▶ Antibiotics to treat infection of the diverticula

▶ Meperidine (Demerol) to control pain and to relax smooth muscle

▶ Antispasmodics to control muscle spasms

▶ Colon resection with removal of involved segment to correct cases refractory to medical treatment

▶ Temporary colostomy if necessary to drain abscesses and rest the colon in diverticulitis accompanied by perforation, peritonitis, obstruction, or fistula

▶ Blood transfusions if necessary to treat blood loss from hemorrhage.

Gastroesophageal reflux disease

Popularly known as heartburn, gastroesophageal reflux disease (GERD) refers to backflow of gastric or duodenal contents or both into the esophagus and past the lower esophageal sphincter (LES), without associated belching or vomiting. The reflux of gastric contents causes acute epigastric pain, usually after a meal. The pain may radiate to the chest or arms. It commonly occurs in pregnant or obese persons. Lying down after a meal also contributes to reflux.

Causes

Causes of GERD include:

▶ weakened esophageal sphincter

▶ increased abdominal pressure, such as with obesity or pregnancy

▶ hiatal hernia

▶ medications such as morphine, diazepam, calcium channel blockers, meperidine, and anticholinergic agents

▶ food, alcohol, or cigarettes that lower LES pressure

HOW HEARTBURN OCCURS

Hormonal fluctuations, mechanical stress, and the effects of certain foods and drugs can lower esophageal sphincter (LES) pressure. When LES pressure falls and intraabdominal or intragastric pressure rises, the normally contracted LES relaxes inappropriately and allows reflux of gastric acid or bile secretions into the lower esophagus. There, the reflux irritates and inflames the esophageal mucosa, causing pyrosis.

Persistent inflammation can cause LES pressure to decrease even more and may trigger a recurrent cycle of reflux and pyrosis.

Pyrosis

Esophagus

Esophageal mucosa

Lower esophageal sphincter

Increased intraabdominal pressure

Increased intragastric pressure

▶ nasogastric intubation for more than 4 days.

Pathophysiology

Normally, the LES maintains enough pressure around the lower end of the esophagus to close it and prevent reflux. Typically the sphincter relaxes after each swallow to allow food into the stomach. In GERD, the sphincter does not remain closed (usually due to deficient LES pressure or pressure within the stomach exceeding LES pressure) and the pressure in the stomach pushes stomach contents into the esophagus. The high acidity of the stomach contents causes pain and irritation when it enters the esophagus. (See *How heartburn occurs*.)

Signs and symptoms

▶ Burning pain in the epigastric area, possibly radiating to the arms and chest, from the reflux of gastric contents into the

esophagus causing irritation and also esophageal spasm

▶ Pain, usually after a meal or when lying down, secondary to increased abdominal pressure causing reflux

▶ Feeling of fluid accumulation in the throat without a sour or bitter taste due to hypersecretion of saliva.

Complications

▶ Reflux esophagitis
▶ Esophageal stricture
▶ Esophageal ulceration
▶ Chronic pulmonary disease from aspiration of gastric contents in throat.

Diagnosis

Diagnostic tests are aimed at determining the underlying cause of GERD:

▶ Esophageal acidity test evaluates the competence of the lower esophageal sphincter and provides objective measure of reflux.

▶ Acid perfusion test confirms esophagitis and distinguishes it from cardiac disorders.

▶ Esophagoscopy allows visual examination of the lining of the esophagus to reveal extent of disease and confirm pathologic changes in mucosa.

▶ Barium swallow identifies hiatal hernia as the cause.

▶ Upper GI series detects hiatal hernia or motility problems.

▶ Esophageal manometry evaluates resting pressure of LES and determines sphincter competence.

Treatment

▶ Diet therapy with frequent, small meals and avoidance of eating before going to bed to reduce abdominal pressure and reduce the incidence of reflux

▶ Positioning — sitting up during and after meals and sleeping with head of bed elevated — to reduce abdominal pressure and prevent reflux

▶ Increased fluid intake to wash gastric contents out of the esophagus

▶ Antacids to neutralize acidic content of the stomach and minimize irritation

▶ Histamine H_2 antagonists to inhibit gastric acid secretion

▶ Proton pump inhibitors to reduce gastric acidity

▶ Cholinergic agents to increase LES pressure

▶ Smoking cessation to improve LES pressure (nicotine lowers LES pressure)

▶ Surgery if hiatal hernia is the cause or patient has refractory symptoms.

Hemorrhoids

Hemorrhoids are varicosities in the superior or inferior hemorrhoidal venous plexus. Dilation and enlargement of the superior plexus of the superior hemorrhoidal veins above the dentate line cause internal hemorrhoids. Enlargement of the plexus of the inferior hemorrhoidal veins below the dentate line causes external hemorrhoids, which may protrude from the rectum. Hemorrhoids occur in both sexes. Incidence is generally highest between ages 20 and 50.

Causes

Causes may include:
▶ prolonged sitting
▶ straining at defecation
▶ constipation, low-fiber diet
▶ pregnancy
▶ obesity.

Pathophysiology

Hemorrhoids result from activities that increase intravenous pressure, causing distention and engorgement. Predisposing factors include prolonged sitting, straining at defecation, constipation, low-fiber diet, pregnancy, and obesity. Other factors include hepatic disease, such as cirrhosis, amebic abscesses, or hepatitis; alcoholism; and anorectal infections.

Hemorrhoids are classified as first, second, third, or fourth degree, depending on their severity. First-degree hemorrhoids are confined to the anal canal. Second-degree hemorrhoids prolapse during straining but reduce spontaneously. Third-degree hemorrhoids are prolapsed hemorrhoids that require manual reduction after each bowel movement. Fourth-degree hemorrhoids are irreducible. Signs and symptoms vary accordingly.

Signs and symptoms

▶ Painless, intermittent bleeding during defecation from irritation and injury to the hemorrhoid mucosa

▶ Bright red blood on stool or toilet tissue due to injury to hemorrhoid mucosa

▶ Anal itching from poor anal hygiene

▶ Vague feeling of anal discomfort when bleeding occurs

▶ Prolapse of rectal mucosa from straining

▶ Pain from thrombosis of external hemorrhoids.

Complications

▶ Constipation
▶ Local infection

▶ Thrombosis of hemorrhoids
▶ Secondary anemia from severe or recurrent bleeding.

Diagnosis
▶ Physical examination confirms external hemorrhoids.
▶ Anoscopy and flexible sigmoidoscopy visualizes internal hemorrhoids.

Treatment
Treatment depends on the type and severity of the hemorrhoids:
▶ high fiber diet, increased fluid intake, and bulking agents to relieve constipation
▶ avoidance of prolonged sitting on the toilet to prevent venous congestion
▶ local anesthetic agents to decrease local swelling and pain
▶ hydrocortisone cream and suppositories to reduce edematous, prolapsed hemorrhoids, and itching
▶ warm sitz baths to relieve pain
▶ injection sclerotherapy or rubber band ligation to reduce prolapsed hemorrhoids
▶ hemorrhoidectomy by cauterization or excision.

Hepatitis, nonviral

Nonviral hepatitis is an inflammation of the liver that usually results from exposure to certain chemicals or drugs. Most patients recover from this illness, although a few develop fulminating hepatitis or cirrhosis.

Causes
▶ Hepatotoxic chemicals
▶ Hepatotoxic drugs.

Pathophysiology
Various hepatotoxins — such as carbon tetrachloride, acetaminophen, trichloroethylene, poisonous mushrooms, vinyl chloride — can cause hepatitis. After exposure to these agents, hepatic cellular necrosis, scarring, Kupffer cell hyperplasia, and infiltration by mononuclear phagocytes occur with varying severity. Alcohol, anox-

ia, and preexisting liver disease exacerbate the effects of some toxins.

Drug-induced (idiosyncratic) hepatitis may begin with a hypersensitivity reaction unique to the individual, unlike toxic hepatitis, which appears to affect all exposed people indiscriminately. Among possible causes are niacin, halothane, sulfonamides, isoniazid, acetaminophen, methyldopa, and phenothiazines (cholestasis-induced hepatitis). Symptoms of hepatic dysfunction may appear at any time during or after exposure to these drugs, but it usually manifests after 2 to 5 weeks of therapy.

Signs and symptoms
Signs and symptoms include:
▶ anorexia, nausea and vomiting due to systemic effects of liver inflammation
▶ jaundice from decreased bilirubin metabolism, leading to hyperbilirubinemia
▶ dark urine from elevated urobilinogen
▶ hepatomegaly due to inflammation
▶ possible abdominal pain from liver inflammation
▶ clay colored stool secondary to decreased bile in the gastrointestinal tract from liver necrosis
▶ pruritus secondary to jaundice and hyperbilirubinemia.

Complications
▶ Cirrhosis
▶ Hepatic failure.

Diagnosis
▶ Liver enzymes, such as serum aspartate aminotransferase and alanine aminotransferase levels, are elevated.
▶ Total and direct bilirubin levels are elevated.
▶ Alkaline phosphatase levels are elevated.
▶ White blood cell count and eosinophil count are elevated.
▶ Liver biopsy identifies underlying pathology, especially infiltration with white blood cells and eosinophils.

Treatment

Treatment includes:

▶ lavage, catharsis, or hyperventilation, depending on the route of exposure, to remove the causative agent

▶ acetylcysteine as an antidote for acetaminophen poisoning

▶ corticosteroid to relieve symptoms of drug-induced nonviral hepatitis.

Hepatitis, viral

Viral hepatitis is a common infection of the liver, resulting in hepatic cell destruction, necrosis, and autolysis. In most patients, hepatic cells eventually regenerate with little or no residual damage. However, old age and serious underlying disorders make complications more likely. The prognosis is poor if edema and hepatic encephalopathy develop.

Five major forms of hepatitis are currently recognized:

▶ Type A (infectious or short-incubation hepatitis) is most common among male homosexuals and in people with human immunodeficiency virus (HIV) infection.

▶ Type B (serum or long-incubation hepatitis) also is most common among HIV-positive individuals. Routine screening of donor blood for the hepatitis B surface antigen has reduced the incidence of post-transfusion cases, but transmission by needles shared by drug abusers remains a major problem.

▶ Type C accounts for about 20% of all viral hepatitis cases and for most post-transfusion cases.

▶ Type D (delta hepatitis) is responsible for about 50% of all cases of fulminant hepatitis, which has a high mortality. Developing in 1% of patients with viral hepatitis, fulminant hepatitis causes unremitting liver failure with encephalopathy. It progresses to coma and commonly leads to death within 2 weeks. In the United States, type D occurs only in people who are frequently exposed to blood and blood products, such as I.V. drug users and hemophilia patients.

▶ Type E (formerly grouped with types C and D under the name non-A, non-B hepatitis) occurs primarily among patients who have recently returned from an endemic area (such as India, Africa, Asia, or Central America). It's more common in young adults and more severe in pregnant woman. (See *Viral hepatitis from A to E*, pages 322 and 323.)

▶ Other types continue to be identified with growing patient population and sophisticated laboratory identification techniques.

Causes

▶ The five major forms of viral hepatitis result from infection with the causative viruses: A, B, C, D, or E.

Pathophysiology

Hepatic damage is usually similar in all types of viral hepatitis. Varying degrees of cell injury and necrosis occur.

On entering the body, the virus causes hepatocyte injury and death, either by directly killing the cells or by activating inflammatory and immune reactions. The inflammatory and immune reactions will, in turn, injure or destroy hepatocytes by lysing the infected or neighboring cells. Later, direct antibody attack against the viral antigens causes further destruction of the infected cells. Edema and swelling of the interstitium lead to collapse of capillaries and decreased blood flow, tissue hypoxia, and scarring and fibrosis.

Signs and symptoms

Signs and symptoms reflect the stage of the disease.

Prodromal stage. Signs and symptoms of the prodromal stage include:

▶ easy fatigue and generalized malaise due to systemic effects of liver inflammation

▶ anorexia and mild weight loss due to systemic effects of liver inflammation

▶ arthralgia and myalgia due to systemic effects of liver inflammation

▶ nausea and vomiting from gastrointestinal effects of liver inflammation

VIRAL HEPATITIS FROM A TO E

This chart compares the features of each type of viral hepatitis that has been characterized. Other types are emerging.

FEATURE	HEPATITIS A	HEPATITIS B
Incubation	15 to 45 days	30 to 180 days
Onset	Acute	Insidious
Age group most affected	Children, young adults	Any age
Transmission	Fecal-oral, sexual (especially oral-anal contact), nonpercutaneous (sexual, maternal-neonatal), percutaneous (rare)	Blood-borne; parenteral route, sexual, maternal-neonatal; virus is shed in all body fluids
Severity	Mild	Often severe
Prognosis	Generally good	Worsens with age and debility
Progression to chronicity	None	Occasional

▶ changes in the senses of taste and smell related to liver inflammation
▶ fever secondary to inflammatory process
▶ right upper quadrant tenderness from liver inflammation and irritation of area nerve fibers
▶ dark colored urine from urobilinogen
▶ clay colored stools from decreased bile in the gastrointestinal tract.

Clinical stage. Signs and symptoms of the clinical stage include:
▶ worsening of all symptoms of prodromal stage

▶ itching from increased bilirubin in the blood
▶ abdominal pain or tenderness from continued liver inflammation
▶ jaundice from elevated bilirubin in the blood.

Recovery stage. The patient's symptoms subside and appetite returns.

Complications
▶ Chronic persistent hepatitis, which may prolong recovery up to 8 months
▶ Chronic active hepatitis

HEPATITIS C	HEPATITIS D	HEPATITIS E
15 to 160 days	14 to 64 days	14 to 60 days
Insidious	Acute and chronic	Acute
More common in adults	Any age	Ages 20 to 40
Blood-borne; parenteral route	Parenteral route; most people infected with hepatitis D are also infected with hepatitis B	Primarily fecal-oral
Moderate	Can be severe and lead to fulminant hepatitis	Highly virulent with common progression to fulminant hepatitis and hepatic failure, especially in pregnant patients
Moderate	Fair, worsens in chronic cases; can lead to chronic hepatitis D and chronic liver disease	Good unless pregnant
10% to 50% of cases	Occasional	None

- Cirrhosis
- Hepatic failure and death
- Primary hepatocellular carcinoma.

Diagnosis

- Hepatitis profile study identifies antibodies specific to the causative virus, establishing the type of hepatitis.
- Serum aspartate aminotransferase and serum alanine aminotransferase levels are increased in the prodromal stage.
- Serum alkaline phosphatase is slightly increased.

- Serum bilirubin may remain high into late disease, especially in severe cases.
- Prothrombin time prolonged (greater than 3 seconds longer than normal indicates severe liver damage).
- White blood cell counts reveal transient neutropenia and lymphopenia followed by lymphocytosis.
- Liver biopsy to confirm suspicion of chronic hepatitis.

Treatment

- Rest to minimize energy demands

▶ Avoidance of alcohol or other drugs to prevent further hepatic damage

▶ Diet therapy with small, high-calorie meals to combat anorexia

▶ Parental nutrition if patient can't eat due to persistent vomiting

▶ Vaccination against hepatitis A and B to provide immunity to these viruses before transmission occurs.

Hirschsprung's disease

Hirschsprung's disease, also called congenital megacolon and congenital aganglionic megacolon, is a congenital disorder of the large intestine, characterized by absence or marked reduction of parasympathetic ganglion cells in the colorectal wall. Hirschsprung's disease appears to be a familial, congenital defect, occurring in 1 in 2,000 to 1 in 5,000 live births. It's up to seven times more common in males than in females (although the aganglionic segment is usually shorter in males) and is most prevalent in whites. Total aganglionosis affects both sexes equally. Females with Hirschsprung's disease are at higher risk for having affected children. This disease usually coexists with other congenital anomalies, particularly trisomy 21 and anomalies of the urinary tract such as megaloureter.

Without prompt treatment, an infant with colonic obstruction may die within 24 hours from enterocolitis that leads to severe diarrhea and hypovolemic shock. With prompt treatment, prognosis is good.

Cause

▶ Familial congenital defect.

Pathophysiology

In Hirschsprung's disease, parasympathetic ganglion cells in the colorectal wall are absent or markedly reduced in number. The aganglionic bowel segment contracts without the reciprocal relaxation needed to propel feces forward. Impaired intestinal motility causes severe, intractable constipation. Colonic obstruction can ensue, causing bowel dilation and subsequent occlusion of surrounding blood and lymphatics. The ensuing mucosal edema, ischemia, and infarction draw large amounts of fluid into the bowel, causing copious amounts of liquid stool. Continued infarction and destruction of the mucosa can lead to infection and sepsis.

Signs and symptoms

In the newborn, signs and symptoms include:

▶ failure to pass meconium within 24 to 48 hours due to inability to propel intestinal contents forward

▶ bile stained or fecal vomiting as a result of bowel obstruction

▶ abdominal distention secondary to retention of intestinal contents and bowel obstruction

▶ irritability due to resultant abdominal distention

▶ feeding difficulties and failure to thrive related to retention of intestinal contents and abdominal distention

▶ dehydration related to subsequent feeding difficulties and inability to ingest adequate fluids

▶ overflow diarrhea secondary to increased water secretion into bowel with bowel obstruction.

In children, signs and symptoms include:

▶ intractable constipation due to decreased gastrointestinal motility

▶ abdominal distention from retention of stool

▶ easily palpated fecal masses from retention of stool

▶ wasted extremities (in severe cases) secondary to impaired intestinal motility and its effects on nutrition and intake

▶ loss of subcutaneous tissue (in severe cases) secondary to malnutrition

▶ large protuberant abdomen due to retention of stool and consequent changes in fluid and electrolyte homeostasis.

In adults, signs and symptoms (occurs rarely and is more prevalent in men) include:

abdominal distention from decreased bowel motility and constipation

chronic intermittent constipation secondary to impaired intestinal motility.

Complications
- Bowel perforation
- Electrolyte imbalances
- Nutritional deficiencies
- Enterocolitis
- Hypovolemic shock
- Sepsis.

Diagnosis
Rectal biopsy confirms diagnosis by showing absence of ganglion cells.

Barium enema, used in older infants, reveals a narrowed segment of distal colon with a sawtoothed appearance and a funnel-shaped segment above it. This confirms the diagnosis and assesses the extent of intestinal involvement.

Rectal manometry detects failure of the internal anal sphincter to relax and contract.

Upright plain abdominal X-rays show marked colonic distention.

Treatment
Treatment may include:

corrective surgery to pull the normal ganglionic segment through to the anus (usually delayed until the infant is at least 10 months old)

daily colonic lavage to empty the bowel of the infant until the time of surgery

temporary colostomy or ileostomy to compress the colon in instances of total bowel obstruction.

Hyperbilirubinemia
Hyperbilirubinemia, also called neonatal jaundice, is the result of hemolytic processes in the newborn marked by elevated serum bilirubin levels and mild jaundice. It can be physiologic (with jaundice the only symptom) or pathologic (resulting from an underlying disease).

Physiologic jaundice generally develops 24 to 48 hours after birth and disappears by day 7 in full term babies and by day 9 or 10 in premature infants. Serum unconjugated bilirubin levels do not exceed 12mg/dl. Pathologic jaundice may appear anytime after the first day of life and persist beyond 7 days. Serum bilirubin levels are greater than 12 mg/dl in a term infant, 15 mg/dl in a premature infant, or increase more than 5 mg/dl in 24 hours. Physiologic jaundice is self-limiting; pathologic jaundice varies, depending on the cause.

CULTURAL DIVERSITY Hyperbilirubinia tends to be more common and more severe in Chinese, Japanese, Koreans, and Native Americans, whose mean peak of unconjugated bilirubin is approximately twice that of the rest of the population.

Causes
Causes may include:
- blood type incompatibility
- intrauterine infection (rubella, cytomegalic inclusion body disease, toxoplasmosis, syphilis and, occasionally, bacteria such as *Escherichia coli*, Staphylococcus, Pseudomonas, Klebsiella, Proteus, and Streptococcus)
- infection (gram-negative bacteria)
- polycythemia
- enclosed hemorrhages (bruises, subdural hematoma)
- respiratory distress syndrome (hyaline membrane disease)
- heinz body anemia from drugs and toxins (vitamin K_3, sodium nitrate)
- transient neonatal hyperbilirubinemia
- abnormal red blood cell morphology
- deficiencies of red blood cell enzymes (glucose 6-phosphate dehydrogenase, hexokinase)
- physiologic jaundice
- breast feeding
- maternal diabetes
- crigler-Najjar syndrome
- gilbert syndrome
- herpes simplex
- pyloric stenosis
- hypothyroidism

- neonatal giant cell hepatitis
- bile duct atresia
- galactosemia
- choledochal cyst.

Pathophysiology

As erythrocytes break down at the end of their neonatal life cycle, hemoglobin separates into globin (protein) and heme (iron) fragments. Heme fragments form unconjugated (indirect) bilirubin, which binds to albumin for transport to liver cells to conjugate with glucuronide, forming direct bilirubin. Because unconjugated bilirubin is fat-soluble and can't be excreted in the urine or bile, it may escape to extravascular tissue, especially fatty tissue and the brain, causing hyperbilirubinemia.

Certain drugs (such as aspirin, tranquilizers, and sulfonamides) and conditions (such as hypothermia, anoxia, hypoglycemia, and hypoalbuminemia) can disrupt conjugation and usurp albumin-binding sites.

Decreased hepatic function also reduces bilirubin conjugation. Biliary obstruction or hepatitis can cause hyperbilirubinemia by blocking normal bile flow.

Increased erythrocyte production or breakdown in hemolytic disorders, or in Rh or ABO incompatibility can cause hyperbilirubinemia. Lysis releases bilirubin and stimulates cell agglutination. As a result, the liver's capacity to conjugate bilirubin becomes overloaded.

Finally, maternal enzymes in breast milk inhibit the infant's glucuronyl-transferase conjugating activity.

Signs and symptoms

- Jaundice from the escape of unconjugated bilirubin to extra vascular tissue (primary sign of hyperbilirubinemia).

Complications

- Kernicterus
- Cerebral palsy, epilepsy, or mental retardation
- Perceptual-motor handicaps and learning disorders.

Diagnosis

- Bilirubin levels greater than 0.5 mg/dl.

Treatment

- Phototherapy (treatment of choice for physiologic jaundice, and pathologic jaundice due to erythroblastosis fetalis, after the initial exchange transfusion) with fluorescent lights to decompose bilirubin in the skin by oxidation (usually discontinued after bilirubin levels fall below 10mg/dl and continue to decrease for 24 hours)
- Exchange transfusion to replace the infant's blood with fresh blood (less then 48 hours old), removing some of the unconjugated bilirubin in serum; indicated for conditions such as hydrops fetalis, polycythemia, erythroblastosis fetalis; marked reticulocytosis, drug toxicity, and jaundice that develops within the first 6 hours after birth
- Albumin administration to provide additional albumin for binding unconjugated bilirubin
- Phenobarbital administration (rare) to the mother before delivery and to the newborn several days after delivery to stimulate the hepatic glucuronide-conjugating system

Irritable bowel syndrome

Also referred to as spastic colon or spastic colitis, irritable bowel syndrome (IBS) is marked by chronic symptoms of abdominal pain, alternating constipation and diarrhea, excess flatus, a sense of incomplete evacuation, and abdominal distention. Irritable bowel syndrome is a common, stress-related disorder. However, 20% of patients never seek medical attention. IBS is a benign condition that has no anatomical abnormality or inflammatory component. It occurs in women twice as often as men.

Causes

- Psychological stress (most common)
- Ingestion of irritants (coffee, raw fruit, or vegetables)

WHAT HAPPENS IN IRRITABLE BOWEL SYNDROME

Typically, the patient with irritable bowel syndrome has a normal-appearing GI tract. However, careful examination of the colon may reveal functional irritability — an abnormality in colonic smooth-muscle function marked by excessive peristalsis and spasms, even during remission.

INTESTINAL FUNCTION

To understand what happens in irritable bowel syndrome, consider how smooth muscle controls bowel function. Normally, segmental muscle contractions mix intestinal contents while peristalsis propels the contents through the GI tract. Motor activity is most propulsive in the proximal (stomach) and the distal (sigmoid) portions of the intestine. Activity in the rest of the intestines is slower, permitting nutrient and water absorption.

In irritable bowel syndrome, the autonomic nervous system, which innervates the large intestine, doesn't cause the alternating contractions and relaxations that propel stools smoothly toward the rectum.

The result is constipation or diarrhea or both.

CONSTIPATION

Some patients have spasmodic intestinal contractions that set up a partial obstruction by trapping gas and stools. This causes distention, bloating, gas pain, and constipation.

DIARRHEA

Other patients have dramatically increased intestinal motility. Usually eating or cholinergic stimulation triggers the small intestine's contents to rush into the large intestine, dumping watery stools and irritating the mucosa. The result is diarrhea.

MIXED SYMPTOMS

If further spasms trap liquid stools, the intestinal mucosa absorbs water from the stools, leaving them dry, hard, and difficult to pass. The result: a pattern of alternating diarrhea and constipation.

▶ Lactose intolerance
▶ Abuse of laxatives
▶ Hormonal changes (menstruation).

Pathophysiology

Irritable bowel syndrome appears to reflect motor disturbances of the entire colon in response to stimuli. Some muscles of the small bowel are particularly sensitive to motor abnormalities and distention; others are particularly sensitive to certain foods and drugs. The patient may be hypersensitive to the hormones gastrin and cholecystokinin. The pain of IBS seems to be caused by abnormally strong contractions of the intestinal smooth muscle as it reacts to distention, irritants, or stress.

(See *What happens in irritable bowel syndrome.*)

Signs and symptoms

▶ Crampy lower abdominal pain secondary to muscle contraction; usually occurring during the day and relieved by defecation or passage of flatus
▶ Pain that intensifies 1 to 2 hours after a meal from irritation of nerve fibers by causative stimulus
▶ Constipation alternating with diarrhea, with one dominant; secondary to motor disturbances from causative stimulus
▶ Mucus passed through the rectum from altered secretion in intestinal lumen due to motor abnormalities

▶ Abdominal distention and bloating caused by flatus and constipation.

Diagnosis

▶ Physical examination reveals contributing psychological factors, such as a recent stressful life change.

▶ Stool samples for ova, parasites, bacteria, and blood rule out infection.

▶ Lactose intolerance test rules out lactose intolerance.

▶ Barium enema may reveal colon spasm and tubular appearance of descending colon without evidence of cancers and diverticulosis.

▶ Sigmoidoscopy or colonoscopy may reveal spastic contractions without evidence of colon cancer or inflammatory bowel disease.

▶ Rectal biopsy rules out malignancy.

Treatment

Treatment may include:

▶ stress relief measures, including counseling or mild anti-anxiety agents

▶ investigation and avoidance of food irritants

▶ application of heat to abdomen

▶ bulking agents to reduce episodes of diarrhea and minimize effect of nonpropulsive colonic contractions

▶ antispasmodics (dicyclomine, hycomine sulfate) for pain

▶ loperamide possibly to reduce urgency and fecal soiling in patients with persistent diarrhea

▶ bowel training (if the cause of IBS is chronic laxative abuse) to regain muscle control.

Liver failure

Liver failure can be the end result of any liver disease. The liver performs over 100 separate functions in the body. When it fails, a complex syndrome involving the impairment of many different organs and body functions ensues. (See *Functions of the liver.*) Hepatic encephalopathy and hepatorenal syndrome are two conditions

occurring in liver failure. The only cure for liver failure is a liver transplant.

Causes

▶ Viral hepatitis
▶ Nonviral hepatitis
▶ Cirrhosis
▶ Liver cancer.

Pathophysiology

Manifestations of liver failure include hepatic encephalopathy and hepatorenal syndrome.

Hepatic encephalopathy, a set of central nervous system disorders, results when the liver can no longer detoxify the blood. Liver dysfunction and collateral vessels that shunt blood around the liver to the systemic circulation permit toxins absorbed from the GI tract to circulate freely to the brain. Ammonia is one of the main toxins causing hepatic encephalopathy. Ammonia is a byproduct of protein metabolism. The normal liver transforms ammonia to urea, which the kidneys excrete. When the liver fails and is no longer able to transform ammonia to urea, ammonia blood levels rise and the ammonia is delivered to the brain. Short-chain fatty acids, serotonin, tryptophan, and false neurotransmitters may also accumulate in the blood and contribute to hepatic encephalopathy.

Hepatorenal syndrome is renal failure concurrent with liver disease; the kidneys appear to be normal but abruptly cease functioning. It causes expanded blood volume, accumulation of hydrogen ions, and electrolyte disturbances. It is most common in patients with alcoholic cirrhosis or fulminating hepatitis. The cause may be the accumulation of vasoactive substances that cause inappropriate constriction of renal arterioles, leading to decreased glomerular filtration and oliguria. The vasoconstriction may also be a compensatory response to portal hypertension and the pooling of blood in the splenic circulation.

Signs and symptoms

▶ Jaundice from the failure of liver to conjugate bilirubin

▶ Abdominal pain or tenderness from liver inflammation

▶ Nausea and anorexia from systemic effects of inflammation

▶ Fatigue and weight loss from failure of hepatic metabolism

▶ Pruritus due to the accumulation of bilirubin in the skin

▶ Oliguria from intrarenal vasoconstriction

▶ Splenomegaly secondary to portal hypertension

▶ Ascites due to portal hypertension and decreased plasma proteins

▶ Peripheral edema from accumulation of fluid retained because of decreased plasma protein production and loss of albumin with ascites

▶ Varices of the esophagus, rectum, and abdominal wall secondary to portal hypertension

▶ Bleeding tendencies from thrombocytopenia (secondary to blood accumulation in the spleen) and prolonged prothrombin time (from the impaired production of coagulation factors)

▶ Petechia resulting from thrombocytopenia

▶ Amenorrhea secondary to altered steroid hormone production and metabolism

▶ Gynecomastia in males from estrogen buildup due to failure of hepatic biotransformation functions.

Complications

▶ Variceal bleeding

▶ Gastrointestinal hemorrhage

▶ Coma

▶ Death.

Diagnosis

▶ Liver function tests reveal elevated levels of aspartate aminotransferase (AST), alanine aminotrasferase (ALT), alkaline phosphatase, and bilirubin.

▶ Blood studies reveal anemia, impaired red blood cell production, elevated bleed-

---

FUNCTIONS OF THE LIVER

The liver is one of the most essential organs of the body. To understand how liver disease affects the body, it is best to understand its main functions:

▶ Detoxifies poisonous chemicals, including alcohol, beer, wine, and drugs (prescribed and over-the-counter, as well as illegal substances)

▶ Makes bile to help digest food

▶ Stores energy by stockpiling sugar (carbohydrates, glucose, and fat) until needed

▶ Stores iron reserves, as well as vitamins and minerals

▶ Manufactures new proteins

▶ Produces important plasma proteins necessary for blood coagulation, including prothrombin and fibrinogen

▶ Serves as a site for hematopoiesis during fetal development.

---

ing and clotting times, low blood glucose levels, and increased serum ammonia levels.

▶ Urine osmolarity is increased.

Treatment

Treatment may include:

▶ liver transplantation

▶ low protein, high carbohydrate diet to correct nutritional deficiencies and prevent overtaxing liver

▶ lactulose to reduce ammonia blood levels and help alleviate some symptoms of hepatic encephalopathy.

For *ascites*, treatment includes:

▶ salt restriction and potassium-sparing diuretics to increase water excretion

▶ potassium supplements to reverse the effects of high aldosterone

▶ paracentesis to remove ascitic fluid and alleviate abdominal discomfort

▶ shunt placement to aid in removal of ascitic fluid and alleviate abdominal discomfort.

For *portal hypertension*, treament includes:

▶ shunt placement between the portal vein and another systemic vein to divert blood flow and relieve pressure.

For *variceal bleeding*, treatment includes:

▶ vasoconstrictor drugs to decrease blood flow

▶ balloon tamponade to control bleeding by exerting pressure on the varices with the use of a balloon catheter

▶ surgery to tie off bleeding collaterals sprouting from the portal vein

▶ vitamin K to control bleeding by decreasing prothrombin time.

Malabsorption

Malabsorption is failure of the intestinal mucosa to absorb single or multiple nutrients efficiently. Absorption of amino acids, fat, sugar, or vitamins may be impaired. The result is inadequate movement of nutrients from the small intestine to the bloodstream or lymphatic system. Manifestations depend primarily on what is not being absorbed.

Causes

A wide variety of disorders result in malabsorption. (See *Causes of malabsorption*.)

▶ prior gastric surgery
▶ pancreatic disorders
▶ hepatobiliary disease
▶ disease of the small intestine, such as celiac disease
▶ hereditary disorder
▶ drug toxicity.

Pathophysiology

The small intestine's inability to absorb nutrients efficiently may result from a variety of disease processes. The mechanism of malabsorption depends on the cause. Some common causes of malabsorption syndrome include celiac disease, lactase

deficiency, gastrectomy, Zollinger-Ellison syndrome, and bacterial overgrowth in the duodenal stump.

In celiac sprue, dietary gluten — a product of wheat, barley, rye, and oats — is toxic to the patient, causing injury to the mucosal villi. The mucosa appear flat and have lost absorptive surface. Symptoms generally disappear when gluten is removed from the diet.

Lactase deficiency is a disaccharide deficiency syndrome. Lactase is an intestinal enzyme that splits nonabsorbable lactose (a disaccharide) into the absorbable monosaccharides glucose and galactose. Production may be deficient, or another intestinal disease may inhibit the enzyme.

Malabsorption may occur after gastrectomy. Poor mixing of chyme with gastric secretions is the cause of postsurgical malabsorption.

In Zollinger-Ellison syndrome, increased acidity in the duodenum inhibits release of cholecystokinin, which stimulates pancreatic enzyme secretion. Pancreatic enzyme deficiency leads to decreased breakdown of nutrients and malabsorption.

Bacterial overgrowth in the duodenal stump (loop created in the Billroth II procedure) causes malabsorption of vitamin B_{12}.

Signs and symptoms

Signs and symptoms include:

▶ weight loss and generalized malnutrition from impaired absorption of carbohydrate, fat, and protein

▶ diarrhea from decreased absorption of fluids, electrolytes, bile acids and fatty acids in colon

▶ steatorrhea from excess fat in stool

▶ flatulence and abdominal distention secondary to fermentation of undigested lactose

▶ nocturia from delayed absorption of water

▶ weakness and fatigue from anemia and electrolyte depletion from diarrhea

CAUSES OF MALABSORPTION

Many disorders — from systemic to organ-specific diseases — may give rise to malabsorption.

DISEASES OF THE SMALL INTESTINE

Primary small bowel disease
▶ Bacterial overgrowth due to stasis in afferent loop after Billroth II gastrectomy
▶ Massive bowel resection
▶ Nontropical sprue (celiac disease)
▶ Regional enteritis
▶ Tropical sprue

Ischemic small bowel disease
▶ Chronic congestive heart failure
▶ Mesenteric atherosclerosis

Small bowel infections and infestations
▶ Acute enteritis
▶ Giardiasis

Systemic disease involving small bowel
▶ Amyloidosis
▶ Lymphoma
▶ Sarcoidosis

▶ Scleroderma
▶ Whipple's disease

DRUG-INDUCED MALABSORPTION
▶ Calcium carbonate
▶ Neomycin

HEPATOBILIARY DISEASE
▶ Biliary fistula
▶ Biliary tract obstruction
▶ Cirrhosis and hepatitis

HEREDITARY DISORDER
▶ Primary lactase deficiency

PANCREATIC DISORDERS
▶ Chronic pancreatitis
▶ Cystic fibrosis
▶ Pancreatic cancer
▶ Pancreatic resection
▶ Zollinger-Ellison syndrome

PREVIOUS GASTRIC SURGERY
▶ Billroth II gastrectomy
▶ Pyloroplasty
▶ Total gastrectomy
▶ Vagotomy

▶ edema from impaired absorption of amino acids, resulting in protein depletion and hypoproteinemia
▶ amenorrhea from protein depletion leading to hypopituitarism
▶ anemia from the impaired absorption of iron, folic acid, and vitamin B_{12}
▶ glossitis, cheilosis secondary to a deficiency of iron, folic acid, vitamin B_{12}, and other vitamins
▶ peripheral neuropathy from a deficiency of vitamin B_{12} and thiamine
▶ bruising, bleeding tendency from vitamin K malabsorption and hypoprothrombinemia
▶ bone pain, skeletal deformities, fractures from calcium malabsorption that leads to hypocalcemia; protein depletion leading to osteoporosis and vitamin D malabsorption causing impaired calcium absorption
▶ tetany, paresthesias resulting from calcium malabsorption leading to hypocalcemia and magnesium malabsorption, leading to hypomagnesemia and hypokalemia.

Complications
▶ Fractures
▶ Anemias
▶ Bleeding disorders
▶ Tetany
▶ Malnutrition.

Diagnosis
▶ Stool specimen for fat reveals excretion of greater than 6 g of fat per day.
▶ D-Xylose absorption test shows less than 20% of 25 g of D-Xylose in the urine af-

ter 5 hours, reflects disorders of proximal bowel.

◗ Schilling test reveals deficiency of vitamin B_{12} absorption.

◗ Culture of duodenal and jejunal contents confirms bacterial overgrowth in the proximal bowel.

◗ Gastrointestinal barium studies show characteristic features of the small intestine.

◗ Small intestine biopsy reveals the atrophy of mucosal villi.

Treatment

◗ Identification of cause and appropriate correction

◗ Gluten-free diet to stop progression of celiac disease and malabsorption

◗ Lactose-free diet to treat lactase deficiency

◗ Dietary supplementation to replace nutrient deficiencies

◗ Vitamin B_{12} injections to treat vitamin B_{12} deficiency.

Pancreatitis

Pancreatitis, inflammation of the pancreas, occurs in acute and chronic forms and may be due to edema, necrosis, or hemorrhage. In men, this disease is commonly associated with alcoholism, trauma, or peptic ulcer; in women, with biliary tract disease. The prognosis is good in pancreatitis associated with biliary tract disease, but poor when associated with alcoholism. Mortality is as high as 60% when pancreatitis is associated with necrosis and hemorrhage.

Causes

◗ Biliary tract disease

◗ Alcoholism

◗ Abnormal organ structure

◗ Metabolic or endocrine disorders, such as high cholesterol levels or overactive thyroid

◗ Pancreatic cysts or tumors

◗ Penetrating peptic ulcers

◗ Blunt trauma or surgical trauma

◗ Drugs, such as glucocorticoids, sulfonamides, thiazides, and oral contraceptives, and NSAIDS

◗ Kidney failure or transplantation

◗ Endoscopic examination of the bile ducts and pancreas.

Pathophysiology

Acute pancreatitis occurs in two forms: edematous (interstitial) and necrotizing. Edematous pancreatitis causes fluid accumulation and swelling. Necrotizing pancreatitis causes cell death and tissue damage. The inflammation that occurs with both types is caused by premature activation of enzymes, which causes tissue damage.

Normally, the acini in the pancreas secrete enzymes in an inactive form. Two theories explain why enzymes become prematurely activated.

In one view, a toxic agent such as alcohol alters the way the pancreas secretes enzymes. Alcohol probably increases pancreatic secretion, alters the metabolism of the acinar cells, and encourages duct obstruction by causing pancreatic secretory proteins to precipitate.

Another theory is that a reflux of duodenal contents containing activated enzymes enters the pancreatic duct, activating other enzymes and setting up a cycle of more pancreatic damage.

In chronic pancreatitis, persistent inflammation produces irreversible changes in the structure and function of the pancreas. It sometimes follows an episode of acute pancreatitis. Protein precipitates block the pancreatic duct and eventually harden or calcify. Structural changes lead to fibrosis and atrophy of the glands. Growths called pseudocysts contain pancreatic enzymes and tissue debris. An abscess results if pseudocysts become infected.

If pancreatitis damages the islets of Langerhans, diabetes mellitus may result. Sudden severe pancreatitis causes massive hemorrhage and total destruction of the pancreas, manifested as diabetic acidosis, shock, or coma.

Signs and symptoms
▶ Pain caused by the escape of inflammatory exudate and enzymes into the back of the peritoneum, edema and distention of the pancreatic capsule, and obstruction of the biliary tract
▶ Persistent vomiting (in a severe attack) from hypermotility or paralytic ileus secondary to pancreatitis or peritonitis
▶ Abdominal distention (in a severe attack) from bowel hypermotility and the accumulation of fluids in the abdominal cavity
▶ Diminished bowel activity (in severe attack) suggesting altered motility secondary to peritonitis
▶ Crackles at lung bases (in a severe attack) secondary to heart failure
▶ Left pleural effusion (in a severe attack) from circulating pancreatic enzymes
▶ Mottled skin from hemorrhagic necrosis of the pancreas
▶ Tachycardia secondary to dehydration and possible hypovolemia
▶ Low-grade fever resulting from the inflammatory response
▶ Cold, sweaty extremities secondary to cardiovascular collapse
▶ Restlessness related to pain associated with acute pancreatitis
▶ Extreme malaise (in chronic pancreatitis) related to malabsorption or diabetes.

Complications
▶ Massive hemorrhage/shock
▶ Pseudocysts
▶ Biliary and duodenal obstruction
▶ Portal and splenic vein thrombosis
▶ Diabetes mellitus
▶ Respiratory failure.

Diagnosis
▶ Elevated serum amylase and lipase confirm diagnosis.
▶ Blood and urine glucose tests reveal transient glucose in urine and hyperglycemia.

In chronic pancreatitis, serum glucose levels may be transiently elevated.
▶ White blood cell count is elevated.
▶ Serum bilirubin levels are elevated in both acute and chronic pancreatitis.
▶ Blood calcium levels may be decreased.
▶ Stool analysis shows elevated lipid and trypsin levels in chronic pancreatitis.
▶ Abdominal and chest X-rays detect pleural effusions and differentiate pancreatitis from diseases that cause similar symptoms; may detect pancreatic calculi.
▶ Computed tomography and ultrasonography show enlarged pancreas with cysts and pseudocysts.
▶ Endoscopic retrograde cholangiopancreatography identifies ductal system abnormalities, such as calcification or strictures; helps differentiate pancreatitis from other disorders such as pancreatic cancer.

Treatment
Treatment may include:
▶ intravenous replacement of fluids, protein, and electrolytes to treat shock
▶ fluid volume replacement to help correct metabolic acidosis
▶ blood transfusions to replace blood loss from hemorrhage
▶ withholding of food and fluids to rest the pancreas and reduce pancreatic enzyme secretion
▶ nasogastric tube suctioning to decrease stomach distention and suppress pancreatic secretions
▶ meperidine to relieve abdominal pain
▶ antacids to neutralize gastric secretions
▶ histamine antagonists to decrease hydrochloric acid production
▶ antibiotics to fight bacterial infections
▶ anticholinergics to reduce vagal stimulation, decrease gastrointestinal motility, and inhibit pancreatic enzyme secretion
▶ insulin to correct hyperglycemia
▶ surgical drainage for a pancreatic abscess or a pseudocyst
▶ laparotomy (if biliary tract obstruction causes acute pancreatitis) to remove obstruction.

A CLOSER LOOK AT PEPTIC ULCERS

A gastrointestinal lesion is not necessarily an ulcer. Lesions that don't extend below the mucosal lining (epithelium) are called erosions. Lesions of both acute and chronic ulcers can extend through the epithelium and perforate the stomach wall. Chronic ulcers also have scar tissue at the base.

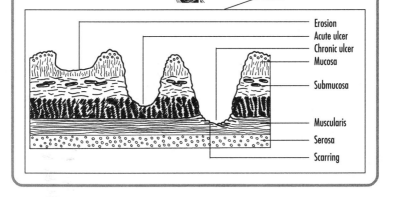

Erosion
Acute ulcer
Chronic ulcer
Mucosa

Submucosa

Muscularis

Serosa

Scarring

Peptic ulcer

Peptic ulcers, circumscribed lesions in the mucosal membrane extending below the epithelium, can develop in the lower esophagus, stomach, pylorus, duodenum, or jejunum. Although erosions are often referred to as ulcers, erosions are breaks in the mucosal membranes that do not extend below the epithelium. Ulcers may be acute or chronic in nature. Chronic ulcers are identified by scar tissue at their base. (See *A closer look at peptic ulcers.*) About 80% of all peptic ulcers are duodenal ulcers, which affect the proximal part of the small intestine and occur most commonly in men between ages 20 and 50. Duodenal ulcers usually follow a chronic course with remissions and exacerbations; 5% to 10% of patients develop complications that necessitate surgery.

Gastric ulcers are most common in middle-aged and elderly men, especially in chronic users of nonsteroidal anti-inflammatory drugs, alcohol, or tobacco.

Causes

▶ Infection with *Helicobacter pylori*
▶ Use of nonsteroidal anti-inflammatory drugs (NSAIDs)
▶ Pathologic hypersecretory disorders.

Pathophysiology

Although the stomach contains acidic secretions that can digest substances, intrinsic defenses protect the gastric mucosal membrane from injury. A thick, tenacious layer of gastric mucus protects the stomach from autodigestion, mechanical trauma, and chemical trauma. Prostaglandins provide another line of defense. Gastric ulcers may be a result of destruction of the mucosal barrier.

The duodenum is protected from ulceration by the function of Brunner's glands. These glands produce a viscid, mucoid, alkaline secretion that neutralizes the acid chyme. Duodenal ulcers appear to result from excessive acid protection.

Helicobacter pylori releases a toxin that destroys the gastric and duodenal mucosa, reducing the epithelium's resistance to acid digestion and causing gastritis and ulcer disease.

Salicylates and other NSAIDs inhibit the secretion of prostaglandins (substances that block ulceration). Certain illnesses, such as pancreatitis, hepatic disease, Crohn's disease, preexisting gastritis, and Zollinger-Ellison syndrome, also contribute to ulceration.

Besides peptic ulcer's main causes, several predisposing factors are acknowledged. They include blood type (gastric ulcers and type A; duodenal ulcers and type O) and other genetic factors. Exposure to irritants, such as alcohol, coffee, and tobacco, may contribute by accelerating gastric acid emptying and promoting mucosal breakdown. Emotional stress also contributes to ulcer formation because of the increased stimulation of acid and pepsin secretion and decreased mucosal defense. Physical trauma and normal aging are additional predisposing conditions.

Signs and symptoms
Symptoms vary by the type of ulcer.
A gastric ulcer produces the following signs and symptoms:
▶ pain that worsens with eating due to stretching of the mucosa by food
▶ nausea and anorexia secondary to mucosal stretching.

A duodenal ulcer produces the following signs and symptoms:
▶ epigastric pain that is gnawing, dull, aching, or "hunger like" due to excessive acid production
▶ pain relieved by food or antacids, but usually recurring 2 to 4 hours later secondary to food acting as a buffer for acid.

Complications
▶ Hemorrhage

▶ Shock
▶ Gastric perforation
▶ Gastric outlet obstruction.

Diagnosis
▶ Barium swallow or upper GI and small bowel series may reveal the presence of the ulcer. This is the first test performed on a patient when symptoms aren't severe.
▶ Endenoscopy confirms the presence of an ulcer and permits cytologic studies and biopsy to rule out *H. pylori* or cancer.
▶ Upper GI tract X-rays reveal mucosal abnormalities.
▶ Stool analysis may reveal occult blood.
▶ Serologic testing may disclose clinical signs of infection, such as elevated white blood cell count.
▶ Gastric secretory studies show hyperchlorhydria.
▶ Carbon[13] urea breath test results reflect activity of *H. pylori*.

Treatment
▶ Antimicrobial agents (tetracycline, bismuth subsalicylate, and metronidazole) to eradicate *H. pylori* infection (See *Treating peptic ulcer*, pages 336 and 337.)
▶ Misoprostol (a prostaglandin analog) to inhibit gastric acid secretion and increase carbonate and mucus production, to protect the stomach lining
▶ Antacids to neutralize acid gastric contents by elevating the gastric pH, thus protecting the mucosa and relieving pain
▶ Avoidance of caffeine and alcohol to avoid stimulation of gastric acid secretion
▶ Anticholinergic drugs to inhibit the effect of the vagal nerve on acid-secreting cells
▶ H_2 blockers to reduce acid secretion
▶ Sucralfate, mucosal protectant to form an acid-impermeable membrane that adheres to the mucous membrane and also accelerates mucus production
▶ Dietary therapy with small infrequent meals and avoidance of eating before bedtime to neutralize gastric contents

(Text continues on page 338.)

TREATING PEPTIC ULCER

Peptic ulcers can result from factors that increase gastric acid production or from factors that impair mucosal barrier protection. This illustration highlights the actions of the major treatments used for peptic ulcer and where they interfere with the pathophysiologic chain of events.

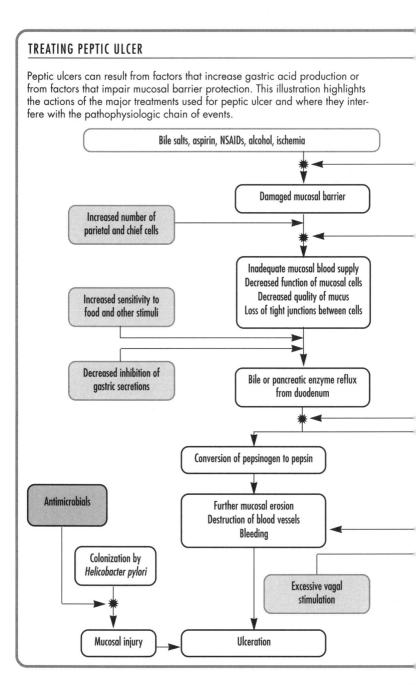

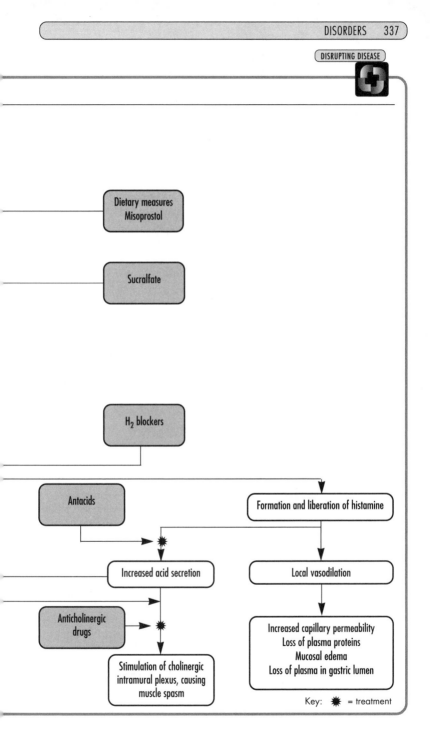

Dietary measures
Misoprostol

Sucralfate

H₂ blockers

Antacids

Formation and liberation of histamine

Increased acid secretion

Local vasodilation

Anticholinergic drugs

Stimulation of cholinergic intramural plexus, causing muscle spasm

Increased capillary permeability
Loss of plasma proteins
Mucosal edema
Loss of plasma in gastric lumen

Key: ✳ = treatment

◗ Insertion of a nasogastric tube (in instances of gastrointestinal bleeding) for gastric decompression and rest, and also to permit iced saline lavage that may also contain norepinephrine

◗ Gastroscopy to allow visualization of the bleeding site and coagulation by laser or cautery to control bleeding

◗ Surgery to repair perforation or treat unresponsiveness to conservative treatment, and suspected malignancy.

Ulcerative colitis

Ulcerative colitis is an inflammatory, often chronic, disease that affects the mucosa of the colon. It invariably begins in the rectum and sigmoid colon, and commonly extends upward into the entire colon, rarely affecting the small intestine, except for the terminal ileum. Ulcerative colitis produces edema (leading to mucosal friability) and ulcerations. Severity ranges from a mild, localized disorder to a fulminant disease that may cause a perforated colon, progressing to potentially fatal peritonitis and toxemia. The disease cycles between exacerbation and remission.

Ulcerative colitis occurs primarily in young adults, especially women. It's more prevalent among Ashkenazic Jews and in higher socioeconomic groups, and there seems to be a familial tendency. The prevalence is unknown; however, some studies suggest as many as 100 of 100,000 persons have the disease. Onset of symptoms seems to peak between ages 15 and 30 and between ages 55 and 65.

Causes

Specific causes of ulcerative colitis are unknown but may be related to abnormal immune response in the GI tract, possibly associated with food or bacteria such as *Escherichia coli.*

Pathophysiology

Ulcerative colitis usually begins as inflammation in the base of the mucosal layer of the large intestine. The colon's mucosal surface becomes dark, red, and velvety. Inflammation leads to erosions that coalesce and form ulcers. The mucosa becomes diffusely ulcerated, with hemorrhage, congestion, edema, and exudative inflammation. Ulcerations are continuous. Abscesses in the mucosa drain purulent pus, become necrotic, and ulcerate. Sloughing causes bloody, mucus filled stools. As abscesses heal, scarring and thickening may appear in the bowel's inner muscle layer. As granulation tissue replaces the muscle layer, the colon narrows, shortens, and loses its characteristic pouches (haustral folds).

Signs and symptoms

Signs and symptoms may include:

◗ recurrent bloody diarrhea, often containing pus and mucus (hallmark sign), from accumulated blood and mucus in the bowel

◗ abdominal cramping and rectal urgency from accumulated blood and mucus

◗ weight loss secondary to malabsorption

◗ weakness related to possible malabsorption and subsequent anemia.

Complications

Complications may include:

◗ perforation
◗ toxic megacolon
◗ liver disease
◗ stricture formation
◗ colon cancer
◗ anemia.

Diagnosis

◗ Sigmoidoscopy confirms rectal involvement: specifically, mucosal friability and flattening and thick, inflammatory exudate.

◗ Colonoscopy reveals extent of the disease, stricture areas, and pseudopolyps (not performed when the patient has active signs and symptoms).

◗ Biopsy with colonoscopy confirms the diagnosis.

▶ Barium enema reveals the extent of the disease, detects complications, and identifies cancer (not performed when the patient has active signs and symptoms).

▶ Stool specimen analysis reveals blood, pus, and mucus but no disease-causing organisms.

▶ Serology shows decreased serum potassium, magnesium, and albumin levels; decreased WBC count; decreased hemoglobin; and prolonged prothrombin time. Elevated erythrocyte sedimentation rate correlates with severity of the attack.

Treatment

Treatment includes:

▶ corticotropin and adrenal corticosteroids to control inflammation

▶ sulfasalazine for its anti-inflammatory and antimicrobial effects

▶ antidiarrheals to relieve frequent, troublesome diarrhea in patients whose ulcerative colitis is otherwise under control

▶ iron supplements to correct anemia

▶ total parental nutrition and nothing by mouth for patients with severe disease, to rest the intestinal tract, decrease stool volume, and restore nitrogen balance

▶ supplemental drinks to supplement nutrition in patients with moderate symptoms

▶ intravenous hydration to replace fluid loss from diarrhea

▶ surgery to correct massive dilation of the colon and to treat patients with symptoms that are unbearable or unresponsive to drugs and supportive measures

▶ proctocolectomy with ileostomy to divert stool and to allow rectal anastomosis to heal, removing all the potentially malignant epithelia of the rectum and colon.

The musculoskeletal system is a complex system of bones, muscles, ligaments, tendons, and other tissues that gives the body form and shape. It also protects vital organs, makes movement possible, stores calcium and other minerals in the bony matrix for mobilization if deficiency occurs, and provides sites for hematopoiesis (blood cell production) in the marrow.

BONES

The human skeleton contains 206 bones, which are composed of inorganic salts (primarily calcium and phosphate), embedded in a framework of collagen fibers.

Bone shape and structure

Bones are classified by shape as either long, short, flat, or irregular. Long bones are found in the extremities and include the humerus, radius, and ulna of the arm; the femur, tibia, and fibula of the leg; and the phalanges, metacarpals, and metatarsals of the hands and feet (See *Structure of long bones.*) Short bones include the tarsal and carpal bones of the feet and hands, respectively. Flat bones include the frontal and parietal bones of the cranium, ribs, sternum, scapulae, ilium, and pubis. Irregular bones include the bones of the spine (vertebrae, sacrum, coccyx) and certain bones of the skull — the temporal, sphenoid, ethmoid, and mandible.

Classified according to structure, bone is either cortical (compact) or cancellous (spongy or trabecular). Adult cortical bone consists of networks of interconnecting canals, or canaliculi. Each of these networks, or haversian systems, runs parallel to the bone's long axis and consists of a central haversian canal surrounded by layers (lamellae) of bone. Between adjacent lamellae are small openings called lacunae, which contain bone cells or osteocytes. The canaliculi, each containing one or more capillaries, provide a route for tissue fluids transport; they connect all the lacunae.

Cancellous bone consists of thin plates (trabeculae) that form the interior meshwork of bone. These trabeculae are arranged in various directions to correspond with the lines of maximum stress or pressure. This gives the bone added structural strength. Chemically, inorganic salts (calcium and phosphate, with small amounts of sodium, potassium carbonate, and magnesium ions) comprise 70% of the mature bone. The salts give bone its elasticity and ability to withstand compression.

Bone growth

Bone formation is ongoing and is determined by hormonal stimulation, dietary factors, and the amount of stress put on the bone. It is accomplished by the continual actions of bone-forming osteoblasts and bone-reabsorbing cells osteoclasts. Osteoblasts are present on the outer surface of and within bones. They respond to various stimuli to produce the bony matrix, or osteoid. As calcium salts precipitate on the organic matrix, the bone hardens. As the bone forms, a system of microscopic canals in the bone form around the osteocytes. Osteoclasts are

STRUCTURE OF LONG BONES

Long bones are the weight-bearing bones of the body. Their structure provides maximal strength and minimal weight. Structure of a long bone in an adult is shown below.

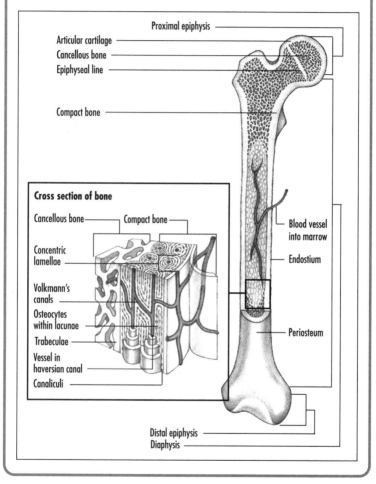

Proximal epiphysis
Articular cartilage
Cancellous bone
Epiphyseal line
Compact bone

Cross section of bone

Cancellous bone — Compact bone
Concentric lamellae
Volkmann's canals
Osteocytes within lacunae
Trabeculae
Vessel in haversian canal
Canaliculi

Blood vessel into marrow
Endostium
Periosteum

Distal epiphysis
Diaphysis

phagocytic cells that digest old, weakened bone section by section. As they finish, osteoblasts simultaneously replace the cleared section with new, stronger bone.

Vitamin D supports bone calcification by stimulating osteoblast activity and calcium absorption from the gut to make it available for bone building. When serum calcium levels fall, the parathyroid gland releases parathyroid hormone, which then stimulates osteoclast activity and bone breakdown, freeing calcium into the blood. Parathyroid hormone also increases serum calcium by decreasing renal excretion of

STRUCTURE OF A SYNOVIAL JOINT

The metacarpophalangeal joint depicted here, permits angular motion between the finger and the hand. A synovial joint is characterized by a synovial pouch full of fluid that lubricates the two articulating bones.

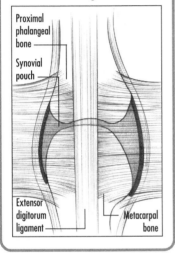

Proximal phalangeal bone

Synovial pouch

Extensor digitorum ligament

Metacarpal bone

calcium and increasing renal excretion of phosphate ion.

Phosphates are essential to bone formation; about 85% of the body's phosphates are found in bone. The intestine absorbs a considerable amount of phosphates from dietary sources, but adequate levels of vitamin D are necessary for their absorption. Because calcium and phosphates interact in a reciprocal relationship, renal excretion of phosphates increases or decreases in inverse proportion to serum calcium levels. Alkaline phosphatase (ALP) influences bone calcification and lipid and metabolite transport. Osteoblasts contain an abundance of ALP. A rise in serum ALP levels can identify skeletal diseases primarily those characterized by marked osteoblastic activity such as bone metastases

or Paget's disease. It can also identify biliary obstruction or hyperparathyroidism, or excessive ingestion of Vitamin D.

In children and young adults, bone growth occurs in the epiphyseal plate, a layer of cartilage between the diaphysis and epiphysis of long bones.

Osteoblasts deposit new bone in the area just beneath the epiphysis, making the bone longer, and osteoclasts model the new bone's shape by resorbing previously deposited bone. These remodeling activities promote longitudinal bone growth, which continues until the epiphyseal growth plates, located at both ends, close during adolescence. In adults, bone growth is complete, and this cartilage is replaced by bone, becoming the epiphyseal line.

JOINTS

The tendons, ligaments, cartilage, and other tissues that connect two bones comprise a joint. Depending on their structure, joints either predominantly permit motion or provide stability. Joints, like bones, are classified according to structure and function.

Classification of joints

The three structural types of joints are fibrous, cartilaginous, and synovial.

▶ Fibrous joints, or *synarthroses*, have only minute motion and provide stability when tight union is necessary, as in the sutures that join the cranial bones.

▶ Cartilaginous joints, or *amphiarthroses*, allow limited motion, as between vertebrae.

▶ Synovial joints, or *diarthroses*, are the most common and permit the greatest degree of movement. These joints include the elbows and knees. (See *Structure of a synovial joint*.)

Synovial joints have distinguishing characteristics:

▶ The two articulating surfaces of the bones have a smooth hyaline covering (articular cartilage) that is resilient to pressure.

▶ Their opposing surfaces are congruous and glide smoothly on each other.

▶ A fibrous (articular) capsule holds them together.

▶ Beneath the capsule, lining the joint cavity, is the synovial membrane, which secretes a clear viscous fluid called synovial fluid. This fluid lubricates the two opposing surfaces during motion and also nourishes the articular cartilage.

▶ Surrounding a synovial joint are ligaments, muscles, and tendons, which strengthen and stabilize the joint but allow free movement.

Joint movement

The two types of synovial joint movement are angular and circular.

Angular movement

Joints of the knees, elbows, and phalanges permit the following angular movements:

▶ flexion (closing of the joint angle)

▶ extension (opening of the joint angle)

▶ hyperextension (extension of the angle beyond the usual arc).

Other joints, including the shoulders and hips, permit:

▶ abduction (movement away from the body's midline)

▶ adduction (movement toward the midline).

Circular movements

Circular movements include:

▶ rotation (motion around a central axis), as in the ball and socket joints of the hips and shoulders

▶ pronation (downward wrist or ankle motion)

▶ supination (upward wrist motion to begging position).

Other kinds of movement are inversion (inward turning, as of foot), eversion (outward turning, as of foot), protraction (as in forward motion of the mandible), and retraction (returning protracted part into place).

MUSCLES

The most specialized feature of muscle tissue — contractility — makes the movement of bones and joints possible. Normal skeletal muscles contract in response to neural impulses. Appropriate contraction of muscle usually applies force to one or more tendons. The force pulls one bone toward, away from, or around a second bone, depending on the type of muscle contraction and the type of joint involved. Abnormal metabolism in the muscle may result in inappropriate contractility. For example, when stored glycogen or lipids cannot be used because of the lack of an enzyme necessary to convert energy for contraction, the result may be cramps, fatigue, and exercise intolerance.

Muscles permit and maintain body positions, such as sitting and standing. Muscles also pump blood through the body (cardiac contraction and vessel compression), move food through the intestines (peristalsis), and make breathing possible. Skeletal muscular activity produces heat; it is an important component in temperature regulation. Deep body temperature regulators are found in the abdominal viscera, spinal cord, and great veins. These receptors detect changes in the body core temperature and stimulate the hypothalamus to institute appropriate temperature changing responses, such as shivering in response to cold. Muscle mass accounts for about 40% of an average man's weight.

Muscle classification

Muscles are classified according to structure, anatomic location, and function:

▶ Skeletal muscles are attached to bone and have a striped (striated) appearance that reflects their cellular structure.

▶ Visceral muscles move contents through internal organs and are smooth (nonstriated).

▶ Cardiac muscles (smooth) comprise the heart wall.

When muscles are classified according to activity, they are called either voluntary or involuntary. (See Chapter 8, "Nervous system.") Voluntary muscles can be controlled at will and are under the influence of the somatic nervous system; these are the skeletal muscles. Involuntary muscles, controlled by the autonomic nervous system, include the cardiac and visceral muscles. Some organs contain both voluntary and involuntary muscles.

Muscle contraction

Each skeletal muscle consists of many elongated muscle cells, called muscle fibers, through which run slender threads of protein, called myofibrils. Muscle fibers are held together in bundles by sheaths of fibrous tissue, called fascia. Blood vessels and nerves pass into muscles through the fascia to reach the individual muscle fibers. Motor neurons synapse with the motor nerve fibers of voluntary muscles. These fibers reach the membranes of skeletal muscle cells at neuromuscular (myoneural) junctions. When an impulse reaches the myoneural junction, the junction releases the neurotransmitter, acetylcholine, which releases calcium from the sarcoplasmic reticulum, a membranous network in the muscle fiber, which, in turn, triggers muscle contraction. The energy source for this contraction is adenosine triphosphate (ATP). ATP release is also triggered by the impulse at the myoneural junction. Relaxation of a muscle is believed to take place by reversal of these mechanisms.

Muscle fatigue results when the sources of ATP in a muscle are depleted. If a muscle is deprived of oxygen, fatigue occurs rapidly. As the muscle fatigues, it switches to anaerobic metabolism of glycogen stores, in which the stored glycogen is split into glucose (glycolysis) without the use of oxygen. Lactic acid is a by-product of anaerobic glycolysis and may accumulate in the muscle and blood with intense or prolonged muscle contraction.

TENDONS AND LIGAMENTS

Skeletal muscles are attached to bone directly or indirectly by fibrous cords known as tendons. The least movable end of the muscle attachment (generally proximal) is called the point of origin; the most movable end (generally distal) is the point of insertion.

Ligaments are fibrous connections that control joint movement between two bones or cartilages. Their purpose is to support and strengthen joints.

PATHOPHYSIOLOGIC MANIFESTATIONS

Alterations of the normal functioning of bones and muscles are described next. Most musculoskeletal disorders are caused by or profoundly affect other body systems.

Alterations in bone

Disease may alter density, growth, or strength of bone.

Density

In healthy young adults, the resorption and formation phases are tightly coupled to maintain bone mass in a steady state. Bone loss occurs when the two phases become uncoupled, and resorption exceeds formation. Estrogen not only regulates calcium uptake and release, it also regulates osteoblastic activity. Decreased estrogen levels may lead to diminished osteoblastic activity and loss of bone mass, called osteoporosis. In children, vitamin D deficiency prevents normal bone growth and leads to rickets.

AGE ALERT Bone density and structural integrity decrease after the age of 30 years in women and after the age of 45 years in men. The relatively steady loss of bone matrix can be partially offset by exercise.

CULTURAL DIVERSITY Age, race, and sex affect bone mass, structural integrity (ability to withstand stress), and bone loss. For example, blacks commonly have denser bones than whites, and men typically have denser bones than women.

Growth

The *osteochondroses* are a group of disorders characterized by avascular necrosis of the epiphyseal growth plates in growing children and adolescents. In these disorders, a lack of blood supply to the femoral head leads to septic necrosis, with softening and resorption of bone. Revascularization then initiates new bone formation in the femoral head or tibial tubercle, which leads to a malformed femoral head.

Bone strength

Both cortical and trabecular bone contribute to skeletal strength. Any loss of the inorganic salts that comprise the chemical structure of bone will weaken bone. Cancellous bone is more sensitive to metabolic influences, so conditions that produce rapid bone loss tend to affect cancellous bone more quickly than cortical bone.

Alterations of muscle

Pathologic effects on muscle include atrophy, fatigue, weakness, myotonia, and spasticity.

Atrophy

Atrophy is a decrease in the size of a tissue or cell. In muscles, the myofibrils atrophy after prolonged inactivity from bed rest or trauma (casting), when local nerve damage makes movement impossible, or when illness removes needed nutrients from muscles. The effects of muscular deconditioning associated with lack of physical activity may be apparent in a matter of days. An individual on bed rest loses muscle strength, as well as muscle mass, from baseline levels at a rate of 3% a day.

Conditioning and stretching exercises may help prevent atrophy. If reuse isn't restored within 1 year, regeneration of muscle fibers is unlikely.

 AGE ALERT Some degree of muscle atrophy is normal with aging.

Fatigue

Pathologic muscle fatigue may be the result of impaired neural stimulation of muscle or energy metabolism or disruption of calcium flux. See Chapter 5, Fluid and electrolytes, for a detailed discussion of these events.

Weakness

AGE ALERT Muscle mass and muscle strength decrease in the elderly, usually as a result of disuse. This can be reversed with moderate, regular, weight-bearing exercise.

Periodic paralysis is a disorder that can be triggered by exercise or a process or chemical (such as medication) that increases serum potassium levels. This hyperkalemic periodic paralysis may be caused by a high-carbohydrate diet, emotional stress, prolonged bed rest, or hyperthyroidism. During an attack of periodic paralysis, the muscle membrane is unresponsive to neural stimuli, and the electrical charge needed to initiate the impulse (resting membrane potential) is reduced from –90 mV to –45 mV.

Myotonia and spasticity

Myotonia is delayed relaxation after a voluntary muscle contraction — such as grip, eye closure, or muscle percussion — accompanied by prolonged depolarization of the muscle membrane. Depolarization is the reversal of the resting potential in stimulated cell membranes. It is the process by which the cell membrane "resets" its positive charge with respect to the negative charge outside the cell. Myotonia occurs in myotonic muscular dys-

MANAGING MUSCULOSKELETAL PAIN

A patient with a musculoskeletal disorder that causes chronic, non-malignant pain should be assessed and treated in a stepped approach. Measures include:
▶ nonpharmacologic methods, such as heat, ice, elevation, and rest
▶ acetaminophen (Tylenol)
▶ nonsteroidal anti-inflammatories such as ibuprofen (Motrin)
▶ other nonnarcotic analgesics such as tramadol (Ultram) or topical capsaicin cream (Zostrix)
▶ tricyclic antidepressants such as amitriptyline hydrochloride (Elavil) may decrease the pain signal at the neurosynaptic junctions
▶ opioid analgesics alone or with a tricyclic antidepressant.

trophy and some forms of periodic paralysis.

Stress-induced muscle tension, or spasticity, is presumably caused by increased activity in the reticular activating system and gamma loop in the muscle fiber. The reticular activating system consists of multiple diffuse pathways in the brain that control wakefulness and response to stimuli. A pathologic contracture is permanent muscle shortening caused by muscle spasticity, seen in central nervous system injury or severe muscle weakness.

DISORDERS

AGE ALERT Patients with musculoskeletal disorders are often elderly, have other concurrent medical conditions, or are victims of trauma. Generally, they face prolonged immobilization. (See *Managing musculoskeletal pain*.)

Bone fracture

When a force exceeds the compressive or tensile strength (the ability of the bone to hold together) of the bone, a fracture will occur. (For an explanation of the terms used to identify fractures, see *Classifying fractures*.)

An estimated 25% of the population has traumatic musculoskeletal injury each year, and a significant number of these involve fractures.

The prognosis varies with the extent of disablement or deformity, amount of tissue and vascular damage, adequacy of reduction and immobilization, and patient's age, health, and nutritional status.

AGE ALERT Children's bones usually heal rapidly and without deformity. However, epiphyseal plate fractures in children are likely to cause deformity because they interfere with normal bone growth. In the elderly, underlying systemic illness, impaired circulation, or poor nutrition may cause slow or poor healing.

Causes

Risk factors for fractures are those that involve force to bone, such as:
▶ falls
▶ motor vehicle accidents
▶ sports
▶ use of drugs that impair judgment or mobility
▶ young age (immaturity of bone)
▶ bone tumors
▶ metabolic illnesses(such as hypoparathyroidism or hyperparathyroidism)
▶ medications that cause iatrogenic osteoporosis, such as steroids.

AGE ALERT The highest incidence of fractures occurs in young males between the ages of 15 and 24 years (tibia, clavicle, and lower humerus) and are usually the result of trauma. In the elderly, upper femur, upper humerus, vertebrae, and pelvis fractures are often associated with osteoporosis.

CLASSIFYING FRACTURES

One of the best-known systems for classifying fractures uses a combination of terms that describe general classification, fragment position, and fracture line — such as simple, nondisplaced, and oblique — to describe fractures.

GENERAL CLASSIFICATION OF FRACTURES
▶ Simple (closed) — Bone fragments don't penetrate the skin.
▶ Compound (open) — Bone fragments penetrate the skin.
▶ Incomplete (partial) — Bone continuity isn't completely interrupted.
▶ Complete — Bone continuity is completely interrupted.

CLASSIFICATION BY FRAGMENT POSITION
▶ Comminuted — The bone breaks into small pieces.
▶ Impacted — One bone fragment is forced into another.
▶ Angulated — Fragments lie at an angle to each other.
▶ Displaced — Fracture fragments separate and are deformed.

▶ Nondisplaced — The two sections of bone maintain essentially normal alignment.
▶ Overriding — Fragments overlap, shortening the total bone length.
▶ Segmental — Fractures occur in two adjacent areas with an isolated central segment.
▶ Avulsed — Fragments are pulled from the normal position by muscle contractions or ligament resistance.

CLASSIFICATION BY FRACTURE LINE
▶ Linear — The fracture line runs parallel to the bone's axis.
▶ Longitudinal — The fracture line extends in a longitudinal (but not parallel) direction along the bone's axis.
▶ Oblique — The fracture line crosses the bone at about a 45-degree angle to the bone's axis.
▶ Spiral — The fracture line crosses the bone at an oblique angle, creating a spiral pattern.
▶ Transverse — The fracture line forms a right angle with the bone's axis.

Pathophysiology
When a bone is fractured, the periosteum and blood vessels in the cortex, marrow, and surrounding soft tissue are disrupted. A hematoma forms between the broken ends of the bone and beneath the periosteum, and granulation tissue eventually replaces the hematoma.

Damage to bone tissue triggers an intense inflammatory response in which cells from surrounding soft tissue and the marrow cavity invade the fracture area, and blood flow to the entire bone is increased. Osteoblasts in the periosteum, endosteum, and marrow produce osteoid (collagenous, young bone that has not yet calcified, also called callus), which hardens along the outer surface of the shaft and over the broken ends of the bone. Osteoclasts reabsorb material from previously formed bones and osteoblasts to rebuild bone. Osteoblasts then transform into osteocytes (mature bone cells).

Signs and symptoms
Signs and symptoms of bone fracture may include:
▶ deformity due to unnatural alignment
▶ swelling due to vasodilation and infiltration by inflammatory leukocytes and mast cells
▶ muscle spasm
▶ tenderness

RECOGNIZING COMPARTMENT SYNDROME

Compartment syndrome occurs when edema or bleeding increases pressure within a muscle compartment (a smaller section of a muscle), to the point of interfering with circulation. Crush injuries, burns, bites, and fractures requiring casts or dressings may cause this syndrome. Compartment syndrome most commonly occurs in the lower arm, hand, lower leg, or foot.

Symptoms include:
▶ increased pain
▶ decreased touch sensation
▶ increased weakness of the affected part
▶ increased swelling and pallor
▶ decreased pulses and capillary refill.

Treatment of compartment syndrome consists of:
▶ placing the limb at heart level
▶ removing constricting forces
▶ monitoring neurovascular status
▶ subfascial injection of hyaluronidase (Wydase)
▶ emergency fasciotomy.

▶ impaired sensation distal to the fracture site due to pinching or severing of neurovascular elements by the trauma or by bone fragments
▶ limited range of motion
▶ crepitus, or "clicking" sounds on movement caused by shifting bone fragments.

Complications

Possible complications of fracture are:
▶ permanent deformity and dysfunction if bones fail to heal (nonunion) or heal improperly (malunion)
▶ aseptic (not caused by infection) necrosis of bone segments due to impaired circulation

▶ hypovolemic shock as a result of blood vessel damage (especially with a fractured femur)
▶ muscle contractures
▶ compartment syndrome (See *Recognizing compartment syndrome*.)
▶ renal calculi from decalcification due to prolonged immobility
▶ fat embolism due to disruption of marrow or activation of the sympathetic nervous system after the trauma (may lead to respiratory or central nervous system distress).

Diagnosis

Diagnosis of bone fracture includes:
▶ history of traumatic injury and results of the physical examination, including gentle palpation and a cautious attempt by the patient to move parts distal to the injury
▶ X-rays of the suspected fracture and the joints above and below (confirm the diagnosis; after reduction, confirm bone alignment).

Treatment

For arm or leg fractures, emergency treatment consists of:
▶ splinting the limb above and below the suspected fracture to immobilize it
▶ applying a cold pack to reduce pain and edema
▶ elevating the limb to reduce pain and edema.

Treatment in severe fractures that cause blood loss includes:
▶ direct pressure to control bleeding
▶ fluid replacement as soon as possible to prevent or treat hypovolemic shock.

After confirming a fracture, treatment begins with a reduction. Closed reduction involves:
▶ manual manipulation
▶ local anesthetic (such as lidocaine [Xylocaine])
▶ analgesic (such as morphine IM)
▶ muscle relaxant (such as diazepam [Valium] I.V.) or a sedative (such as midazo-

lam [Versed]) to facilitate the muscle stretching necessary to realign the bone.

When closed reduction is impossible, open reduction by surgery involves:

▶ immobilization of the fracture by means of rods, plates, or screws and application of a plaster cast

▶ tetanus prophylaxis

▶ prophylactic antibiotics

▶ surgery to repair soft tissue damage

▶ thorough wound debridement

▶ physical therapy after cast removal to restore limb mobility.

When a splint or cast fails to maintain the reduction, immobilization requires skin or skeletal traction, using a series of weights and pulleys. This may involve:

▶ elastic bandages and sheepskin coverings to attach traction devices to the patient's skin (skin traction)

▶ pin or wire inserted through the bone distal to the fracture and attached to a weight to allow more prolonged traction (skeletal traction).

Clubfoot

Clubfoot, also called talipes, is the most common congenital disorder of the lower extremities. It is marked primarily by a deformed talus and shortened Achilles tendon, which give the foot a characteristic club-like appearance. In talipes equinovarus, the foot points downward (equinus) and turns inward (varus), while the front of the foot curls toward the heel (forefoot adduction).

Clubfoot occurs in about 1 per 1,000 live births, is usually bilateral and is twice as common in boys as girls. It may be associated with other birth defects, such as myelomeningocele, spina bifida, and arthrogryposis. Clubfoot is correctable with prompt treatment.

Causes

A combination of genetic and environmental factors in utero appears to cause clubfoot, including:

▶ heredity (mechanism of transmission is undetermined; the sibling of a child born with clubfoot has 1 chance in 35 of being born with the same anomaly, and a child of a parent with clubfoot has 1 chance in 10)

▶ arrested development during the 9th and 10th weeks of embryonic life when the feet are formed (children without a family history of clubfoot)

▶ muscle abnormalities leading to variations in length and tendon insertions

▶ secondary to paralysis, poliomyelitis, or cerebral palsy (older children), in which case treatment includes management of the underlying disease.

Pathophysiology

Abnormal development of the foot during fetal growth leads to abnormal muscles and joints and contracture of soft tissue. The condition called apparent clubfoot results when a fetus maintains a position in utero that gives his feet a clubfoot appearance at birth; it can usually be corrected manually. Another form of apparent clubfoot is inversion of the feet, resulting from the denervation type of progressive muscular atrophy and progressive muscular dystrophy.

Signs and symptoms

Talipes equinovarus varies greatly in severity. Deformity may be so extreme that the toes touch the inside of the ankle, or it may be only vaguely apparent.

Every case includes:

▶ deformed talus

▶ shortened Achilles tendon

▶ shortened and flattened calcaneus bone of the heel

▶ shortened, underdeveloped calf muscles and soft-tissue contractures at the site of the deformity (depending on degree of the varus deformity)

▶ foot is tight in its deformed position, resisting manual efforts to push it back into normal position

▶ no pain, except in elderly, arthritic patients with secondary deformity.

Complications

Possible complications of talipes equino-varius are:

▶ chronic impairment (neglected clubfoot)
▶ rarely totally correctable (when severe enough to require surgery).

Diagnosis

Early diagnosis of clubfoot is usually no problem because the deformity is obvious. In subtle deformity, however, true clubfoot must be distinguished from apparent clubfoot (metatarsus varus or pigeon toe), usually by:

▶ X-rays showing superimposition of the talus and calcaneus and a ladder-like appearance of the metatarsals (true clubfoot).

Treatment

Treatment for clubfoot is done in three stages: correcting the deformity, maintaining the correction until the foot regains normal muscle balance, and observing the foot closely for several years to prevent the deformity from recurring.

Clubfoot deformities are usually corrected in sequential order: forefoot adduction first, then varus (or inversion), then equinus (or plantar flexion). Trying to correct all three deformities at once only results in a misshapen, rocker-bottomed foot.

Other essential parts of management are:

▶ stressing to parents importance of prompt treatment and orthopedic supervision until growth is completed
▶ teaching parents cast care and how to recognize circulatory impairment before a child in a clubfoot cast is discharged
▶ explaining to older child and his parents that surgery can improve clubfoot with good function, but can't totally correct it; the affected calf muscle will remain slightly underdeveloped
▶ emphasizing the need for long-term orthopedic care to maintain correction; correcting this defect permanently takes time and patience.

Developmental hip dysplasia

Developmental hip dysplasia (DHD), an abnormality of the hip joint present from birth, is the most common disorder affecting the hip joints of children younger than 3 years. About 85% of affected infants are females.

DHD can be unilateral or bilateral. This abnormality occurs in three forms of varying severity:

▶ Unstable dysplasia: the hip is positioned normally but can be dislocated by manipulation.
▶ Subluxation or incomplete dislocation: the femoral head rides on the edge of the acetabulum.
▶ Complete dislocation: the femoral head is totally outside the acetabulum.

Causes

Although the causes of DHD are not clear, it's more likely to occur in the following circumstances:

▶ dislocation after breech delivery (malposition in utero, 10 times more common than after cephalic delivery)
▶ elevated maternal relaxin, hormone secreted by the corpus luteum during pregnancy that causes relaxation of pubic symphysis and cervical dilation (may promote relaxation of the joint ligaments, predisposing the infant to DHD)
▶ large neonates and twins (more common).

Pathophysiology

The precise cause of congenital dislocation is unknown. Excessive or abnormal movement of the joint during a traumatic birth may cause dislocation. Displacement of bones within the joint may damage joint structures, including articulating surfaces, blood vessels, tendons, ligaments, and nerves. This may lead to ischemic necrosis because of the disruption of blood flow to the joint.

Signs and symptoms

Signs and symptoms of hip dysplasia vary with age and include:

▶ no gross deformity or pain (in newborns)
▶ the hip rides above the acetabulum, causing the level of the knees to be uneven (complete dysplasia)
▶ limited abduction on the dislocated side (as the child grows older and begins to walk)
▶ swaying from side to side ("duck waddle" due to uncorrected bilateral dysplasia)
▶ limp due to uncorrected unilateral dysplasia.

Complications

If corrective treatment isn't begun until after the age of 2 years, DHD may cause:
▶ degenerative hip changes
▶ abnormal acetabular development
▶ lordosis (abnormally increased concave curvature of the lumbar and cervical spine)
▶ joint malformation
▶ soft tissue damage
▶ permanent disability.

Diagnosis

Diagnostic measures may include:
▶ X-rays to show the location of the femur head and a shallow acetabulum (also to monitor disease or treatment progress)
▶ sonography and magnetic resonance imaging to assess reduction.

Observations during physical examination of the relaxed child that strongly suggest DHD include:
▶ the number of folds of skin over the thighs on each side when the child is placed on his back (a child in this position usually has an equal number of folds, but a child with subluxation or dislocation may have an extra fold on the affected side, which is also apparent when the child lies prone)
▶ buttock fold on the affected side higher with the child lying prone (also restricted abduction of the affected hip). (See *Ortolani's and Trendelenburg's signs of DHD*.)

ORTOLANI'S AND TRENDELENBURG'S SIGNS OF DHD

A positive Ortolani's or Trendelenburg's sign confirms developmental hip dysplasia (DHD).

Ortolani's sign

▶ Place infant on his back, with hip flexed and in abduction. Adduct the hip while pressing the femur downward. This will dislocate the hip.
▶ Then, abduct the hip while moving the femur upward. A click or a jerk (produced by the femoral head moving over the acetabular rim) indicates subluxation in an infant younger than 1 month. The sign indicates subluxation or complete dislocation in an older infant.

Trendelenburg's sign

▶ When the child rests his weight on the side of the dislocation and lifts his other knee, the pelvis drops on the normal side because abductor muscles in the affected hip are weak.
▶ However, when the child stands with his weight on the normal side and lifts the other knee, the pelvis remains horizontal.

Treatment

The earlier an infant receives treatment, the better the chances are for normal development. Treatment varies with the patient's age.

In infants younger than 3 months, treatment includes:
▶ gentle manipulation to reduce the dislocation, followed by splint-brace or harness to hold the hips in a flexed and abducted position to maintain the reduction
▶ splint-brace or harness worn continuously for 2 to 3 months, then a night splint

for another month to tighten and stabilize the joint capsule in correct alignment.

If treatment doesn't begin until after the age of 3 months, it may include:

▶ bilateral skin traction (in infants) or skeletal traction (in children who have started walking) to try to reduce the dislocation by gradually abducting the hips

▶ Bryant's traction or divarication traction (both extremities placed in traction, even if only one is affected, to help maintain immobilization) for children younger than 3 years and weighing less than 35 lb (16 kg) for 2 to 3 weeks

▶ gentle closed reduction under general anesthesia to further abduct the hips, followed by a spica cast for 4 to 6 months (if traction fails)

▶ if closed treatment fails, open reduction, followed by immobilization in a spica cast for an average of 6 months, or surgical division and realignment of bone (osteotomy).

In a child aged 2 to 5 years, treatment is difficult and includes:

▶ skeletal traction and subcutaneous adductor tenotomy (surgical cutting of the tendon).

Treatment begun after the age of 5 years rarely restores satisfactory hip function.

Gout

Gout, also called gouty arthritis, is a metabolic disease marked by urate deposits that cause painfully arthritic joints. It's found mostly in the foot, especially the great toe, ankle, and midfoot, but may affect any joint. Gout follows an intermittent course, and patients may be totally free of symptoms for years between attacks. The prognosis is good with treatment.

 AGE ALERT Primary gout usually occurs in men after age 30 (95% of cases) and in postmenopausal women; secondary gout occurs in the elderly.

Causes

Although the exact cause of primary gout remains unknown, it may be caused by:

▶ genetic defect in purine metabolism, causing overproduction of uric acid (hyperuricemia), retention of uric acid, or both.

In secondary gout, which develops during the course of another disease (such as obesity, diabetes mellitus, hypertension, sickle cell anemia, and renal disease), the cause may be:

▶ breakdown of nucleic acid causing hyperuricemia

▶ result of drug therapy, especially after the use of hydrochlorothiazide or pyrazinamide (Zinamide), which decrease urate excretion (ionic form of uric acid).

Pathophysiology

When uric acid becomes supersaturated in blood and other body fluids, it crystallizes and forms a precipitate of urate salts that accumulate in connective tissue throughout the body; these deposits are called tophi. The presence of the crystals triggers an acute inflammatory response when neutrophils begin to ingest the crystals. Tissue damage begins when the neutrophils release their lysosomes (see Chapter 12, Immune System). The lysosomes not only damage the tissues, but also perpetuate the inflammation.

In asymptomatic gout, serum urate levels increase but don't crystallize or produce symptoms. As the disease progresses, it may cause hypertension or urate kidney stones may form.

The first acute attack strikes suddenly and peaks quickly. Although it generally involves only one or a few joints, this initial attack is extremely painful. Affected joints appear hot, tender, inflamed, dusky red, or cyanotic. The metatarsophalangeal joint of the great toe usually becomes inflamed first (podagra), then the instep, ankle, heel, knee, or wrist joints. Sometimes a low-grade fever is present. Mild acute attacks often subside quickly but tend to recur at irregular intervals. Severe attacks may persist for days or weeks.

Intercritical periods are the symptom-free intervals between gout attacks. Most patients have a second attack within 6 months to 2 years, but some attacks, common in those who are untreated, tend to be longer and more severe than initial attacks. Such attacks are also polyarticular, invariably affecting joints in the feet and legs, and sometimes accompanied by fever. A migratory attack sequentially strikes various joints and the Achilles tendon and is associated with either subdeltoid or olecranon bursitis.

Eventually, chronic polyarticular gout sets in. This final, unremitting stage of the disease is marked by persistent painful polyarthritis, with large tophi in cartilage, synovial membranes, tendons, and soft tissue. Tophi form in fingers, hands, knees, feet, ulnar sides of the forearms, helix of the ear, Achilles tendons and, rarely, in internal organs, such as the kidneys and myocardium. The skin over the tophus may ulcerate and release a chalky, white exudate that is composed primarily of uric acid crystals.

Signs and symptoms

Possible signs and symptoms of gout include:
◗ joint pain due to uric acid deposits and inflammation
◗ redness and swelling in joints due to uric acid deposits and irritation
◗ tophi in the great toe, ankle, and pinna of ear due to urate deposits
◗ elevated skin temperature from inflammation.

Complications

Complications of gout may include:
◗ eventual erosions, deformity, and disability due to chronic inflammation and tophi that cause secondary joint degeneration
◗ hypertension and albuminuria (in some patients)
◗ kidney involvement, with tubular damage from aggregates of urate crystals; progressively poorer excretion of uric acid and chronic renal dysfunction.

Diagnosis

The following test results help diagnose gout:
◗ needle-like monosodium urate crystals in synovial fluid (shown by needle aspiration) or tissue sections of tophaceous deposits
◗ hyperuricemia (uric acid greater than 420 μmol/mmol of creatinine)
◗ elevated 24-hour urine uric acid (usually higher in secondary than in primary gout)
◗ X-rays initially normal; in chronic gout, damage of articular cartilage and subchondral bone. Outward displacement of the overhanging margin from the bone contour characterizes gout.

Treatment

The goals of treatment are to end the acute attack as quickly as possible, prevent recurring attacks, and prevent or reverse complications. Treatment for acute gout consists of:
◗ immobilization and protection of the inflamed, painful joints
◗ local application of heat or cold
◗ increased fluid intake (to 3 L/day if not contradicted by other conditions to prevent kidney stone formation)
◗ concomitant treatment with colchicine (oral or I.V.) every hour for 8 hours to inhibit phagocytosis of uric acid crystals by neutrophils, until the pain subsides or nausea, vomiting, cramping, or diarrhea develops (in acute inflammation)
◗ nonsteroidal anti-inflammatory drugs for pain and inflammation.

AGE ALERT Older patients are at risk for GI bleeding associated with nonsteroidal anti-inflammatory drug use. Encourage the patient to take these drugs with meals, and monitor the patient's stools for occult blood.

Treatment for chronic gout aims to decrease serum uric acid levels, including:
◗ maintenance dosage of allopurinol (Zyloprim) to suppress uric acid formation or control uric acid levels, preventing further

attacks (use cautiously in patients with renal failure)
▶ colchicine to prevent recurrent acute attacks until uric acid returns to its normal level (doesn't affect uric acid level)
▶ uricosuric agents (probenecid [Benemid] and sulfinpyrazon [Anturane]) to promote uric acid excretion and inhibit uric acid accumulation (of limited value in patients with renal impairment)
▶ dietary restrictions, primarily avoiding alcohol and purine-rich foods (shellfish, liver, sardines, anchovies, and kidneys) that increase urate levels (adjunctive therapy).

Muscular dystrophy

Muscular dystrophy is a group of congenital disorders characterized by progressive symmetric wasting of skeletal muscles without neural or sensory defects. Paradoxically, some wasted muscles tend to enlarge (pseudohypertrophy) because connective tissue and fat replace muscle tissue, giving a false impression of increased muscle strength.

The four main types of muscular dystrophy are:
▶ Duchenne, or pseudohypertrophic; 50% of all cases
▶ Becker, or benign pseudohypertrophic
▶ facioscapulohumeral, or Landouzy-Dejerine, dystrophy
▶ limb-girdle dystrophy.

The prognosis varies with the form of disease. Duchenne muscular dystrophy strikes during early childhood and is usually fatal during the second decade of life. It mostly affects males, 13 to 33 per 100,000 persons. Patients with Becker muscular dystrophy can live into their 40s, and it mostly affects males, 1 to 3 per 100,000 persons. Facioscapulohumeral and limb-girdle dystrophies usually don't shorten life expectancy, and they affect both sexes equally.

Causes

Causes of muscular dystrophy include:

▶ various genetic mechanisms typically involving an enzymatic or metabolic defect
▶ X-linked recessive disorders due to defects in the gene coding, mapped genetically to the Xp21 locus, for the muscle protein dystrophin, which is essential for maintaining muscle cell membrane; muscle cells deteriorate or die without it. (Duchenne and Becker dystrophy)
▶ autosomal dominant disorder (facioscapulohumeral dystrophy)
▶ autosomal recessive disorder (limb-girdle dystrophy).

Pathophysiology

Abnormally permeable cell membranes allow leakage of a variety of muscle enzymes, particularly creatine kinase. This metabolic defect that causes the muscle cells to die is present from fetal life onward. The absence of progressive muscle wasting at birth suggests that other factors compound the effect of dystrophin deficiency. The specific trigger is unknown, but phagocytosis of the muscle cells by inflammatory cells causes scarring and loss of muscle function.

As the disease progresses, skeletal muscle becomes almost totally replaced by fat and connective tissue. The skeleton eventually becomes deformed, causing progressive immobility. Cardiac and smooth muscle of the GI tract often become fibrotic. No consistent structural abnormalities are seen in the brain.

Signs and symptoms

Signs and symptoms of Duchenne muscular dystrophy include:
▶ insidious onset between the ages of 3 and 5 years
▶ initial effect on legs, pelvis, and shoulders
▶ waddling gait, toe-walking, and lumbar lordosis due to muscle weakness
▶ difficulty climbing stairs, frequent falls
▶ enlarged, firm calf muscles
▶ confined to wheelchair (usually by 9 to 12 years of age).

Signs and symptoms of Becker (benign pseudohypertrophic) muscular dystrophy are:

▶ similar to those of Duchenne muscular dystrophy but with slower progression.

Signs of facioscapulohumeral (Landouzy-Dejerine) dystrophy include:

▶ weakened face, shoulder, and upper arm muscles (initial sign)

▶ pendulous lip and absent nasolabial fold

▶ inability to pucker mouth or whistle

▶ abnormal facial movements and absence of facial movements when laughing or crying

▶ diffuse facial flattening leading to a masklike expression

▶ inability to raise arms above the head.

Signs and symptoms of limb-girdle dystrophy include:

▶ weakness in upper arms and pelvis first

▶ lumbar lordosis with abdominal protrusion

▶ winging of the scapulae

▶ waddling gait

▶ poor balance

▶ inability to raise the arms.

Complications

Possible complications of Duchenne muscular dystrophy are:

▶ weakened cardiac and respiratory muscles leading to tachycardia, electrocardiographic abnormalities, and pulmonary complications

▶ death commonly due to sudden heart failure, respiratory failure, or infection.

Diagnosis

Diagnosis depends on typical clinical findings, family history, and diagnostic test findings. If another family member has muscular dystrophy, its clinical characteristics can suggest the type of dystrophy the patient has and how he may be affected. The following tests may help in the diagnosis:

▶ electromyography showing short, weak bursts of electrical activity in affected muscles

▶ muscle biopsy showing a combination of muscle cell degeneration and regeneration (in later stages, showing fat and connective tissue deposits)

▶ immunologic and molecular biological techniques (now available in specialized medical centers) to facilitate accurate prenatal and postnatal diagnosis of Duchenne and Becker dystrophies (replacing muscle biopsy and elevated serum creatine kinase levels in diagnosis).

Treatment

No treatment can stop the progressive muscle impairment. Supportive treatments include:

▶ coughing and deep-breathing exercises and diaphragmatic breathing

▶ teaching parents to recognize early signs of respiratory complications

▶ orthopedic appliances, exercise, physical therapy, and surgery to correct contractures (to help preserve mobility and independence)

▶ genetic counseling regarding risk for transmitting disease for family members who are carriers

▶ adequate fluid intake, increased dietary bulk, and stool softener for constipation due to inactivity

▶ low-calorie, high-protein, high-fiber diet (physical inactivity predisposes to obesity).

Osteoarthritis

Osteoarthritis, the most common form of arthritis, is a chronic condition causing the deterioration of joint cartilage and the formation of reactive new bone at the margins and subchondral areas of the joints. It usually affects weight-bearing joints (knees, feet, hips, lumbar vertebrae). Osteoarthritis is widespread (affecting more than 60 million persons in the United States) and is most common in women. Typically, its earliest symptoms manifest in middle age and progress from there.

Disability depends on the site and severity of involvement and can range from minor limitation of finger movement to severe disability in persons with hip or knee involvement. The rate of progression varies, and joints may remain stable for years in an early stage of deterioration.

Causes
The primary defect in both idiopathic and secondary osteoarthritis is loss of articular cartilage due to functional changes in chondrocytes (cells responsible for the formation of the proteoglycans, glycoproteins that act as cementing material in the cartilage, and collagen).

Idiopathic osteoarthritis, a normal part of aging, results from many factors, including:
▶ metabolic (endocrine disorders such as hyperparathyroidism) and genetic (decreased collagen synthesis)
▶ chemical (drugs that stimulate the collagen-digesting enzymes in the synovial membrane, such as steroids)
▶ mechanical factors (repeated stress on the joint).

Secondary osteoarthritis usually follows an identifiable predisposing event that leads to degenerative changes, such as:
▶ trauma (most common cause)
▶ congenital deformity
▶ obesity.

Pathophysiology
Osteoarthritis occurs in synovial joints. The joint cartilage deteriorates, and reactive new bone forms at the margins and subchondral areas of the joints. The degeneration results from damage to the chondrocytes. Cartilage softens with age, narrowing the joint space. Mechanical injury erodes articular cartilage, leaving the underlying bone unprotected. This causes sclerosis, or thickening and hardening of the bone underneath the cartilage.

Cartilage flakes irritate the synovial lining, which becomes fibrotic and limits joint movement. Synovial fluid may be forced into defects in the bone, causing cysts. New bone, called osteophyte (bone spur), forms at joint margins as the articular cartilage erodes, causing gross alteration of the bony contours and enlargement of the joint.

Signs and symptoms
Symptoms, which increase with poor posture, obesity, and occupational stress, include:
▶ deep, aching joint pain due to degradation of the cartilage, inflammation, and bone stress, particularly after exercise or weight bearing (the most common symptom, usually relieved by rest)
▶ stiffness in the morning and after exercise (relieved by rest)
▶ crepitus, or "grating" of the joint during motion due to cartilage damage
▶ Heberden's nodules (bony enlargements of the distal interphalangeal joints) due to repeated inflammation
▶ altered gait from contractures due to overcompensation of the muscles supporting the joint
▶ decreased range of motion due to pain and stiffness
▶ joint enlargement due to stress on the bone and disordered bone growth
▶ localized headaches (may be a direct result of cervical spine arthritis).

Complications
Complications of osteoarthritis include:
▶ irreversible joint changes and node formation (nodes eventually becoming red, swollen, and tender, causing numbness and loss of finger dexterity)
▶ subluxation of the joint.

Diagnosis
Findings that help diagnose osteoarthritis include:
▶ absence of systemic symptoms (ruling out inflammatory joint disorder)
▶ arthroscopy showing bone spurs, narrowing of joint space
▶ increased erythrocyte sedimentation rate (with extensive synovitis).

X-rays of the affected joint help confirm the diagnosis but may be normal in

SPECIFIC CARE FOR ARTHRITIC JOINTS

Specific care depends on the affected joint:
▶ Hand: hot soaks and paraffin dips to relieve pain, as ordered.
▶ Lumbar and sacral spine: a firm mattress or bed board to decrease morning pain.
▶ Cervical spine: cervical collar; check for constriction; watch for redness with prolonged use.
▶ Hip: moist heat pads to relieve pain and antispasmodic drugs, as ordered. Assist with range-of-motion and strengthening exercises, always making sure the patient gets the proper rest afterward. Check crutches, cane, braces, and walker for proper fit, and teach the patient to use them correctly. For example, the patient with unilateral joint involvement should use an orthopedic appliance (such as a cane or walker) on the unaffected side. Advise use of cushions when sitting and use of an elevated toilet seat .
▶ Knee: assist with prescribed range-of-motion exercises, exercises to

maintain muscle tone, and progressive resistance exercises to increase muscle strength. Provide elastic supports or braces if needed.

To minimize the long-term effects of osteoarthritis, teach the patient to:
▶ plan for adequate rest during the day, after exertion, and at night
▶ take medication exactly as prescribed and report adverse effects immediately
▶ avoid overexertion, take care to stand and walk correctly, minimize weight-bearing activities, and be especially careful when stooping or picking up objects
▶ always wear well-fitting supportive shoes and not let the heels become too worn down
▶ install safety devices at home, such as guard rails in the bathroom
▶ do range-of-motion exercises as gently as possible
▶ maintain proper body weight to lessen strain on joints
▶ avoid percussive activities.

the early stages. X-rays may require many views and typically show:
▶ narrowing of joint space or margin
▶ cyst-like bony deposits in joint space and margins, sclerosis of the subchondral space
▶ joint deformity due to degeneration or articular damage
▶ bony growths at weight-bearing areas
▶ joint fusion.

Treatment

The goal of treatment is to relieve pain, maintain or improve mobility, and minimize disability. Treatment may include:
▶ weight loss to reduce stress on the joint
▶ balance of rest and exercise
▶ medications, including aspirin, phenylbutazone, indomethacin (Indocin), fenopro-

fen (Nalfon), ibuprofen (Motrin), and other nonsteroidal anti-inflammatory drugs; propoxyphene (Darvon), celecoxib (Celebrex), and glucosamine (See *Specific care for arthritic joints.*)
▶ support or stabilization of joint with crutches, braces, cane, walker, cervical collar, or traction to reduce stress
▶ intra-articular injections of corticosteroids (every 4 to 6 months) to possibly delay node development in the hands (if used too frequently, may accelerate arthritic progression by depleting the normal ground substance of the cartilage).

Surgical treatment, reserved for patients with severe disability or uncontrollable pain, may include:

▶ arthroplasty (partial or total replacement of deteriorated part of joint with prosthetic appliance)

▶ arthrodesis (surgical fusion of bones, primarily in spine [laminectomy])

▶ osteoplasty (scraping and lavage of deteriorated bone from joint)

▶ osteotomy (change in alignment of bone to relieve stress by excision of wedge of bone or cutting of bone).

Osteogenesis imperfecta

Osteogenesis imperfecta is a genetic disease in which bones are thin, poorly developed, and fracture easily.

The expression of the disease varies, depending on whether the defect is carried as a trait or is clinically obvious. (See Chapter 4, "Genetics.") If it's inherited as an autosomal dominant disorder, a heterozygote may eventually express the disease, which occurs in about 1 in 30,000 people. If inheritance is as an autosomal recessive, the homozygous child will likely die before, during, or soon after birth from multiple fractures sustained in utero or during delivery.

Causes

Causes of osteogenesis imperfecta include:
▶ genetic disease, typically autosomal dominant (characterized by a defect in the synthesis of connective tissue)
▶ autosomal recessive carriage of gene defects that produce osteogenesis imperfecta in homozygotes (osteoporosis in some).

Pathophysiology

Most forms of the disease appear to be caused by mutations in the genes that determine the structure of collagen. Possible mutations in other genes may cause variations in the assembly and maintenance of bone and other connective tissues. Collectively or alone, these mutated genes lead to pathologic fractures and impaired healing.

Signs and symptoms

In the autosomal dominant disorder, the following symptoms may not be apparent until the child's mobility increases:
▶ frequent fractures and poor healing due to falls as toddlers begin to walk
▶ short stature due to multiple fractures caused by minor physical stress
▶ deformed cranial structure and limbs from multiple fractures
▶ thin skin and bluish sclera of the eyes; thin collagen fibers of the sclera allowing the choroid layer to be seen
▶ abnormal tooth and enamel development due to improper deposition of dentine.

Complications

Possible complications of osteogenesis imperfecta include:
▶ deafness due to bone deformity and scarring of the middle and inner ear
▶ stillbirth or death within the first year of life (autosomal-recessive disorder).

Diagnosis

Diagnosis involves:
▶ fractures early in life, hearing loss, and blue sclerae, showing that mutation is expressed in more than one connective tissue
▶ elevated serum alkaline phosphatase levels (during periods of rapid bone formation and cellular injury)
▶ skin culture showing reduced quantity of fibroblasts
▶ echocardiography, possibly showing mitral regurgitation or floppy mitral valves.

Treatment

Possible treatments are:
▶ prevention of fractures
▶ internal fixation of fractures to ensure stabilization.

Osteomalacia and rickets

In Vitamin D deficiency, bone cannot calcify normally; the result is called *rickets* in infants and young children and *osteomalacia* in adults.

Once a common childhood disease, rickets is now rare in the United States. It does appear occasionally in breast-fed infants who don't receive a vitamin D supplement or in infants fed a formula with a nonfortified milk base. Rickets also occurs in overcrowded, urban areas where smog limits sunlight penetration.

 CULTURAL DIVERSITY Incidence of rickets is highest in children with black or dark brown skin, who, because of their pigmentation, absorb less sunlight. In urban Asia, osteomalacia is most prevalent in young women who have had several children, eat a cereal-based diet, and have minimal exposure to sunlight.

With treatment, the prognosis is good. In osteomalacia, bone deformities may disappear; however they usually persist in children with rickets.

Causes

Causes of osteomalacia and rickets include:

▶ inadequate dietary intake of preformed vitamin D
▶ malabsorption of vitamin D
▶ inadequate exposure to sunlight (solar ultraviolet rays irradiate 7-dehydrocholesterol, a precursor of vitamin D, to form calciferol)
▶ inherited impairment of renal tubular reabsorption of phosphate (from vitamin D insensitivity) in vitamin D-resistant rickets (refractory rickets, familial hypophosphatemia)
▶ conditions reducing the absorption of fat-soluble vitamin D (such as chronic pancreatitis, celiac disease, Crohn's disease, cystic fibrosis, gastric or small-bowel resections, fistulas, colitis, and biliary obstruction)
▶ hepatic or renal disease (interfering with hydroxylated calciferol formation, needed to form a calcium-binding protein in intestinal absorption sites)
▶ malfunctioning parathyroid gland (decreased secretion of parathyroid hormone), contributing to calcium deficiency (nor-

mally, vitamin D controls absorption of calcium and phosphorus through the intestine) and interfering with activation of vitamin D in the kidneys.

Pathophysiology

Vitamin D regulates the absorption of calcium ions from the intestine. When vitamin D is lacking, falling serum calcium concentration stimulates synthesis and secretion of parathyroid hormone, causing release of calcium from bone, decreasing renal calcium excretion, and increasing renal phosphate excretion. When the concentration of phosphate in the bone decreases, osteoid may be produced, but mineralization can't proceed normally. Large quantities of osteoid accumulate, coating the trabeculae and linings of the haversian canals and areas beneath the periosteum.

When mineralization of bone matrix is delayed or inadequate, bone is disorganized in structure and lacks density. The result is gross deformity of both spongy and compact bone.

Signs and symptoms

Osteomalacia may be asymptomatic until a fracture occurs. Chronic vitamin D deficiency induces numerous bone malformations due to bone softening. Possible signs and symptoms include:

▶ pain in the legs and lower back due to vertebral collapse
▶ bow legs
▶ knock knees
▶ rachitic rosary (beading of ends of ribs)
▶ enlarged wrists and ankles
▶ pigeon breast (protruding ribs and sternum)
▶ delayed closing of fontanels
▶ softening skull
▶ bulging forehead
▶ poorly developed muscles (pot belly)
▶ difficulty walking and climbing stairs.

Complications

Complications of osteomalacia and rickets may include:

• spontaneous multiple fractures
• tetany in infants.

Diagnosis

Physical examination, dietary history, and laboratory tests establish the diagnosis. Test results that suggest vitamin D deficiency include:

• serum calcium less than 7.5 mg/dl
• serum inorganic phosphorus less than 3 mg/dl
• serum citrate less than 2.5 mg/dl, and alkaline phosphatase less than 4 Bodansky units/dl
• X-rays showing characteristic bone deformities and abnormalities such as Looser's zones (radiolucent bands perpendicular to the surface of the bones indicating reduced bone ossification; confirms the diagnosis).

Treatment

Possible treatments include:

• massive oral doses of vitamin D or cod liver oil (for osteomalacia and rickets, except when caused by malabsorption)
• teaching patient on prolonged vitamin D supplementation signs of vitamin D toxicity (headache, nausea, constipation, and, after prolonged use, renal calculi)
• 25-hydroxycholecalciferol, 1,25-dihydroxycholecalciferol, or a synthetic analogue of active vitamin (for rickets refractory to vitamin D or rickets accompanied by hepatic or renal disease)
• foods high in vitamin D (fortified milk, fish liver oils, herring, liver, and egg yolks) and sufficient sun exposure (obtain a dietary history to assess the patient's current vitamin D intake)
• supplemental aqueous preparations of vitamin D for chronic fat malabsorption, hydroxylated cholecalciferol for refractory rickets, and supplemental vitamin D for breast-fed infants (to prevent rickets).

Osteomyelitis

Osteomyelitis is a bone infection characterized by progressive inflammatory destruction after formation of new bone. It may be chronic or acute. It commonly results from a combination of local trauma — usually trivial but causing a hematoma — and an acute infection originating elsewhere in the body. Although osteomyelitis often remains localized, it can spread through the bone to the marrow, cortex, and periosteum. Acute osteomyelitis is usually a blood-borne disease and most commonly affects rapidly growing children. Chronic osteomyelitis, which is rare, is characterized by draining sinus tracts and widespread lesions.

AGE ALERT Osteomyelitis occurs more often in children (especially boys) than in adults — usually as a complication of an acute localized infection. The most common sites in children are the lower end of the femur and the upper ends of the tibia, humerus, and radius. The most common sites in adults are the pelvis and vertebrae, generally after surgery or trauma.

The incidence of both chronic and acute osteomyelitis is declining, except in drug abusers.

With prompt treatment, the prognosis for acute osteomyelitis is very good; for chronic osteomyelitis, prognosis remains poor.

Causes

The most common pyogenic organism in osteomyelitis is:

• *Staphylococcus aureus.*
  Others include:
• *Streptococcus pyogenes*
• *Pneumococcus* species
• *Pseudomonas aeruginosa*
• *Escherichia coli*
• *Proteus vulgaris*
• *Pasteurella multocida* (part of the normal mouth flora of cats and dogs).

Pathophysiology

Typically, these organisms find a culture site in a hematoma from recent trauma or in a weakened area, such as the site of local infection (for example, furunculosis), and travel through the bloodstream to the metaphysis, the section of a long bone that is continuous with the epiphysis plates, where the blood flows into sinusoids. (See *Avoiding osteomyelitis,* page 362.)

Signs and symptoms
Clinical features of chronic and acute osteomyelitis are generally the same and may include:
◗ rapid onset of acute osteomyelitis, with sudden pain in the affected bone and tenderness, heat, swelling, and restricted movement
◗ chronic infection persisting intermittently for years, flaring after minor trauma or persisting as drainage of pus from an old pocket in a sinus tract.

Complications
Possible complications of osteomyelitis include:
◗ amputation (of an arm or leg when resistant chronic osteomyelitis causes severe, unrelenting pain and decreases function)
◗ weakened bone cortex, predisposing the bone to pathologic fracture
◗ arrested growth of an extremity (in children with severe disease).

Diagnosis
Diagnosis must rule out septicemia, foreign bodies, poliomyelitis (rare), rheumatic fever, myositis (inflammation of voluntary muscle), and bone fracture. History, physical examination, and laboratory tests that help confirm osteomyelitis may include:
◗ history of a urinary tract, respiratory tract, ear, or skin infection; human or animal bite; or other penetrating trauma
◗ white blood cell count showing leukocytosis
◗ elevated erythrocyte sedimentation rate
◗ blood cultures showing causative organism

◗ magnetic resonance imaging to delineate bone marrow from soft tissue (facilitates diagnosis)
◗ X-rays (may not show bone involvement until the disease has been active for 2 to 3 weeks)
◗ bone scans to detect early infection.

Treatment
Treatment for acute osteomyelitis should begin before definitive diagnosis and includes:
◗ large doses of antibiotics I.V. (usually a penicillinase-resistant penicillin, such as nafcillin [Nafcil]or oxacillin [Bactocill]) after blood cultures are taken
◗ early surgical drainage to relieve pressure and abscess formation
◗ immobilization of the affected body part by cast, traction, or bed rest to prevent failure to heal or recurrence
◗ supportive measures, such as analgesics for pain and I.V. fluids to maintain hydration
◗ incision and drainage, followed by a culture of the drainage (if an abscess or sinus tract forms).

Antibiotic therapy to control infection may include:
◗ systemic antibiotics
◗ intracavitary instillation of antibiotics through closed-system continuous irrigation with low intermittent suction
◗ limited irrigation with blood drainage system with suction (Hemovac)
◗ packed, wet, antibiotic-soaked dressings.

Chronic osteomyelitis care may include:
◗ surgery, usually required to remove dead bone and promote drainage (prognosis remains poor even after surgery)
◗ hyperbaric oxygen to stimulate normal immune mechanisms
◗ skin, bone, and muscle grafts to fill in dead space and increase blood supply
◗ teach patient to avoid jerky movements and falls (may threaten bone integrity); report sudden pain, crepitus, or deformity immediately; watch for sudden malposition of the limb (may indicate fracture).

AVOIDING OSTEOMYELITIS

Bones are essentially isolated from the body's natural defense system once an organism gets through the periosteum. Bones are limited in their ability to replace necrotic tissue caused by infection, which may lead to chronic osteomyelitis.

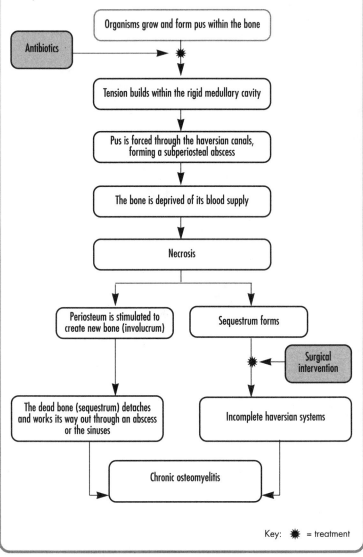

Key: ✴ = treatment

Osteoporosis

Osteoporosis is a metabolic bone disorder in which the rate of bone resorption accelerates while the rate of bone formation slows, causing a loss of bone mass. Bones affected by this disease lose calcium and phosphate salts and become porous, brittle, and abnormally vulnerable to fractures. Osteoporosis may be primary or secondary to an underlying disease, such as Cushing syndrome or hyperthyroidism. It primarily affects the weight-bearing vertebrae. Only when the condition is advanced or severe, as in secondary disease, do similar changes occur in the skull, ribs, and long bones. Often, the femoral heads and pelvic acetabula are selectively affected.

Primary osteoporosis is often called senile or postmenopausal osteoporosis because it most commonly develops in elderly, postmenopausal women.

 CULTURAL DIVERSITY Persons of African origin have a much lower incidence of osteoporosis than those of European or Asian origin.

Causes

The cause of primary osteoporosis is unknown, but contributing factors include:
▶ mild but prolonged negative calcium balance due to inadequate dietary intake of calcium (may be an important contributing factor)
▶ declining gonadal and adrenal function
▶ faulty protein metabolism due to relative or progressive estrogen deficiency (estrogen stimulates osteoblastic activity and limits the osteoclastic-stimulating effects of parathyroid hormones)
▶ sedentary lifestyle.

The many causes of secondary osteoporosis include:
▶ prolonged therapy with steroids or heparin (heparin promotes bone resorption by inhibiting collagen synthesis or enhancing collagen breakdown)
▶ total immobilization or disuse of a bone (as in hemiplegia)
▶ alcoholism

▶ malnutrition
▶ malabsorption
▶ scurvy
▶ lactose intolerance
▶ endocrine disorders such as hyperthyroidism, hyperparathyroidism, Cushing's syndrome, diabetes mellitus (plasma calcium and phosphate concentrations are maintained by the endocrine system)
▶ osteogenesis imperfecta
▶ Sudeck's atrophy (localized to hands and feet, with recurring attacks)
▶ medications   (aluminum-containing antacids, corticosteroids, anticonvulsants)
▶ cigarette smoking.

Pathophysiology

In normal bone, the rates of bone formation and resorption are constant; replacement follows resorption immediately, and the amount of bone replaced equals the amount of bone resorbed. Osteoporosis develops when the remodeling cycle is interrupted, and new bone formation falls behind resorption.

When bone is resorbed faster than it forms, the bone becomes less dense. Men have approximately 30% greater bone mass than women, which may explain why osteoporosis develops later in men.

Signs and symptoms

Osteoporosis is typically discovered suddenly, such as:
▶ a postmenopausal woman bends to lift something, hears a snapping sound, then feels a sudden pain in her lower back
▶ vertebral collapse causes back pain that radiates around the trunk (most common presenting feature) and is aggravated by movement or jarring.

In another common pattern, osteoporosis can develop insidiously, showing:
▶ increasing deformity, kyphosis, loss of height, decreased exercise intolerance, and a markedly aged appearance
▶ spontaneous wedge fractures, pathologic fractures of the neck and femur, Colles' fractures of the distal radius after a minor

fall, and hip fractures (common as bone is lost from the femoral neck).

Complications
Possible complications of osteoporosis include:
▶ spontaneous fractures as the bones lose volume and become brittle and weak
▶ shock, hemorrhage, or fat embolism (fatal complications of fractures).

Diagnosis
Differential diagnosis must exclude other causes of bone loss, especially those affecting the spine, such as metastatic cancer or advanced multiple myeloma. History is the key to identify the specific cause of osteoporosis. Diagnosis may include:
▶ serial height measurements
▶ dual- or single-photon absorptiometry to measure bone mass of the extremities, hips, and spine
▶ X-rays showing typical degeneration in the lower thoracic and lumbar vertebrae (vertebral bodies may appear flattened and may look denser than normal; bone mineral loss is evident in only in later stages)
▶ computed tomography scan to assess spinal bone loss
▶ normal serum calcium, phosphorus, and alkaline phosphatase, possibly elevated parathyroid hormone.

Since the advent of readily available bone density measurement, the following studies are seldom done:
▶ bone biopsy showing thin, porous, but otherwise normal-looking bone
▶ radionuclide bone scans showing diseased areas as darker portions.

Treatment
Treatment to control bone loss, prevent fractures, and control pain may include:
▶ physical therapy emphasizing gentle exercise and activity and regular, moderate weight-bearing exercise to slow bone loss and possibly reverse demineralization (the mechanical stress of exercise stimulates bone formation)

▶ supportive devices, such as a back brace
▶ surgery, if indicated, for hip fracture
▶ estrogen during 2 years after menopause to slow bone loss
▶ analgesics and local heat to relieve pain.
 Other medications include:
▶ calcium and vitamin D supplements to support normal bone metabolism
▶ calcitonin (Calcimar) to reduce bone resorption and slow the decline in bone mass
▶ biophosphonates (such as etidronate [Didronel] to increase bone density and restore lost bone
▶ fluoride (such as alendronate [Fosamax]) to stimulate bone formation; requires strict dosage precautions, and can cause gastric distress
▶ vitamin C, calcium, and protein to support skeletal metabolism (through a balanced diet rich in nutrients).
 Other measures include:
▶ early mobilization after surgery or trauma
▶ decreased alcohol and tobacco consumption
▶ careful observation for signs of malabsorption, (fatty stools, chronic diarrhea)
▶ prompt, effective treatment of the underlying disorder (to prevent secondary osteoporosis)
▶ safety precautions for frail patients (keeping side rails up, moving the patient gently and carefully at all times, explaining to the patient's family and ancillary health care personnel how easily an osteoporotic patient's bones can fracture)
▶ advising patient to report new pain immediately, especially after trauma, no matter how slight
▶ advising patient to sleep on a firm mattress and avoid excessive bed rest
▶ teaching the patient good body mechanics, such as stooping before lifting and avoiding twisting movements and prolonged bending
▶ instructing a female patient taking estrogen in the proper technique for breast self-examination, to perform self-examination at least monthly and to report lumps

immediately; to have regular gynecologic exams; and to report abnormal bleeding promptly.

Paget's disease

Paget's disease, also called osteitis deformans, is a slowly progressive metabolic bone disease characterized by accelerated patterns of bone remodeling. An initial phase of excessive bone resorption (osteoclastic phase) is followed by a reactive phase of excessive abnormal bone formation (osteoblastic phase). Chronic accelerated remodeling eventually enlarges and softens the affected bones. The new bone structure, which is chaotic, fragile, and weak, causes painful deformities of both external contour and internal structure. Paget's disease usually localizes in one or several areas of the skeleton (most frequently the lumbosacral spine, skull, pelvis, femur, and tibia are affected), but occasionally skeletal deformity is widely distributed.

CULTURAL DIVERSITY Paget's disease occurs worldwide but is extremely rare in Asia, the Middle East, Africa, and Scandinavia.

In the United States, Paget's disease affects about 2.5 million people older than 40 years (mostly men). It can be fatal, particularly when it's associated with heart failure (widespread disease creates a continuous need for high cardiac output), bone sarcoma, or giant-cell tumors.

Causes

Although the exact cause of Paget's disease is unknown, one theory is that early viral infection causes a dormant skeletal infection that erupts many years later as Paget's disease.

Other possible causes include:
▶ benign or malignant bone tumors
▶ vitamin D deficiency during the bone-developing years of childhood
▶ autoimmune disease
▶ estrogen deficiency

Pathophysiology

Repeated episodes of accelerated osteoclastic resorption of spongy bone occur. The trabeculae diminish, and vascular fibrous tissue replaces marrow. This is followed by short periods of rapid, abnormal bone formation. The collagen fibers in this new bone are disorganized, and glycoprotein levels in the matrix decrease. The partially resorbed trabeculae thicken and enlarge because of excessive bone formation, and the bone becomes soft and weak.

Signs and symptoms

Clinical effects of Paget's disease vary. Early stages may be asymptomatic. When signs and symptoms appear, they may include:
▶ usually severe and persistent pain intensifying with weight bearing, possibly with impaired movement due to impingement of abnormal bone on the spinal cord or sensory nerve root (pain may also result from the constant inflammation accompanying cell breakdown)
▶ characteristic cranial enlargement over frontal and occipital areas (hat size may increase) and possibly headaches, sensory abnormalities, and impaired motor function (with skull involvement).

Other deformities include:
▶ kyphosis (spinal curvature due to compression fractures of vertebrae)
▶ barrel chest
▶ asymmetric bowing of the tibia and femur (often reduces height)
▶ waddling gait (from softening of pelvic bones)
▶ warm and tender disease sites susceptible to pathologic fractures after minor trauma
▶ slow and often incomplete healing of pathologic fractures.

Complications

Possible complications of Paget's disease include:

- blindness and hearing loss with tinnitus and vertigo due to bony impingement on the cranial nerves
- pathologic fractures
- hypertension
- renal calculi
- hypercalcemia
- gout
- heart failure due to high blood flow demands of remodeling bones
- respiratory failure due to deformed thoracic bones
- malignant changes in involved bone (1% of the patients).

Diagnosis

Diagnosis of Paget's disease may include:
- X-rays, computed tomography scan, and magnetic resonance imaging taken before overt symptoms develop showing increased bone expansion and density
- radionuclide bone scan (more sensitive than X-rays) clearly showing early Paget's lesions (radioisotope concentrates in areas of active disease)
- bone biopsy showing characteristic mosaic pattern.

Other laboratory findings include:
- anemia
- elevated serum alkaline phosphatase (an index of osteoblastic activity and bone formation)
- elevated 24-hour urine levels for hydroxyproline (amino acid excreted by kidneys and an index of osteoclastic hyperactivity) and pyridinolines.

Treatment

Primary treatment consists of drug therapy and includes one of the following medications:
- bisphosphonate (alendronate [Fosamax], etidronate [Didronel]) to inhibit osteoclast-mediated bone resorption
- calcitonin (Calcimar, a hormone, and etidronate [Didronel]) to retard bone resorption and reduce serum alkaline phosphate and urinary hydroxyproline secretion (Calcitonin requires long-term maintenance therapy, but improvement is

noticeable after the first few weeks of treatment; etidronate produces improvement after 1 to 3 months)
- mithramycin (Mithracin), a cytotoxic antibiotic, to decrease serum calcium, urinary hydroxyproline, and serum alkaline phosphatase levels; produces remission of symptoms within 2 weeks and biochemical improvement in 1 to 2 months, but may destroy platelets or compromise renal function.

Other treatment varies according to symptoms:
- surgery to reduce or prevent pathologic fractures, correct secondary deformities, and relieve neurologic impairment; drug therapy with calcitonin and etidronate or mithramycin must precede surgery to decrease the risk for excessive bleeding from hypervascular bone
- joint replacement (difficult because bonding material [methyl methacrylate] doesn't set properly on pagetic bone)
- aspirin, indomethacin (Indocin), or ibuprofen (Motrin) to control pain
- monitoring for new areas of pain or restricted movements (may indicate new fracture sites) and sensory or motor disturbances, such as difficulty in hearing, seeing, or walking
- instructing the patient to follow specific instructions when taking etidronate or alendronate, to minimize adverse affects and avoid neutralizing effects of drug, and to watch for and report stomach cramps, diarrhea, fractures, and increasing or new bone pain
- instructing the patient taking mithramycin to watch for signs of infection, easy bruising, bleeding, and temperature elevation, and to report for regular follow-up laboratory tests.

Rhabdomyolysis

Rhabdomyolysis, the breakdown of muscle tissue, may cause myoglobinuria, in which varying amounts of muscle protein (myoglobin) appear in the urine. Rhabdomyolysis usually follows major muscle trauma, especially a muscle crush injury.

Long-distance running, certain severe infections, and exposure to electric shock can cause extensive muscle damage and excessive release of myoglobin. Prognosis is good if contributing causes are stopped or disease is checked before damage has progressed to an irreversible stage. Unchecked, it can cause renal failure.

Causes

Possible causes of rhabdomyolysis include:
▶ familial tendency
▶ strenuous exertion
▶ infection
▶ anesthetic agents (halothane) causing intraoperative rigidity
▶ heat stroke
▶ electrolyte disturbances
▶ cardiac arrhythmias
▶ excessive muscular activity associated with status epilepticus, electroconvulsive therapy, or high-voltage electrical shock.

Pathophysiology

Muscle trauma that compresses tissue causes ischemia and necrosis. The ensuing local edema further increases compartment pressure and tamponade; pressure from severe swelling causes blood vessels to collapse, leading to tissue hypoxia, muscle infarction, and neural damage in the area of the fracture, and release of myoglobin from the necrotic muscle fibers into the circulation.

Signs and symptoms

Signs and symptoms of rhabdomyolysis include:
▶ tenderness, swelling, and muscle weakness due to muscle trauma and pressure
▶ dark, reddish-brown urine from myoglobin.

Complications

Possible complications of rhabdomyolysis are:
▶ renal failure as myoglobin is trapped in renal capillaries or tubules

▶ amputation if muscle necrosis is substantial.

Diagnosis

Diagnosis may include:
▶ urine myoglobin greater than 0.5 μg/dl (evident with only 200 g of muscle damage)
▶ elevated creatinine kinase (0.5 to 0.95 mg/dl) due to muscle damage
▶ elevated serum potassium, phosphate, creatinine, and creatine levels
▶ hypocalcemia in early stages, hypercalcemia in later stages
▶ computed tomography, magnetic resonance imaging, and bone scintigraphy to detect muscle necrosis
▶ intracompartmental venous pressure measurements using a wick catheter, needle, or slit catheter inserted into the muscle.

Treatment

Treatment of rhabdomyolysis may include:
▶ treating the underlying disorder
▶ preventing renal failure
▶ bed rest
▶ anti-inflammatory agents
▶ corticosteroids (in extreme cases)
▶ analgesics for pain
▶ immediate fasciotomy and debridement (if compartment venous pressure is greater than 25 mm Hg).

Scoliosis

Scoliosis is a lateral curvature of the thoracic, lumbar, or thoracolumbar spine. The curve may be convex to the right (more common in thoracic curves) or to the left (more common in lumbar curves). Rotation of the vertebral column around its axis may cause rib cage deformity. Scoliosis is often associated with kyphosis (humpback) and lordosis (swayback).

About 2% to 3% of adolescents have scoliosis. In general, the greater the magnitude of the curve and the younger the child at the time of diagnosis, the greater the risk for progression of the spinal abnormality. Favorable outcomes are usually achieved with optimal treatment.

Types of structural scoliosis are:
▶ congenital, such as wedge vertebrae, fused ribs or vertebrae, or hemivertebrae
▶ paralytic or musculoskeletal, developing several months after asymmetric paralysis of the trunk muscles due to polio, cerebral palsy, or muscular dystrophy
▶ idiopathic (most common), may be transmitted as an autosomal dominant or multifactorial trait (appears in a previously straight spine during the growing years).

Idiopathic scoliosis can be further classified according to age at onset:
▶ infantile (affects mostly male infants between birth and 3 years and causes left thoracic and right lumbar curves)
▶ juvenile (affects both sexes between the ages of 4 and 10 years and causes varying types of curvature)
▶ adolescent (generally affecting girls from the age of 10 years until skeletal maturity and causing varying types of curvature).

Causes

Scoliosis may be functional or structural. Possible causes include:
▶ functional: poor posture or a discrepancy in leg lengths, not fixed deformity of the spinal column (postural scoliosis)
▶ structural: deformity of the vertebral bodies leading to curvature.

Pathophysiology

Differential stress on vertebral bone causes an imbalance of osteoblastic activity; thus the curve progresses rapidly during adolescent growth spurt. Without treatment, the imbalance continues into adulthood.

Signs and symptoms

Scoliosis rarely produces subjective symptoms until it's well established. When symptoms occur, they include:
▶ backache
▶ fatigue
▶ dyspnea.

The most common curve in functional or structural scoliosis arises in the thoracic segment, with convexity to the right and compensatory curves (S curves) in the cervical and lumbar segments, both with convexity to the left. As the spine curves laterally, compensatory curves develop to maintain body balance. Subtle signs include:
▶ uneven hemlines or pant legs that appear unequal in length
▶ one hip that appears higher than the other.

Physical examination shows:
▶ unequal shoulder heights, elbow levels, and heights of iliac crests
▶ asymmetric thoracic cage and misalignment of the spinal vertebrae when the patient bends over
▶ asymmetric paraspinal muscles, rounded on the convex side of the curve and flattened on the concave side
▶ asymmetric gait.

Complications

Without treatment, curves greater than 40 degrees progress. Untreated scoliosis may result in:
▶ pulmonary insufficiency (curvature may decrease lung capacity)
▶ back pain
▶ degenerative arthritis of the spine
▶ vertebral disk disease
▶ sciatica.

Diagnosis

Diagnosis of scoliosis includes:
▶ anterior, posterior, and lateral spinal X-rays, taken with the patient standing upright and bending (confirm scoliosis and determine the degree of curvature [Cobb method] and flexibility of the spine)
▶ scoliometer to measure the angle of trunk rotation.

Treatment

The severity of the deformity and potential spine growth determine appropriate treatment, which may include:

▶ close observation
▶ exercise
▶ brace
▶ surgery
▶ a combination of these.

To be most effective, treatment should begin early, when spinal deformity is still subtle. For a curve less than 25 degrees, or mild scoliosis, treatment includes:
▶ X-rays to monitor curve
▶ examination every 3 months
▶ exercise program to strengthen torso muscles and prevent curve progression.

For a curve of 30 to 50 degrees:
▶ spinal exercises and a brace (may halt progression but doesn't reverse the established curvature); braces can be adjusted as the patient grows and worn until bone growth is complete)
▶ transcutaneous electrical stimulation (alternative therapy).

A lateral curve continues to progress at the rate of 1 degree a year even after skeletal maturity. For a curve of 40 degrees or more, treatment includes:
▶ surgery (supportive instrumentation, with spinal fusion in severe cases)
▶ periodic postoperative checkups for several months to monitor stability of the correction.

Sprains

A sprain is a complete or incomplete tear of the supporting ligaments surrounding a joint. It usually follows a sharp twist. An immobilized sprain may heal in 2 to 3 weeks without surgical repair, after which the patient can gradually resume normal activities. A sprained ankle is the most common joint injury, followed by sprains of the wrist, elbow, and knee.

Causes
Causes of sprains include:
▶ sharply twisting with force stronger than that of the ligament, inducing joint movement beyond normal range of motion
▶ concurrent fractures or dislocations.

Pathophysiology
When a ligament is torn, an inflammatory exudate develops in the hematoma between the torn ends. Granulation tissue grows inward from the surrounding soft tissue and cartilage. Collagen formation begins 4 to 5 days after the injury, eventually organizing fibers parallel to the lines of stress. With the aid of vascular fibrous tissue, the new tissue eventually fuses with surrounding tissues. As further reorganization takes place, the new ligament separates from the surrounding tissue, and eventually becomes strong enough to withstand normal muscle tension.

Signs and symptoms
Possible signs and symptoms of sprain are:
▶ localized pain (especially during joint movement)
▶ swelling and heat due to inflammation
▶ loss of mobility due to pain (may not occur until several hours after the injury)
▶ skin discoloration from blood extravasating into surrounding tissues.

Complications
Possible complications of sprain include:
▶ recurring dislocation due to torn ligaments that don't heal properly, requiring surgical repair (occasionally)
▶ loss of function in a ligament (if a strong muscle pull occurs before it heals and stretches it, it may heal in a lengthened shape with an excessive amount of scar tissue).

Diagnosis
Sprain may be diagnosed by:
▶ history of recent injury or chronic overuse
▶ X-ray to rule out fractures
▶ stress radiography to visualize the injury in motion
▶ arthroscopy
▶ arthrography.

MUSCLE-TENDON RUPTURES

Perhaps the most serious muscle-tendon injury is a rupture of the muscle-tendon junction. This type of rupture may occur at any such junction, but it's most common at the Achilles tendon, which extends from the posterior calf muscle to the foot. An Achilles tendon rupture produces a sudden, sharp pain and, until swelling begins, a palpable defect. This rupture typically occurs in men between the ages of 35 and 40 years, especially during physical activities such as jogging or tennis.

To distinguish an Achilles tendon rupture from other ankle injuries, perform this simple test: With the patient prone and his feet hanging off the foot of the table, squeeze the calf muscle. The response establishes the diagnosis:

▶ plantar flexion, the tendon is intact

▶ ankle dorsiflexion, it's partially intact

▶ no flexion of any kind, the tendon is ruptured.

An Achilles tendon rupture usually requires surgical repair, followed by a long leg cast for 4 weeks, and then a short cast for an additional 4 weeks.

Treatment

Treatment to control pain and swelling includes:

▶ immobilizing the injured joint to promote healing

▶ elevating the joint above the level of the heart for 48 to 72 hours (immediately after the injury)

▶ intermittently applying ice for 12 to 48 hours to control swelling (place a towel between the ice pack and the skin to prevent a cold injury)

▶ an elastic bandage or cast, or if the sprain is severe, a soft cast or splint to immobilize the joint

▶ codeine or another analgesic (if injury is severe)

▶ crutch and gait training (sprained ankle)

▶ immediate surgical repair to hasten healing, including suturing the ligament ends in close approximation (some athletes)

▶ tape wrists or ankles before sports activities to prevent sprains (athletes).

Strains

A strain is an injury to a muscle or tendinous attachment usually seen after traumatic or sports injuries. Strain is a general term for muscle or tendon damage that often results from sudden, forced motion causing it to be stretched beyond normal capacity. Injury ranges from excessive stretch (muscle pull) to muscle rupture. (See *Muscle-tendon ruptures.*) If the muscle ruptures, the body of the muscle protrudes through the fascia. A strained muscle can usually heal without complications.

 AGE ALERT Tendon rupture is more common in the elderly; muscle rupture, in the young.

Causes

Possible causes of strain include:

▶ vigorous muscle overuse or overstress, causing the muscle to become stretched beyond normal capacity, especially when the muscle isn't adequately stretched before the activity (acute strain)

▶ knife or gunshot wound causing a traumatic rupture (acute strain)

▶ repeated overuse (chronic strain).

Pathophysiology

Bleeding into the muscle and surrounding tissue occurs if the muscle is torn. When a tendon or muscle is torn, an inflammatory exudate develops between the torn ends. Granulation tissue grows inward from the surrounding soft tissue and cartilage. Collagen formation begins 4 to 5 days after the injury, eventually organizing fibers parallel to the lines of stress. With the aid of vascular fibrous tissue, the new tissue eventually fuses with surrounding tissues. As further reorganization takes place, the new tendon or muscle separates from the surrounding tissue and eventually becomes strong enough to withstand normal muscle strain. If a muscle is chronically strained, calcium may deposit into a muscle, limiting movement by causing stiffness, and muscle fatigue.

Signs and symptoms

Signs and symptoms of acute strain include:
◗ sharp, transient pain (myalgia)
◗ snapping noise
◗ rapid swelling that may continue for 72 hours
◗ limited function
◗ tender muscle (when severe pain subsides)
◗ ecchymoses (after several days).
   Signs and symptoms of chronic strain include:
◗ stiffness
◗ soreness
◗ generalized tenderness.

Complications

Possible complications of strain include:
◗ complete muscle rupture requiring surgical repair
◗ myositis ossificans (chronic inflammation with bony deposits) due to scar tissue calcification (late complication).

Diagnosis

Diagnosis of strain may include:
◗ history of a recent injury or chronic overuse
◗ X-ray to rule out fracture
◗ stress radiography to visualize the injury in motion
◗ biopsy showing muscle regeneration and connective tissue repair (rarely done).

Treatment

Possible treatments for acute strain includes:

◗ compression wrap to immobilize the affected area
◗ elevating the injured part above the level of the heart to reduce swelling
◗ analgesics
◗ application of ice for up to 48 hours, then application of heat to enhance blood flow, reduce cramping, and promote healing
◗ surgery to suture the tendon or muscle ends in close approximation.
   Chronic strains usually don't need treatment. Discomfort may be relieved by:
◗ heat application
◗ nonsteroidal anti-inflammatory drugs (such as ibuprofen [Motrin])
◗ analgesic muscle relaxant.

11

Blood, although a fluid, is one of the body's major tissues. It continuously circulates through the heart and blood vessels, carrying vital elements to every part of the body.

Blood performs several vital functions through its special components: the liquid protein (plasma) and the formed constituents (erythrocytes, leukocytes, and thrombocytes) suspended in it. Erythrocytes (red blood cells) carry oxygen to the tissues and remove carbon dioxide. Leukocytes (white blood cells) act in inflammatory and immune responses. Plasma (a clear, straw-colored fluid) carries antibodies and nutrients to tissues and carries waste away. Plasma coagulation factors and thrombocytes (platelets) control clotting.

Hematopoiesis, the process of blood formation, occurs primarily in the marrow. There primitive blood cells (stem cells) differentiate into the precursors of erythrocytes (normoblasts), leukocytes, and thrombocytes.

The average person has 5 to 6 L of circulating blood, which comprises 5% to 7% of body weight (as much as 10% in premature newborns). Blood is three to five times more viscous than water, has an arterial pH of 7.35 to 7.45, and is either bright red (arterial blood) or dark red (venous blood), depending on the degree of oxygen saturation and the hemoglobin level.

PATHOPHYSIOLOGIC MANIFESTATIONS

Bone marrow cells reproduce rapidly and have a short life span, and the storage of circulating cells in the marrow is minimal. Thus, bone marrow cells and their precursors are particularly vulnerable to physiologic changes that affect cell production. Disease can affect the structure or concentration of any hematologic cell.

Hemoglobin

The protein hemoglobin is the major component of the red blood cell. Hemoglobin consists of an iron-containing molecule (heme) bound to the protein globulin. Oxygen binds to the heme component and is transported throughout the body and released to the cells. The hemoglobin picks up carbon dioxide and hydrogen ions from the cells and delivers them to the lungs, where they are released.

A variety of mutations or abnormalities in the hemoglobin protein can cause abnormal oxygen transport.

Red blood cells

Red blood cell (RBC) disorders may be quantitative or qualitative. A deficiency of RBCs (anemia) can follow a condition that destroys or inhibits the formation of these cells. (See *Erythropoiesis.*)

Common factors leading to anemia include:
▶ drugs, toxins, ionizing radiation
▶ congenital or acquired defects that cause bone marrow to stop producing new RBCs

cells (aplasia) and generally suppress production of all blood cells (hematopoiesis, aplastic anemia)

◗ metabolic abnormalities (sideroblastic anemia)

◗ deficiency of vitamins (vitamin B_{12} deficiency, or pernicious anemia) or minerals (iron, folic acid, copper, and cobalt deficiency anemias) leading to inadequate erythropoiesis

◗ excessive chronic or acute blood loss (posthemorrhagic anemia)

◗ chronic illnesses, such as renal disease, cancer, and chronic infections

◗ intrinsically (sickle cell anemia) or extrinsically (hemolytic transfusion reaction) defective RBCs.

Decreased plasma volume can cause a relative excess of RBCs. The few conditions characterized by excessive production of RBCs include:

◗ abnormal proliferation of all bone marrow cells (polycythemia vera)

◗ abnormality of a single element (such as erythropoietin excess caused by hypoxemia or pulmonary disease).

Leukocytosis

Leukocytosis is an elevation in the number of white blood cells (WBCs). All types of WBCs may be increased, or only one type. (See *WBC types and functions*, page 374.) Leukocytosis is a normal physiologic response to infection or inflammation. Other factors, such as temperature changes, emotional disturbances, anesthesia, surgery, strenuous exercise, pregnancy, and some drugs, hormones, and toxins can also cause leukocytosis. Abnormal leukocytosis occurs in malignancies and bone marrow disorders.

Leukopenia

Leukopenia is a deficiency of WBCs — all types or only one type. It can be caused by a number of conditions or diseases, such as human immunodeficiency virus (HIV) infection, prolonged stress, bone marrow disease or destruction, radiation

ERYTHROPOIESIS

The tissues' demand for oxygen and the blood cells' ability to deliver it regulate red blood cell (RBC) production. Lack of oxygen in the tissues (hypoxia) stimulates RBC production, which triggers the formation and release of the hormone erythropoietin. In turn, erythropioetin, 90% of which is produced by the kidneys and 10% by the liver, activates bone marrow to produce RBCs. Androgens may also stimulate erythropoiesis, which accounts for higher RBC counts in men.

The formation of an erythrocyte (RBC) begins with an uncommitted stem cell that may eventually develop into an RBC or white blood cell. Such formation requires certain vitamins — B_{12} and folic acid — and minerals — copper, cobalt, and especially iron, which is vital to hemoglobin's oxygen-carrying capacity. Iron is obtained from various foods and is absorbed in the duodenum and jejunum. An excess of iron is temporarily stored in reticuloendothelial cells, especially those in the liver, as ferritin and hemosiderin until it's released for use in the bone marrow to form new RBCs.

or chemotherapy, lupus erythematosus, leukemia, thyroid disease, or Cushing syndrome. Because WBCs fight infection, leukopenia increases the risk of infectious illness.

Thrombocytosis

Thrombocytosis is an excess of circulating platelets to greater than 400,000/μl. Thrombocytosis may be primary or secondary.

WBC TYPES AND FUNCTIONS

White blood cells (WBCs), or leukocytes, protect the body against harmful bacteria and infection. WBCs are classified as granular leukocytes (basophils, neutrophils, and eosinophils) or nongranular leukocytes (lymphocytes, monocytes, and plasma cells). WBCs are usually produced in bone marrow; lymphocytes and plasma cells are produced in lymphoid tissue as well. Neutrophils have a circulating half-life of less than 6 hours, while some lymphocytes may survive for weeks or months. Normally, WBCs number between 5,000 and 10,000 μl. There are six types of WBCs:

▶ Neutrophils — The predominant form of granulocyte, they make up about 60% of WBCs and help devour invading organisms by phagocytosis.

▶ Eosinophils — Minor granulocytes, they may defend against parasites and lung and skin infections and act in allergic reactions. They account for 1% to 5% of the total WBC count.

▶ Basophils — Minor granulocytes, they may release heparin and histamine into the blood and participate in delayed hypersensitivity reactions. They account for 0% to 1% of the total WBC count.

▶ Monocytes — Along with neutrophils, they help devour invading organisms by phagocytosis. Monocytes help process antigens for lymphocytes and form macrophages in the tissues. They account for 1% to 6% of the total WBC count.

▶ Lymphocytes — They occur as B cells and T cells. B cells form lymphoid follicles, produce humoral antibodies, and help T cells mediated delayed hypersensitivity reactions and the rejection of foreign cells or cell products. Lymphocytes account for 20% to 40% of the total WBC count.

▶ Plasma cells — They develop from lymphoblasts, reside in the tissue, and produce antibodies.

Primary thrombocytosis

In primary thrombocytosis, the number of platelet precursor cells, called megakaryocytes, is increased and the platelet count is greater than 1 million/μl. The condition may result from an intrinsic abnormality of platelet function and increased platelet mass. It may accompany polycythemia vera or chronic granulocytic leukemia. In the presence of thrombocytosis, both hemorrhage and thrombosis may occur. This paradox occurs because accelerated clotting results in a generalized activation of prothrombin and a consequent excess of thrombin clots in the microcirculation. This process consumes exorbitant amounts of coagulation factors and thereby increases the risk of hemorrhage.

Secondary thrombocytosis

Secondary thrombocytosis is a result of an underlying cause, such as stress, exercise, hemorrhage, or hemolytic anemia. Stress and exercise release stored platelets from the spleen. Hemorrhage or hemolytic anemia signal the bone marrow to produce more megakaryocytes.

Thrombocytosis may also occur after a splenectomy. Because the spleen is the primary site of platelet storage and destruction, platelet count may rise after its removal until the bone marrow begins producing fewer platelets.

DISORDERS

Specific causes of hematologic disorders include trauma, chronic disease, surgery, malnutrition, drug, exposure to toxins or radiation, and genetic or congenital defects that disrupt production or function of blood cells.

Aplastic anemias

Aplastic or hypoplastic, anemias result from injury to or destruction of stem cells in bone marrow or the bone marrow matrix, causing pancytopenia (anemia, leukopenia, and thrombocytopenia) and bone marrow hypoplasia. Although commonly used interchangeably with other terms for bone marrow failure, aplastic anemia properly refers to pancytopenia resulting from the decreased functional capacity of a hypoplastic, fatty bone marrow.

These disorders generally produce fatal bleeding or infection, especially when they're idiopathic or caused by chloramphenicol (Chloromycetin) use or infectious hepatitis. The death rate for severe aplastic anemia is 80% to 90%.

Causes

Possible causes of aplastic anemia are:
▶ radiation (about half of such anemias)
▶ drugs (antibiotics, anticonvulsants), or toxic agents (such as benzene or chloramphenicol [Chloromycetin])
▶ autoimmune reactions (unconfirmed), severe disease (especially hepatitis), or preleukemic and neoplastic infiltration of bone marrow
▶ congenital (idiopathic anemias): two identified forms of aplastic anemia are congenital — hypoplastic or Blackfan-Diamond anemia (develops between ages 2 and 3 months); and Fanconi syndrome (develops between birth and 10 years of age).

Pathophysiology

Aplastic anemia usually develops when damaged or destroyed stem cells inhibit blood cell production. Less commonly,

they develop when damaged bone marrow microvasculature creates an unfavorable environment for cell growth and maturation.

Signs and symptoms

Signs and symptoms of aplastic anemia vary with the severity of pancytopenia, but develop insidiously in many cases. They may include:
▶ progressive weakness and fatigue, shortness of breath, headache, pallor, and ultimately tachycardia and heart failure due to hypoxia and increased venous return
▶ ecchymosis, petechiae, and hemorrhage, especially from the mucous membranes (nose, gums, rectum, vagina) or into the retina or central nervous system due to thrombocytopenia
▶ infection (fever, oral and rectal ulcers, sore throat) without characteristic inflammation due to neutropenia (neutrophil deficiency).

Complications

A possible complication of aplastic anemia is:
▶ life-threatening hemorrhage from the mucous membranes.

Diagnosis

The following test results help diagnose aplastic anemia:
▶ 1 million/μl or fewer RBC of normal color and size (normochromic and normocytic).

RBCs may be macrocytic (larger than normal) and anisocytotic (excessive variation in size), with:
▶ very low absolute reticulocyte count
▶ elevated serum iron (unless bleeding occurs), normal or slightly reduced total iron-binding capacity, presence of hemosiderin (a derivative of hemoglobin), and microscopically visible tissue iron storage
▶ decreased platelet, neutrophil, and lymphocyte counts
▶ abnormal coagulation test results (bleeding time) reflecting decreased platelet count

▶ "dry tap" (no cells) from bone marrow aspiration at several sites

▶ biopsy showing severely hypocellular or aplastic marrow, with varied amounts of fat, fibrous tissue, or gelatinous replacement; absence of tagged iron (because iron is deposited in the liver rather than bone marrow) and megakaryocytes (platelet precursors); and depression of RBCs and precursors (erythroid elements).

Differential diagnosis must rule out paroxysmal nocturnal hemoglobinuria and other diseases in which pancytopenia is common.

Treatment

Effective treatment must eliminate an identifiable cause and provide vigorous supportive measures, including:

▶ packed RBC or platelet transfusion; experimental histocompatibility locus antigen-matched leukocyte transfusions

▶ bone marrow transplantation (treatment of choice for anemia due to severe aplasia and for patients who need constant RBC transfusions)

▶ for patients with leukopenia, special measures to prevent infection (avoidance of exposure to communicable diseases, diligent handwashing, etc.)

▶ specific antibiotics for infection (not given prophylactically because they encourage resistant strains of organisms)

▶ respiratory support with oxygen in addition to blood transfusions (for patients with low hemoglobin levels)

▶ corticosteroids to stimulate erythropoiesis; marrow-stimulating agents, such as androgens (controversial); antilymphocyte globulin (experimental); immunosuppressive agents (if the patient doesn't respond to other therapy); and colony-stimulating factors to encourage growth of specific cellular components.

Iron deficiency anemia

Iron deficiency anemia is a disorder of oxygen transport in which hemoglobin synthesis is deficient. A common disease worldwide, iron deficiency anemia affects 10% to 30% of the adult population of the United States. Iron deficiency anemia occurs most commonly in premenopausal women, infants (particularly premature or low-birth-weight infants), children, and adolescents (especially girls). The prognosis after replacement therapy is favorable.

Causes

Possible causes of iron deficiency anemia are:

▶ inadequate dietary intake of iron (less than 1 to 2 mg/day), as in prolonged nonsupplemented breast-feeding or bottle-feeding of infants or during periods of stress, such as rapid growth, in children and adolescents

▶ iron malabsorption, as in chronic diarrhea, partial or total gastrectomy, and malabsorption syndromes, such as celiac disease and pernicious anemia

▶ blood loss due to drug-induced GI bleeding (from anticoagulants, aspirin, steroids) or heavy menses, hemorrhage from trauma, peptic ulcers, cancer, or varices

▶ pregnancy, which diverts maternal iron to the fetus for erythropoiesis

▶ intravascular hemolysis-induced hemoglobinuria or paroxysmal nocturnal hemoglobinuria

▶ mechanical trauma to RBCs caused by a prosthetic heart valve or vena cava filters.

Pathophysiology

Iron deficiency anemia occurs when the supply of iron is inadequate for optimal formation of RBCs, resulting in smaller (microcytic) cells with less color (hypochromic) on staining. Body stores of iron, including plasma iron, become depleted, and the concentration of serum transferrin, which binds with and transports iron, decreases. Insufficient iron stores lead to a depleted RBC mass with subnormal hemoglobin concentration, and, in turn, subnormal oxygen-carrying capacity of the blood.

Signs and symptoms

Because iron deficiency anemia progresses gradually, many patients exhibit only symptoms of an underlying condition. They tend not to seek medical treatment until anemia is severe.

At advanced stages, signs and symptoms include:

▶ dyspnea on exertion, fatigue, listlessness, pallor, inability to concentrate, irritability, headache, and a susceptibility to infection due to decreased oxygen-carrying capacity of the blood caused by decreased hemoglobin levels

▶ increased cardiac output and tachycardia due to decreased oxygen perfusion

▶ coarsely ridged, spoon-shaped (koilonchyia), brittle, and thin nails due to decreased capillary circulation

▶ sore, red, and burning tongue due to papillae atrophy

▶ sore, dry skin in the corners of the mouth due to epithelial changes.

Complications

Possible complications include:

▶ infection and pneumonia

▶ pica, compulsive eating of nonfood materials, such as starch or dirt

▶ bleeding

▶ overdosage of oral or IM iron supplements.

Diagnosis

Blood studies (serum iron, total iron-binding capacity, ferritin levels) and iron stores in bone marrow may confirm iron deficiency anemia. However, the results of these tests can be misleading because of complicating factors, such as infection, pneumonia, blood transfusion, or iron supplements. Characteristic blood test results include:

▶ low hemoglobin (males, less than 12 g/dl; females, less than 10 g/dl)

▶ low hematocrit (males, less than 47; females, less than 42)

▶ low serum iron with high binding capacity

▶ low serum ferritin

▶ low RBC count, with microcytic and hypochromic cells (in early stages, RBC count may be normal, except in infants and children)

▶ decreased mean corpuscular hemoglobin in severe anemia

▶ depleted or absent iron stores (by specific staining) and hyperplasia of normal precursor cells (by bone marrow studies).

Diagnosis must also include:

▶ exclusion of other causes of anemia, such as thalassemia minor, cancer, and chronic inflammatory, hepatic, or renal disease.

Treatment

The first priority of treatment is to determine the underlying cause of anemia. Only then can iron replacement therapy begin. Possible treatments are:

▶ oral preparation of iron (treatment of choice) or a combination of iron and ascorbic acid (enhances iron absorption)

▶ parenteral iron (for patient noncompliant with oral dose, needing more iron than can be given orally, with malabsorption preventing adequate iron absorption, or for a maximum rate of hemoglobin regeneration).

Because total-dose I.V. infusion of supplemental iron is painless and requires fewer injections, it's usually preferred to IM administration. Considerations include:

▶ total-dose infusion of iron dextran (IN-FeD) in normal saline solution given over 1 to 8 hours (pregnant patients and geriatric patients with severe anemia)

▶ I.V. test dose of 0.5 ml given first (to minimize the risk for an allergic reaction).

Pernicious anemia

Pernicious anemia, the most common type of megaloblastic anemia, is caused by malabsorption of vitamin B_{12}.

 AGE ALERT Onset typically occurs between the ages of 50 and 60 years, and incidence increases with age. It's rare in children.

CULTURAL DIVERSITY Pernicious anemia primarily affects people of northern European ancestry. In the United States, it's most common in New England and the Great Lakes region because of ethnic distribution.

If not treated, pernicious anemia is fatal. Its manifestations subside with treatment, but some neurologic deficits may be permanent.

Pathophysiology

Pernicious anemia is characterized by decreased production of hydrochloric acid in the stomach, and a deficiency of intrinsic factor, which is normally secreted by the parietal cells of the gastric mucosa and is essential for vitamin B_{12} absorption in the ileum. The resulting vitamin B_{12} deficiency inhibits cell growth, particularly of RBCs, leading to production of few, deformed RBCs with poor oxygen-carrying capacity. It also causes neurologic damage by impairing myelin formation.

Causes

Possible causes of pernicious anemia include:
▶ genetic predisposition (suggested by familial incidence)
▶ immunologically related diseases, such as thyroiditis, myxedema, and Graves' disease (significantly higher incidence in these patients)
▶ partial gastrectomy (iatrogenic induction)
▶ older age (progressive loss of vitamin B_{12} absorption).

AGE ALERT The elderly often have a dietary deficiency of B_{12} in addition to or instead of poor absorption.

Signs and symptoms

Characteristically, pernicious anemia has an insidious onset but eventually causes an unmistakable triad of symptoms:
▶ weakness due to tissue hypoxia

▶ sore tongue due to atrophy of the papillae
▶ numbness and tingling in the extremities as a result of interference with impulse transmission from demyelination.

Other common manifestations include:
▶ pale appearance of lips and gums
▶ faintly jaundiced sclera and pale to bright yellow skin due to hemolysis-induced hyperbilirubinemia
▶ high susceptibility to infection, especially of the genitourinary tract.

Pernicious anemia may also have gastrointestinal, neurologic, and cardiovascular effects.

Gastrointestinal symptoms include:
▶ nausea, vomiting, anorexia, weight loss, flatulence, diarrhea, and constipation from disturbed digestion due to gastric mucosal atrophy and decreased hydrochloric acid production
▶ gingival bleeding and tongue inflammation (may hinder eating and intensify anorexia).

Neurologic symptoms include:
▶ neuritis; weakness in extremities
▶ peripheral numbness and paresthesia
▶ disturbed position sense
▶ lack of coordination; ataxia; impaired fine finger movement
▶ positive Babinski and Romberg signs
▶ light-headedness
▶ altered vision (diplopia, blurred vision), taste, and hearing (tinnitus); optic muscle atrophy
▶ loss of bowel and bladder control; and, in males, impotence, due to demyelination (initially affects peripheral nerves but gradually extends to the spinal cord) caused by vitamin B_{12} deficiency
▶ irritability, poor memory, headache, depression, and delirium (some symptoms are temporary, but irreversible central nervous system [CNS] changes may have occurred before treatment).

Cardiovascular symptoms include:
▶ low hemoglobin levels due to widespread destruction of RBCs caused by increasingly fragile cell membranes

▶ palpitations, wide pulse pressure, dyspnea, orthopnea, tachycardia, premature beats, and, eventually, heart failure due to compensatory increased cardiac output.

Complications

Possible complications include:
▶ hypokalemia (first week of treatment)
▶ permanent CNS symptoms (if the patient is not treated within 6 months of appearance of symptoms)
▶ gastric polyps
▶ stomach cancer.

Diagnosis

Laboratory screening must rule out other anemias with similar symptoms but different treatments, such as:
▶ folic acid deficiency anemia
▶ vitamin B_{12} deficiency resulting from malabsorption due to GI disorders, gastric surgery, radiation, or drug therapy.

Decreased hemoglobin levels by 1 to 2 g/dl in elderly men and slightly decreased hematocrit in both men and women reflect decreased bone marrow and hematopoiesis and, in men, decreased androgen levels; they aren't an indicator of pernicious anemia. Diagnosis of pernicious anemia is established by:
▶ positive family history
▶ hemoglobin 4 to 5 g/dl
▶ low RBC count
▶ mean corpuscular volume greater than 120 μl due to increased amounts of hemoglobin in larger-than-normal RBCs
▶ serum vitamin B_{12} less than than 0.1 μg/ml
▶ bone marrow aspiration showing erythroid hyperplasia (crowded red bone marrow), with increased numbers of megaloblasts but few normally developing RBCs
▶ gastric analysis showing absence of free hydrochloric acid after histamine or pentagastrin injection
▶ Schilling test for excretion of radiolabeled vitamin B_{12} (definitive test for pernicious anemia)

▶ serologic findings including intrinsic factor antibodies and antiparietal cell antibodies.

Treatment

Treatment for pernicious anemia is:
▶ early parenteral vitamin B_{12} replacement (can reverse pernicious anemia, minimize complications, and possibly prevent permanent neurologic damage)
▶ concomitant iron and folic acid replacement to prevent iron deficiency anemia (rapid cell regeneration increases the patient's iron and folate requirements)
▶ after initial response, decrease vitamin B_{12} dosage to monthly self-administered maintenance dose (treatment must be given for life)
▶ bed rest for extreme fatigue until hemoglobin rises
▶ blood transfusions for dangerously low hemoglobin
▶ digoxin (Lanoxin), diuretic, low-sodium diet (if patient is in heart failure)
▶ antibiotics to combat infections.

Sideroblastic anemias

Sideroblastic anemias are a group of heterogenous disorders with a common defect: they fail to use iron in hemoglobin synthesis, despite the availability of adequate iron stores. These anemias may be hereditary or acquired. The acquired form can be primary or secondary. Hereditary sideroblastic anemia commonly responds to treatment with pyridoxine (vitamin B_6). The primary acquired (idiopathic) form, known as refractory anemia with ringed sideroblasts, resists treatment and is usually fatal within 10 years of the onset of complications or a concomitant disease. This form is most common in the elderly. It's commonly associated with thrombocytopenia or leukopenia as part of a myelodysplastic syndrome. Correction of the secondary acquired form depends on the cause.

RINGED SIDEROBLAST

Electron microscopy shows large iron deposits in the mitochondria that surround the nucleus, forming the characteristic ringed sideroblast of hemochromatosis.

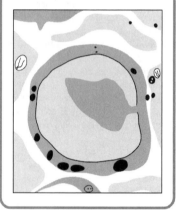

Cause

Hereditary sideroblastic anemia appears to be transmitted by:

▸ X-linked inheritance, occurring mostly in young males (female carriers usually show no signs of this disorder).

The acquired form may be secondary to:

▸ ingestion of or exposure to toxins (such as alcohol and lead) or drugs (such as isoniazid [Laniazid] and chloramphenicol [Chloromycetin])

▸ other diseases, such as rheumatoid arthritis, lupus erythematosus, multiple myeloma, tuberculosis, and severe infections.

Pathophysiology

In sideroblastic anemia, normoblasts fail to use iron to synthesize hemoglobin. As a result, iron is deposited in the mitochondria of normoblasts, which are then termed ringed sideroblasts. Iron toxicity can cause organ damage; untreated, it can damage the nuclei of RBC precursors.

Signs and symptoms

Possible signs and symptoms of sideroblastic anemia include:

▸ anorexia, fatigue, weakness, dizziness, pale skin and mucous membranes, and, occasionally, enlarged lymph nodes due to iron toxicity

▸ dyspnea, exertional angina, slight jaundice, and hepatosplenomegaly due to heart and liver failure caused by excessive iron accumulation in these organs

▸ increased GI absorption of iron, causing signs of hemosiderosis (hereditary sideroblastic anemia)

▸ other symptoms depend on the underlying cause (secondary sideroblastic anemia).

Complications

Possible complications are:
▸ heart, liver, and pancreatic disease
▸ respiratory complications
▸ acute myelogenous leukemia.

Diagnosis

Diagnosis is confirmed by:

▸ ringed sideroblasts on microscopic examination of bone marrow aspirate stained with Prussian blue or alizarin red dye. (See *Ringed sideroblast.*)

▸ hypochromic or normochromic and slightly macrocytic RBCs on microscopic examination; RBC precursors may be megaloblastic, with anisocytosis and poikilocytosis (abnormal variation in shape)

▸ low hemoglobin with high serum iron, transferrin, urobilinogen, and bilirubin levels due to RBC lysis

▸ normal platelet and leukocyte counts (occasional thrombocytopenia or leukopenia).

Treatment

Treatment of sideroblastic anemias depends on the underlying cause and includes:

▸ several weeks of treatment with high doses of pyridoxine (vitamin B_6)for hereditary form

▶ removal of the causative drug or toxin or treatment of the underlying condition (symptoms usually subside in acquired secondary form)

▶ folic acid supplements (may be beneficial when concomitant megaloblastic nuclear changes in RBC precursors are present)

▶ deferoxamine (Desferal) to treat chronic iron overload as needed

▶ blood transfusions (providing hemoglobin) or high doses of androgens (effective palliative measures for some patients with primary acquired form)

▶ phlebotomy to prevent hemochromatosis (the accumulation of iron in body tissues) increases the rate of erythropoiesis and uses up excess iron stores, reducing serum and total-body iron levels.

Thalassemia

Thalassemia, a hereditary group of hemolytic anemias, is characterized by defective synthesis in the polypeptide chains of the protein component of hemoglobin. Consequently, RBC synthesis is also impaired.

CULTURAL DIVERSITY Thalassemia is most common in people of Mediterranean ancestry (especially Italian and Greek), but also occurs in people whose ancestors originated in Africa, southern China, southeast Asia, and India.

In β-thalassemia, the most common form of this disorder, synthesis of the beta polypeptide chain is defective. It occurs in three clinical forms: major, intermedia, and minor. The severity of the resulting anemia depends on whether the patient is homozygous or heterozygous for the thalassemic trait. The prognosis varies:

▶ *thalassemia major:* patients seldom survive to adulthood

▶ *thalassemia intermedia:* children develop normally into adulthood, although puberty is usually delayed

▶ *thalassemia minor:* normal life span.

Causes
Causes of thalassemia are:

▶ homozygous inheritance of the partially dominant autosomal gene (thalassemia major or thalassemia intermedia)

▶ heterozygous inheritance of the same gene (thalassemia minor).

Pathophysiology
Total or partial deficiency of beta polypeptide chain production impairs hemoglobin synthesis and results in continual production of fetal hemoglobin, lasting even past the neonatal period. Normally, immunoglobulin synthesis switches from gamma- to beta-polypeptides at the time of birth. This conversion doesn't happen in thalassemic infants. Their red cells are hypochromic and microcytic.

Signs and symptoms
Possible signs and symptoms of thalassemia major (also known as Cooley's anemia, Mediterranean disease, and erythroblastic anemia) are:

▶ healthy infant at birth, during second 6 months if life develops severe anemia, bone abnormalities, failure to thrive, and life-threatening complications

▶ pallor and yellow skin and sclera in 3- to 6-month-old infants

▶ splenomegaly or hepatomegaly, with abdominal enlargement; frequent infections; bleeding tendencies (especially nose bleeds); anorexia

▶ small body, large head (characteristic features), and possible mental retardation

▶ possible features similar to Down syndrome in infants, due to thickened bone at the base of the nose from bone marrow hyperactivity.

Signs and symptoms of thalassemia intermedia are:

▶ some degree of anemia, jaundice, and splenomegaly

▶ possibly signs of hemosiderosis due to increased intestinal absorption of iron.

Signs of thalassemia minor are:

▶ mild anemia (usually produces no symptoms and is often overlooked; it should be differentiated from iron deficiency anemia).

Complications

Possible complications of thalassemia include:

▶ pathologic fractures due to expansion of the marrow cavities with thinning of the long bones
▶ cardiac arrhythmias
▶ heart failure.

Diagnosis

Diagnosis of thalassemia major includes:
▶ low RBC and hemoglobin, microcytosis, and high reticulocyte count
▶ elevated bilirubin and urinary and fecal urobilinogen levels
▶ low serum folate reflects increased folate use by hypertrophied bone marrow
▶ peripheral blood smear showing target cells, microcytes, pale nucleated RBCs, and marked anisocytosis
▶ thinning and widening of the marrow space on skull and long bone X-rays due to overactive bone marrow
▶ granular appearance of bones of skull and vertebrae, areas of osteoporosis in long bones, deformed (rectangular or biconvex) phalanges
▶ significantly increased fetal hemoglobin and slightly increased hemoglobin A_2 quantitative hemoglobin studies
▶ excluding iron deficiency anemia (also produces hypochromic microcytic RBCs).

Diagnosis of thalassemia intermedia includes:
▶ hypochromic microcytic RBCs (less severe than in thalassemia major).

Diagnosis of thalassemia minor includes:
▶ hypochromic microcytic RBCs
▶ significantly increased hemoglobin A^2 and moderately increased fetal hemoglobin on quantitative hemoglobin studies.

Treatment

Treatment of thalassemia major is essentially supportive and includes:

▶ prompt treatment with appropriate antibiotics for infections
▶ folic acid supplements to help maintain folic acid levels despite increased requirements
▶ transfusions of packed RBCs to increase hemoglobin levels (used judiciously to minimize iron overload)
▶ splenectomy and bone marrow transplantation (effectiveness has not been confirmed)
▶ no treatment for thalassemia intermedia and thalassemia minor
▶ no iron supplements (contraindicated in all forms of thalassemia).

Disseminated intravascular coagulation

Disseminated intravascular coagulation (DIC) occurs as a complication of diseases and conditions that accelerate clotting, causing small blood vessel occlusion, organ necrosis, depletion of circulating clotting factors and platelets, activation of the fibrinolytic system, and consequent severe hemorrhage. Clotting in the microcirculation usually affects the kidneys and extremities but may occur in the brain, lungs, pituitary and adrenal glands, and GI mucosa. DIC, also called consumption coagulopathy or defibrination syndrome, is generally an acute condition but may be chronic in cancer patients. Prognosis depends on early detection and treatment, the severity of the hemorrhage, and treatment of the underlying disease.

Causes

Causes of DIC include:
▶ infection, including gram-negative or gram-positive septicemia and viral, fungal, rickettsial, or protozoal infection
▶ obstetric complications, including abruption placentae, amniotic fluid embolism, retained dead fetus, septic abortion, eclampsia
▶ neoplastic disease, including acute leukemia, metastatic carcinoma, aplastic anemia

▶ disorders that produce necrosis, including extensive burns and trauma, brain tissue destruction, transplant rejection, hepatic necrosis

▶ other conditions, including heatstroke, shock, poisonous snakebite, cirrhosis, fat embolism, incompatible blood transfusion, cardiac arrest, surgery requiring cardiopulmonary bypass, giant hemangioma, severe venous thrombosis, and purpura fulminans.

Pathophysiology

It isn't clear why certain disorders lead to DIC or whether they use a common mechanism. In many patients, the triggering mechanisms may be the entrance of foreign protein into the circulation and vascular endothelial injury.

Regardless of how DIC begins, the typical accelerated clotting results in generalized activation of prothrombin and a consequent excess of thrombin. The thrombin converts fibrinogen to fibrin, producing fibrin clots in the microcirculation. This process uses huge amounts of coagulation factors (especially fibrinogen, prothrombin, platelets, and factors V and VIII), causing hypofibrinogenemia, hypoprothrombinemia, thrombocytopenia, and deficiencies in factors V and VIII. Circulating thrombin also activates the fibrinolytic system, which dissolves fibrin clots into fibrin degradation products. Hemorrhage may be mostly the result of the anticoagulant activity of fibrin degradation products as well as depletion of plasma coagulation factors.

Signs and symptoms

Signs and symptoms of DIC caused by the anticoagulant activity of fibrin degradation products and depletion of plasma coagulation factors include:
▶ abnormal bleeding
▶ cutaneous oozing of serum
▶ petechiae or blood blisters
▶ bleeding from surgical or IV sites
▶ bleeding from the GI tract
▶ epistaxis

▶ hemoptysis.
Other signs and symptoms are:
▶ cyanotic, cold, mottled fingers and toes, due to fibrin clots in the microcirculation resulting in tissue ischemia
▶ severe muscle, back, abdominal, and chest pain from tissue hypoxia
▶ nausea and vomiting (may be a manifestation of GI bleeding)
▶ shock due to hemorrhage
▶ confusion, possibly due to cerebral thrombus and decreased cerebral perfusion
▶ dyspnea due to poor tissue perfusion and oxygenation
▶ oliguria due to decreased renal perfusion.

Complications

Complications of DIC include:
▶ acute tubular necrosis
▶ shock
▶ multiple organ failure.

Diagnosis

Diagnosis of DIC is based on:
▶ decreased platelet count, usually less than 100,00/µl, because platelets are consumed during thrombosis
▶ fibrinogen less than150 mg/dl because fibrinogen is consumed in clot formation (levels may be normal if elevated by hepatitis or pregnancy)
▶ prothrombin time greater than15 seconds
▶ partial thromboplastin time greater than 60 seconds
▶ increased fibrin degradation products, often greater than 45 mcg/ml, due to excess fibrinolysis by plasmin
▶ D-dimer test (presence of an asymmetrical carbon compound fragment formed in the presence of fibrin split products) positive at less than 1:8 dilution
▶ positive fibrin monomers, diminished levels of factors V and VIII, fragmentation of RBCs, and hemoglobin less than 10 g/dl
▶ reduced urine output (less than 30 ml/hour), elevated blood urea nitrogen

(greater than 25 mg/dl), and elevated serum creatinine (greater than 1.3mg/dl).

Treatment
Treatment includes:
▶ prompt recognition and treatment of underlying disorder
▶ blood, fresh frozen plasma, platelet, or packed RBC transfusions to support hemostasis in active bleeding
▶ heparin in early stages to prevent microclotting and as a last resort in hemorrhage (controversial in acute DIC after sepsis). (See *Understanding DIC and its treatment.*)

Erythroblastosis fetalis
Erythroblastosis fetalis, a hemolytic disease of the fetus and newborn, stems from an incompatibility of fetal and maternal blood; that is, mother and fetus have different ABO blood types or the fetus is Rh positive and the mother is Rh negative. The mother's immune system generates antibodies against fetal red cells.

The effects of hemolytic disease are more severe in Rh incompatibility than ABO incompatibility. ABO incompatibility may resolve after birth without life-threatening complications. ABO incompatibility occurs in about 25% of all pregnancies, but only 1 in 10 cases results in hemolytic disease. Rh incompatibility occurs in less than 10% of pregnancies and rarely causes hemolytic disease in the first pregnancy.

In severe, untreated erythroblastosis fetalis, the prognosis is poor, especially if brain and spinal cord become infiltrated with bilirubin (kernicterus). About 70% of these infants die, usually within the first week of life; survivors inevitably have severe neurologic damage, including sensory impairment, mental deficiencies, and cerebral palsy. Most fetuses with hydrops fetalis (the most severe form of this disorder, associated with profound anemia and edema) are stillborn; the few who are delivered alive rarely survive longer than a few hours.

Causes
Erythroblastosis fetalis is caused by:
▶ ABO incompatibility
▶ Rh isoimmunization. (See *What happens in Rh isoimmunization,* page 386.)

Pathophysiology
The pathophysiologies of ABO and Rh incompatibility are different.

ABO incompatibility. Each blood group has specific antigens on RBCs and specific antibodies in the serum. As in transfusion, the maternal immune system forms antibodies against fetal cells when blood groups differ. Most commonly, the mother has blood type O and the fetus has type A or B. Of course, a mother with type A or B will not form antibodies against a type O fetus, who has no fetal blood type antigens. Because the blood of most adults already contains anti-A or anti-B antibodies, ABO incompatibility can cause hemolytic disease even if fetal erythrocytes don't escape into the maternal circulation during pregnancy.

Rh incompatibility. During her first pregnancy, an Rh-negative female becomes sensitized (during delivery or abortion) by exposure to Rh-positive fetal blood antigens inherited from the father. A female may also become sensitized from receiving blood transfusions with alien Rh antigens; from inadequate doses of Rh_o (D) (RhoGAM); or from failure to receive Rh_o (D) after significant fetal-maternal leakage during abruption placentae (premature detachment of the placenta).

A subsequent pregnancy with an Rh-positive fetus provokes maternal production of agglutinating antibodies, which cross the placental barrier, attach to Rh-positive cells in the fetus, and cause hemolysis and anemia. To compensate, the fetal blood forming organs step up the production of RBCs, and erythroblasts (immature RBCs) appear in the fetal circulation. Extensive hemolysis releases more unconjugated bilirubin than the liver can

DISRUPTING DISEASE

UNDERSTANDING DIC AND ITS TREATMENT

Key: ✳ = treatment

WHAT HAPPENS IN RH ISOIMMUNIZATION

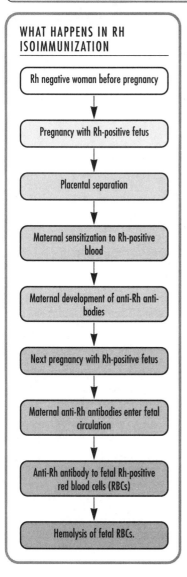

Rh negative woman before pregnancy

↓

Pregnancy with Rh-positive fetus

↓

Placental separation

↓

Maternal sensitization to Rh-positive blood

↓

Maternal development of anti-Rh antibodies

↓

Next pregnancy with Rh-positive fetus

↓

Maternal anti-Rh antibodies enter fetal circulation

↓

Anti-Rh antibody to fetal Rh-positive red blood cells (RBCs)

↓

Hemolysis of fetal RBCs.

conjugate and excrete, causing hyper-bilirubinemia and hemolytic anemia.

Signs and symptoms
Signs and symptoms of erythroblastosis fetalis include:

▶ jaundice due to large amounts of unconjugated bilirubin released by hemolysis
▶ anemia due to hemolysis
▶ hepatosplenomegaly.

Complications
Complications of erythroblastosis fetalis include:
▶ fetal death in utero
▶ severe anemia
▶ heart failure
▶ kernicterus.

Diagnosis
Diagnosis considers both prenatal and neonatal findings. Prenatal findings include:
▶ maternal history (for erythroblastotic stillbirths, abortions, previously affected children, previous anti-Rh titers)
▶ blood typing and screening (should be done frequently to determine changes in the degree of maternal immunization)
▶ paternal blood typing for ABO and Rh
▶ history of blood transfusion
▶ amniotic fluid analysis showing increased bilirubin and anti-Rh titers
▶ radiologic studies showing edema and, in hydrops fetalis, the halo sign (edematous, elevated, subcutaneous fat layers) and the Buddha position (fetus's legs are crossed).

Neonatal findings indicating erythroblastosis fetalis include:
▶ direct Coombs' test of umbilical cord blood to measure RBC (Rh-positive) antibodies in the newborn (positive only when the mother is Rh negative and the fetus is Rh positive)
▶ cord hemoglobin level less than 10 g, indicating severe disease
▶ many nucleated peripheral RBCs.

Treatment
Treatment depends on the degree of maternal sensitization and the effects of hemolytic disease on the fetus or newborn. It may include:

• intrauterine-intraperitoneal transfusion (if amniotic fluid analysis suggests the fetus is severely affected and is not mature enough to deliver)
• planned delivery (usually 2 to 4 weeks before term date, depending on maternal history, serologic test results, and amniocentesis)
• exchange transfusion to remove antibody-coated RBCs and prevent hyperbilirubinemia by replacing the infant's blood with fresh group O, Rh-negative blood
• albumin infusion to bind bilirubin
• phototherapy (exposure to ultraviolet light to reduce bilirubin levels)
• gamma globulin containing anti-Rh antibody (Rh_o [D]) to prevent Rh isoimmunization in Rh-negative females (ineffective if a previous pregnancy, abortion, or transfusion has already sensitized the mother).

Neonatal therapy for hydrops fetalis includes:
• intubation to maintain ventilation
• removal of excess fluid to relieve ascites and respiratory distress
• exchange transfusion
• maintaining body temperature.

Idiopathic thrombocytopenic purpura

Idiopathic thrombocytopenic purpura (ITP) is a deficiency of platelets that occurs when the immune system destroys the body's own platelets. ITP may be acute, as in postviral thrombocytopenia, or chronic, as in essential thrombocytopenia or autoimmune thrombocytopenia.

 AGE ALERT Acute ITP usually affects children between the ages of 2 and 6 years; chronic ITP mainly affects adults younger than age 50, especially women between the ages of 20 and 40.

The prognosis for acute ITP is excellent; nearly four of five patients recover without treatment. The prognosis for chronic ITP is good; remissions lasting weeks or years are common, especially among women.

Causes
Causes of ITP include:
• viral infection
• immunization with a live virus vaccine
• immunologic disorders
• drug reactions.

Pathophysiology
ITP occurs when circulating immunoglobulin G (IgG) molecules react with host platelets, which are then destroyed in the spleen and, to a lesser degree, in the liver. Normally, the life span of platelets in circulation is 7 to 10 days. In ITP, platelets survive 1 to 3 days or less.

Signs and symptoms
Signs and symptoms of ITP are caused by decreased levels of platelets and may include:
• nose bleeds
• oral bleeding
• hemorrhages into the skin, mucous membranes, and other tissues causing red discoloration of skin (purpura)
• small purplish hemorrhagic spots on skin (petechiae)
• excessive menstrual bleeding.

Complications
Possible complications of ITP are:
• hemorrhage
• cerebral hemorrhage
• purpuric lesions of vital organs (such as the brain and kidney).

Diagnosis
Diagnosis of ITP includes:
• platelet count less than 20,000 μl
• prolonged bleeding time
• abnormal size and appearance of platelets
• decreased hemoglobin level (if bleeding occurred)
• bone marrow studies showing abundant megakaryocytes (platelet precursor cells)

and a circulating platelet survival time of only several hours to a few days
▶ humoral tests that measure platelet-associated IgG (may help establish the diagnosis; half the patients have elevated IgG).

Treatment

Treatment for acute ITP includes:
▶ glucocorticoids to prevent further platelet destruction
▶ immunoglobulin to prevent platelet destruction
▶ plasmapheresis
▶ platelet pheresis.

Treatment for chronic ITP includes:
▶ corticosteroids to suppress phagocytic activity and enhance platelet production
▶ splenectomy (when splenomegaly accompanies the initial thrombocytopenia)
▶ blood and blood component transfusions and vitamin K to correct anemia and coagulation defects.

Alternative treatments include:
▶ immunosuppressants to help stop platelet destruction
▶ high-dose I.V. immunoglobulin
▶ immunoabsorption apheresis using staphylococcal protein-A columns.

Polycythemia vera

Polycythemia vera is a chronic disorder characterized by increased RBC mass, erythrocytosis, leukocytosis, thrombocytosis, and increased hemoglobin level, with normal or increased plasma volume. This disease is also known as primary polycythemia, erythremia, polycythemia rubra vera, splenomegalic polycythemia, or Vaquez-Osler disease. It usually occurs between the ages of 40 and 60, most commonly among Jewish males of European ancestry. It seldom affects children and doesn't appear to be familial.

The prognosis depends on age at diagnosis, the type of treatment used, and complications. Mortality is high if polycythemia is untreated, associated with leukemia, or associated with myeloid metaplasia (presence of marrow-like tissue and ectopic hematopoiesis in extramedullary sites, such as liver and spleen, and nucleated erythrocytes in blood).

Causes

The cause of polycythemia vera is unknown, but is probably related to:
▶ multipotential stem cell defect.

Pathophysiology

In polycythemia vera, uncontrolled and rapid cellular reproduction and maturation cause proliferation or hyperplasia of all bone marrow cells (panmyelosis).

Increased RBC mass makes the blood abnormally viscous and inhibits blood flow to microcirculation. Diminished blood flow and thrombocytosis set the stage for intravascular thrombosis.

Signs and symptoms

Possible signs and symptoms of polycythemia vera include:
▶ feeling of fullness in the head or headache due to altered hypervolemia and hyperviscosity
▶ dizziness due to hypervolemia and hyperviscosity
▶ ruddy cyanosis (plethora) of the nose and clubbing of the digits due to thrombosis in smaller vessels
▶ painful pruritus due to abnormally high concentrations of mast cells in the skin and their release of heparin and histamine.

Complications

Possible complications include:
▶ hemorrhage
▶ vascular thromboses
▶ uric acid stones.

Diagnosis

The following test results help diagnose polycythemia vera:
▶ increased RBC mass
▶ normal arterial oxygen saturation in association with splenomegaly
▶ increased uric acid
▶ increased blood histamine
▶ decreased serum iron

▶ decreased or absent urinary erythropoietin

▶ bone marrow biopsy showing excess production of myeloid stem cells.

Treatment

Treatment may include:
▶ phlebotomy to reduce RBC mass
▶ myelosuppressive therapy with radioactive phosphorus to suppress erythropoiesis (may increase the risk for leukemia).

Secondary polycythemia

Secondary polycythemia, also called reactive polycythemia, is excessive production of circulating RBCs due to hypoxia, tumor, or disease. It occurs in approximately 2 of every 100,000 people living at or near sea level; the incidence increases among those living at high altitudes.

Causes

Secondary polycythemia may be caused by:
▶ increased production of erythropoietin.

Pathophysiology

Secondary polycythemia may result from increased production of the hormone erythropoietin — which stimulates bone marrow to produce RBCs — in a compensatory response to several conditions. These include hypoxemia caused by such conditions as chronic obstructive pulmonary disease, hemoglobin abnormalities (such as carboxyhemoglobinemia in heavy smokers), heart failure (causing a decreased ventilation-perfusion ratio), right-to-left shunting of blood in the heart (as in transposition of the great vessels), central or peripheral alveolar hypoventilation (as in barbiturate intoxication), and low oxygen content at high altitudes.

Increased production of erythropoietin may also be an inappropriate (pathologic) response to renal, central nervous system, or endocrine disorders or to certain neoplasms (such as renal tumors, uterine myoma, or cerebellar hemangiomas).

Signs and symptoms

Possible signs and symptoms are:
▶ ruddy cyanotic skin, emphysema, and hypoxemia without hepatomegaly or hypertension (in the hypoxic patient)
▶ clubbing of the fingers (when the underlying cause is cardiovascular).

Diagnosis

Diagnosis of secondary polycythemia is based on the following test results:
▶ high hematocrit and hemoglobin
▶ high mean corpuscular volume and mean corpuscular hemoglobin
▶ high urinary erythropoietin
▶ high blood histamine
▶ normal or low arterial oxygen saturation
▶ bone marrow biopsy showing hyperplasia or erythroid precursors.

Treatment

The goal of treatment is to correct the underlying disease or environmental condition, and may include:
▶ phlebotomy or pheresis to reduce blood volume (to correct hazardous hyperviscosity or if the patient doesn't respond to treatment of the primary disease)
▶ continuous low-flow oxygen therapy to correct severe hypoxia.

Spurious polycythemia

Spurious polycythemia is characterized by an increased hematocrit and a normal or low RBC total mass. It results from diminished plasma volume and subsequent hemoconcentration. It is also known as relative polycythemia, stress erythrocytosis, stress polycythemia, benign polycythemia, Gaisböck's syndrome, or pseudopolycythemia. It usually affects middle-aged people and is more common in men than in women.

Causes

Causes of spurious polycythemia include:
▶ dehydration

- hemoconcentration due to stress
- high-normal RBC mass and low-normal plasma volume
- hypertension
- thromboembolic disease
- elevated serum cholesterol and uric acid
- familial tendency.

Pathophysiology

Conditions that promote severe fluid loss decrease plasma volume and lead to hemoconcentration. Such conditions include persistent vomiting or diarrhea, burns, adrenocortical insufficiency, aggressive diuretic therapy, decreased fluid intake, diabetic acidosis, and renal disease.

Nervous stress causes hemoconcentration by some unknown mechanism. This form of erythrocytosis (chronically elevated hematocrit) is particularly common in the middle-aged man who is a chronic smoker and has a type A personality (tense, hard driving, and anxious).

In many patients, an increased hematocrit merely reflects a normally high RBC mass and low plasma volume. This is particularly common in patients who don't smoke, aren't obese, and have no history of hypertension.

Signs and symptoms

Signs and symptoms of spurious polycythemia may include:
- headaches or dizziness due to altered circulation secondary to hypervolemia and hyperviscosity
- ruddy appearance caused by cyanosis
- slight hypertension from increased blood volume
- tendency to hyperventilate when recumbent
- cardiac or pulmonary disease.

Diagnosis

The following test results help diagnose spurious polycythemia:
- high hemoglobin and hematocrit
- high RBC count
- normal RBC mass

- normal arterial oxygen saturation
- normal bone marrow
- low or normal plasma volume
- possibly hyperlipidemia
- possibly uricosuria.

Treatment

Treatment includes:
- appropriate fluids and electrolytes to correct dehydration
- measures to prevent further fluid loss, such as antidiarrheals if needed, avoiding dietary diuretics (e.g., caffeine), preventing excessive perspiration, remaining hydrated.

Thrombocytopenia

Thrombocytopenia, the most common cause of hemorrhagic disorders, is a deficiency of circulating platelets. It may be congenital or acquired; the acquired form is more common. Because platelets are needed for coagulation, this disease poses a serious threat to hemostasis. The prognosis is excellent in drug-induced thrombocytopenia if the offending drug — usually carbamazepine (Tegretol) or heparin — is withdrawn; in such cases, recovery may be immediate. In other types, the prognosis depends on the patient's response to treatment of the underlying cause.

Causes

Possible causes of thrombocytopenia include:
- decreased or defective platelet production in the bone marrow (as in leukemia, aplastic anemia, or drug toxicity)
- increased platelet destruction outside the marrow due to an underlying disorder (such as cirrhosis of the liver, disseminated intravascular coagulation, or severe infection)
- sequestration (increased amount of blood in a limited vascular area, such as the spleen)
- blood loss.

Pathophysiology

In thrombocytopenia, lack of platelets can cause inadequate hemostasis. Four mechanisms are responsible: decreased platelet production, decreased platelet survival, pooling of blood in the spleen, and intravascular dilution of circulating platelets. Megakaryocytes, giant cells in the bone marrow, produce platelets. Platelet production decreases when the number of megakaryocytes is reduced or when platelet production becomes dysfunctional. (See *What happens in thrombocytopenia,* pages 392 and 393.)

Signs and symptoms

Possible signs and symptoms of thrombocytopenia are:
▶ petechiae or blood blisters caused by bleeding into the skin
▶ bleeding into the mucous membrane
▶ malaise, fatigue, and general weakness
▶ large blood-filled blisters in the mouth (in adults).

Complications

Complications include:
▶ hemorrhage
▶ death.

Diagnosis

The following tests help diagnose thrombocytopenia:
▶ platelet count usually less than 100,000/µl in adults
▶ prolonged bleeding time
▶ platelet antibody studies to help determine why the platelet count is low (also used to select treatment)
▶ platelet survival studies to help differentiate between ineffective platelet production and platelet destruction as causes of thrombocytopenia
▶ bone marrow studies to determine the number, size, and maturity of megakaryocytes in severe disease, helping identify ineffective platelet production as the cause and ruling out malignant disease.

Treatment

Treatment of thrombocytopenia may include:
▶ withdrawing the offending drug or treating the underlying cause
▶ corticosteroids to increase platelet production
▶ lithium carbonate (Eskalith) or folate to stimulate bone marrow production
▶ I.V. gamma globulin (experimental use for severe or refractory thrombocytopenia)
▶ platelet transfusion to stop episodic abnormal bleeding due to low platelet count
▶ splenectomy to correct disease caused by platelet destruction (because the spleen is the primary site of platelet removal and antibody production).

Von Willebrand's disease

Von Willebrand's disease is a hereditary bleeding disorder, occurring more often in females and characterized by prolonged bleeding time, moderate deficiency of clotting factor VIII (antihemophilic factor), and impaired platelet function. This disease commonly causes bleeding from the skin or mucosal surfaces and, in females, excessive uterine bleeding. Bleeding may range from mild and asymptomatic to severe, potentially fatal, hemorrhage. The prognosis is usually good.

Causes

Von Willebrand's disease is caused by:
▶ inherited autosomal dominant trait.
Recently, an acquired form has been identified in patients with cancer and immune disorders.
.

Pathophysiology

A possible mechanism is that mild to moderate deficiency of factor VIII and defective platelet adhesion prolong coagulation time. Specifically, this results from a deficiency of von Willebrand's factor (VWF), which stabilizes the factor VIII molecule and is needed for proper platelet function.

WHAT HAPPENS IN THROMBOCYTOPENIA

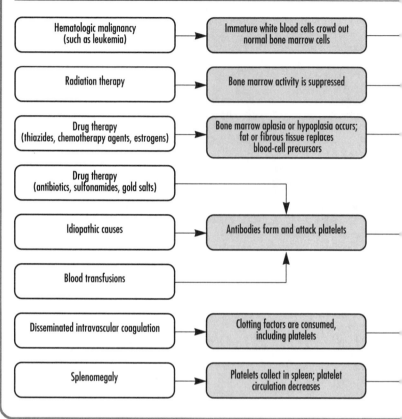

Hematologic malignancy (such as leukemia)	Immature white blood cells crowd out normal bone marrow cells
Radiation therapy	Bone marrow activity is suppressed
Drug therapy (thiazides, chemotherapy agents, estrogens)	Bone marrow aplasia or hypoplasia occurs; fat or fibrous tissue replaces blood-cell precursors
Drug therapy (antibiotics, sulfonamides, gold salts)	
Idiopathic causes	Antibodies form and attack platelets
Blood transfusions	
Disseminated intravascular coagulation	Clotting factors are consumed, including platelets
Splenomegaly	Platelets collect in spleen; platelet circulation decreases

Defective platelet function is characterized in vivo by decreased agglutination and adhesion at the bleeding site and in vitro by reduced platelet retention when blood is filtered through a column of packed glass beads, and diminished ristocetin-induced platelet aggregation.

Signs and symptoms

Prolonged coagulation time may cause:
▶ easy bruising
▶ epistaxis (nose bleed)
▶ bleeding from the gums
▶ petechiae (rarely)
▶ hemorrhage after laceration or surgery (in severe forms)
▶ menorrhagia (in severe forms)
▶ GI bleeding (in severe forms)
▶ excessive postpartum bleeding (uncommon)
▶ massive soft tissue hemorrhage and bleeding into joints (rare).

Complications

A complication of von Willebrand's disease is:
▶ hemorrhage.

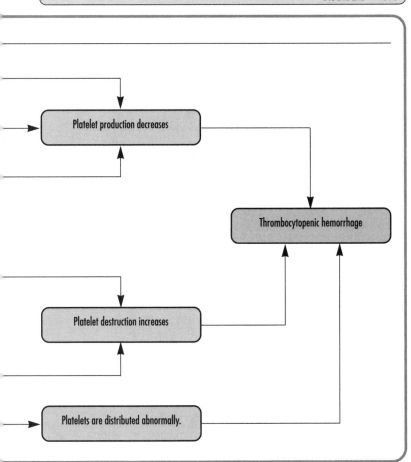

Diagnosis
The following test results help diagnose von Willebrand's disease:

▶ prolonged bleeding time (greater than 6 minutes)

▶ slightly prolonged partial thromboplastin time (greater than 45 seconds)

▶ absent or low factor VIII

▶ absent or low factor VIII-related antigens

▶ low factor VIII activity

▶ ristocetin coagulation factor assay showing defective in vitro platelet aggregation

▶ normal platelet count and clot retraction.

Treatment
Treatment includes:

▶ infusion of cryoprecipitate or blood fractions rich in factor VIII to shorten bleeding time and replace factor VIII

▶ parenteral or intranasal desmopressin (DDAVP) to increase serum levels of VWF.

12

The immune system is responsible for safeguarding the body from disease-causing microorganisms. It is part of a complex system of host defenses.

Host defenses may be innate or acquired. Innate defenses include physical and chemical barriers, the complement complex, and cells such as phagocytes (cells programmed to destroy foreign cells, such as bacteria) and natural killer lymphocytes.

Physical barriers, such as the skin and mucous membranes, prevent invasion by most organisms. Chemical barriers include lysozymes (found in such body secretions as tears, mucus, and saliva) and hydrochloric acid in the stomach. Lysozymes destroy bacteria by removing cell walls. Hydrochloric acid breaks down foods and destroys pathogens carried by food or swallowed mucus.

Organisms that penetrate this first line of defense simultaneously trigger the inflammatory and immune responses, some innate and others acquired.

Acquired immunity comes into play when the body encounters a cell or cell product that it recognizes as foreign, such as a bacterium or a virus. The two types of cell-mediated immunity are humoral (provided by B lymphocytes) and cell-mediated (provided by T lymphocytes). All cells involved in the inflammatory and immune responses arrive from a single type of stem cell in the bone marrow. B cells mature in the marrow, and T cells migrate to the thymus, where they mature.

The inflammatory response is the immediate local response to tissue injury, whether from trauma or infection. It involves the action of polymorphonuclear leukocytes, basophils and mast cells, platelets, and, to some extent, monocytes and macrophages. Each of these cells is described in a later section.

IMMUNE RESPONSE

The immune response primarily involves the interaction of antigens (foreign proteins), B lymphocytes, T lymphocytes, macrophages, cytokines, complement, and polymorphonuclear leukocytes. Some immunoactive cells circulate constantly; others remain in the tissues and organs of the immune system, such as the thymus, lymph nodes, bone marrow, spleen, and tonsils. In the thymus, the T lymphocytes, which are involved in cell-mediated immunity, become able to differentiate self (host) from nonself (foreign) substances (antigens). In contrast, B lymphocytes, which are involved in humoral immunity, mature in the bone marrow. The key mechanism in humoral immunity is the production of immunoglobulin by B cells and the subsequent activation of the complement cascade. The lymph nodes, spleen, liver, and intestinal lymphoid tissue help remove and destroy circulating antigens in the blood and lymph.

Antigens
An antigen is a substance that can induce an immune response. T and B lymphocytes have specific receptors that respond to specific antigen molecular shapes, called

epitopes. In B cells, this receptor is an immunoglobulin, also called an antibody.

Major histocompatibility complex

The T-cell antigen receptor recognizes antigens only in association with specific cell-surface molecules known as the major histocompatibility complex (MHC).

The MHC, also known as the human leukocyte antigen (HLA) locus, is a cluster of genes on human chromosome 6 that has a pivotal role in the immune response. Every person receives one set of MHC genes from each parent, and both sets of genes are expressed on the individual's cells. These genes produce MHC molecules, which participate in:

▶ the recognition of self versus nonself
▶ the interaction of immunologically active cells by coding for cell-surface proteins.

MHC molecules differ among individuals. Slightly different antigen receptors can recognize a large number of distinct antigens, coded by distinct, variable region genes.

Groups or clones of lymphocytes exist that have identical receptors for a specific antigen. The clone of a lymphocyte rapidly proliferates when exposed to the specific antigen. Some lymphocytes further differentiate, while others become memory cells, which allow a more rapid response — the memory or anamnestic response — to subsequent challenge by the antigen.

Haptens

Most antigens are large molecules, such as proteins or polysaccharides. Smaller molecules, such as drugs, that aren't antigenic by themselves are known as haptens. They can bind with larger molecules, or carriers, and become antigenic or immunogenic.

Antigenicity

Many factors influence the intensity of a foreign substance's interaction with the host's immune system (antigenicity):

▶ physical and chemical characteristics of the antigen
▶ its relative foreignness; for example, little or no immune response may follow the transfusion of serum proteins between humans, but a vigorous immune response (serum sickness) commonly follows transfusion of horse serum proteins to a human
▶ the host's genetic makeup, especially the MHC molecules.

Humoral immunity

The humoral immune response is one of two types of immune responses that can occur when foreign substances invade the body. The other is the cell-mediated response. The humoral response is also called an antibody-mediated response.

B lymphocytes

B lymphocytes and their products, immunoglobulins, are the basis of humoral immunity. A soluble antigen binds with the B-cell antigen receptor, initiating the humoral immune response. The activated B cells differentiate into plasma cells, which secrete immunoglobulins, also called antibodies. This response is regulated by T lymphocytes and their products — lymphokines, such as interleukin-2 (IL-2), IL-4, and IL-5, and interferon-8 — which determine which class of immunoglobulins a B cell will manufacture.

Immunoglobulins

The immunoglobulins secreted by plasma cells are four-chain molecules with two heavy and two light chains. Each chain has a variable (V) region and one or more constant (C) regions, which are coded by separate genes. The V regions of both light and heavy chains participate in antigen binding. The C regions of the heavy chain provide a binding site for Fc receptors on cells and govern other mechanisms. (See

STRUCTURE OF THE IMMUNOGLOBULIN MOLECULE

The immunoglobulin molecule consists of four polypeptide chains: two heavy (H) and two light (L) chains held together by disulfide bonds. The H chain has one variable (V) and at least three constant (C) regions. The L chain has one V and one C region. Together, the V regions form a pocket known as the antigen-binding site. This site is located within the antigen-binding fragment (Fab) region of the molecule. Part of the C region of the H chains forms the crystallizable fragment (Fc) region of the molecule. This region mediates effector mechanisms, such as complement activation, and is the portion of the immunoglobulin molecule bound by Fc receptors on phagocytic cells, mast cells, and basophils. Each immunoglobulin molecule also has two antibody-combining sites (except for the immunoglobulin M [IgM] molecule, which has ten, and IgA, which may have two or more).

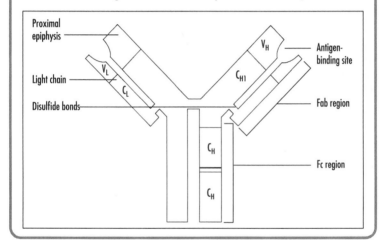

Structure of the immunoglobulin molecule.)

There are five known classes of immunoglobulins: IgG, IgM, IgA, IgE, and IgD. These are distinguished by the constant portions of their heavy chains. However, each class has a kappa or lambda light chain, which gives rise to many subtypes and provides almost limitless combinations of light and heavy chains that give immunoglobulins their specificity. (See *Classification of immunoglobulins.*)

A clone of B cells is specific for only one antigen, and the V regions of its Ig light chains determines that specificity. However, the class of immunoglobulin can change if the association between the cell's V region genes and heavy chain C region genes changes through a process known as isotype switching. For example, a clone of B cells genetically programmed to recognize tetanus toxoid will first make an IgM antibody against tetanus toxoid and later an IgG or other antibody against it.

Cell-mediated immunity

The cell-mediated immune response protects the body against bacterial, viral, and fungal infections and defends against transplanted cells and tumor cells. T lymphocytes and macrophages are the chief participants in the cell-mediated immune response. A macrophage processes the

CLASSIFICATION OF IMMUNOGLOBULINS

The following chart shows the five classifications of immunoglobulins.

CLASSIFICATION	DESCRIPTION
IgA	▶ Secretory immunoglobulin (monomer in serum, dimer in secretory form) ▶ Found in colostrum, saliva, tears, nasal fluids, and respiratory, GI, and genitourinary secretions ▶ Accounts for 20% of total serum immunoglobulins ▶ Important role in preventing antigenic agents from attaching to epithelial surfaces
IgD	▶ Minute amounts found in serum (monomer) ▶ Predominant on surface of B lymphocytes ▶ Primarily an antigen receptor ▶ Possible function in controlling lymphocyte activation or suppression
IgE	▶ Found only in trace amounts ▶ Involved in release of vasoactive amines stored in basophils and tissue mast cell granules that cause the allergic effects
IgG	▶ Smallest immunoglobulin (monomer) ▶ Found in all body fluids ▶ Can cross membranes as a single structural unit ▶ Accounts for 75% of total serum immunoglobulins ▶ Produced mainly in secondary immune response ▶ Classic antibody reactions, including precipitation, agglutination, neutralization, and complement fixation Major antibacterial and antiviral antibody
IgM	▶ Largest immunoglobulin (pentamer) ▶ Usually found only in the vascular system ▶ Cannot readily cross membrane barriers because of its size ▶ Accounts for 5% of total serum immunoglobulins ▶ Dominant activity in primary or initial immune response ▶ Classic antibody reactions, including precipitation, agglutination, neutralization, and complement fixation

antigen and then presents it to T lymphocytes.

Macrophages

Macrophages influence both immune and inflammatory responses. Macrophage precursors circulate in the blood. When they collect in various tissues and organs, they differentiate into different types of macrophages. Unlike B and T lymphocytes, macrophages lack surface receptors for specific antigens. Instead, they have receptors for the C region of the heavy chain (Fc region) of immunoglobulin, for fragments of the third component of complement (C3), and for nonimmunologic substances such as carbohydrate molecules.

One of the most important functions of macrophages is presentation of antigen to T lymphocytes. Macrophages ingest and process the antigen, then deposit it on their own surfaces in association with HLA antigen. T lymphocytes become activated when they recognize the antigen-HLA complex. Macrophages also function in the inflammatory response by producing IL-1, which generates fever, and by synthesizing complement proteins and other mediators that have phagocytic, microbicidal, and tumoricidal effects.

T lymphocytes

Immature T lymphocytes are derived from the bone marrow and migrate to the thymus, where they mature. In maturation, the products of the MCH genes "teach" T cells to distinguish between self and nonself.

Five types of T cells exist with specific functions:
▶ memory cells, sensitized cells that remain dormant until second exposure to antigen, also known as secondary immune response
▶ lymphokine-producing cells, delayed hypersensitivity reactions
▶ cytotoxic T cells, direct destruction of antigen or the cells carrying the antigen
▶ helper T cells, also known as T4 cells, facilitate the humoral and cell-mediated responses
▶ suppressor T cells, also known as T8 cells, inhibit humoral and cell-mediated responses.

T cells acquire specific surface molecules (markers) that identify their potential role when needed in the immune response. These markers and the T cell antigen receptor together promote the particular activation of each type of T cell. T-cell activation requires presentation of antigens in the context of a specific HLA antigen: class II HLA for helper T cells; class I for cytotoxic T cells. T cell activation also requires IL-1, produced by macrophages, and IL-2, produced by T cells.

Natural killer cells. This is a discrete population of large lymphocytes, some of which resemble T cells. Natural killer cells recognize surface changes on body cells infected with a virus. They bind to and, in many cases, kill the infected cells.

Cytokines

Cytokines are low-molecular-weight proteins involved in the communication among macrophages and the lymphocytes. They induce or regulate a variety of immune or inflammatory responses. Cytokines include colony-stimulating factors, interferons, interleukins, tumor necrosis factors, and transforming growth factor.

Complement system

The chief humoral effector of the inflammatory response, the complement system includes more than 20 serum proteins. When activated, these proteins interact in a cascade-like process that has profound biological effects. Complement activation takes place through one of two pathways.

Classic pathway

In the classic pathway, IgM or IgG binds with the antigen to form antigen-antibody complexes that activate the first complement component, C1. This in turn activates C4, C2, and C3.

Alternate pathway

In the alternate pathway, activating surfaces such as bacterial cell membranes directly amplify spontaneous cleavage of C3. Once C3 is activated in either pathway, activation of the terminal components, C5 to C9, follows.

The major biological effects of complement activation include chemotaxis (phagocyte attraction), phagocyte activation, histamine release, viral neutralization, promotion of phagocytosis by opsonization (making the bacteria susceptible to phagocytosis), and lysis of cells and bacteria. Kinins (peptides that cause vasodilation and enhance vascular permeability and smooth muscle contraction) and other mediators of inflammation derived from the kinin and coagulation pathways interact with the complement system.

Polymorphonuclear leukocytes

Other key factors in the inflammatory response are the polymorphonuclear leukocytes: neutrophils, eosinophils, basophils, and mast cells.

Neutrophils

Neutrophils, the most numerous of these leukocytes, derive from bone marrow and increase dramatically in number in response to infection and inflammation. They're the first to respond in acute infection. Neutrophils are highly mobile cells attracted to areas of inflammation and are the main constituent of pus.

Neutrophils have surface receptors for immunoglobulins and complement fragments, and they avidly ingest bacteria or other particles that are coated with target-identifying antibodies (opsons). Toxic oxygen metabolites and enzymes such as lyzozyme promptly kill the ingested organisms. Unfortunately, in addition to killing invading organisms, neutrophils also damage host tissues.

Eosinophils

Eosinophils, also derived from bone marrow, multiply in allergic and parasitic disorders. Although their phagocytic function isn't clearly understood, evidence suggests that they participate in host defense against parasites. Their products may also diminish inflammatory response in allergic disorders.

Basophils and mast cells

Basophils and mast cells also function in immune disorders. Mast cells, unlike basophils, aren't blood cells. Basophils circulate in peripheral blood, whereas mast cells accumulate in connective tissue, particularly in the lungs, intestines, and skin. Both types of cells have surface receptors for IgE. When their receptors are cross-linked by an IgE antigen complex, they release mediators characteristic of the allergic response.

PATHOPHYSIOLOGIC MANIFESTATIONS

The host defense system and the immune response are highly complex processes, subject to malfunction at any point along the sequence of events. This malfunction may involve exaggeration, misdirection, or an absence or depression of activity leading to an immune disorder.

Immune response malfunction

When the immune system responds inappropriately, three basic categories of reactions may occur: hypersensitivity, autoimmune response, and alloimmune response. The type of reaction is determined by the source of the antigen, such as environmental, self, or other person, to which the immune system is responding.

Hypersensitivity

Hypersensitivity is an exaggerated or inappropriate response that occurs on second exposure to an antigen. The result is inflammation and the destruction of healthy tissue. *Allergy* refers to the harmful effects resulting from a hypersensitivity to antigens, also called *allergens*.

Hypersensitivity reactions may be *immediate*, occurring within minutes to hours of re-exposure, or *delayed*, occurring several hours after re-exposure. A delayed hypersensitivity reaction typically is most severe days after the re-exposure.

Generally, hypersensitivity reactions are classified as one of four types: type I (mediated by IgE), type II (tissue-specific), type III (immune-complex-mediated), type IV (cell-mediated). (See *Classification of hypersensitivity reactions,* pages 400 and 401.)

Type I hypersensitivity. Allergens activate T cells, which induce B-cell production of IgE, which binds to the Fc
(Text continues on page 402.)

CLASSIFICATION OF HYPERSENSITIVITY REACTIONS

TYPE I

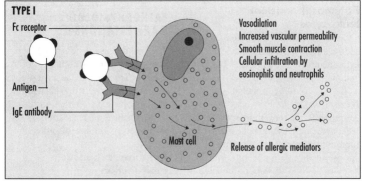

Fc receptor

Antigen

IgE antibody

Mast cell

Vasodilation
Increased vascular permeability
Smooth muscle contraction
Cellular infiltration by
eosinophils and neutrophils

Release of allergic mediators

Reactions
Anaphylactic (immediate, atopic, mediated by immunoglobulin E)

Pathophysiology
Binding of antigens to IgE antibodies on mast cell surfaces releases allergic mediators, causing vasodilation, increased capillary permeability, smooth muscle contraction, and eosinophilia

Clinical examples
Extrinsic asthma, seasonal allergic rhinitis, systemic anaphylaxis, reactions to insect stings, some food and drug reactions, some cases of urticaria, infantile eczema

TYPE II

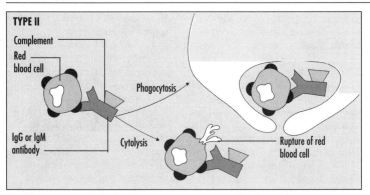

Complement

Red blood cell

Phagocytosis

IgG or IgM antibody

Cytolysis

Rupture of red blood cell

Reactions
Cytotoxic (cytolytic, complement-dependent)

Pathophysiology
Binding of IgG or IgM antibodies to cellular or exogenous antigens activates the complement cascade, resulting in phagocytosis or cytolysis

Clinical examples
Goodpasture's syndrome, pernicious anemia, autoimmune hemolytic anemia, thrombocytopenia, some drug reactions, hyperacute renal allograft rejection, and hemolytic disease of the newborn

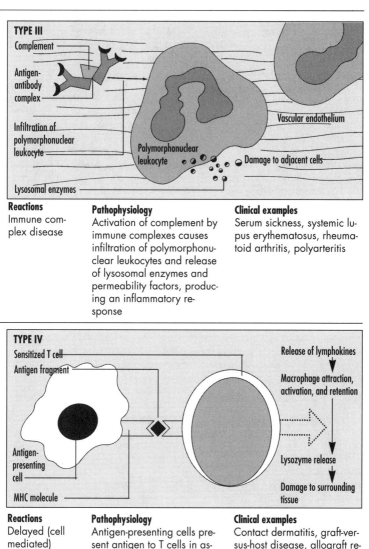

TYPE III

Complement

Antigen-antibody complex

Infiltration of polymorphonuclear leukocyte

Polymorphonuclear leukocyte

Lysosomal enzymes

Vascular endothelium

Damage to adjacent cells

Reactions	Pathophysiology	Clinical examples
Immune complex disease	Activation of complement by immune complexes causes infiltration of polymorphonuclear leukocytes and release of lysosomal enzymes and permeability factors, producing an inflammatory response	Serum sickness, systemic lupus erythematosus, rheumatoid arthritis, polyarteritis

TYPE IV

Sensitized T cell

Antigen fragment

Antigen-presenting cell

MHC molecule

Release of lymphokines

Macrophage attraction, activation, and retention

Lysozyme release

Damage to surrounding tissue

Reactions	Pathophysiology	Clinical examples
Delayed (cell mediated)	Antigen-presenting cells present antigen to T cells in association with major histocompatibility complex (MCH). The sensitized T cells release lymphokines that stimulate macrophages. Lysozymes are released and surrounding tissue is damaged	Contact dermatitis, graft-versus-host disease, allograft rejection, some drug sensitivities, Hashimoto's thyroiditis, sarcoidosis

receptors on the surface of mast cells. Repeated exposure to relatively large doses of the allergen is usually necessary to cause this response. When enough IgE has been produced, the person is *sensitized* to the allergen. At the next exposure to the same antigen, the antigen binds with the surface IgE, cross-links the Fc receptors, and causes mast cells to degranulate and release various mediators. Degranulation also may be triggered by complement-driven anaphylatoxins — C3a and C5a — or by certain drugs such as morphine.

Some of the mediators released are preformed, whereas others are newly synthesized on activation of the mast cells. Preformed mediators include heparin, histamine, proteolytic (protein-splitting) and other enzymes, and chemotactic factors for eosinophils and neutrophils. Newly synthesized mediators include prostaglandins and leukotrienes. Mast cells also produce a variety of cytokines, which initiate smooth muscle contraction, vasodilation, bronchospasm, edema, increased vascular permeability, mucus secretion, and cellular infiltration by eosinophils and neutrophils. These effects result in the some of the classic associated signs and symptoms, such as hypotension, wheezing, swelling, urticaria, and rhinorrhea.

Type II hypersensitivity. Type II hypersensitivity, a tissue-specific reaction, generally involves the destruction of a target cell by an antibody directed against cell-surface antigens. Alternatively, the antibody may be directed against small molecules adsorbed to cells or against cell-surface receptors, rather than against the cell constituents themselves. Tissue damage occurs through several mechanisms:

▸ binding of antigen and antibody activates complement, which ultimately disrupts cellular membranes — complement-mediated lysis

▸ various phagocytic cells with receptors for immunoglobulin (Fc region) and complement fragments envelop and destroy opsonized targets, such as red blood cells, leukocytes, and platelets

▸ cytotoxic T cells and natural killer cells, although not antigen specific, also contribute to tissue damage by releasing toxic substances that destroy the cells

▸ antibody binding causes the target cell to malfunction rather than causing its destruction.

Type III hypersensitivity. Circulating antigen-antibody complexes (immune complexes) accumulate and are deposited in the tissues. The most common tissues involved are the kidneys, joints, skin, and blood vessels. Normally, they clear excess immune complexes from the circulation. However, immune complexes deposited in the tissues activate the complement cascade, causing local inflammation, and trigger platelet release of vasoactive amines that increase vascular permeability, so that more immune complexes accumulate in the vessel walls.

Probably the most harmful effects result from the generation of complement fragments that attract neutrophils. The neutrophils attempt to ingest the immune complexes. They are generally unsuccessful, but in the attempt, the neutrophils release lysosomal enzymes, which exacerbate the tissue damage.

The formation of immune complexes is dynamic and ever-changing. The complexes that form in children may be totally different from those formed in later years. Also, more than one type of immune complex may be present at one time.

Type IV hypersensitivity. These cell-mediated reactions involve the processing of the antigen by the macrophages. Once processed, the antigen is presented to the T cells. Cytotoxic T cells, if activated, attack and destroy the target cells directly. When lymphokine T cells are activated, they release lymphokines, which recruit and activate other lymphocytes, monocytes, macrophages, and polymorphonuclear leukocytes. The coagulation, kinin, and complement cascades also contribute to tissue damage in this type of reaction.

Autoimmune reactions

In autoimmune reactions, the body's normal defenses become self-destructive, recognizing self-antigens as foreign. What causes this misdirected response is not clearly understood. For example, drugs or viruses have been implicated as causing some autoimmune reactions, but in diseases such as rheumatoid arthritis and systemic lupus erythematosus, the mechanism for misdirection is unclear.

Autoimmune reactions are believed to result from a combination of factors, including genetic, hormonal, and environmental influences. Many are characterized by B-cell hyperactivity and by hypergammaglobulinemia. B-cell hyperactivity may be related to T-cell abnormalities. Hormonal and genetic factors strongly influence the onset of some autoimmune disorders.

 AGE ALERT Immune function starts declining at sexual maturity and continues declining with age. During this decline, the immune system begins losing its ability to differentiate between self and nonself, leading to an increase in the incidence of autoimmune disorders.

Alloimmune reactions

Alloimmune reactions are directed at antigens from the tissues of others of the same species. Alloimmune reactions commonly occur in transplant and transfusion reactions, in which the recipient reacts to antigens, primarily HLA, on the donor cells. This immune response is also seen in infants with erythroblastosis fetalis (see Chapter 11). This type of response is commonly associated with a type II hypersensitivity reaction.

Immunodeficiency

An absent or depressed immune response increases susceptibility to infection. Immunodeficiency may be primary, reflecting a defect involving T cells, B cells, or lymphoid tissues, or secondary, resulting from an underlying disease or factor that depresses or blocks the immune response.

The most common forms of immunodeficiency are caused by viral infection or are iatrogenic reactions to therapeutic drugs.

DISORDERS

The environment contains thousands of pathogenic microorganisms. Normally, our host defense system protects us from these harmful invaders. When this network of safeguards breaks down, the result is an altered immune response or immune system failure.

Acquired immunodeficiency syndrome

Human immunodeficiency virus (HIV) infection may cause acquired immunodeficiency syndrome (AIDS). Although it's characterized by gradual destruction of cell-mediated (T cell) immunity, it also affects humoral immunity and even autoimmunity because of the central role of the CD4+ (helper) T lymphocyte in immune reactions. The resulting immunodeficiency makes the patient susceptible to opportunistic infections, cancers, and other abnormalities that define AIDS.

This syndrome was first described by the Centers for Disease Control and Prevention (CDC) in 1981. Because transmission is similar, AIDS shares epidemiologic patterns with hepatitis B and sexually transmitted diseases.

As of June 1997, there were 612,078 reported cases of AIDS and 379,258 deaths from AIDS in adults, adolescents, and children in the United States.

AIDS is more prevalent in large urban areas with a high incidence of I.V. drug use and high-risk sexual practices. HIV is predominantly an infection of young people, with most cases involving persons between the ages of 17 and 55 years. However, it has also been reported in elderly men and women. In the United States, AIDS is the leading cause of death among women aged 25 to 44 years. The incidence is increasing faster among women than men, and heterosexual transmission of HIV is the major mode of transmission. The majority of women with heterosexually transmitted HIV infection report having had sexual contact with an I.V. drug user, often during adolescence. An increase of AIDS in this childbearing age group is expected to cause an increase in the number of children with HIV infection.

Depending on individual variations and the presence of cofactors that influence disease progression, the time from acute HIV infection to the appearance of symptoms (mild to severe) to the diagnosis of AIDS and, eventually, to death varies greatly. The average duration between HIV exposure and diagnosis is 8 to 10 years, but shorter and longer incubation periods have been reported. Current combination drug therapy in conjunction with treatment and prophylaxis of common opportunistic infections can delay the natural progression and prolong survival.

Causes
The HIV-I retrovirus is the primary etiologic agent. Transmission occurs by contact with infected blood or body fluids and is associated with identifiable high-risk behaviors. It's disproportionately represented in:
▶ homosexual and bisexual men
▶ I.V. drug users
▶ neonates of infected women
▶ recipients of contaminated blood or blood products (dramatically decreased since mid-1985)
▶ heterosexual partners of persons in the former groups.

Pathophysiology
The natural history of AIDS begins with infection by the HIV retrovirus, which is detectable only by laboratory tests, and ends with death. Twenty years of data strongly suggests that HIV isn't transmitted by casual household or social contact. The HIV virus may enter the body by any of several routes involving the transmission of blood or body fluids, for example:

▶ direct inoculation during intimate sexual contact, especially associated with the mucosal trauma of receptive rectal intercourse
▶ transfusion of contaminated blood or blood products (a risk diminished by routine testing of all blood products)
▶ sharing contaminated needles
▶ transplacental or postpartum transmission from infected mother to fetus (by cervical or blood contact at delivery and in breast milk).

HIV strikes helper T cells bearing the CD4+ antigen. Normally a receptor for MHC molecules, the antigen serves as a receptor for the retrovirus and allows it to enter the cell. Viral binding also requires the presence of a coreceptor (believed to be the chemokine receptor CCR5) on the cell surface. The virus also may infect CD4+ antigen-bearing cells of the GI tract, uterine cervix, and neuroglia.

Like other retroviruses, HIV copies its genetic material in a reverse manner compared with other viruses and cells. Through the action of reverse transcriptase, HIV produces DNA from its viral RNA. Transcription is often poor, leading to mutations, and some such mutations make HIV resistant to antiviral drugs. The viral DNA enters the nucleus of the cell and is incorporated into the host cell's DNA, where it is transcribed into more viral RNA. If the host cell reproduces, it duplicates the HIV DNA along with its own and passes it on to the daughter cells. Thus, if activated, the host cell carries this information and, if activated, replicates the virus. Viral enzymes, proteases, arrange the structural components and RNA into viral particles that move out to the periphery of the host cell, where the virus buds and emerges from the host cell. Thus, the virus is now free to travel and infect other cells.

HIV replication may lead to cell death or it may become latent. HIV infection leads to profound pathology, either directly through destruction of CD4+ cells, other immune cells, and neuroglial cells, or indirectly through the secondary effects of CD4+ T-cell dysfunction and resulting immunosuppression.

The HIV infectious process takes three forms:

▶ immunodeficiency (opportunistic infections and unusual cancers)

▶ autoimmunity (lymphoid interstitial pneumonitis, arthritis, hypergammaglobulinemia, and production of autoimmune antibodies)

▶ neurologic dysfunction (AIDS dementia complex, HIV encephalopathy, and peripheral neuropathies).

Signs and symptoms

HIV infection manifests in many ways. After a high-risk exposure and inoculation, the infected person usually experiences a mononucleosis-like syndrome, which may be attributed to flu or another virus and then may remain asymptomatic for years. In this latent stage, the only sign of HIV infection is laboratory evidence of seroconversion.

When symptoms appear, they may take many forms, including:

▶ persistent generalized lymphadenopathy secondary to impaired function of CD4+ cells

▶ nonspecific symptoms, including weight loss, fatigue, night sweats, fevers related to altered function of CD4+ cells, immunodeficiency, and infection of other CD4+ antigen-bearing cells

▶ neurologic symptoms resulting from HIV encephalopathy and infection of neuroglial cells

▶ opportunistic infection or cancer related to immunodeficiency.

 AGE ALERT In children, HIV infection has a mean incubation time of 17 months. Signs and symptoms resemble those in adults, except for findings related to sexually transmitted diseases. Children have a high incidence of opportunistic bacterial infections: otitis media, sepsis, chronic salivary gland enlargement, lymphoid interstitial pneumonia, *Mycobacterium avium* complex function, and pneumonias, including *Pneumocystis carinii*.

Complications

Complications of AIDS are:

▶ repeated opportunistic infections. (See *Opportunistic infections in AIDS,* page 406.)

Diagnosis

The CDC has developed an HIV/AIDS classification matrix defining AIDS as an illness characterized by one or more indicator diseases, coexisting with laboratory evidence of HIV infection and other possible causes of immunosuppression. Diagnosis of AIDS includes one or more of the following:

▶ confirmed presence of HIV infection

▶ CD4+ T-cell count of less than 200 cells/μl

▶ the presence of one or more conditions specified by the CDC as Categories A, B, or C. (See *Conditions associated with AIDS,* page 407.)

Treatment

No cure has yet been found for AIDS. Primary therapy includes the use of various combinations of three different types of antiretroviral agents to try to gain the maximum benefit of inhibiting HIV viral replication with the fewest adverse reactions. Current recommendations include the use of two nucleosides plus one protease inhibitor, or two nucleosides and one nonnucleoside to help inhibit the production of resistant, mutant strains. The drugs include:

▶ protease inhibitors to block replication of virus particle formed through the action of viral protease (reducing the number of new virus particles produced)

▶ nucleoside reverse-transcriptase inhibitors to interfere with the copying of viral RNA into DNA by the enzyme reverse transcriptase

▶ nonnucleoside reverse-transcriptase inhibitors to interfere with the action of reverse transcriptase.

Additional treatment may include:

▶ immunomodulatory agents to boost the immune system weakened by AIDS and retroviral therapy

OPPORTUNISTIC INFECTIONS IN AIDS

The following chart shows the complicating infections that may occur in acquired immunodeficiency syndrome (AIDS).

MICROBIOLOGICAL AGENT	ORGANISM	CONDITION
Protozoa	Pneumocystis carinii	Pneumocystis carinii pneumonia
	Cryptosporidium	Cryptosporidiosis
	Toxoplasmosis gondii	Toxoplasmosis
	Histoplasma	Histoplasmosis
Fungi	Candida albicans	Candidiasis
	Cryptococcus neoformans	Cryptococcosis
Viruses	Herpes	Herpes simplex 1 and 2
	Cytomegalovirus	Cytomegalovirus retinitis
Bacteria	Mycobacteria tuberculosis	Tuberculosis
	Mycobacteria avium	Mycobacteria avium complex

Other opportunistic conditions include:
- Kaposi's sarcoma
- Wasting syndrome
- AIDS dementia complex.

- human granulocyte colony-stimulating growth factor to stimulate neutrophil production (retroviral therapy causes anemia, so patients may receive epoetin alfa)
- anti-infective and antineoplastic agents to combat opportunistic infections and associated cancers (some prophylactically to help resist opportunistic infections)
- supportive therapy, including nutritional support, fluid and electrolyte replacement therapy, pain relief, and psychological support.

Anaphylaxis

Anaphylaxis is an acute, potentially life-threatening type I (immediate) hypersensitivity reaction marked by the sudden onset of rapidly progressive urticaria (vascular swelling in skin accompanied by itching) and respiratory distress. With prompt recognition and treatment, the prognosis is good. However, a severe reaction may precipitate vascular collapse, leading to systemic shock and, sometimes, death. The reaction typically occurs within minutes, but can occur up to 1 hour after re-exposure to the antigen.

Causes

The cause of anaphylaxis is usually the ingestion of or other systemic exposure to sensitizing drugs or other substances. Such substances may include:
- serums (usually horse serum)
- vaccines
- allergen extracts
- enzymes such L-asparginase
- hormones
- penicillin or other antibiotics (induce anaphylaxis in 1 to 4 of every 10,000 pa-

CONDITIONS ASSOCIATED WITH AIDS

The Centers for Disease Control and Prevention (CDC) lists associated diseases under three categories. From time to time the CDC, adds to these lists.

CATEGORY A

▶ persistent generalized lymph node enlargement
▶ acute primary HIV infection with accompanying illness
▶ history of acute HIV infection

CATEGORY B

▶ bacillary angiomatosis
▶ oropharyngeal or persistent vulvo-vaginal candidiasis, fever or diarrhea lasting longer than 1 month
▶ idiopathic thrombocytopenic purpura
▶ pelvic inflammatory disease, especially with a tubulo-ovarian abscess
▶ peripheral neuropathy

CATEGORY C

▶ candidiasis of the bronchi, trachea, lungs, or esophagus
▶ invasive cervical cancer
▶ disseminated or extrapulmonary coccoidiomycosis
▶ extrapulmonary cryptococcosis
▶ chronic interstitial cryptosporidiosis

▶ cytomegalovirus (CMV) disease affecting organs other than the liver, spleen, or lymph nodes
▶ CMV retinitis with vision loss
▶ encephalopathy related to HIV
▶ herpes simplex infection with chronic ulcers or herpetic bronchitis, pneumonitis, or exophagitis
▶ disseminated or extrapulmonary histoplasmosis
▶ chronic intestinal isopsoriasis
▶ Kaposi's sarcoma
▶ Burkitt's lymphoma or its equivalent
▶ immunoblastic lymphoma or its equivalent
▶ primary brain lymphoma
▶ disseminated or extrapulmonary *Mycobacterium avium* complex or *M. kansasii*
▶ pulmonary or extrapulmonary *M. tuberculosis*
▶ disseminated or extrapulmonary infection with any other species of *Mycobacterium*
▶ *Pneumocystis carinii* pneumonia
▶ recurrent pneumonia
▶ progressive multifocal leukoencephalopathy
▶ recurrent *Salmonella* septicemia
▶ toxoplasmosis of the brain
▶ wasting syndrome caused by HIV

tients treated; most likely after parenteral administration or prolonged therapy and in patients with an inherited tendency to food or drug allergy, or atopy)
▶ sulfonamides
▶ local anesthetics
▶ salicylates
▶ polysaccharides
▶ diagnostic chemicals, such as sulfobromophthalein, sodium dehydrocholate, and radiographic contrast media
▶ food proteins, such as those in legumes, nuts, berries, seafood, and egg albumin
▶ food additives containing sulfite
▶ insect venom.

Pathophysiology

Anaphylaxis requires previous sensitization or exposure to the specific antigen, resulting in IgE production by plasma cells in the lymph nodes and enhancement by helper T cells. IgE antibodies then bind to membrane receptors on mast cells in connective tissue, and basophils.

On re-exposure, the antigen binds to adjacent IgE antibodies or cross-linked IgE receptors, activating a series of cellular reactions that trigger mast cell degranulation. With degranulation, powerful chemical mediators, such as histamine, eosinophil chemotactic factor of anaphylaxis, and platelet-activating factor, are released from the mast cells. IgG or IgM enters into the reaction and activates the complement cascade, leading to the release of the complement fractions.

At the same time, two other chemical mediators, bradykinin and leukotrienes, induce vascular collapse by stimulating contraction of certain groups of smooth muscles and increasing vascular permeability. These substances, together with the other chemical mediators, cause vasodilation, smooth muscle contraction, enhanced vascular permeability, and increased mucus production. Continued release, along with the spread of these mediators through the body by way of the basophils in the circulation, triggers the systemic responses. Also, increased vascular permeability leads to decreased peripheral resistance and plasma leakage from the circulation to the extravascular tissues. Consequent reduction of blood volume causes hypotension, hypovolemic shock, and cardiac dysfunction. (See *Understanding anaphylaxis,* pages 409 to 411.)

Signs and symptoms

An anaphylactic reaction produces sudden physical distress within seconds or minutes after exposure to an allergen. A delayed or persistent reaction may occur up to 24 hours later. The severity of the reaction is inversely related to the interval between exposure to the allergen and the onset of symptoms. Usually, the first symptoms include:

▶ feeling of impending doom or fright due to activation of IgE and subsequent release of chemical mediators

▶ sweating due to release of histamine and vasodilation

▶ sneezing, shortness of breath, nasal pruritus, urticaria, and angioedema (swelling of nerves and blood vessels) secondary to histamine release and increased capillary permeability.

Systemic manifestations may include:
▶ hypotension, shock, and sometimes cardiac arrhythmias due to increased vascular permeability and subsequent decrease in peripheral resistance and leakage of plasma fluids

▶ nasal mucosal edema, profuse watery rhinorrhea, itching, nasal congestion, and sudden sneezing attacks due to histamine release, vasodilation, and increased capillary permeability

▶ edema of the upper respiratory tract, resulting in hypopharyngeal and laryngeal obstruction, due to increased capillary permeability and mast cell degranulation

▶ hoarseness, stridor, wheezing, and accessory muscle use secondary to bronchiole smooth muscle contraction and increased mucus production

▶ severe stomach cramps, nausea, diarrhea, and urinary urgency and incontinence resulting from smooth muscle contraction of the intestines and bladder.

Complications

Complications of anaphylaxis include:
▶ respiratory obstruction
▶ systemic vascular collapse
▶ death.

Diagnosis

No single diagnostic test can identify anaphylaxis. Anaphylaxis can be diagnosed by the rapid onset of severe respiratory or cardiovascular symptoms after ingestion or injection of a drug, vaccine, diagnostic agent, food, or food additive, or after an insect sting. If these symptoms occur without a known allergic stimulus, other possible causes of shock (such as acute myocardial infarction, status asthmaticus, or heart failure) must be ruled out.

The following test results may provide some clues to the patient's risk for anaphylaxis:
▶ skin tests showing hypersensitivity to a specific allergen
▶ elevated serum IgE levels.

Treatment

Treatment includes:
▶ immediate administration of epinephrine 1:1000 aqueous solution to reverse
(Text continues on page 412.)

UNDERSTANDING ANAPHYLAXIS

An anaphylactic reaction requires previous sensitization or exposure to the specific antigen. What happens in anaphylaxis is described next.

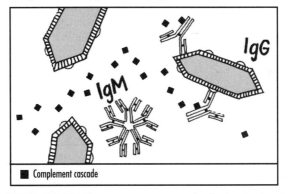

■ Complement cascade

1. RESPONSE TO THE ANTIGEN

Immunoglobulin M (IgM) and IgG recognize the antigen as a foreign substance and attach to it.

Destruction of the antigen by the complement cascade begins but remains unfinished, either because of insufficient amounts of the protein catalyst or because the antigen inhibits certain complement enzymes. The patient has no signs and symptoms at this stage.

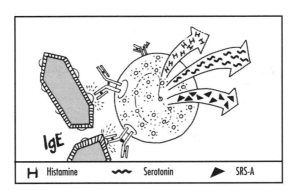

H   Histamine   〜 Serotonin   ▶ SRS-A

2. RELEASED CHEMICAL MEDIATORS

The antigen's continued presence activates IgE on basophils. The activated IgE promotes the release of mediators, including histamine, serotonin, and slow-reacting substance of anaphylaxis (SRS-A). The sudden release of histamine causes vasodilation and increases capillary permeability. The patient begins to have signs and symptoms, including sudden nasal congestion, itchy and watery eyes, flushing, sweating, weakness, and anxiety.

(continued)

UNDERSTANDING ANAPHYLAXIS *(continued)*

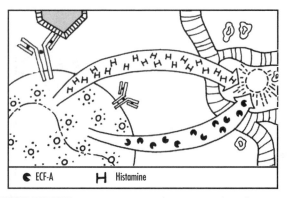

C ECF-A   **H** Histamine

3. INTENSIFIED RESPONSE

The activated IgE also stimulates mast cells in connective tissue along the venule walls to release more histamine and eosinophil chemotactic factor of anaphylaxis (ECF-A). These substances produce disruptive lesions that weaken the venules. Now, red and itchy skin, wheals, and swelling appear, and signs and symptoms worsen.

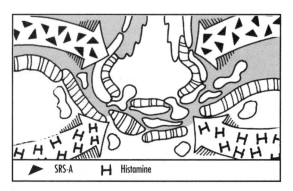

▶ SRS-A   **H** Histamine

4. DISTRESS

In the lungs, histamine causes endothelial cells to burst and endothelial tissue to tear away from surrounding tissue. Fluids leak into the alveoli, and SRS-A prevents the alveoli from expanding, thus reducing pulmonary compliance. Tachypnea, crowing, use of accessory muscles, and cyanosis signal respiratory distress. Resulting neurologic signs and symptoms include changes in level of consciousness, severe anxiety, and possibly seizures.

UNDERSTANDING ANAPHYLAXIS

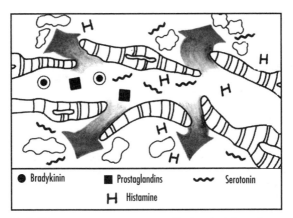

● Bradykinin   ■ Prostaglandins   ∿ Serotonin

H Histamine

5. DETERIORATION

Meanwhile, basophils and mast cells begin to release prostaglandins and bradykinin along with histamine and serotonin. These substances increase vascular permeability, caus-ing fluids to leak from the vessels. Shock, confusion, cool and pale skin, generalized edema, tachycardia, and hypotension signal rapid vascular collapse.

▶ SRS-A   ◗ Hepari

6. FAILED COMPENSATORY MECHANISMS

Damage to the endothelial cells causes basophils and mast cells to release heparin. Additional substances are also released to neutralize the other mediators. Eosinophils release arylsulfatase B to neutralize SRS-A, phospholipase D to neutral-ize heparin, and cyclic adenosine monophosphate and the prostaglandins E_1 and E_2 to increase the metabolic rate. But these events can't reverse anaphylaxis. Hemorrhage, disseminated intravascular coagulation, and cardiopulmonary arrest result.

bronchoconstriction and cause vasoconstriction, IM or subcutaneously if the patient has not lost consciousness and is normotensive, or I.V. if the reaction is severe
(repeating dosage every 5 to 20 minutes
as needed)
▶ tracheostomy or endotracheal intubation
and mechanical ventilation to maintain a
patent airway
▶ oxygen therapy to increase tissue perfusion
▶ longer-acting epinephrine, corticosteroids, and diphenhydramine (Benadryl)
to reduce the allergic response (long-term
management)
▶ albuterol mini-nebulizer treatment
▶ tagamet or another histamine-2 blocker
▶ aminophylline to reverse bronchospasm
▶ volume expanders to maintain and restore circulating plasma volume
▶ I.V. vasopressors such as norepinephrine (Levophed) and dopamine (Intropin)
to stabilize blood pressure
▶ cardiopulmonary resuscitation to treat
cardiac arrest.

Latex allergy

Latex allergy is a hypersensitivity reaction to products that contain natural latex,
a substance found in an increasing number of products at home and at work, that
is derived from the sap of a rubber tree,
not synthetic latex. The hypersensitivity
reactions can range from local dermatitis
to life-threatening anaphylactic reaction.

Until 1980, there were few reports of
latex allergy. However, since the 1987 rec-
ommendation of the Centers for Disease
Control and Prevention for universal precautions and the Occupational Safety and
Health Administration requirement that
employers provide gloves and other protective measures for their employees, the
number of hypersensitivity reactions to
latex has increased. Currently more than
40,000 products on the market are made
with natural rubber latex.

The exact incidence of latex allergy isn't
known. As of 1997, the Food and Drug

Administration had received slightly more
than 1,000 reports of reactions to latex
products, including 16 deaths (attributed
to the use of latex catheters for barium en-
emas).

The term "allergy" has been used loosely to describe any reaction that occurred
after exposure to latex. The National Institute for Occupational Safety and Health
developed a classification system to distinguish a true latex allergy from other
types of reactions, as follows:
▶ irritant contact dermatitis, the most common response to latex exposure and largely associated with glove use, is the result
of a direct skin injury, not an allergic reaction, possibly from the glove itself, the
glove powder, or perspiration during glove
use
▶ chemical sensitivity dermatitis (delayed
hypersensitivity) results from a hypersensitivity to the chemicals added to latex during harvesting, processing, or manufacturing; the rash appears 24 to 48 hours
after latex exposure
▶ latex allergy (immediate hypersensitivity) is the result of a response to latex proteins. A number of exposures are necessary before allergy occurs, and the symptoms appear within minutes of exposure,
often mild at first, possibly progressing to
anaphylaxis.

Causes

Exposure to latex proteins found in natural rubber products produces a true latex
allergy. Those in frequent contact with la-
tex-containing products are at risk for developing a latex allergy. More frequent
exposure leads to a higher risk.

The populations at highest risk are:
▶ medical and dental professionals
▶ workers in latex companies
▶ patients with spina bifida, or other conditions that require multiple surgeries involving latex material.

Other individuals at risk include patients with a history of the following conditions:

▶ asthma or other allergies, especially to bananas, avocados, tropical fruits, or chestnuts

▶ multiple intra-abdominal or genitourinary surgeries

▶ frequent intermittent urinary catheterization.

Pathophysiology

A true latex allergy is an IgE-mediated immediate hypersensitivity reaction. Mast cells release histamine and other secretory products. Vascular permeability increases and vasodilation and bronchoconstriction occur.

Chemical sensitivity dermatitis is a type IV delayed hypersensitivity reaction to the chemicals used in processing rather than the latex itself. In a cell-mediated allergic reaction, sensitized T lymphocytes are triggered, stimulating the proliferation of other lymphocytes and mononuclear cells. This results in tissue inflammation and contact dermatitis.

Signs and symptoms

With a true latex allergy, the patient shows signs and symptoms of anaphylaxis, including:

▶ hypotension due to vasodilation and increased vascular permeability

▶ tachycardia secondary to hypotension

▶ urticaria and pruritus due to histamine release

▶ difficulty breathing, bronchospasm, wheezing, and stridor secondary to bronchoconstriction

▶ angioedema from increased vascular permeability and loss of water to tissues.

Complications

Like anaphylaxis, a true latex allergy may lead to:

▶ respiratory obstruction

▶ systemic vascular collapse

▶ death.

Diagnosis

Diagnosis of latex allergy may include:

▶ radioallergosorbent test showing specific IgE antibodies to latex (safest for use in patients with history of type I hypersensitivity)

▶ skin prick test showing positive response as an indicator of IgE sensitivity

▶ patch test resulting in hives with itching or redness as a positive response.

Treatment

Treatment includes:

▶ prevention of exposure, including use of latex-free products to decrease possible exacerbation of hypersensitivity

▶ drug therapy, such as corticosteroids, antihistamines, and histamine$_2$ receptor blockers before and after possible exposure to latex to depress immune response and block histamine release.

If the patient is experiencing an acute emergency, treatment includes:

▶ immediate administration of epinephrine 1:1000 aqueous solution to reverse bronchoconstriction and cause vasoconstriction, IM or subcutaneously if the patient has not lost consciousness and is normotensive, or I.V. if the reaction is severe (repeating dosage every 5 to 20 minutes as needed)

▶ tracheostomy or endotracheal intubation and mechanical ventilation to maintain a patent airway

▶ oxygen therapy to increase tissue perfusion

▶ volume expanders to maintain and restore circulating plasma volume

▶ I.V. vasopressors such as norepinephrine (Levophed) and dopamine (Intropin) to stabilize blood pressure

▶ cardiopulmonary resuscitation to treat cardiac arrest

▶ longer acting epinephrine, corticosteroids, and diphenhydramine (Benadryl) to reduce the allergic response (long-term management)

▶ drugs to reverse bronchospasm, including aminophylline, histamine-2 blockers, albuterol.

Lupus erythematosus

Lupus erythematosus is a chronic inflammatory disorder of the connective tissues that appears in two forms: discoid lupus erythematosus, which affects only the skin, and systemic lupus erythematosus (SLE), which affects multiple organ systems as well as the skin and can be fatal. SLE is characterized by recurring remissions and exacerbations, which are especially common during the spring and summer.

The annual incidence of SLE averages 27.5 cases per 1 million whites and 75.4 cases per 1 million blacks.

 CULTURAL DIVERSITY SLE strikes women 8 times as often as men, increasing to 15 times as often during childbearing years. It occurs worldwide but is most prevalent among people of Asian, Hispanic, or African origin.

The prognosis improves with early detection and treatment but remains poor for patients who develop cardiovascular, renal, or neurologic complications, or severe bacterial infections.

Causes

The exact cause of SLE remains a mystery, but available evidence points to interrelated immunologic, environmental, hormonal, and genetic factors. These may include:

▶ physical or mental stress
▶ streptococcal or viral infections
▶ exposure to sunlight or ultraviolet light
▶ immunization
▶ pregnancy
▶ abnormal estrogen metabolism
▶ treatment with certain drugs, such as procainamide (Pronestyl), hydralazine (Apresoline), anticonvulsants, and, less frequently, penicillins, sulfa drugs, and oral contraceptives.

Pathophysiology

Autoimmunity is believed to be the prime mechanism involved with SLE. The body produces antibodies against components of its own cells, such as the antinuclear antibody (ANA), and immune complex disease follows. Patients with SLE may produce antibodies against many different tissue components, such as red blood cells, neutrophils, platelets, lymphocytes, or almost any organ or tissue in the body.

Signs and symptoms

The onset of SLE may be acute or insidious and produces no characteristic clinical pattern. (See *Signs of systemic lupus erythematosis.*)

Although SLE may involve any organ system, symptoms all relate to tissue injury and subsequent inflammation and necrosis resulting from the invasion by immune complexes. They commonly include:

▶ fever
▶ weight loss
▶ malaise
▶ fatigue
▶ rashes
▶ polyarthralgia.

Additional signs and symptoms may include:

▶ joint involvement, similar to rheumatoid arthritis (although the arthritis of lupus is usually nonerosive)
▶ skin lesions, most commonly an erythematous rash in areas exposed to light (the classic butterfly rash over the nose and cheeks occurs in less than 50% of the patients) or a scaly, papular rash (mimics psoriasis), especially in sun-exposed areas
▶ vasculitis (especially in the digits), possibly leading to infarctive lesions, necrotic leg ulcers, or digital gangrene
▶ Raynaud's phenomenon (about 20% of patients)
▶ patchy alopecia and painless ulcers of the mucous membranes
▶ pulmonary abnormalities, such as pleurisy, pleural effusions, pneumonitis, pulmonary hypertension, and, rarely, pulmonary hemorrhage
▶ cardiac involvement, such as pericarditis, myocarditis, endocarditis, and early coronary atherosclerosis
▶ microscopic hematuria, pyuria, and urine sediment with cellular casts due to glo-

merulonephritis, possibly progressing to kidney failure (particularly when untreated)

▶ urinary tract infections, possibly due to heightened susceptibility to infection
▶ seizure disorders and mental dysfunction
▶ central nervous system (CNS) involvement, such as emotional instability, psychosis, and organic brain syndrome
▶ headaches, irritability, and depression (common).

Constitutional symptoms of SLE include:
▶ aching, malaise, fatigue
▶ low-grade or spiking fever and chills
▶ anorexia and weight loss
▶ lymph node enlargement (diffuse or local, and nontender)
▶ abdominal pain
▶ nausea, vomiting, diarrhea, constipation
▶ irregular menstrual periods or amenorrhea during the active phase of SLE.

Complications
Possible complications of SLE include:
▶ concomitant infections
▶ urinary tract infections
▶ renal failure
▶ osteonecrosis of hip from long-term steroid use.

Diagnosis
Test results that may indicate SLE include:
▶ complete blood count with differential possibly showing anemia and a decreased white blood cell count
▶ platelet count, which may be decreased
▶ erythrocyte sedimentation rate, which is often elevated
▶ serum electrophoresis, which may show hypergammaglobulinemia.

Other diagnostic tests include:
▶ Antinuclear antibodies (ANA), and lupus erythematosus cell tests showing positive results in active SLE
▶ anti–double-stranded deoxyribonucleic acid antibody (anti-dsDNA); most specific test for SLE, correlates with disease activity, especially renal involvement, and

SIGNS OF SYSTEMIC LUPUS ERYTHEMATOSUS

Diagnosing systemic lupus erythematosus (SLE) is difficult because it often mimics other diseases; symptoms may be vague and vary greatly among patients.

For these reasons, the American Rheumatism Association issued a list of criteria for classifying SLE to be used primarily for consistency in epidemiologic surveys. Usually, four or more of these signs are present at some time during the course of the disease:
▶ malar or discoid rash
▶ photosensitivity
▶ oral or nasopharyngeal ulcerations
▶ nonerosive arthritis (of two or more peripheral joints)
▶ pleuritis or pericarditis
▶ profuse proteinuria (more than 0.5 g/day) or excessive cellular casts in the urine
▶ seizures or psychoses
▶ hemolytic anemia, leukopenia, lymphopenia, or thrombocytopenia
▶ anti–double-stranded deoxyribonucleic acid or anti-Smith antibody test or positive findings of antiphospholipid antibodies (elevated immunoglobulin G [IgG] or IgM anticardiolipin antibodies, positive test result for lupus anticoagulant, or false-positive serologic test results for syphilis)
▶ abnormal antinuclear antibody titer.

helps monitor response to therapy; may be low or absent in remission
▶ additional autoantibody testing — such as Smith antigen (anti-SM, highly specific for SLE), Sjögren's syndrome (anti-SSA and anti-SSB [Sjögren's syndrome antigen B]), and anti-ribonucleoprotein (RNP, antibodies to nuclear antigens) —

to differentiate autoimmune disorders from those with similar signs and symptoms

▶ urine studies possibly showing red blood cells and white blood cells, urine casts and sediment, and significant protein loss (more than 0.5 g/24 hours)

▶ serum complement blood studies showing decreased serum complement (C3 and C4) levels indicating active disease

▶ chest X-ray possibly showing pleurisy or lupus pneumonitis

▶ electrocardiography possibly showing a conduction defect with cardiac involvement or pericarditis

▶ kidney biopsy to determine disease stage and extent of renal involvement

▶ lupus anticoagulant and anticardiolipin tests possibly positive in some patients (usually in patients prone to antiphospholipid syndrome of thrombosis, abortion, and thrombocytopenia).

Treatment

Treatment for SLE may include:

▶ nonsteroidal anti-inflammatory compounds, including aspirin, to control arthritis symptoms

▶ topical corticosteroid creams such as hydrocortisone (Acticort) or triamcinolone (Aristocort) for acute skin lesions

▶ intralesional corticosteroids or antimalarials such as hydroxychloroquine (Plaquenil Sulfate) to treat refractory skin lesions

▶ systemic corticosteroids to reduce systemic symptoms of SLE, for acute generalized exacerbations, or for serious disease related to vital organ systems, such as pleuritis, pericarditis, lupus nephritis, vasculitis, and CNS involvement

▶ high-dose steroids and cytotoxic therapy (such as cyclophosphamide [Cytoxin]) to treat diffuse proliferative glomerulonephritis

▶ dialysis or kidney transplant for renal failure

▶ antihypertensive drugs and dietary changes to minimize effects of renal involvement.

Rheumatoid arthritis

Rheumatoid arthritis (RA) is a chronic, systemic inflammatory disease that primarily attacks peripheral joints and the surrounding muscles, tendons, ligaments, and blood vessels. Partial remissions and unpredictable exacerbations mark the course of this potentially crippling disease. Rheumatoid arthritis strikes women three times more often than men.

Rheumatoid arthritis occurs worldwide, affecting more than 6.5 million people in the United States alone.

 AGE ALERT RA can occur at any age, but 80% of the patients develop rheumatoid arthritis between the ages of 35 and 50 years.

This disease usually requires lifelong treatment and, sometimes, surgery. (See *Drug therapy for rheumatoid arthritis*, pages 418 and 419.) In most patients, it follows an intermittent course and allows normal activity between flares, although 10% of affected people have total disability from severe joint deformity, associated extra-articular symptoms, such as vasculitis, or both. The prognosis worsens with the development of nodules, vasculitis, and high titers of rheumatoid factor (RF).

Causes

The cause of the chronic inflammation characteristic of rheumatoid arthritis isn't known. Possible theories include:

▶ abnormal immune activation (occurring in a genetically susceptible individual) leading to inflammation, complement activation, and cell proliferation within joints and tendon sheaths

▶ possible infection (viral or bacterial), hormone action, or lifestyle factors influencing onset

▶ development of an IgM antibody against the body's own IgG (also called rheumatoid factor, RF); RF aggregates into complexes, generates inflammation, causing eventual cartilage damage and triggering other immune responses.

Pathophysiology

If not arrested, the inflammatory process in the joints occurs in four stages:

◗ synovitis develops from congestion and edema of the synovial membrane and joint capsule. Infiltration by lymphocytes, macrophages, and neutrophils continues the local inflammatory response. These cells, as well as fibroblast-like synovial cells, produce enzymes that help to degrade bone and cartilage

◗ pannus — thickened layers of granulation tissue — covers and invades cartilage and eventually destroys the joint capsule and bone

◗ fibrous ankylosis — fibrous invasion of the pannus and scar formation— occludes the joint space. Bone atrophy and misalignment cause visible deformities and disrupt the articulation of opposing bones, causing muscle atrophy and imbalance and, possibly, partial dislocations (subluxations).

◗ fibrous tissue calcifies, resulting in bony ankylosis and total immobility.

Signs and symptoms

Rheumatoid arthritis usually develops insidiously and initially causes nonspecific symptoms, most likely related to the initial inflammatory reactions before the inflammation of the synovium, including:

◗ fatigue

◗ malaise

◗ anorexia and weight loss

◗ persistent low-grade fever

◗ lymphadenopathy

◗ vague articular symptoms.

As the disease progresses, signs and symptoms include:

◗ specific localized, bilateral, and symmetric articular symptoms, frequently in the fingers at the proximal interphalangeal, metacarpophalangeal, and metatarsophalangeal joints, possibly extending to the wrists, knees, elbows, and ankles from inflammation of the synovium

◗ stiffening of affected joints after inactivity, especially on arising in the morn-

ing, due to progressive synovial inflammation and destruction

◗ spindle-shaped fingers from marked edema and congestion in the joints

◗ joint pain and tenderness, at first only with movement but eventually even at rest, due to prostaglandin release, edema, and synovial inflammation and destruction

◗ feeling of warmth at joint from inflammation

◗ diminished joint function and deformities as synovial destruction continues

◗ flexion deformities or hyperextension of metacarpophalangeal joints, subluxation of the wrist, and stretching of tendons pulling the fingers to the ulnar side (ulnar drift), or characteristic "swan's neck" appearance or "boutonnière" deformity from joint swelling and loss of joint space

◗ carpal tunnel syndrome from synovial pressure on the median nerve causing paresthesia in the fingers.

Extra-articular findings may include:

◗ gradual appearance of rheumatoid nodules — subcutaneous, round or oval, nontender masses (20% of RF-positive patients), usually on elbows, hands, or Achilles tendon from destruction of the synovium

◗ vasculitis possibly leading to skin lesions, leg ulcers, and multiple systemic complications from infiltration of immune complexes and subsequent tissue damage and necrosis in the vasculature

◗ pericarditis, pulmonary nodules or fibrosis, pleuritis, or inflammation of the sclera and overlying tissues of the eye from immune complex invasion and subsequent tissue damage and necrosis

◗ peripheral neuropathy with numbness or tingling in the feet or weakness and loss of sensation in the fingers from infiltration of the nerve fibers

◗ stiff, weak, or painful muscles secondary to limited mobility and decreased use.

Complications

Complications of rheumatoid arthritis include:

◗ fibrosis and ankylosis

DRUG THERAPY FOR RHEUMATOID ARTHRITIS

The following flow chart identifies the major pathophysiologic events in rheumatoid arthritis and shows where in this chain of events the major drug therapies act to control the disease.

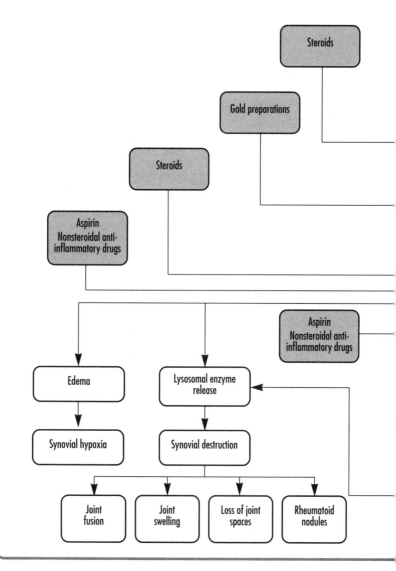

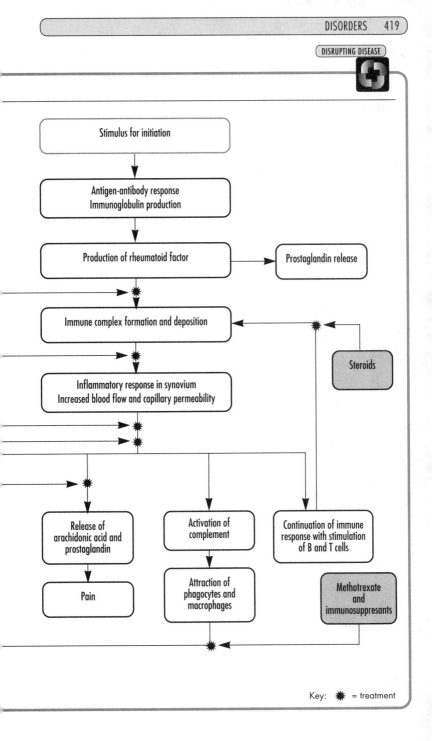

- soft tissue contractures
- pain
- joint deformities
- Sjögren's syndrome
- destruction of second cervical vertebra
- spinal cord compression
- temporomandibular joint disease
- infection
- osteoporosis
- myositis (inflammation of voluntary muscles)
- cardiopulmonary lesions
- lymphadenopathy
- peripheral neuritis.

Diagnosis

Test results indicating rheumatoid arthritis include:

- X-rays showing bone demineralization and soft-tissue swelling (early stages), cartilage loss and narrowed joint spaces, and, finally, cartilage and bone destruction and erosion, subluxations, and deformities (later stages)
- RF titer positive in 75% to 80% of patients (titer of 1:160 or higher)
- synovial fluid analysis showing increased volume and turbidity but decreased viscosity and elevated white blood cell counts (often greater than 10,000/µl)
- serum protein electrophoresis possibly showing elevated serum globulin levels
- erythrocyte sedimentation rate and C-reactive protein levels showing elevations in 85% to 90% of patients (may be useful to monitor response to therapy because elevation frequently parallels disease activity)
- complete blood count usually showing moderate anemia (hemoglobin less than 12 g/dl, hematocrit less than 38%), slight leukocytosis (11,000 to 20,000/µl), and thrombocytosis (alkaline phosphatase greater than 2.0 U/dl).

Treatment

Treatment for rheumatoid arthritis involves pharmacologic therapy and supportive measures, including:

- salicylates, particularly aspirin (mainstay of therapy) to decrease inflammation and relieve joint pain
- nonsteroidal anti-inflammatory agents such as indomethacin (Indocin), fenoprofen (Nalfon), and ibuprofen (Motrin) to relieve inflammation and pain
- antimalarials such as hydroxychloroquine (Plaquenil Sulfate), sulfasalazine (Azulfidine), gold salts, and penicillamine (Cupramine) to reduce acute and chronic inflammation
- corticosteroids (prednisone) in low doses for anti-inflammatory effects, in higher doses for immunosuppressive effect on T cells
- methotrexate (Folex), cyclosporine (Neoral), and azathioprine (Imuran) in early disease for immunosuppression by suppressing T and B lymphocyte proliferation causing destruction of the synovium
- supportive measures, including rest, splinting to rest inflamed joints, range-of-motion exercises, physical therapy, heat applications for chronic disease and ice application for acute episodes
- synovectomy (removal of destructive, proliferating synovium, usually in the wrists, knees, and fingers) to possibly halt or delay the course of the disease
- osteotomy (cutting of bone or excision of a wedge of bone) to realign joint surfaces and redistribute stress
- tendon transfers to prevent deformities or relieve contractures
- joint reconstruction or total joint arthroplasty, including metatarsal head and distal ulnar resectional arthroplasty, insertion of a Silastic prosthesis between metacarpophalangeal and proximal interphalangeal joints (severe disease)
- arthrodesis (joint fusion) for stability and relief of pain (sacrifices joint mobility).

13

The endocrine system consists of glands, specialized cell clusters, hormones, and target tissues. The glands and cell clusters secrete hormones in response to stimulation from the nervous system and other sites. Together with the nervous system, the endocrine system regulates and integrates the body's metabolic activities and maintains internal homeostasis. Each target tissue has receptors for specific hormones. Hormones connect with the receptors, and the resulting hormone-receptor complex triggers the response of the target cell.

HORMONAL REGULATION

The hypothalamus, the main integrative center for the endocrine and nervous systems, helps control some endocrine glands by neural and hormonal pathways. Neural pathways connect the hypothalamus to the posterior pituitary gland, or neurohypophysis. Neural stimulation of the posterior pituitary causes the secretion of two effector hormones: antidiuretic hormone (ADH, also known as vasopressin) and oxytocin.

The hypothalamus also exerts hormonal control at the anterior pituitary gland, or adenohypophysis, by releasing and inhibiting hormones and factors, which arrive by a portal system. Hypothalamic hormones stimulate the pituitary gland to synthesize and release trophic hormones, such as corticotropin (ACTH, also called adrenocorticostimulating hormone), thyroid-stimulating hormone (TSH), and gonadotropins, such as luteinizing hormone

(LH) and follicle-stimulating hormone (FSH). Secretion of trophic hormones stimulates the adrenal cortex, thyroid gland, and gonads. Hypothalamic hormones also stimulate the pituitary gland to release or inhibit the release of effector hormones, such as growth hormone (GH) and prolactin.

In a patient with a possible endocrine disorder, this complex hormonal sequence requires careful assessment to identify the dysfunction, which may result from defects in the gland; defects of releasing, trophic, or effector hormones; or defects of the target tissue. Hyperthyroidism, for example, may result from excessive thyrotropin-releasing hormone (TSH), or thyroid hormones, or excessive response of the thyroid gland.

Besides hormonal and neural controls, a feedback system regulates the endocrine system. (See *Feedback mechanism of the endocrine system*, page 422.) The feedback mechanism may be simple or complex. Simple feedback occurs when the level of one substance regulates secretion of a hormone. For example, a low serum calcium level stimulates the parathyroid glands to secrete parathyroid hormone (PTH), and a high serum calcium level inhibits PTH secretion.

One example of complex feedback occurs through the hypothalamic-pituitary target organ axis. Secretion of the hypothalamic corticotropin-releasing hormone releases pituitary ACTH, which in turn stimulates adrenal cortisol secretion. Subsequently, an increase in serum cortisol levels inhibits ACTH by decreasing cor-

FEEDBACK MECHANISM OF THE ENDOCRINE SYSTEM

The hypothalamus receives regulatory information (feedback) from its own circulating hormones (simple loop) and also from target glands (complex loop).

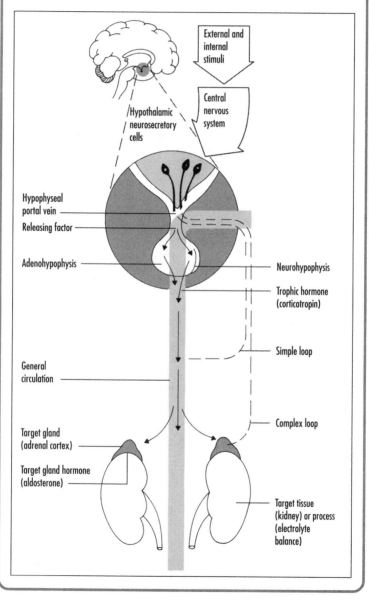

ticotropin-releasing hormone secretion or ACTH directly. Corticosteroid therapy disrupts the hypothalamic-pituitary-adrenal axis by suppressing the hypothalamic-pituitary secretion mechanism. Because abrupt withdrawal of steroids doesn't allow time for recovery of the hypothalamic-pituitary-adrenal axis to stimulate cortisol secretion, it can induce life-threatening adrenal crisis (hypocortisolism).

Rhythms

The endocrine system is also controlled by rhythms, many of which last 24 hours (circadian). Circadian rhythm control of ACTH and cortisol increases levels of these hormones in the early morning hours and decreases them in the late afternoon. The menstrual cycle is an example of an infradian rhythm — in this case, 28 days.

Hormonal effects

The posterior pituitary gland secretes oxytocin and ADH. Oxytocin stimulates contraction of the uterus and is responsible for the milk-letdown reflex in lactating women. ADH controls the concentration of body fluids by altering the permeability of the distal and collecting tubules of the kidneys to conserve water. ADH secretion depends on plasma osmolality, the characteristic of a solution determined by the ionic concentration of the dissolved substance and the solution, which is monitored by hypothalamic neurons. Hypovolemia and hypotension are the most powerful stimulators of ADH release. Other stimulators include trauma, nausea, morphine, tranquilizers, certain anesthetics, and positive-pressure breathing.

In addition to the trophic hormones, the anterior pituitary secretes prolactin, which stimulates milk secretion, and GH. GH affects most body tissues. It triggers growth by stimulating protein synthesis and fat mobilization, and by decreasing carbohydrate use by muscle and fat tissue. The thyroid gland synthesizes and secretes the iodinated hormones, thyroxine and triiodothyronine. Thyroid hormones are necessary for normal growth and development, and act on many tissues to increase metabolic activity and protein synthesis.

The parathyroid glands secrete PTH, which regulates calcium and phosphate metabolism. PTH elevates serum calcium levels by stimulating resorption of calcium and excretion of phosphate and — by stimulating the conversion of vitamin D to its most active form — enhances absorption of calcium from the GI tract. Calcitonin, another hormone secreted by the thyroid gland, affects calcium metabolism, although its precise role in humans is unknown.

The pancreas produces glucagon from the alpha cells and insulin from the beta cells. Glucagon, the hormone of the fasting state, releases stored glucose from the liver to increase blood glucose levels. Insulin, the hormone of the postprandial state, facilitates glucose transport into the cells, promotes glucose storage, stimulates protein synthesis, and enhances free fatty acid uptake and storage.

The adrenal cortex secretes mineralocorticoids, glucocorticoids, and sex steroid hormones (androgens). Aldosterone, a mineralocorticoid, regulates the reabsorption of sodium and the excretion of potassium by the kidneys. Although affected by ACTH, aldosterone is mainly regulated by the renin-angiotension system. Together, aldosterone, angiotensin II, and renin may be implicated in the pathogenesis of hypertension.

Cortisol, a glucocorticoid, stimulates gluconeogenesis, increases protein breakdown and free fatty acid mobilization, suppresses the immune response, and facilitates an appropriate response to stress.

The adrenal medulla is an aggregate of nervous tissue that produces the catecholamines, epinephrine and norepinephrine, which cause vasoconstriction. In addition, epinephrine stimulates the fight-or-flight response — dilation of bronchioles and increased blood pressure, blood glucose level, and heart rate. The adrenal cortex as well as the gonads secretes androgens, which are steroid sex hormones. In males and premenopausal females, the contribution of adrenal androgens is very small, but in postmenopausal females, the adrenals are the major source of sex hormones.

The testes synthesize and secrete testosterone in response to gonadotropic hormones, especially LH, from the anterior pituitary gland; spermatogenesis occurs in response to FSH. The ovaries produce sex steroid hormones (primarily estrogen and progesterone) in response to anterior pituitary trophic hormones.

PATHOPHYSIOLOGIC MANIFESTATIONS

Alterations in hormone levels, either significantly high or low, may result from various causes. Feedback systems may fail to function properly or may respond to the wrong signals. Dysfunction of an endocrine gland may manifest as either failure to produce adequate amounts of active hormone or excessive synthesis or release. Once the hormones are released, they may be degraded at an altered rate or inactivated by antibodies before reaching the target cell. Abnormal target cell responses include receptor-associated alterations and intracellular alterations.

Receptor-associated alterations

These alterations have been associated with water-soluble hormones (peptides) and involve:
▶ decreased number of receptors, resulting in diminished or defective hormone-receptor binding
▶ impaired receptor function, resulting in insensitivity to the hormone
▶ presence of antibodies against specific receptors, either reducing available binding sites or mimicking hormone action and suppressing or exaggerating target cell response
▶ unusual expression of receptor function.

Intracellular alterations

These involve the inadequate synthesis of the second messenger needed to convert the hormonal signal into intracellular events. The two different mechanisms that may be involved are:
▶ faulty response of target cells for water-soluble hormones to hormone-receptor binding and failure to generate the required second messenger
▶ abnormal response of the target cell to the second messenger and failure to express the usual hormonal effect.

Pathophysiologic aberrations affecting target cells for lipid-soluble (steroid hormones) hormones occur less frequently or may be recognized less frequently.

DISORDERS

Common dysfunctions of the endocrine system are classified as hypofunction and hyperfunction, inflammation, and tumor.

Adrenal hypofunction

Adrenal hypofunction is classified as primary or secondary. Primary adrenal hypofunction or insufficiency (Addison's disease) originates within the adrenal gland and is characterized by the decreased secretion of mineralocorticoids, glucocorticoids, and androgens. Secondary adrenal hypofunction is due to impaired pituitary secretion of corticotropin (ACTH) and is characterized by decreased glucocorticoid secretion. The secretion of aldosterone, the major mineralocorticoid, is often unaffected.

Addison's disease is relatively uncommon and can occur at any age and in both sexes. Secondary adrenal hypofunction occurs when a patient abruptly stops long-term exogenous steroid therapy or when the pituitary is injured by a tumor or by infiltrative or autoimmune processes — these occur when circulating antibodies react specifically against adrenal tissue, causing inflammation and infiltration of the cells by lymphocytes. With early diagnosis and adequate replacement therapy, the prognosis for adrenal hypofunction is good.

Adrenal crisis (Addisonian crisis), a critical deficiency of mineralocorticoids and glucocorticoids, generally follows acute stress, sepsis, trauma, surgery, or the omission of steroid therapy in patients who have chronic adrenal insufficiency. Adrenal crisis is a medical emergency that needs immediate, vigorous treatment.

CULTURAL DIVERSITY Autoimmune Addison's disease is most common in white females, and a genetic predisposition is likely. It's more common in patients with a familial predisposition to autoimmune endocrine diseases. Most persons with Addison's disease are diagnosed in their third to fifth decades.

Causes

Primary and secondary adrenal hypofunction and adrenal crisis have different causes. The most common cause of primary hypofunction is:

◗ Addison's disease (destruction of more than 90% of both adrenal glands, usually due to an autoimmune process in which circulating antibodies react specifically against the adrenal tissue).

Other causes include:

◗ tuberculosis (once the chief cause, now responsible for less than 20% of adult cases)
◗ bilateral adrenalectomy
◗ hemorrhage into the adrenal gland
◗ neoplasms
◗ infections (histoplasmosis, cytomegalovirus [CMV])
◗ family history of autoimmune disease (may predispose the patient to Addison's disease and other endocrinopathies).

Causes of secondary hypofunction (glucocorticoid deficiency) include:

◗ hypopituitarism (causing decreased ACTH secretion)
◗ abrupt withdrawal of long-term corticosteroid therapy (long-term exogenous corticosteroid stimulation suppresses pituitary ACTH secretion, resulting in adrenal gland atrophy)
◗ removal of an ACTH-secreting tumor.

Adrenal crisis is usually caused by:
◗ exhausted body stores of glucocorticoids in a person with adrenal hypofunction after trauma, surgery, or other physiologic stress.

Pathophysiology

Addison's disease is a chronic condition that results from the partial or complete destruction of the adrenal cortex. It manifests as a clinical syndrome in which the symptoms are associated with deficient production of the adrenocortical hormones, cortisol, aldosterone, and androgens. High levels of ACTH and corticotropin-releasing hormone accompany the low glucocorticoid levels.

ACTH acts primarily to regulate the adrenal release of glucocorticoids (primarily cortisol); mineralocorticoids, including aldosterone; and sex steroids that supplement those produced by the gonads. ACTH secretion is controlled by corticotropin-releasing hormone from the hypothalamus and by negative feedback control by the glucocorticoids.

Addison's disease involves all zones of the cortex, causing deficiencies of the adrenocortical secretions, glucocorticoids, androgens, and mineralocorticoids.

Manifestations of adrenocortical hormone deficiency become apparent when 90% of the functional cells in both glands are lost. In most cases, cellular atrophy is limited to the cortex, although medullary involvement may occur, resulting in catecholamine deficiency. Cortisol deficiency causes decreased liver gluconeogenesis (the formation of glucose from molucules that are not carbohydrates). The resulting low blood glucose levels can become dangerously low in patients who take insulin on a routine basis.

Aldosterone deficiency causes increased renal sodium loss and enhances potassium reabsorption. Sodium excretion causes a reduction in water volume that leads to hypotension. Patients with Addison's disease may have normal blood pressure when supine, but show marked hypotension and tachycardia after standing for several minutes. Low plasma volume and arteriolar pressure stimulate renin release and a resulting increased production of angiotensin II.

Androgen deficiency may decrease hair growth in axillary and pubic areas as well as on the extremities of women. The metabolic effects of testicular androgens make such hair growth less noticeable in men.

Addison's disease is a decrease in the biosynthesis, storage, or release of adrenocortical hormones. In about 80% of the patients, an autoimmune process causes partial or complete destruction of both adrenal glands. Autoimmune antibodies can block the ACTH receptor or bind with ACTH, preventing it from stimulating adrenal cells. Infection is the second most common cause of Addison's disease, specifically tuberculosis, which causes about 20% of the cases. Other diseases that can cause Addison's disease include acquired immunodeficiency syndrome, systemic fungal infections, CMV, adrenal tumor, and metastatic cancers. Infection can impair cellular function and affect ACTH at any stage of regulation.

Signs and symptoms

Clinical features vary with the type of adrenal hypofunction. Signs and symptoms of primary hypofunction include:
▶ weakness
▶ fatigue
▶ weight loss
▶ nausea, vomiting, and anorexia
▶ conspicuous bronze color of the skin, especially in the creases of the hands and over the metacarpophalangeal joints (hand/finger), elbows, and knees
▶ darkening of scars, areas of vitiligo (absence of pigmentation), and increased pigmentation of the mucous membranes, especially the buccal mucosa, due to decreased secretion of cortisol, causing simultaneous secretion of excessive

amounts of ACTH and melanocyte-stimulating hormone by the pituitary gland
▶ associated cardiovascular abnormalities, including orthostatic hypotension, decreased cardiac size and output, and weak, irregular pulse
▶ decreased tolerance for even minor stress
▶ fasting hypoglycemia due to decreased gluconeogenesis
▶ craving for salty food due to decreased mineralocorticoid secretion (which normally causes salt retention).

Signs and symptoms of secondary hypofunction are:
▶ similar to primary hypofunction, but without hyperpigmentation due to low ACTH and melanocyte-stimulating hormone levels
▶ possibly no hypotension and electrolyte abnormalities due to fairly normal aldosterone secretion
▶ usually normal androgen secretion.

Signs and symptoms of Addisonian crisis may include:
▶ profound weakness and fatigue
▶ nausea, vomiting, and dehydration
▶ hypotension
▶ high fever followed by hypothermia (occasionally).

Complications

Possible complications of adrenal hypofunction include:
▶ hyperpyrexia
▶ psychotic reactions
▶ deficient or excessive steroid treatment
▶ ultimate vascular collapse, renal shutdown, coma, and death (if untreated).

Diagnosis

Diagnosis of adrenal hypofunction is based on:
▶ plasma cortisol levels confirming adrenal insufficiency (ACTH stimulation test to differentiate between primary and secondary adrenal hypofunction)
▶ metyrapone test for suspicion of secondary adrenal hypofunction (oral or I.V. metyrapone blocks cortisol production and should stimulate the release of ACTH from

the hypothalamic-pituitary system; in Addison's disease, the hypothalamic-pituitary system responds normally and plasma ACTH levels are high, but because the adrenal glands are destroyed, plasma concentrations of the cortisol precursor 11-deoxycortisol increase, as do urinary 17-hydroxycorticosteroids)

▶ rapid ACTH stimulation test by I.V. or I.M. administration of cosyntropin (Cortrosyn) after baseline sampling for cortisol and ACTH (samples drawn for cortisol 30 and 60 minutes after injection), to differentiate between primary and secondary adrenal hypofunction.

In a patient with typical Addisonian symptoms, the following laboratory findings strongly suggest acute adrenal insufficiency:

▶ decreased plasma cortisol level (less than10 mcg/dl in the morning; less in the evening)

▶ decreased serum sodium and fasting blood glucose levels

▶ increased serum potassium and blood urea nitrogen levels

▶ decreased hematocrit; increased lymphocyte and eosinophil counts

▶ X-rays showing adrenal calcification if the cause is infectious.

Treatment

Treatment for adrenal hypofunction may include:

▶ lifelong corticosteroid replacement, usually with cortisone or hydrocortisone, both of which have a mineralocorticoid effect (primary or secondary adrenal hypofunction)

▶ oral fludrocortisone (Florinef), a synthetic mineralocorticoid, to prevent dangerous dehydration, hypotension, hyponatremia, and hyperkalemia (Addison's disease)

▶ I.V. bolus of hydrocortisone, 100 mg every 6 hours for 24 hours; then, 50 to 100 mg I.M. or diluted with dextrose in saline solution and given I.V. until the patient's condition stabilizes; up to 300 mg/day of hydrocortisone and 3 to 5 L of I.V. saline

and glucose solutions may be needed (adrenal crisis).

With proper treatment, adrenal crisis usually subsides quickly; blood pressure stabilizes, and water and sodium levels return to normal. After the crisis, maintenance doses of hydrocortisone preserve physiologic stability.

Congenital adrenal hyperplasia

Congenital adrenal hyperplasia (CAH) encompasses a group of genetic disorders resulting in the deficiency or absence of one of five enzymes needed for the biosynthesis of glucocorticoids and mineralocorticoids. Manifestations are usually present at birth or during early childhood, but symptoms may appear later in life in nonclassic CAH. CAH is uncommon and often has an autosomal recessive mode of inheritance. When successfully treated, sexual functioning and fertility aren't affected.

 AGE ALERT Salt-losing CAH may cause fatal adrenal crisis in newborns.

The prevalent adrenal disorder in infants and children, simple virilizing CAH and salt-losing CAH, are the most common forms. Acquired adrenal virilism is rare and affects twice as many females as males. With successful treatment, a normal quality of life and life span are expected. In older patients, androgen excess may be part of the syndrome of polycystic ovaries or may be secondary to an adrenal carcinoma.

Causes

The cause of CAH is:

▶ genetic, as an autosomal recessive trait.

Pathophysiology

Cortisol levels are regulated by a negative-feedback mechanism. Corticotropin (ACTH) in the blood stimulates the release of cortisol precursors and, consequently, of cortisol, aldosterone, and androgens. In turn, cortisol suppresses ACTH secretion. With a deficiency of the enzyme 21-hydroxylase, cortical secretion of cortisol is impaired and pituitary secretion of ACTH is increased. ACTH stimulates the adrenal cortex, which in turn stimulates both aldosterone and androgen biosynthesis and release.

Signs and symptoms
Signs and symptoms of CAH may include:
◗ ambiguous genitalia (enlarged clitoris with urethral opening at the base and a combination of the labia and scrotum), normal genital tract and gonads (newborn females)
◗ pubic and axillary hair at an earlier age, a deep voice, acne, and facial hair, but no menarche (female approaching puberty)
◗ no apparent manifestations (newborn males)
◗ accentuated masculine characteristics, including a deepened voice, acne, enlarged phallus with small testes, and frequent erections (male approaching puberty)
◗ high androgen levels causing rapid bone and muscle growth (in children)
◗ short stature due to premature epiphyseal closure and high androgen levels (adults)
◗ more severe changes, including development of a penis in females (salt-losing CAH).

Because males have no external abnormalities, diagnosis is more difficult and commonly delayed until other symptoms occur. In the second week of life, symptoms of a salt-wasting crisis include apathy, failure to eat, diarrhea, and adrenal crisis (vomiting, dehydration from hyponatremia, and hyperkalemia). If adrenal crisis is not treated promptly, dehydration and electrolyte imbalance cause cardiovascular collapse and cardiac arrest.

Complications
Possible complications of CAH are:
◗ death (salt-wasting crisis)
◗ precocious puberty
◗ menstrual irregularities
◗ sexual dysfunction and infertility.

Diagnosis
Diagnosis of CAH may include:
◗ elevated urine 17-ketosteroid levels (can be suppressed by dexamethasone [Decadron])
◗ elevated serum 17-hydroxyprogesterone level after I.V. bolus of ACTH
◗ serum hyperkalemia, hyponatremia, and hypochloremia (present but not diagnostic)
◗ elevated 24-hour urine pregnanetriol level
◗ normal or decreased 24-hour urine 17-hydroxycorticosteroid levels.

Treatment
Treatment of CAH includes:
◗ daily cortisone (Cortone) or hydrocortisone (Cortef) to stop the excessive output of ACTH and subsequent excessive androgen production (initial and subsequent doses guided by urinary 17-ketosteroids levels) given I.M. until the infant is old enough to tolerate pills (usually about 18 months)
◗ I.V. sodium chloride and glucose to reestablish fluid and electrolyte balance, with desoxycorticosterone I.M. and hydrocortisone I.V. as needed (adrenal crisis); glucocorticoid (cortisone or hydrocortisone) and perhaps mineralocorticoids (desoxycorticosterone, fludrocortisone, or both after stabilization)
◗ sex chromatin and karyotype studies to determine genetic sex (with ambiguous external genitalia); possible reconstructive surgery for females between the ages of 1 and 3 years.

Cushing syndrome
Cushing syndrome is a cluster of clinical abnormalities caused by excessive adrenocortical hormones (particularly cortisol) or related corticosteroids and, to a lesser extent, androgens and aldosterone. Cushing's disease (pituitary corticotropin [ACTH] excess) accounts for about 70% of the cases of Cushing syndrome. Cushing's disease occurs most commonly be-

tween 20 and 40 years of age and is eight times more common in females.

AGE ALERT Cushing syndrome caused by ectopic corticotropin secretion is more common in adult men, with the peak incidence between 40 and 60 years of age. In 30% of patients, Cushing syndrome results from a cortisol-secreting tumor. Adrenal tumors, rather than pituitary tumors, are more common in children, especially girls.

The annual incidence of endogenous cortisol excess in the United States is 2 to 4 cases per 1 million people per year. The incidence of Cushing syndrome resulting from exogenous administration of cortisol is uncertain, but it is known to be much greater than that of endogenous types. The prognosis for endogenous Cushing syndrome is guardedly favorable with surgery, but morbidity and mortality are high without treatment. About 50% of the individuals with untreated Cushing syndrome die within 5 years of onset as a result of overwhelming infection, suicide, complications from generalized arteriosclerosis (coronary artery disease), and severe hypertensive disease.

Causes
Causes of Cushing syndrome include:
▶ anterior pituitary hormone (ACTH) excess
▶ autonomous, ectopic ACTH secretion by a tumor outside the pituitary (usually malignant, frequently oat cell carcinoma of the lung).

Pathophysiology
Cortisol excess results in anti-inflammatory effects and excessive catabolism of protein and peripheral fat to support hepatic glucose production. The mechanism may be ACTH dependent, in which elevated plasma ACTH levels stimulate the adrenal cortex to produce excess cortisol, or ACTH independent, in which excess cortisol is produced by the adrenal cortex or exogenously administered. This suppresses the hypothalamic-pituitary-adrenal axis, also present in ectopic ACTH-secreting tumors.

Signs and symptoms
Like other endocrine disorders, Cushing syndrome induces changes in many body systems. Specific clinical effects vary with the system affected, and include:
▶ diabetes mellitus, with decreased glucose tolerance, fasting hyperglycemia, and glucosuria due to cortisol-induced insulin resistance and increased glycogenolysis and glucogenesis in the liver (endocrine and metabolic systems)
▶ muscle weakness due to hypokalemia or loss of muscle mass from increased catabolism, pathologic fractures due to decreased bone mineral ionization, and skeletal growth retardation in children (musculoskeletal system)
▶ purple striae; facial plethora (edema and blood vessel distention); acne; fat pads above the clavicles, over the upper back (buffalo hump), on the face (moon facies), and throughout the trunk (truncal obesity), with slender arms and legs; little or no scar formation; poor wound healing due to decreased collagen and weakened tissues (skin)
▶ peptic ulcer due to increased gastric secretions and pepsin production and decreased gastric mucus (GI system)
▶ irritability and emotional lability, ranging from euphoric behavior to depression or psychosis; insomnia due to the role of cortisol in neurotransmission (central nervous system)
▶ hypertension due to sodium and secondary fluid retention; left ventricular hypertrophy; capillary weakness from protein loss, which leads to bleeding and ecchymosis; dyslipidemia (cardiovascular system)
▶ increased susceptibility to infection due to decreased lymphocyte production and suppressed antibody formation; decreased resistance to stress; suppressed inflammatory response masking even severe infection (immunologic system)

sodium and secondary fluid retention, increased potassium excretion, ureteral calculi from increased bone demineralization with hypercalciuria (renal and urologic systems)

increased androgen production, with clitoral hypertrophy, mild virilism, hirsutism, and amenorrhea or oligomenorrhea in women, and sexual dysfunction (reproductive system).

Complications

Complications of Cushing syndrome include:

osteoporosis
increased susceptibility to infections
hirsutism
ureteral calculi
metastases of malignant tumors.

Diagnosis

Diagnosis is based on the following laboratory test results:

hyperglycemia, hypernatremia, glucosuria, hypokalemia, and metabolic alkalosis

urinary free cortisol levels more than 150 μg/24 hours

dexamethasone suppression test to confirm the diagnosis and determine the cause, possibly an adrenal tumor or a nonendocrine, corticotropin-secreting tumor

blood levels of corticotropin-releasing hormone, ACTH, and different glucocorticoids to diagnose and localize cause to pituitary or adrenal gland.

Treatment

Differentiation among pituitary, adrenal, and ectopic causes of hypercortisolism is essential for effective treatment, which is specific for the cause of cortisol excess and includes medication, radiation, and surgery. Possible treatments are:

surgery for tumors of the adrenal and pituitary glands or other tissue (such as the lung)

radiation therapy (tumor)

drugs, such as mitotane (Lysodren) or aminoglutethimide (Cytadren), to block steroid synthesis for inoperable tumor.

Diabetes insipidus

A disorder of water metabolism, diabetes insipidus results from a deficiency of circulating vasopressin (also called antidiuretic hormone, or ADH) or from renal resistance to this hormone. Pituitary diabetes insipidus is caused by a deficiency of vasopressin, and nephrogenic diabetes insipidus is caused by the resistance of renal tubules to vasopressin. Diabetes insipidus is characterized by excessive fluid intake and hypotonic polyuria. A decrease in ADH levels leads to altered intracellular and extracellular fluid control, causing renal excretion of a large amount of urine.

The disorder may start at any age and is slightly more common in men than in women. The incidence is slightly greater today than in the past.

In uncomplicated diabetes insipidus, the prognosis is good with adequate water replacement, and patients usually lead normal lives.

Causes

The cause of diabetes insipidus may be:

acquired, familial, idiopathic, neurogenic, or nephrogenic

associated with stroke, hypothalamic or pituitary tumors, and cranial trauma or surgery (neurogenic diabetes insipidus)

X-linked recessive trait or end-stage renal failure (nephrogenic diabetes insipidus, less common)

certain drugs, such as lithium (Duralith), phenytoin (Dilantin), or alcohol (transient diabetes insipidus).

Pathophysiology

Diabetes insipidus is related to an insufficiency of ADH, leading to polyuria and polydipsia. The three forms of diabetes insipidus are neurogenic, nephrogenic, and psychogenic.

Neurogenic, or central, diabetes insipidus is an inadequate response of ADH to plasma osmolarity, which occurs when an organic lesion of the hypothalamus, infundibular stem, or posterior pituitary partially or completely blocks ADH synthesis, transport, or release. The many organic lesions that can cause diabetes insipidus include brain tumors, hypophysectomy, aneurysms, thrombosis, infections, and immunologic disorders. Neurogenic diabetes insipidus has an acute onset. A three-phase syndrome can occur, which involves:

◗ progressive loss of nerve tissue and increased diuresis
◗ normal diuresis
◗ polyuria and polydipsia, the manifestation of permanent loss of the ability to secrete adequate ADH.

Nephrogenic diabetes insipidus is caused by an inadequate renal response to ADH. The collecting duct permeability to water does not increase in response to ADH. Nephrogenic diabetes insipidus is generally related to disorders and drugs that damage the renal tubules or inhibit the generation of cyclic adenosine monophosphate in the tubules, preventing activation of the second messenger. Causative disorders include pyelonephritis, amyloidosis, destructive uropathies, polycystic disease, and intrinsic renal disease. Drugs include lithium carbonate (Eskalith), general anesthetics such as methoxyflurane, and demeclocycline (Declomycin). In addition, hypokalemia or hypercalcemia impairs the renal response to ADH. A rare genetic form of nephrogenic diabetes insipidus is an X-linked recessive trait.

Psychogenic diabetes insipidus is caused by an extremely large fluid intake, which may be idiopathic or related to psychosis or sarcoidosis. The polydipsia and resultant polyuria wash out ADH more quickly than it can be replaced. Chronic polyuria may overwhelm the renal medullary concentration gradient, rendering patients partially or totally unable to concentrate urine.

Regardless of the cause, insufficient ADH causes the immediate excretion of large volumes of dilute urine and consequent plasma hyperosmolality. In conscious individuals, the thirst mechanism is stimulated, usually for cold liquids. With severe ADH deficiency, urine output may be greater than 12 L/day, with a low specific gravity. Dehydration develops rapidly if fluids aren't replaced.

Signs and symptoms

Signs and symptoms of diabetes insipidus include:

◗ polydipsia and polyuria up to 12 L/day (cardinal symptoms)
◗ sleep disturbance and fatigue due to nocturia
◗ headache and visual disturbance due to electrolyte disturbance and dehydration
◗ abdominal fullness, anorexia, and weight loss due to almost continuous fluid consumption.

Complications

Possible complications of diabetes insipidus are:
◗ dilatation of the urinary tract
◗ severe dehydration
◗ shock and renal failure if dehydration is severe.

Diagnosis

Diagnosis is based on:
◗ urinalysis showing almost colorless urine of low osmolality (50 to 200 mOsm/kg, less than that of plasma) and low specific gravity (less than 1.005)
◗ water deprivation test to identify vasopressin deficiency, resulting in renal inability to concentrate urine
◗ hyponatremia.

Treatment

Until the cause of diabetes insipidus can be identified and eliminated, the administration of vasopressin (Pitressin Syn-

thetic) can control fluid balance and prevent dehydration. Medications include:
▶ vasopressin aqueous preparation S.C. or I.M. several times daily, effective for only 2 to 6 hours (used as a diagnostic agent and, rarely, in acute disease)
▶ desmopressin acetate (DDAVP) orally (not available in the United States), by nasal spray absorbed through the mucous membranes, or S.C. or I.V. injection, effective for 8 to 20 hours depending on the dosage.

Diabetes mellitus

Diabetes mellitus is a metabolic disorder characterized by hyperglycemia (elevated serum glucose level) resulting from lack of insulin, lack of insulin effect, or both. Three general classifications are recognized:
▶ type 1, absolute insulin insufficiency
▶ type 2, insulin resistance with varying degrees of insulin secretory defects
▶ gestational diabetes, which emerges during pregnancy.

Onset of type 1 usually occurs before the age of 30 years (although it may occur at any age); the patient is usually thin and requires exogenous insulin and dietary management to achieve control. Conversely, type 2 usually occurs in obese adults after the age of 40 years and is treated with diet and exercise in combination with various antidiabetic drugs, although treatment may include insulin therapy.

CULTURAL DIVERSITY More than 8% of all adults in the United States have diabetes, and 93% of these have type 2. The prevalence of type 1 diabetes is higher in white populations in the United States. Type 2 diabetes is more prevalent in persons of African, American Indian, Asian, Hispanic, and Pacific Islander descent. Although whites usually develop type 2 diabetes after the age of 40, type 2 diabetes tends to occur at an earlier age in nonwhite populations, and about 25% of the diabetes that occurs in youth in nonwhite populations is type 2.

Medical advances permit increased longevity and improved quality of life if the patient carefully monitors blood glucose levels, uses the data to make pharmacologic and life-style changes, and uses new insulin delivery systems, such as subcutaneous insulin pumps. In addition, medications now available enhance the body's own glucose metabolism and insulin sensitivity to optimize glycemic control and prevent progression to long-term complications.

Causes

Evidence indicates that diabetes mellitus has diverse causes, including:
▶ heredity
▶ environment (infection, diet, toxins, stress)
▶ life-style changes in genetically susceptible persons.

Pathophysiology

In persons genetically susceptible to type 1 diabetes, a triggering event, possibly a viral infection, causes production of autoantibodies against the beta cells of the pancreas. The resultant destruction of the beta cells leads to a decline in and ultimate lack of insulin secretion. Insulin deficiency leads to hyperglycemia, enhanced lipolysis (decomposition of fat), and protein catabolism. These characteristics occur when more than 90% of the beta cells have been destroyed.

Type 2 diabetes mellitus is a chronic disease caused by one or more of the following factors: impaired insulin production, inappropriate hepatic glucose production, or peripheral insulin receptor insensitivity. Genetic factors are significant, and onset is accelerated by obesity and a sedentary lifestyle. Again, added stress can be a pivotal factor.

Gestational diabetes mellitus occurs when a woman not previously diagnosed with diabetes shows glucose intolerance during pregnancy. This may occur if placental hormones counteract insulin, causing insulin resistance. Gestational diabetes mellitus is a significant risk factor for the future occurrence of type 2 diabetes mellitus.

Signs and symptoms
Signs and symptoms of diabetes mellitus include:
▶ polyuria and polydipsia due to high serum osmolality caused by high serum glucose levels
▶ anorexia (common) or polyphagia (occasional)
▶ weight loss (usually 10% to 30%; persons with type 1 diabetes often have almost no body fat at time of diagnosis) due to prevention of normal metabolism of carbohydrates, fats, and proteins caused by impaired or absent insulin function
▶ headaches, fatigue, lethargy, reduced energy levels, and impaired school and work performance due to low intracellular glucose levels
▶ muscle cramps, irritability, and emotional lability due to electrolyte imbalance
▶ vision changes, such as blurring, due to glucose-induced swelling
▶ numbness and tingling due to neural tissue damage
▶ abdominal discomfort and pain due to autonomic neuropathy, causing gastroparesis and constipation
▶ nausea, diarrhea, or constipation due to dehydration and electrolyte imbalances or autonomic neuropathy.

Complications
Complications of diabetes mellitus include:
▶ microvascular disease, including retinopathy, nephropathy, and neuropathy
▶ dyslipidemia
▶ macrovascular disease, including coronary, peripheral, and cerebral artery disease
▶ hypoglycemia
▶ diabetic ketoacidosis
▶ hyperglycemic, hyperosmolar, nonketotic syndrome
▶ excessive weight gain

▶ skin ulcerations
▶ chronic renal failure.

Diagnosis
In adult men and nonpregnant women, diabetes mellitus is diagnosed by two of the following criteria obtained more than 24 hours apart, using the same test twice or any combination:
▶ fasting plasma glucose level of 126 mg/dl or more on at least two occasions
▶ typical symptoms of uncontrolled diabetes and random blood glucose level of 200 mg/dl or more
▶ blood glucose level of 200 mg/dl or more 2 hours after ingesting 75 g of oral dextrose.

Diagnosis may also be based on:
▶ diabetic retinopathy on ophthalmologic examination
▶ other diagnostic and monitoring tests, including urinalysis for acetone and glycosylated hemoglobin (reflects glycemic control over the past 2 to 3 months).

Treatment
Effective treatment for all types of diabetes optimizes blood glucose control and decreases complications. Treatment for type 1 diabetes includes:
▶ insulin replacement, meal planning, and exercise (current forms of insulin replacement include mixed-dose, split mixed-dose, and multiple daily injection regimens and continuous subcutaneous insulin infusions)
▶ pancreas transplantation (currently requires chronic immunosuppression). (See *Treatment of type 1 diabetes mellitus*, pages 434 and 435.)

Treatment of type 2 diabetes mellitus includes:
▶ oral antidiabetic drugs to stimulate endogenous insulin production, increase insulin sensitivity at the cellular level, suppress hepatic gluconeogenesis, and delay GI absorption of carbohydrates (drug combinations may be used)

(Text continues on page 436.)

TREATMENT OF TYPE 1 DIABETES MELLITUS

The following algorithm shows the pathophysiologic process of diabetes and points for treatment intervention.

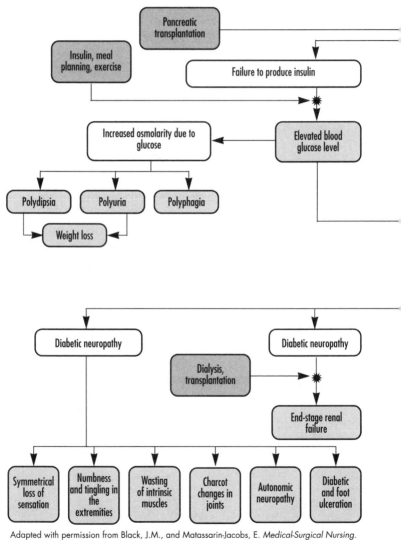

Adapted with permission from Black, J.M., and Matassarin-Jacobs, E. *Medical-Surgical Nursing.* 5th ed. Philadelphia: W.B. Saunders Company, 1997.

DISRUPTING DISEASE

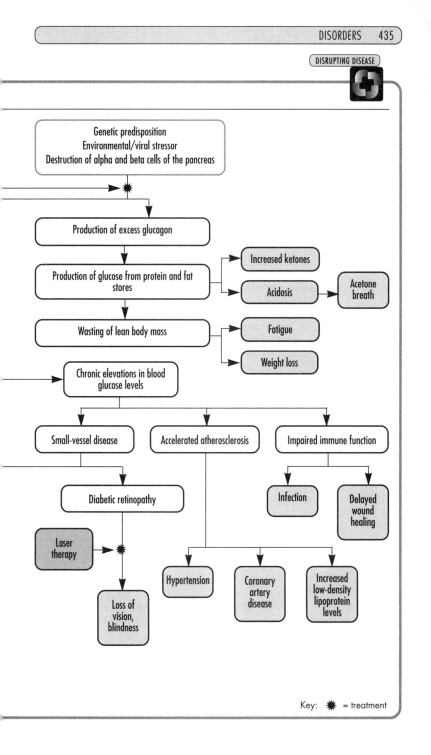

Genetic predisposition
Environmental/viral stressor
Destruction of alpha and beta cells of the pancreas

Production of excess glucagon

Production of glucose from protein and fat stores

Increased ketones

Acidosis → Acetone breath

Wasting of lean body mass

Fatigue

Weight loss

Chronic elevations in blood glucose levels

Small-vessel disease

Accelerated atherosclerosis

Impaired immune function

Diabetic retinopathy

Infection

Delayed wound healing

Laser therapy → Loss of vision, blindness

Hypertension

Coronary artery disease

Increased low-density lipoprotein levels

Key: ✹ = treatment

▶ exogenous insulin, alone or with oral antidiabetic drugs, to optimize glycemic control.

Treatment of both types of diabetes mellitus includes:

▶ individualized meal plan designed to meet nutritional needs, control blood glucose and lipid levels, and reach and maintain appropriate body weight (plan to be followed consistently with meals eaten at regular times)

▶ weight reduction (obese patient with type 2 diabetes mellitus) or high calorie allotment, depending on growth stage and activity level (type 1 diabetes mellitus).

Treatment of gestational diabetes involves:

▶ medical nutrition therapy and exercise

▶ alpha glucosidase inhibitors, injected insulin, or both (if euglycemia not achieved); postpartum counseling to address the high risk for gestational diabetes in subsequent pregnancies and type 2 diabetes later in life; regular exercise and prevention of weight gain to help prevent type 2 diabetes.

Gonadotropin deficiency

Gonadotropin deficiency is a lack of hormones (follicle-stimulating hormone [FSH] and luteinizing hormone [LH]) that stimulate the sex glands, primarily the testes and ovaries. Chronic gonadotropin deficiency, if not treated, can cause infertility and osteopenia (decreased bone mass). A decrease in testosterone results in decreased bone cell formation.

Causes

Causes of gonadotropin deficiency include:

▶ pituitary tumor or hemorrhage

▶ oversecretion of target gland hormone, such as estrogen, progesterone, or testosterone

▶ prolactin-secreting tumor

▶ hypothalamic suppression of gonadotropin-releasing hormone (GnRH) during periods of physical or emotional stress, obesity, and starvation

▶ genetics.

Pathophysiology

GnRH is secreted by the hypothalamus and causes the anterior pituitary to secrete the gonadotropins — testosterone, estrogen, FSH, and LH. Estrogen, progesterone, and testosterone, produced by the gonads, function in a negative-feedback loop that regulates GnRH secretion.

Testosterone, which is responsible for masculine sex characteristics and sperm production, also functions in bone, muscle, and red blood cell formation, as well as having a role in neural signaling. Estrogen serves many functions, among them cognitive, bone, and vaginal maintenance. FSH and LH function to maintain the corpus luteum and pregnancy.

Several mechanisms can cause GnRH deficiency, including:

▶ pituitary tumor producing another hormone that impinges on the gonadotropin-producing cells and physically impairs GnRH biosynthesis

▶ medical treatments such as radiation (impairs GnRH-producing cells)

▶ oversecretion of estrogen, progesterone, or testosterone by dysfunctional target glands, causing GnRH inhibition through the negative-feedback loop

▶ prolactin (inhibits pituitary secretion of GnRH; prolactin-secreting tumors can cause GnRH deficiency)

▶ reduced GnRH secretion due to response of hypothalamus to physical stress, obesity, or starvation (for example, females in competitive athletics may not enter menarche or may cease menstruation for extended periods of time).

Signs and symptoms

Many symptoms are directly related to a reduction in the differentiating sexual characteristics that are caused and maintained by the gonadotropins (FSH and LH) and the hormones they stimulate, androgens and estrogen. These signs and symptoms vary with the degree and length of GnRH deficiency, and may include:

▶ decreased libido, strength, and body hair, and fine wrinkles around the eyes and lips (adults)

▶ amenorrhea; vaginal, uterine, and breast atrophy; clitoral enlargement; voice deepening; and beard growth (women)

▶ testicular atrophy, reduction in beard growth, and erectile dysfunction (men)

▶ decreased red blood cells and loss of bone and muscle mass due to low testosterone levels

▶ mood and behavior changes due to changes in testosterone levels

▶ anosmia, which is the absence of a sense of smell (genetic cases).

The age of onset of GnRH deficiency affects the presentation in children:

▶ inadequate sexual differentiation shown by ambiguity, pseudohermaphroditism (individual showing one or more contraindications of the morphologic sex criteria), or normal-appearing female genitalia with male genetic coding (first trimester)

▶ microphallus and partial or complete lack of testicular descent (second and third trimesters)

▶ poor secondary sex characteristics and muscle development, lack of deepening voice in males, sparse body hair, gynecomastia (enlarged breast tissue), delayed fusion of epiphyseal plates, and continued long bone growth (childhood through puberty).

Complications

A complication of gonadotropin deficiency is:

▶ infertility.

Diagnosis

Diagnosis of gonadotropin deficiency is based on:

▶ serum estrogen, testosterone, and GnRH levels to differentiate between dysfunction of the hypothalamus or of the ovaries or testicles

▶ low testosterone and high GnRH levels (primary testicular failure)

▶ low estrogen and high GnRH levels (primary ovarian failure)

▶ low GnRH and testosterone or estrogen levels (hypothalamic or pituitary dysfunction)

▶ human chorionic gonadotropin (HCG) stimulation test (HCG, 500 IU/1.7 m^2 or 100 IU/kg in children, given after measuring baseline testosterone; after 3 to 4 days, testosterone levels should increase by 50% to 200% because HCG and LH stimulate the Leydig cells to stimulate testicular function)

▶ clomiphene citrate test (normal response, 30% to 200% increase in FSH and 0% to 65% increase in testosterone) with impaired or absent increase in hypothalamic or pituitary disorders

▶ GnRH stimulation test (rapid I.V. injection of GnRH stimulates the pituitary to secrete LH and FSH), with insufficient elevation of LH or FSH levels indicating pituitary or hypothalamus dysfunction.

Treatment

Treatment of gonadotropin deficiency includes:

▶ surgery to remove tumors

▶ gonadotropin, estrogen, or testosterone replacement

▶ stress reduction and weight gain or loss.

Growth hormone deficiency

Growth hormone (GH) deficiency results from hypofunction of the anterior pituitary gland with a resulting decreased secretion of GH. GH deficiency includes a group of childhood disorders characterized by subnormal growth velocity, delayed bone age, and a subnormal response to at least two stimuli for release of the hormone. (See *GH deficiency in children*, page 438.) GH deficiency in adults is characterized by general weakness and increased mortality.

Causes

Possible causes of GH deficiency are:

▶ autosomal recessive, autosomal dominant, or X-linked trait

▶ pituitary or central nervous system tumor

GH DEFICIENCY IN CHILDREN

SYNDROME	DESCRIPTION
Isolated growth hormone (GH) deficiency	Genetic disorder causing dwarfism if not treated with GH replacement
Neurosecretory failure	Idiopathic disorder with subnormal growth velocity and delayed bone age, but GH levels exceed the limits for deficiency; variable response to GH replacement
Panhypopituitarism	GH and other trophic hormones deficient; all requiring replacement
Turner syndrome	Chromosomal defect causing multiple abnormalities, decreased GH response to provocative stimuli, and possibly increased target-organ resistance to GH or insulin-like growth factors; GH in high doses produces modest increases in height
Down syndrome	Chromosomal defect causing multiple abnormalities, including short stature; GH replacement stimulates growth but doesn't improve quality of life
Intrauterine growth retardation	Low birth weight; in absence of catch-up growth during the first year, adults have short stature; catch-up growth may occur; GH may stimulate growth
Growth retardation	Short children have normal growth velocity and bone age; adult height at least 10 cm less than predicted from parents' heights; modest and inconsistent response to GH therapy
Familial short stature	Short children have normal growth velocity and bone age; adult height within range predicted from parents' heights; modest and inconsistent response to GH therapy
Constitutional delay of growth and development	Short children have delayed bone age but normal growth velocity and rates of bone age advancement
Chronic renal failure and renal transplantation	Children grow slowly but have normal GH response and levels of insulin-like growth factor 1; GH therapy may increase height but is associated with increased risk for malignancy in patients taking immunosuppressive drugs

Adapted with permission from a report by the Drug and Therapeutics Committee of the Lawson Wilkins Pediatric Endocrine Society."Guidelines for the use of growth hormone in children with short stature," *Journal of Pediatrics* 127(6):857-867, 1994.

- pituitary hypoxic necrosis
- pituitary inflammation
- hypothalamic failure
- GH receptor insensitivity
- biologically inactive GH
- hematologic disorders
- idiopathic causes
- trauma
- pituitary irradiation.

Pathophysiology
The absence or deficiency of GH synthesis causes growth failure in children. In adults, metabolic derangements decrease GH response to stimulation.

Signs and symptoms
Signs and symptoms of GH deficiency are:
- short stature (2 standard deviations less than the predicted mean for age and gender)
- reduced muscle mass and increased subcutaneous fat due to decreased protein synthesis and insufficient muscle anabolism.

Complications
Complications of GH deficiency may include:
- short stature and possible related psychosocial difficulties (if untreated)
- fatal convulsions, especially during periods of stress, due to fasting hypoglycemia
- gonadotropin deficiency
- multiple pituitary hormone deficiencies
- increased cardiovascular mortality (adults).

Diagnosis
Diagnosis of GH deficiency is based on:
- decreased serum GH and somatomedin C levels.

Treatment
Treatment of GH deficiency includes:
- exogenous GH given S.C. up to several times weekly during puberty.

Growth hormone excess
Growth hormone (GH) excess that begins in adulthood (after epiphyseal closure) is called acromegaly. GH excess that is present before closure of the epiphyseal growth plates of the long bones causes pituitary gigantism. In both cases, the result is increased growth of bone, cartilage, and other tissues, as well as increased catabolism of carbohydrates and protein synthesis. Acromegaly is rare, with a prevalence of about 70 people per million in the United States. Most cases are diagnosed in the fourth and fifth decades, but acromegaly is usually present for years before diagnosis. GH excess is a slow but progressive disease that decreases longevity if untreated. Morbidity and mortality tend to be related to coronary artery disease and hypertension.

The earliest clinical sign of acromegaly is soft-tissue swelling of the extremities, which causes coarsening of the facial features.

In gigantism, a proportional overgrowth of all body tissues starts before epiphyseal closure. This causes remarkable height increases — as much as 6 inches (15 cm) a year. Gigantism affects infants and children, causing them to reach as much as three times the normal height for their age. As adults, they may reach a height of more than 7.5 feet (203 cm).

Causes
GH excess is caused by:
- eosinophilic or mixed-cell adenomas of the anterior pituitary gland.

Pathophysiology
A GH-secreting tumor creates an unpredictable GH secretion pattern, which replaces the usual peaks that occur 1 to 4 hours after the onset of sleep. Elevated GH and somatomedin levels stimulate tissue growth. In pituitary gigantism, because the epiphyseal plates aren't closed, the excess GH stimulates linear growth. It also increases the bulk of bones and joints and causes enlargement of internal organs and metabolic abnormalities. In acromegaly, the excess GH increases bone density and

width, and the proliferation of connective and soft tissues.

Signs and symptoms

Acromegaly develops slowly, and gigantism is characterized by rapid growth. Other signs and symptoms of acromegaly include:

▶ diaphoresis, oily skin, hypermetabolism, hypertrichosis (excessive hair growth), weakness, arthralgias, malocclusion of the teeth, and new skin tags (typical)

▶ severe headache, central nervous system impairment, bitemporal hemianopia (defective vision), loss of visual acuity, and blindness (if the intrasellar tumor compresses the optic chiasm or nerves)

▶ cartilaginous and connective tissue overgrowth, causing the characteristic hulking appearance, with an enlarged supraorbital ridge and thickened ears and nose

▶ marked prognathism (projection of the jaw) that may interfere with chewing

▶ laryngeal hypertrophy, paranasal sinus enlargement, and thickening of the tongue causing the voice to sound deep and hollow

▶ arrowhead appearance of distal phalanges on X-rays, thickened fingers

▶ irritability, hostility, and various psychological disturbances

▶ bow legs, barrel chest, arthritis, osteoporosis, kyphosis, hypertension, and arteriosclerosis (prolonged effects of excessive GH secretion)

▶ glucose intolerance and clinical diabetes mellitus due to action of GH as an insulin antagonist.

Signs and symptoms of gigantism include:

▶ backache, arthralgia, and arthritis due to rapid bone growth

▶ excessive height due to rapid growth before epiphyseal plate closure

▶ headache, vomiting, seizure activity, visual disturbances, and papilledema (edema wherein the optic nerve enters the eye chamber) due to tumor compressing nerves and tissue in surrounding structures

▶ deficiencies of other hormone systems (if GH-producing tumor destroys other hormone-secreting cells)

▶ glucose intolerance and diabetes mellitus due to insulin-antagonistic actions of GH.

Complications

Possible complications of GH excess are:
▶ cardiomegaly
▶ hypertension
▶ diabetes mellitus.

Diagnosis

Diagnosis of GH excess is based on:
▶ elevated plasma GH level measured by radioimmunoassay (results of random sampling may be misleading owing to pulsatile GH secretion)

▶ somatomedin-C, a metabolite of GH (a better diagnostic alternative)

▶ glucose suppression test (glucose normally suppresses GH secretion; if glucose infusion doesn't suppress GH to less than 2 ng/ml and the patient has characteristic clinical features, hyperpituitarism is likely)

▶ skull X-rays, computed tomography scan, or magnetic resonance imaging to show the presence and extent of pituitary lesion

▶ bone X-rays showing a thickening of the cranium (especially frontal, occipital, and parietal bones) and long bones, and osteoarthritis in the spine (support the diagnosis)

▶ elevated blood glucose levels.

Treatment

Treatment may involve:
▶ tumor removal by cranial or transsphenoidal hypophysectomy or pituitary radiation therapy

▶ mandatory surgery for a tumor causing blindness or other severe neurologic disturbances (acromegaly)

▶ replacement of thyroid, cortisone, and gonadal hormones (postoperative therapy)

▶ bromocriptine (Parlodel) and octreotide (Sandostatin) to inhibit GH synthesis (adjunctive treatment).

Hyperparathyroidism

Hyperparathyroidism results from excessive secretion of parathyroid hormone (PTH) from one or more of the four parathyroid glands. PTH promotes bone resorption, and hypersecretion leads to hypercalcemia and hypophosphatemia. Renal and GI absorption of calcium increase.

Primary hyperparathyroidism is commonly diagnosed based on elevated calcium levels found on laboratory test results in asymptomatic patients. It affects women two to three times more frequently than men.

Causes

Hyperparathyroidism may be primary or secondary. In primary hyperparathyroidism:

▶ one or more parathyroid glands enlarge and increase PTH secretion, most commonly caused by a single adenoma, but this may be a component of multiple endocrine neoplasia (all four glands usually involved).

In secondary hyperparathyroidism, a hypocalcemia-producing abnormality outside the parathyroids causes excessive compensatory production of PTH. Causes include:

▶ rickets, vitamin D deficiency, chronic renal failure, and osteomalacia due to phenytoin (Dilantin).

Pathophysiology

Overproduction of PTH by a tumor or hyperplastic tissue increases intestinal calcium absorption, reduces renal calcium clearance, and increases bone calcium release. Response to this excess varies for each patient for an unknown reason.

Hypophosphatemia results when excessive PTH inhibits renal tubular phosphate reabsorption. The hypophosphatemia aggravates hypercalcemia by increasing the sensitivity of the bone to PTH.

Signs and symptoms

Signs and symptoms of primary hyperparathyroidism result from hypercalcemia and are typically present in several body systems. Signs and symptoms may include:

▶ polyuria, nephrocalcinosis, or recurring nephrolithiasis and consequent renal insufficiency (renal system)

▶ chronic low back pain and easy fracturing due to bone degeneration; bone tenderness; chondrocalcinosis (decreased bone mass); osteopenia and osteoporosis, especially on the vertebrae; erosions of the juxta-articular (adjoining joint) surface; subchondral fractures; traumatic synovitis; and pseudogout (skeletal and articular systems)

▶ pancreatitis causing constant, severe epigastric pain that radiates to the back; peptic ulcers, causing abdominal pain, anorexia, nausea, and vomiting (GI system)

▶ muscle weakness and atrophy, particularly in the legs (neuromuscular system)

▶ psychomotor and personality disturbances, depression, overt psychosis, stupor, and possibly coma (central nervous system)

▶ skin necrosis, cataracts, calcium microthrombi to lungs and pancreas, anemia, and subcutaneous calcification (other systems).

Secondary hyperparathyroidism may produce the same features of calcium imbalance with skeletal deformities of the long bones (such as rickets) as well as symptoms of the underlying disease.

Complications

Complications of hyperparathyroidism include:

▶ pathologic fractures
▶ renal damage
▶ urinary tract infections
▶ hypertension.

Diagnosis

Findings differ in primary and secondary disease. In primary disease, diagnosis is based on:

▶ hypercalcemia and high concentrations of serum PTH on radioimmunoassay (confirms the diagnosis)

▶ X-rays showing diffuse demineralization of bones, bone cysts, outer cortical bone absorption, and subperiosteal erosion of the phalanges and distal clavicles

▶ microscopic bone examination by X-ray spectrophotometry typically showing increased bone turnover

▶ elevated urine and serum calcium, chloride, and alkaline phosphatase levels; decreased serum phosphorus levels

▶ elevated uric acid and creatinine levels, which may also increase basal gastric acid secretion and serum immunoreactive gastrin

▶ increased serum amylase levels (may indicate acute pancreatitis).

Diagnosis of secondary disease is based on:

▶ normal or slightly decreased serum calcium level, variable serum phosphorus level, especially when the cause is rickets, osteomalacia, or kidney disease

▶ patient history possibly showing familial kidney disease, seizure disorders, or drug ingestion.

Treatment

Effective treatment varies, depending on the cause of the disease. In primary hyperparathyroidism, surgery is the only definitive therapy. The only effective long-term medical therapy is maintaining hydration in mild hyperparathyroidism.

Treatment of primary disease includes:

▶ surgery to remove the adenoma or, depending on the extent of hyperplasia, all but half of one gland, to provide normal PTH levels (may relieve bone pain within 3 days, but renal damage may be irreversible)

▶ treatments to decrease calcium levels, such as forcing fluids, limiting dietary intake of calcium, and promoting sodium and calcium excretion through forced diuresis (using as much as 6 L of urine output in life-threatening circumstances), and use of furosemide (Lasix) or ethacrynic

acid (Edecrin) (preoperatively or if surgery isn't feasible or necessary)

▶ oral sodium or potassium phosphate; S.C. calcitonin (Calcimar); I.V. mithramycin or biphosphonate

▶ I.V. magnesium and phosphate or sodium phosphate solution by mouth or retention enema (for potential postoperative magnesium and phosphate deficiencies), possibly supplemental calcium, vitamin D, or calcitriol (Calcijex) (serum calcium level decreases to low normal range during the first 4 to 5 days after surgery).

Treatment of secondary disease includes:

▶ vitamin D to correct the underlying cause of parathyroid hyperplasia; oral calcium preparation to correct hyperphosphatemia in the patient with kidney disease

▶ dialysis in the patient with renal failure to decrease phosphorus levels (may be lifelong)

▶ enlarged glands may not revert to normal size and function even after calcium levels have been controlled in the patient with chronic secondary hyperparathyroidism.

Hypoparathyroidism

Hypoparathyroidism is caused by disease, injury, or congenital malfunction of the parathyroid glands. Because the parathyroid glands primarily regulate calcium balance, hypoparathyroidism causes hypocalcemia and consequent neuromuscular symptoms ranging from paresthesia to tetany.

The clinical effects of hypoparathyroidism are usually correctable with replacement therapy. Some complications of long-term hypocalcemia, such as cataracts and basal ganglion calcifications, are irreversible.

Causes

Hypoparathyroidism may be acute or chronic and is classified as idiopathic or acquired. Possible causes include:

▶ autoimmune genetic disorder or congenital absence of the parathyroid glands (idiopathic)

▶ accidental removal of or injury to the parathyroid glands during thyroidectomy or other neck surgery or, rarely, from massive thyroid irradiation (acquired)

▶ ischemic infarction of the parathyroid glands during surgery, amyloidosis, neoplasms, or trauma (acquired)

▶ impairment of hormone synthesis and release due to hypomagnesemia, suppression of normal gland function due to hypercalcemia, and delayed maturation of parathyroid function (acquired, reversible).

Parathyroid hormone (PTH) is regulated directly by serum calcium levels, not by the pituitary or hypothalamus. It normally maintains normocalcemia by regulating bone resorption and GI absorption of calcium. It also maintains an inverse relationship between serum calcium and phosphate levels by inhibiting phosphate reabsorption in the renal tubules.

AGE ALERT The incidence of the idiopathic and reversible forms is greatest in children; the incidence of the irreversible acquired form is greatest in adults who have undergone surgery for hyperthyroidism or other head and neck conditions.

Pathophysiology
Underproduction of PTH causes hypocalcemia and hyperphosphatemia. Surgical manipulation of the neck may damage the parathyroid glands, possibly by causing ischemia. The degree of hypoparathyroidism can vary from decreased reserve to frank tetany. Hypomagnesemia can prevent PTH secretion in patients with chronic GI magnesium losses, nutritional deficiencies, and renal magnesium wasting.

Signs and symptoms
Mild hypoparathyroidism may be asymptomatic but usually causes:

▶ hypocalcemia and high serum phosphate levels affecting the central nervous system (CNS) and other systems.

Signs and symptoms of chronic hypoparathyroidism include:

▶ neuromuscular irritability, increased deep tendon reflexes, Chvostek's sign (spasm of the hyperirritable facial nerve when it's tapped), dysphagia, organic brain syndrome, psychosis, mental deficiency in children, and tetany

▶ difficulty walking and a tendency to fall (chronic tetany).

Signs and symptoms of acute hypoparathyroidism include:

▶ tingling in the fingertips, around the mouth, and occasionally in the feet (first symptom); spreading and becoming more severe, producing muscle tension and spasms and consequent adduction of the thumbs, wrists, and elbows; pain varying with the degree of muscle tension but seldom affecting the face, legs, and feet (acute overt tetany)

▶ laryngospasm, stridor, cyanosis, and seizures (CNS abnormalities); worst during hyperventilation, pregnancy, infection, withdrawal of thyroid hormone, or administration of diuretics and before menstruation (acute tetany)

▶ abdominal pain; dry, lusterless hair; spontaneous hair loss; brittle fingernails developing ridges or falling out; dry, scaly skin; cataracts; and weakened tooth enamel, causing teeth to stain, crack, and decay easily (effects of hypocalcemia).

Complications
Possible complications are:
▶ cardiac arrhythmias, heart failure
▶ cataracts
▶ basal ganglia calcifications
▶ stunted growth, teeth malformation, and mental retardation
▶ Parkinson's symptoms
▶ hypothyroidism.

Diagnosis
The following test results confirm the diagnosis of hypoparathyroidism:
▶ radioimmunoassay for PTH showing decreased PTH level
▶ decreased serum calcium level

OTHER FORMS OF HYPERTHYROIDISM

▶ *Toxic adenoma* — a small, benign nodule in the thyroid gland that secretes thyroid hormone — is the second most common cause of hyperthyroidism. The cause of toxic adenoma is unknown; incidence is highest in the elderly. Clinical effects are essentially similar to those of Graves' disease, except that toxic adenoma doesn't induce ophthalmopathy, pretibial myxedema, or acropachy. Presence of adenoma is confirmed by radioactive iodine (^{131}I) uptake and thyroid scan, which show a single hyperfunctioning nodule suppressing the rest of the gland. Treatment includes ^{131}I therapy, or surgery to remove adenoma after antithyroid drugs achieve a euthyroid state.

▶ *Thyrotoxicosis factitia* results from chronic ingestion of thyroid hormone for thyrotropin suppression in patients with thyroid carcinoma, or from thyroid hormone abuse by people who are trying to lose weight.

▶ *Functioning metastatic thyroid carcinoma* is a rare disease that causes excess production of thyroid hormone.

▶ *Thyroid-stimulating hormone-secreting pituitary tumor* causes overproduction of thyroid hormone.

▶ *Subacute thyroiditis* is a virus-induced granulomatous inflammation of the thyroid, producing transient hyperthyroidism associated with fever, pain, pharyngitis, and tenderness in the thyroid gland.

▶ *Silent thyroiditis* is a self-limiting, transient form of hyperthyroidism, with histologic thyroiditis but no inflammatory symptoms.

▶ increased serum phosphorus level

▶ electrocardiography showing prolonged QT and ST intervals due to hypocalcemia

▶ inflating a blood pressure cuff on the upper arm to between diastolic and systolic blood pressure and maintaining this inflation for 3 minutes, eliciting Trousseau's sign (carpal spasm), to show clinical evidence of hypoparathyroidism.

Treatment

Treatment of hypothyroidism includes:

▶ immediate I.V. calcium salts, such as 10% calcium gluconate, to increase serum calcium levels (acute, life-threatening tetany)

▶ breathing into a paper bag and inhaling one's own carbon dioxide causes a mild respiratory acidosis that increases serum calcium levels (awake patient able to cooperate)

▶ sedatives and anticonvulsants to control spasms until calcium levels increase

▶ maintenance therapy with oral calcium and vitamin D supplements (chronic tetany)

▶ vitamin D and calcium supplements because of calcium absorption from the small intestine needing the presence of vitamin D (treatment of reversible disease, usually lifelong)

▶ calcitriol (Calcijex) if hepatic or renal problems make the patient unable to tolerate vitamin D.

Hyperthyroidism

Hyperthyroidism, or thyrotoxicosis, is a metabolic imbalance that results from the overproduction of thyroid hormone. The most common form is Graves' disease, which increases thyroxine (T_4) production, enlarges the thyroid gland (goiter), and causes multiple system changes. (See *Other forms of hyperthyroidism*.)

AGE ALERT The incidence of Graves' disease is greatest in women between the ages of 30 and 60 years, especially those with a family history of thyroid abnormalities; only 5% of the patients are younger than 15 years.

With treatment, most patients can lead normal lives. However, thyroid storm — an acute, severe exacerbation of thyrotoxicosis — is a medical emergency that may have life-threatening cardiac, hepatic, or renal consequences.

Causes

Thyrotoxicosis may result from both genetic and immunologic factors, including:

▶ increased incidence in monozygotic twins, pointing to an inherited factor, probably autosomal recessive gene

▶ occasional coexistence with other endocrine abnormalities, such as type 1 diabetes mellitus, thyroiditis, and hyperparathyroidism

▶ defect in suppressor T-lymphocyte function permitting production of autoantibodies (thyroid-stimulating immunoglobulin and thyroid-stimulating hormone [TSH]-binding inhibitory immunoglobulin)

▶ clinical thyrotoxicosis precipitated by excessive dietary intake of iodine or possibly stress (patients with latent disease)

▶ stress, such as surgery, infection, toxemia of pregnancy, or diabetic ketoacidosis, can precipitate thyroid storm (inadequately treated thyrotoxicosis).

Pathophysiology

The thyroid gland secretes the thyroid precursor, T_4, thyroid hormone or triiodothyronine (T_3), and calcitonin. T_4 and T_3 stimulate protein, lipid, and carbohydrate metabolism primarily through catabolic pathways. Calcitonin removes calcium from the blood and incorporates it into bone.

Biosynthesis, storage, and release of thyroid hormones are controlled by the hypothalamic-pituitary axis through a negative-feedback loop. Thyrotropin-releasing hormone (TRH) from the hypothalamus stimulates the release of TSH by the pituitary. Circulating T_3 levels provide negative feedback through the hypothalamus to decrease TRH levels, and through the pituitary to decrease TSH levels.

Although the exact mechanism isn't understood, hyperthyroidism has a hereditary component, and it is frequently associated with other autoimmune endocrinopathies.

Graves' disease is an autoimmune disorder characterized by the production of autoantibodies that attach to and then stimulate TSH receptors on the thyroid gland. A goiter is an enlarged thyroid gland, either the result of increased stimulation or a response to increased metabolic demand. The latter occurs in iodine-deficient areas of the world, where the incidence of goiter increases during puberty (a time of increased metabolic demand). These goiters often regress to normal size after puberty in males, but not in females. Sporadic goiter in non–iodine-deficient areas is of unknown origin. Endemic and sporadic goiters are nontoxic and may be diffuse or nodular. Toxic goiters may be uninodular or multinodular and may secrete excess thyroid hormone.

Pituitary tumors with TSH-producing cells are rare, as is hypothalamic disease causing TRH excess.

Signs and symptoms

Signs and symptoms of hyperthyroidism include:

▶ enlarged thyroid (goiter)
▶ nervousness
▶ heat intolerance and sweating
▶ weight loss despite increased appetite
▶ frequent bowel movements
▶ tremor and palpitations
▶ exophthalmos (characteristic, but absent in many patients with thyrotoxicosis).

Other signs and symptoms, common because thyrotoxicosis profoundly affects virtually every body system, include:

◗ difficulty concentrating due to accelerated cerebral function; excitability or nervousness caused by increased basal metabolic rate from T_4; fine tremor, shaky handwriting, and clumsiness from increased activity in the spinal cord area that controls muscle tone; emotional instability and mood swings ranging from occasional outbursts to overt psychosis (central nervous system)

◗ moist, smooth, warm, flushed skin (patient sleeps with minimal covers and little clothing); fine, soft hair; premature patchy graying and increased hair loss in both sexes; friable nails and onycholysis (distal nail separated from the bed); pretibial myxedema (nonpitting edema of the anterior surface of the legs, dermopathy), producing thickened skin; accentuated hair follicles; sometimes itchy or painful raised red patches of skin with occasional nodule formation; microscopic examination showing increased mucin deposits (skin, hair, and nails)

◗ systolic hypertension, tachycardia, full bounding pulse, wide pulse pressure, cardiomegaly, increased cardiac output and blood volume, visible point of maximal impulse, paroxysmal supraventricular tachycardia and atrial fibrillation (especially in elderly people), and occasional systolic murmur at the left sternal border (cardiovascular system)

◗ increased respiratory rate, dyspnea on exertion and at rest, possibly due to cardiac decompensation and increased cellular oxygen use (respiratory system)

◗ excessive oral intake with weight loss; nausea and vomiting due to increased GI motility and peristalsis; increased defecation; soft stools or, in severe disease, diarrhea; liver enlargement (GI system)

◗ weakness, fatigue, and muscle atrophy; rare coexistence with myasthenia gravis; possibly generalized or localized paralysis associated with hypokalemia; and, rarely, acropachy (soft-tissue swelling accompanied by underlying bone changes where new bone formation occurs) (musculoskeletal system)

◗ oligomenorrhea or amenorrhea, decreased fertility, increased incidence of spontaneous abortion (females), gynecomastia due to increased estrogen levels (males), diminished libido (both sexes) (reproductive system)

◗ exophthalmos due to combined effects of accumulated mucopolysaccharides and fluids in the retro-orbital tissues, forcing the eyeball outward and lid retraction, thereby producing characteristic staring gaze; occasional inflammation of conjunctivae, corneas, or eye muscles; diplopia; and increased tearing (eyes).

When thyrotoxicosis escalates to thyroid storm, these symptoms may occur:
◗ extreme irritability, hypertension, tachycardia, vomiting, temperature up to 106° F (41.1° C), delirium, and coma.

AGE ALERT Consider apathetic thyrotoxicosis, a morbid condition resulting from overactive thyroid, in elderly patients with atrial fibrillation or depression.

Complications

Possible complications include:
◗ muscle wasting
◗ visual loss or diplopia
◗ cardiac failure
◗ hypoparathyroidism after surgical removal of thyroid
◗ hypothyroidism after radioiodine treatment.

Diagnosis

The diagnosis of thyrotoxicosis is usually straightforward. It depends on a careful clinical history and physical examination, a high index of suspicion, and routine hormone determinations. The following tests confirm the disorder:
◗ radioimmunoassay showing increased serum T_4 and T_3 levels
◗ low TSH levels
◗ thyroid scan showing increased uptake of radioactive iodine 131 (^{131}I) in Graves' disease and, usually, in toxic multinodular goiter and toxic adenoma; low radioactive uptake in thyroiditis and thyro-

toxic factitia (test contraindicated in pregnancy)
▶ ultrasonography confirming subclinical ophthalmopathy.

Treatment
The primary forms of therapy include:
▶ antithyroid drugs
▶ single oral dose of ^{131}I
▶ surgery.
　Appropriate treatment depends on:
▶ severity of thyrotoxicosis
▶ causes
▶ patient age and parity
▶ how long surgery will be delayed (if patient is appropriate candidate for surgery).

　Antithyroid therapy includes antithyroid drugs for children, young adults, pregnant women, and patients who refuse surgery or ^{131}I treatment. Antithyroid drugs are preferred in patients with new-onset Graves' disease because of spontaneous remission in many of these patients; they are also used to correct the thyrotoxic state in preparation for ^{131}I treatment or surgery. Treatment options include:
▶ thyroid hormone antagonists, including propylthiouracil (PTU) and methimazole (Tapazole), to block thyroid hormone synthesis (hypermetabolic symptoms subside within 4 to 8 weeks after therapy begins, but remission of Graves' disease requires continued therapy for 6 months to 2 years)
▶ propranolol (Inderal) until antithyroid drugs reach their full effect, to manage tachycardia and other peripheral effects of excessive hypersympathetic activity resulting from blocking the conversion of T_4 to the active T_3 hormone
▶ minimum dosage needed to keep maternal thyroid function within the high-normal range until delivery, and to minimize the risk for fetal hypothyroidism; propylthiouracil (PTU) is preferred agent (during pregnancy)
▶ possibly antithyroid medications and propranolol for neonates for 2 to 3 months because most infants of hyperthyroid mothers are born with mild and transient thyrotoxicosis caused by placental trans-

fer of thyroid-stimulating immunoglobulins (neonatal thyrotoxicosis)
▶ continuous control of maternal thyroid function because thyrotoxicosis is sometimes exacerbated in the puerperal period; antithyroid drugs gradually tapered and thyroid function reassessed after 3 to 6 months postpartum
▶ periodic checks of infant's thyroid function with a breast-feeding mother on low-dose antithyroid treatment due to possible presence of small amounts of the drug in breast milk, which can rapidly lead to thyrotoxicity in the neonate
▶ single oral dose of ^{131}I (treatment of choice for patients not planning to have children; patients of reproductive age must give informed consent for this treatment, because ^{131}I concentrates in the gonads).

　During treatment with ^{131}I, the thyroid gland picks up the radioactive element as it would regular iodine. The radioactivity destroys some of the cells that normally concentrate iodine and produce T_4, thus decreasing thyroid hormone production and normalizing thyroid size and function.

　In most patients, hypermetabolic symptoms diminish 6 to 8 weeks after such treatment. However, some patients may require a second dose. Almost all patients treated with ^{131}I eventually become hypothyroid.

　Treatment with surgery includes:
▶ subtotal thyroidectomy to decrease the thyroid gland's capacity for hormone production (patients who refuse or aren't candidates for ^{131}I treatment)
▶ iodides (Lugol's solution or saturated solution of potassium iodide), antithyroid drugs, and propranolol to relieve hyperthyroidism preoperatively (if patient doesn't become euthyroid, surgery should be delayed, and antithyroid drugs and propranolol given to decrease the systemic effects [cardiac arrhythmias] of thyrotoxicosis)
▶ lifelong regular medical supervision because most patients become hypothyroid, sometimes as long as several years after surgery.

　Treatment for ophthalmopathy includes:

- local application of topical medications, such as prednisone acetate suspension, but may require high doses of corticosteroids
- external-beam radiation therapy or surgical decompression (severe exophthalmos causing pressure on optic nerve and orbital contents).

Treatment for thyroid storm includes:
- antithyroid drug to stop conversion of T_4 to T_3 and to block sympathetic effect; corticosteroids to inhibit the conversion of T_4 to T_3; and iodide to block the release of thyroid hormone
- supportive measures, including the administration of nutrients, vitamins, fluids, and sedatives.

Hypopituitarism

Hypopituitarism, also known as panhypopituitarism, is a complex syndrome marked by metabolic dysfunction, sexual immaturity, and growth retardation (when it occurs in childhood). The cause is a deficiency of the hormones secreted by the anterior pituitary gland. Panhypopituitarism is a partial or total failure of all six of this gland's vital hormones — corticotropin (ACTH), thyroid-stimulating hormone (TSH), luteinizing hormone (LH), follicle-stimulating hormone (FSH), human growth hormone, and prolactin. Partial and complete forms of hypopituitarism affect adults and children; in children, these diseases may cause dwarfism and delayed puberty. The prognosis may be good with adequate replacement therapy and correction of the underlying causes.

Primary hypopituitarism usually develops in a predictable pattern. It generally starts with decreased gonadotropin (FSH and LH) levels and consequent hypogonadism, reflected by cessation of menses in women and impotence in men. Growth hormone deficiency follows, causing short stature, delayed growth, and delayed puberty in children. Subsequent decreased TSH levels cause hypothyroidism, and, finally, decreased ACTH levels result in adrenal insufficiency. When hypopituitarism follows surgical ablation or trauma, the pattern of hormonal events may not necessarily follow that sequence. Damage to the hypothalamus or neurohypophysis may cause diabetes insipidus.

Causes

Hypopituitarism may be primary or secondary. Primary hypopituitarism may be caused by:
- tumor of the pituitary gland
- congenital defects (hypoplasia or aplasia of the pituitary gland)
- pituitary infarction (most often from postpartum hemorrhage)
- partial or total hypophysectomy by surgery, irradiation, or chemical agents
- granulomatous disease, such as tuberculosis
- idiopathic or autoimmune origin (occasionally).

Secondary hypopituitarism is caused by:
- deficiency of releasing hormones produced by the hypothalamus, either idiopathic or resulting from infection, trauma, or a tumor.

Pathophysiology

Hypopituitarism describes the low secretion of an anterior pituitary hormone, and panhypopituitarism describes the low secretion of all anterior pituitary hormones. Both can result from malfunction of the pituitary gland or the hypothalamus. The result is a lack of stimulation of target endocrine organs and some degree of deficiency of the target organ hormone, which may not be discovered until the body is stressed and the expected increases in secretions from the target organs don't occur.

Signs and symptoms

Signs and symptoms of hypopituitarism include:
- ACTH deficiency, causing weakness, fatigue, weight loss, fasting hypoglycemia, and altered mental function due to hypocortisolism; loss of axillary and pu-

bic hair due to androgen deficiency in females; orthostatic hypotension and hyponatremia due to aldosterone deficiency
▶ TSH deficiency, causing weight gain, constipation, cold intolerance, fatigue, and coarse hair
▶ gonadotropin deficiency, causing sexual dysfunction and infertility
▶ antidiuretic hormone deficiency, causing diabetes insipidus
▶ prolactin deficiency, causing lactation dysfunction or gynecomastia.

Complications
Possible complications include:
▶ blindness
▶ adrenal crisis.

Diagnosis
Diagnosis of hypopituitarism includes:
▶ hormonal deficiency of the tropic and target organ hormone(s) affected, chosen after evaluation of clinical picture
▶ computed tomography or magnetic resonance imaging of pituitary and target glands, showing destruction of the anterior pituitary or atrophy of target glands (adrenal cortex, thyroid, or gonads).

Treatment
The most effective treatment for hypopituitarism is:
▶ replacement of hormones secreted by the target glands (cortisol, thyroxine, and androgen or cyclic estrogen); prolactin not replaced
▶ clomiphene or cyclic gonadotropin-releasing hormone to induce ovulation in the patient of reproductive age.

Hypothyroidism in adults
Hypothyroidism results from hypothalamic, pituitary, or thyroid insufficiency or resistance to thyroid hormone. The disorder can progress to life-threatening myxedema coma. Hypothyroidism is more prevalent in women than men; in the United States, the incidence is increasing significantly in people ages 40 to 50.

AGE ALERT Hypothyroidism occurs primarily after the age of 40. After 65 years of age, the prevalence increases to as much as 10% in females and 3% in males.

Pathophysiology
Hypothyroidism may reflect a malfunction of the hypothalamus, pituitary, or thyroid gland, all of which are part of the same negative-feedback mechanism. However, disorders of the hypothalamus and pituitary rarely cause hypothyroidism. Primary hypothyroidism is most common.

Chronic autoimmune thyroiditis, also called chronic lymphocytic thyroiditis, occurs when autoantibodies destroy thyroid gland tissue. Chronic autoimmune thyroiditis associated with goiter is called Hashimoto's thyroiditis. The cause of this autoimmune process is unknown, although heredity has a role, and specific human leukocyte antigen subtypes are associated with greater risk.

Outside the thyroid, antibodies can reduce the effect of thyroid hormone in two ways. First, antibodies can block the thyroid-stimulating hormone (TSH) receptor and prevent the production of TSH. Second, cytotoxic antithyroid antibodies may attack thyroid cells.

Subacute thyroiditis, painless thyroiditis, and postpartum thyroiditis are self-limited conditions that usually follow an episode of hyperthyroidism. Untreated subclinical hypothyroidism in adults is likely to become overt at a rate of 5% to 20% per year.

Causes
Causes of hypothyroidism in adults include:
▶ inadequate production of thyroid hormone, usually after thyroidectomy or radiation therapy (particularly with iodine 131 [^{131}I]), or due to inflammation, chronic autoimmune thyroiditis (Hashimoto's disease), or such conditions as amyloidosis and sarcoidosis (rare)

CLINICAL FINDINGS IN ACQUIRED HYPOTHYROIDISM

Typical findings in acquired hypothyroidism are listed below:

HISTORY	PHYSICAL EXAMINATION
Arthritis	Anemia
Constipation	Bradycardia
Decreased sociability	Brittle hair
Drowsiness	Cool skin
Dry skin	Delayed relaxation of reflexes
Fatigue	Dementia
Intolerance to cold	Dry skin
Lethargy	Gravelly voice
Memory impairment	Hypothermia
Menstrual disorders	Large tongue
Muscle cramps	Loss of lateral third of eyebrow
Psychosis	Puffy face and hands
Somnolence	Slow speech
Weakness	Weight changes

Adapted with permission from Martinez, M., et al. "Making sense of hypothyroidism: An approach to testing and treatment," *Postgraduate Medicine* 93(6);143, 1993.

pituitary failure to produce TSH, hypothalamic failure to produce thyrotropin-releasing hormone (TRH), inborn errors of thyroid hormone synthesis, iodine deficiency (usually dietary), or use of such antithyroid medications as propylthiouracil (PTU).

Signs and symptoms

Signs and symptoms of hypothyroidism include:

weakness, fatigue, forgetfulness, sensitivity to cold, unexplained weight gain, and constipation (typical, vague, early clinical features) (See *Clinical findings in acquired hypothyroidism.*)

characteristic myxedematous signs and symptoms of decreasing mental stability; coarse, dry, flaky, inelastic skin; puffy face, hands, and feet; hoarseness; periorbital edema; upper eyelid droop; dry, sparse hair; and thick, brittle nails (as disorder progresses)

cardiovascular involvement, including decreased cardiac output, slow pulse rate, signs of poor peripheral circulation, and, occasionally, an enlarged heart.

Other common effects include:

anorexia, abdominal distention, menorrhagia, decreased libido, infertility, ataxia, and nystagmus; reflexes with delayed relaxation time (especially in the Achilles' tendon)

progression to myxedema coma, usually gradual but may develop abruptly, with stress aggravating severe or prolonged hypothyroidism, including progressive stupor, hypoventilation, hypoglycemia, hyponatremia, hypotension, and hypothermia.

Diagnosis

Diagnosis of hypothyroidism is based on:

radioimmunoassay showing low triiodothyronine (T_3) and thyroxine (T_4) levels

increased TSH level with cause of thyroid disorder; decreased with hypothalamic or pituitary disorder cause

thyroid panel differentiating primary hypothyroidism (thyroid gland hypofunction), secondary hypothyroidism (pituitary hyposecretion of TSH), tertiary hypothyroidism (hypothalamic hyposecretion of TRH), and euthyroid sick syndrome (impaired peripheral conversion of thyroid hormone due to a suprathyroidal illness,

THYROID TEST RESULTS IN HYPOTHYROIDISM

DYSFUNCTION INVOLVES	THYROTROPIN-RELEASING HORMONE	THYROID-STIMULATING HORMONE	TH (T₃ AND T₄)
Hypothalamus	Low	Low	Low
Pituitary	High	Low	Low
Thyroid gland	High	High	Low
Peripheral conversion of thyroid hormone (TH)	High	Low or normal	T_3 and T_4 low, but reverse T_3 elevated

such as severe infection) (See *Thyroid test results in hypothyroidism.*)
▶ elevated serum cholesterol, alkaline phosphatase, and triglyceride levels
▶ normocytic, normochromic anemia
▶ low serum sodium levels, decreased pH, and increased partial pressure of carbon dioxide, indicating respiratory acidosis (myxedema coma).

Complications
Possible complications are:
▶ heart failure
▶ myxedema coma
▶ infection
▶ megacolon
▶ organic psychosis
▶ infertility.

Treatment
Treatment includes:
▶ gradual thyroid hormone replacement with T_4 and, occasionally, T_3
▶ surgical excision, chemotherapy, or radiation for tumors.

 AGE ALERT Elderly patients should be started on a very low dose of T_4 to avoid cardiac problems; TSH levels guide gradual increases in dosage.

Hypothyroidism in children
A deficiency of thyroid hormone secretion during fetal development and early infancy results in infantile cretinism (congenital hypothyroidism). Hypothyroidism in infants is seen as respiratory difficulties, cyanosis, persistent jaundice, lethargy, somnolence, large tongue, abdominal distention, poor feeding, and hoarse crying. Prompt treatment of hypothyroidism in infants prevents physical and mental retardation. Older children who become hypothyroid have similar symptoms to those of adults, plus poor skeletal growth and late epiphyseal maturation and dental development. Sexual maturation may be accelerated in younger children and delayed in older children.

Cretinism is three times more common in girls than boys. Early diagnosis and treatment allow the best prognosis; infants treated before the age of 3 months usually grow and develop normally. Athyroid children who remain untreated beyond the age of 3 months, and children with acquired hypothyroidism who remain untreated beyond the age of 2 years, have irreversible mental retardation; their skeletal abnormalities are reversible with treatment.

Causes
Causes include:
▶ defective embryonic development (most common cause), causing congenital absence or underdevelopment of the thyroid gland (cretinism in infants)

▶ inherited autosomal recessive defect in the synthesis of thyroxine (next most common cause)

▶ antithyroid drugs taken during pregnancy, causing cretinism in infants (less frequently)

▶ chronic autoimmune thyroiditis (cretinism after age 2 years).

Pathophysiology

Hypothyroidism in infants and children is related to decreased thyroid hormone production or secretion. Loss of functional thyroid tissue can be caused by an autoimmune process. Defective thyroid synthesis may be related to congenital defects, with thyroid dysgenesis (defective development) the most common. Iodine deficiency or antithyroid drugs used by the mother during pregnancy can also contribute. Hypothyroidism may also be related to decreased thyroid-stimulating hormone (TSH) secretion or resistance to TSH.

Signs and symptoms

Signs and symptoms include:

▶ infant with infantile cretinism will have normal weight and length at birth, with characteristic signs developing within 3 to 6 months; delayed onset of most symptoms until weaning from breast-feeding due to small amounts of thyroid hormone in breast milk

▶ typically, an infant with cretinism sleeps excessively, seldom cries (except for occasional hoarse crying), and is inactive; parents may describe a "good baby — no trouble at all" (behavior actually due to reduced metabolism and progressive mental impairment)

▶ abnormal deep tendon reflexes, hypotonic abdominal muscles, protruding abdomen, and slow, awkward movements

▶ feeding difficulties, constipation, and jaundice because the immature liver can't conjugate bilirubin

▶ large, protruding tongue obstructing respiration; loud and noisy breathing through open mouth; dyspnea on exertion; anemia; abnormal facial features, such as a short forehead, puffy wide-set eyes (periorbital edema), wrinkled eyelids, a broad short and upturned nose, and a dull expression reflecting mental retardation

▶ cold, mottled skin due to poor circulation; and dry, brittle, and dull hair

▶ teeth erupting late and decaying early, below-normal body temperature, and slow pulse rate

▶ growth retardation shown as short stature, due to delayed epiphyseal maturation, particularly in the legs; obesity; and head appearing abnormally large due to stunted arms and legs; delayed or accelerated sexual development; mental retardation can be prevented by appropriate treatment if child acquires hypothyroidism after the age of 2 years.

Complications

Complications include:

▶ irreversible mental retardation (for hypothyroid infant not treated by the age of 3 months; early treatment helps prevent retardation)

▶ learning disabilities

▶ short stature

▶ accelerated or delayed sexual maturation.

Diagnosis

Diagnosis is based on:

▶ elevated TSH level associated with low T_3 and T_4 levels pointing to cretinism (because early detection and treatment can minimize the effects of cretinism, many states require measurement of infant thyroid hormone levels at birth)

▶ thyroid scan and ^{131}I uptake tests showing decreased uptake and confirming the absence of thyroid tissue in athyroid children

▶ increased gonadotropin levels compatible with sexual precocity in older children may coexist with hypothyroidism

▶ electrocardiogram showing bradycardia and flat or inverted T waves in untreated infants

▶ hip, knee, and thigh X-rays showing absence of the femoral or tibial epiphyseal line and markedly delayed skeletal development relative to chronological age
▶ low T_4 and normal TSH levels suggesting hypothyroidism secondary to hypothalamic or pituitary disease (rare).

Treatment
Early detection is mandatory to prevent irreversible mental retardation and permit normal physical development. Treatment includes:
▶ oral levothyroxine (Synthroid), beginning with moderate doses and gradually increasing to levels sufficient for lifelong maintenance (rapid increase in dosage may precipitate thyrotoxicity); proportionately higher doses in children than in adults because children metabolize thyroid hormone more quickly (infants younger than age 1).

Syndrome of inappropriate antidiuretic hormone
The syndrome of inappropriate antidiuretic hormone secretion (SIADH) results when excessive ADH secretion is triggered by stimuli other than increased extracellular fluid osmolarity and decreased extracellular fluid volume, reflected by hypotension. SIADH is a relatively common complication of surgery or critical illness. The prognosis varies with the degree of disease and the speed at which it develops. SIADH usually resolves within 3 days of effective treatment.

Causes
The most common cause of SIADH is small-cell carcinoma of the lung, which secretes excessive levels of ADH or vasopressin-like substances. Other neoplastic diseases — such as pancreatic and prostatic cancer, Hodgkin's disease, and thymoma (tumor on the thymus) — may also trigger SIADH.
    Less common causes include:
▶ central nervous system disorders, including brain tumor or abscess, cerebrovascular accident, head injury, and Guillain-Barré syndrome
▶ pulmonary disorders, including pneumonia, tuberculosis, lung abscess, and positive-pressure ventilation
▶ drugs, including chlorpropamide (Diabinase), vincristine (Oncovin), cyclophosphamide (Cytoxin), carbamazepine (Tegretol), clofibrate (Atromid-S), metoclopramide (Reglan), and morphine
▶ miscellaneous conditions, including psychosis and myxedema.

Pathophysiology
In the presence of excessive ADH, excessive water reabsorption from the distal convoluted tubule and collecting ducts causes hyponatremia and normal to slightly increased extracellular fluid volume.

Signs and symptoms
Signs and symptoms of SIADH include:
▶ thirst, anorexia, fatigue, and lethargy (first signs), followed by vomiting and intestinal cramping due to hyponatremia and electrolyte imbalance manifestations
▶ water retention and decreased urinary output due to hyponatremia
▶ additional neurologic symptoms, such as restlessness, confusion, anorexia, headache, irritability, decreasing reflexes, seizures, and coma, due to electrolyte imbalances, worsening with the degree of water intoxication.

Complications
Complications of SIADH include:
▶ cerebral edema
▶ brain herniation
▶ central pontine myelinosis.

Diagnosis
SIADH is diagnosed by the following laboratory results:
▶ serum osmolarity less than 280 mOsm/kg of water
▶ hyponatremia (serum sodium less than 135 mEq/L); lower values indicating worse condition

▶ elevated urinary sodium level (more than 20 mEq/day)

▶ elevated serum ADH level.

Treatment

Treatment of SIADH includes:

▶ restricted water intake (500 to 1,000 ml/day) (symptomatic treatment)

▶ administration of 200 to 300 ml of 3% saline solution to increase serum sodium level (severe water intoxication)

▶ correction of underlying cause of SIADH when possible

▶ surgical resection, irradiation, or chemotherapy to alleviate water retention for SIADH resulting from cancer

▶ demeclocycline (Declomycin) to block the renal response to ADH (if fluid restriction is ineffective)

▶ furosemide (Lasix) with normal or hypertonic saline to maintain urine output and block ADH secretion.

14

The components of the renal system are the kidneys, ureters, bladder, and urethra. The kidneys, located retroperitoneally in the lumbar area, produce and excrete urine to maintain homeostasis. They regulate the volume, electrolyte concentration, and acid-base balance of body fluids; detoxify the blood and eliminate wastes; regulate blood pressure; and support red blood cell production (erythropoiesis). The ureters are tubes that extend from the kidneys to the bladder; their only function is to transport urine to the bladder. The bladder is a muscular bag that serves as reservoir for urine until it leaves the body through the urethra.

PATHOPHYSIOLOGIC MANIFESTATIONS

Wastes are eliminated from the body by urine formation — glomerular filtration, tubular reabsorption, and tubular secretion — and excretion. Glomerular filtration is the process of filtering the blood as it flows through the kidneys. The glomerulus of the renal tubule filters plasma and then reabsorbs the filtrate. Glomerular function depends on the permeability of the capillary walls, vascular pressure, and filtration pressure. The normal glomerular filtration rate (GFR) is about 120 ml/min. To prevent too much fluid from leaving the vascular system, tubular reabsorption opposes capillary filtration. Reabsorption takes place as capillary filtration progresses. When fluid filters through the capillaries, albumin, which doesn't pass through capillary walls, remains behind.

As the albumin concentration inside the capillaries increases, the capillaries begin to draw water back in by osmosis. This osmotic force controls the quantities of water and diffusible solutes that enter and leave the capillaries.

Anything that affects filtration or reabsorption affects total filtration effort. Capillary pressure and interstitial fluid colloid osmotic pressure affect filtration. Interstitial fluid pressure and plasma colloid osmotic pressure affect reabsorption.

Altered renal perfusion; renal disease affecting the vessels, glomeruli, or tubules; or obstruction to urine flow can slow the GFR. The results are retention of nitrogenous wastes (azotemia), such as blood urea nitrogen and creatinine, which are consequent to acute renal failure.

Capillary pressure

The renal arteries branch into five segmental arteries, which supply different areas of the kidneys. The segmental arteries then branch into several divisions from which the afferent arterioles and vasa recta arise. Renal veins follow a similar branching pattern — characterized by stellate vessels and segmental branches — and empty into the inferior vena cava. The tubular system receives its blood supply from a peritubular capillary network. The ureteral veins follow the arteries and drain into the renal vein. The bladder receives blood through vesical arteries. Vesical veins unite to form the pudendal plexus, which empties into the iliac veins. A rich lymphatic system drains the renal cortex, kidneys, ureters, and bladder.

Capillary pressure reflects mean arterial pressure (MAP). Increased MAP increases capillary pressure, which in turn increases the GFR. When MAP decreases, so do capillary pressure and GFR. Autoregulation of afferent and efferent arterioles minimizes and controls changes in capillary pressure, unless MAP exceeds 180 mm Hg or is less than 80 mm Hg.

Sympathetic branches from the celiac plexus, upper lumbar splanchnic and thoracic nerves, and intermesenteric and superior hypogastric plexuses, which surround the kidneys, innervate the kidneys. Similar numbers of sympathetic and parasympathetic nerves from the renal plexus, superior hypogastric plexus, and intermesenteric plexus innervate the ureters. Nerves that arise from the inferior hypogastric plexus innervate the bladder. The parasympathetic nerve supply to the bladder controls urination.

Increased sympathetic activity and angiotensin II constrict afferent and efferent arterioles, decreasing the capillary pressure. Because these changes affect both the afferent and efferent arterioles, they have no net effect on GFR.

Inadequate renal perfusion accounts for 40% to 80% of acute renal failure. Volume loss (as with GI hemorrhage, burns, diarrhea, and diuretic use), volume sequestration (as in pancreatitis, peritonitis, and rhabdomyolysis), or decreased effective circulating volume (as in cardiogenic shock and sepsis) may reduce circulating blood volume. Decreased cardiac output due to peripheral vasodilatation (by sepsis or drugs) or profound renal vasoconstriction (as in severe cardiac failure, hepatorenal syndrome, or with such drugs as nonsteroidal anti-inflammatories [NSAIDs]) also diminish renal perfusion.

Hypovolemia causes a decrease in MAP that triggers a series of neural and humoral responses: activation of the sympathetic nervous system and renin-angiotensin-aldosterone system, and release of arginine vasopressin. Prostaglandin-mediated relaxation of afferent arterioles and angiotensin II–mediated constriction of efferent arterioles maintain GFR. GRF decreases steeply if MAP decreases to less than 80 mm Hg. Drugs that block prostaglandin production (such as NSAIDs) can cause severe vasoconstriction and acute renal failure during hypotension.

Prolonged renal hypoperfusion causes acute tubular necrosis. Processes involving large renal vessels, microvasculature, glomeruli, or tubular interstitium cause intrinsic renal disease. Emboli or thrombi, aortic dissection, or vasculitis can occlude renal arteries. Cholesterol-rich atheroemboli can occur spontaneously or follow aortic instrumentation. If they lodge in medium and small renal arteries, they trigger an eosinophil-rich inflammatory reaction.

Interstitial fluid colloid osmotic pressure

Few plasma proteins and red blood cells are filtered out of the glomeruli, so interstitial fluid colloid osmotic pressure (the force of albumin in the interstitial fluid) remains low. Large quantities of plasma protein flow through glomerular capillaries. Size and surface charge keep albumin, globulin, and other large proteins from crossing the glomerular wall. Smaller proteins leave the glomerulus but are absorbed by the proximal tubule.

Injury to the glomeruli or peritubular capillaries can increase interstitial fluid colloid osmotic pressure, drawing fluid out of the glomerulus and the peritubular capillaries. Swelling and edema occur in Bowman's space and the interstitial space surrounding the tubule. Increased interstitial fluid pressure opposes glomerular filtration, causes collapse of the surrounding nephrons and peritubular capillaries, and leads to hypoxia and renal cell injury or death. When cells die, intracellular enzymes are released that stimulate immune and inflammatory reactions. This further contributes to swelling and edema.

A CLOSER LOOK AT THE GLOMERULUS

The normal internal structures separating the capillary lumen and the urinary space in the glomerulus are shown.

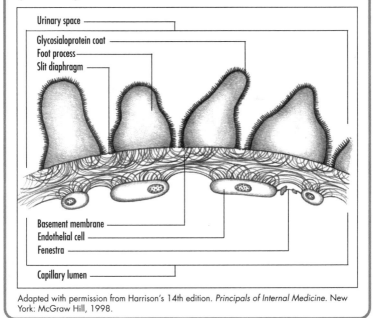

Urinary space

Glycosialoprotein coat
Foot process
Slit diaphragm

Basement membrane
Endothelial cell
Fenestra

Capillary lumen

Adapted with permission from Harrison's 14th edition. *Principals of Internal Medicine.* New York: McGraw Hill, 1998.

The resulting increase in interstitial fluid pressure can interfere with glomerular filtration and tubular reabsorption. Loss of glomerular filtration renders the kidney incapable of regulating blood volume and electrolyte composition. Diseases that damage the tubules cause tubular proteinuria because small proteins can move from capillaries into tubules.

Normal glomerular cells, which are endothelial in nature, form a barrier that holds cells and other particles back. The basement membrane typically traps larger proteins. The channels of the basement membrane are coated with glycoproteins that are rich in glutamate, aspartate, and sialic acid. This produces a negative charge barrier that impedes the passage of such anionic molecules as albumin. (See *A closer look at the glomerulus*).

Glomerular disease disrupts the basement membrane, allowing large proteins to leak out. Damage to epithelial cells permits albumin leakage. Hypoalbuminemia, as in nephrotic syndrome, is the result of excessive urine loss, increased renal catabolism, and inadequate hepatic synthesis. Plasma oncotic pressure decreases, and edema results as fluid moves from capillaries into the interstitium. Consequent activation of the renin-angiotensin system, AVP, and sympathetic nervous system increases renal salt and water reabsorption, which further contributes to edema. The severity of edema is directly related to the degree of hypoalbuminemia, and is exacerbated by heart disease or peripheral vascular disease.

CONGENITAL NEPHROPATHIES AND UROPATHIES

The following congenital conditions can affect kidney function:
▶ Renal hypoplasia — the kidney is small due to a reduction in the number of normally developed nephrons, and it may be unilateral or bilateral
▶ Renal dysplasia — the kidney is abnormally shaped, and the involved areas are nonfunctional
▶ Renal malrotation — the kidney is positioned abnormally
▶ Ectopic kidney — the kidney is located in the pelvic or thoracic area, causing reflux from the bladder into the ureters
▶ Horseshoe kidney — the lower poles of the kidneys are fused by an isthmus
▶ Obstructive uropathy — renal outlet obstruction due to abnormal vasculature, adhesions, kinks, or masses, usually causing hydronephrosis
▶ Ureterocele — a cystic dilation of the intravesicular ureter associated with duplication of the ureter.

From Hansen, M. *Pathophysiology: Foundations of Disease and Clinical Interventions.* Philadelphia: W.B. Saunders Company, 1998.

Plasma colloid pressure

Protein concentration of the plasma determines the plasma colloid pressure (the pulling force of albumin in the intravascular fluid), the major force on reabsorption of fluid into the capillaries. Plasma protein levels can decrease as a result of liver disease, protein loss in the urine, and protein malnutrition.

As oncotic pressure decreases, less fluid moves back into the capillaries and fluid begins to accumulate in the tubular and peritubular areas. Swelling around the tubule causes collapse of the tubule and peritubular capillaries, hypoxia, and death of the nephrons.

Diminished plasma oncotic pressure and urinary protein loss stimulate hepatic lipoprotein synthesis, and the resulting hyperlipedemia manifests as lipid bodies (fatty casts, oval fat bodies) in the urine. Metabolic disturbances result as other proteins are lost in the urine, including thyroxine-binding globulin, cholecalciferol-binding protein, transferrin, and metal-binding proteins. Urine losses of antithrombin III, decreased serum levels of proteins S and C, hyperfibrinogenemia, and enhanced platelet aggregation lead to a hypercoagulable state, as in nephrotic syndrome. Some patients also develop severe immunoglobulin G deficiency, which increases susceptibility to infection.

Structural variations

Variations in normal anatomic structure of the urinary tract occur in 10% to 15% of the total population and range from minor and easily correctable to lethal. Ectopic kidneys, which result if the embryonic kidneys do not ascend from the pelvis to the abdomen, function normally. If the embryonic kidneys fuse as they ascend, the single, U-shaped or horseshoe kidney causes no symptoms in about one third of affected persons. The most common problems associated with horseshoe kidneys include hydronephrosis, infection, and stone formation.

AGE ALERT Structural abnormalities of the renal system account for about 45% of the renal failure in children.

Urinary tract malformations are commonly associated with certain nonrenal anomalies. These characteristics include low-set and malformed ears, chromosomal disorders (especially trisomies 13 and 18), absent abdominal muscles, spinal cord and lower extremity anomalies, imperforate anus or genital deviation, Wilms' tumor, congenital ascites, cystic disease of the liver, and positive family history of renal disease (hereditary nephritis or cystic disease). (See *Congenital nephropathies and uropathies.*)

SOURCES OF URINARY FLOW OBSTRUCTION

Shown are the major sites of urinary tract obstruction.

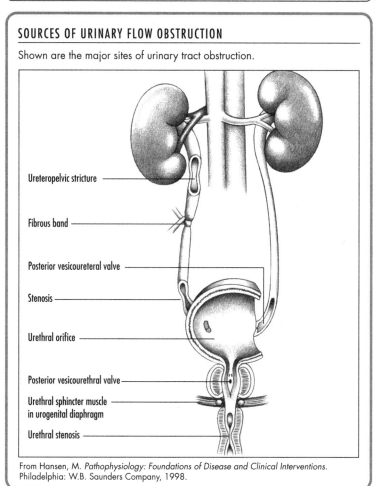

Ureteropelvic stricture

Fibrous band

Posterior vesicoureteral valve

Stenosis

Urethral orifice

Posterior vesicourethral valve

Urethral sphincter muscle
in urogenital diaphragm

Urethral stenosis

From Hansen, M. *Pathophysiology: Foundations of Disease and Clinical Interventions.*
Philadelphia: W.B. Saunders Company, 1998.

Obstruction

Obstruction along the urinary tract causes urine to accumulate behind the source of interference, leading to infection or damage. (See *Sources of urinary flow obstruction.*) Obstructions may be congenital or acquired. Causes include tumors, stones (calculi), trauma, edema, pregnancy, benign prostatic hyperplasia or carcinoma, inflammation of the GI tract, and loss of ureteral peristaltic activity or bladder muscle function.

Consequences of obstruction depend on the location and whether it is unilateral or bilateral, partial or complete, and acute or chronic, as well as the cause. For example, obstruction of a ureter causes hydroureter, or an accumulation of urine within the ureter, which increases retrograde pressure to the renal pelvis and calyces. As urine accumulates in the renal collection system, hydronephrosis results. If the obstruction is complete and acute in nature, increasing pressure transmitted to the proximal tubule inhibits glomerular filtration. If GFR declines to zero, the result is renal failure.

Chronic partial obstruction compresses structures as urine accumulates and the papilla and medulla infarct. The kidneys initially increase in size, but progressive atrophy follows, with eventual loss of renal mass. The underlying tubular damage decreases the kidney's ability to conserve sodium and water and excrete hydrogen ions and potassium; sodium and bicarbonate are wasted. Urine volume is excessive, even though GFR has declined. The result is an increased risk for dehydration and metabolic acidosis.

Tubular obstruction, caused by renal calculi or scarring from repeated infection, can increase interstitial fluid pressure. As fluid accumulates in the nephron, it backs up into Bowman's capsule and space. If the obstruction is unrelieved, nephrons and capillaries collapse, and renal damage is irreversible. The papillae, which are the final site of urine concentration, are particularly affected.

Relief of the obstruction is usually followed by copious diuresis of sodium and water retained during the period of obstruction, and a return to normal GFR. Excessive loss of sodium and water (more than 10 L/day) is uncommon. If GFR doesn't recover quickly, diuresis may not be significant after relief of the obstruction.

Unresolved obstruction can result in infection or even renal failure. Obstructions below the bladder cause urine to accumulate, forming a medium for bacterial growth.

 AGE ALERT Urinary tract infections are most common in girls aged 7 to 11 years. This is a result of bacteria ascending the urethra.

Cystitis is an infection of the bladder that results in mucosal inflammation and congestion. The detrusor muscle becomes hyperactive, decreasing bladder capacity and leading to reflux into the ureters. This transient reflux can cause acute or chronic pyelonephritis if bacteria ascend to the kidney.

Bilateral obstruction not relieved within 1 week of onset causes acute or chronic renal failure. Chronic renal failure progresses over weeks to months without symptoms until 90% of renal function is lost.

DISORDERS

Renal disorders include acute and chronic renal failure, glomerulonephritis, hypospadias and epispadias, nephrotic syndrome, neurogenic bladder, polycystic kidney, renal agenesis, renal calculi, and vesicoureteral reflux.

Acute renal failure

Acute renal failure, the sudden interruption of renal function, can be caused by obstruction, poor circulation, or underlying kidney disease. Whether prerenal, intrarenal, or postrenal, it usually passes through three distinct phases: oliguric, diuretic, and recovery. About 5% of all hospitalized patients develop acute renal failure. The condition is usually reversible with treatment, but if not treated, it may progress to end-stage renal disease, prerenal azotemia, and death.

Causes

Acute renal failure may be prerenal, intrarenal, or postrenal. Causes of prerenal failure include:
▶ arrhythmias
▶ cardiac tamponade
▶ cardiogenic shock
▶ heart failure
▶ myocardial infarction
▶ burns
▶ dehydration
▶ diuretic overuse
▶ hemorrhage
▶ hypovolemic shock
▶ trauma
▶ antihypertensive drugs
▶ sepsis
▶ arterial embolism
▶ arterial or venous thrombosis
▶ tumor
▶ disseminated intravascular coagulation
▶ eclampsia

- malignant hypertension
- vasculitis.
    Causes of intrarenal failure include:
- poorly treated prerenal failure
- nephrotoxins
- obstetric complications
- crush injuries
- myopathy
- transfusion reaction
- acute glomerulonephritis
- acute interstitial nephritis
- acute pyelonephritis
- bilateral renal vein thrombosis
- malignant nephrosclerosis
- papillary necrosis
- polyarteritis nodosa
- renal myeloma
- sickle cell disease
- systemic lupus erythematosus
- vasculitis.
    Causes of postrenal failure include:
- bladder obstruction
- ureteral obstruction
- urethral obstruction.

Pathophysiology
The pathophysiology of prerenal, intrarenal, and postrenal failure differ.

Prerenal failure. Prerenal failure ensues when a condition that diminishes blood flow to the kidneys leads to hypoperfusion. Examples include hypovolemia, hypotension, vasoconstriction, or inadequate cardiac output. Azotemia (excess nitrogenous waste products in the blood) develops in 40% to 80% of all cases of acute renal failure.

When renal blood flow is interrupted, so is oxygen delivery. The ensuing hypoxemia and ischemia can rapidly and irreversibly damage the kidney. The tubules are most susceptible to the effects of hypoxemia.

Azotemia is a consequence of renal hypoperfusion. The impaired blood flow results in decreased glomerular filtration rate (GFR) and increased tubular reabsorption of sodium and water. A decrease in GFR causes electrolyte imbalance and metabolic acidosis. Usually, restoring renal blood flow and glomerular filtration reverses azotemia.

Intrarenal failure. Intrarenal failure, also called intrinsic or parenchymal renal failure, results from damage to the filtering structures of the kidneys. Causes of intrarenal failure are classified as nephrotoxic, inflammatory, or ischemic. When the damage is caused by nephrotoxicity or inflammation, the delicate layer under the epithelium (the basement membrane) becomes irreparably damaged, often leading to chronic renal failure. Severe or prolonged lack of blood flow by ischemia may lead to renal damage (ischemic parenchymal injury) and excess nitrogen in the blood (intrinsic renal azotemia).

Acute tubular necrosis, the precursor to intrarenal failure, can result from ischemic damage to renal parenchyma during unrecognized or poorly treated prerenal failure; or from obstetric complications, such as eclampsia, postpartum renal failure, septic abortion, or uterine hemorrhage.

The fluid loss causes hypotension, which leads to ischemia. The ischemic tissue generates toxic oxygen-free radicals, which cause swelling, injury, and necrosis.

Another cause of acute failure is the use of nephrotoxins, including analgesics, anesthetics, heavy metals, radiographic contrast media, organic solvents, and antimicrobials, particularly aminoglycoside antibiotics. These drugs accumulate in the renal cortex, causing renal failure that manifests well after treatment or other toxin exposure. The necrosis caused by nephrotoxins tends to be uniform and limited to the proximal tubules, whereas ischemia necrosis tends to be patchy and distributed along various parts of the nephron.

Postrenal failure. Bilateral obstruction of urine outflow leads to postrenal failure. The cause may be in the bladder, ureters, or urethra.

Bladder obstruction can result from:

▶ anticholinergic drugs
▶ autonomic nerve dysfunction
▶ infection
▶ tumors.

Ureteral obstructions, which restrict blood flow from kidneys to bladder, can result from:

▶ blood clots
▶ calculi
▶ edema or inflammation
▶ necrotic renal papillae
▶ retroperitoneal fibrosis or hemorrhage
▶ surgery (accidental ligation)
▶ tumor or uric acid crystals.

Urethral obstruction can be the result of prostatic hyperplasia, tumor, or strictures.

The three types of acute renal failure (prerenal, intrarenal, or postrenal) usually pass through three distinct phases: oliguric, diuretic, and recovery.

Oliguric phase. Oliguria may be the result of one or several factors. Necrosis of the tubules can cause sloughing of cells, cast formations, and ischemic edema. The resulting tubular obstruction causes a retrograde increase in pressure and a decrease in GFR. Renal failure can occur within 24 hours from this effect. Glomerular filtration may remain normal in some cases of renal failure, but tubular reabsorption of filtrate may be accelerated. In this instance, ischemia may increase tubular permeability and cause backleak. Another concept is that intrarenal release of angiotensin II or redistribution of blood flow from the cortex to the medulla may constrict the afferent arterioles, increasing glomerular permeability and decreasing GFR.

Urine output may remain at less than 30 mL/hour or 400 mL/day for a few days to weeks. Before damage occurs, the kidneys respond to decreased blood flow by conserving sodium and water.

Damage impairs the kidney's ability to conserve sodium. Fluid (water) volume excess, azotemia (elevated serum levels of urea, creatinine, and uric acid), and electrolyte imbalance occur. Ischemic or toxic injury leads to the release of mediators and intrarenal vasoconstriction. Medullary hypoxia results in the swelling of tubular and endothelial cells, adherence of neutrophils to capillaries and venules, and inappropriate platelet activation. Increasing ischemia and vasoconstriction further limit perfusion.

Injured cells lose polarity, and the ensuing disruption of tight junctions between the cells promotes backleak of filtrate. Ischemia impairs the function of energy-dependent membrane pumps, and calcium accumulates in the cells. This excess calcium further stimulates vasoconstriction and activates proteases and other enzymes. Untreated prerenal oliguria may lead to acute tubular necrosis.

Diuretic phase. As the kidneys become unable to conserve sodium and water, the diuretic phase, marked by increased urine secretion of more than 400 ml/24 hours, ensues. GFR may be normal or increased, but tubular support mechanisms are abnormal. Excretion of dilute urine causes dehydration and electrolyte imbalances. High blood urea nitrogen (BUN) levels produce osmotic diuresis and consequent deficits of potassium, sodium, and water.

Recovery phase. If the cause of the diuresis is corrected, azotemia gradually disappears and recovery occurs. The diuretic phase may last days or weeks. The recovery phase is a gradual return to normal or near-normal renal function over 3 to 12 months.

AGE ALERT Even with treatment, the elderly patient is particularly susceptible to volume overload, precipitating acute pulmonary edema, hypertensive crisis, hyperkalemia, and infection.

Renal failure affects many of the body processes. Metabolic acidosis may be the result of decreased excretion of hydrogen ions. Anemia occurs from erythropoietinemia, glomerular filtration of erythrocytes, or bleeding associated with platelet dysfunction. Sepsis is also common because of decreased white blood cell–mediated immunity. Heart failure can result because of fluid overload and anemia, which cause additional workload to the heart. Anemia also causes tissue hypoxia, which then stimulates increased ventilation and work of breathing. Respiratory compensation for metabolic acidosis has a similar effect on the respiratory system. Abnormalities in quantities or function of anticoagulant proteins, coagulation factor, platelet, or endothelial mediators result in a hypercoagulable state. This results in bleeding or clotting difficulties. Altered mental status and peripheral sensation are believed to be due to effects on the highly sensitive cells of nerves secondary to retained toxins, hypoxia, electrolyte imbalance, and acidosis. The hypermetabolic state induced by this critical illness promotes tissue catabolism.

Signs and symptoms

Signs and symptoms of acute renal failure include:
▶ oliguria due to decreased GFR
▶ tachycardia due to hypotension
▶ hypotension due to hypovolemia
▶ dry mucous membranes due to stimulation of the sympathetic nervous system
▶ flat neck veins due to hypovolemia
▶ lethargy due to altered cerebral perfusion
▶ cool, clammy skin due to decreased cardiac output and heart failure.
    Progressive symptoms include:
▶ edema related to fluid retention
▶ confusion due to altered cerebral perfusion and azotemia
▶ GI symptoms due to altered metabolic status
▶ crackles on auscultation due to fluid in the lungs
▶ infection due to altered immune response
▶ seizures and coma related to alteration in consciousness
▶ hematuria, petechiae, and ecchymosis related to bleeding abnormalities.

Complications

Complications of acute renal failure may include:
▶ chronic renal failure
▶ ischemic parenchymal injury
▶ intrinsic renal azotemia
▶ electrolyte imbalance
▶ metabolic acidosis
▶ pulmonary edema
▶ hypertensive crisis
▶ infection.

Diagnosis

Diagnosis of acute renal failure is based on the following results:
▶ blood studies showing elevated BUN, serum creatinine, and potassium levels; decreased bicarbonate level, hematocrit, and hemoglobin; and acid pH
▶ urine studies showing casts, cellular debris, and decreased specific gravity; in glomerular diseases, proteinuria and urine osmolality close to serum osmolality; sodium level less than 20 mEq/L if oliguria results from decreased perfusion, and more than 40 mEq/L if cause is intrarenal
▶ creatinine clearance test measuring GFR and reflecting the number of remaining functioning nephrons
▶ electrocardiogram (ECG) showing tall, peaked T waves; widening QRS complex; and disappearing P waves if hyperkalemia is present
▶ ultrasonography, plain films of the abdomen, kidney-ureter-bladder radiography, excretory urography, renal scan, retrograde pyelography, computed tomographic scans, and nephrotomography.

Treatment

Treatment for acute renal failure includes:
▶ high-calorie diet that's low in protein, sodium, and potassium to meet metabolic needs
▶ I.V. therapy to maintain and correct fluid and electrolyte balance
▶ fluid restriction to minimize edema
▶ diuretic therapy to treat oliguric phase
▶ sodium polystyrene sulfonate (Kayexalate) by mouth or enema to reverse hy-

perkalemia with mild hyperkalemic symptoms (malaise, loss of appetite, muscle weakness)

❱ hypertonic glucose, insulin, and sodium bicarbonate I.V.— for more severe hyperkalemic symptoms (numbness and tingling and ECG changes)

❱ hemodialysis to correct electrolyte and fluid imbalances

❱ peritoneal dialysis to correct electrolyte and fluid imbalances.

Chronic renal failure

Chronic renal failure is usually the end result of gradual tissue destruction and loss of kidney function. It can also result from a rapidly progressing disease of sudden onset that destroys the nephrons and causes irreversible kidney damage.

Few symptoms develop until less than 25% of glomerular filtration remains. The normal parenchyma then deteriorates rapidly, and symptoms worsen as renal function decreases. This syndrome is fatal without treatment, but maintenance on dialysis or a kidney transplant can sustain life.

Causes

Chronic renal failure may be caused by:
❱ chronic glomerular disease (glomerulonephritis)
❱ chronic infection (such as chronic pyelonephritis and tuberculosis)
❱ congenital anomalies (polycystic kidney disease)
❱ vascular disease (hypertension, nephrosclerosis)
❱ obstruction (kidney stones)
❱ collagen disease (lupus erythematosus)
❱ nephrotoxic agents (long-term aminoglycoside therapy)
❱ endocrine disease (diabetic neuropathy).

Pathophysiology

Chronic renal failure often progresses through four stages. Reduced renal reserve shows a glomerular filtration rate (GFR) of 35% to 50% of normal; renal insufficiency has a GFR of 20% to 35% of normal; renal failure has a GFR of 20% to 25% of normal; and end-stage renal disease has a GFR less than 20% of normal.

Nephron damage is progressive; damaged nephrons can't function and don't recover. The kidneys can maintain relatively normal function until about 75% of the nephrons are nonfunctional. Surviving nephrons hypertrophy and increase their rate of filtration, reabsorption, and secretion. Compensatory excretion continues as GFR diminishes.

Urine may contain abnormal amounts of protein and red (RBCs) and white blood cells or casts, the major end products of excretion remain essentially normal, and nephron loss becomes significant. As GFR decreases, plasma creatinine levels increase proportionately without regulatory adjustment. As sodium delivery to the nephron increases, less is reabsorbed, and sodium deficits and volume depletion follow. The kidney becomes incapable of concentrating and diluting urine.

If tubular interstitial disease is the cause of chronic renal failure, primary damage to the tubules — the medullary portion of the nephron — precedes failure, as do such problems as renal tubular acidosis, salt wasting, and difficulty diluting and concentrating urine. If vascular or glomerular damage is the primary cause, proteinuria, hematuria, and nephrotic syndrome are more prominent.

Changes in acid-base balance affect phosphorus and calcium balance. Renal phosphate excretion and $1,25(OH)_2$ vitamin D_3 synthesis are diminished. Hypocalcemia results in secondary hypoparathyroidism, diminished GFR, and progressive hyperphosphatemia, hypocalcemia, and dissolution of bone. In early renal insufficiency, acid excretion and phosphate reabsorption increase to maintain normal pH. When GFR decreases by 30% to 40%, progressive metabolic acidosis ensues and tubular secretion of potassium increases. Total-body potassium levels may increase to life-threatening levels requiring dialysis.

In glomerulosclerosis, distortion of filtration slits and erosion of the glomerular epithelial cells lead to increased fluid transport across the glomerular wall. Large proteins traverse the slits but become trapped in glomerular basement membranes, obstructing the glomerular capillaries. Epithelial and endothelial injury cause proteinuria. Mesangial-cell proliferation, increased production of extracellular matrix, and intraglomerular coagulation cause the sclerosis.

Tubulointerstitial injury occurs from toxic or ischemic tubular damage, as with acute tubular necrosis. Debris and calcium deposits obstruct the tubules. The resulting defective tubular transport is associated with interstitial edema, leukocyte infiltration, and tubular necrosis. Vascular injury causes diffuse or focal ischemia of renal parenchyma, associated with thickening, fibrosis, or focal lesions of renal blood vessels. Decreased blood flow then leads to tubular atrophy, interstitial fibrosis, and functional disruption of glomerular filtration, medullary gradients, and concentration.

The structural changes trigger an inflammatory response. Fibrin deposits begin to form around the interstitium. Microaneurysms result from vascular wall damage and increased pressure secondary to obstruction or hypertension. Eventual loss of the nephron triggers compensatory hyperfunction of uninjured nephrons, which initiates a positive-feedback loop of increasing vulnerability.

Eventually, the healthy glomeruli are so overburdened that they become sclerotic, stiff, and necrotic. Toxins accumulate and potentially fatal changes ensue in all major organ systems.

Extrarenal consequences. Physiologic changes affect more than one system, and the presence and severity of manifestations depend on the duration of renal failure and its response to treatment. In some fluid and electrolyte imbalances, the kidneys can't retain salt, and hyponatremia results. Dry mouth, fatigue, nausea, hypotension, loss of skin turgor, and listlessness can progress to somnolence and confusion. Later, as the number of functioning nephrons decreases, so does the capacity to excrete sodium and potassium. Sodium retention leads to fluid overload and edema; the potassium overload leads to muscle irritability and weakness, and life-threatening cardiac arrhythmias.

As the cardiovascular system becomes involved, hypertension occurs, and distant heart sounds may be auscultated if pericardial effusion occurs. Bibasilar crackles and peripheral edema reflect cardiac failure.

Pulmonary changes include reduced macrophage activity and increasing susceptibility to infection. Decreased lung sounds in areas of consolidation reflect the presence of pneumonia. As the pleurae become more involved, the patient may experience pleuritic pain and friction rubs.

The GI mucosa becomes inflamed and ulcerated, and gums may also be ulcerated and bleeding. Stomatitis, uremic fetor (an ammonia smell to the breath), hiccups, peptic ulcer, and pancreatitis in end-stage renal failure are believed to be due to retention of metabolic acids and other metabolic waste products. Malnutrition may be secondary to anorexia, malaise, and reduced dietary intake of protein. The reduced protein intake also affects capillary fragility, and results in decreased immune functioning and poor wound healing.

Normochromic normocytic anemia and platelet disorders with prolonged bleeding time ensue as diminished erythropoietin secretion leads to reduced RBC production in the bone marrow. Uremic toxins associated with chronic renal failure shorten RBC survival time. The patient experiences lethargy and dizziness.

Demineralization of the bone (renal osteodystrophy) manifested by bone pain and pathologic fractures is due to several factors:
◗ decreased renal activation of vitamin D, decreasing absorption of dietary calcium

▶ retention of phosphate, increasing urinary loss of calcium

▶ increased circulation of parathyroid hormone due to decreased urinary excretion.

The skin acquires a grayish-yellow tint as urine pigments (urochromes) accumulate. Inflammatory mediators released by retained toxins in the skin cause pruritus. Uric acid and other substances in the sweat crystallize and accumulate on the skin as uremic frost. High plasma calcium levels are also associated with pruritus.

Restless leg syndrome (abnormal sensation and spontaneous movement of the feet and lower legs), muscle weakness, and decreased deep tendon reflexes are believed to result from the effect of toxins on the nervous system.

Chronic renal failure increases the risk for death from infection. This is related to suppression of cell-mediated immunity and a reduction in the number and function of lymphocytes and phagocytes.

All hormone levels are impaired, in both excretion and activation. Females may be anovulatory, amenorrheic, or unable to carry pregnancy to full term. Males tend to have decreased sperm counts and impotence.

Signs and symptoms

Signs and symptoms of chronic renal failure include:

▶ hypervolemia due to sodium retention

▶ hypocalcemia and hyperkalemia due to electrolyte imbalance

▶ azotemia due to retention of nitrogenous wastes

▶ metabolic acidosis due to loss of bicarbonate

▶ bone and muscle pain and fractures caused by calcium-phosphorus imbalance and consequent parathyroid malfunction

▶ peripheral neuropathy due to accumulation of toxins

▶ dry mouth, fatigue, and nausea due to hyponatremia

▶ hypotension due to sodium loss

▶ altered mental state due to hyponatremia and toxin accumulation

▶ irregular pulses due to hyperkalemia

▶ hypertension due to fluid overload

▶ gum sores and bleeding due to coagulopathies

▶ yellow-bronze skin due to altered metabolic processes

▶ dry, scaly skin and severe itching due to uremic frost

▶ muscle cramps and twitching, including cardiac irritability, due to hyperkalemia

▶ Kussmaul's respirations due to metabolic acidosis

AGE ALERT Growth retardation in children occurs from endocrine abnormalities induced by renal failure. Impaired bone growth and bowlegs in children are also due to rickets.

▶ infertility, decreased libido, amenorrhea, and impotence due to endocrine disturbances

▶ GI bleeding, hemorrhage, and bruising due to thrombocytopenia and platelet defects

▶ pain, burning, and itching in legs and feet associated with peripheral neuropathy

▶ infection related to decreased macrophage activity.

Complications

Possible complications of chronic renal failure include:

▶ anemia

▶ peripheral neuropathy

▶ cardiopulmonary complications

▶ GI complications

▶ sexual dysfunction

▶ skeletal defects

▶ paresthesias

▶ motor nerve dysfunction, such as foot drop and flaccid paralysis

▶ pathologic fractures.

Diagnosis

Blood study results that help diagnose chronic renal failure include:

▶ decreased arterial pH and bicarbonate, low hemoglobin and hematocrit

▶ decreased RBC survival time, mild thrombocytopenia, platelet defects

- elevated blood urea nitrogen, serum creatinine, sodium, and potassium levels
- increased aldosterone secretion related to increased renin production
- hyperglycemia (a sign of impaired carbohydrate metabolism)
- hypertriglyceridemia and low levels of high-density lipoprotein.

Urinalysis results aiding in diagnosis include:

- specific gravity fixed at 1.010
- proteinuria, glycosuria, RBCs, leukocytes, casts, or crystals, depending on the cause.

Other study results used to diagnose chronic renal failure include:

- reduced kidney size on kidney-ureter-bladder radiography, excretory urography, nephrotomography, renal scan, or renal arteriography
- renal biopsy to identify underlying disease
- EEG to identify metabolic encephalopathy.

Treatment

Treatment of chronic renal failure involves:

- low-protein diet, to limit accumulation of end products of protein metabolism that the kidneys can't excrete
- high-protein diet for patients on continuous peritoneal dialysis
- high-calorie diet, to prevent ketoacidosis and tissue atrophy
- sodium and potassium restrictions, to prevent elevated levels
- fluid restrictions, to maintain fluid balance
- loop diuretics, such as furosemide (Lasix), to maintain fluid balance
- digitalis glycosides, such as digoxin, to mobilize fluids causing edema
- calcium carbonate (Caltrate) or calcium acetate (PhosLo), to treat renal osteodystrophy by binding phosphate and supplementing calcium
- transfusions, to treat anemia
- antihypertensives, to control blood pressure and edema

- antiemetics, to relieve nausea and vomiting
- cimetidine (Tagamet) or rantidine (Zantac), to decrease gastric irritation
- methylcellulose or docusate, to prevent constipation
- iron and folate supplements or RBC transfusion for anemia
- synthetic erythropoietin, to stimulate the bone marrow to produce RBCs; supplemental iron, conjugated estrogens, and 1-desamino-8-D-arginine vasopressin (DDAVP), to combat hematologic effects
- antipruritics, such as trimeprazine (Temaril) or diphenhydramine (Benadryl), to relieve itching
- aluminum hydroxide gel (AlaGel), to reduce serum phosphate levels
- supplementary vitamins, particularly B and D, and essential amino acids
- dialysis for hyperkalemia and fluid imbalances
- oral or rectal administration of cation exchange resins, such as sodium polystyrene sulfonate (Kayexalate), and I.V. administration of calcium gluconate, sodium bicarbonate, 50% hypertonic glucose, and regular insulin, to reverse hyperkalemia
- emergency pericardiocentesis or surgery for cardiac tamponade
- intensive dialysis and thoracentesis, to relieve pulmonary edema and pleural effusion
- peritoneal or hemodialysis, to help control end-stage renal disease
- renal transplantation (often the treatment of choice if a donor is available).

Glomerulonephritis

Glomerulonephritis is a bilateral inflammation of the glomeruli, often following a streptococcal infection. Acute glomerulonephritis is also called *acute poststreptococcal glomerulonephritis.*

CHARACTERISTICS OF GLOMERULAR LESIONS

The types of glomerular lesions and their characteristics are:
▶ Diffuse lesions: relatively uniform, involve most or all glomeruli (for example, glomerulonephritis)
▶ Focal lesions: involve only some glomeruli; others normal
▶ Segmental-local: involve only one part of the glomerulus
▶ Mesangial: deposits of immunoglobulins in mesangial matrix
▶ Membranous: thickening of glomerular capillary wall
▶ Proliferative lesions: increased number of glomerular cells
▶ Sclerotic lesions: glomerular scarring from previous glomerular injury
▶ Crescent lesions: accumulation of proliferating cells in Bowman's space.

From Huether, S. *Understanding Pathophysiology.* St. Louis: Mosby, 1996.

AGE ALERT Acute glomerulonephritis is most common in boys aged 3 to 7 years, but it can occur at any age. Up to 95% of children and 70% of adults recover fully; the rest, especially elderly patients, may progress to chronic renal failure within months.

Rapidly progressive glomerulonephritis (RPGN) — also called subacute, crescentic, or extracapillary glomerulonephritis — most commonly occurs between the ages of 50 and 60. It may be idiopathic or associated with a proliferative glomerular disease, such as poststreptococcal glomerulonephritis.

AGE ALERT Goodpasture's syndrome, a type of rapidly progressive glomerulonephritis, is rare, but occurs most frequently in men aged 20 to 30 years.

Chronic glomerulonephritis is a slowly progressive disease characterized by inflammation, sclerosis, scarring, and, eventually, renal failure. It usually remains undetected until the progressive phase, which is usually irreversible.

Causes
Causes of acute and RPGN include:
▶ streptococcal infection of the respiratory tract
▶ impetigo
▶ immunoglobulin A (IgA) nephropathy (Berger's disease)
▶ lipoid nephrosis.
    Chronic glomerulonephritis is caused by:
▶ membranoproliferative glomerulonephritis
▶ membranous glomerulopathy
▶ focal glomerulosclerosis
▶ RPGN
▶ poststreptococcal glomerulonephritis
▶ systemic lupus erythematosus
▶ Goodpasture's syndrome
▶ hemolytic uremic syndrome.

Pathophysiology
In nearly all types of glomerulonephritis, the epithelial or podocyte layer of the glomerular membrane is disturbed. This results in a loss of negative charge. (See *Characteristics of glomerular lesions.*)

Acute poststreptococcal glomerulonephritis results from the entrapment and collection of antigen-antibody complexes in the glomerular capillary membranes, after infection with a group A beta-hemolytic streptococcus. The antigens, which are endogenous or exogenous, stimulate the formation of antibodies. Circulating antigen-antibody complexes become lodged in the glomerular capillaries. (See *Glomerulonephritis.*) Glomerular injury occurs when the complexes initiate complement activation and the release of immunologic substances that lyse cells and increase membrane permeability. Antibody damage to basement membranes causes crescent formation. The severity of glomerular damage and renal insufficiency is related to the size, number, lo-

GLOMERULONEPHRITIS

The immune complex depositions that occur in glomerulonephritis are shown.

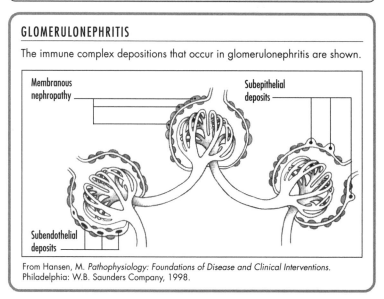

Membranous nephropathy

Subepithelial deposits

Subendothelial deposits

From Hansen, M. *Pathophysiology: Foundations of Disease and Clinical Interventions.*
Philadelphia: W.B. Saunders Company, 1998.

cation (focal or diffuse), duration of exposure, and type of antigen-antibody complexes.

Antibody or antigen-antibody complexes in the glomerular capillary wall activate biochemical mediators of inflammation — complement, leukocytes, and fibrin. Activated complement attracts neutrophils and monocytes, which release lysosomal enzymes that damage the glomerular cell walls and cause a proliferation of the extracellular matrix, affecting glomerular blood flow. Those events increase membrane permeability, which causes a loss of negative charge across the glomerular membrane as well as enhanced protein filtration.

Membrane damage leads to platelet aggregation, and platelet degranulation releases substances that increase glomerular permeability. Protein molecules and red blood cells (RBCs) can now pass into the urine, resulting in proteinuria or hematuria. Activation of the coagulation system leads to fibrin deposits in Bowman's space. The result is crescent formation and diminished renal blood flow and glomerular filtration rate (GFR). Glomerular bleed-

ing causes acidic urine, which transforms hemoglobin to methemoglobin and results in brown urine without clots.

The inflammatory response decreases GFR, which causes fluid retention and decreased urine output, extracellular fluid volume expansion, and hypertension. Gross proteinuria is associated with nephrotic syndrome. After 10 to 20 years, renal insufficiency develops and is followed by nephrotic syndrome and end-stage renal failure.

Goodpasture's syndrome is an RPGN in which antibodies are produced against the pulmonary capillaries and glomerular basement membrane. Diffuse intracellular antibody proliferation in Bowman's space leads to a crescent-shaped structure that obliterates the space. The crescent is composed of fibrin and endothelial, mesangial, and phagocytic cells, which compress the glomerular capillaries, diminish blood flow, and cause extensive scarring of the glomeruli. GFR is reduced, and renal failure occurs within weeks or months.

IgA nephropathy, or Berger's disease, is usually idiopathic. Plasma IgA level is elevated, and IgA and inflammatory cells are deposited into Bowman's space. The result is sclerosis and fibrosis of the glomerulus and a reduced GFR.

Lipid nephrosis causes disruption of the capillary filtration membrane and loss of its negative charge. This increased permeability with resultant loss of protein leads to nephrotic syndrome.

Systemic diseases, such as hepatitis B virus, systemic lupus erythematosus, or solid malignant tumors, cause a membranous nephropathy. An inflammatory process causes thickening of the glomerular capillary wall. Increased permeability and proteinuria lead to nephrotic syndrome.

Sometimes the immune complement further damages the glomerular membrane. The damaged and inflamed glomeruli lose the ability to be selectively permeable so that RBCs and proteins filter through as GFR decreases. Uremic poisoning may result. Renal function may deteriorate, especially in adults with sporadic acute post-streptococcal glomerulonephritis, often in the form of glomerulosclerosis accompanied by hypertension. The more severe the disorder, the more likely the occurrence of complications. Hypervolemia leads to hypertension, resulting from either sodium and water retention (caused by the decreased GFR) or inappropriate renin release. The patient develops pulmonary edema and heart failure. (See *Averting renal failure in glomerulonephritis.*)

Signs and symptoms

Possible signs and symptoms of glomerulonephritis include:
- decreased urination or oliguria due to decreased GFR
- smoky or coffee-colored urine due to hematuria
- shortness of breath due to pulmonary edema
- dyspnea due to pulmonary edema
- orthopnea due to hypervolemia
- periorbital edema due to hypervolemia
- mild to severe hypertension due to sodium or water retention

- bibasilar crackles due to heart failure.

 AGE ALERT An elderly patient with glomerulonephritis may report vague, nonspecific symptoms such as nausea, malaise, and arthralgia.

Complications

Possible complications of glomerulonephritis are:
- pulmonary edema
- heart failure
- sepsis
- renal failure
- severe hypertension
- cardiac hypertrophy.

Diagnosis

Blood study results that aid in diagnosis include:
- elevated electrolyte, blood urea nitrogen, and creatinine levels
- decreased serum protein level
- decreased hemoglobin in chronic glomerulonephritis
- elevated antistreptolysin-O titers in 80% of patients, elevated streptozyme and anti-DNAase B titers, low serum complement levels indicating recent streptococcal infection.

Urinalysis results that help diagnose glomerulonephritis include:
- RBCs, white blood cells, mixed cell casts, and protein indicating renal failure
- fibrin-degradation products and C3 protein.

 AGE ALERT Significant proteinuria is not a common finding in an elderly patient.

Other results that help diagnose glomerulonephritis are:
- throat culture showing group A beta-hemolytic streptococcus
- bilateral kidney enlargement on kidney-ureter-bladder X-ray (acute glomerulonephritis)
- symmetric contraction with normal pelves and calyces (chronic glomerulonephritis) as seen on X-ray
- renal biopsy confirming the diagnosis or assessing renal tissue status.

AVERTING RENAL FAILURE IN GLOMERULONEPHRITIS

Shown is a flowchart of pathophysiologic occurrences and the treatments that can alter the course of glomerulonephritis.

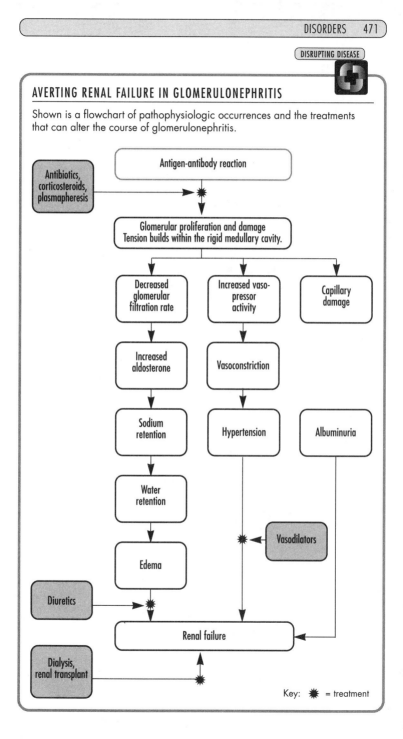

Treatment

Treatment involves:
- treating the primary disease to alter immunologic cascade
- antibiotics for 7 to 10 days to treat infections contributing to ongoing antigen-antibody response
- anticoagulants to control fibrin crescent formation in RPGN
- bed rest to reduce metabolic demands
- fluid restrictions to decrease edema
- dietary sodium restriction to prevent fluid retention
- correction of electrolyte imbalances
- loop diuretics such as metolazone (Zaroxolyn) or furosemide (Lasix) to reduce extracellular fluid overload
- vasodilators such as hydralazine (Apresoline) or nifedipine (Procardia) to decrease hypertension
- dialysis or kidney transplantation for chronic glomerulonephritis
- corticosteroids to decrease antibody synthesis and suppress inflammatory response
- plasmapheresis in RPGN to suppress rebound antibody production, possibly combined with corticosteroids and cyclophosphamide (Cytoxan).

Hypospadias and epispadias

Among the most common birth defects, congenital anomalies of the ureter, bladder, and urethra occur in about 5% of all births. The abnormality may be obvious at birth or may go unrecognized until symptoms appear. Hypospadias is a congenital abnormality in which the opening of the urethra is misplaced to the perineal or scrotal region. The defect may be slight or extreme in nature, and it occurs in 1 of 300 live male births.

Epispadias occurs in 1 in 200,000 infant boys and 1 in 400,000 infant girls, and it is expressed in differing degrees. In males, the urethral opening is on the dorsal aspect of the penis. In females, a cleft along the ventral urethral opening extends to the bladder neck.

Causes

Hypospadias and epispadias may be caused by:
- congenital defect
- genetic factors.

Pathophysiology

In hypospadias, the urethral opening is on the ventral surface of the penis. (See *Comparing hypospadias and epispadias*.) A genetic factor is suspected in less severe cases. It's usually associated with a downward bowing of the penis (chordee), making normal urination with the penis elevated impossible. The ventral prepuce may be absent or defective, and the genitalia may be ambiguous. In the rare case of hypospadias in a female, the urethral opening is in the vagina, and vaginal discharge may be present.

Epispadias occurs more commonly in males than females and often accompanies bladder exstrophy, which is the absence of a portion of the lower abdominal and anterior bladder wall, with a portion of the posterior bladder wall through the deficit. In mild cases, the orifice is on the dorsum of the glans; in severe cases, on the dorsum of the penis. Affected females have a bifid (cleft into two parts) clitoris and a short, wide urethra. Total urinary incontinence occurs when the urethral opening is proximal to the sphincter.

Signs and symptoms

Signs and symptoms of hypospadias and epispadias include:
- displaced urethral opening
- altered voiding patterns due to displaced opening of the urethra
- chordee, or bending of the penis (in hypospadias)
- ejaculatory dysfunction due to displaced penile opening.

Complications

Possible complications are:
- UTI
- urinary obstruction.

Diagnosis

If sexual identification is questionable, diagnosis is based on:

❯ buccal smears and karotyping.

Treatment

Treatment may include:

❯ no treatment (for mild hypospadias in a asymptomatic patient)

❯ surgery, preferably before the child reaches school age (severe hypospadias)

❯ surgical repair in several stages, which is almost always necessary (epispadias).

Nephrotic syndrome

Marked proteinuria, hypoalbuminemia, hyperlipidemia, and edema characterize nephrotic syndrome. It results from a defect in the permeability of glomerular vessels. About 75% of the cases result from primary (idiopathic) glomerulonephritis. The prognosis is highly variable, depending on the underlying cause.

 AGE ALERT Age has no part in the progression or prognosis of nephrotic syndrome. Primary nephrotic syndrome is found predominantly in the preschool child. The incidence peaks at ages 2 and 3 years, and it is rare after the age of 8.

Boys are more frequently affected with primary nephrotic syndrome than girls; the incidence is 3 per 100,000 children per year. Some forms of nephrotic syndrome may eventually progress to end-stage renal failure.

Causes

Causes of nephrotic syndrome include:

❯ lipid nephrosis (nil lesions)

 AGE ALERT Lipid nephrosis is the main cause of nephrotic syndrome in children younger than 8 years.

❯ membranous glomerulonephritis

 AGE ALERT Membranous glomerulonephritis is the most common lesion in adult idiopathic nephrotic syndrome.

❯ focal glomerulosclerosis

COMPARING HYPOSPADIAS AND EPISPADIAS

In hypospadias, the urethral opening is on the ventral surface of the penis or within the vagina.

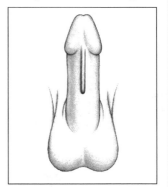

In males with epispadias, a urethral opening occurs on the dorsal surface of the penis; in females, a fissure occurs on the upper wall of the urethra.

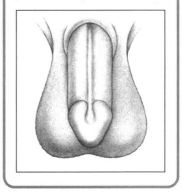

❯ membranoproliferative glomerulonephritis

❯ metabolic diseases, such as diabetes mellitus

❯ collagen-vascular disorders, such as systemic lupus erythematosus and periarteritis nodosa

▶ circulatory diseases, such as heart failure, sickle cell anemia, and renal vein thrombosis

▶ nephrotoxins, such as mercury, gold, and bismuth

▶ infections, such as tuberculosis and enteritis

▶ allergic reactions

▶ pregnancy

▶ hereditary nephritis

▶ neoplastic diseases, such as multiple myeloma.

Pathophysiology

In lipid nephrosis, the glomeruli appear normal by light microscopy, and some tubules may contain increased lipid deposits. Membranous glomerulonephritis is characterized by the appearance of immune complexes, seen as dense deposits in the glomerular basement membrane, and by the uniform thickening of the basement membrane. It eventually progresses to renal failure.

Focal glomerulosclerosis can develop spontaneously at any age, can occur after kidney transplantation, or may result from heroin injection. Ten percent of children and up to 20% of adults with nephrotic syndrome develop this condition. Lesions initially affect some of the deeper glomeruli, causing hyaline sclerosis. Involvement of the superficial glomeruli occurs later. These lesions usually cause slowly progressive deterioration in renal function, although remission may occur in children.

Membranoproliferative glomerulonephritis causes slowly progressive lesions in the subendothelial region of the basement membrane. This disorder may follow infection, particularly streptococcal infection, and occurs primarily in children and young adults.

Regardless of the cause, the injured glomerular filtration membrane allows the loss of plasma proteins, especially albumin and immunoglobulin. In addition, metabolic, biochemical, or physiochemical disturbances in the glomerular basement membrane result in the loss of negative charge as well as increased permeability to protein. Hypoalbuminemia results not only from urinary loss, but also from decreased hepatic synthesis of replacement albumin. Increased plasma concentration and low molecular weight accentuate albumin loss. Hypoalbuminemia stimulates the liver to synthesize lipoprotein, with consequent hyperlipidemia, and clotting factors. Decreased dietary intake, as with anorexia, malnutrition, or concomitant disease, further contributes to decreased levels of plasma albumin. Loss of immunoglobulin also increases susceptibility to infections.

Extensive proteinuria (more than 3.5 g/day) and a low serum albumin level, secondary to renal loss, lead to low serum colloid osmotic pressure and edema. The low serum albumin level also leads to hypovolemia and compensatory salt and water retention. Consequent hypertension may precipitate heart failure in compromised patients.

Signs and symptoms

Possible signs and symptoms of nephrotic syndrome include:

▶ periorbital edema, due to fluid overload

▶ mild to severe dependent edema of the ankles or sacrum

▶ orthostatic hypotension, due to fluid imbalance

▶ ascites, due to fluid imbalance

▶ swollen external genitalia, due to edema in dependent areas

▶ respiratory difficulty, due to pleural effusion

▶ anorexia, due to edema of intestinal mucosa

▶ pallor and shiny skin with prominent veins

▶ diarrhea, due to edema of intestinal mucosa

▶ frothy urine in children

▶ change in quality of hair, related to protein deficiency

▶ pneumonia, due to susceptibility of infections.

Complications

Possible complications include:

▶ malnutrition
▶ infection
▶ coagulation disorders
▶ thromboembolic vascular occlusion (especially in the lungs and legs)
▶ accelerated atherosclerosis
▶ hypochromic anemia, due to excessive urinary excretion of transferrin
▶ acute renal failure.

Diagnosis

Diagnosis is based on:

▶ consistent heavy proteinuria (24-hour protein more than 3.5 mg/dl)
▶ urinalysis showing hyaline, granular and waxy fatty casts, and oval fat bodies
▶ increased serum cholesterol, phospholipid (especially low-density and very low-density lipoproteins), and triglyceride levels, and decreased albumin levels
▶ renal biopsy for histologic identification of the lesion.

Treatment

Treatment includes:

▶ correction of underlying cause, if possible
▶ nutritious diet, including 0.6 g of protein/kg of body weight
▶ restricted sodium intake, to reduce edema
▶ diuretics, to diminish edema
▶ antibiotics, to treat infection
▶ 8-week course of a corticosteroid, such as prednisone (Deltasone), followed by maintenance therapy or a combination of prednisone and azathioprine (Imuran) or cyclophosphamide (Cytoxan)
▶ treatment for hyperlipidemia (frequently unsuccessful)
▶ paracentesis, for acites.

Neurogenic bladder

All types of bladder dysfunction caused by an interruption of normal bladder innervation are referred to as neurogenic bladder. Other names for this disorder include neuromuscular dysfunction of the lower urinary tract, neurologic bladder dysfunction, and neuropathic bladder. Neurogenic bladder can be hyperreflexic (hypertonic, spastic, or automatic) or flaccid (hypotonic, atonic, or autonomous).

Causes

Many factors can interrupt bladder innervation. Cerebral disorders causing neurogenic bladder include:

▶ cerebrovascular accident
▶ brain tumor (meningioma and glioma)
▶ Parkinson's disease
▶ multiple sclerosis
▶ dementia
▶ incontinence associated with aging.

Spinal cord disease or trauma can also cause neurogenic bladder, including:

▶ spinal stenosis causing cord compression
▶ arachnoiditis (inflammation of the membrane between the dura and pia mater) causing adhesions between membranes covering the cord
▶ cervical spondylosis
▶ spina bifida
▶ poliomyelitis
▶ myelopathies from hereditary or nutritional deficiencies
▶ tabes dorsalis (degeneration of the dorsal columns of the spinal cord)
▶ disorders of peripheral innervation, including autonomic neuropathies, due to endocrine disturbances such as diabetes mellitus (most common).

Other causes include:

▶ metabolic disturbances, such as hypothyroidism or uremia
▶ acute infectious diseases, such as Guillain-Barré syndrome or transverse myelitis (pathologic changes extending across the spinal cord)
▶ heavy metal toxicity
▶ chronic alcoholism
▶ collagen diseases, such as systemic lupus erythematosus
▶ vascular diseases, such as atherosclerosis
▶ distant effects of certain cancers, such as primary oat cell carcinoma of the lung

TYPES OF NEUROGENIC BLADDER

NEURAL LESION	TYPE	CAUSE
Upper motor	Uninhibited	▶ Lack of voluntary control in infancy ▶ Multiple sclerosis
	Reflex or automatic	▶ Spinal cord transection ▶ Cord tumors ▶ Multiple sclerosis
Lower motor	Autonomous	▶ Sacral cord trauma ▶ Tumors ▶ Herniated disk ▶ Abdominal surgery with transection of pelvic parasympathetic nerves
	Motor paralysis	▶ Lesions at levels S2, S3, S4 ▶ Poliomyelitis ▶ Trauma ▶ Tumors
	Sensory paralysis	▶ Posterior lumbar nerve roots ▶ Diabetes mellitus ▶ Tabes dorsalis

From Huether, S. *Understanding Pathophysiology.* St. Louis: Mosby, 1996.

▶ herpes zoster
▶ sacral agenesis (absence of a completely formed sacrum).

Pathophysiology

An upper motor neuron lesion (at or above T12) causes spastic neurogenic bladder, with spontaneous contractions of detrusor muscles, increased intravesical voiding pressure, bladder wall hypertrophy with trabeculation, and urinary sphincter spasms. The patient may experience small urine volume, incomplete emptying, and loss of voluntary control of voiding. Urinary retention also sets the stage for infection.

A lower motor neuron lesion (at or below S2 to S4) affects the spinal reflex that controls micturition. The result is a flaccid neurogenic bladder with decreased intravesical pressure, and bladder capacity, residual urine retention, and poor detrusor contraction. The bladder may not empty spontaneously. The patient experiences loss of both voluntary and involuntary control of urination. Lower motor neuron lesions lead to overflow incontinence. When sensory neurons are interrupted, the patient can't perceive the need to void.

Interruption of the efferent nerves at the cortical, or upper motor neuron, level results in loss of voluntary control. Higher centers also control micturition, and voiding may be incomplete. Sensory neuron interruption leads to dribbling and overflow incontinence. (See *Types of neurogenic bladder*.) Altered bladder sensation often makes symptoms difficult to discern.

Retention of urine contributes to renal calculi, as well as infection. Neurogenic bladder can lead to deterioration of renal function if not promptly diagnosed and treated.

Signs and symptoms

Possible signs and symptoms of neurogenic bladder include:

▶ some degree of incontinence, changes in initiation or interruption of micturition, inability to completely empty the bladder

▶ frequent urinary tract infections, due to urine retention

▶ hyperactive autonomic reflexes (autonomic dysreflexia) when the bladder is distended and the lesion is at upper thoracic or cervical level

▶ severe hypertension, bradycardia, and vasodilation (blotchy skin) above the level of the lesion

▶ piloerection and profuse sweating above the level of the lesion

▶ involuntary or frequent scanty urination without a feeling of bladder fullness, due to hyperreflexic neurogenic bladder

▶ spontaneous spasms (caused by voiding) of the arms and legs, due to hyperreflexic neurogenic bladder

▶ increased anal sphincter tone, due to hyperreflexic neurogenic bladder

▶ voiding and spontaneous contractions of the arms and legs, due to tactile stimulation of the abdomen, thighs, or genitalia

▶ overflow incontinence and diminished anal sphincter tone, due to flaccid neurogenic bladder

▶ greatly distended bladder without feeling of bladder fullness, due to sensory impairment.

Complications

Complications of neurogenic bladder may include:

▶ incontinence
▶ residual urine retention
▶ urinary tract infection
▶ calculus formation
▶ renal failure.

Diagnosis

The following studies may help diagnose neurogenic bladder:

▶ voiding cystourethrography, to evaluate bladder neck function, vesicoureteral reflux, and incontinence

▶ urodynamic studies, to evaluate how urine is stored in the bladder, how well the bladder empties urine, and the rate of movement of urine out of the bladder during voiding

▶ urine flow study (uroflow), to show diminished or impaired urine flow

▶ cystometry, to evaluate bladder nerve supply, detrusor muscle tone, and intravesical pressures during bladder filling and contraction

▶ urethral pressure profile, to determine urethral function with respect to length of the urethra and outlet pressure resistance

▶ sphincter electromyelography, to correlate neuromuscular function of the external sphincter with bladder muscle function during bladder filling and contraction, and to evaluate how well the bladder and urinary sphincter muscles work together

▶ videourodynamic studies, to correlate visual documentation of bladder function with pressure studies

▶ retrograde urethrography, to show strictures and diverticula.

Treatment

Treatment includes:

▶ intermittent self-catheterization, to empty the bladder

▶ anticholinergics and alpha-adrenergic stimulators for the patient with hyperreflexic neurogenic bladder, until intermittent self-catheterization is performed

▶ terazosin (Hytrin) and doxazosin (Cardura), to facilitate bladder emptying in neurogenic bladder

▶ propantheline (Pro-Banthine), methantheline, flavoxate (Urispas), dicyclomine (Bentyl), imipramine (Tofranil), and pseudoephedrine (Sudafed), to facilitate urine storage

▶ surgery, to correct structural impairment through transurethral resection of the blad-

POLYCYSTIC KIDNEY

This cross-sectional drawing shows multiple areas of cystic damage. Each indentation depicts a cyst.

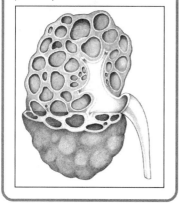

der neck, urethral dilation, external sphincterotomy, or urinary diversion procedures
▶ implantation of an artificial urinary sphincter may be necessary if permanent incontinence follows surgery.

Polycystic kidney disease

Polycystic kidney disease is an inherited disorder characterized by multiple, bilateral, grapelike clusters of fluid-filled cysts that enlarge the kidneys, compressing and eventually replacing functioning renal tissue. (See *Polycystic kidney*.) The disease affects males and females equally and appears in two distinct forms. Autosomal dominant polycystic kidney disease (ADPKD) occurs in 1 in 1,000 to 1 in 3,000 persons and accounts for about 10% of end-stage renal disease in the United States. The rare infantile form causes stillbirth or early neonatal death. The adult form has an insidious onset but usually becomes obvious between the ages of 30 and 50; rarely, it remains asymptomatic until the patient is in his 70s.

AGE ALERT Renal deterioration is more gradual in adults than infants, but in both age groups, the disease progresses relentlessly to fatal uremia.

The prognosis in adults is extremely variable. Progression may be slow, even after symptoms of renal insufficiency appear. Once uremia symptoms develop, polycystic disease usually is fatal within 4 years, unless the patient receives dialysis. Three genetic variants of the autosomal dominant form have been identified (see below).

Causes

Polycystic kidney disease is inherited as:
▶ autosomal dominant trait (adult type)
▶ autosomal recessive trait (infantile type).

Pathophysiology

ADPKD occurs as ADPKD-1, mapped to the short arm of chromosome 16 and encoded for a 4,300–amino acid protein; ADPKD-2, mapped to the short arm of chromosome 4 with later onset of symptoms; and a third variety not yet mapped. Autosomal recessive polycystic kidney disease occurs in 1 in 10,000 to 1 in 40,000 live births, and has been localized to chromosome 6.

Grossly enlarged kidneys are caused by multiple spherical cysts, a few millimeters to centimeters in diameter, that contain straw-colored or hemorrhagic fluid. The cysts are distributed evenly throughout the cortex and medulla. Hyperplastic polyps and renal adenomas are common. Renal parenchyma may have varying degrees of tubular atrophy, interstitial fibrosis, and nephrosclerosis. The cysts cause elongation of the pelvis, flattening of the calyces, and indentations in the kidney.

Characteristically, an affected infant shows signs of respiratory distress, heart failure, and, eventually, uremia and renal failure. Accompanying hepatic fibrosis and intrahepatic bile duct abnormalities may cause portal hypertension and bleeding varices.

In most cases, about 10 years after symptoms appear, progressive compression of kidney structures by the enlarging mass causes renal failure.

Cysts also form elsewhere — such as on the liver, spleen, pancreas, and ovaries. Intracranial aneurysms, colonic diverticula, and mitral valve prolapse also occur.

In the autosomal recessive form, death in the neonatal period is most commonly due to pulmonary hypoplasia.

Signs and symptoms
Signs and symptoms in neonates include:
▶ pronounced epicanthic folds (vertical fold of skin on either side of the nose); a pointed nose; small chin; and floppy, low-set ears (Potter facies), due to genetic abnormalities
▶ huge, bilateral, symmetrical masses on the flanks that are tense and can't be transilluminated, due to kidney enlargement
▶ uremia, due to renal failure.

Signs and symptoms in adults include:
▶ hypertension, due to activation of the renin-angiotensin system
▶ lumbar pain, due to enlarging kidney mass
▶ widening abdominal girth, due to enlarged kidneys
▶ swollen or tender abdomen caused by the enlarging kidney mass, worsened by exertion and relieved by lying down
▶ grossly enlarged kidneys on palpation.

Complications
 AGE ALERT A few infants with this disease survive for 2 years, and then die of hepatic complications or renal, heart, or respiratory failure.
Possible complications in adults include:
▶ pyelonephritis
▶ recurrent hematuria
▶ life-threatening retroperitoneal bleeding from cyst rupture
▶ proteinuria
▶ colicky abdominal pain from ureteral passage of clots or calculi
▶ renal failure.

Diagnosis
Diagnosis is based on the following test results:
▶ excretory or retrograde urography showing enlarged kidneys, with elongation of the pelvis, flattening of the calyces, and indentations in the kidney caused by cysts
▶ excretory urography of the neonate showing poor excretion of contrast medium
▶ ultrasonography, tomography, and radioisotope scans showing kidney enlargement and cysts; tomography, computed tomography, and magnetic resonance imaging showing multiple areas of cystic damage
▶ urinalysis and creatinine clearance tests showing nonspecific results indicating abnormalities.

Treatment
Treatment includes:
▶ antibiotics for infections
▶ adequate hydration to maintain fluid balance
▶ surgical drainage of cystic abscess or retroperitoneal bleeding
▶ surgery for intractable pain (uncommon symptom) or analgesics for abdominal pain
▶ dialysis or kidney transplantation for progressive renal failure
▶ nephrectomy not recommended (polycystic kidney disease occurs bilaterally, and the infection could recur in the remaining kidney).

Renal agenesis
Renal agenesis is the failure of a kidney to grow or develop. The kidney is usually polycystic and dysplastic. The disease may be unilateral or bilateral, random or hereditary, and occur in isolation or associated with other disorders. Unilateral renal agenesis occurs in 1 in 1,000 live births, and more commonly in males than females.

Bilateral renal agenesis is also called Potter syndrome. It occurs in 1 of every 3,000 live births, and 75% of the cases are in males. Bilateral renal agenesis is not compatible with life, and most affected infants die in utero.

Causes
The causes of renal agenesis are:
▶ unknown, but suspected to be hereditary.

Pathophysiology
In unilateral renal agenesis, the left kidney is usually absent. The remaining kidney may be completely normal. During the first years of life, this kidney hypertrophies to functionally compensate for the missing kidney. If the kidney has abnormalities of the collecting system, compensation is virtually impossible. Extrarenal congenital abnormalities are common with this type of agenesis.

 AGE ALERT Infants with Potter syndrome rarely live longer than a few hours.

Signs and symptoms
There are no symptoms of unilateral renal agenesis if the kidney is functioning appropriately. Signs and symptoms of Potter syndrome are due to a congenital defect and include:
▶ wide-set eyes
▶ parrot-beak nose
▶ low-set ears
▶ receding chin
▶ pulmonary pathophysiology.

Complications
Renal agenesis may be complicated by:
▶ renal failure.

Diagnosis
Diagnosis is based on:
▶ prenatal ultrasound.

Treatment
Treatment includes:
▶ surgery for structural or functional defects in the remaining kidney.

Renal calculi
Renal calculi, or stones (nephrolithiasis), can form anywhere in the urinary tract, although they most commonly develop on the renal pelves or calyces. They may vary in size and may be solitary or multiple. (See *Renal calculi*.)

Renal calculi are more common in men than in women and rarely occur in children. Calcium stones generally occur in middle-age men with a familial history of stone formation.

CULTURAL DIVERSITY Renal calculi rarely occur in blacks. They are prevalent in certain geographic areas, such as the southeastern United States (called the "stone belt"), possibly because a hot climate promotes dehydration and concentrates calculus-forming substances, or because of regional dietary habits.

Causes
Although the exact cause is unknown, predisposing factors of renal calculi include:
▶ dehydration
▶ infection
▶ changes in urine pH (calcium carbonate stones, high pH; uric acid stones, lower pH)
▶ obstruction to urine flow leading to stasis in the urinary tract
▶ immobilization causing bone reabsorption
▶ metabolic factors
▶ dietary factors
▶ renal disease
▶ gout (a disease of increased uric acid production or decreased excretion).

Pathophysiology
The major types of renal stones are calcium oxalate and calcium phosphate, accounting for 75% to 80% of stones; struvite (magnesium, ammonium, and phosphate), 15%; and uric acid, 7%. Cystine stones are relatively rare, making up 1% of all renal stones.

Calculi form when substances that are normally dissolved in the urine, such as

RENAL CALCULI

Renal calculi vary in size and type. Small calculi may remain in the renal pelvis or pass down the ureter. A staghorn calculus (a cast of the calyceal and pelvic collecting system) may develop from a stone that stays in the kidney.

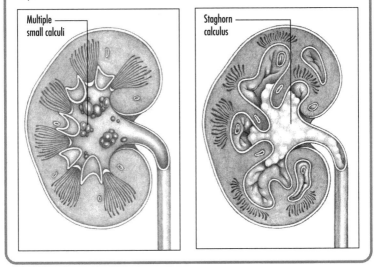

Multiple small calculi

Staghorn calculus

calcium oxalate and calcium phosphate, precipitate. Dehydration may lead to renal calculi as calculus-forming substances concentrate in urine.

Stones form around a nucleus or nidus in the appropriate environment. A crystal evolves in the presence of stone-forming substances (calcium oxalate, calcium carbonate, magnesium, ammonium, phosphate, or uric acid) and becomes trapped in the urinary tract, where it attracts other crystals to form a stone. A high urine saturation of these substances encourages crystal formation and results in stone growth.

Stones may be composed of different substances, and the pH of the urine affects the solubility of many stone-forming substances. Formation of calcium oxalate and cystine stones is independent of urine pH.

Stones may occur on the papillae, renal tubules, calyces, renal pelves, ureter, or bladder. Many stones are less than 5 mm in diameter and are usually passed in the urine. Staghorn calculi can continue to grow in the pelvis, extending to the calyces, forming a branching stone, and ultimately resulting in renal failure if not surgically removed.

Calcium stones are the smallest. Most are calcium oxalate or a combination of oxalate and phosphate. Although 80% are idiopathic, they frequently occur with hyperuricosuria (a high level of uric acid in the urine). Prolonged immobilization can lead to bone demineralization, hypercalciuria, and stone formation. In addition, hyperparathyroidism, renal tubular acidosis, and excessive intake of vitamin D or dietary calcium may predispose to renal calculi.

Struvite stones are often precipitated by an infection, particularly with *Pseudomonas* or *Proteus* species. These urea-splitting organisms are more common in women. Struvite calculi can destroy renal parenchyma.

Gout results in a high uric acid production, hyperuricosuria, and uric acid stones. Diets high in purine (such as meat, fish, and poultry) elevate levels of uric acid in the body. Regional enteritis and ulcerative colitis can precipitate the formation of uric acid stones. These diseases often result in fluid loss and loss of bicarbonate, leading to metabolic acidosis. Acidic urine enhances the formation of uric acid stones.

Cystinuria is a rare hereditary disorder in which a metabolic error causes decreased tubular reabsorption of cystine. This causes an increased amount of cystine in the urine. Because cystine is a relatively insoluble substance, its presence contributes to stone formation.

Infected, scarred tissue may be an ideal site for calculus development. In addition, infected calculi (usually magnesium ammonium phosphate or staghorn calculi) may develop if bacteria serve as the nucleus in calculus formation.

Urinary stasis allows calculus constituents to collect and adhere and also encourages infection, which compounds the obstruction.

Calculi may either enter the ureter or remain in the renal pelvis, where they damage or destroy renal parenchyma and may cause pressure necrosis.

In ureters, calculi cause obstruction with resulting hydronephrosis and tend to recur. Intractable pain and serious bleeding also can result from calculi and the damage they cause. Large, rough calculi occlude the opening to the ureteropelvic junction and increase the frequency and force of peristaltic contractions, causing hematuria from trauma. The patient usually reports pain travelling from the costovertebral angle to the flank and then to the suprapubic region and external genitalia (classic renal colic pain). Pain intensity fluctuates and may be excruciating at its peak. The patient with calculi in the renal pelvis and calyces may report a constant dull pain. He may also report back pain if calculi are causing obstruction within a kidney and severe abdominal pain from calculi traveling down a ureter. Infection can develop in static urine or after trauma as the stone abrades surfaces. If the stone lodges and blocks urine, hydronephrosis can occur.

Signs and symptoms
Possible signs and symptoms of renal calculi include:
▶ severe pain resulting from obstruction
▶ nausea and vomiting
▶ fever and chills from infection
▶ hematuria when calculi abrade a ureter
▶ abdominal distention
▶ anuria from bilateral obstruction, or obstruction of a patient's only kidney.

Complications
Complications include:
▶ damage or destruction of renal parenchyma
▶ pressure necrosis
▶ obstruction by the stone
▶ hydronephrosis
▶ bleeding
▶ pain
▶ infection.

Diagnosis
The following tests may be used to diagnose renal calculi:
▶ kidney-ureter-bladder (KUB) radiography, to show most renal calculi
▶ excretory urography, to help confirm the diagnosis and determine the size and location of calculi
▶ kidney ultrasonography, to detect obstructive changes, such as unilateral or bilateral hydronephrosis and radiolucent calculi not seen on KUB radiography
▶ urine culture showing pyuria, a sign of urinary tract infection
▶ 24-hour urine collection, for calcium oxalate, phosphorus, and uric acid excretion levels
▶ calculus analysis, for mineral content

serial blood calcium and phosphorus levels diagnose hyperparathyroidism and increased calcium, relative to normal serum protein

blood protein levels, to determine the level of free calcium unbound to protein.

Treatment

Treatment may include:

increasing fluid intake to more than 3 L/day to promote hydration

antimicrobial agents to treat infection, varying with the cultured organism

analgesics such as meperidine (Demerol) or morphine for pain

diuretics to prevent urinary stasis and further calculus formation; thiazides to decrease calcium excretion into the urine

methenamine mandelate to suppress calculus formation when infection is present

low-calcium diet to prevent recurrence

oxalate-binding cholestyramine for absorptive hypercalciuria

parathyroidectomy for hyperparathyroidism

allopurinol (Zyloprim) for uric acid calculi

daily small doses of ascorbic acid to acidify urine

cystoscope with manipulation of the calculus to remove renal calculi too large for natural passage

percutaneous ultrasonic lithotripsy and extracorporeal shock wave lithotripsy or laser therapy to shatter the calculus into fragments for removal by suction or natural passage

surgical removal of cystine calculi or large stones or placement of urinary diversion around the stone to relieve obstruction.

Vesicoureteral reflux

In vesicoureteral reflux, urine flows from the bladder back into the ureters and eventually into the renal pelvis or the parenchyma. When the bladder empties only part of the stored urine, urinary tract infection may result. This disorder is most common during infancy in boys and during early childhood (ages 3 to 7) in girls. Primary vesicoureteral reflux that results from congenital anomalies is more common in females.

CULTURAL DIVERSITY There is a much lower incidence of vesicoureteral reflux in blacks. Indeed, it's extremely rare.

Up to 25% of asymptomatic siblings of children with diagnosed primary vesicoureteral reflux also have the disorder. Secondary vesicoureteral reflux occurs in adults.

Causes

Primary vesicoureteral reflux is caused by congenital anomalies of the ureters or bladder, including:

short or absent intravesical ureters

ureteral ectopia lateralis (ureter that opens more laterally in the bladder wall)

ureteral duplication

ureterocele

gaping or golf-hole ureteral orifice.

Secondary vesicoureteral reflux is due to damage by:

bladder outlet obstruction

iatrogenic injury

trauma

inadequate detrusor muscle buttress in the bladder

cystitis, repeated infections

neurogenic bladder

Pathophysiology

Incompetence of the ureterovesical junction and shortening of intravesical ureteral musculature allow backflow of urine into the ureters when the bladder contracts during voiding. Congenital paraureteral bladder diverticulum, acquired diverticulum (from outlet obstruction), flaccid neurogenic bladder, and high intravesical pressure may cause inadequate detrusor muscle contraction in the bladder from outlet obstruction or an unknown cause.

Vesicoureteral reflux also may result from cystitis; inflammation of the intravesical ureter causes edema and intramural ureter fixation. This usually leads to reflux in people with congenital ureteral or bladder anomalies or other predisposing conditions. Recurrent urinary tract infections can lead to acute or chronic pyelonephritis and renal damage due to renal scarring, hypertension, or calculi.

Signs and symptoms

Signs and symptoms of vesicoureteral reflux include:

▶ urinary frequency and urgency due to urinary tract infection
▶ burning on urination
▶ hematuria
▶ foul-smelling urine
▶ high fever and chills due to urinary tract infection
▶ flank pain
▶ painful urination
▶ vomiting
▶ malaise
▶ palpation showing a hard, thickened bladder if posterior urethral valves are causing an obstruction in males.

 AGE ALERT Infants with vesicoureteral reflux may have dark, concentrated urine due to retention. In children, fever, nonspecific abdominal pain, and diarrhea may be the only clinical effects. In children younger than 5 years, repeated urinary tract infections are suggestive of reflux.

Complications

Possible complications include:

▶ recurrent urinary tract infections
▶ pyelonephritis
▶ anemia
▶ hypertension
▶ renal obstruction
▶ renal failure.

Diagnosis

The following test results help diagnose vesicoureteral reflux:

▶ clean-catch urinalysis showing bacterial count more than 100,000/ml, sometimes without pyuria; microscopic examination showing red and white blood cells and increased urine pH (active infection); spe-

cific gravity less than 1.010 due to inability to concentrate urine

▶ elevated serum creatinine (more than 1.2 mg/dl) and blood urea nitrogen (more than 18 mg/dl) levels due to advanced renal dysfunction

▶ voiding cystourethrography to identify and determine the degree of reflux and show when reflux occurs

▶ catheterization of the bladder after the patient voids to determine the amount of residual urine

▶ I.V. pyelogram to diagnose vesicoureteral reflux by visualization

▶ excretory urography to show a dilated lower ureter, a ureter visible for its entire length, hydronephrosis, calyceal distortion, and renal scarring

▶ cystoscopy (may confirm diagnosis)

▶ radioisotope scanning and renal ultrasonography to detect reflux and screen the upper urinary tract for damage secondary to infection and other renal abnormalities.

Treatment

Treatment of vesicoureteral reflux includes:

▶ antibiotics to treat reflux secondary to infection or related to neurogenic bladder and, in children, a short intravesical ureter (disappears spontaneously with growth)

▶ long-term prophylactic antibiotic therapy for recurrent infection

▶ vesicoureteral reimplantation to treat recurrent infection despite prophylactic antibiotic therapy

▶ transurethral sphincterotomy to relieve obstructed outlet

▶ bladder augmentation to decrease intravesical pressure.

Through the sensory system, a person receives stimuli that facilitate interaction with the surrounding world. Afferent pathways connect specialized sensory receptors in the eyes, ears, nose, and mouth to the brain — the final station for continual processing of sensory stimuli. Alterations in sensory function may lead to dysfunctions of sight and hearing, as well as smell, taste, balance, and coordination.

PATHOPHYSIOLOGIC MANIFESTATIONS

Alterations can occur in all the senses.

Vision

Disorders of vision include alterations in ocular movement, visual acuity, accommodation, refraction, and color vision.

Ocular movement

The eyes constantly move to keep objects being viewed on the fovea, which is a small area of the retina that contains only cones and is responsible for the best peripheral visual acuity. The six extraocular muscles that move each eye are innervated by the oculomotor (III), trochlear (IV), and abducens (VI) cranial nerves. (See *Extraocular control of eye movement*, page 486.) Alterations in ocular movement include strabismus, diplopia, and nystagmus.

Strabismus. Strabismus occurs when one eye deviates from its normal position due to the absence of normal, parallel, or coordinated movement. The eyes may have

an uncoordinated appearance, and the person may experience diplopia.

In children, types of strabismus are:
▶ concomitant, in which the degree of deviation doesn't vary with the direction of gaze
▶ nonconcomitant, in which the degree of deviation varies with the direction of gaze
▶ congenital (present at birth or during the first 6 months)
▶ acquired (present during the first 2½ years)
▶ latent (phoria; apparent only when the child is tired or sick)
▶ constant.

Tropias are categorized into four types: esotropia (inward deviation), exotropia (outward deviation), hypertropia (upward deviation), and hypotropia (downward deviation).

Strabismic amblyopia is characterized by the loss of central vision in one eye; it typically results in esotropia (due to fixation in the dominant eye and suppression of images in the deviating eye). Strabismic amblyopia may result from hyperopia (farsightedness) or anisometropia (unequal refractive power).

Esotropia may result from muscle imbalance and may be congenital or acquired. In accommodative esotropia, the child's attempt to compensate for the farsightedness affects the convergent reflex, and the eyes cross.

Strabismus is frequently inherited, but its cause is unknown. In adults, strabismus may result from trauma. The incidence of strabismus is higher in patients with central nervous system disorders,

EXTRAOCULAR CONTROL OF EYE MOVEMENT

The six muscles that control the movement of each eye are innervated by three cranial nerves.

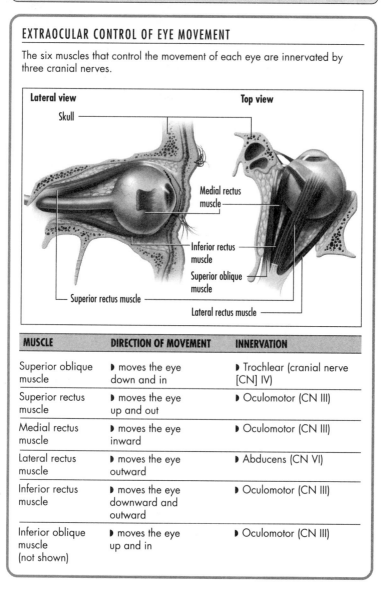

Lateral view   Top view

Skull

Medial rectus muscle

Inferior rectus muscle

Superior oblique muscle

Superior rectus muscle

Lateral rectus muscle

MUSCLE	DIRECTION OF MOVEMENT	INNERVATION
Superior oblique muscle	▸ moves the eye down and in	▸ Trochlear (cranial nerve [CN] IV)
Superior rectus muscle	▸ moves the eye up and out	▸ Oculomotor (CN III)
Medial rectus muscle	▸ moves the eye inward	▸ Oculomotor (CN III)
Lateral rectus muscle	▸ moves the eye outward	▸ Abducens (CN VI)
Inferior rectus muscle	▸ moves the eye downward and outward	▸ Oculomotor (CN III)
Inferior oblique muscle (not shown)	▸ moves the eye up and in	▸ Oculomotor (CN III)

such as cerebral palsy, mental retardation, and Down syndrome.

Muscle imbalances may be corrected by glasses, patching, or surgery, depending on the cause. However, residual defects in vision and extraocular muscle alignment may persist even after treatment.

 AGE ALERT In the absence of early intervention, children with strabismus may develop amblyopia as a result of cerebral suppression of visual stimuli. Deviation of an eye is the second most common symptom in a child with retinoblastoma. Therefore, acquired strabismus should always be checked.

Diplopia. Diplopia, or double vision, results when the extraocular muscles fail to work together and images fall on noncorresponding parts of the retinas. Diplopia usually begins intermittently or affects near or far vision exclusively. It can be classified as monocular (persisting when one eye is covered) or, more commonly, binocular (clearing when one eye is covered). Monocular diplopia may result from an early cataract, retinal edema or scarring, iridolysis (surgical lysis of adhesions of the iris), subluxated lens (partial dislocation of the lens of the eye), poorly fitting contact lens, or uncorrected refractive error. Binocular diplopia may result from ocular deviation or displacement, extraocular muscle palsies, or psychoneurosis, or after retinal surgery. Other causes of binocular diplopia include infection, neoplastic disease, metabolic disorders, degenerative disease, inflammatory disorders, and vascular disease.

Nystagmus. Nystagmus refers to involuntary oscillations or alternating movements of one or both eyes. These oscillations are usually rhythmic and may be horizontal, vertical, or rotary. They may be transient or sustained and may occur spontaneously or on deviation or fixation. Nystagmus may be classified as pendular or jerk. Nystagmoid movements usually have a fast and a slow component. The direction of the nystagmus is given by the fast component. (See *Classifying nystagmus,* page 488.)

Nystagmus is a supranuclear ocular palsy resulting from pathology in the visual perceptual area, vestibular system, or cerebellum. Causes of nystagmus include brain stem or cerebellar lesions, labyrinthine disease, stroke, encephalitis, Meniere's disease, multiple sclerosis, and alcohol and drug toxicity, including barbiturate,

phenytoin (Dilantin), or carbamazepine (Tegretol) toxicity.

AGE ALERT In children, pendular nystagmus may be idiopathic or may result from early impairment of vision associated with such disorders as optic atrophy, albinism, congenital cataracts, and severe astigmatism.

Visual acuity

Visual acuity refers to the ability to see clearly. A lack of visual acuity is often associated with refractive errors. In nearsightedness, or myopia, the eye focuses the visual image in front of the retina, causing objects in close view to be seen clearly and those at a distance to be blurry. In farsightedness, or hyperopia, the eye focuses the visual image behind the retina, causing objects in close view to be blurry and those at a distance to be clear. Both these problems are caused by an alteration in the shape of the eyeball. Other causes of reduced visual acuity include aging, amblyopia, cataracts, glaucoma, papilledema, dark adaptation, and scotoma (an area of diminished visual acuity surrounded by an area of normal vision within the visual field).

Amblyopia is severely decreased visual acuity or virtual blindness in a structurally intact eye. It may be caused by toxins (including alcohol and tobacco) or may accompany such systemic diseases as diabetes mellitus or renal failure.

AGE ALERT With age, the pupil becomes smaller, which decreases the amount of light that reaches the retina. Older adults need about three times as much light as a younger person to see objects clearly.

Accommodation

Accommodation occurs as the thickness of the lens of the eye changes to maintain visual acuity. For near vision, the ciliary body contracts and relaxes the zonules, the lens becomes spherical, the pupil constricts, and the eyes converge. For far vision, the ciliary body relaxes, the zonules

CLASSIFYING NYSTAGMUS

Nystagmus is classified as pendular or jerk. Each type has further classifications.

PENDULAR NYSTAGMUS

Oscillating: slow, steady oscillations of equal velocity around a center point; caused by congenital loss of visual acuity or multiple sclerosis.

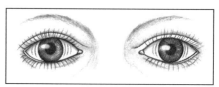

Vertical or seesaw: rapid, seesaw movement in which one eye appears to rise while the other appears to fall; suggests an optic chiasm lesion.

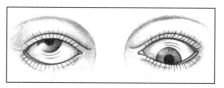

JERK NYSTAGMUS

Convergence-retraction: irregular jerking of the eyes back into the orbit during upward gaze; can reflect midbrain tegmental (roof of the midbrain) damage.

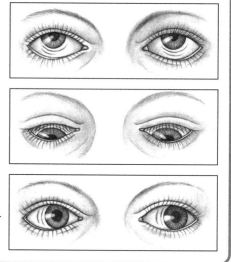

Downbeat: irregular downward jerking of the eyes during downward gaze; can signal lower medullary damage.

Vestibular: horizontal or rotary movements of the eyes; suggests vestibular disease or cochlear dysfunction.

tighten, the lens becomes flatter, the eyes straighten, and the pupils dilate. The oculomotor nerve and coordinated brain stem pathways account for accommodation. Alterations in accommodation may be caused by pressure, inflammation, the aging process, or disorders affecting the oculomotor nerve. The result of impaired ac-

commodation may be diplopia, blurred vision, or headache.

Alterations in refraction

Refraction is the process of bending light rays so that they fall on the retina. As rays of light reach the surface of the cornea from all directions, the cornea directs them toward the lens. The lens further bends the light and directs the light rays to one spot on the retina. The greater the refractive power, the more the light rays are bent. Emmetropia is the condition in which light rays fall exactly on the retina. Alterations in refraction occur when light is not properly focused. Causes include abnormalities in curvature of the cornea, focusing of the lens, and eye length. Results include myopia, hyperopia, and astigmatism. (See *Refractive errors*, page 490.)

Myopia. Myopia, or nearsightedness, occurs when light rays are focused in front of the retina. Near objects can be seen clearly, and distant objects appear blurry. This condition may occur if the eye is too long or the refractive power of the cornea or lens is too great. Myopia may also occur if hyperglycemia in uncontrolled diabetes causes lens swelling. A concave lens that bends light rays outward is used to correct myopia.

Hyperopia. Hyperopia, or farsightedness, occurs when light rays are focused behind the retina. Distant objects appear clear, and nearby objects are blurred. This condition occurs when the eye is too short or the refractive power of the cornea or lens is too low, and may be corrected with a convex lens that bends light rays inward.

 AGE ALERT Presbyopia is a form of hyperopia that begins in middle age as the lens becomes firm and loses its elasticity. As a result, the refractive power of the lens is reduced, the eye loses its ability to accommodate, and near objects appear blurred. This condition is treated with a convex lens that bends light rays in different directions so they focus in a single point.

Astigmatism. Astigmatism occurs when unequal curvature of the cornea or eyeball causes light rays to focus on different points on the retina, resulting in distorted images. Astigmatism may occur in conjunction with other refractive disorders.

Color vision
The cones of the retina are responsible for color vision. Each cone contains one of three different visual pigments (red, green, or blue) that absorb light waves of different wavelengths.

Color blindness is inherited on the X chromosome and therefore usually affects males. Acquired color blindness may also be due to diabetes, bilateral strokes affecting the ventral portion of the occipital lobe, or disease of the macula or optic nerve.

 AGE ALERT Older adults often experience impaired color vision, especially in the blue and green ranges, because cones in the retina deteriorate. Yellowing of the aging lens also impairs color vision.

Hearing
Sound waves normally enter the external auditory canal, then travel to the tympanic membrane in the middle ear, causing it to vibrate. This vibration causes the malleus to move, setting in motion the incus and, in turn, the stapes. The malleus, incus, and stapes are collectively referred to as the ossicles. The stapes presses on the oval window of the inner ear, setting in motion the fluid of the cochlea and stimulating hair cells. The hair cells carry impulses through the cochlear division of the auditory cranial nerve (VIII) to the brain. This type of sound transmission to the inner ear, called air conduction, is normally better than transmission through bone (bone conduction).

Alterations in hearing are classified as conductive or sensorineural. Mixed hearing loss combines aspects of conductive and sensorineural hearing loss.

Conductive hearing loss
Conductive hearing loss results from disorders of the external and middle ear that block sound transmission. Causes include

REFRACTIVE ERRORS

Normal refraction
In the normal eye, light rays are focused exactly on the retina.

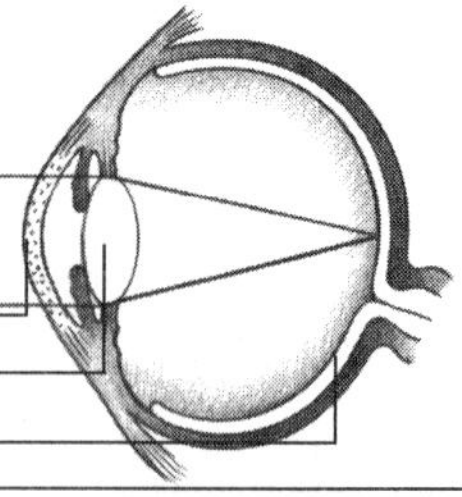

Light rays ⟶

Cornea ———
Lens ———
Retina ———

Myopia
The myopic eye is elongated, so light rays focus in front of the retina.

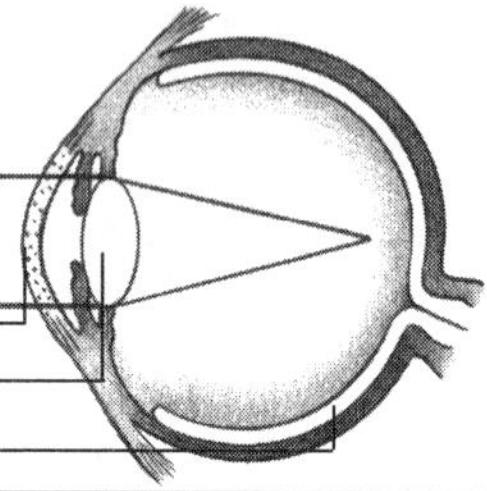

Light rays ⟶

Cornea ———
Lens ———
Retina ———

Hyperopia
In hyperopia, the light rays focus behind the retina.

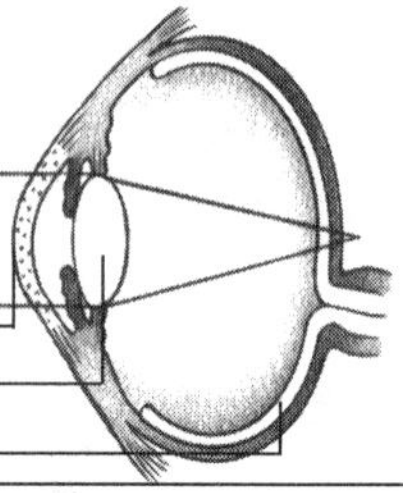

Light rays ⟶

Cornea ———
Lens ———
Retina ———

Astigmatism
In astigmatism, light rays don't come to a single focus on the retina.

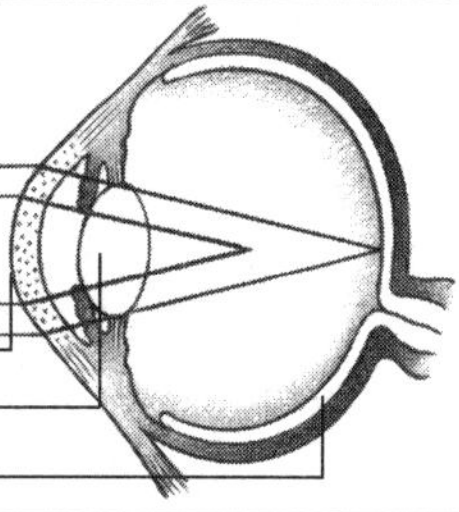

Light rays ⟶

Cornea ———
Lens ———
Retina ———

obstruction of the external auditory canal, tumors or fluid in the middle ear, perforation of the tympanic membrane, trauma or infection that affects the ossicular chain, and fixation of the ossicles. Treatment includes hearing aids, tympanoplasty for chronic otitis media and trauma, and stapedectomy for otosclerosis.

Sensorineural hearing loss

Sensorineural hearing loss results from damage to the hair cells of the organ of Corti or cranial nerve VIII by very loud noises; infection; such ototoxic drugs as aminoglycoside antibiotics, aspirin, diuretics including ethacrynic acid (Edecrin) and furosemide (Lasix); tobacco and alcohol; meningitis; cochlear otosclerosis; Meniere's disease; or aging. Sensorineural hearing loss may be helped by hearing aids and cochlear implantation.

 AGE ALERT The most common sensorineural hearing loss in older adults, presbycusis affects the cochlear hair cells and nerve fibers. It begins with the loss of high-frequency sounds and may progress to middle and low frequencies. (See *Types of presbycusis.*)

AGE ALERT Peripheral or central hearing disorders in children may lead to speech, language, and learning problems. Early identification and treatment of hearing loss is crucial so that early intervention may be started. (See *Congenital hearing loss,* page 492.)

Taste and smell

The senses of taste and smell are also subject to alterations.

Taste

The sensory receptors for taste are the taste buds, concentrated over the surface of the tongue and scattered over the palate, pharynx, and larynx. These buds can differentiate among sweet, salty, sour, and bitter stimuli. Taste and olfactory receptors together perceive more complex flavors. Much of what is considered taste is actually smell; food odors typically stimulate the olfactory system more strongly than related food tastes stimulate the taste buds.

A factor interrupting the transmission of taste stimuli to the brain may cause taste abnormalities. (See *Taste pathways to the brain,* page 493.) Taste abnormalities may result from trauma, infection, vitamin or mineral deficiencies, neurologic or oral disorders, and the effects of drugs. More-

---

TYPES OF PRESBYCUSIS

Presbycusis, sensorineural hearing loss affecting cochlear hair cells and nerve fibers, occurs in four known types: sensory, neural, metabolic, and cochlear conduction.

▶ Sensory presbycusis begins in middle age and progresses slowly; loss of cochlear neurons parallels loss in the organ of Corti.

▶ Neural presbycusis develops later in life; loss of cochlear neurons occurs without loss in the organ of Corti. It progresses rapidly and may be accompanied by other signs of central nervous system decline, including intellectual deterioration, memory loss, and loss of motor coordination.

▶ Metabolic presbycusis tends to run in families, usually begins in middle age, and progresses slowly.

▶ Cochlear-conduction presbycusis also starts in middle age and reflects changes in the motion mechanism of the cochlear duct.

---

over, because tastes are most accurately perceived in a fluid medium, mouth dryness may interfere with taste. Two major pathologic causes of impaired taste are aging, which normally reduces the number of taste buds, and heavy smoking (especially pipe smoking), which dries the tongue.

Alterations in taste may include:
▶ ageusia, a complete loss of taste
▶ hypogeusia, a partial loss of taste
▶ dysgeusia, a distorted sense of taste
▶ cacogeusia, an unpleasant or revolting taste of food.

 AGE ALERT Young children are frequently unable to differentiate between an abnormal taste sensation and a simple taste dislike.

CONGENITAL HEARING LOSS

Hearing loss may be transmitted genetically as an autosomal dominant, autosomal recessive, or X-linked recessive trait. In neonates, it may result from trauma, toxicity, or infection during pregnancy or delivery.

Predisposing factors include:
▶ family history of hearing loss or hereditary disorders (such as otosclerosis)
▶ maternal exposure to rubella or syphilis during pregnancy
▶ exposure to ototoxic drugs during pregnancy
▶ prolonged fetal anoxia during delivery
▶ congenital abnormality of ears, nose, or throat.

Premature or low-birth-weight infants are likely to have structural or functional hearing impairments. Infants with a serum bilirubin level more than 20 mg/dl also risk hearing impairment from bilirubin toxicity to the brain.

In addition, trauma during delivery may cause intracranial hemorrhage and damage to the cochlea or acoustic nerve.

Smell

As air travels between the septum and the turbinates of the nose, it touches sensory hairs (cilia) and olfactory nerve endings in the mucosal surface. The resultant stimulation of cranial nerve I sends impulses to the olfactory-receiving area, primarily in the frontal cortex. (See *Olfactory perception,* page 494.) Temporary impairment in the sense of smell can result from a condition that irritates and causes swelling of the nasal mucosa and obstructs the olfactory area in the nose, such as heavy smoking, rhinitis, or sinusitis. Permanent alterations in the sense of smell usually result when the olfactory neuroepithelium or a part of the olfactory nerve is destroyed. Permanent or temporary loss can also result from inhaling irritants, such as cocaine or acid fumes, that paralyze nasal cilia. Conditions such as aging, Parkinson's disease, Alzheimer's disease, or Kallmann syndrome (a congenital disorder) may also alter the sense of smell. Because combined stimulation of taste buds and olfactory cells produces the sense of taste, the loss of the sense of smell is usually accompanied by the loss of the sense of taste.

Alterations in smell include:
▶ anosmia, a total loss of the sense of smell
▶ hyposmia, an impaired sense of smell
▶ parosmia, an abnormal sense of smell.

DISORDERS

The most common disorders of vision are cataract, glaucoma, and macular degeneration. Other common sensory disorders include Meniere's disease and otosclerosis.

Cataract

A cataract is a gradually developing opacity of the lens or lens capsule. Light shining through the cornea is blocked by this opacity, and a blurred image is cast onto the retina. As a result, the brain interprets a hazy image. Cataracts commonly occur bilaterally, and each progresses independently. Exceptions are traumatic cataracts, which are usually unilateral, and congenital cataracts, which may remain stationary. Cataracts are most prevalent in people aged older than 70 as part of the aging process. The prognosis is generally good; surgery improves vision in 95% of affected people.

Causes

Causes of cataracts include:
▶ aging (senile cataracts)
▶ congenital disorders
▶ genetic abnormalities
▶ maternal rubella during the first trimester of pregnancy

TASTE PATHWAYS TO THE BRAIN

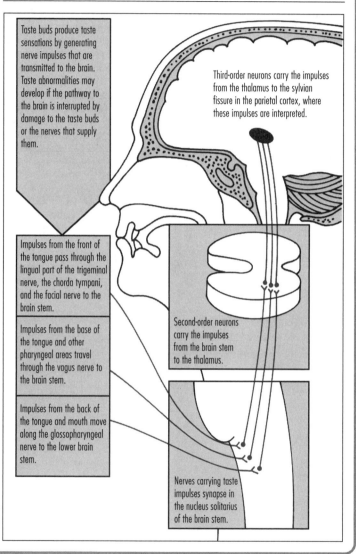

Taste buds produce taste sensations by generating nerve impulses that are transmitted to the brain. Taste abnormalities may develop if the pathway to the brain is interrupted by damage to the taste buds or the nerves that supply them.

Third-order neurons carry the impulses from the thalamus to the sylvian fissure in the parietal cortex, where these impulses are interpreted.

Impulses from the front of the tongue pass through the lingual part of the trigeminal nerve, the chorda tympani, and the facial nerve to the brain stem.

Impulses from the base of the tongue and other pharyngeal areas travel through the vagus nerve to the brain stem.

Impulses from the back of the tongue and mouth move along the glossopharyngeal nerve to the lower brain stem.

Second-order neurons carry the impulses from the brain stem to the thalamus.

Nerves carrying taste impulses synapse in the nucleus solitarius of the brain stem.

- traumatic cataracts
- foreign body injury
- complicated cataracts
- uveitis
- glaucoma
- retinitis pigmentosa
- retinal detachment
- diabetes mellitus
- hypoparathyroidism
- myotonic dystrophy

OLFACTORY PERCEPTION

The exact mechanism of olfactory perception remains unknown. The most likely theory suggests that the sticky mucus covering the olfactory cells traps airborne odorous molecules. As the molecules fit into appropriate receptors on the cell surface, the opposite end of the cell transmits an electrical impulse to the brain by way of the olfactory nerve (cranial nerve I).

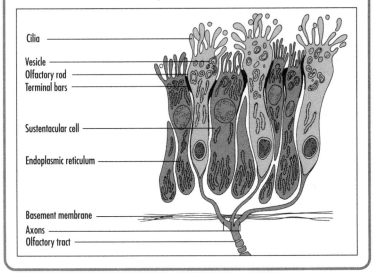

Cilia

Vesicle
Olfactory rod
Terminal bars

Sustentacular cell

Endoplasmic reticulum

Basement membrane
Axons
Olfactory tract

▶ atopic dermatitis
▶ exposure to ionizing radiation or infrared rays
▶ drugs that are toxic to the lens—see below
▶ prednisone (Deltasone)
▶ ergot alkaloids
▶ dinitrophenol
▶ naphthalene
▶ phenothiazines
▶ pilocarpine
▶ exposure to ultraviolet rays.

Pathophysiology

Pathophysiology may vary with each form of cataract. Congenital cataracts are particularly challenging. (See *Congenital cataracts*.) Senile cataracts show evidence of protein aggregation, oxidative injury, and increased pigmentation in the center of the lens. In traumatic cataracts, phagocytosis of the lens or inflammation may occur when a lens ruptures. The mechanism of a complicated cataract varies with the disease process; for example, in diabetes, increased glucose in the lens causes it to absorb water.

Typically, cataract development goes through four stages:
▶ immature: the lens is not totally opaque
▶ mature: the lens is completely opaque and vision loss is significant
▶ tumescent: the lens is filled with water; may lead to glaucoma
▶ hypermature: the lens proteins deteriorate, causing peptides to leak through the lens capsule; glaucoma may develop if intraocular fluid outflow is obstructed.

Signs and symptoms

Possible signs and symptoms of cataracts include:

▶ gradual painless blurring and loss of vision due to lens opacity

▶ milky white pupil due to lens opacity

▶ blinding glare from headlights at night due to the inefficient reflection of light rays by the opacities

▶ poor reading vision caused by reduced clarity of images

▶ better vision in dim light than in bright light in patients with central opacity; as pupils dilate, patients can see around the opacity.

 AGE ALERT Elderly patients with reduced vision may become depressed and withdraw from social activities rather than complain about reduced vision.

Complications

Complications of cataracts include:

▶ blindness

▶ glaucoma.

Surgical complications may include:

▶ loss of vitreous humor

▶ wound dehiscence from loosening of sutures and flat anterior chamber or iris prolapse into the wound

▶ hyphema, which is a hemorrhage into the anterior chamber of the eye

▶ vitreous-block glaucoma

▶ retinal detachment

▶ infection.

Diagnosis

Diagnosis is based on the following tests:

▶ physical examination (shining a penlight on the pupil to show the white area behind the pupil, which remains unnoticeable until the cataract is advanced)

▶ indirect ophthalmoscopy and slit-lamp examination to show a dark area in the normally homogeneous red reflex

▶ visual acuity test to confirm vision loss.

Treatment

Cataract treatment may include:

CONGENITAL CATARACTS

Congenital cataracts may be caused by:

▶ chromosomal abnormalities

▶ metabolic disease (such as galactosemia)

▶ intrauterine nutritional deficiencies

▶ infection during pregnancy (such as rubella).

Congenital cataracts may not be apparent at birth unless the eye is examined by funduscope.

If the cataract is removed within a few months of birth, the infant will be able to develop proper retinal fixation and cortical visual responses. After surgery, the child is likely to favor the normal eye; the brain suppresses the poor image from the affected eye, leading to underdeveloped vision (amblyopia) in that eye. Postoperatively, in the child with bilateral cataracts, vision develops equally in both eyes.

▶ extracapsular cataract extraction to remove the anterior lens capsule, and cortex and intraocular lens implant in the posterior chamber, typically performed by using phacoemulsification to fragment the lens with ultrasonic vibrations, then aspirating the pieces (See *Comparing methods of cataract removal,* page 496.)

▶ intracapsular cataract extraction to remove the entire lens within the intact capsule by cryoextraction (the moist lens sticks to an extremely cold metal probe for easy and safe extraction; rarely performed today)

▶ laser surgery after an extracapsular cataract extraction to restore visual acuity when a secondary membrane forms in the posterior lens capsule that has been left intact

COMPARING METHODS OF CATARACT REMOVAL

Cataracts can be removed by extracapsular or intracapsular techniques.

EXTRACAPSULAR CATARACT EXTRACTION

The surgeon may use irrigation and aspiration or phacoemulsification. To irrigate and aspirate, he makes an incision at the limbus, opens the anterior lens capsule with a cystotome, and exerts pressure from below to express the lens. He then irrigates and suctions the remaining lens cortex.

In phacoemulsification, he uses an ultrasonic probe to break the lens into minute particles and aspirates the particles.

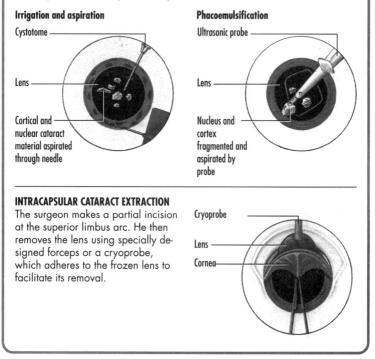

Irrigation and aspiration

Cystotome

Lens

Cortical and nuclear cataract material aspirated through needle

Phacoemulsification

Ultrasonic probe

Lens

Nucleus and cortex fragmented and aspirated by probe

INTRACAPSULAR CATARACT EXTRACTION

The surgeon makes a partial incision at the superior limbus arc. He then removes the lens using specially designed forceps or a cryoprobe, which adheres to the frozen lens to facilitate its removal.

Cryoprobe

Lens

Cornea

♦ discission (an incision) and aspiration may still be used in children with soft cataracts

♦ contact lenses or lens implantation after surgery to improve visual acuity, binocular vision, and depth perception.

Glaucoma

Glaucoma is a group of disorders characterized by an abnormally high intraocular pressure (IOP) that damages the optic nerve and other intraocular structures. Untreated, it leads to a gradual loss of vision and, ultimately, blindness. Glaucoma occurs in several forms: chronic open-angle (primary), acute angle-closure, congenital (inherited as an autosomal recessive trait), and secondary to other causes. Chronic open-angle glaucoma is usually bilateral, with insidious onset and a slowly progressive course. Acute angle-closure glaucoma typically has a rapid onset, constituting an ophthalmic emergency. Unless treated promptly, this acute form of glaucoma causes blindness in 3 to 5 days.

In the United States, approximately 2.5 million people have been diagnosed with glaucoma; another 1 million people have the disease but are undiagnosed. Glaucoma accounts for 12% of all new cases of blindness in the United States. The prognosis is good with early treatment.

 CULTURAL DIVERSITY Blacks have the highest incidence of glaucoma, and it's the single most common cause of blindness in this group.

Causes

Risk factors for chronic open-angle glaucoma include:
▶ genetics
▶ hypertension
▶ diabetes mellitus
▶ aging
▶ black ethnicity
▶ severe myopia.

Precipitating factors for acute angle-closure glaucoma include:
▶ drug-induced mydriasis (extreme dilation of the pupil)
▶ emotional excitement, which can lead to hypertension

 CULTURAL DIVERSITY The risk for acute angle-closure glaucoma is greater in persons with narrow iridic angles, such as those of Asian and Eskimo descent.

Secondary glaucoma may result from:
▶ uveitis
▶ trauma
▶ steroids
▶ diabetes
▶ infections
▶ surgery.

Pathophysiology

Chronic open-angle glaucoma results from overproduction or obstruction of the outflow of aqueous humor through the tra-

NORMAL FLOW OF AQUEOUS HUMOR

Aqueous humor, a transparent fluid produced by the ciliary epithelium of the ciliary body, flows from the posterior chamber through the pupil to the anterior chamber. It then flows peripherally and filters through the trabecular meshwork to the canal of Schlemm, through which the fluid ultimately enters venous circulation.

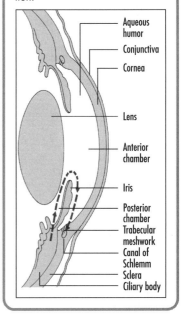

Aqueous humor
Conjunctiva
Cornea
Lens
Anterior chamber
Iris
Posterior chamber
Trabecular meshwork
Canal of Schlemm
Sclera
Ciliary body

becular meshwork or the canal of Schlemm, causing increased IOP and damage to the optic nerve. (See *Normal flow of aqueous humor.*) In secondary glaucoma, conditions such as trauma and surgery increase the risk for obstruction of intraocular fluid outflow caused by edema or other abnormal processes.

Acute angle-closure glaucoma results from obstruction to the outflow of aqueous humor. Obstruction may be caused by

CONGENITAL GLAUCOMA

Congenital glaucoma, a rare disease, occurs when a congenital defect in the angle of the anterior chamber obstructs the outflow of aqueous humor. Congenital glaucoma is usually bilateral, with an enlarged cornea that may be cloudy and bulging. Symptoms in a newborn, although difficult to assess, may include tearing, pain, and photophobia.

Untreated, congenital glaucoma causes damage to the optic nerve and blindness. Surgical intervention (such as goniotomy, goniopuncture, trabeculotomy, or trabeculectomy) is necessary to reduce intraocular pressure and prevent loss of vision.

anatomically narrow angles between the anterior iris and the posterior corneal surface, shallow anterior chambers, a thickened iris that causes angle closure on pupil dilation, or a bulging iris that presses on the trabeculae, closing the angle (peripheral anterior synechiae). Any of these may cause IOP to increase suddenly. (See *Congenital glaucoma.*)

 AGE ALERT In older patients, partial closure of the angle may also occur, so that two forms of glaucoma may coexist.

Signs and symptoms

Clinical manifestations of chronic open-angle glaucoma typically are bilateral and include:
▶ mild aching in the eyes caused by increased IOP
▶ loss of peripheral vision due to compression of retinal rods and nerve fibers
▶ halos around lights as a result of corneal edema
▶ reduced visual acuity, especially at night, not correctable with glasses.

Clinical manifestations of acute angle-closure glaucoma have a rapid onset, are usually unilateral, and include:
▶ inflammation
▶ red, painful eye caused by an abrupt elevation of IOP
▶ sensation of pressure over the eye due to increased IOP
▶ moderate papillary dilation nonreactive to light
▶ cloudy cornea due to compression of intraocular components
▶ blurring and decreased visual acuity due to aberrant neural conduction
▶ photophobia due to abnormal intraocular pressures
▶ halos around lights due to corneal edema
▶ nausea and vomiting caused by increased IOP.

Complications

Glaucoma may be complicated by:
▶ blindness.

Diagnosis

Glaucoma may be diagnosed using the following tests:
▶ pressure measurement tonometry using an applanation, Schiotz, or pneumatic tonometer; fingertip tension to estimate IOP (on gentle palpation of closed eyelids, one eye feels harder than the other in acute angle-closure glaucoma)
▶ slit-lamp examination of anterior structures of the eye, including the cornea, iris, and lens
▶ gonioscopy, to determine the angle of the anterior chamber of the eye, enabling differentiation between chronic open-angle glaucoma and acute angle-closure glaucoma (normal angle in chronic open-angle glaucoma and abnormal angle in acute angle-closure glaucoma (See *Optic disk changes in chronic glaucoma.*)
▶ ophthalmoscopy, to show cupping of the optic disk in chronic open-angle glaucoma; pale disk suggesting acute angle-closure glaucoma

▶ perimetry or visual field tests, to detect loss of peripheral vision due to chronic open-angle glaucoma

▶ fundus photography, to monitor the disk for changes.

Treatment

Treatment of chronic open-angle glaucoma may include:

▶ beta-blockers, such as timolol or betaxolol (a beta 1-receptor antagonist), to decrease aqueous humor production

▶ alpha agonists, such as brimonidine or apraclonidine, to reduce intraocular pressure

▶ carbonic anhydrase inhibitors, such as dorzolamide or acetazolamide, to decrease the formation and secretion of aqueous humor

▶ epinephrine, to reduce IOP by improving aqueous outflow

▶ prostaglandins, such as latanoprost, to reduce intraocular pressure

▶ miotic eye drops, such as pilocarpine, to reduce IOP by facilitating the outflow of aqueous humor.

When medical therapy fails to reduce intraocular pressure, the following surgical procedures may be performed:

▶ argon laser trabeculoplasty of the trabecular meshwork of an open angle, to produce a thermal burn that changes the surface of the meshwork and increases the outflow of aqueous humor

▶ trabeculectomy, to remove scleral tissue, followed by a peripheral iridectomy, to produce an opening for aqueous outflow under the conjunctiva, creating a filtering bleb.

Treatment for acute angle-closure glaucoma is an ocular emergency requiring immediate intervention to reduce high IOP, including:

▶ I.V. mannitol (20%) or oral glycerin (50%), to reduce IOP by creating an osmotic pressure gradient between the blood and intraocular fluid

▶ steroid drops, to reduce inflammation

▶ acetazolamide, a carbonic anhydrase inhibitor, to reduce IOP by decreasing the

OPTIC DISK CHANGES IN CHRONIC GLAUCOMA

Ophthalmoscopy and slit-lamp examination show cupping of the optic disk, which is characteristic of chronic glaucoma.

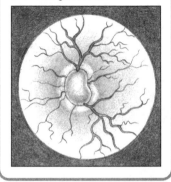

formation and secretion of aqueous humor

▶ pilocarpine, to constrict the pupil, forcing the iris away from the trabeculae and allowing fluid to escape

▶ timolol, a beta-blocker, to decrease IOP

▶ narcotic analgesics, to reduce pain if necessary.

▶ laser iridotomy or surgical peripheral iridectomy, if drug therapy does not reduce IOP, to relieve pressure and preserve vision

▶ cycloplegic drops, such as apraclonidine, in the affected eye (only after laser peripheral iridectomy), to relax the ciliary muscle and reduce inflammation to prevent adhesions.

Macular degeneration

Macular degeneration — atrophy or degeneration of the macular disk — is the most common cause of legal blindness in adults. Commonly affecting both eyes, it accounts for about 12% of blindness in the United States and for about 17% of new blindness. It's also one of the causes of severe irreversible and unpreventable loss of central vision in the elderly.

Two types of age-related macular degeneration occur. The dry, or atrophic, form is characterized by atrophic pigment epithelial changes and most often gradually causes mild visual loss. The wet, exudative form rapidly causes severe vision loss. It's characterized by the subretinal formation of new blood vessels (neovascularization) that cause leakage, hemorrhage, and fibrovascular scar formation.

Causes

The causes of macular degeneration are unknown but may include:
◗ aging
◗ inflammation
◗ injury
◗ infection
◗ nutrition.

Pathophysiology

Age-related macular degeneration results from hardening and obstruction of retinal arteries, which probably reflect normal degenerative changes. The formation of new blood vessels in the macular area obscures central vision. Underlying pathologic changes occur primarily in the retinal pigment epithelium, Bruch's membrane, and choriocapillaris in the macular region.

The dry form develops as yellow extracellular deposits, or drusen, accumulate beneath the pigment epithelium of the retina; they may be prominent in the macula. Drusen are common in the elderly. Over time, drusen grow and become more numerous. Visual loss occurs as the retinal pigment epithelium detaches and becomes atrophic.

Exudative macular degeneration develops as new blood vessels in the choroid project through abnormalities in Bruch's membrane and invade the potential space underneath the retinal pigment epithelium. As these vessels leak, fluid in the retinal pigment epithelium is increased, resulting in blurry vision.

Signs and symptoms

Signs and symptoms of macular degeneration include:
◗ changes in central vision due to neovascularization, such as a blank spot (scotoma) in the center of a page when reading
◗ distorted appearance of straight lines caused by relocation of retinal receptors.

Complications

Possible complications are:
◗ visual impairment progressing to blindness
◗ nystagmus.

Diagnosis

Diagnosis is based on the following test results:
◗ indirect ophthalmoscopy, to show gross macular changes, opacities, hemorrhage, neovascularization, retinal pallor, or retinal detachment
◗ I.V. fluorescein angiography sequential photographs, to show leaking vessels as fluorescein dye flows into the tissues from the subretinal neovascular net
◗ Amsler's grid test, to show central visual field loss.

Treatment

Treatment includes:
◗ laser photocoagulation, to reduce the incidence of severe visual loss in patients with subretinal neovascularization (exudative form)
◗ currently no cure for the atrophic form.

Meniere's disease

Meniere's disease, a labyrinthine dysfunction also known as endolymphatic hydrops, causes severe vertigo, sensorineural hearing loss, and tinnitus. It usually affects adults between the ages of 30 and 60, men slightly more often than women, and rarely occurs in children. Usually, only one ear is involved. After multiple attacks over several years, residual tinnitus and hearing loss can be incapacitating.

Causes

The cause of Meniere's disease is un-known. It may be associated with:

▶ family history
▶ immune disorder
▶ migraine headaches
▶ middle ear infection
▶ head trauma
▶ autonomic nervous system dysfunction
▶ premenstrual edema.

Pathophysiology

Meniere's disease may result from over-production or decreased absorption of en-dolymph—the fluid contained in the labyrinth of the ear. Accumulated en-dolymph dilates the semicircular canals, utricle, and saccule and causes degenera-tion of the vestibular and cochlear hair cells. Overstimulation of the vestibular branch of cranial nerve VIII impairs pos-tural reflexes and stimulates the vomiting reflex. (See *Normal vestibular function*.) Perception of sound is impaired as a re-sult of this excessive cranial nerve stimu-lation, and injury to sensory receptors for hearing may affect auditory acuity.

Signs and symptoms

Signs and symptoms of Meniere's disease include:

▶ sudden severe spinning, whirling verti-go, lasting from 10 minutes to several hours, due to increased endolymph (at-tacks may occur several times a year, or remissions may last as long as several years)
▶ tinnitus caused by altered firing of sen-sory auditory neurons (may have residual tinnitus between attacks)
▶ hearing impairment due to sensorineu-ral loss (hearing may be normal between attacks, but repeated attacks may pro-gressively cause permanent hearing loss)
▶ feeling of fullness or blockage in the ear preceding an attack, a result of changing sensitivity of pressure receptors
▶ severe nausea, vomiting, sweating, and pallor during an acute attack due to auto-nomic dysfunction

NORMAL VESTIBULAR FUNCTION

The semicircular canals and vestibule of the inner ear are re-sponsible for equilibrium and bal-ance. Each of the three semicircu-lar canals lies at a 90-degree an-gle to the others. When the head is moved, endolymph inside each semicircular canal moves in an opposite direction. The movement stimulates hair cells, which send electrical impulses to the brain through the vestibular portion of cranial nerve VIII. Head move-ment also causes movement of the vestibular otoliths (crystals of cal-cium salts) in their gel medium, which tugs on hair cells, initiating the transmission of electrical im-pulses to the brain through the vestibular nerve. Together, these two organs help detect the body's present position as well as a change in direction or motion.

▶ nystagmus due to asymmetry and in-tensity of impulses reaching the brain stem
▶ loss of balance and falling to the affect-ed side due to vertigo.

Complications

A complication is:

▶ hearing loss.

Diagnosis

Diagnosis of Meniere's disease is based on:

▶ patient history of signs and symptoms
▶ audiometric testing showing a sen-sorineural hearing loss and loss of dis-crimination and recruitment
▶ electronystagmography showing normal or reduced vestibular response on the af-fected side
▶ cold caloric testing showing impairment of oculovestibular reflex

- electrocochleography showing increased ratio of summating potential to action potential
- brain stem evoked response audiometry test, to rule out acoustic neurinoma, brain tumor, and vascular lesions in the brain stem
- computed tomography scan and magnetic resonance imaging, to rule out acoustic neurinoma as a cause of symptoms.

Treatment

During an acute attack, treatment may include:

- lying down to minimize head movement, and avoiding sudden movements and glaring lights to reduce dizziness
- promethazine (Phenergan) or prochlorperazine (Compazine), to relieve nausea and vomiting
- atropine, to control an attack by reducing autonomic nervous system function
- dimenhydrinate (Dramamine), to control vertigo and nausea
- central nervous system depressants, such as lorazepam (Ativan) or diazepam (Valium) during an acute attack, to reduce excitability of vestibular nuclei
- antihistamines, such as meclizine (Antivert) or diphenhydramine (Benadryl), to reduce dizziness and vomiting.

Long-term management may include:

- diuretics, such as triamterene (Dyrenium) or acetazolamide (Diamox), to reduce endolymph pressure
- betahistine, to alleviate vertigo, hearing loss, and tinnitus
- vasodilators, to dilate blood vessels supplying the inner ear
- sodium restriction, to reduce endolymphatic hydrops
- antihistamines or mild sedatives, to prevent attacks
- systemic streptomycin, to produce chemical ablation of the sensory neuroepithelium of the inner ear, and thereby control vertigo in patients with bilateral disease for whom no other treatment can be considered.

In Meniere's disease that persists despite medical treatment or produces incapacitating vertigo, the following surgical procedures may be performed:

- endolymphatic drainage and shunt procedures, to reduce pressure on the hair cells of the cochlea and prevent further sensorineural hearing loss
- vestibular nerve resection in patients with intact hearing, to reduce vertigo and prevent further hearing loss
- labyrinthectomy for relief of vertigo in patients with incapacitating symptoms and poor or no hearing, because destruction of the cochlea results in a total loss of hearing in the affected ear
- cochlear implantation, to improve hearing in patients with profound deafness due to Meniere's disease.

Otosclerosis

The most common cause of chronic, progressive, conductive hearing loss, otosclerosis is the slow formation of spongy bone in the otic capsule, particularly at the oval window. It occurs in at least 10% of persons of European descent and is three times as prevalent in females as in males; onset is usually between the ages of 15 and 30. Occurring unilaterally at first, the disorder may progress to bilateral conductive hearing loss. With surgery, the prognosis is good.

CULTURAL DIVERSITY Otosclerosis occurs less frequently in Asians and blacks, but it is a common disorder in southern India.

Causes

Causes include:

- autosomal dominant trait
- pregnancy.

AGE ALERT Children with osteogenesis imperfecta, an inherited condition characterized by brittle bones, may also have otosclerosis.

Pathophysiology

In otosclerosis, the normal bone of the otic capsule is gradually replaced with a highly vascular spongy bone. This spongy bone immobilizes the footplate of the normally mobile stapes, disrupting the conduction of vibrations from the tympanic membrane to the cochlea. Because the sound pressure vibrations aren't transmitted to the fluid of the inner ear, the result is conductive hearing loss. If the inner ear becomes involved, sensorineural hearing loss may develop.

Signs and symptoms

Signs and symptoms of otosclerosis include:

▶ bilateral conductive hearing loss due to the disruption of the conduction of vibrations from the tympanic membrane to the cochlea

▶ tinnitus due to overstimulation of cranial nerve VIII afferents

▶ ability to hear a conversation better in a noisy environment than in a quiet one (paracusis of Willis) as a result of masking effects.

Complications

A complication of otosclerosis is:
▶ deafness.

Diagnosis

Diagnosis is based on:

▶ otoscopic examination showing a normal-appearing tympanic membrane; occasionally, the tympanic membrane may appear pinkish-orange (Schwartz's sign) as a result of vascular and bony changes in the middle ear

▶ Rinne test showing bone conduction lasting longer than air conduction (normally, the reverse is true); as otosclerosis progresses, bone conduction also deteriorates

▶ audiometric testing showing hearing loss ranging from 60 dB in early stages to total loss

▶ Weber's test to detect sounds lateralizing to the more affected ear.

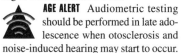 **AGE ALERT** Audiometric testing should be performed in late adolescence when otosclerosis and noise-induced hearing may start to occur.

Treatment

Treatment of otosclerosis may include:

▶ stapedectomy (removal of the stapes) and insertion of a prosthesis to restore partial or total hearing

▶ stapedotomy (creation of a small hole in the footplate of the stapes) and insertion of a wire and piston as a prosthesis to help restore hearing

▶ hearing aid (air conduction aid with molded ear insertion receiver) if surgery is not possible to permit hearing of conversation in normal surroundings.

The integumentary system, the largest and heaviest body system, includes the skin—the integument, or external covering of the body—and the epidermal appendages, including the hair, nails, and sebaceous, eccrine, and apocrine glands. It protects against injury and invasion of microorganisms, harmful substances, and radiation; regulates body temperature; serves as a reservoir for food and water; and synthesizes vitamin D. Emotional well-being, including one's responses to the daily stresses of life, is reflected in the skin.

SKIN

The skin is composed of three layers: the epidermis, dermis, and subcutaneous tissues. The epidermis is the outermost layer. It's thin and contains sensory receptors for pain, temperature, touch, and vibration. The epidermal layer has no blood vessels and relies on the dermal layer for nutrition. The dermis contains connective tissue, the sebaceous glands, and some hair follicles. The subcutaneous tissue lies beneath the dermis; it contains fat and sweat glands and the rest of the hair follicles. The subcutaneous layer is able to store calories for future use in the body. (See *Close-up view of the skin.*)

HAIR AND NAILS

The hair and nails are considered appendages of the skin. Both have protective functions in addition to their cosmetic appeal. The cuticle of the nail, for example, functions as a seal, protecting the area between two portions of the nail from external hazards. (See *Nail structure,* page 506.)

GLANDS

The sebaceous glands, found on all areas of the skin except the palms and soles, produce sebum, a semifluid material composed of fat and epithelial cells. Sebum is secreted into the hair follicle and exits to the skin surface. It helps waterproof the hair and skin and promotes the absorption of fat-soluble substances into the dermis.

The eccrine glands produce sweat, an odorless, watery fluid. Glands in the palms and soles secrete sweat primarily in response to emotional stress. The other remaining eccrine glands respond mainly to thermal stress, effectively regulating temperature.

Located mainly in the axillary and anogenital areas, apocrine glands have a coiled secretory portion that lies deeper in the dermis than the eccrine glands. These glands begin to function at puberty and have no known biological function. Bacterial decomposition of the apocrine fluid produced by these glands causes body odor.

Skin color depends on four pigments: melanin, carotene, oxyhemoglobin, and deoxyhemoglobin. Each pigment is unique in its function and effect on the skin. For example, melanin, the brownish pigment of the skin, is genetically determined, though it can be altered by sunlight exposure. Excessive dietary carotene (from carrots, sweet potatoes, and leafy vegeta-

CLOSE-UP VIEW OF THE SKIN

The skin is composed of two major layers — the epidermis and dermis. The epidermis consists of five strata, shown below. Subcutaneous tissue lying beneath the dermis consists of loose connective tissue that attaches the skin to underlying structures.

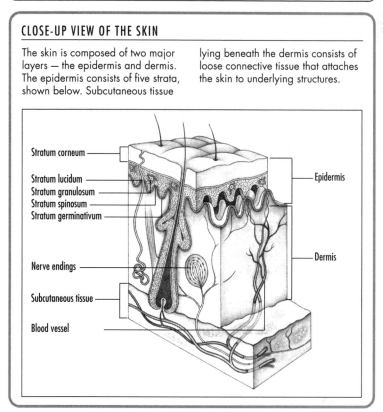

Stratum corneum
Stratum lucidum
Stratum granulosum
Stratum spinosum
Stratum germinativum
Epidermis
Dermis
Nerve endings
Subcutaneous tissue
Blood vessel

bles) causes a yellowing of the skin. Excessive oxyhemoglobin in the blood causes a reddening of the skin, and excessive deoxyhemoglobin (not bound to oxygen) causes a bluish discoloration.

PATHOPHYSIOLOGIC MANIFESTATIONS

Clinical manifestations of skin dysfunction include the inflammatory reaction of the skin and the formation of lesions.

Inflammatory reaction of the skin

An inflammatory reaction occurs with injury to the skin. The reaction can only occur in living organisms. Although a beneficial response, it's usually accompanied by some degree of discomfort at the site. Irritation changes the epidermal structure, and consequent increase of immunoglobulin E (IgG) activity. Other classic signs of inflammatory skin responses are redness, edema, and warmth, due to bioamines released from the granules of tissue mast cells and basophils.

Formation of lesions

Primary skin lesions appear on previously healthy skin in response to disease or external irritation. They're classified by their appearance as macules, papules, plaques, patches, nodules, tumors, wheals, cysts, vesicles, bullae, or pustules. (See *Recognizing primary skin lesions,* pages 507 and 508.)

NAIL STRUCTURE

The following illustration shows the anatomic components of a fingernail.

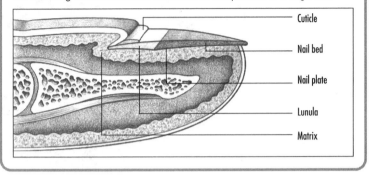

- Cuticle
- Nail bed
- Nail plate
- Lunula
- Matrix

Modified lesions are described as secondary skin lesions. These lesions occur as a result of rupture, mechanical irritation, extension, invasion, or normal or abnormal healing of primary lesions. These include atrophy, erosions, ulcers, fissures, crusts, scales, lichenification, excoriation, and scars. (See *Recognizing secondary skin lesions*, pages 509 and 510.)

DISORDERS

Trauma, abnormal cellular function, infection, and systemic disease may cause disruptions in skin integrity.

Acne

Acne is a chronic inflammatory disease of the sebaceous glands. It's usually associated with a high rate of sebum secretion and occurs on areas of the body that have sebaceous glands, such as the face, neck, chest, back, and shoulders. There are two types of acne: *inflammatory,* in which the hair follicle is blocked by sebum, causing bacteria to grow and eventual rupture of the follicle; and *noninflammatory,* in which the follicle doesn't rupture but remains dilated.

AGE ALERT Acne occurs in both males and females. Acne vulgaris develops in 80% to 90% of adolescents or young adults, primarily between ages 15 and 18. Although the lesions can appear as early as age 8, acne primarily affects adolescents.

Although the severity and overall incidence of acne is usually greater in males, it tends to start at an earlier age lasts longer in females.

The prognosis varies and depends on the severity and underlying cause(s); with treatment, the prognosis is usually good.

Causes

The cause of acne is multifactorial. Diet isn't believed to be a precipitating factor. Possible causes of acne include increased activity of sebaceous glands and blockage of the pilosebaceous ducts (hair follicles).

Factors that may predispose to acne include:
- heredity
- androgen stimulation
- certain drugs, including corticosteroids, corticotropin (ACTH), androgens, iodides, bromides, trimethadione, phenytoin (Dilantin), isoniazid (Laniozid), lithium (Eskalith), and halothane
- cobalt irradiation

(Text continues on page 510.)

RECOGNIZING PRIMARY SKIN LESIONS

The most common primary lesions are illustrated below.

BULLA
Fluid-filled lesion more than ¾" (2 cm) in diameter (also called a blister) (e.g., severe poison oak or ivy dermatitis, bullous pemphigoid, second-degree burn)

COMEDO
Plugged pilosebaceous duct, exfoliative, formed from sebum and keratin (e.g., blackhead [open comedo], whitehead [closed comedo])

CYST
Semisolid or fluid-filled encapsulated mass extending deep into the dermis (e.g., acne)

MACULE
Flat, pigmented, circumscribed area less than ⅜" (1 cm) in diameter (e.g., freckle, rubella)

NODULE
Firm, raised lesion; deeper than a papule, extending into dermal layer; ¼" to ¾" (0.5 to 2 cm) in diameter (e.g., intradermal nevus)

PAPULE
Firm, inflammatory, raised lesion up to ¼" (0.5 cm) in diameter, may be same color as skin or pigmented (e.g., acne papule, lichen planus)

(continued)

RECOGNIZING PRIMARY SKIN LESIONS *(continued)*

PATCH
Flat, pigmented, circumscribed area more than ⅜" (1 cm) in diameter (e.g., herald patch [pityriasis rosea])

PLAQUE
Circumscribed, solid, elevated lesion more than ⅜" (1 cm) in diameter; elevation above skin surface occupies larger surface area compared with height (e.g., psoriasis)

PUSTULE
Raised, circumscribed lesion usually less than ⅜" (1 cm) in diameter; containing purulent material, making it a yellow-white color (e.g., acne pustule, impetigo, furuncle)

TUMOR
Elevated solid lesion more than ¾" (2 cm) in diameter, extending into dermal and subcutaneous layers (e.g., dermatofibroma)

VESICLE
Raised, circumscribed, fluid-filled lesion less than ¼" (0.5 cm) in diameter (e.g., chicken pox, herpes simplex)

WHEAL
Raised, firm lesion with intense localized skin edema, varying in size and shape; color ranging from pale pink to red, disappears in hours (e.g., hive [urticaria], insect bite)

RECOGNIZING SECONDARY SKIN LESIONS

The most common secondary lesions are illustrated below.

ATROPHY
Thinning of skin surface at site of disorder (e.g., striae, aging skin)

CRUST
Dried sebum, serous, sanguineous, or purulent exudate overlying an erosion or weeping vesicle, bulla, or pustule (e.g., impetigo)

EROSION
Circumscribed lesion involving loss of superficial epidermis (e.g., rug burn, abrasion)

EXCORIATION
Linear scratched or abraded areas, often self-induced (e.g., abraded acne, eczema)

FISSURE
Linear cracking of the skin extending into the dermal layer (e.g., hand dermatitis [chapped skin])

LICHENIFICATION
Thickened, prominent skin markings by constant rubbing (e.g., chronic atopic dermatitis)

(continued)

RECOGNIZING SECONDARY SKIN LESIONS *(continued)*

SCALE
Thin, dry flakes of shedding skin (e.g., psoriasis, dry skin, newborn desquamation)

SCAR
Fibrous tissue caused by trauma, deep inflammation, or surgical incision; red and raised (recent), pink and flat (6 weeks), and depressed (old) (e.g., on a healed surgical incision)

ULCER
Epidermal and dermal destruction may extend into subcutaneous tissue; usually heals with scarring (e.g., pressure sore or ulcer)

▶ hyperalimentation
▶ exposure to heavy oils, greases, or tars
▶ trauma or rubbing from tight clothing
▶ cosmetics
▶ emotional stress
▶ unfavorable climate
▶ oral contraceptive use. (Many females experience acne flare-up during their first few menses after starting or discontinuing oral contraceptives.)

Pathophysiology
Androgens stimulate sebaceous gland growth and the production of sebum, which is secreted into dilated hair follicles that contain bacteria. The bacteria, usually *Propionibacterium acne* and *Staphylococcus epidermis,* are normal skin flora that secrete lipase. This enzyme interacts with sebum to produce free fatty acids, which provoke inflammation. Hair follicles also produce more keratin, which joins with the sebum to form a plug in the dilated follicle.

Signs and symptoms
The acne plug may appear as:
▶ a closed comedo, or whitehead (not protruding from the follicle and covered by the epidermis)
▶ an open comedo, or blackhead (protruding from the follicle and not covered by the epidermis; melanin or pigment of the follicle causes the black color).

Rupture or leakage of an enlarged plug into the epidermis produces inflammation, characteristic acne pustules, papules, or, in severe forms, acne cysts or abscesses (chronic, recurring lesions producing acne scars).

In women, signs and symptoms may include increased severity just before or during menstruation when estrogen levels are at the lowest.

Complications

Complications of acne may include:
▶ acne conglobata
▶ scarring (when acne is severe)
▶ impaired self-esteem (mostly adolescents afflicted).

Diagnosis

Diagnosis of acne vulgaris is confirmed by characteristic acne lesions, especially in adolescents.

Treatment

Topical treatments of acne include the application of antibacterial agents, such as benzyl peroxide (Benzac 5 or 10), clindamycin (Cleocin), or benzyl peroxide plus erythromycin (Benzamycin) antibacterial agents. These may be applied alone or with tretinoin (Retin-A; retinoic acid), which is a keratolytic. Keratolytic agents, such as benzyl peroxide and tretinoin, dry and peel the skin in order to help open blocked follicles, moving the sebum up to the skin level.

Systemic therapy consists primarily of:
▶ antibiotics, usually tetracycline, to decrease bacterial growth (reduced dosage for long-term maintenance when the patient is in remission)
▶ culture to identify a possible secondary bacterial infection (for exacerbation of pustules or abscesses while on tetracycline or erythromycin drug therapy)
▶ oral isotretinoin (Accutane) to inhibit sebaceous gland function and abnormal keratinization (16- to 20-week course of isotretinoin limited to patients with severe papulopustular or cystic acne not responding to conventional therapy due to its severe adverse effects)
▶ for females only, antiandrogens: birth control pills, such as norgestimate/ethinyl estradiol (Ortho TriCyclen) or spironolactone
▶ cleansing with an abrasive sponge in order to dislodge superficial comedones

▶ surgery to remove comedones and to open and drain pustules (usually on an outpatient basis)
▶ dermabrasion (for severe acne scarring) with a high-speed metal brush to smooth the skin (performed only by a well-trained dermatologist or plastic surgeon)
▶ bovine collagen injections into the dermis beneath the scarred area to fill in affected areas and even out the skin surface (not recommended by all dermatologists).

Atopic dermatitis

Atopic (allergic) dermatitis (also called atopic or infantile eczema) is a chronic or recurrent inflammatory response often associated with other atopic diseases, such as bronchial asthma and allergic rhinitis. It usually develops in infants and toddlers between ages 1 month and 1 year, usually in those with a strong family history of atopic disease. These children often develop other atopic disorders as they grow older.

Typically, this form of dermatitis flares and subsides repeatedly before finally resolving during adolescence, but it can persist into adulthood. Atopic dermatitis affects about 9 of every 1,000 persons.

Causes

Possible causes of atopic dermatitis include food allergy, infection, irritating chemicals, extremes of temperature and humidity, psychological stress or strong emotions (flare ups). These causes may be exacerbated by genetic predisposition.

AGE ALERT About 10% of childhood cases of atopic dermatitis are caused by allergy to certain foods, especially eggs, peanuts, milk, and wheat.

Pathophysiology

The allergic mechanism of hypersensitivity results in a release of inflammatory mediators through sensitized antibodies of the immunoglobulin E (IgE) class. Histamine and other cytokines induce an inflammatory response that results in edema and skin breakdown, along with pruritus.

Signs and symptoms
Possible signs and symptoms of atopic dermatitis are:
▶ erythematous areas on excessively dry skin; in children, typically on the forehead, cheeks, and extensor surfaces of the arms and legs; in adults, at flexion points (antecubital fossa, popliteal area, and neck)
▶ edema, crusting, and scaling due to pruritus and scratching
▶ multiple areas of dry, scaly skin, with white dermatographia, blanching, and lichenification with chronic atrophic lesions
▶ pink pigmentation and swelling of upper eyelid with a double fold under the lower lid (Morgan's, Dennie's, or mongolian fold) due to severe pruritus
▶ viral, fungal, or bacterial infections and ocular disorders (common secondary conditions).

Complications
Possible complications include:
▶ cataracts developing between ages 20 and 40
▶ Kaposi's varicelliform eruption (eczema herpeticum), a potentially serious widespread cutaneous viral infection (may develop if the patient comes in contact with a person infected with herpes simplex)
▶ subclinical (not requiring treatment) skin infection that may progress to cellulitis.

Diagnosis
Diagnosis of atopic dermatitis may involve:
▶ family history of atopic disorders (helpful in diagnosis)
▶ typical distribution of skin lesions
▶ ruling out other inflammatory skin lesions, such as diaper rash (lesions confined to the diapered area), seborrheic dermatitis (moist or greasy scaling with yellow-crusted patches), and chronic contact dermatitis (lesions affect hands and forearms, not antecubital and popliteal areas)

▶ serum IgE levels (often elevated but not diagnostic).

Treatment
Treatments include:
▶ eliminating allergens and avoiding irritants (strong soaps, cleansers, and other chemicals), extreme temperature changes, and other precipitating factors
▶ preventing excessive dryness of the skin (critical to successful therapy)
▶ topical application of a corticosteroid ointment, especially after bathing, to alleviate inflammation (moisturizing cream between steroid doses to help retain moisture); systemic antihistamines, such as Benadryl (diphenhydramine)

 AGE ALERT Chronic use of potent fluorinated corticosteroids may cause striae or atrophy in children.
▶ administering systemic corticosteroid therapy (during extreme exacerbations)
▶ applying weak tar preparations and ultraviolet B light therapy to increase thickness of stratum corneum
▶ administering antibiotics (for positive culture for bacterial agent).

Burns
Burns are classified as first degree, second-degree superficial, second-degree deep partial thickness, third-degree full thickness, and fourth degree. A first-degree burn is limited to the epidermis. The most common example of a first-degree burn is sunburn, which results from exposure to the sun. In a second-degree burn, the epidermis and part of the dermis are damaged. A third-degree burn damages the epidermis and dermis, and vessels and tissue are visible. In fourth-degree burns, the damage extends through deeply charred subcutaneous tissue to muscle and bone. A major burn is a horrifying injury needing painful treatment and a long period of rehabilitation.

Each year in the United States, about 2 million persons receive burn injuries. Of these, 300,000 are burned seriously, and more than 6,000 die, making burns this nation's third leading cause of accidental death. About 60,000 people are hospitalized each year for burns. Most significant burns occur in the home; home fires account for the highest burn fatality rate.

In victims younger than 4 years and older than 60 years, there's a higher incidence of complications and thus a higher mortality rate. Immediate, aggressive burn treatment increases the patient's chance for survival. Later, supportive measures and strict aseptic technique can minimize infection. Meticulous, comprehensive burn care can make the difference between life and death. Survival and recovery from a major burn are more likely once the burn wound is reduced to less than 20% of the total body surface area (BSA).

Causes
Thermal burns, the most common type, frequently result from:
▶ residential fires
▶ automobile accidents
▶ playing with matches
▶ improper handling of firecrackers
▶ scalding accidents and kitchen accidents (such as a child climbing on top of a stove or grabbing a hot iron)
▶ parental abuse of (in children or elders)
▶ clothes that have caught on fire.

Chemical burns result from contact, ingestion, inhalation, or injection of acids, alkalis, or vesicants.

Electrical burns usually result from contact with faulty electrical wiring or high-voltage power lines. Sometimes young children chew electrical cords.

Friction or abrasion burns occur when the skin rubs harshly against a coarse surface.

Sunburn results from excessive exposure to sunlight.

Pathophysiology
The injuring agent denatures cellular proteins. Some cells die because of traumatic or ischemic necrosis. Loss of collagen cross-linking also occurs with denatura-

tion, creating abnormal osmotic and hydrostatic pressure gradients that cause the movement of intravascular fluid into interstitial spaces. Cellular injury triggers the release of mediators of inflammation, contributing to local and, in the case of major burns, systemic increases in capillary permeability. Specific pathophysiologic events depend on the cause and classification of the burn. (See *Classifications of burns,* page 514.)

First-degree burns. A first-degree burn causes localized injury or destruction to the skin (epidermis only) by direct (such as chemical spill) or indirect (such as sunlight) contact. The barrier function of the skin remains intact, and these burns aren't life threatening.

Second-degree superficial partial-thickness burns. These burns involve destruction to the epidermis and some dermis. Thin-walled, fluid-filled blisters develop within a few minutes of the injury. As these blisters break, the nerve endings become exposed to the air. Because pain and tactile responses remain intact, subsequent treatments are very painful. The barrier function of the skin is lost.

Second-degree deep partial-thickness burns. These burns involve destruction of the epidermis and dermis, producing blisters and mild to moderate edema and pain. The hair follicles are still intact, so hair will grow again. Compared with second-degree superficial partial-thickness burns, there's less pain sensation with this burn because the sensory neurons have undergone extensive destruction. The areas around the burn injury remain very sensitive to pain. The barrier function of the skin is lost.

Third- and fourth-degree burns. A major burn affects every body system and organ. A third-degree burn extends through the epidermis and dermis and into the subcutaneous tissue layer. A fourth-degree

CLASSIFICATIONS OF BURNS

The depth of skin and tissue damage determines burn classification. The following illustration shows the four degrees of burn classifications.

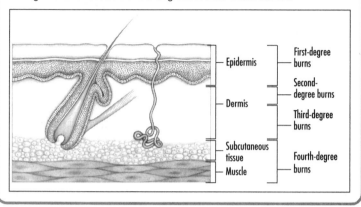

burn involves muscle, bone, and interstitial tissues. Within only hours, fluids and protein shift from capillary to interstitial spaces, causing edema. There's an immediate immunologic response to a burn injury, making burn wound sepsis a potential threat. Finally, an increase in calorie demand after a burn injury increases the metabolic rate.

Signs and symptoms
Signs and symptoms depend on the type of burn and may include:
▶ localized pain and erythema, usually without blisters in the first 24 hours (first-degree burn)
▶ chills, headache, localized edema, and nausea and vomiting (more severe first-degree burn)
▶ thin-walled, fluid-filled blisters appearing within minutes of the injury, with mild to moderate edema and pain (second-degree superficial partial-thickness burn)
▶ white, waxy appearance to damaged area (second-degree deep partial-thickness burn)
▶ white, brown, or black leathery tissue and visible thrombosed vessels due to destruction of skin elasticity (dorsum of hand

most common site of thrombosed veins), without blisters (third-degree burn)
▶ silver-colored, raised area, usually at the site of electrical contact (electrical burn)
▶ singed nasal hairs, mucosal burns, voice changes, coughing, wheezing, soot in mouth or nose, and darkened sputum (with smoke inhalation and pulmonary damage).

Complications
Possible complications of burns include:
▶ loss of function (burns to face, hands, feet, and genitalia)
▶ total occlusion of circulation in extremity (due to edema from circumferential burns)
▶ airway obstruction (neck burns) or restricted respiratory expansion (chest burns)
▶ pulmonary injury (from smoke inhalation or pulmonary embolism)
▶ adult respiratory distress syndrome (due to left-sided heart failure or myocardial infarction)
▶ greater damage than indicated by the surface burn (electrical and chemical burns) or internal tissue damage along the conduction pathway (electrical burns)
▶ cardiac arrhythmias (due to electrical shock)

◗ infected burn wound

◗ stroke, heart attack, or pulmonary embolism (due to formation of blood clots resulting from slower blood flow)

◗ burn shock (due to fluid shifts out of the vascular compartments, possibly leading to kidney damage and renal failure)

◗ peptic ulcer disease (due to decreased blood supply in the abdominal area)

◗ disseminated intravascular coagulation (more severe burn states)

◗ added pain, depression, and financial burden (due to psychological component of disfigurement).

Diagnosis

Diagnosis involves determining the size and classifying the wound. The following methods are used to determine size:

◗ percentage of BSA covered by the burn using the Rule of Nines chart

◗ Lund-Browder chart (more accurate because it allows BSA changes with age); correlation of the burn's depth and size to estimate its severity. (See *Using the Rule of Nines and the Lund and Browder chart,* page 516 and 517.)

Major burns are classified as:

◗ third-degree burns on more than 10% of BSA

◗ second-degree burns on more than 25% of adult BSA (over 20% in children)

◗ burns of hands, face, feet, or genitalia

◗ burns complicated by fractures or respiratory damage

◗ electrical burns

◗ all burns in poor-risk patients.

Moderate burns are classified as:

◗ third-degree burns on 2% to 10% of BSA

◗ second-degree burns on 15% to 25% of adult BSA (10% to 20% in children).

Minor burns are classified as:

◗ third-degree burns on less than 2% of BSA

◗ second-degree burns on less than 15% of adult BSA (10% in children).

Treatment

Initial burn treatments are based on the type of burn and may include:

◗ immersing the burned area in cool water (55°F [12.8°C]) or applying cool compresses (minor burns)

◗ pain medication as needed or anti-inflammatory medications

◗ covering the area with an antimicrobial agent and a nonstick bulky dressing (after debridement); prophylactic tetanus injection as needed

◗ maintaining an open airway; assessing airway, breathing, and circulation; checking for smoke inhalation immediately on receipt of the patient; assisting with endotracheal intubation; and giving 100% oxygen (first immediate treatment for moderate and major burns)

◗ controlling active bleeding

◗ covering partial-thickness burns over 30% of BSA or full-thickness burns over 5% of BSA with a clean, dry, sterile bed sheet (because of drastic reduction in body temperature, *do not* cover large burns with saline-soaked dressings)

◗ removing smoldering clothing (first soaking in saline solution if clothing is stuck to the patient's skin), rings, and other constricting items

◗ immediate I.V. therapy to prevent hypovolemic shock and maintain cardiac output (lactated Ringer's solution or a fluid replacement formula; additional I.V. lines may be needed)

◗ antimicrobial therapy (all patients with major burns)

◗ complete blood count, electrolyte, glucose, blood urea nitrogen, and serum creatinine levels; arterial blood gas analysis; typing and cross-matching; urinalysis for myoglobinuria and hemoglobinuria

◗ closely monitoring intake and output, frequently checking vital signs (every 15 minutes), possibly inserting indwelling urinary catheter

◗ nasogastric tube to decompress the stomach and avoid aspiration of stomach contents

USING THE RULE OF NINES AND THE LUND AND BROWDER CHART

You can quickly estimate the extent of an adult patient's burn by using the Rule of Nines. This method divides an adult's body surface area into percentages. To use this method, mentally transfer your patient's burns to the body chart shown below, then add up the corresponding percentages for each burned body section. The total, an estimate of the extent of your patient's burn, enters into the formula to determine his initial fluid replacement needs.

You can't use the Rule of Nines for infants and children because their body section percentages differ from those of adults. For example, an infant's head accounts for about 17% at the total body surface area compared with 7% for an adult. Instead, use the Lund and Browder chart.

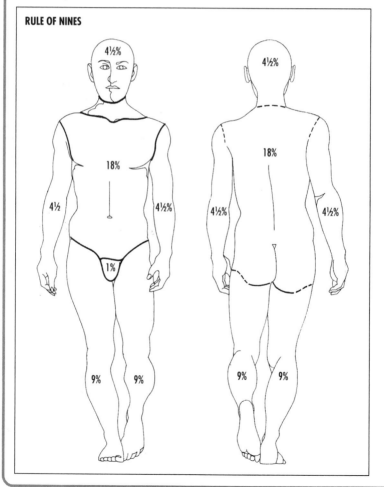

RULE OF NINES

4½% 4½%

18% 18%

4½ 4½% 4½% 4½%

1%

9% 9% 9% 9%

LUND AND BROWDER CHART

To determine the extent of an infant's or child's burns, use the Lund and Browder chart shown here.

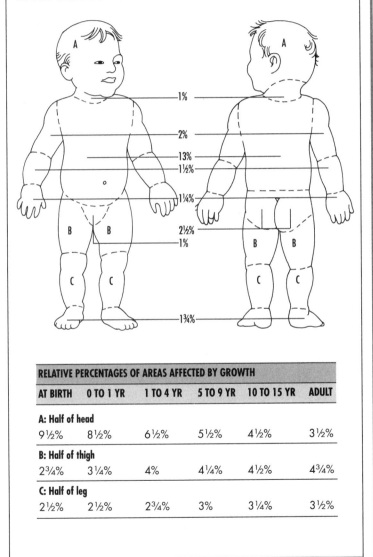

| 1% |
| 2% |
| 13% |
| 1½% |
| 1¼% |
| 2½% |
| 1% |
| 1¾% |

RELATIVE PERCENTAGES OF AREAS AFFECTED BY GROWTH					
AT BIRTH	0 TO 1 YR	1 TO 4 YR	5 TO 9 YR	10 TO 15 YR	ADULT
A: Half of head					
9½%	8½%	6½%	5½%	4½%	3½%
B: Half of thigh					
2¾%	3¼%	4%	4¼%	4½%	4¾%
C: Half of leg					
2½%	2½%	2¾%	3%	3¼%	3½%

▶ irrigating the wound with copious amounts of normal saline solution (chemical burns)

▶ surgical intervention, including skin grafts and more thorough surgical cleansing (major burns)

▶ not treating the burn wound itself for a patient being transferred to a specialty hospital within 4 hours, but wrapping the patient in a sterile sheet and blanket for warmth, elevating the burned extremity, and preparing the patient for transport.

Upon discharge or during prolonged care:

▶ increased caloric intake due to increased metabolic rate to promote healing and recovery

▶ teaching the patient and giving complete discharge instructions for home care; stressing importance of keeping the dressing clean and dry, elevating the burned extremity for the first 24 hours, and having the wound rechecked in 1 to 2 days.

Cellulitis

Cellulitis is an infection of the dermis or subcutaneous layer of the skin. It may follow damage to the skin, such as a bite or wound. As the cellulitis spreads, fever, erythema, and lymphangitis may occur.

A cellulitis infection may occur after a person sustains surface skin damage, such as from a bite or wound injury.

If treated in a timely manner, the prognosis is usually good. Persons with other contributing health factors, such as diabetes, immunodeficiency, and impaired circulation, have an increased risk for developing or spreading cellulitis.

Causes

Possible causes of cellulitis are bacterial and fungal infections, commonly group A streptococcus and *Staphylococcus aureus.*

Pathophysiology

As the offending organism invades the compromised area, it overwhelms the defensive cells (neutrophils, eosinophils, basophils, and mast cells) that break down

the cellular components that normally contain and localize the inflammation. As cellulitis progresses, the organism invades tissue around the initial wound site.

Signs and symptoms

Signs and symptoms of cellulitis are:

▶ erythema and edema due to inflammatory responses to the injury (classic signs)

▶ pain at the site and possibly surrounding area

▶ fever and warmth due to temperature increase caused by infection.

Complications

Possible complications of cellulitis include:

▶ sepsis (untreated cellulitis)

▶ progression of cellulitis to involve more tissue area

▶ local abscesses

▶ thrombophlebitis

▶ lymphangitis in recurrent cellulitis.

Diagnosis

Diagnosis is based on:

▶ visual examination and inspection of the affected area

▶ white blood cell count showing mild leukocytosis with a left shift

▶ mildly elevated erythrocyte sedimentation rate

▶ culture and gram stain results of fluid from abscesses and bulla positive for the offending organism.

Treatment

Treatment of cellulitis may include:

▶ oral or I.V. penicillin (drug of choice for initial treatment) unless patient has known penicillin allergy

▶ warm soaks to the site to help relieve pain and decrease edema by increasing vasodilation

▶ pain medication as needed to promote comfort

▶ elevation of infected extremity to promote comfort and decrease edema.

AGE ALERT Cellulitis of the lower extremity is more likely to develop into thrombophlebitis in an elderly patient.

Dermatitis

Dermatitis is an inflammation of the skin that occurs in several forms: atopic (see "Atopic dermatitis"), seborrheic, nummular, contact, chronic, localized neurodermatitis (lichen simplex chronicus), exfoliative, and stasis. (See *Types of dermatitis,* pages 520 to 523.)

Folliculitis, furuncles, and carbuncles

Folliculitis is a bacterial infection of a hair follicle that causes a pustule to form. The infection can be superficial (follicular impetigo or Bockhart's impetigo) or deep (sycosis barbae).

Furuncles, also known as boils, are another form of deep folliculitis. Carbuncles are a group of interconnected furuncles. (See *Forms of bacterial skin infection,* page 524.)

The incidence of folliculitis in the general population is difficult to determine because many affected people never seek treatment.

With appropriate treatment, the prognosis for patients with folliculitis is good. The disorder usually resolves within 2 to 3 weeks. The prognosis for patients with carbuncles depends on the severity of the infection and the patient's physical condition and ability to resist infection.

Causes

The most common cause of folliculitis, furuncles, and carbuncles is coagulase-positive *Staphylococcus aureus.*

Other causes may include:
▶ *Klebsiella, Enterobacter*, or *Proteus* organisms (causing gram-negative folliculitis in patients on long-term antibiotic therapy, such as for acne)
▶ *Pseudomonas aeruginosa* (thriving in a warm environment with a high pH and

low chlorine content — "hot tub folliculitis").

Predisposing risk factors include:
▶ infected wound
▶ poor hygiene
▶ debilitation
▶ tight clothes
▶ friction
▶ immunosuppressive therapy
▶ exposure to certain solvents.

Pathophysiology

The affecting organism enters the body, usually at a break in the skin barrier (such as a wound site). The organism then causes an inflammatory reaction within the hair follicle.

Signs and symptoms

Folliculitis, furuncles, and carbuncles have different signs and symptoms.
▶ Folliculitis shows as pustules on the scalp, arms, and legs in children and the trunk, buttocks, and legs in adults.
▶ Furuncles show as hard, painful nodules, commonly on the neck, face, axillae, and buttocks. The nodules enlarge for several days, then rupture, discharging pus and necrotic material; after the nodules rupture, subsiding pain but erythema and edema persisting for days or weeks.
▶ Carbuncles show as extremely painful, deep abscesses draining through multiple openings onto the skin surface, usually around several hair follicles; with accompanying fever and malaise. Carbuncles are now rare.

Complications

Possible complications are:
▶ scarring
▶ bacteremia
▶ metastatic seeding of a cardiac valve defect or arthritic joint.

Diagnosis

Diagnosis is based on:
▶ patient history showing preexistent furuncles (carbuncles)

(Text continues on page 522.)

TYPES OF DERMATITIS

TYPE	CAUSES
Seborrheic dermatitis A subacute skin disease affecting the scalp, face, and occasionally other areas that's characterized by lesions covered with yellow or brownish-gray scales.	▶ Unknown; stress, immunodeficiency, and neurologic conditions may be predisposing factors; related to the yeast *Pityrosporum ovale* (normal flora)
Nummular dermatitis A subacute form of dermatitis characterized by inflammation in coin-shaped, scaling, or vesicular patches, usually pruritic.	▶ Possibly precipitated by stress, dry skin, irritants, or scratching
Contact dermatitis Often sharply demarcated inflammation of the skin resulting from contact with an irritating chemical or atopic allergen (a substance producing an allergic reaction in the skin) and irritation of the skin resulting from contact with concentrated substances to which the skin is sensitive, such as perfumes, soaps, or chemicals.	▶ Mild irritants: chronic exposure to detergents or solvents ▶ Strong irritants: damage on contact with acids or alkalis ▶ Allergens: sensitization after repeated exposure
Hand or foot dermatitis A skin disease characterized by inflammatory eruptions of the hands or feet.	▶ In many cases unknown, but may result from irritant or allergic contact ▶ Excessively dry skin often a contributing factor ▶ 50% of patients are atopic

SIGNS AND SYMPTOMS	TREATMENT AND INTERVENTIONS
▶ Eruptions in areas with many sebaceous glands (usually scalp, face, chest, axillae, and groin) and in skin folds ▶ Itching, redness, and inflammation of affected areas; lesions may appear greasy; fissures may occur ▶ Indistinct, occasionally yellowish scaly patches from excess stratum corneum (dandruff may be a mild seborrheic dermatitis)	▶ Removal of scales with frequent washing and shampooing with selenium sulfide suspension (most effective), zinc pyrithione, or tar and salicylic acid shampoo ▶ Application of topical corticosteroids and antifungals to involved area
▶ Round, nummular (coin-shaped), red lesions, usually on arms and legs, with distinct borders of crusts and scales ▶ Possible oozing and severe itching ▶ Summertime remissions common, with wintertime recurrence	▶ Elimination of known irritants ▶ Measures to relieve dry skin: increased humidification, limited frequency of baths, use of bland soap and bath oils, and application of emollients ▶ Application of wet dressings in acute phase ▶ Topical corticosteroids (occlusive dressings or intralesional injections) for persistent lesions ▶ Tar preparations and antihistamines to control itching ▶ Antibiotics for secondary infection ▶ Same as for atopic dermatitis
▶ Mild irritants and allergens: erythema and small vesicles that ooze, scale, and itch ▶ Strong irritants: blisters and ulcerations ▶ Classic allergic response: clearly defined lesions, with straight lines following points of contact ▶ Severe allergic reaction: marked erythema, blistering, and edema of affected areas	▶ Elimination of known allergens and decreased exposure to irritants, wearing protective clothing such as gloves, and washing immediately after contact with irritants or allergens ▶ Topical anti-inflammatory agents (including corticosteroids), systemic corticosteroids for edema and bullae, antihistamines, and local applications of Burow's solution (for blisters) ▶ Same as for atopic dermatitis
▶ Redness and scaling of the palms or soles ▶ May produce painful fissures ▶ Some cases present with blisters (dyshidrotic eczema)	▶ Same as for nummular dermatitis ▶ Severe cases may require systemic steroids

(continued)

TYPES OF DERMATITIS (continued)

TYPE	CAUSES
Localized neurodermatitis (lichen simplex chronicus, essential pruritus) Superficial inflammation of the skin characterized by itching and papular eruptions that appear on thickened, hyperpigmented skin.	▶ Chronic scratching or rubbing of a primary lesion or insect bite or other skin irritation ▶ May be psychogenic
Exfoliative dermatitis Severe skin inflammation characterized by redness and widespread erythema and scaling, covering virtually the entire skin surface.	▶ Preexisting skin lesions progressing to exfoliative stage, such as in contact dermatitis, drug reaction, lymphoma, leukemia, or atopic dermatitis ▶ May be idiopathic
Stasis dermatitis A condition usually caused by impaired circulation and characterized by eczema of the legs with edema, hyperpigmentation, and persistent inflammation.	▶ Secondary to peripheral vascular diseases affecting the legs, such as recurrent thrombophlebitis and resultant chronic venous insufficiency

▶ physical examination showing the presence of the skin lesion to diagnose either folliculitis or carbuncle
▶ wound cultures of the infected site (usually showing *S. aureus*)
▶ possibly elevated white blood cell count (leukocytosis).

Treatment
Appropriate treatments include:
▶ cleaning the infected area thoroughly with antibacterial soap and water
▶ applying warm, wet compresses to promote vasodilation and drainage from the lesions
▶ applying topical antibiotics, such as mupirocin ointment or clindamycin or erythromycin solution.

SIGNS AND SYMPTOMS	TREATMENT AND INTERVENTIONS
▶ Intense, sometimes continual scratching ▶ Thick, sharp-bordered, possibly dry, scaly lesions with raised papules and accentuated skin lines (lichenification) ▶ Usually affects easily reached areas, such as ankles, lower legs, anogenital area, back of neck, and ears ▶ One or several lesions may be present; asymmetric distribution	▶ Scratching must stop; then lesions disappear in about 2 weeks ▶ Fixed dressings or Unna's boot to cover affected areas ▶ Topical corticosteroids under occlusion or by intralesional injection ▶ Antihistamines and open wet dressings ▶ Emollients ▶ Patient informed about underlying cause
▶ Generalized dermatitis, with acute loss of stratum corneum, erythema, and scaling ▶ Sensation of tight skin ▶ Hair loss ▶ Possible fever, sensitivity to cold, shivering, gynecomastia, and lymphadenopathy	▶ Hospitalization, with protective isolation and hygienic measures to prevent secondary bacterial infection ▶ Open wet dressings, with colloidal baths ▶ Bland lotions over topical corticosteroids ▶ Maintenance of constant environmental temperature to prevent chilling or overheating ▶ Careful monitoring of renal and cardiac status ▶ Systemic antibiotics and steroids ▶ Same as for atopic dermatitis
▶ Varicosities and edema common, but obvious vascular insufficiency not always present ▶ Usually affects the lower leg just above internal malleolus or sites of trauma or irritation ▶ Early signs: dusky-red deposits of hemosiderin in skin, with itching and dimpling of subcutaneous tissue ▶ Later signs: edema, redness, and scaling of large areas of legs ▶ Possible fissures, crusts, and ulcers	▶ Measures to prevent venous stasis: avoidance of prolonged sitting or standing, use of support stockings, weight reduction in obesity, and leg elevation ▶ Corrective surgery for underlying cause ▶ After ulcer develops, encourage rest periods with legs elevated, open wet dressings, Unna's boot (zinc gelatin dressing provides continuous pressure to affected areas), and antibiotics for secondary infection after wound culture

Specific treatments include:
▶ folliculitis (extensive infection) — giving systemic antibiotics, such as a cephalosporin (Ancef) or dicloxacillin (Diclocil)
▶ furuncles — incision and drainage of ripe lesions after applying warm, wet compresses, then giving systemic antibiotics
▶ carbuncles — systemic antibiotic therapy and incision and drainage.

Fungal infections

Fungal infections of the skin are often regarded as superficial infections affecting the hair, nails, and dermatophytes (the dead top layer of the skin). They are unique in that they infect and survive on the keratin within these structures. The most common fungal infections are tinea and candidiasis.

FORMS OF BACTERIAL SKIN INFECTION

The degree of hair follicle involvement in bacterial skin infection ranges from superficial erythema and pustule of a single follicle to deep abscesses (carbuncles) involving several follicles.

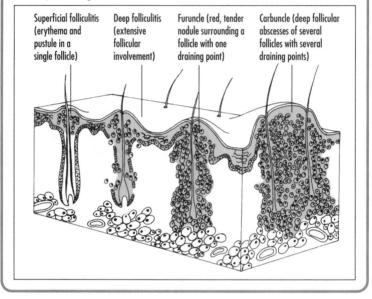

Superficial folliculitis (erythema and pustule in a single follicle)

Deep folliculitis (extensive follicular involvement)

Furuncle (red, tender nodule surrounding a follicle with one draining point)

Carbuncle (deep follicular abscesses of several follicles with several draining points)

Tinea infections are classified by the body location in which they occur. (See *Common sites of tinea infections.*) Tinea commonly infects children and adolescents. Obese patients are also at greater risk for these infections. Some forms, such as tinea cruris (a fungal infection of the groin), infect one gender more than the other. Tinea crusis occurs more commonly in males.

Candidiasis can infect the skin or mucous membranes. These infections are also classified according to the infected site or area. *Candida* organisms are the normal flora found in some people (on the skin, in the mouth, GI tract, and genitalia).

Candidiasis usually occurs in children and immunosuppressed individuals. There are also higher rates of candidiasis in pregnant women and in patients with diabetes mellitus, as well as those with indwelling catheters and I.V. lines.

The prognosis for tinea and candidiasis is very good. It usually responds well to appropriate drug therapy and resolves completely. It's also important to reduce risk factors to obtain a good outcome from the infection. Antifungal therapy usually resolves candidiasis, but if risk factors aren't avoided, a chronic condition can develop.

Causes

Tinea infections are caused by:
▶ *Microsporum, Trichophyton,* or *Epidermophyton* organisms
▶ contact with contaminated objects or surfaces.

Risk factors for tinea include:
▶ obesity
▶ exposure to the causative organisms

▶ antibiotic therapy with suppression of normal flora

▶ softened skin from prolonged water contact, such as with water sports or diaphoresis.

Causes of candidiasis include:

▶ overgrowth of *Candida* organisms and infection due to depletion of the normal flora (such as with antibiotic therapy)

▶ neutropenia and bone marrow suppression in immunocompromised patients (at greater risk for the disseminating form)

▶ *Candida albicans,* normal GI flora (cause candidiasis in susceptible patients)

▶ *Candida* overgrowth in the mouth (thrush).

Pathophysiology

In tinea infections, the tinea fungi attack the outer, dead skin layers. These fungi prefer a dark, warm, moist environment. Tinea infections can be spread from human to human, animal to human, or soil to human. Tinea corpis, for example, can be contracted from animals infected with *Microsporium canis* or *Trichophyton mentagrophytes*, and also from humans infected with *Trichophyton rubrum*.

In candidiasis, the *Candida* organism penetrates the epidermis after it binds to integrin receptors and adhesion molecules. The secretion of proteolytic enzymes facilitates tissue invasion. An inflammatory response results from the attraction of neutrophils to the area and from activation of the complement cascade.

Signs and symptoms

Signs and symptoms of tinea include:

▶ erythema and pustules in a ring-like formation

▶ itching, often severe.

Signs and symptoms of candidiasis are:

▶ superficial papules and pustules

▶ erythematous and edematous areas of the infected epidermis or mucous membrane (with progression of inflammation, a white-yellow, curd-like crust covering the infected area)

COMMON SITES OF TINEA INFECTIONS

TYPE	SITE
Tinea capitis	▶ Scalp
Tinea corporis	▶ Skin areas excluding scalp, face, hands, feet, and groin
Tinea cruris	▶ Groin
Tinea pedis	▶ Foot
Tinea manus	▶ Hand
Tinea unguium or onychomycosis	▶ Nails

▶ severe pruritus and pain at the lesion sites (common)

▶ white coating of the tongue and possibly lesions in the mouth (thrush).

Complications

Possible complications include:

▶ secondary bacterial infections of wounds opened by scratching

▶ ulcers with chronic forms (candidiasis lesions)

▶ candidal meningitis, endocarditis, or septicemia due to systemic disseminating candidiasis.

Diagnosis

Diagnosis is based on:

▶ fungal culture to determine the causative organism and suggest the mode of infection transmission (tinea infection)

▶ microscopic examination of a potassium hydroxide-treated skin scraping and culture (candida infection).

Treatment

Treatment includes:

STAGING PRESSURE ULCERS

The staging system described below is based on the recommendations of the National Pressure Ulcer Advisory Panel (NPUAP) (Consensus Conference, 1991) and the Agency for Health Care Policy and Research (Clinical Practice Guidelines for Treatment of Pressure Ulcers, 1992). The stage 1 definition was updated by the NPUAP in 1997.

STAGE 1

A stage 1 pressure ulcer is an observable pressure-related alteration of intact skin. The indicators, compared with the adjacent or opposite area on the body, may include changes in one or more of the following factors: skin temperature (warmth or coolness), tissue consistency (firm or boggy feel), or sensation (pain or itching). The ulcer appears as a defined area of persistent redness in lightly pigmented skin; in darker skin, the ulcer may appear with persistent red, blue, or purple hues.

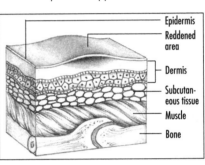

STAGE 2

A stage 2 pressure ulcer is characterized by partial-thickness skin loss involving the epidermis or dermis. The ulcer is superficial and appears as an abrasion, blister, or shallow crater.

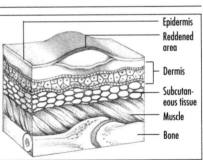

STAGE 3

A stage 3 pressure ulcer is characterized by full-thickness skin loss involving damage or necrosis of subcutaneous tissue, which may extend down to, but not through, the underlying fascia. The ulcer appears as a deep crater with or without undermining of adjacent tissue.

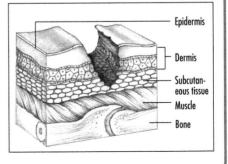

STAGING PRESSURE ULCERS *(continued)*

STAGE 4
Full-thickness skin loss with extensive destruction, tissue necrosis, or damage to muscle, bone, or support structures (for example, tendon or joint capsule) characterize a stage 4 pressure ulcer. Tunneling and sinus tracts also may be associated with stage 4 pressure ulcers.

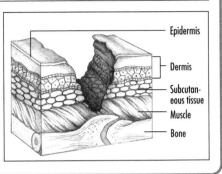

Epidermis

Dermis

Subcutaneous tissue

Muscle

Bone

▶ topical antifungal agents such as keto-conazole (Nizoral), believed to inhibit yeast growth by altering cell membrane permeability (tinea pedis, tinea cruis, and tinea corpuris)

▶ oral therapy with griseofulvin (Fulvicin) (drug of choice) if no response to topical treatment, to arrest fungal cell activity by disrupting its miotic spindle structure (tinea)

▶ oral ketoconazole (Nizoral; second choice) for infections resistant to griseofulvin therapy, to make the fungus more susceptible to osmotic pressure

▶ eliminating risk factors (tinea and candidiasis)

▶ oral nystatin (Mycostatin) and topical antifungals such as miconazole (Monistat) (candidiasis)

▶ I.V. amphotericin B (Fungizone) or oral ketonazole (Nizoral) (systemic infections).

Pressure ulcer
Pressure ulcers, commonly called pressure sores or bedsores, are localized areas of cellular necrosis that occur most often in the skin and subcutaneous tissue over bony prominences. These ulcers may be superficial, caused by local skin irritation with subsequent surface maceration, or deep, originating in underlying tissue. Deep lesions often go undetected until they penetrate the skin, but by then, they've

usually caused subcutaneous damage. (See *Staging pressure ulcers.*)

Most pressure ulcers develop over five body locations: sacral area, greater trochanter, ischial tuberosity, heel, and lateral malleolus. Collectively, these areas account for 95% of all pressure ulcer sites. Patients who have contractures are at an increased risk for developing pressure ulcers due to the added pressure on the tissue and the alignment of the bones.

AGE ALERT Age also has a role in the incidence of pressure ulcers. Muscle is lost with aging, and skin elasticity decreases. Both these factors increase the risk for developing pressure ulcers.

Partial-thickness ulcers usually involve the dermis and epidermis; these wounds heal within a few weeks. Full-thickness ulcers also involve the dermis and epidermis, but in these wounds, the damage is more severe and complete. There may also be damage to the deeper tissue layers. Ulcers of the subcutaneous tissue and muscle may require several months to heal. If the damage has affected the bone in addition to the skin layers, osteomyelitis may occur, which will prolong healing times.

Causes
Possible causes of pressure ulcers include:

▶ immobility and decreased level of activity

▶ friction causing damage to the epidermal and upper dermal skin layers

▶ constant moisture on the skin causing tissue maceration

▶ impaired hygiene status, such as with fecal incontinence, leading to skin breakdown

▶ malnutrition (associated with pressure ulcer development)

▶ medical conditions such as diabetes and orthopedic injuries (predispose to pressure ulcer development)

▶ psychological factors, such as depression and chronic emotional stresses (may have a role in pressure ulcer development).

Pathophysiology

A pressure ulcer is caused by an injury to the skin and its underlying tissues. The pressure exerted on the area causes ischemia and hypoxemia to the affected tissues because of decreased blood flow to the site. As the capillaries collapse, thrombosis occurs, which subsequently leads to tissue edema and progression to tissue necrosis. Ischemia also adds to an accumulation of waste products at the site, which in turn leads to the production of toxins. The toxins further break down the tissue and eventually lead to the death of the cells.

Signs and symptoms

Signs and symptoms of pressure ulcers may include:

▶ blanching erythema, varying from pink to bright red depending on the patient's skin color; in dark-skinned people, purple discoloration or a darkening of normal skin color (first clinical sign); when the examiner presses a finger on the reddened area, the "pressed on" area whitens and color returns within 1 to 3 seconds if capillary refill is good

▶ pain at the site and surrounding area

▶ localized edema due to the inflammatory response

▶ increased body temperature due to initial inflammatory response (in more severe cases, cool skin due to more severe damage or necrosis)

▶ nonblanching erythema (more severe cases) ranging from dark red to purple or cyanotic; indicates deeper dermal involvement

▶ blisters, crusts, or scaling as the skin deteriorates and the ulcer progresses

▶ usually dusky-red appearance, doesn't bleed easily, warm to the touch, and possibly mottled (deep ulcer originating at the bony prominence below the skin surface).

Complications

Possible complications of pressure ulcers include:

▶ progression of the pressure ulcer to a more severe state (greatest risk)

▶ secondary infections such as sepsis

▶ loss of limb from bone involvement.

Diagnosis

Diagnosis is based on:

▶ physical examination showing presence of the ulcer

▶ wound culture with exudate or evidence of infection

▶ elevated white blood cell count with infection

▶ possibly elevated erythrocyte sedimentation rate

▶ total serum protein and serum albumin levels showing severe hypoproteinemia.

Treatment

Treatment for pressure ulcers includes:

▶ repositioning by the caregiver every 2 hours or more often if indicated, with support of pillows for immobile patients; a pillow and encouragement to change position for those able to move

▶ foam, gel, or air mattress to aid in healing by reducing pressure on the ulcer site and reducing the risk for more ulcers

▶ foam, gel, or air mattress on chairs and wheelchairs as indicated

nutritional assessment and dietary consult as indicated; nutritional supplements, such as vitamin C and zinc, for the malnourished patient; monitoring serum albumin and protein markers and body weight

adequate fluid intake (I.V. if indicated) and increased fluids for a dehydrated patient

good skin care and hygiene practices (for example, meticulous hygiene and skin care for the incontinent patient to prevent breakdown of the affected tissue and skin)

stage II, cover ulcer with transparent film, polyurethane foam, or hydrocolloid dressing

stage II or IV, loosely fill wound with saline- or gel-moistened gauze, manage exudate with absorbent dressing (moist gauze or foam) and cover with secondary dressing

clean, bulky dressing for certain types of ulcers, such as decubiti

surgical debridement for deeper wounds stage III or IV as indicated.

Psoriasis

Psoriasis is a chronic, recurrent disease marked by epidermal proliferation and characterized by recurring partial remissions and exacerbations. Flare-ups are often related to specific systemic and environmental factors, but may be unpredictable. Widespread involvement is called exfoliative or erythrodermic psoriasis.

Psoriasis affects about 21% of the population in the United States. Although this disorder often affects young adults, it may strike at any age, including infancy. Genetic factors predetermine the incidence of psoriasis; researchers have discovered a significantly greater incidence of certain human leukocyte antigens (HLA) in families with psoriasis.

Flare-ups can usually be controlled with therapy. Appropriate treatment depends on the type of psoriasis, extent of the disease, the patient's response, and the effect of the disease on the patient's lifestyle. No permanent cure exists, and all methods of treatment are palliative.

Causes

Causes of psoriasis include:

genetically determined tendency to develop psoriasis

possible immune disorder, as shown by in the HLA type in families

environmental factors

isomorphic effect or Koebner's phenomenon, in which lesions develop at sites of injury due to trauma

flare-up of guttate (drop-shaped) lesions due to infections, especially beta-hemolytic streptococci.

Other contributing factors include:

pregnancy

endocrine changes

climate (cold weather tends to exacerbate psoriasis)

emotional stress.

Pathophysiology

A skin cell normally takes 14 days to move from the basal layer to the stratum corneum, where it's sloughed off after 14 days of normal wear and tear. Thus, the life-cycle of a normal skin cell is 28 days compared with only 4 days for a psoriatic skin cell. This markedly shortened cycle doesn't allow time for the cell to mature. Consequently, the stratum corneum becomes thick and flaky, producing the cardinal manifestations of psoriasis.

Signs and symptoms

Possible signs and symptoms include:

itching and occasional pain from dry, cracked, encrusted lesions (most common)

erythematous and usually well-defined plaques, sometimes covering large areas of the body (psoriatic lesions)

lesions most commonly on the scalp, chest, elbows, knees, back, and buttocks

plaques with characteristic silver scales that either flake off easily or thicken, covering the lesion; scale removal can produce fine bleeding

▶ occasional small guttate lesions (usually thin and erythematous, with few scales), either alone or with plaques.

Complications

Possible complications of psoriasis include:

▶ spread to fingernails, producing small indentations or pits and yellow or brown discoloration (about 60% of patients)

▶ accumulation of thick, crumbly debris under the nail, causing it to separate from the nailbed (onycholysis).

Rarely, psoriasis becomes pustular, taking one of two forms:

▶ localized pustular psoriasis, with pustules on the palms and soles that remain sterile until opened

▶ generalized pustular (Von Zumbusch) psoriasis, often occurring with fever, leukocytosis, and malaise, with groups of pustules coalescing to form lakes of pus on red skin (also remain sterile until opened), commonly involving the tongue and oral mucosa

▶ arthritic symptoms, usually in one or more joints of the fingers or toes, the larger joints, or sometimes the sacroiliac joints, which may progress to spondylitis, and morning stiffness (some patients).

Diagnosis

Diagnosis is based on the following factors:

▶ patient history, appearance of the lesions, and, if needed, the results of skin biopsy

▶ serum uric acid level (usually elevated in severe cases due to accelerated nucleic acid degradation), but without indications of gout

▶ HLA-Cw6, -B13, and -Bw57 (may be present in early-onset familial psoriasis).

Treatment

Treatment may include:

▶ ultraviolet B (UVB) or natural sunlight exposure to retard rapid cell production to the point of minimal erythema

▶ tar preparations or crude coal tar applications to the affected areas about 15 minutes before exposure to ultraviolet B, or left on overnight and wiped off the next morning

▶ gradually increasing exposure to UVB (outpatient treatment or day treatment with UVB avoids long hospitalizations and prolongs remission)

▶ steroid creams and ointments applied twice daily, preferably after bathing to facilitate absorption, and overnight use of occlusive dressings to control symptoms of psoriasis

▶ intralesional steroid injection for small, stubborn plaques

▶ anthralin ointment (Anthra-Derm) or paste mixture for well-defined plaques (not applied to unaffected areas due to injury and staining of normal skin); petroleum jelly around affected skin before applying anthralin

▶ anthralin (Anthra-Derm) and steroids (anthralin application at night and steroid use during the day)

▶ calcipotriene ointment (Dovonex), a vitamin D analogue (best when alternated with a topical steroid)

▶ Goeckerman regimen (combines tar baths and UVB treatments) to help achieve remission and clear the skin in 3 to 5 weeks (severe chronic psoriasis)

▶ Ingram technique (variation of the Goeckerman regimen) using anthralin (Anthra-Derm) instead of tar

▶ administration of psoralens (plant extracts that accelearte exfoliation) with exposure to high intensity UVA (PUVA therapy)

▶ cytotoxin, usually methotrexate (Folex) (last-resort treatment for refractory psoriasis)

▶ acitretin (Soriatant), a retinoid compound (extensive psoriasis)

▶ cyclosporine (Neoral), an immunosuppressant (in resistive cases)

▶ low-dose antihistamines, oatmeal baths, emollients, and open wet dressings to help relieve pruritus

▶ aspirin and local heat to help alleviate the pain of psoriatic arthritis; nonsteroidal anti-inflammatory drugs in severe cases

▶ tar shampoo followed by a steroid lotion (psoriasis of the scalp)
▶ no effective topical treatment for psoriasis of the nails.

Scleroderma

Scleroderma (also known as systemic sclerosis) is an uncommon disease of diffuse connective tissue disease characterized by inflammatory and then degenerative and fibrotic changes in the skin, blood vessels, synovial membranes, skeletal muscles, and internal organs (especially the esophagus, intestinal tract, thyroid, heart, lungs, and kidneys). There are several forms of scleroderma, including diffuse systemic sclerosis, localized, linear, chemically induced localized, eosinophilia myalgia syndrome, toxic oil syndrome, and graft-versus-host disease.

 AGE ALERT Scleroderma is an uncommon disease. It affects women three to four times more often than men, especially between ages 30 and 50 years. The peak incidence of occurrence is in 50- to 60-year-olds.

Scleroderma is usually a slowly progressing disorder. When the condition is limited to the skin, the prognosis is usually favorable. However, approximately 30% of patients with scleroderma die within 5 years of onset. Death is usually caused by infection or renal or cardiac failure.

Causes

The cause of scleroderma is unknown, but some possible causes include:
▶ systemic exposure to silica dust or polyvinyl chloride
▶ anticancer agents such as bleomycin (Blenoxane) or nonnarcotic analgesics such as pentazocine hydrochloride (Talwin)
▶ fibrosis due to an abnormal immune system response
▶ underlying vascular cause with tissue changes initiated by a persistent perfusion.

Pathophysiology

Scleroderma usually begins in the fingers and extends proximally to the upper arms, shoulders, neck, and face. The skin atrophies, edema and infiltrates containing CD4+ T cells surround the blood vessels, and inflamed collagen fibers become edematous, losing strength and elasticity, and degenerative. The dermis becomes tightly bound to the underlying structures, resulting in atrophy of the affected dermal appendages and destruction of the distal phalanges by osteoporosis. As the disease progresses, this atrophy can affect other areas. For example, in some patients, muscles and joints become fibrotic.

Signs and symptoms

Possible signs and symptoms of scleroderma are:
▶ skin thickening, commonly limited to the distal extremities and face, but which can also involve internal organs (limited systemic sclerosis)
▶ CREST syndrome (a benign subtype of limited systemic sclerosis): Calcinosis, Raynaud's phenomenon, Esophageal dysfunction, Sclerodactyly, and Telangiectasia
▶ generalized skin thickening and involvement of internal organs (diffuse systemic sclerosis)
▶ patchy skin changes with a teardrop-like appearance known as morphea (localized scleroderma)
▶ band of thickened skin on the face or extremities that severely damages underlying tissues, causing atrophy and deformity (linear scleroderma)

 AGE ALERT Atrophy and deformity with scleroderma are most common in childhood.
▶ Raynaud's phenomenon (blanching, cyanosis, and erythema of the fingers and toes); progressive phalangeal resorption may shorten the fingers (early symptoms)
▶ pain, stiffness, and swelling of fingers and joints (later symptoms)
▶ taut, shiny skin over the entire hand and forearm due to skin thickening

• tight and inelastic facial skin, causing a mask-like appearance and "pinching" of the mouth; contractures with progressive tightening

• thickened skin over proximal limbs and trunk (diffuse systemic sclerosis)

• frequent reflux, heartburn, dysphagia, and bloating after meals due to GI dysfunction

• abdominal distention, diarrhea, constipation, and malodorous floating stool.

Complications

Complications of scleroderma include:

• compromised circulation due to abnormal thickening of the arterial intima, possibly causing slowly healing ulcerations on fingertips or toes leading to gangrene

• decreased food intake and weight loss due to GI symptoms

• arrhythmias and dyspnea due to cardiac and pulmonary fibrosis; malignant hypertension due to renal involvement, called renal crisis (may be fatal if untreated; advanced disease).

Diagnosis

Diagnosis of scleroderma may include:

• typical cutaneous changes (the first clue to diagnosis)

• slightly elevated erythrocyte sedimentation rate, positive rheumatoid factor in 25% to 35% of patients, and positive antinuclear antibody test results

• urinalysis showing proteinuria, microscopic hematuria, and casts (with renal involvement)

• hand X-rays showing terminal phalangeal tuft resorption, subcutaneous calcification, and joint space narrowing and erosion

• chest X-rays showing bilateral basilar pulmonary fibrosis

• GI X-rays showing distal esophageal hypomotility and stricture, duodenal loop dilation, small-bowel malabsorption pattern, and large diverticula

• pulmonary function studies showing decreased diffusion and vital capacity

• electrocardiogram showing nonspecific abnormalities related to myocardial fibrosis

• skin biopsy showing changes consistent with disease progression, such as marked thickening of the dermis and occlusive vessel changes.

Treatment

There's no cure for scleroderma. Treatment aims to preserve normal body functions and minimize complications and may include:

• immunosuppressants, such as cyclosporine (Neoral) or chlorambucil (Leukeran) (common palliative medications)

• vasodilators and antihypertensives; such as nifedipine (Adalat), prazosin (Minipress), or topical nitroglycerin (Nitrol); digital sympathectomy; or, rarely, cervical sympathetic blockade to treat Raynaud's phenomenon

• digital plaster cast to immobilize the area, minimize trauma, and maintain cleanliness; possible surgical debridement for chronic digital ulceration

• antacids (to reduce total acid level in GI tract), omeprazole (Prilosec; antiulcer drug to block the formation of gastric acid), periodic dilation, and a soft, bland diet for esophagitis with stricture

• broad-spectrum antibiotics to treat small-bowel involvement with erythromycin or tetracycline (preferred drugs) to counteract the bacterial overgrowth in the duodenum and jejunum related to hypomotility

• short-term benefit from vasodilators, such as nifedipine (Adalat) or hydralazine (Apresoline), to decrease contractility and oxygen demand, and cause vasodilation (for pulmonary hypertension)

• angiotensin-converting enzyme inhibitor to preserve renal function (early intervention in renal crisis)

• physical therapy to maintain function and promote muscle strength, heat therapy to relieve joint stiffness, and patient teaching to make performance of daily activities easier (for hand debilitation).

Warts

Warts, also known as verrucae, are common, benign, viral infections of the skin and adjacent mucous membranes. Although their incidence is greatest in children and young adults, warts occur at any age. The prognosis varies; many types of warts resolve spontaneously, whereas others need more vigorous and prolonged treatment. Most persons eventually develop an immune response to the papillomavirus that causes the warts to disappear spontaneously. An immune response may develop to some types of warts, but this immune response can be delayed for many years.

Causes

Warts are caused by:
▶ human papillomavirus (HPV)
▶ transmission by touch and skin-to-skin contact
▶ spread on the affected person by autoinoculation.

Pathophysiology

HPV replicates in the epidermal cells, causing irregular thickening of the stratum corneum in the infected areas. People who lack the virus-specific immunity are susceptible to the virus.

Signs and symptoms

Signs and symptoms depend on the type of wart and its location and may include:
▶ rough, elevated, rounded surface, most frequently occurring on extremities, particularly hands and fingers; most prevalent in children and young adults (common warts [verruca vulgaris])
▶ single, thin, threadlike projection; commonly occurring around the face and neck (filiform)
▶ rough, irregularly shaped, elevated surface, occurring around edges of fingernails and toenails (when severe, extending under the nail and lifting it off the nailbed, causing pain [periungual])
▶ multiple groupings of up to several hundred slightly raised lesions with smooth, flat, or slightly rounded tops, common on the face, neck, chest, knees, dorsa of hands, wrists, and flexor surfaces of the forearms (usually occur in children but can affect adults); often linear distribution due to spread by scratching or shaving (flat or juvenile)
▶ slightly elevated or flat; occurring singly or in large clusters (mosaic warts), primarily at pressure points of the feet (plantar)
▶ fingerlike, horny projection arising from a pea-shaped base, occurring on scalp or near hairline (digitale)
▶ usually small, pink to red, moist, and soft, occurring singly or in large cauliflower-like clusters on the penis, scrotum, vulva, and anus; although transmitted through sexual contact, it's not always venereal in origin (condyloma acuminatum [moist wart]).

Complications

Possible complications are:
▶ autoinoculation
▶ scar formation
▶ chronic pain after plantar wart removal and scar formation
▶ nail deformity after injury to nail matrix
▶ cervical cancer, with increased risk if a woman smokes (certain strains of HPV)
▶ esophageal warts (in newborn exposed to genital warts).

AGE ALERT The presence of perianal warts in children may be a sign of sexual abuse.

Diagnosis

Diagnosis is based on:
▶ physical examination of the patient
▶ sigmoidoscopy with recurrent anal warts to rule out internal involvement necessitating surgery.

Treatment

If immunity develops, warts resolve by themselves. Treatments may include:
▶ skin irritants, such as salicylic acid or formaldehyde, applied to the wart to try to stimulate an immune response

▶ curettage or cryosurgery

▶ carbon dioxide laser treatment for recalcitrant warts of on the feet, groin, or nail bed

▶ abstinence or condom use until warts are eradicated; partners should also be examined and treated as needed (genital warts).

The reproductive system must function properly to ensure survival of the species. The male reproductive system produces sperm and delivers them to the female reproductive tract. The female reproductive system produces the ovum. If a sperm fertilizes an ovum, this system also nurtures and protects the embryo and developing fetus and delivers it at birth. The functioning of the reproductive system is determined not only by anatomic structure, but also by complex hormonal, neurologic, vascular, and psychogenic factors.

Anatomically, the main distinction between the male and female is the presence of conspicuous external genitalia in the male. In contrast, the major reproductive organs of the female lie within the pelvic cavity.

MALE REPRODUCTIVE SYSTEM

The male reproductive system consists of the organs that produce sperm, transfer mature sperm from the testes, and introduce them into the female reproductive tract, where fertilization occurs.

Besides supplying male sex cells (in a process called spermatogenesis), the male reproductive system plays a part in the secretion of male sex hormones. The penis also functions in urine elimination.

Reproductive organs

The male reproductive organs include the penis, scrotum, testes, duct system, and accessory reproductive glands.

Penis

The penis consists of three cylinders of erectile tissue: two corpora cavernosa, and the corpus spongiosum, which contains the urethra. The glans or tip of the penis contains the urethral meatus, through which urine and semen pass to the exterior, and many nerve endings for sexual sensation.

Scrotum

The scrotum, which contains the testes, epididymis, and lower spermatic cords, maintains the proper testicular temperature for spermatogenesis through relaxation and contraction. This is important because excessive heat reduces the sperm count.

Testes

The testes (also called gonads, which is a term for any reproductive organ, or testicles) produce sperm in the seminiferous tubules.

 AGE ALERT Complete spermatogenesis develops in most males by age 15 or 16 years.

The testes form in the abdominal cavity of the fetus and descend into the scrotum during the seventh month of gestation.

Duct system

The vas deferens connects the epididymis, in which sperm mature and ripen for up to 6 weeks, and the ejaculatory ducts. The seminal vesicles — two convoluted membranous pouches — secrete a viscous liquid of fructose-rich semen and prostaglandins that probably facilitates fertilization.

Accessory reproductive glands

The prostate gland secretes the thin alkaline substance that comprises most of the seminal fluid; this fluid also protects sperm from acidity in the male urethra and in the vagina, thus increasing sperm motility.

The bulbourethral (Cowper's) glands secrete an alkaline ejaculatory fluid, probably similar in function to that produced by the prostate gland. The spermatic cords are cylindrical fibrous coverings in the inguinal canal containing the vas deferens, blood vessels, and nerves.

Testosterone

The testes produce and secrete hormones, especially testosterone, in their interstitial cells (Leydig's cells). Testosterone affects the development and maintenance of secondary sex characteristics and sex drive. It also regulates metabolism, stimulates protein anabolism (encouraging skeletal growth and muscular development), inhibits pituitary secretion of the gonadotropins (follicle-stimulating hormone and interstitial cell-stimulating hormone), promotes potassium excretion, and mildly influences renal sodium reabsorption.

In males, the reproductive and urinary systems are structurally integrated; most disorders, therefore, affect both systems. Congenital abnormalities or prostate enlargement may impair both sexual and urinary function. Abnormal findings in the pelvic area may result from pathologic changes in other organ systems, such as the upper urinary and GI tracts, endocrine glands, and neuromusculoskeletal system.

FEMALE REPRODUCTIVE SYSTEM

Female reproductive structures include the mammary glands, external genitalia, and internal genitalia. Hormonal influences determine the development and function of these structures and affect fertility, childbearing, and the ability to experience sexual pleasure.

In no other part of the body do so many interrelated physiologic functions occur in such proximity as in the area of the female reproductive tract. Besides the internal genitalia, the female pelvis contains the organs of the urinary and GI systems (bladder, ureters, urethra, sigmoid colon, and rectum). The reproductive tract and its surrounding area are thus the site of urination, defecation, menstruation, ovulation, copulation, impregnation, and parturition.

Mammary glands

Located in the breasts, the mammary glands are specialized accessory glands that secrete milk. Although present in both sexes, they normally function only in females.

External structures

Female genitalia include the following external structures, collectively known as the vulva: mons pubis (or mons veneris), labia majora, labia minora, clitoris, and the vestibule. The perineum is the external region between the vulva and the anus. The size, shape, and color of these structures — as well as pubic hair distribution and skin texture and pigmentation — vary greatly among individuals. Furthermore, these external structures undergo distinct changes during the life cycle.

Mons pubis

The mons pubis is the pad of fat over the symphysis pubis (pubic bone), which is usually covered by the base of the inverted triangular patch of pubic hair that grows over the vulva after puberty.

Labia majora

The labia majora are the two thick, longitudinal folds of fatty tissue that extend from the mons pubis to the posterior aspect of the perineum. The labia majora protect the perineum and contain large sebaceous glands that help maintain lubrication. Virtually absent in the young child, their development is a characteristic sign of onset of puberty. The skin of the more prominent parts of the labia majora is pigmented and darkens after puberty.

Labia minora

The labia minora are the two thin, longitudinal folds of skin that border the vestibule. Firmer than the labia majora, they extend from the clitoris to the posterior fourchette.

Clitoris

The clitoris is the small, protuberant organ located just beneath the arch of the mons pubis. The clitoris contains erectile tissue, venous cavernous spaces, and specialized sensory corpuscles that are stimulated during coitus. It's homologous to the male penis.

Vestibule

The vestibule is the oval space bordered by the clitoris, labia minora, and fourchette. The urethral meatus is located in the anterior portion of the vestibule, and the vaginal meatus is in the posterior portion. The hymen is the elastic membrane that partially obstructs the vaginal meatus in virgins. Its absence doesn't necessarily imply a history of coitus, nor does its presence obstruct menstrual blood flow.

Several glands lubricate the vestibule. Skene's glands (also known as the paraurethral glands) open on both sides of the urethral meatus; Bartholin's glands, on both sides of the vaginal meatus.

The fourchette is the posterior junction of the labia majora and labia minora. The perineum, which includes the underlying muscles and fascia, is the external surface of the floor of the pelvis, extending from the fourchette to the anus.

Internal structures

The internal structures of the female genitalia include the vagina, cervix, uterus, Fallopian tubes (or oviducts), and ovaries.

Vagina

The vagina occupies the space between the bladder and the rectum. A muscular, membranous tube approximately 2 to 3 inches (5 to 7.5 cm) long, the vagina connects the uterus with the vestibule of the external genitalia. It serves as a passageway for sperm to the Fallopian tubes, a conduit for the discharge of menstrual fluid, and the birth canal during parturition.

Cervix

The cervix is the most inferior part of the vagina, protruding into the vaginal canal. The cervix provides a passageway between the vagina and the uterine cavity.

Uterus

The uterus is the hollow, pear-shaped organ in which the conceptus grows during pregnancy. The thick uterine wall consists of mucosal, muscular, and serous layers. The inner mucosal lining (the endometrium) undergoes cyclic changes to facilitate and maintain pregnancy.

The smooth muscular middle layer (the myometrium) interlaces the uterine and ovarian arteries and veins that circulate blood through the uterus. During pregnancy, this vascular system expands dramatically. After abortion or childbirth, the myometrium contracts to constrict the vasculature and control loss of blood.

The outer serous layer (the parietal peritoneum) covers all of the fundus, part of the corpus, but none of the cervix. This incompleteness allows surgical entry into the uterus without incision of the peritoneum, thus reducing the risk for peritonitis in the days before effective antibiotic therapy.

Fallopian tubes

The Fallopian tubes extend from the sides of the fundus and terminate near the ovaries. Each tube has a fimbriated (fringelike) end adjacent to the ovary that serves to capture an oocyte after ovulation. Through ciliary and muscular action, these small tubes carry ova from the ovaries to the uterus and facilitate the movement of sperm from the uterus toward the ovaries. The same ciliary and muscular action helps move a zygote (fertilized ovum) down to the uterus, where it may implant in the blood-rich inner uterine lining, the endometrium.

Ovaries

The ovaries are two almond-shaped organs, one on either side of the pelvis, situated behind and below the Fallopian tubes. The ovaries produce ova and two primary hormones — estrogen and progesterone — in addition to small amounts of androgen. These hormones in turn produce and maintain secondary sex characteristics, prepare the uterus for pregnancy, and stimulate mammary gland development. The ovaries are connected to the uterus by the utero-ovarian ligament.

In the 30th week of gestation, the fetus has about 7 million follicles, which degenerate, leaving about 2 million present at birth. By puberty, only 400,000 remain, and these ova precursors become Graafian follicles in response to the effects of pituitary gonadotropic hormones (follicle-stimulating hormone [FSH] and luteinizing hormone [LH]). Fewer than 500 of each woman's ova mature and become potentially fertile.

The menstrual cycle

Maturation of the hypothalamus and the resultant increase in hormone levels initiate puberty. In the young girl, the appearance of pubic and axillary hair (pubarche) and the characteristic adolescent growth spurt follow breast development (thelarche) — the first sign of puberty. The reproductive system begins to undergo a series of hormone-induced changes that result in menarche, or the onset of menstruation (or menses).

The menstrual cycle consists of three different phases: menstrual, proliferative (estrogen dominated), and secretory (progesterone dominated). (See *Understanding the menstrual cycle*.)

At the end of the secretory phase, the uterine lining is ready to receive and nourish a zygote. If fertilization doesn't occur, increasing estrogen and progesterone levels decrease LH and FSH production. Because LH is needed to maintain the corpus luteum, a decrease in LH production causes the corpus luteum to atrophy and halt the secretion of estrogen and progesterone. The thickened uterine lining then begins to slough off, and menstruation begins again.

In the nonpregnant female, LH controls the secretions of the corpus luteum, thereby increasing progesterone levels in the bloodstream. In the pregnant woman, human chorionic gonadotropin (HCG), produced by the nascent placenta, controls these secretions.

If fertilization and pregnancy occur, the endometrium grows even thicker and vascular ingrowth occurs. After implantation of the zygote (about 5 or 6 days after fertilization), the endometrium becomes the decidua. Trophoblastic cells produce HCG soon after implantation, stimulating the corpus luteum to continue secreting estrogen and progesterone, which prevents further ovulation and menstruation.

HCG continues to stimulate the corpus luteum until the placenta (the vascular organ that develops to transport materials to and from the fetus) forms and starts producing its own estrogen and progesterone. After the placenta takes over hormonal production, secretions of the corpus luteum are no longer needed to maintain the pregnancy, and the corpus luteum gradually decreases its function and begins to degenerate. This is termed the luteoplacental shift and commonly occurs by the end of the first trimester.

PATHOPHYSIOLOGIC MANIFESTATIONS

Alterations may occur in the structure, process, or function of both the male and female reproductive systems.

UNDERSTANDING THE MENSTRUAL CYCLE

The menstrual cycle is divided into three distinct phases.

▶ During the menstrual phase, which starts on the first day of menstruation, the top layer of the endometrium breaks down and flows out of the body. This flow, the menses, consists of blood, mucous, and unneeded tissue.

▶ During the proliferatve (follicular) phase, the endometrium begins to

thicken, and the level of estrogen in the blood increases.

▶ During the secretory (luteal) phase, the endometrium begins to thicken to nourish an embryo should fertilization occur. Without fertilization, the top layer of the endometrium breaks down and the menstrual phase of the cycle begins again.

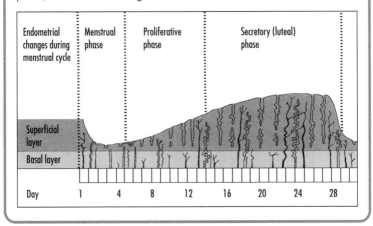

Sexual maturation alteration

Sexual maturation, or puberty, can be affected by various congenital and endocrine disorders. The timing of puberty may be too early (precocious puberty) or too late (delayed puberty). Precocious puberty is the onset of sexual maturation before age 9 years in boys and before age 8 years in girls. It occurs more often in girls than boys. In girls, the cause is most commonly idiopathic, whereas in boys it is more likely to be organic.

In delayed puberty, there's no evidence of the development of secondary sex characteristics in a boy age 14 years or a girl age 13 years. There's usually no evidence of hormonal abnormalities. The hypothalamic-pituitary-ovarian axis, a system that stimulates and regulates the pro-

duction of hormones necessary for normal sexual development and function, is intact, but maturation is slow. The cause is unknown.

Hormonal alterations

Complex hormonal interactions determine the normal function of the female reproductive tract and require an intact hypothalamic-pituitary-ovarian axis. A defect or malfunction of this system can cause infertility due to insufficient gonadotropin secretions (both LH and FSH). The ovary controls, and is controlled by, the hypothalamus through a system of negative and positive feedback mediated by estrogen production. Insufficient gonadotropin levels may result from infections, tumors, or neurologic disease of the hypothalamus or pituitary gland. A mild hormonal imbalance in gonadotropin production and regulation, possibly caused by polycystic disease of the ovary or abnormalities in the adrenal or thyroid gland that adversely affect hypothalamic-pituitary functioning, may sporadically inhibit ovulation. Because gonadotropins are released in a pulsatile fashion, a significant disturbance in this pulsatility will adversely affect ovulatory function.

Male hypogonadism, or an abnormal decrease in gonad size and function, results from decreased androgen production in males, which may impair spermatogenesis (causing infertility) and inhibit the development of normal secondary sex characteristics. The clinical effects of androgen deficiency depend on age at onset. Primary hypogonadism results directly from interstitial (Leydig's cell) cellular or seminiferous tubular damage due to faulty development or mechanical damage. Androgen deficiency causes increased secretion of gonadotropins by the pituitary in an attempt to increase the testicular functional state and is therefore termed hypergonadotropic hypogonadism. This form of hypogonadism includes Klinefelter's syndrome (47 XXY), Reifenstein's syndrome, male Turner's syndrome, Sertoli-cell-only syndrome, anorchism, orchitis, and sequelae of irradiation.

Secondary hypogonadism is due to faulty interaction within the hypothalamic-pituitary axis, resulting in failure to secrete normal levels of gonadotropins, and is therefore termed hypogonadotropic hypogonadism. This form of hypogonadism includes hypopituitarism, isolated FSH deficiency, isolated LH deficiency, Kallmann's syndrome, and Prader-Willi syndrome. Depending on the patient's age at onset, hypogonadism may cause eunuchism (complete gonadal failure) or eunuchoidism (partial failure).

Symptoms vary depending on the specific cause of hypogonadism. Some characteristic findings may include delayed bone maturation, delayed puberty; infan-

tile penis and small, soft testes; less than average muscle development and strength; fine, sparse facial hair; scant or absent axillary, pubic, and body hair; and a high-pitched, effeminate voice. In an adult, hypogonadism diminishes sex drive and potency and causes regression of secondary sex characteristics.

Menstrual alterations

Alterations in menstruation include the absence of menses, abnormal bleeding patterns, or painful menstruation. Menopause is the cessation of menstruation. It results from a complex continuum of physiologic changes — the climacteric — caused by declining ovarian function. The climacteric produces various changes in the body, the most dramatic being the cessation of menses.

AGE ALERT The climacteric, a normal gradual reduction in ovarian function due to aging, begins in most women between ages 40 and 50 years and results in infrequent ovulation, decreased menstrual function and, eventually, cessation of menstruation (usually between ages 45 and 55 years).

Premature menopause, the gradual or abrupt cessation of menstruation before age 35 years, occurs without apparent cause in about 5% of women in the United States. Certain diseases, especially autoimmune diseases such as premature ovarian failure, may cause pathologic menopause. Other factors that may precipitate premature menopause include malnutrition, debilitation, extreme emotional stress, pelvic irradiation, and surgical procedures that impair ovarian blood supply. Artificial menopause may follow radiation therapy or surgical procedures such as removal of both ovaries (bilateral oophorectomy). Hysterectomy decreases the interval before menopause even when the ovaries are not removed. It's speculated that hysterectomy may decrease ovarian blood flow in some fashion.

Ovarian failure, in which no ova are produced, may result from a functional

ABNORMAL PREMENOPAUSAL BLEEDING

Causes of abnormal premenopausal bleeding vary with the type of bleeding:

▶ Oligomenorrhea (infrequent menses) and polymenorrhea (menses occurring too frequently) usually result from anovulation due to an endocrine or systemic disorder.

▶ Hypomenorrhea (decreased amount of menstrual fluid) results from local, endocrine, or systemic disorders or blockage caused by partial obstruction by the hymen or cervical obstruction.

▶ Hypermenorrhea (excessive bleeding occurring at the regular intervals) usually results from local lesions, such as uterine leiomyomas, endometrial polyps, and endometrial hyperplasia. It may also result from endometritis, salpingitis, and anovulation.

▶ Cryptomenorrhea (no external bleeding, although menstrual symptoms are experienced) may result from an imperforate hymen or cervical stenosis.

▶ Metrorrhagia (bleeding occurring at irregular intervals) usually results from slight physiologic bleeding from the endometrium during ovulation but may also result from local disorders, such as uterine malignancy, cervical erosions, polyps (which tend to bleed after intercourse), or inappropriate estrogen therapy.

Complications of pregnancy can also cause premenopausal bleeding, which may be as mild as spotting or as severe as hypermenorrhea.

ovarian disorder from premature menopause. Amenorrhea is a natural consequence of ovarian failure.

Pain is often associated with the menstrual cycle; in many common diseases of the female reproductive tract, such pain may follow a cyclic pattern. A patient with endometriosis, for example, may report increasing premenstrual pain that decreases at the end of menstruation. For a description of the types of abnormal menstrual bleeding, see *Abnormal premenopausal bleeding.*

Sexual dysfunction

Sexual dysfunction includes arousal problems, orgasmic problems, and sexual pain (dyspareunia, vaginismus). Dysfunction may be caused by a general medical condition, psychological condition, substance use or abuse, or a combination of these factors.

Arousal disorder is an inability to experience sexual pleasure. According to the Diagnostic and Statistical Manual of Mental Disorders, 4th edition (DSM-IV), the essential feature is a persistent or recurrent inability to attain or to maintain an adequate lubrication-swelling response of sexual excitement until completion of the sexual act. Orgasmic disorder, according to DSM-IV, is a persistent or recurrent delay in or absence of orgasm after a normal sexual excitement phase.

Both arousal and orgasmic disorders are considered primary if they exist in a female who has never experienced sexual arousal or orgasm; they are secondary when a physical, mental, or situational condition has inhibited or obliterated a previously normal sexual function. The prognosis is good for temporary or mild disorders resulting from misinformation or situational stress but is guarded for disorders that result from intense anxiety, chronically discordant relationships, psychological disturbances, or drug or alcohol abuse in either partner.

The following factors, alone or in combination, may cause arousal or orgasmic disorder:

▶ certain drugs, including central nervous system depressants, alcohol, street drugs, and, rarely, oral contraceptives
▶ general systemic illnesses, diseases of the endocrine or nervous system, or diseases that impair muscle tone or contractility
▶ gynecologic factors, such as chronic vaginal or pelvic infection or pain, congenital anomalies, and genital cancers
▶ stress and fatigue
▶ inadequate or ineffective stimulation
▶ psychological factors, such as performance anxiety, guilt, depression, or unconscious conflicts about sexuality
▶ relationship problems, such as poor communication, hostility, or ambivalence toward the partner, fear of abandonment or independence, or boredom with sex.

All these factors may contribute to involuntary inhibition of the orgasmic reflex. Another crucial factor is the fear of losing control of feelings or behavior. Whether these factors produce sexual dysfunction and the type of dysfunction depend on how well the woman copes with the resulting pressures. Physical factors may also cause arousal or orgasmic disorder.

Female sexual function and responses decline, along with estrogen levels, in the perimenopausal period. The decrease in estradiol levels during menopause affects nerve transmission and response in the peripheral vascular system. As a result, the timing and degree of vasoconstriction during the sexual response is affected, vasocongestion decreases, muscle tension decreases, and contractions are fewer and less intense during orgasm.

A female with arousal disorder has limited or absent sexual desire and experiences little or no pleasure from sexual stimulation. Physical signs of this disorder include lack of vaginal lubrication or absence of signs of genital vasocongestion.

Dyspareunia is genital pain associated with intercourse. Insufficient lubrication is the most common cause. Other physical causes of dyspareunia include:

▶ endometriosis
▶ genital, rectal, or pelvic scar tissue
▶ acute or chronic infections of the genitourinary tract
▶ disorders of the surrounding viscera (including residual effects of pelvic inflammatory disease or disease of the adnexal and broad ligaments).

Among the many other possible physical causes are:

▶ deformities or lesions of the introitus or vagina
▶ benign and malignant growths and tumors
▶ intact hymen
▶ radiation to the pelvis
▶ allergic reactions to diaphragms, condoms, or other contraceptives.

Psychological causes include:

▶ fear of pain or injury during intercourse
▶ previous painful experience, including sexual abuse
▶ guilt feelings about sex
▶ fear of pregnancy or injury to the fetus during pregnancy
▶ anxiety caused by a new sexual partner or technique
▶ mental or physical fatigue.

Vaginismus is an involuntary spastic constriction of the lower vaginal muscles, usually from fear of vaginal penetration. This disorder may coexist with dyspareunia and, if severe, may prevent intercourse (a common cause of unconsummated marriages). Vaginismus may be physical or psychological in origin. It may occur spontaneously as a protective reflex to pain or result from organic causes, such as hymenal abnormalities, genital herpes, obstetric trauma, and atrophic vaginitis.

Psychological causes may include:

▶ childhood and adolescent exposure to rigid, punitive, and guilt-ridden attitudes toward sex
▶ fear resulting from painful or traumatic sexual experiences, such as incest or rape

▶ early traumatic experience with pelvic examinations

▶ fear of pregnancy, sexually transmitted disease, or cancer.

In males, the normal sexual response involves erection, emission, and ejaculation. Sexual dysfunction is the impairment of one or all of these processes.

Erectile disorder, or impotence, refers to inability to attain or maintain penile erection sufficient to complete intercourse. Transient periods of impotence aren't considered dysfunction and probably occur in half the adult males. Erectile disorder affects all age groups but increases in frequency with age.

Psychogenic factors are responsible for approximately 50% to 60% of the cases of erectile dysfunction; organic factors, for the rest. In some patients, psychogenic and organic factors coexist, making isolation of the primary cause difficult.

Most problems with emission and ejaculation usually have structural causes.

Male structural alterations

Structural defects of the male reproductive system may be congenital or acquired. Testicular disorders such as cryptorchidism or torsion may result in infertility.

In cryptorchidism, a congenital disorder, one or both testes fail to descend into the scrotum, remaining in the abdomen or inguinal canal or at the external ring. If bilateral cryptorchidism persists untreated into adolescence, it may result in sterility, make the testes more vulnerable to trauma, and significantly increase the risk for testicular cancer, particularly germ cell tumors. In about 80% of affected infants, the testes descend spontaneously during the first year; in the rest, the testes may descend later.

AGE ALERT The testes of an older male may be slightly smaller than those of a younger male, but they should be equal in size, smooth, freely moveable, and soft, without nodules. The left testis is often lower than the right. Benign prostatic hyperplasia is a disorder of prostate enlargement due to androgen-induced growth of prostate cells. It's more prevalent with aging and may result in urinary obstructive symptoms.

Hypospadias is the most common penile structural abnormality. The midline fusion of the urethral folds is incomplete, so the urethral meatus opens on the ventral (anterior, or "belly") side of the penis. In epispadias, the urethral meatus is located on the dorsal (posterior, or "back") side of the penis.

Priapism is prolonged, painful erection in the absence of sexual stimulation. It results from arteriovenous shunting within the corpus cavernosum that leads to obstructed venous outflow from the penis. In adults, it's usually idiopathic due to trauma. In children, it may be associated with sickle cell disease. Without prompt treatment, it can lead to ischemic fibrosis and infertility.

A urethral stricture is a narrowing of the urethra caused by scarring. It may result from trauma or infection. Common complications include prostatitis and secondary infection.

During the first 3 years of life, congenital adhesions between the foreskin and the glans penis separate naturally with penile erections. Phimosis is a condition in which the foreskin can't be retracted over the glans penis; poor hygiene and chronic infection can cause it. Paraphimosis is a condition in which the foreskin is retracted and can't be reduced to cover the glans; the penis becomes constricted, causing edema of the glans. Severe paraphimosis is a surgical emergency.

To prevent threatened spontaneous abortion, millions of women took diethylstilbestrol (DES) between 1946 and 1971. Men whose mothers took DES during their eighth to sixteenth weeks of pregnancy have experienced structural abnormalities, such as urethral meatal stenosis, hypospadias, epididymal cysts, varicoceles, cryptorchidism, and decreased fertility.

DISORDERS

Disorders of the reproductive system may affect sexual, reproductive, or urinary function.

Abnormal uterine bleeding

Abnormal uterine bleeding refers to abnormal endometrial bleeding without recognizable organic lesions. Abnormal uterine bleeding is the indication for almost 25% of gynecologic surgical procedures. The prognosis varies with the cause. Correction of hormonal imbalance or structural abnormality yields a good prognosis.

Causes

Abnormal uterine bleeding usually results from an imbalance in the hormonal-endometrial relationship in which persistent and unopposed stimulation of the endometrium by estrogen occurs. Disorders that cause sustained high estrogen levels are:

▶ polycystic ovary syndrome
▶ obesity (because enzymes present in peripheral adipose tissue convert the androgen androstenedione to estrogen precursors)
▶ immaturity of the hypothalamic-pituitary-ovarian mechanism (postpubertal teenagers)
▶ anovulation (women in their late thirties or early forties).

   Other causes include:
▶ trauma (foreign object insertion or direct trauma)
▶ endometriosis
▶ coagulopathy such as thrombocytopenia or leukemia (rare).

Pathophysiology

Irregular bleeding is associated with hormonal imbalance and anovulation (failure of ovulation to occur). When progesterone secretion is absent but estrogen secretion continues, the endometrium proliferates and become hypervascular. When ovulation does not occur, the endometrium is randomly broken down, and exposed vascular channels cause prolonged and excessive bleeding. In most cases of abnormal uterine bleeding, the endometrium shows no pathologic changes. However, in chronic unopposed estrogen stimulation (as from a hormone-producing ovarian tumor), the endometrium may show hyperplastic or malignant changes.

Signs and symptoms

Abnormal uterine bleeding usually occurs as:

▶ metrorrhagia (episodes of vaginal bleeding between menses)
▶ hypermenorrhea (heavy or prolonged menses, longer than 8 days, also incorrectly termed menorrhagia)
▶ chronic polymenorrhea (menstrual cycle less than 18 days) or oligomenorrhea (infrequent menses)
▶ fatigue due to anemia
▶ oligomenorrhea and infertility due to anovulation.

Complications

Possible complications are:

▶ iron-deficiency anemia (blood loss of more than 1.6 L over a short time) and hemorrhagic shock or right-sided cardiac failure (rare)
▶ endometrial adenocarcinoma due to chronic estrogen stimulation.

Diagnosis

Abnormal uterine bleeding may be caused by anovulation. Diagnosis of anovulation is based on:

▶ history of abnormal bleeding, bleeding in response to a brief course of progesterone, absence of ovulatory cycle body temperature changes, and low serum progesterone levels
▶ diagnostic studies ruling out other causes of excessive vaginal bleeding, such as organic, systemic, psychogenic, and endocrine causes, including certain cancers, polyps, pregnancy, and infection

• dilatation and curettage (D&C) or office endometrial biopsy to rule out endometrial hyperplasia and cancer

• hemoglobin levels and hematocrit to determine the need for blood or iron replacement.

Treatment

Possible treatment of abnormal uterine bleeding includes:

• high-dose estrogen-progestogen combination therapy (oral contraceptives) to control endometrial growth and reestablish a normal cyclic pattern of menstruation (usually given four times daily for 5 to 7 days even though bleeding usually stops in 12 to 24 hours; drug choice and dosage determined by patient's age and cause of bleeding); maintenance therapy with lower dose combination oral contraceptives

• endometrial biopsy to rule out endometrial adenocarcinoma (patients over age 35 years)

• progestogen therapy (alternative in many women, such as those susceptible to such adverse effects of estrogen as thrombophlebitis)

• I.V. estrogen followed by progesterone or combination oral contraceptives if the patient is young (more likely to be anovulatory) and severely anemic (if oral drug therapy is ineffective)

• D&C (short-lived treatment and not clinically useful, but an important diagnostic tool)

• iron replacement or transfusions of packed cells or whole blood, as indicated, due to anemia caused by recurrent bleeding

• explaining the importance of following the prescribed hormonal therapy; explaining D&C or endometrial biopsy procedure and purpose (if ordered)

• stressing the need for regular checkups to assess the effectiveness of treatment.

Amenorrhea

Amenorrhea is the abnormal absence or suppression of menstruation. Absence of menstruation is normal before puberty, after menopause, or during pregnancy and lactation; it's abnormal, and therefore pathologic, at any other time. Primary amenorrhea is the absence of menarche in an adolescent (after age 16 years). Secondary amenorrhea is the failure of menstruation for at least 3 months after the normal onset of menarche. Primary amenorrhea occurs in 0.3% of women; secondary amenorrhea is seen in 1% to 3% of women. Prognosis is variable, depending on the specific cause. Surgical correction of outflow tract obstruction is usually curative.

Causes

Amenorrhea usually results from:

• anovulation due to hormonal abnormalities, such as decreased secretion of estrogen, gonadotropins, luteinizing hormone, and follicle-stimulating hormone (FSH)

• lack of ovarian response to gonadotropins

• constant presence of progesterone or other endocrine abnormalities.

Amenorrhea may also result from:

• absence of a uterus

• endometrial damage

• ovarian, adrenal, or pituitary tumors

• emotional disorders (common in patients with severe disorders such as depression and anorexia nervosa); mild emotional disturbances tending to distort the ovulatory cycle; severe psychic trauma abruptly changing the bleeding pattern or completely suppressing one or more full ovulatory cycles

• malnutrition and intense exercise, causing an inadequate hypothalamic response.

Pathophysiology

The mechanism varies depending on the cause and whether the defect is structural, hormonal, or both. Women who have adequate estrogen levels but a progesterone deficiency don't ovulate and are thus infertile. In primary amenorrhea, the hypothalamic-pituitary-ovarian axis is dysfunctional. Because of anatomic defects of the central nervous system, the ovary doesn't receive the hormonal signals that normally initiate the development of secondary sex characteristics and the beginning of menstruation.

Secondary amenorrhea can result from several central factors (hypogonadotropic hypoestrogenic anovulation), uterine factors (as with Asherman syndrome, in which the endometrium is sufficiently scarred that no functional endometrium exists), cervical stenosis, premature ovarian failure, and others.

Signs and symptoms

Amenorrhea is a symptom of many disorders; signs and symptoms depend on the specific cause, and include:
▶ absence of menstruation
▶ vasomotor flushes, vaginal atrophy, hirsutism (abnormal hairiness), and acne (secondary amenorrhea).

Complications

Complications of amenorrhea include:
▶ infertility
▶ endometrial adenocarcinoma (amenorrhea associated with anovulation that gives rise to unopposed estrogen stimulation of the endometrium).

Diagnosis

Diagnosis of amenorrhea is based on:
▶ history of failure to menstruate in a female over age 16 years, if consistent with bone age (confirms primary amenorrhea)
▶ absence of menstruation for 3 months in a previously established menstrual pattern (secondary amenorrhea)
▶ physical and pelvic examination and sensitive pregnancy test ruling out pregnancy, as well as anatomic abnormalities (such as cervical stenosis) that may cause false amenorrhea (cryptomenorrhea), in which menstruation occurs without external bleeding
▶ onset of menstruation (spotting) within 1 week after giving pure progestational agents such as medroxyprogesterone

(Provera), indicating enough estrogen to stimulate the lining of the uterus (if menstruation doesn't occur, special diagnostic studies such as gonadotropin levels are indicated)
▶ blood and urine studies showing hormonal imbalances, such as lack of ovarian response to gonadotropins (elevated pituitary gonadotropin levels), failure of gonadotropin secretion (low pituitary gonadotropin levels), and abnormal thyroid levels (without suspicion of premature ovarian failure or central hypogonadotropism, gonadotropin levels aren't clinically meaningful because they're released in a pulsatile fashion; at a given time of day, levels may be elevated, low, or average)
▶ complete medical workup, including appropriate X-rays, laparoscopy, and a biopsy, to identify ovarian, adrenal, and pituitary tumors.

Tests to identify dominant or missing hormones include:
▶ "ferning" of cervical mucus on microscopic examination (an estrogen effect)
▶ vaginal cytologic examination
▶ endometrial biopsy
▶ serum progesterone level
▶ serum androgen levels
▶ elevated urinary 17-ketosteroid levels with excessive androgen secretions
▶ plasma FSH level more than 50 IU/L, depending on the laboratory (suggests primary ovarian failure); or normal or low FSH level (possible hypothalamic or pituitary abnormality, depending on the clinical situation).

Treatment

Treatment of amenorrhea may include:
▶ appropriate hormone replacement to reestablish menstruation
▶ treatment of the cause of amenorrhea not related to hormone deficiency (for example, surgery for amenorrhea due to a tumor)
▶ inducing ovulation; with intact pituitary gland, clomiphene citrate (Clomid) may induce ovulation in women with secondary amenorrhea due to gonadotropin defi-

ciency, polycystic ovarian disease, or excessive weight loss or gain if it's reversed

▶ FSH and human menopausal gonadotropins (Pergonal) for women with pituitary disease

▶ providing reassurance and emotional support (psychiatric counseling if amenorrhea results from emotional disturbances)

▶ teaching the patient how to keep an accurate record of her menstrual cycles to aid in early detection of recurrent amenorrhea (after treatment).

Benign prostatic hyperplasia

Although most men over age 50 years have some prostatic enlargement, in benign prostatic hyperplasia (BPH, also known as benign prostatic hypertrophy), the prostate gland enlarges enough to compress the urethra and cause overt urinary obstruction. Depending on the size of the enlarged prostate, the age and health of the patient, and the extent of obstruction, BPH is treated symptomatically or surgically.

 AGE ALERT BPH is common, affecting up to 50% of men over age 50 years and 75% of men over age 80 years.

Causes

The main cause of BPH may be age-associated changes in hormone activity. Androgenic hormone production decreases with age, causing imbalance in androgen and estrogen levels and high levels of dihydrotestosterone, the main prostatic intracellular androgen.

Other causes include:
▶ arteriosclerosis
▶ inflammation
▶ metabolic or nutritional disturbances.

Pathophysiology

Regardless of the cause, BPH begins with nonmalignant changes in periurethral glandular tissue. The growth of the fibroadenomatous nodules (masses of fibrous glandular tissue) progresses to compress the remaining normal gland (nodular hyperplasia). The hyperplastic tissue is mostly glandular, with some fibrous stroma and smooth muscle. As the prostate enlarges, it may extend into the bladder and obstruct urinary outflow by compressing or distorting the prostatic urethra. There are periodic increases in sympathetic stimulation of the smooth muscle of the prostatic urethra and bladder neck. Progressive bladder distention may also cause a pouch to form in the bladder that retains urine when the rest of the bladder empties. This retained urine may lead to calculus formation or cystitis.

Signs and symptoms

Clinical features of BPH depend on the extent of prostatic enlargement and the lobes affected. Characteristically, the condition starts with a group of symptoms known as prostatism, which include:
▶ reduced urinary stream caliber and force
▶ urinary hesitancy
▶ difficulty starting micturition (resulting in straining, feeling of incomplete voiding, and an interrupted stream).

As the obstruction increases, it causes:
▶ frequent urination with nocturia
▶ sense of urgency
▶ dribbling
▶ urine retention
▶ incontinence
▶ possible hematuria.

Complications

As BPH worsens, a common complication is complete urinary obstruction after infection or while using decongestants, tranquilizers, alcohol, antidepressants, or anticholinergics.

Other complications include:
▶ infection
▶ renal insufficiency and, if untreated, renal failure
▶ urinary calculi
▶ hemorrhage
▶ shock.

Diagnosis

Diagnosis includes physical examination showing:

▶ visible midline mass above the symphysis pubis (sign of an incompletely emptied bladder)

▶ enlarged prostate with rectal palpation.

Clinical features and a rectal examination are usually sufficient for diagnosis. Other findings that help confirm BPH may include:

▶ excretory urography to rule out urinary tract obstruction, hydronephrosis (distention of the renal pelvis and calices due to obstruction of the ureter and consequent retention of urine), calculi or tumors, and filling and emptying defects in the bladder

▶ alternatively, if patient is not cooperative, cystoscopy to rule out other causes of urinary tract obstruction (neoplasm, stones)

▶ elevated blood urea nitrogen and serum creatinine levels (suggest renal dysfunction)

▶ elevated prostate-specific antigen (PSA) (prostatic carcinoma must be ruled out)

▶ urinalysis and urine cultures showing hematuria, pyuria, and, with bacterial count more than $100,000/\mu l$, urinary tract infection (UTI)

▶ cystourethroscopy for severe symptoms (definitive diagnosis) showing prostate enlargement, bladder wall changes, and a raised bladder (only done immediately before surgery to help determine the best procedure).

Treatment

Conservative therapy includes:
▶ prostate massages
▶ sitz baths
▶ fluid restriction for bladder distention
▶ antimicrobials for infection
▶ regular ejaculation to help relieve prostatic congestion
▶ alpha-adrenergic blockers, such as terazosin (Hytrin) and prazosin (Minipress), to improve urine flow rates to relieve bladder outlet obstruction by preventing contractions of the prostatic capsule and bladder neck

▶ finasteride (Proscar) to possibly reduce the size of the prostate in some patients

▶ continuous drainage with an indwelling urinary catheter to alleviate urine retention (high-risk patients).

Surgery is the only effective therapy to relieve acute urine retention, hydronephrosis, severe hematuria, recurrent UTIs, and other intolerable symptoms. The following procedures involve open surgical removal:

▶ transurethral resection (if prostate weighs less than 2 oz [56.7 g]); tissue removed with a wire loop and electric current using a resectoscope

▶ suprapubic (transvesical) resection (most common and useful for prostatic enlargement remaining within the bladder)

▶ retropubic (extravesical) resection allowing direct visualization (potency and continence usually maintained)

▶ monitoring and recording the patient's vital signs, intake and output, and daily weight; watching closely for signs of postobstructive diuresis (such as increased urine output and hypotension) that may lead to serious dehydration, reduced blood volume, shock, electrolyte loss, and anuria

▶ indwelling urinary catheter for urine retention (usually difficult in a patient with BPH)

▶ suprapubic cystostomy under local anesthetic if indwelling urinary catheter can't be passed transurethrally (watching for rapid bladder decompression)

▶ balloon dilation of the urethra and prostatic stents to maintain urethral patency (occasionally)

▶ laser excision to relieve prostatic enlargement

▶ nerve-sparing surgical techniques to reduce common complications such as erectile dysfunction.

After prostatic surgery, interventions may include:

▶ maintaining patient comfort; watching for and preventing postoperative complications; observing for immediate dangers of prostatic bleeding (shock, hemorrhage); checking the catheter often (every 15 minutes for the first 2 to 3 hours) for patency and urine color; checking dressings for bleeding

▶ three-way catheter with continuous bladder irrigation (inserted postoperatively by many urologists); involves keeping the catheter open at a rate sufficient to maintain clear, light-pink returns; watching for fluid overload from absorption of the irrigating fluid into systemic circulation; observing an indwelling regular catheter closely (if used); irrigating a catheter with stopped drainage due to clots with 80 to 100 ml of normal saline solution, as ordered, maintaining strict aseptic technique

▶ watching for septic shock (most serious complication of prostatic surgery); immediately reporting severe chills, sudden fever, tachycardia, hypotension, or other signs of shock; starting rapid infusion of I.V. antibiotics, as ordered; watching for pulmonary embolus, heart failure, and renal shutdown; monitoring vital signs, central venous pressure, and arterial pressure continuously (supportive care in the intensive care unit may be needed)

▶ belladonna and opium suppositories or other anticholinergics, as ordered, to relieve painful bladder spasms that often occur after transurethral resection

▶ after an open procedure, patient comfort measures, such as providing suppositories (except after perineal prostatectomy), analgesic medication to control incisional pain, and frequent dressing changes

▶ I.V. fluids until the patient can drink sufficient fluids (2 to 3 L/day) to maintain adequate hydration

▶ stool softeners and laxatives, as ordered, to prevent straining (don't check for fecal impaction because a rectal examination may precipitate bleeding)

▶ reassuring patient that temporary frequency, dribbling, and occasional hematuria will likely occur after the catheter is removed

▶ reinforcing prescribed limits on activity, such as lifting, strenuous exercise, and long automobile rides that increase bleeding tendency; cautioning patient to restrict sexual activity for several weeks after discharge

▶ instructing the patient about the prescribed oral antibiotic drug regimen and indications for using gentle laxatives; urging him to seek medical care immediately if he can't void, passes bloody urine, or develops a fever

▶ encouraging annual digital rectal exams and screening for PSA to identify a possible malignancy.

Cryptorchidism

Cryptorchidism is a congenital disorder in which one or both testes fail to descend into the scrotum, remaining in the abdomen or inguinal canal or at the external ring. Although this condition may be bilateral, it more commonly affects the right testis. True undescended testes remain along the path of normal descent, while ectopic testis deviate from that path.

Cryptorchidism occurs in 30% of premature male newborns but in only 3% of those born at term. In about 80% of affected infants, the testes descend spontaneously during the first year; in the rest, the testes may descend later. If indicated, surgical therapy is successful in up to 95% of the cases if the infant is treated early enough.

Causes

The mechanism by which the testes descend into the scrotum is still unexplained. Possible causes of cryptorchidism include:

▶ hormonal factors, most likely androgenic hormones from the placenta, maternal or fetal adrenals, or the immature fetal testis and possibly maternal progesterone or gonadotropic hormones from the maternal pituitary

▶ testosterone deficiency resulting in a defect in the hypothalamic-pituitary-gonadal axis, causing failure of gonadal differentiation and gonadal descent

▶ structural factors impeding gonadal descent, such as ectopic location of the testis or short spermatic cord

▶ genetic predisposition in a small number of cases (greater incidence of cryptorchidism in infants with neural tube defects)

▶ premature newborns most commonly affected due to normal descent of testes into the scrotum during the seventh month of gestation.

Pathophysiology

A prevalent but still unsubstantiated theory links undescended testes to the development of the gubernaculum, a fibromuscular band that connects the testes to the scrotal floor. Normally in the male fetus, testosterone stimulates the formation of the gubernaculum. This band probably helps pull the testes into the scrotum by shortening as the fetus grows. Thus, cryptorchidism may result from inadequate testosterone levels or a defect in the testes or the gubernaculum. Because the testis is maintained at a higher temperature, spermatogenesis is impaired, leading to reduced fertility.

Signs and symptoms

Possible signs and symptoms of cryptorchidism include:

▶ testis on the affected side not palpable in the scrotum; underdeveloped scrotum (unilateral cryptorchidism)

▶ scrotum enlarged on the unaffected side due to compensatory hypertrophy (occasionally)

▶ infertility after puberty due to prevention of spermatogenesis (uncorrected bilateral cryptorchidism) despite normal testosterone levels.

Complications

Bilateral cryptorchidism that is untreated into adolescence may result in:

▶ sterility because of testicular temperature higher than optimal for spermatogenesis

▶ significantly increased risk for testicular cancer because the higher tempertures can cause abnormal division of germ cells.

▶ increased vulnerability of the testes to trauma.

Diagnosis

Physical examination confirms after sex is determined by the following laboratory tests:

▶ buccal smear (cells from oral mucosa) to determine genetic sex (a male sex chromatin pattern)

▶ serum gonadotropin to confirm the presence of testes by showing presence of circulating hormone.

Treatment

If the testes don't descend spontaneously by age 1 year, surgical correction is generally indicated. Surgery should be performed by age 2 years, because by this time about 40% of undescended testes can no longer produce viable sperm. Treatment includes:

▶ orchiopexy to secure the testes in the scrotum and to prevent sterility, excessive trauma from abnormal positioning, and harmful psychological effects (usually before age 4 years; optimum age, 1 to 2 years)

▶ human chorionic gonadotropin I.M. to stimulate descent (rarely); ineffective for testes located in the abdomen

▶ providing information on causes, available treatments, and effect on reproduction; emphasizing that testes may descend spontaneously (especially in premature infants).

After orchiopexy, treatment includes:

▶ monitoring vital signs, intake, and output; checking dressings; encouraging coughing and deep breathing; watching for urine retention

▶ keeping the operative site clean, telling the child to wipe from front to back after defecating

▶ maintaining tension on rubber band applied to keep the testis in place but checking that it isn't too tight

▶ encouraging parents to participate in postoperative care, such as bathing or feeding the child; urging the child to do as much for himself as possible.

Dysmenorrhea

Dysmenorrhea is painful menstruation associated with ovulation that isn't related to pelvic disease. It's the most common gynecologic complaint and a leading cause of absenteeism from school (affecting 10% of high school girls each month) and work (estimated 140 million work hours lost annually).

 AGE ALERT The incidence peaks in women in their early twenties and then slowly decreases.

Dysmenorrhea can occur as a primary disorder or secondary to an underlying disease. Because primary dysmenorrhea is self-limiting, the prognosis is generally good. The prognosis for secondary dysmenorrhea depends on the underlying disorder.

Causes

Although primary dysmenorrhea is unrelated to an identifiable cause, possible contributing factors include:
▶ hormonal imbalance
▶ psychogenic factors.

Dysmenorrhea may also be secondary to such gynecologic disorders as:
▶ endometriosis
▶ cervical stenosis
▶ uterine leiomyomas (benign fibroid tumors)
▶ pelvic inflammatory disease
▶ pelvic tumors.

Pathophysiology

The pain of dysmenorrhea probably results from increased prostaglandin secretion in menstrual blood, which intensifies normal uterine contractions. Prostaglandins intensify myometrial smooth muscle contraction and uterine blood vessel constriction, thereby worsening the uterine hypoxia normally associated with menstruation. This combination of intense muscle contractions and hypoxia causes the intense pain of dysmenorrhea. Prostaglandins and their metabolites can also cause GI disturbances, headache, and syncope.

Because dysmenorrhea almost always follows an ovulatory cycle, both the primary and secondary forms are rare during the anovulatory cycle of menses. After age 20 years, dysmenorrhea is generally secondary.

Signs and symptoms

Possible signs and symptoms of dysmenorrhea include sharp, intermittent, cramping, lower abdominal pain, usually radiating to the back, thighs, groin, and vulva, and typically starting with or immediately before menstrual flow and peaking within 24 hours.

Dysmenorrhea may also be associated with signs and symptoms suggestive of premenstrual syndrome, including:
▶ urinary frequency
▶ nausea
▶ vomiting
▶ diarrhea
▶ headache
▶ backache
▶ chills
▶ abdominal bloating
▶ painful breasts
▶ depression
▶ irritability.

Complications

A possible but rare complication is dehydration due to nausea, vomiting, and diarrhea.

Diagnosis

Diagnosis of dysmenorrhea may include:
▶ pelvic examination and a detailed patient history to help suggest the cause
▶ ruling out secondary causes for menses painful since menarche (primary dysmenorrhea)
▶ tests such as laparoscopy, D&C, hysteroscopy, and pelvic ultrasound to diag-

nose underlying disorders in secondary dysmenorrhea.

Treatment

Initial treatment aims to relieve pain and may include:

▶ analgesics such as nonsteroidal anti-inflammatory drugs for mild to moderate pain (most effective when taken 24 to 48 hours before onset of menses), especially effective due to inhibition of prostaglandin synthesis through inhibition of the enzyme cyclooxygenase

▶ narcotics for severe pain (infrequently used)

▶ heat applied locally to the lower abdomen (may relieve discomfort in mature women), used cautiously in young adolescents because appendicitis may mimic dysmenorrhea.

For primary dysmenorrhea:

▶ sex steroids (effective alternative to treatment with antiprostaglandins or analgesics), such as oral contraceptives to relieve pain by suppressing ovulation and inhibiting endometrial prostaglandin synthesis (patients attempting pregnancy should rely on antiprostaglandin therapy)

▶ psychological evaluation and appropriate counseling due to possible psychogenic cause of persistently severe dysmenorrhea.

Treatment of secondary dysmenorrhea is designed to identify and correct the underlying cause and may include surgical treatment of underlying disorders, such as endometriosis or uterine leiomyomas (after conservative therapy fails).

Effective management of the patient with dysmenorrhea focuses on relief of symptoms, emotional support, and appropriate patient teaching, especially for the adolescent, and includes:

▶ complete history focusing on the patient's gynecologic complaints, including detailed information on symptoms of pelvic disease, such as excessive bleeding, changes in bleeding pattern, vaginal discharge, and dyspareunia (painful intercourse)

▶ patient teaching, including explanation of normal female anatomy and physiology as well as the nature of dysmenorrhea (depending on circumstances, providing the adolescent patient with information on pregnancy and contraception)

▶ encouraging the patient to keep a detailed record of her menstrual symptoms and to seek medical care if symptoms persist.

Endometriosis

Endometriosis is the presence of endometrial tissue outside the lining of the uterine cavity. Ectopic tissue is generally confined to the pelvic area, usually around the ovaries, uterovesical peritoneum, uterosacral ligaments, and cul de sac, but it can appear anywhere in the body.

Active endometriosis may occur at any age, including adolescence. As many as 50% of infertile women may have endometriosis, although the true incidence in both fertile and infertile women remains unknown.

Severe symptoms of endometriosis may have an abrupt onset or may develop over many years. Thirty percent to 40% of women with endometriosis become infertile. Endometriosis usually manifests during the menstrual years; after menopause, it tends to subside. Hormonal treatment of endometriosis (continuous use of oral contraceptives, danazol [Danocrine], and gonadotropin-releasing hormone [GnRH] antagonists) is potentially effective in relieving discomfort, although treatment for advanced stages of endometriosis is usually not as successful because of impaired follicular development. However, nonsurgical treatment of endometriosis generally remains inadequate. Surgery appears to be the more effective way to enhance fertility, although definitive class I evidence doesn't currently exist. Pharmacologic and surgical treatment of endometriosis may be beneficial for managing chronic pelvic pain.

Causes

The cause of endometriosis remains unknown. The main theories to explain this disorder (one or more are perhaps true for certain populations of women) are:

▶ retrograde menstruation with implantation at ectopic sites (retrograde menstruation alone may not be sufficient for endometriosis to occur because it occurs in women with no clinical evidence of endometriosis)

▶ genetic predisposition and depressed immune system (may predispose to endometriosis)

▶ coelomic metaplasia (repeated inflammation inducing metaplasia of mesothelial cells to the endometrial epithelium)

▶ lymphatic or hematogenous spread (extraperitoneal disease).

Pathophysiology

The ectopic endometrial tissue responds to normal stimulation in the same way as the endometrium, but more unpredictably. The endometrial cells respond to estrogen and progesterone with proliferation and secretion. During menstruation, the ectopic tissue bleeds, which causes inflammation of the surrounding tissues. This inflammation causes fibrosis, leading to adhesions that produce pain and infertility.

Signs and symptoms

Signs and symptoms of endometriosis include:

▶ dysmenorrhea, abnormal uterine bleeding, and infertility (classic symptoms)

▶ pain that begins 5 to 7 days before menses peaks and lasts for 2 to 3 days (varies between patients); severity of pain not indicative of extent of disease.

Other signs and symptoms depend on the location of the ectopic tissue and may include:

▶ infertility and profuse menses (ovaries and oviducts)

▶ deep-thrust dyspareunia (ovaries or cul de sac)

▶ suprapubic pain, dysuria, and hematuria (bladder)

▶ abdominal cramps, pain on defecation, constipation; bloody stools due to bleeding of ectopic endometrium in the rectosigmoid musculature (large bowel and appendix)

▶ bleeding from endometrial deposits in these areas during menses; pain on intercourse (cervix, vagina, and perineum).

Complications

Complications of endometriosis include:

▶ infertility due to fibrosis, scarring, and adhesions (major complication)

▶ chronic pelvic pain

▶ ovarian carcinoma (rare).

Diagnosis

The only definitive way to diagnose endometriosis is through laparoscopy or laparotomy. Pelvic examination may suggest endometriosis or be unremarkable. Findings suggestive of endometriosis include:

▶ multiple tender nodules on uterosacral ligaments or in the rectovaginal septum (in one-third of the patients)

▶ ovarian enlargement in the presence of endometrial cysts on the ovaries.

Although laparoscopy is recommended to diagnose and determine the extent of disease, some clinicians recommend:

▶ empiric trial of GnRH agonist therapy to confirm or refute the impression of endometriosis before resorting to laparoscopy (controversial, but may be cost-effective)

▶ biopsy at the time of laparoscopy (helpful to confirm the diagnosis), although in some instances, diagnosis is confirmed by visual inspection.

Treatment

Treatment of endometriosis varies according to the stage of the disease and the patient's age and desire to have children. Conservative therapy for young women who want to have children includes:

▶ androgens such as danazol (Danocrine)

▶ progestins and continuous combined oral contraceptives (pseudopregnancy regimen) to relieve symptoms by causing a regression of endometrial tissue

▶ GnRH agonists to induce pseudomenopause (medical oophorectomy), causing remission of the disease (commonly used).

No pharmacologic treatment has been shown to cure the disease or be effective in all women. Some disadvantages of nonsurgical therapy include:

▶ adverse reaction to drug-induced menopause (including osteoporosis if used for more than 6 months), high expense of use for an extended duration, and possible recurrence of endometriosis after discontinuation of GnRH agonists

▶ high expense and weight gain when using danazol (Danocrine)

▶ lowest fertility rates of any medical treatment for endometriosis when using continuous oral contraceptive pills

▶ weight gain and depressive symptoms when using progestin (but as effective as GnRH antagonists).

When ovarian masses are present, surgery must rule out cancer. Conservative surgery includes:

▶ laparoscopic removal of endometrial implants with conventional or laser techniques (no benefit shown for laser laparoscopy over electrocautery or suture methods)

▶ presacral neurectomy or laparoscopic uterosacral nerve ablation (LUNA) for central pelvic pain; effective in about 50% or less of appropriate candidates (clinical studies of both presacral neurectomy and LUNA use different surgical techniques, degrees of resection, and definitions of success)

▶ advising a patient who wants children not to postpone childbearing, as she may become infertile (pregnancy may temporarily improve endometriosis–associated chronic pelvic pain)

▶total abdominal hysterectomy with or without bilateral salpingo-oophorectomy; success rates vary; unclear whether ovarian conservation is appropriate (treatment

of last resort for women who don't want to bear children or for extensive disease)

▶ annual pelvic examination and Papanicolaou test (all patients).

Erectile dysfunction

Erectile dysfunction, or impotence, refers to a male's inability to attain or maintain penile erection sufficient to complete intercourse. The patient with primary impotence has never achieved a sufficient erection. Secondary impotence is more common but no less disturbing than the primary form, and implies that the patient has succeeded in completing intercourse in the past.

Transient periods of impotence aren't considered dysfunction and probably occur in half of adult males.

 AGE ALERT Erectile disorder affects all age groups but increases in frequency with age.

The prognosis for erectile dysfunction patients depends on the severity and duration of their impotence and the underlying causes.

Causes

Causes of erectile dysfunction include psychogenic factors (50% to 60% of cases), organic causes, or both psychogenic and organic factors in some patients. This complexity makes the isolation of the primary cause difficult.

Psychogenic causes of erectile dysfunction include:

▶ intrapersonal psychogenic causes reflecting personal sexual anxieties and generally involving guilt, fear, depression, or feelings of inadequacy resulting from previous traumatic sexual experience, rejection by parents or peers, exaggerated religious orthodoxy, abnormal mother-son intimacy, or homosexual experiences

▶ psychogenic factors reflecting a disturbed sexual relationship, possibly stemming from differences in sexual preferences between partners, lack of communication, insufficient knowledge of sexual function, or nonsexual personal conflicts

▶ situational impotence, a temporary condition in response to stress.

Organic causes include:

▶ chronic diseases that cause neurologic and vascular impairment, such as cardiopulmonary disease, diabetes, multiple sclerosis, or renal failure

▶ liver cirrhosis causing increased circulating estrogen due to reduced hepatic inactivation

▶ spinal cord trauma

▶ complications of surgery, particularly radical prostatectomy

▶ drug- or alcohol-induced dysfunction

▶ genital anomalies or central nervous system defects (rare).

Pathophysiology

Neurologic dysfunction results in lack of the autonomic signal and, in combination with vascular disease, interferes with arteriolar dilation. The blood is shunted around the sacs of the corpus cavernosum into medium-sized veins, which prevents the sacs from filling completely. Also, perfusion of the corpus cavernosum is initially compromised because of partial obstruction of small arteries, leading to loss of erection before ejaculation.

Psychogenic causes may exacerbate emotional problems in a circular pattern; anxiety causes fear of erectile dysfunction, which causes further emotional problems.

Signs and symptoms

Secondary erectile disorder is classified as:

▶ partial; inability to achieve a full erection

▶ intermittent; sometimes potent with the same partner

▶ selective; potent only with certain partners.

Some men lose erectile function suddenly, and others lose it gradually. If the cause isn't organic, erection may still be achieved through masturbation.

Immediately before a sexual encounter, patients with psychogenic impotence may:

▶ feel anxious

▶ perspire

▶ have palpitations

▶ lose interest in sexual activity.

Complications

A complication of erectile dysfunction is severe depression (patients with psychogenic or organic drug-induced erectile dysfunction), causing the impotence or resulting from it.

Diagnosis

A detailed sexual history helps differentiate between organic and psychogenic factors and primary and secondary impotence. Questions should include:

▶ Does the patient have intermittent, selective nocturnal or early morning erections?

▶ Can he achieve erections through other sexual activity?

▶ When did his dysfunction begin, and what was his life situation at that time?

▶ Did erectile problems occur suddenly or gradually?

▶ What prescription or nonprescription drugs is he taking?

Diagnosis also includes:

▶ ruling out such chronic diseases as diabetes and other vascular, neurologic, or urogenital problems

▶ fulfilling the diagnostic criteria for a DSM-IV diagnosis (when the disorder causes marked distress or interpersonal difficulty).

Treatment

Treatment for psychogenic impotence includes:

▶ sex therapy including both partners (course and content of therapy depend on the specific cause of dysfunction and nature of the partner relationship)

▶ teaching or helping the patient to improve verbal communication skills, eliminate unreasonable guilt, or reevaluate attitudes toward sex and sexual roles.

Treatment for organic impotence includes:

▶ reversing the cause, if possible
▶ psychological counseling to help the couple deal realistically with their situation and explore alternatives for sexual expression if reversing the cause is not possible
▶ sildenafil citrate (Viagra) to cause vasodilatation within the penis (may effectively manage erectile dysfunction in appropriate patients)
▶ adrenergic antagonist, yohimbine, to enhance parasympathetic neurotransmission
▶ testosterone supplementation for hypogonadal men (not given to men with prostate cancer)
▶ prostaglandin E injected directly into the corpus cavernosum (may induce an erection for 30 to 60 minutes in some men)
▶ surgically inserted inflatable or noninflatable penile implants (some patients with organic impotence); patient should be instructed to avoid intercourse until the incision heals (usually in 6 weeks) after penile implant surgery.

Measures to help prevent impotence include providing information on resuming sexual activity as part of discharge instructions for a patient with a condition that requires modification of daily activities, including those with cardiac disease, diabetes, hypertension, and chronic obstructive pulmonary disease, and all postoperative patients.

Fibrocystic change of the breast

Also incorrectly known as fibrocystic disease of the breast, this is a disorder of benign breast tissue alterations. The condition is usually bilateral.

AGE ALERT Fibrocystic change is the most common benign breast disorder, affecting an estimated 10% of women ages 21 years and younger, 25% of women ages 22 years and older, and 50% of postmenopausal women.

Although most lesions are benign, some may proliferate and show atypical cellular growth. Fibrocystic change by itself is not a precursor to breast cancer, but if atypical cells are present, the risk for breast carcinoma increases.

Causes

The precise cause is unknown. Some theories include:

▶ imbalance between estrogen (excess) and progesterone (deficiency) during the luteal phase of the menstrual cycle
▶ altered prolactin levels
▶ enzymatic alteration of breast tissue caused by methylxanthines (found in caffeinated food and drinks), tyramine (found in cheese, wine, and nuts), and tobacco, which inhibit cyclic guanosine monophosphate enzymes.

Pathophysiology

Changes in the breast tissue appear to respond to hormonal stimulation, although the exact mechanism is unknown.

The first type of breast tissue alteration is cyst formation. Cysts may form within lobular or subareolar areas. Cysts are classified as microcysts (smaller than 1 mm) and grow to macrocysts (3 mm or larger).

Ductal epithelial proliferation (hyperplasia) is a condition in which dilations of the ductal system occur below the areola and the nipple. The ductal epithelium may undergo metaplastic changes. Fibrotic areas may form as a result of inflammation caused by either the cysts or ductal hyperplasia.

Signs and symptoms

Signs and symptoms of fibrocystic change of the breast include:

▶ breast pain due to inflammation and nerve root stimulation (most common symptom), beginning 4 to 7 days into the luteal phase of the menstrual cycle and continuing until the onset of menstruation

◗ pain in the upper outer quadrant of both breasts (common site)

◗ palpable lumps that increase in size premenstrually and are freely moveable (about 50% of all menstruating women)

◗ granular feeling of breasts on palpation

◗ occasional greenish-brown to black nipple discharge that contains fat, proteins, ductal cells, and erythrocytes (ductal hyperplasia).

Complications

A possible complication is the greater risk for developing breast cancer in women with proliferative lesions that have atypical cells and who also have a family history of malignancy.

Diagnosis

Diagnosis of fibrocystic change of the breast includes:

◗ ultrasound to distinguish cystic (fluid-filled) from solid masses

◗ tissue biopsy to distinguish benign from malignant changes.

Treatment

Treatment is often symptomatic and may include:

◗ diet low in caffeine and fat and high in fruits and vegetables to help alleviate pain

◗ support bra to reduce pain

◗ draining painful cysts under local anesthesia (if the aspirated fluid is bloody, it should be sent for cytologic analysis to rule out possible malignancy)

◗ synthetic androgens (danazol [Danocrine]) for severe pain (occasionally).

Fibroid disease of the uterus

Uterine leiomyomas, the most common benign tumors in women, are also known as myomas, fibromyomas, or fibroids. They're tumors composed of smooth muscle and fibrous connective tissue that usually occur in the uterine corpus, although they may appear on the cervix or on the round or broad ligament.

CULTURAL DIVERSITY Uterine leiomyomas occur in 20% to 25% of women of reproductive age and may affect three times as many blacks as whites, although the true incidence in either population is unknown.

The tumors become malignant (leiomyosarcoma) in less than 0.1% of patients, which should serve to comfort women concerned with the possibility of a uterine malignancy in association with a fibroid.

Causes

The cause of uterine leiomyomas is unknown, but some factors implicated as regulators of leiomyoma growth include:

◗ several growth factors, including epidermal growth factor

◗ steroid hormones, including estrogen and progesterone (leiomyomas typically arise after menarche and regress after menopause, implicating estrogen as a promoter of leiomyoma growth).

Pathophysiology

Leiomyomas are classified according to location. They may be located within the uterine wall (intramural) or protrude into the endometrial cavity (submucous) or from the serosal surface of the uterus (subserous). Their size varies greatly. They're usually firm and surrounded by a pseudocapsule composed of compressed but otherwise normal uterine myometrium. The uterine cavity may become larger, increasing the endometrial surface area. This can cause increased uterine bleeding.

Signs and symptoms

Most leiomyomas are asymptomatic. Signs and symptoms of leiomyoma include:

◗ abnormal bleeding, typically menorrhagia with disrupted submucosal vessels (most common symptom)

◗ pain only associated with torsion of a pedunculated (stemmed) subserous tumor or leiomyomas undergoing degeneration

(fibroid outgrows its blood supply and shrinks down in size; can be artificially induced through myolysis, a laparoscopic procedure to shrink fibroids, or uterine artery embolization)

◗ pelvic pressure and impingement on adjacent viscera (indications for treatment, depending on severity) resulting in mild hydronephrosis (not believed to be an indication for treatment because renal failure rarely, if ever, results).

Complications
Various disorders have been attributed to uterine leiomyomas, including:
◗ recurrent spontaneous abortion
◗ preterm labor
◗ malposition of the fetus
◗ anemia secondary to excessive bleeding
◗ bladder compression
◗ infection (if tumor protrudes out of the vaginal opening)
◗ secondary infertility (rare).

Diagnosis
Diagnosis of leiomyoma may be based on:
◗ clinical findings (enlarged uterus) and patient history suggesting uterine leiomyomas
◗ blood studies showing anemia from abnormal bleeding (may support the diagnosis)
◗ bimanual examination showing enlarged, firm, nontender, and irregularly contoured uterus (also seen with adenomyosis and other pelvic abnormalities)
◗ ultrasound for accurate assessment of the dimension, number, and location of tumors
◗ magnetic resonance imaging (especially sensitive with regard to fibroid imaging).

Other diagnostic procedures include:
◗ hysterosalpingography
◗ hysteroscopy
◗ endometrial biopsy (to rule out endometrial cancer in patients over age 35 years with abnormal uterine bleeding)
◗ laparoscopy.

Treatment
Treatment depends on the severity of symptoms, size and location of the tumors, and the patient's age, parity, pregnancy status, desire to have children, and general health.

Treatment options include nonsurgical as well as surgical procedures. Pharmacologic treatment generally isn't effective in the long term for fibroids. Although often prescribed by gynecologists, progestational agents are ineffective as primary treatment for fibroids.

Besides observation, nonsurgical methods include:
◗ gonadotropin-releasing hormone (GnRH) agonists to rapidly suppress pituitary gonadotropin release, which leads to profound hypoestrogenemia, a 50% reduction in uterine volume (peak effects occurring in the 12th week of therapy), and consequent benefit of reductions in tumor size before surgery and blood loss during surgery, and an increase in preoperative hematocrit (This treatment is not a cure, as tumors increase in size after cessation of therapy. [Increases in tumor size *during* therapy can indicate uterine sarcoma.] The treatment is best used preoperatively or for up to 6 months in a perimenopausal woman who might soon experience a natural menopause and thus avoid surgery.)
◗ nonsteroidal anti-inflammatory drugs for dysmenorrhea or pelvic discomfort.

Surgical procedures include:
◗ abdominal, laparoscopic, or hysteroscopic myomectomy (removal of tumors in the uterine muscle) for patients of any age who want to preserve their uterus
◗ myolysis (a laparoscopic procedure to treat fibroids without hysterectomy or major surgery, performed on an outpatient basis) to coagulate the fibroids and preserve the uterus and childbearing potential
◗ uterine artery embolization (radiologic procedure) to block uterine arteries using small pieces of polyvinyl chloride (This is a promising alternative to surgery in

many women, but no existing long-term studies confirm if this procedure is appropriate in women desiring future child-bearing or establish long-term success or side effects. Recent anecdotal data suggest decreased time to menopause after embolization.)

▶ hysterectomy (Though this is the definitive treatment for symptomatic women who have completed childbearing, it is critical to inform women of all their choices because hysterectomy usually isn't the only available option.)

▶ blood transfusions (with severe anemia due to excessive bleeding).

Prior to undergoing surgery, patients should be helped to understand the effects of hysterectomy or oophorectomy, if indicated, on menstruation, menopause, and sexual activity. Patients should also understand that pregnancy is still possible if multiple myomectomy is necessary, though cesarean delivery may be necessary. Extensive scar tissue may rupture during the contractions of vaginal delivery. (Violation of the endometrial cavity is the classic indication for cesarean section in such patients, but it's unclear why a cell layer 1 to 2 cells thick should be protective against uterine dehiscence in subsequent pregnancy.)

Gynecomastia

Gynecomastia is the enlargement of breast tissue in males.

 AGE ALERT Gynecomastia is usually bilateral, but in men over age 50 years, it's usually unilateral.

Usually the cause is physiologic. Gynecomastia often resolves spontaneously in 6 to 12 months.

Causes

Excessive estrogen production from conditions including:
▶ testicular tumors
▶ obesity
▶ pituitary tumors
▶ some hypogonadism syndromes

Systemic disorders associated with gynecomastia that may alter the estrogen-testosterone ratio include:
▶ liver disease causing inability to break down normal male estrogen secretions
▶ chronic renal failure
▶ chronic obstructive lung disease.

Pharmacologic agents that may cause gynecomastia include marijuana and exogenous estrogen, as given for prostatic malignancy.

Pathophysiology

A disturbance in the normal ratio of active androgen to estrogen results in proliferation of the fibroblastic stroma and the duct system of the breast.

Signs and symptoms

Signs and symptoms of gynecomastia include:
▶ enlarged breast tissue (at least ¾" [2 cm] in diameter), either unilateral or bilateral, beneath the areola
▶ bilateral enlargement (hormone-induced gynecomastia).

Complications

A possible complication of gynecomastia is malignant changes in the breast tissue.

Diagnosis

Diagnosis depends on:
▶ biopsy to rule out malignancy
▶ excessively high estrogen levels and normal testosterone levels (in drug- and tumor-induced hyperestrogenism)
▶ very low testosterone levels and normal estrogen levels (hypergonadism).

Treatment

Gynecomastia usually resolves spontaneously without treatment. If indicated, treatments include:
▶ treatment of the cause to reduce excess breast tissue
▶ resection of extra breast tissue for cosmetic reasons.

Hydrocele

A hydrocele is a collection of fluid between the visceral and parietal layers of the tunica vaginalis of the testicle or along the spermatic cord. It's the most common cause of scrotal swelling.

Causes

Possible causes of hydrocele include:
▶ congenital malformation (infants)
▶ trauma to the testes or epididymis
▶ infection of the testes or epididymis
▶ testicular tumor.

Pathophysiology

Congenital hydrocele occurs because of a patency between the scrotal sac and the peritoneal cavity, allowing peritoneal fluids to collect in the scrotum. The exact mechanism of congenital hydrocele is unknown.

In adults, the fluid accumulation may be caused by infection, trauma, tumor, an imbalance between the secreting and absorptive capacities of scrotal tissue, or an obstruction of lymphatic or venous drainage in the spermatic cord. This leads to a displacement of fluid in the scrotum, outside the testes. Subsequent swelling results, leading to reduced blood flow to the testes.

Signs and symptoms

Possible signs and symptoms of hydrocele include:
▶ scrotal swelling and feeling of heaviness
▶ inguinal hernia (often present in congenital hydrocele)
▶ size from slightly larger than the testes to the size of a grapefruit or larger
▶ fluid collection with either flaccid or tense mass
▶ pain with acute epididymal infection or testicular torsion
▶ scrotal tenderness due to severe swelling.

Complications

Complications may include:
▶ epididymitis
▶ testicular atrophy.

Diagnosis

Diagnosis of hydrocele may include:
▶ transillumination to distinguish fluid-filled from solid mass (a tumor doesn't transilluminate)
▶ ultrasound to visualize the testes and determine the presence of a tumor
▶ fluid biopsy to determine the cause and differentiate between normal cells and malignancy.

Treatment

Usually, no treatment of congenital hydrocele is indicated, as this condition frequently resolves spontaneously by age 1 year. Otherwise, possible treatments for hydrocele include:
▶ surgical repair to avoid strangulation of the bowel (inguinal hernia with bowel present in the sac)
▶ aspiration of fluid and injection of sclerosing drug into the scrotal sac for a tense hydrocele that impedes blood circulation or causes pain
▶ excision of the tunica vaginalis for recurrent hydroceles
▶ suprainguinal excision for testicular tumor detected by ultrasound.

Ovarian cysts

Ovarian cysts are usually nonneoplastic sacs on an ovary that contain fluid or semisolid material. Although these cysts are usually small and produce no symptoms, they may require thorough investigation as possible sites of malignant change. Cysts may be single or multiple (polycystic ovarian disease). Common physiologic ovarian cysts include follicular cysts, theca-lutein cysts, and corpus luteum cysts. Ovarian cysts can develop any time between puberty and menopause, including during pregnancy. The prognosis for nonneoplastic ovarian cysts is excellent. The risk for ovarian malignancy isn't increased with a functional (physiologic) ovarian cyst.

Causes

Possible causes of ovarian cyst are:

▶ granulosa-lutein cysts, which occur within the corpus luteum, are functional (arising during some variation of the ovulatory process), non-neoplastic enlargements of the ovaries caused by excessive accumulation of blood during the hemorrhagic phase of the menstrual cycle

▶ Theca-lutein cysts are commonly bilateral and filled with clear, straw-colored liquid; they are often associated with hydatidiform mole, choriocarcinoma, or hormone therapy (with human chorionic gonadotropin [HCG] or clomiphene citrate).

Pathophysiology

Follicular cysts are generally very small and arise from follicles that overdistend, either because they haven't ruptured or have ruptured and resealed before their fluid is reabsorbed. (See *Follicular cyst,* page 562.) Luteal cysts develop if a mature corpus luteum persists abnormally and continues to secrete progesterone. They consist of blood or fluid that accumulates in the cavity of the corpus luteum and are typically more symptomatic than follicular cysts. When such cysts persist into menopause, they secrete excessive amounts of estrogen in response to the hypersecretion of follicle-stimulating hormone and luteinizing hormone that normally occurs during menopause.

Signs and symptoms

Possible signs and symptoms of ovarian cysts are:

▶ no symptoms (small ovarian cysts such as follicular cysts)

▶ mild pelvic discomfort, low back pain, dyspareunia, or abnormal uterine bleeding, secondary to a disturbed ovulatory pattern (large or multiple cysts)

▶ acute abdominal pain similar to that of appendicitis (ovarian cysts with torsion)

▶ unilateral pelvic discomfort (from granulosa-lutein cysts appearing early in pregnancy and growing as large as 2 to 2½" [5 to 6 cm] in diameter), delayed menses, followed by prolonged or irregular bleeding (granulosa-lutein cysts in nonpregnant women).

Complications

Complications of ovarian cysts may include torsion or rupture causing signs of an acute abdomen (abdominal tenderness, distention, and rigidity) due to massive intraperitoneal hemorrhage or peritonitis.

Diagnosis

Generally, characteristic clinical features suggest ovarian cysts. They're confirmed by visualization of the ovary through ultrasound, laparoscopy, or surgery (often for another condition).

Treatment

Treatment depends on the type of cyst and symptoms and may include:

▶ no treatment due to tendency of cyst to disappear spontaneously within one to two menstrual cycles; persisting cyst indicates excision to rule out malignancy (follicular cysts)

▶ hormonal treatment (common, but no proven benefit in ovarian cyst management)

▶ analgesics to relieve symptoms (functional cysts that occur during pregnancy); cysts usually diminish during the third trimester (rarely needing surgery)

▶ elimination of the hydatidiform mole, destruction of choriocarcinoma, or discontinuation of HCG or clomiphene citrate (Clomid) therapy (theca-lutein cysts)

▶ laparoscopy or exploratory laparotomy with possible ovarian cystectomy or oophorectomy for persistent or suspicious ovarian cyst (These may be performed during pregnancy, if necessary. Optimal timing for surgery is the second trimester; laparoscopic management during pregnancy is promising.)

▶ surgery for ongoing hemorrhage from ruptured corpus luteum cyst. (Otherwise, ruptured ovarian cysts may be treated by draining intraperitoneal fluid through culdocentesis in the emergency room or office setting.)

FOLLICULAR CYST

A common type of ovarian cyst, a follicular cyst is usually semitransparent and overdistended, with watery fluid visible through its thin walls.

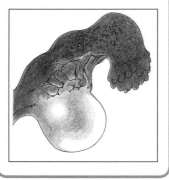

Counseling and support required by surgical cyst patients may include:
▶ teaching to explain the nature of the cyst; type of discomfort, if any; and how long condition may last
▶ preoperatively, watching for signs of cyst rupture, such as increasing abdominal pain, distention, and rigidity; monitoring vital signs for fever, tachypnea, or hypotension (possibly indicating peritonitis or intraperitoneal hemorrhage)
▶ postoperatively, encouraging frequent movement in bed and early ambulation as ordered to prevent pulmonary embolism
▶ providing emotional support; offering appropriate reassurance if patient fears cancer or infertility
▶ advising the patient to gradually increase her activities at home over 4 to 6 weeks after surgery.

Precocious puberty

Precocious puberty may occur in males or females. Males begin to mature sexually before age 9 years. This disorder occurs most commonly as true precocious puberty — early maturation of the hypothalamic-pituitary-gonadal axis, develop-

ment of secondary sex characteristics, gonadal development, and spermatogenesis — or as pseudoprecocious puberty, marked by development of secondary sex characteristics without gonadal development. Males with true precocious puberty are reported to have fathered children as early as age 7 years.

In most males with precocious puberty, sexual characteristics develop in essentially normal sequence; these children function normally when they reach adulthood.

In females, precocious puberty is the early onset of pubertal changes, such as breast development, pubic and axillary hair development, and menarche, before the age of 8 years. The occurrence is 5 times more common in females than males. Normally, the mean age for menarche is 13 years. In true precocious puberty, the ovaries mature and pubertal changes progress in an orderly manner.

In pseudoprecocious puberty, pubertal changes occur without corresponding ovarian maturation. (See *Precocious puberty*, pages 564 to 569.) In many cases, precocious puberty can be reversed.

Polycystic ovarian syndrome

Polycystic ovarian syndrome is a metabolic disorder characterized by multiple ovarian cysts. About 22% of the women in the United States have the disorder, and obesity is present in 50% to 80% of these women. Among those who seek treatment for infertility, more than 75% have some degree of polycystic ovarian syndrome, usually manifested by anovulation alone. Prognosis is very good for ovulation and fertility with appropriate treatment.

Causes

The precise cause of polycystic ovarian syndrome is unknown. Theories include:
▶ abnormal enzyme activity triggering excess androgen secretion from the ovaries and adrenal glands
▶ endocrine abnormalities causing the full spectrum of polycystic ovarian disease;

amenorrhea, polycystic ovaries on ultrasound, hyperandrogenism (part of the Stein-Leventhal syndrome).

Pathophysiology

A general feature of all anovulation syndromes is a lack of pulsatile release of gonadotropin-releasing hormone. Initial ovarian follicle development is normal. Many small follicles begin to accumulate because there's no selection of a dominant follicle. These follicles may respond abnormally to the hormonal stimulation, causing an abnormal pattern of estrogen secretion during the menstrual cycle. Endocrine abnormalities may be the cause of polycystic ovarian syndrome or cystic abnormalities; muscle and adipose tissue are resistant to the effects of insulin, and lipid metabolism is abnormal.

Signs and symptoms

Signs and symptoms of classic polycystic ovarian syndrome (Stein-Leventhal syndrome) include:
▶ mild pelvic discomfort
▶ low back pain
▶ dyspareunia
▶ abnormal uterine bleeding secondary to disturbed ovulatory pattern
▶ hirsutism
▶ acne
▶ male-pattern hair loss.

Complications

Possible complications of polycystic ovarian syndrome include:
▶ malignancy due to sustained estrogenic stimulation of the endometrium
▶ increased risk for cardiovascular disease and type 2 diabetes mellitus due to insulin resistance.

Polycystic ovarian disease may produce:
▶ secondary amenorrhea
▶ oligomenorrhea
▶ infertility.

Diagnosis

Diagnosis of polycystic ovarian syndrome includes:
▶ history and physical examination showing bilaterally enlarged polycystic ovaries and menstrual disturbance, usually dating back to menarche
▶ visualization of the ovary through ultrasound, laparoscopy, or surgery, often for another condition (may confirm ovarian cysts)
▶ slightly elevated urinary 17-ketosteroid levels and anovulation (shown by basal body temperature graphs and endometrial biopsy)
▶ elevated ratio of luteinizing hormone to follicle-stimulating hormone (usually 3:1 or greater), and elevated levels of testosterone, and androstenedione
▶ unopposed estrogen action during the menstrual cycle due to anovulation
▶ direct visualization by laparoscopy to rule out paraovarian cysts of the broad ligament, salpingitis, endometriosis, and neoplastic cysts.

Treatment

Treatment of polycystic ovarian syndrome includes monitoring patient's weight to maintain a normal body mass index in order to reduce risks associated with insulin resistance, which may cause spontaneous ovulation in some women.

Treatment of polycystic ovarian disease may include the administration of drugs such as:
▶ clomiphene citrate (Clomid) to induce ovulation
▶ medroxyprogesterone acetate (Provera) for 10 days each month for a patient wanting to become pregnant
▶ low-dose oral contraceptives to treat abnormal bleeding for the patient needing reliable contraception.

Other treatments inlcude:
▶ preoperatively, watching for signs of cyst rupture, such as increasing abdominal pain, distention, and rigidity; monitoring vital signs for fever, tachypnea, or hypotension

(Text continues on page 568.)

PRECOCIOUS PUBERTY

Although precocious puberty occurs in both male and female children, the origins of the disorder are different, so treatment varies.

	FEMALE

Causes	About 85% of the cases of true precocious puberty in females are idiopathic. Other causes of true precocious puberty are pathologic, including central nervous system (CNS) disorders resulting from tumors, trauma, infection, or other lesions, such as:
	▶ hypothalamic tumors
	▶ intracranial tumors (pinealoma, granuloma, hamartoma)
	▶ hydrocephaly
	▶ degenerative encephalopathy
	▶ tuberous sclerosis
	▶ neurofibromatosis
	▶ encephalitis
	▶ skull injuries
	▶ meningitis
	▶ peptic arachnoiditis.
	  Other conditions associated with precocity include:
	▶ Albright's syndrome
	▶ Silver's syndrome
	▶ juvenile hypothyroidism.
	  Pseudoprecocious puberty may result from:
	▶ increased levels of sex hormones due to ovarian and adrenocortical tumors
	▶ adrenal cortical virilizing hyperplasia
	▶ ingestion of estrogens or androgens
	▶ increased end-organ sensitivity to low levels of circulating sex hormones, with estrogens promoting premature breast development and androgens promoting premature pubic and axillary hair growth.

Pathophysiology	Idiopathic precocious puberty results from early development and activation of the endocrine glands without corresponding abnormality.
Signs and symptoms	Changes that may occur independently or simultaneously before the age of 8 years include: ▶ rapid growth spurt ▶ thelarche (breast development) ▶ pubarche (pubic hair development) ▶ menarche.

MALE

True precocious puberty may be:
- idiopathic and genetically transmitted as a dominant trait
- cerebral (neurogenic).
  Pseudoprecocious puberty may result from:
- testicular tumors (hyperplasia, adenoma, or carcinoma) that produce excessive testosterone levels
- congenital adrenogenital syndrome, producing high levels of adrenocortical steroids.

Idiopathic precocious puberty results from pituitary or hypothalamic intracranial lesions that cause excessive secretion of gonadotropin.

All boys with precocious puberty experience:
- early bone development, causing an initial growth spurt
- early muscle development
- premature closure of the epiphyses, resulting in stunted adult stature
- adult hair pattern
- penile growth
- bilateral enlarged testes.
  Symptoms of precocity due to cerebral lesions include:
- nausea
- vomiting
- headache
- visual disturbances
- internal hydrocephalus.

(continued)

PRECOCIOUS PUBERTY (continued)

	FEMALE

Signs and symptoms
(continued)

Complications	▶ Development of ovarian or adrenal malignancy.
Diagnosis	Diagnosis requires: ▶ complete patient history ▶ thorough physical examination ▶ special tests to differentiate between true and pseudoprecocious puberty and to indicate the necessary treatment ▶ X-rays of the hands, wrists, knees, and hips to determine bone age and possible premature epiphyseal closure ▶ ultrasound, laparoscopy, or exploratory laparotomy to verify a suspected abdominal lesion ▶ EEG, ventriculography, pneumoencephalography, computed axial tomography scan, or angiography to detect CNS disorders. Other tests detect abnormally high hormonal levels for the patient's age and may include: ▶ vaginal smear for estrogen secretion ▶ urinary tests for gonadotropic activity and excretion of 17-ketosteroids ▶ radioimmunoassay for both luteinizing and follicle-stimulating hormones.
Treatments	Treatment of constitutional true precocious puberty may include medroxyprogesterone (Provera) to reduce gonadotropin secretion and prevent menstruation. Other therapy depends on the cause of precocious puberty and its stage of development and includes: ▶ cortical or adrenocortical steroid replacement for adrenogenital syndrome ▶ surgery to remove ovarian and adrenal tumors, resulting in regression of secondary sex characteristics, especially in young children ▶ surgery and chemotherapy for choriocarcinomas ▶ thyroid extract or levothyroxine to decrease gonadotropic secretions in hypothyroidism ▶ discontinuation of medication for drug ingestion ▶ no treatment in precocious thelarche and pubarche. The dramatic physical changes produced by precocious puberty can be upsetting and alarming for the child and her family. Interventions include: ▶ providing a calm, supportive atmosphere

MALE

Symptoms of pseudoprecocity caused by testicular tumors include:
▸ adult hair patterns
▸ acne
▸ discrepancy in testis size; with the enlarged testis feeling hard or containing a palpable, isolated nodule.
   Adrenogenital syndrome produces:
▸ adult skin tone
▸ excessive hair (including beard)
▸ deepened voice
▸ stocky and muscular appearance
▸ penile, scrotal sac, and prostate enlargement (but not the testes).

▸ Development of testicular tumors
▸ In precocious puberty caused by a brain tumor, the outlook is less encouraging, and may be fatal.

Assessing the cause of precocious puberty requires:
▸ complete physical examination
▸ detailed patient history to evaluate the patient's recent growth pattern, behavior changes, family history of precocious puberty, or ingestion of hormones.
   In true precocity, laboratory results include:
▸ elevated serum levels of luteinizing and follicle-stimulating hormones and corticotropin
▸ elevated plasma testosterone levels (equal to those of an adult male)
▸ evaluation of ejaculate showing the presence of live spermatozoa
▸ brain scan, skull X-rays, and EEG to detect possible CNS tumors
▸ skull and hand X-rays showing advanced bone age.
   In pseudoprecocity, diagnosis includes:
▸ chromosomal karyotype analysis showing abnormal pattern of autosomes and sex chromosomes
▸ elevated levels of 24-hour urinary 17-ketosteroids and other steroids.

Boys with idiopathic precocious puberty generally require no medical treatment and have no physical complications in adulthood. Supportive psychological counseling is the most important therapy.
   Interventions for specific conditions include:
▸ reassessing regularly for possible tumors in a child with an initial diagnosis of idiopathic precocious puberty
▸ neurosurgery for brain tumors (they commonly resist treatment)
▸ removing the affected testis (orchiectomy) for testicular tumors; chemotherapy and lymphatic radiation therapy for malignant tumors (poor prognosis)
▸ lifelong therapy with maintenance doses of glucocorticoids (cortisol) to inhibit corticotropin production in adrenogenital syndrome causing precocious puberty.
   Interventions to help the child undergoing these changes and his family include:
▸ emphasizing to parents that the child's social and emotional development should remain consistent with his chronological age, not with his physical development; advising parents not to place unrealistic demands on the child
▸ reassuring the child that although his body is changing more rapidly than those of other boys, they'll eventually experience the same changes
▸ helping him feel less self-conscious about his changing body; suggesting clothing that de-emphasizes sexual development

(continued)

PRECOCIOUS PUBERTY *(continued)*

FEMALE

Treatments
(continued)

- encouraging the patient and family to express their feelings about these changes
- explaining all diagnostic procedures and telling the patient and family that surgery may be necessary
- explaining the condition to the child in terms she can understand to prevent feelings of shame and loss of self-esteem
- providing appropriate sex education, including information on menstruation and related hygiene
- telling parents that although their daughter seems physically mature, she's not psychologically mature, and the discrepancy between physical appearance and psychological and psychosexual maturation may create problems; warning them against expecting more of her than expected of other children her age
- suggesting that parents continue to dress their daughter in clothes appropriate for her age that don't call attention to her physical development
- reassuring parents that precocious puberty doesn't usually precipitate precocious sexual behavior.

(possibly indicating peritonitis or intraperitoneal hemorrhage)

- postoperatively, encouraging frequent movement in bed and early ambulation as ordered to prevent pulmonary embolism
- providing emotional support, offering appropriate reassurance if the patient fears cancer or infertility.

Prostatitis

Prostatitis, or inflammation of the prostate gland, may be acute or chronic. It's usually nonbacterial and idiopathic in origin (95%). The nonbacterial form of the disorder is also known as prostatodynia. Acute prostatitis most often results from gram-negative bacteria and is easy to recognize and treat. However, chronic prostatitis, the most common cause of recurrent urinary tract infections (UTIs) in men, is less easy to recognize.

 AGE ALERT As many as 35% of men aged over 50 years have chronic prostatitis.

Causes

The cause of nonbacterial prostatitis is unknown. Bacterial prostatitis is caused by:

- *Escherichia coli* (80% of cases)
- *Klebsiella, Enterobacter, Proteus, Pseudomonas, Streptococcus,* or *Staphylococcus* organisms (20% of cases).

These organisms probably spread to the prostate by:

- an ascending urethral infection or through the blood stream
- invasion of rectal bacteria through lymphatics
- reflux of infected bladder urine into prostate ducts
- infrequent or excessive sexual intercourse
- procedures such as cystoscopy or catheterization (less commonly)

 AGE ALERT Acute prostatitis is associated with benign prostatic hypertrophy in older men.

- bacterial invasion from the urethra (chronic prostatitis).

Pathophysiology

Spasms in the genitourinary tract or tension in the pelvic floor muscles may cause inflammation and nonbacterial prostatitis.

Bacterial prostatic infections can be the result of a previous or concurrent infec-

▶ providing sex education for the child with true precocity
▶ explaining adverse effects of medication (cushingoid symptoms) to family if a child must take glucocorticoids for the rest of his life.

tion. The bacteria ascend from the infected urethra, bladder, lymphatics, or blood through the prostatic ducts and into the prostate. Infection stimulates an inflammatory response in which the prostate becomes larger, tender, and firm. Inflammation is usually limited to a few of the gland's excretory ducts.

Signs and symptoms

Acute prostatitis begins with:
▶ chills
▶ low back pain, especially when standing, due to compression of the prostate gland
▶ perineal fullness
▶ suprapubic tenderness
▶ frequent and urgent urination
▶ dysuria, nocturia, and urinary obstruction due to blocked urethra by enlarged prostate
▶ cloudy urine.
    Signs of systemic infection include:
▶ fever
▶ myalgia
▶ fatigue
▶ arthralgia.

    Signs and symptoms of chronic bacterial prostatitis may include:
▶ the same urinary symptoms as the acute form but to a lesser degree
▶ recurrent symptomatic cystitis.
    Other possible signs include:
▶ evidence of UTI, such as urinary frequency, burning, cloudy urine
▶ painful ejaculation
▶ bloody semen
▶ persistent urethral discharge
▶ sexual dysfunction.

Complications

Possible complications of prostatitis include:
▶ UTI (common)
▶ infected and abscessed testis (removed surgically).

Diagnosis

Diagnosis of prostatitis may include:
▶ rectal examination finding evidence of acute prostatitis, such as very tender, warm, and enlarged prostate (characteristic)
▶ rectal examination finding firm, irregularly shaped, and slightly enlarged prostate

due to fibrosis (chronic bacterial prostatitis)
- palpation showing normal prostate gland by exclusion (nonbacterial prostatitis)
- pelvic X-ray showing prostatic calculi
- urine culture identifying the causative infectious organism
- urine culture identifying no UTI or causative organism (nonbacterial prostatitis).

Firm diagnosis depends on a comparison of urine cultures of specimens obtained by the Meares and Stamey technique. A significant increase in colony count in the prostatic specimens confirms prostatitis. This test requires four specimens:
- first specimen when the patient starts voiding (voided bladder one [VB1])
- second specimen midstream (VB2)
- third specimen after the patient stops voiding and the doctor massages the prostate to produce secretions (expressed prostate secretions [EPS])
- final voided specimen (VB3).

Treatment
Systemic antibiotic therapy is the treatment of choice for acute prostatitis, and may include:
- co-trimoxazole (Bactrim) orally for 30 days (for culture showing sensitivity)
- I.V. co-trimoxazole (Bactrim) or I.V. gentamicin (Garamycin) plus ampicillin (Unasyn) until sensitivity test results are known (sepsis)
- parenteral therapy for 48 hours to 1 week; then oral agent for 30 more days (with favorable test results and clinical response)
- co-trimoxazole (Bactrim) for at least 6 weeks (chronic prostatitis due to E. coli).
   Supportive therapy includes:
- bed rest
- adequate hydration
- analgesics
- antipyretics
- sitz baths
- stool softeners as necessary.

In symptomatic chronic prostatitis, treatment may include:
- instructing patient to drink at least 8 glasses of water daily
- regular careful massage of the prostate to relieve discomfort (vigorous massage may cause secondary epididymitis or septicemia)
- regular ejaculation to help promote drainage of prostatic secretions
- anticholingerics and analgesics to help relieve nonbacterial prostatitis symptoms
- alpha-adrenergic blockers and muscle relaxants to relieve pain
- continuous low-dose anabolic steroid therapy (effective in some men).

If drug therapy is unsuccessful, surgical treatment may include:
- transurethral resection of the prostate removing all infected tissue (not usually performed on young adults; may cause retrograde ejaculation and sterility)
- total prostatectomy (curative but may cause impotence and incontinence).

Testicular torsion
Testicular torsion is an abnormal twisting of the spermatic cord due to rotation of a testis or the mesorchium (a fold in the area between the testis and epididymis), which causes strangulation and, if left untreated, eventual infarction of the testis. Onset may be spontaneous or may follow physical exertion or trauma. This condition is almost always (90%) unilateral. The greatest risk occurs during the neonatal period and again between ages 12 and 18 years (puberty), but it may occur at any age. Infants with torsion of one testis have a greater incidence of torsion of the other testis later in life than do males in the general population. The prognosis is good with early detection and prompt treatment.

Causes
In intravaginal torsion (the most common type of testicular torsion in adolescents), testicular twisting may result from:
- abnormality of the coverings of the testis and abnormally positioned testis

▶ incomplete attachment of the testis and spermatic fascia to the scrotal wall, leaving the testis free to rotate around its vascular pedicle.

In extravaginal torsion (most common in neonates):

▶ loose attachment of the tunica vaginalis to the scrotal lining causing spermatic cord rotation above the testis

▶ sudden forceful contraction of the cremaster muscle (may precipitate this condition).

Pathophysiology

Normally, the tunica vaginalis envelops the testis and attaches to the epididymis and spermatic cord. Normal contraction of the cremaster muscle causes the left testis to rotate counterclockwise and the right testis to rotate clockwise. In testicular torsion, the testis rotates on its vascular pedicle and twists the arteries and vein in the spermatic cord, causing an interruption of circulation to the testis. Vascular engorgement and ischemia develop, causing scrotal swelling unrelieved by rest or elevation of the scrotum. If manual reduction is unsuccessful, it must be surgically corrected within 6 hours after the onset of symptoms to preserve testicular function (70% salvage rate). After 12 hours, the testis becomes dysfunctional and necrotic. (See *Extravaginal torsion.*)

Signs and symptoms

Signs and symptoms of testicular torsion include:

▶ excruciating pain in the affected testis or iliac fossa of the pelvis.

▶ edematous, elevated, and ecchymotic scrotum with loss of the cremasteric reflex (stimulation of the skin on the inner thigh retracts the testis on the same side) on the affected side.

Associated symptoms include:

▶ abdominal pain

▶ nausea and vomiting.

EXTRAVAGINAL TORSION

In extravaginal torsion, rotation of the spermatic cord above the testis causes strangulation and, eventually, infarction of the testis.

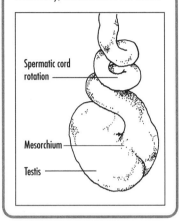

Spermatic cord rotation

Mesorchium

Testis

Complications

Possible complications of testicular torsion are:

▶ testicular infarction and necrosis

▶ infertility.

Diagnosis

Diagnosis of testicular torsion is based on:

▶ physical examination showing tense, tender swelling in the scrotum or inguinal canal; persistent reddening of the overlying skin; possibly palpable twisting of the spermatic cord (when examined before severe edema develops)

▶ Doppler ultrasonography to help distinguish testicular torsion from strangulated hernia, undescended testes, or epididymitis (absent blood flow and avascular testis in torsion).

Treatment

Treatment consists of immediate surgical repair by:

▶ orchiopexy (fixation of a viable testis to the scrotum and prophylactic fixation of the contralateral testis)

▶ orchiectomy (excision of a nonviable testis) to decrease the risk for autoimmune response to necrotic testis and its contents, damage to unaffected testis, and subsequent infertility

▶ manual manipulation of the testis counterclockwise in order to improve blood flow before surgery (not always possible).

Varicocele

A mass of dilated and tortuous varicose veins in the spermatic cord is called a varicocele. It is classically described as a "bag of worms." Thirty percent of all men diagnosed with infertility have a varicocele. Varicocele occurs in the left spermatic cord 95% of the time.

 AGE ALERT It occurs in 10% to 15% of all males, usually between age 13 and 18 years.

Causes

Causes of varicocele include:
▶ incompetent or congenitally absent valves in the spermatic veins
▶ tumor or thrombus obstructing the inferior vena cava (unilateral left-sided varicocele).

Pathophysiology

Because of a valvular disorder in the spermatic vein, blood pools in the pampiniform plexus of veins that drain each testis rather than flowing into the venous system. One function of the pampiniform plexus is to keep the testes slightly cooler than the body temperature, which is the optimum temperature for sperm production. Incomplete blood flow through the testis thus interferes with spermatogenesis. Testicular atrophy also may occur because of the reduced blood flow.

Signs and symptoms

Usually no symptoms are associated with the presence of a varicocele. Occasionally, symptoms may include:
▶ feeling of heaviness on the affected side
▶ testicular pain and tenderness on palpation.

Complications

Possible complications of a varicocele include:
▶ infertility due to the elevated temperature caused by increased blood flow to the testes
▶ metastasis from a renal tumor leading to sudden development of a varicocele in an older man (late sign).

Diagnosis

Physical examination shows:
▶ palpation of "bag of worms" when the patient is upright
▶ drained varicocele can't be felt when the patient is recumbent.

Treatment

Possible treatment includes:
▶ conservative treatment with a scrotal support to relieve discomfort (mild varicocele and fertility not an issue)
▶ surgical repair or removal involving ligation of the spermatic cord at the internal inguinal ring (if infertility is an issue).

APPENDIX
SELECTED REFERENCES
INDEX

Appendix:
Less common disorders

DISEASE AND CAUSES	PATHOPHYSIOLOGY	SIGNS AND SYMPTOMS
Actinomycosis ▶ Bacterial infection due to a variety of gram-positive anaerobic or microaerophilic rods, mostly of genus *Actinomyces* ▶ Occurs most frequently at an oral, cervical, or facial site	A disruption of the mucosal barrier permits actinomyces to invade beyond their endogenous environment in the mouth, lower GI tract, and female genitourinary tract causing local inflammation of the area and progressing to lesions with fibrous walls and sinus tracts.	*Oral, cervical, or facial site:* ▶ Infectious lesion of head and neck ▶ Pain ▶ Fever ▶ Leukocytosis ▶ Otitis, sinusitis, and canaliculitis *Extension to cranium, spine or thorax:* ▶ nuchal rigidity ▶ exaggerated deep tendon reflexes ▶ decreased level of consciousness ▶ cough with occasional hemoptysis ▶ fever
Adenoid hyperplasia ▶ Cause unknown; may be hereditary or due to chronic infections or irritations	Increased mitosis leads to an increase in cell numbers in adenoid tissue.	▶ Respiratory obstruction, especially mouth breathing, snoring at night, and frequent, prolonged nasal congestion ▶ Persistent mouth breathing during the formative years ▶ Frequent otitis media ▶ Sinusitis
Adrenogenital syndrome ▶ Autosomal recessive disorder that causes a 21-hydroxylase deficiency or 11-β hydroxylase deficiency ▶ Tumor of the adrenal glands	Lack of an enzyme needed to synthesize cortisol causes an excessive ACTH response through the negative feedback loop to the pituitary. The continual ACTH message to the adrenal glands results in hyperplasia of adrenal tissue. Excessive androgen production is stimulated because the adrenal pathway to androgen production is not blocked.	▶ Enlarged external genitalia in newborn due to excessive androgen production; female infants may have slightly enlarged clitoris or clitoris with a penile shape and labia fused to appear as a scrotum, whereas males have enlarged genitals ▶ Severe electrolyte imbalance, dehydration, vomiting, wasting, and shock due to adrenal crisis at 5 to 10 days

DISEASE AND CAUSES	PATHOPHYSIOLOGY	SIGNS AND SYMPTOMS
Albinism ▶ Autosomal recessive inheritance	Absence of the enzyme tyrosine results in a defect in melanin formation.	▶ Extremely fair skin color ▶ Fine, white hair ▶ Gray or blue irises of the eye ▶ Strabismus ▶ Nystagmus ▶ Photophobia ▶ Persistent loss of visual acuity
Alpha₁-antitrypsin deficiency ▶ Autosomal recessive inheritance	Reduced or absent levels of alpha₁-antitrypsin, which is needed to inhibit proteolytic enzymes, leads to the unopposed action of protease on the liver and lungs.	*In children:* ▶ Cholestasis ▶ Hepatitis ▶ Portal hypertension *In young adults:* ▶ Emphysema ▶ Cirrhosis
Amblyopia ▶ Strabismus in children ▶ Excessive alcohol or tobacco use in adults	Acuity is reduced due to a toxic reaction in the orbital portion of the optic nerve or by cerebral blockage of the visual stimuli.	▶ Visual dimness, photophobia, and ocular discomfort ▶ Small central or pericentral scotoma enlarges slowly ▶ Temporal disk pallor ▶ Possible blindness
Amyloidosis ▶ Pressure due to accumulation and infiltration of amyloid causes atrophy of nearby cells. Abnormal immunoglobulin synthesis and reticuloendothelial cell dysfunction may occur ▶ Familial inheritance in persons with Portuguese ancestry ▶ May occur with tuberculosis, chronic infection, rheumatoid arthritis, multiple myeloma, Hodgkin's disease, paraplegia, brucellosis, and Alzheimer's disease	A rare, chronic disease of abnormal fibrillar scleroprotein accumulation that infiltrates body organs and soft tissues. Perireticular type affects the inner coats of blood vessels whereas pericollagen type affects the outer coats. Amyloidosis can result in permanent, even life-threatening, organ damage.	▶ Proteinuria, leading to nephrotic syndrome, eventually to renal failure ▶ Heart failure due to cardiomegaly, dysrhythmias, and amyloid deposits in subendocardium, endocardium, and myocardium ▶ Stiffness and enlargement of tongue, decreased intestinal motility, malabsorption, bleeding, abdominal pain, constipation, and diarrhea ▶ Appearance of peripheral neuropathy ▶ Liver enlargement, often with azotemia, anemia, albuminuria and, rarely, jaundice

DISEASE AND CAUSES	PATHOPHYSIOLOGY	SIGNS AND SYMPTOMS
Ankylosing spondylitis ▶ Cause unknown; strongly associated with presence of human leukocyte antigen (HLA)-B27 ▶ Familial inheritance	Fibrous tissue of the joint capsule is infiltrated by inflammatory cells that erode the bone and fibrocartilage. Repair of the cartilaginous structures begins with the proliferation of fibroblasts, which synthesize and secrete collagen. The collagen forms fibrous scar tissue that eventually undergoes calcification and ossification, causing the joint to fuse or lose flexibility.	▶ Intermittent low back pain that is most severe following inactivity or in the morning ▶ Stiffness, limited lumbar spine motion ▶ Pain and limited expansion of chest ▶ Peripheral arthritis in shoulders, hips, and knees ▶ Kyphosis in advanced stages ▶ Hip deformity and limited range of motion ▶ Mild fatigue, fever, and anorexia or weight loss
Anthrax ▶ Infection of the skin, lungs, or GI tract that results from contact with contaminated animals or their hides, bones, fur, hair, or wool by *Bacillus anthracis*	After infection, bacterium produces toxins that enter susceptible cells, leading to cell death; mechanism unknown.	Incubation is 12 hours to 5 days. ▶ Red-brown bump on skin enlarges and swells around edges; a black scab forms after the bump blisters and hardens ▶ Swollen lymph nodes ▶ Muscle ache and headache ▶ Nausea, vomiting, fever *In pulmonary anthrax:* ▶ Respiratory problems ▶ Shock and coma *In GI anthrax (rare):* ▶ Extensive bleeding and kills tissue ▶ Fatal if enters the bloodstream.
Aspergillosis ▶ Fungal infection due to Aspergillus species; transmitted by inhalation of fungal spores or invasion of spores through wounds or injured tissue	*Aspergillus* species produce extracellular enzymes such as proteases and peptidases that contribute to tissue invasion, leading to hemorrhage and necrosis.	Incubation is a few days to weeks; may be asymptomatic or mimic tuberculosis, causing a productive cough and purulent or blood-tinged sputum, dyspnea, empyema, and lung abscesses ▶ Allergic aspergillosis causes wheezing, dyspnea, pleural pain, and fever Aspergillosis endophthalmitis appears 2 to 3 weeks after eye surgery ▶ Cloudy vision, eye pain, and reddened conjunctiva ▶ Purulent exudate on exposure to anterior and posterior chambers of the eye

DISEASE AND CAUSES	PATHOPHYSIOLOGY	SIGNS AND SYMPTOMS

Behçet's syndrome

▶ Cause unknown; environmental factor or unknown virus can initiate process if genetic predisposition exists
▶ Family members may exhibit similar symptoms

An overactive immune system produces sudden inflammation of small blood vessels; symptoms based on location of inflammation. Behcet's syndrome is more apparent in persons with Mediterranean, Middle East, and Far East ancestry. Onset is usually ages 10 to 20; five times more common in males.

▶ Recurrent genital ulcerations
▶ Recurrent oral ulcerations
▶ Eye inflammation and skin lesions
▶ Subcutaneous thrombophlebitis
▶ Epididymitis and deep vein thrombosis
▶ Arterial occlusion and aneurysm
▶ Severe headache and fatigue
▶ Bloating, diarrhea, cramping, and bloody stools
▶ Movement and speech difficulties

Bell's palsy

▶ Considered an idiopathic facial paralysis; infectious cause suggested

Blockage of the seventh cranial nerve due to inflammation around the nerve where it leaves bony tissue leads to unilateral or bilateral facial weakness or paralysis. The blockage may result from hemorrhage, tumor, meningitis or local trauma.

▶ Unilateral facial weakness or paralysis, with aching at the jaw angle
▶ Drooping mouth, causing salivation
▶ Distorted taste
▶ Impaired ability to fully close eye on affected side
▶ Loss of taste and tinnitus

Blastomycosis

▶ Fungal infection due to *Blastomyces dermatitidis;* usually infects the lungs and produces bronchopneumonia.
▶ May disseminate through blood causing osteomyelitis and CNS, skin, and genital disorders

Inhalation of the conidia leads to clearing of the organism by alveolar macrophages that kill conidia. Conidia that are not killed convert to yeast forms that trigger an inflammatory response resulting in the formation of noncaseating granulomas.

Symptoms mimic a viral upper respiratory tract infection:
▶ Dry, hacking, or productive cough
▶ Pleuritic chest pain
▶ Fever, shaking, chills, night sweats, malaise, and anorexia
▶ Small, painless, nonpruritic, and nondistinctive macules or papules on exposed body parts
▶ Painful swelling of testes, epididymis, or prostate; deep perineal pain, pyuria, and hematuria

Botulism

▶ Paralytic illness due to an endotoxin produced by *Clostridium botulinum;* often due to consumption of inadequately cooked, contaminated foods

The endotoxin acts at the neuromuscular junction of skeletal muscle, preventing acetylcholine release and blocking neural transmission, eventually resulting in paralysis.

Appears within 12 to 36 hours after digesting food; severity depends an amount consumed. Initial signs include:
▶ Dry mouth, sore throat, weakness, dizziness, vomiting, and diarrhea
Cardinal signs include:
▶ Acute symmetrical cranial nerve impairment, followed by weakness and muscle paralysis
▶ Mental or sensory processes not affected and not associated with fever

DISEASE AND CAUSES	PATHOPHYSIOLOGY	SIGNS AND SYMPTOMS

Bronchiectasis

▶ Conditions associated with continued damage to bronchial walls and abnormal mucociliary clearance cause tissue breakdown to adjacent airways; such conditions include cystic fibrosis, immunologic disorders, and recurrent bacterial respiratory-tract infections

Inflammation and destruction of the structural components of the bronchial wall lead to chronic abnormal dilatation.

In early stages:
▶ Asymptomatic with complaints of frequent pneumonia or hemoptysis
▶ Chronic cough producing foul-smelling, mucopurulent secretions
▶ Coarse crackles during inspiration
▶ Wheezing, dyspnea, sinusitis, fever, and chills
In advanced stage:
▶ Chronic malnutrition and right-sided heart failure due to hypoxic pulmonary vasoconstriction

Bronchiolitis

▶ No known cause; may be associated with specific diseases or conditions, such as bone marrow, heart or lung transplants, rheumatoid arthritis, lupus erythematosus, and Crohn's disease

Infection or other unknown factors cause necrosis of the bronchial epithelium and destruction of ciliated epithelial cells. As the submucosa becomes edematous, cellular debris and fibrin form plugs in the bronchioles.

Subacute symptoms:
▶ Fever, persistent nonproductive cough, dyspnea, malaise, and anorexia
▶ Physical assessment reveals dry crackles
Less common:
▶ Productive cough, hemoptysis, chest pain, general aches, and night sweats

Brucellosis

▶ Due to gram-negative, aerobic *Brucella bacterium* that is transmitted by consumption of unpasteurized dairy products and meat or contact with infected animals or their secretions or excretions

Nonmotile, nonspore-forming, gram-negative coccobacilli of *Brucella* species cause an acute febrile illness.

Usually insidious
In acute phase:
▶ Fever, chills, profuse sweating, fatigue, headache, backache, and enlarged lymph nodes
In chronic phase:
▶ Depression, sleep disturbances, and sexual impotence

Buerger's disease

▶ Cause unknown; however, link to excessive smoking suggests hypersensitivity to nicotine

Polymorphonuclear leukocytes infiltrate the walls of small and medium-sized arteries and veins. Thrombus develops in the vascular lumen, eventually occluding and obliterating portions of the small vessels.

▶ Intermittent claudication of the instep
▶ With exposure to cold, feet become cold, cyanotic, and numb; later, feet become red, hot, and tingle
▶ Painful ulcerations on fingertips
▶ Decreased peripheral pulses
In later stage:
▶ Muscle atrophy and gangrene

DISEASE AND CAUSES	PATHOPHYSIOLOGY	SIGNS AND SYMPTOMS

Cancer of the vulva

- Cause unknown; affects external female reproductive organs
- Usually occurs after menopause
- Accounts for 3% to 4% of all cancers of the female reproductive system.

Squamous and basal cell cancers (90% squamous, 4% basal, 6% rare cancers) that usually begin on skin surface and are mostly slow-growing. Untreated, cancer spreads to vagina, urethra, anus, or lymph nodes.

- Unusual lumps or sores
- Vulvar pruritus
- Bleeding
- Small painful, infected ulcer
- Groin pain
- Abnormal urination and defecation

Cardiogenic shock

- Can result from conditions that cause significant ventricular dysfunction with reduced cardiac output, such as MI, myocardial ischemia, papillary muscle dysfunction, or end-stage cardiomyopathy

Decreased perfusion triggers baroceptor reflexes and fluid regulation mechanisms. The adaptive mechanisms eventually fail and positive feedback loops develop due to increased cardiac workload and tissue hypoxia. This may lead to extensive ischemic damage to vital organs such that the patient cannot recover.

- Cold, clammy, pale skin
- Decrease in systolic blood pressure to 30 mm Hg below baseline
- Tachycardia
- Rapid, shallow respirations
- Restlessness and mental confusion
- Narrowing pulse pressure and cyanosis

Carpal tunnel syndrome

- Mostly idiopathic or may result from trauma or repetitive motion of the wrist
- Many conditions, including alterations in the endocrine or immune systems, may increase the fluid pressure in the tunnel

Compression of the median nerve from inflammation or fibrosis of the tendon sheaths initially impairs sensory transmission to the thumb, index finger, second finger, and inner aspect of the third finger.

- Weakness, pain, burning, numbness, or tingling in both hands
- Paresthesia affects thumb, forefinger, middle finger, and half of the fourth finger
- Inability to clench fist
- Nails may be atrophic
- Dry and shiny skin

Celiac disease

- Results from a complex interaction involving dietary, genetic, and immunologic factors

Ingestion of gluten causes injury to the villi in the upper small intestine, leading to a decreased surface area and malabsorption of most nutrients. Inflammatory enteritis also results, leading to osmotic diarrhea and secretory diarrhea.

- Recurrent diarrhea, abdominal distention, stomach cramps, weakness, or increased appetite without weight gain
- Normochromic, hypochromic, or macrocytic anemia
- Osteomalacia, osteoporosis, tetany, and bone pain in lower back, rib cage, and pelvis
- Peripheral neuropathy, paresthesia, or seizures
- Dry skin, eczema, psoriasis, dermatitis herpetiformis, and acne rosacea
- Amenorrhea, hypometabolism, and adrenocortical insufficiency
- Mood changes and irritability

DISEASE AND CAUSES	PATHOPHYSIOLOGY	SIGNS AND SYMPTOMS

Cerebral aneurysm
▶ May be due to atherosclerosis, trauma, congenital defect, inflammation, infection, or cocaine use

A single pathologic mechanism does not exist.

Several days before event:
▶ Headache
▶ Stiff legs and back
▶ Nuchal rigidity
▶ Nausea
Onset abrupt without warning:
▶ Severe headache
▶ Vomiting
▶ Altered consciousness
▶ Coma

Cervical cancer
▶ Cause unknown; predisposing factors include intercourse under age 16, multiple sex partners, multiple pregnancies, and venereal infections
▶ Two types of cervical cancer: preinvasive and invasive

Preinvasive cancer is curable with early detection causing minimal cervical dysplasia in the lower third of the epithelium. Invasive cancer penetrates basement membrane to disseminate throughout the body via lymphatic routes. Histologic type is 95% squamous cell carcinoma.

Preinvasive cancer:
▶ Early invasive cancer shows no clinical changes or symptoms
▶ Abnormal vaginal bleeding
▶ Persistent vaginal discharge
▶ Postcoital pain and bleeding
Advanced stages:
▶ Pelvic pain
▶ Vaginal leakage of urine or feces from fistula
▶ Anorexia, weight loss, and anemia

Cervical spondylosis
▶ Caused by narrowing of cervical canal or neural foramina due to degenerative changes in the intervertebral disk and annulus and to bony osteophytes

Progressive myelopathy leads to cord compression, causing a spastic gait.

▶ Arm weakness and atrophy with reflex loss
▶ Hyperreflexia and increased tone
▶ Plantar extensor response in legs

Chalazion
▶ Due to obstruction of the meibomian (sebaceous) gland duct

Blockage of the meibomian gland leads to the formation of granulation tissue.

▶ Local swelling
▶ Mild irritation
▶ Blurred vision
▶ Red-yellow elevation on conjunctival surface under eyelid

Chédiak-Higashi syndrome
▶ Linked to consanguinity and transmitted as an autosomal recessive trait

A genetic defect that manifests in morphologic changes in the granulocytes and causes delayed chemotaxis and impaired intracellular digestion of organisms; diminished inflammatory response results.

▶ Recurrent bacterial infections, primarily in the skin, lungs, and subcutaneous tissue
▶ Fever
▶ Thrombocytopenia, neutropenia, and hepatosplenomegaly
▶ Significant photophobia
▶ Motor and sensory neuropathies
▶ Cellular proliferation of liver, spleen, and bone marrow is fatal

DISEASE AND CAUSES	PATHOPHYSIOLOGY	SIGNS AND SYMPTOMS

Cholera

▶ Acute enterotoxin-mediated GI infection due to gram-negative bacillus, which is transmitted through water and food contamination with fecal material from carriers or people with active infections

Following ingestion of a significant inoculum, colonization of the small intestine occurs. The secretion of a potent enterotoxin results in a massive outpouring of isotonic fluid from the mucosal surface of the small intestine. Profuse diarrhea, vomiting, fluid and electrolyte loss occurs and may lead to hypovolemic shock, metabolic acidosis, and death.

Incubation period is several hours to 5 days
▶ Acute, painless, profuse watery diarrhea, and vomiting
▶ Intense thirst, weakness, and loss of skin tone
▶ Muscle cramps
▶ Cyanosis
▶ Oliguria
▶ Tachycardia
▶ Falling blood pressure, fever, and hypoactive bowel sounds

Chronic fatigue syndrome

▶ Cause unknown; may be found in HHV-6 or other herpesviruses, enteroviruses, or retroviruses

Infectious agents or environmental factors trigger an abnormal immune response and hormonal alterations.

▶ Prolonged, overwhelming fatigue
Centers for Disease Control and Prevention use a "working case definition" to group severity and symptoms

Coccidioidomycosis

▶ Fungal infection due to possible inhalation of *Coccidioides immilis* spores from the soil or in plaster casts or dressing of infected people

C. immilis induces a granulomatous reaction that results in caseous necrosis.

▶ Acute or subacute respiratory symptoms
▶ Dry cough, pleuritic chest pain, and pleural effusion
▶ Fever and sore throat
▶ Chills, malaise, headache, and itchy macular rash
▶ Tender red nodules on legs with joint pain in knees and ankles in Caucasian women
▶ Chronic pulmonary cavitation

Colorado tick fever

▶ Virus is transmitted to human by a hard-shelled wood tick, *Dermacentor andersoni*. Tick acquires the virus when it bites an infected rodent and remains a permanently infected vector

Virus circulates inside of erythropoietic cells, producing typical febrile symptoms.

Incubation is 3 to 6 days; symptoms begin abruptly
▶ Chills, high temperature, severe back, arm and leg aches, and lethargy
▶ Headache with ocular movement
▶ Photophobia, abdominal pain, nausea, and vomiting

DISEASE AND CAUSES	PATHOPHYSIOLOGY	SIGNS AND SYMPTOMS
Complement deficiencies ▶ Primary complement deficiencies inherited as autosomal recessive traits; however, C1 esterase inhibitor is autosomal dominant ▶ Secondary deficiencies may follow complement-fixing immunologic reactions	Series of circulating enzymatic serum proteins with nine functional components labeled C1-C9. This disease may increase susceptibility to infections and certain autoimmune disorders.	Clinical effects vary with specific deficiency *C5 deficiency (familial defect in infants):* ▶ Diarrhea and seborrheic dermatitis *C1 esterase inhibitor deficiency:* ▶ Swelling in face, hands, abdomen, or throat, with possible fatal laryngeal edema *C2 and C3 deficiencies and C5 familial dysfunction:* ▶ Increase susceptibility to bacterial infection *C2 and C4 deficiencies:* ▶ Collagen vascular disease (lupus and chronic renal failure)
Costochondritis ▶ Cause unknown	An inflammatory process of the costochondral or costosternal joints is initiated, causing localized pain and tenderness.	▶ Sharp pain in chest wall ▶ Area is sensitive to touch ▶ Pain may radiate into arm ▶ Pain worsens with movement ▶ Reproducible pain
Creutzfeldt-Jakob disease ▶ Prion infection	Organism infects the CNS, leading to myelin destruction and neuronal loss.	▶ Myoclonic jerking, ataxia, aphasia, visual disturbances, paralysis, and early abnormal electroencephalogram
Cryptococcosis ▶ Fungal infection due to *Cryptococcus neoformans*, which is transmitted in particles of dust contamination by pigeon feces	Transmission is by inhalation of cryptococci. An asymptomatic pulmonary infection disseminates to extrapulmonary sites, usually CNS, but also skin, bones, prostate gland, liver, or kidneys. Left untreated, infection progresses from coma to death due to cerebral edema or hydrocephalus.	▶ Fever, cough with pleuritic pain, and weight loss ▶ Severe frontal and temporal headache, diplopia, blurred vision, dizziness, aphasia, and vomiting ▶ Skin abscesses and painful lesions of the long bones, skull, spine, and joints
Cystic echinococcosis ▶ Infection due to tapeworm larvae, *Echinococcus gramulosus, E. multilocularis,* or *E. vogeli* ▶ Mainly transmitted by dogs that ingested the viscera of infected sheep ▶ Also found in coyotes, wolves, dingoes, and jackals	*E. granulosus* forms cysts in the liver, lungs, kidneys, and spleen; infection can be treated with surgery. *E. multilocularis* (alveolar hydatid disease) forms parasite tumors in the liver, lungs, brain, and other organs; infection can be fatal.	Slow-growing cysts may be asymptomatic for years. Symptoms reflect location and size of cysts.

DISEASE AND CAUSES	PATHOPHYSIOLOGY	SIGNS AND SYMPTOMS

Cystinuria
- Inherited autosomal recessive defect
- Prevalent in short people; cause unknown

Impaired function of membrane carrier proteins essential for transport of cystine and other dibasic amino acids results in excessive amino acid concentration in urine and excessive urinary excretion of cystine.

- Dull flank pain and capsular distention from acute renal colic; hematuria
- Tenderness in the costovertebral angle or over the kidneys
- Urinary obstruction with secondary infection (fever, chills, frequency, and foul-smelling urine)

DiGeorge syndrome
- Lack or partial lack of thymus gland

Partial or total absence of cell-mediated immunity resulting from T-lymphocyte deficiency produces life-threatening hypocalcemia associated with cardiovascular and facial anomalies. Without a fetal thymus transplant, patients do not live beyond age 2.

At birth:
- Low-set ears, notches in ear pinna, fish-shaped mouth, an undersized jaw and abnormally wide-set eyes
- Great vessel anomalies and tetralogy of Fallot
- Hypocalcemia
- CNS and early heart failure

Endocarditis
- Infection due to bacteria, viruses, fungi, rickettsiae, and parasites

Endothelial damage allows microorganisms to adhere to the surface where they proliferate and promote the propagation of endocardial vegetation.

- Weakness and fatigue
- Weight loss, fever, night sweats, and anorexia
- Arthralgia, splenomegaly, and new systolic murmur

Epidermolysis bullosa
- Cause unknown
- Nonscarring forms result from an autosomal dominant inheritance, except for junctional epidermolysis bullosa (recessively inherited) and dystrophic epidermolysis bullosa (results from X-linked recessive inheritance)

Blisters occur from frictional trauma or heat; prognosis depends on severity. Fatal in infant and child, but becomes less severe as patient matures.

- Blisters appear on hands, feet, knees, or elbows
- Sloughing of large areas of newborn skin
- Blistering occurs in GI, respiratory, or genitourinary tracts
- Eyelid blisters, conjunctivitis, adhesions, and corneal opacities

Esophageal varices
- Portal hypertension

Shunting of blood to the venae cavae due to portal hypertension leads to dilatation of esophageal veins.

- Hemorrhage and subsequent hypotension
- Compromised oxygen supply
- Altered level of consciousness

DISEASE AND CAUSES	PATHOPHYSIOLOGY	SIGNS AND SYMPTOMS

Fanconi's syndrome
▶ Inherited renal tubular transport disorder

Changes in the proximal renal tubules due to atrophy of epithelial cells and loss of proximal tube volume results in a shortened connection to glomeruli by an unusually narrow segment. Malfunction of the proximal renal tubules leads to hyperkalemia, hypernatremia, glycosuria, phosphaturia, aminoaciduria, uricosuria, retarded growth, and rickets.

Mostly normal appearance at birth with slightly lower birth weights
▶ After 6 months: weakness, failure to thrive, dehydration, cystine crystals in the corners of the eye, and retinal pigment degeneration
▶ Yellow skin with little pigmentation
▶ Slow linear growth

Galactosemia
▶ Inherited autosomal recessive defects
▶ Inability to metabolize galactose

Galactose-1-phosphate and galactose accumulates in the tissues leading to decreased hepatic output of glucose and hypoglycemia.

▶ Vomiting and diarrhea
▶ Jaundice and hepatomegaly
▶ Mental retardation, malnourishment, progressive hepatic failure, and death

Gallbladder and bile duct carcinoma
▶ Rare cancer in patients with cholecystitis
▶ Rapidly progressive and fatal
▶ Cause of extrahepatic bile duct carcinoma unknown
▶ Presents with ulcerative colitis

Direct extension to liver, cystic and common bile ducts, stomach and colon causes obstructions and consequent progressive, profound jaundice and epigastric and right upper quadrant pain.

Difficult to distinguish from cholecystitis
▶ Pain in epigastrium or upper right quadrant
▶ Weight loss, anorexia
▶ Nausea, vomiting, and jaundice
▶ Pruritus, skin excoriations, chills, and fever

Gas gangrene
▶ Local infection in devitalized tissue due to Clostridium perfringens

Bacteria produce hydrolytic enzymes and toxins that destroy connective tissue and cellular membranes and cause gas bubbles to form in muscle cells. Enzymes also lyse RBC membranes, destroying their oxygen-carrying capacity.

▶ Myositis and soft-tissue anaerobic cellulitis
▶ Crepitus
▶ Severe localized pain, swelling, and distortion
▶ Bullae and necrosis form after 36 hours
▶ Skin over wound may rupture, exposing dark red or black necrotic muscle and a foul-smelling watery or frothy discharge
▶ Intravascular hemolysis
▶ Thrombosis of blood vessels
▶ Toxemia
▶ Hypovolemia

DISEASE AND CAUSES	PATHOPHYSIOLOGY	SIGNS AND SYMPTOMS
Gastric cancer ▶ Cause unknown; often associated with gastritis resulting from gastric cancer ▶ May be genetic in people with type A blood	An increase in nitrosoamines damages the DNA of mucosal cells, promoting metaplasia and neoplasia.	*In early stage:* ▶ Chronic dyspepsia and epigastric discomfort *In later stage:* ▶ Weight loss, anorexia, fullness after eating, anemia, and fatigue ▶ Dysphagia, vomiting, and blood in stool
Gastritis ▶ Chronic ingestion of irritating foods, aspirin, other nonsteroidal anti-inflammatory agents, caffeine, corticosteroids, and poisons and endotoxins released by infecting bacteria ▶ Chronic gastritis associated with peptic ulcer disease or gastrostomy, causing chronic reflux of pancreatic secretions	An acute or chronic inflammation of gastric mucosa produces mucosal bleeding, edema, hemorrhage and erosion.	Onset rapid following ingestion ▶ Epigastric discomfort, indigestion, cramping, anorexia, nausea, vomiting, and hematemesis
Gaucher's disease ▶ Autosomal recessive disease caused by decreased activity of the enzyme glucocerebrosides	The most common lipidosis, causes an abnormal accumulation of glucocerebrosides in reticuloendothelial cells and results in infiltration of the bone marrow, hepatomegaly, and skeletal complications.	Key signs include hepatosplenomegaly and bone lesions. *In type I:* ▶ Thinning of cortices, pathologic fractures, collapse of hip joints, vertebral compression, severe episodic and pain in legs, arms, and back ▶ Fever, abdominal pain, respiratory problems, easy bruising and bleeding, and anemia *In type II:* ▶ Motor dysfunction and spasticity at 6 to 7 months ▶ Strabismus, muscle hypertonicity, retroflexion of head, neck rigidity, dysphagia, laryngeal stridor, hyperreflexia, and seizures

DISEASE AND CAUSES	PATHOPHYSIOLOGY	SIGNS AND SYMPTOMS
Giant cell arteritis ▶ Immune-mediated process ▶ Possible infectious etiology ▶ Genetic factors apparent with HLA-DR4 genotype; twice as common in women; incidence increases sharply with age	Lymphocytes, plasma cells, and multinucleated giant cells infiltrate affected vessels. Patchy or segmental changes overcome the medium and large arteries of the head and neck and may extend into the carotids and aorta. A cell-mediated immune response directed toward antigens in or near the elastic tissue component of the arterial wall may account for this disorder.	▶ Continuous, throbbing temporal headache ▶ Ischemia of masseter muscles, tongue, and pharynx ▶ Necrosis and ulceration of scalp ▶ Ocular or orbital pain ▶ Transient loss of vision, visual field defects, blurring, and hallucinations ▶ Tender, red, swollen, and nodular temporal arteries with diminished pulses ▶ Sudden blindness ▶ Pale, swollen optic disk surrounded by pericapillary hemorrhage ▶ Depression, weight loss, and fever
Gilbert's disease ▶ Genetic defect resulting in reduced bilirubin UDP-glucuronosyltransferase-1	Impaired hepatic bilirubin clearance results in hyperbilirubinemia.	▶ Mild jaundice without dark urine ▶ Nausea and vomiting ▶ Portal hypertension, ascites, and skin or endocrine changes ▶ Abdominal pain, anorexia, and malaise
Globoid cell leukodystrophy ▶ Galactosylceramidase deficiency	Deficiency leads to rapid cerebral demyelination with large globoid bodies in the white matter and CNS.	▶ Irritability, rigidity, blindness, tonic-clonic seizures, deafness, and mental deterioration
Goodpasture's syndrome ▶ Cause unknown; associated with exposure to hydrocarbons or type 2 influenza	Abnormal production of autoantibodies directed against alveolar and glomerular basement membranes leads to immune-mediated inflammation of lung and kidney tissues.	▶ Bloody sputum ▶ Anemia ▶ Cough ▶ Dyspnea ▶ Hematuria ▶ Peripheral edema ▶ Elevated serum creatinine and protein levels
Hand, foot, and mouth disease ▶ Common disease in infants and children due to Coxsackie A 16	RNA virus produces fever and vesicles in the oropharynx and on the hands and feet.	▶ Painful vesicular lesions on the mouth, tongue, hands, and feet

DISEASE AND CAUSES	PATHOPHYSIOLOGY	SIGNS AND SYMPTOMS
Hermansky-Pudlak syndrome ▶ Genetic, autosomal recessive disorder	A lipid-like material accumulates in the cells of the reticuloendothelial system, resulting in platelet dysfunction and pulmonary fibrosis.	▶ Low visual acuity ▶ Bruising and prolonged bleeding ▶ Lung fibrosis ▶ Inflammatory bowel disease ▶ Reduced kidney function ▶ Nystagmus
Herpangina ▶ Acute infection due to group A Coxsackie virus transmitted by fecal-oral route	RNA virus produces fever and vesicles in the posterior portion of the oropharynx.	▶ Sore throat ▶ Pain with swallowing ▶ Fever (100° to 104° F) lasting for 1 to 4 days ▶ Febrile seizures ▶ Anorexia ▶ Vomiting ▶ Malaise ▶ Diarrhea ▶ Gray-white papulovesicles on soft palate
Hiatal hernia ▶ Diaphragmatic malformation or weakening	Weakening of anchors from the gastroesophageal junction to the diaphragm or increased abdominal pressure allow herniation of part of the stomach through the esophageal hiatus in the diaphragm.	▶ Reflux of gastric contents ▶ Dysphagia ▶ Chest pain
Hirsutism ▶ Androgen excess due to hereditary, endocrine (such as Cushing's or acromegaly) causes, and pharmacologic adverse effects	Minoxidil, androgenic steroids, or testosterone ingestion can cause signs of: ▶ masculinization ▶ pituitary dysfunction – acromegaly, precocious puberty ▶ adrenal dysfunction – Cushing's syndrome.	▶ Excessive hair growth in women or children, typically in an adult male distribution pattern
Hyperbilirubinemia ▶ Rh or ABO mother/fetal incompatibility or intrauterine viral infection	Massive destruction of RBCs causes high levels of bilirubin in the blood. Bilirubin is derived from hemoglobin in RBCs.	▶ Elevated levels of serum bilirubin ▶ Jaundice

DISEASE AND CAUSES	PATHOPHYSIOLOGY	SIGNS AND SYMPTOMS

Hypersplenism
▶ Increased activity of the spleen, where all types of blood cells are removed from circulation due to chronic myelogenous leukemia, lymphomas, Gaucher's disease, hairy cell leukemia, and sarcoidosis

Spleen growth may be stimulated by an increase in its workload, such as the trapping and destroying of abnormal RBCs.

▶ Enlarged spleen
▶ Cytopenia

Idiopathic pulmonary fibrosis
▶ Chronic progressive lung disease associated with inflammation and fibrosis.
▶ Cause unknown

Interstitial inflammation consists of an alveolar septal infiltrate of lymphocytes, plasma cells, and histiocytes. Fibrotic areas are composed of dense acellular collagen. Areas of honeycombing that form are composed of cystic fibrotic air spaces, frequently lined with bronchiolar epithelium and filled with mucus. Smooth muscle hyperplasia may occur in areas of fibrosis and honeycombing.

▶ Dyspnea
▶ Nonproductive cough
▶ Chest heaviness
▶ Wheezing
▶ Anorexia
▶ Weight loss

Irritable bowel syndrome
▶ Disordered motor and sensory function of GI tract with unknown cause, although environmental, infectious, genetic, autoimmune, and host factors are suspected

Altered neural stimulation results in nonpropulsive segmentation waves, leading to constipation.

▶ Abdominal pain relieved with defecation
▶ Altered stool frequency
▶ Altered stool form
▶ Altered stool passage
▶ Passage of mucus
▶ Bloating and abdominal distention

Kaposi's sarcoma
▶ AIDS-related cancer

A malignant cancer arising from vascular endothelial cells, Kaposi's sarcoma affects endothelial tissue, which compromises all blood vessels.

▶ Red-purple circular lesions, slightly raised on the face, arms, neck, and legs
▶ Internal lesions, especially in GI tract, identified by biopsy

Keratitis
▶ Inflammation of cornea due to microorganisms, trauma, or autoimmune disorders

Bacterial infection leads to ulceration of the cornea.

▶ Decreased visual acuity
▶ Pain
▶ Photophobia

DISEASE AND CAUSES	PATHOPHYSIOLOGY	SIGNS AND SYMPTOMS
Kidney cancer ▶ Renal cell carcinoma associated with obesity and cigarette smoking	Tumors of various cell types and patterns that are usually aggressive in growth and affect younger patients.	▶ Hematuria ▶ Flank pain ▶ Palpable mass ▶ Weight loss ▶ Anemia ▶ Fever ▶ Hypertension ▶ Increased erythrocyte sedimentation rate
Klinefelter syndrome ▶ Genetic abnormality, affecting males resulting from an extra X chromosome	Seminiferous tubule dysgenesis.	▶ At puberty, testicles fail to mature ▶ Degenerative testicular changes ▶ Infertility
Kyphosis ▶ An excessive curvature of the spine with convexity backward due to a congenital anomaly, tuberculosis, syphilis, malignant or compression fracture, arthritis, or rickets	Pathophysiology is related to causative factor.	▶ Abnormally rounded thoracic curve
Lassa fever ▶ Viral infection due to *Lassa* species	An epidemic hemorrhagic fever transmitted to humans by contact with infected rodent urine, feces, or saliva.	▶ Fever lasts 2 to 3 weeks ▶ Exudative pharyngitis ▶ Oral ulcers ▶ Lymphadenopathy ▶ Swelling of face and neck ▶ Purpura ▶ Conjunctivitis ▶ Bradycardia ▶ Shock ▶ Peripheral collapse
Latex allergy ▶ Hypersensitivity to products containing natural latex	Latex protein allergens trigger release of histamine and other mediators of the systemic allergic cascade in sensitized persons.	▶ Local dermatitis to anaphylactic reaction
Legg-Calve-Perthes disease ▶ Caused by congenital or developmental factors	Ischemic necrosis leads to eventual flattening of the head of the femur due to vascular interruption.	▶ Pain in hip and proximal thigh ▶ Loss of internal rotation, extension, adduction, and flexion

DISEASE AND CAUSES	PATHOPHYSIOLOGY	SIGNS AND SYMPTOMS
Legionnaires' disease ▶ Infection due to gram-negative bacillus, *Legionella pneumophila*	Transmission of disease occurs with inhalation of organism carried in aerosols produced by air-conditioning units, water faucets, shower heads, humidifiers, and contaminated respiratory equipment.	▶ Dry cough ▶ Myalgia ▶ GI distress ▶ Pneumonia ▶ Cardiovascular collapse
Leprosy ▶ Infection due to *Mycobacterium leprae*	Chronic, systemic infection with progressive cutaneous lesions, attacking the peripheral nervous system.	▶ Skin lesions ▶ Anesthesia ▶ Muscle weakness ▶ Paralysis
Lichen planus ▶ Cause unknown	Benign pruritic skin eruption that usually resolves in 2 to 3 years.	▶ Itching ▶ Scaly, purple papules with white lines or spots on skin or mouth
Listeriosis ▶ Infection due to gram-positive bacillus *Listeria monocytogenes*	Infection with febrile illness, transmitted by neonates in utero, by inhalation, consumption, or contact with contaminated unpasteurized milk, infected animals, or contaminated sewage, mud, or soil.	▶ Fever ▶ Meningitis ▶ Spontaneous abortion
Mastocytosis ▶ Cause unknown	Proliferation of mast cells systemically and within skin.	▶ Urticaria pigmentosa ▶ Diarrhea ▶ Abdominal pain ▶ Flushing ▶ Vascular collapse ▶ Headaches ▶ Ascites
Mastoiditis ▶ Bacterial infection of the mastoid antrum.	An inflammation and infection of the air cells of the mastoid antrum.	▶ Dull ache and tenderness around the mastoid process ▶ Low grade fever ▶ Headache ▶ Thick purulent drainage ▶ Meningitis ▶ Facial paralysis ▶ Brain abscess ▶ Labyrinthitis

DISEASE AND CAUSES	PATHOPHYSIOLOGY	SIGNS AND SYMPTOMS
Medullary sponge kidney ▶ Genetic disorder	Collecting ducts in the renal pyramids dilate, forming cavities, clefts, and cysts that produce complications of calcium oxylate stones and infections.	▶ Kidney stones ▶ Hematuria ▶ Infection (fever, chills, and malaise)
Melasma ▶ Hypermelanotic skin disorder associated with increased hormonal levels with pregnancy, oral contraceptive use, and ovarian cancer	Thought to be due to the effects of estrogen and progesterone on melanin production.	▶ Patchy, nonraised, hypermelanotic rash
Mycosis fungoides ▶ Cause unknown	Rare, chronic malignant T-cell lymphoma originating in the reticuloendothelial system of the skin, and affecting lymph nodes and internal organs.	▶ Urticarial, erythematous, or eczematous patches with irregular-shaped, well-defined borders; patches may become hard and form ulcers, then eventually drain a thin pus and serum
Myelodysplastic syndromes ▶ Caused by genetic factors and exposure to chemicals or radiation	Preleukemic disorders that progress to leukemia	▶ Pancytopenia ▶ Weakness ▶ Fatigue ▶ Palpitations ▶ Dizziness ▶ Irritability ▶ Anemia
Myocarditis ▶ Inflammation of the myocardium due to bacterial, fungal, viral, or protozoal infections; heat stroke; ionizing radiation; rheumatic fever; and diphtheria	Initial infection triggers an autoimmune, cellular, and possibly humoral response resulting in myocardial inflammation and necrosis.	▶ Rapid, irregular, and weak pulse ▶ Chest tenderness ▶ First heart sound resembles second heart sound ▶ Fatigue
Narcolepsy ▶ Familial disorder	Chronic disorder of recurrent attacks of drowsiness and sleep during the daytime.	▶ Drowsiness ▶ Daytime sleepiness ▶ Sudden loss of muscle tone ▶ Extreme strong emotions

DISEASE AND CAUSES	PATHOPHYSIOLOGY	SIGNS AND SYMPTOMS
Neurofibromatosis ▶ Inherited disorder	Group of developmental disorders of the nervous system, muscles, bones, and skin that affects the cell growth of neural tissue.	▶ Café-au-lait spots ▶ Multiple, pediculated, soft tumors ▶ Hearing loss
Orbital cellulitis ▶ Bacterial infection typically due to streptococcal, staphylococcal, and pneumococcal organisms	Inflammation and infection of the fatty orbital tissues and eyelids.	▶ Unilateral eyelid edema ▶ Hyperemia ▶ Redden eyelids ▶ Matted lashes
Osgood-Schlatter disease ▶ Cause unknown	Osteochondrosis of the tibia, disease of the growth or ossification centers in children.	▶ Frequent fractures ▶ Pain at inferior aspect of patella
Osteogenesis imperfecta ▶ Genetic disorder	Disease of the bones and connective tissue that causes varying degrees of skeletal fragility; may be fatal if presents within first few days of life.	▶ Recurring fractures ▶ Thin skin ▶ Blue sclerae ▶ Poor teeth ▶ Hypermobility of the joints ▶ Progressive deafness
Paraphilia ▶ Psychosexual disorder	Dependence on unusual behaviors or fantasies to achieve sexual excitement.	▶ Exhibitionism ▶ Fetishism ▶ Frotteurism (rubbing against another person without his consent) ▶ Pedophilia ▶ Sexual masochism ▶ Sexual sadism ▶ Transvestic fetishism ▶ Voyeurism
Pediculosis ▶ Infestation by the lice parasite	Ectoparasite attaches itself to the hair shaft with claws and louse feeds on blood several times daily; resides close to the scalp to maintain its body temperature. Itching may be due to an allergic reaction to louse saliva or irritability.	▶ Itching ▶ Eczematous dermatitis ▶ Inflammation ▶ Tiredness ▶ Irritability ▶ Weakness ▶ Lice present in hair (head, axilla, and pubic)

DISEASE AND CAUSES	PATHOPHYSIOLOGY	SIGNS AND SYMPTOMS
Penile cancer ▶ Preceded by chronic irritation, condylomata acuminata, uncircumcised	Neoplasms may be benign or malignant; latter are usually squamous cell carcinomas.	▶ Painless ulcerations on the glans or foreskin ▶ Discharge ▶ Small, warty plaque
Pheochromocytoma ▶ Polyglandular multiple endocrine neoplasia	Tumor of the chromaffin cells of the adrenal medulla that causes an increased production of catecholamines.	▶ Hypertension ▶ High blood sugar ▶ High lipid levels ▶ Headache ▶ Palpitations ▶ Sweating ▶ Dizziness ▶ Constipation ▶ Anxiety
Pituitary tumor ▶ Cause unknown	Tumors are usually macroadenomas with self-secreting thyroid-stimulating hormone.	▶ Signs of hyperthyroidism without skin and eye manifestations ▶ Goiters ▶ High free thyroxine levels
Plague ▶ Infection due to *Yersinia pestis*	Acute infection transmitted by a flea bite from an infected rodent.	▶ Painful, inflamed buboes ▶ Hemorrhagic or necrotic areas ▶ Sudden high fever ▶ Myalgia ▶ Delirium ▶ Prostration ▶ Restlessness ▶ Toxemia ▶ Staggering gait
Pleurisy ▶ Several causes including lupus, rheumatoid arthritis, and tuberculosis	Inflammation of the pleura with exudation into the cavity and lung surface.	▶ Chilliness ▶ Stabbing chest pain ▶ Fever ▶ Suppressed cough ▶ Pallor ▶ Dyspnea

DISEASE AND CAUSES	PATHOPHYSIOLOGY	SIGNS AND SYMPTOMS
Pneumoconioses ▶ Inhalation of dust particles, usually in an occupational setting	Chronic and permanent disposition of particles in the lungs causes a tissue reaction, which may be harmless or destructive.	▶ Critical exposure ▶ Emphysema ▶ Shortness of breath ▶ Cough ▶ Fatigue ▶ Weakness ▶ Weight loss
Polycythemia vera ▶ Cause unknown; possibly due to a multipotential stem cell defect	Increased production of RBCs, neutrophils, and platelets inhibits blood flow to microcirculation, resulting in intravascular thrombosis.	Usually no symptoms in early stages. In later stages, related to expanded blood volume and system affected: ▶ Weakness, headache, light-headedness, visual disturbances, and fatigue ▶ Hepatomegaly and splenomegaly
Polymyositis ▶ Cause unknown; may be due to viral or autoimmune reaction	Damage of skeletal muscle by an inflammatory process dominated by lymphocytic infiltration leads to progressive muscle weakness.	▶ Proximal muscle weakness, dysphonia, dysphagia, and regurgitation ▶ Polyarthralgias, joint effusions, and Raynaud's phenomenon ▶ Rash associated with muscular pain, tenderness, and induration
Porphyrias ▶ Due to inherited or acquired disorders of specific enzymes in heme biosynthetic pathway	Biosynthesis of heme is affected by metabolic disorders that causes excessive production and excretion of porphyrins or their precursors.	▶ Nonspecific symptoms, generally abdominal pain, neurologic symptoms, tachycardia, hypertension, muscle weakness, and skin lesions
Postherpetic neuralgia ▶ Complication of the chronic phase of herpes zoster	Varicella virus in ganglia of the posterior nerve roots reactivates, multiplies, and spreads down the sensory nerves to the skin.	▶ Intractable neurologic pain lasting over 6 weeks after disappearance of herpes zoster rash
Proctitis ▶ Contributing factors that allow the normal mucosa to break down include trauma, infection, allergies, radiation, stress, and sexually transmitted diseases	Acute or chronic inflammation of the rectal mucosa.	▶ Mild rectal pain, mucous discharge, bleeding, feeling of rectal fullness, and tenesmus

DISEASE AND CAUSES	PATHOPHYSIOLOGY	SIGNS AND SYMPTOMS
Pseudogout ▶ Cause unknown; associated with conditions that cause degenerative or metabolic changes in cartilage	Calcium pyrophosphate crystals deposit in periarticular joint structures.	▶ Sudden joint pain and swelling in larger peripheral joints; mimics other form of arthritis
Ptosis ▶ Due to congenital (autosomal dominant trait or anomaly) or acquired (age, mechanical, myogenic, neurogenic, or nutritional) factors	Stretching of eyelid skin or aponeurotic tendon causes upper eyelid to droop. Lesion affects innervation of either of two muscles that open the eyelid.	▶ Drooping of upper eyelid
Pyloric stenosis ▶ Congenital; cause unknown	Pyloric sphincter muscle fibers thicken and become inelastic, leading to a narrowed opening. The extra peristaltic effort that is necessary leads to hypertrophied muscle layers of the stomach.	▶ Progressive nonbilious vomiting, leading to projectile vomiting at ages 2 to 4 weeks
Rectal prolapse ▶ Protrusion of one or more layers of mucous membrane through the anus due to conditions that affect the pelvic floor or rectum	Increased intra-abdominal pressure triggers the circumferential protrusion of one or more layers of the mucous membrane.	▶ Lower abdominal pain due to ulceration, bloody diarrhea, or tissue protruding from rectum during defecation or walking
Reiter's syndrome ▶ Cause unknown; mostly follows a period of diarrhea or sexual contact ▶ Genetic factor (HLA-B27) increases risk of acquiring disorder	An infection (e.g., *Chlamydia*) is thought to initiate an aberrant and hyperactive immune response that produces inflammation in involved target organs.	*Urogenital tract:* ▶ Burning sensation with urination, penile discharge, and prostatitis in men ▶ Cervicitis, urethritis, and vulvovaginitis in women *Joint symptoms or arthritis:* ▶ Affects knees, ankles and feet ▶ Inflammation where tendon attaches to bone *Eye involvement:* ▶ Conjunctivitis and uveitis

DISEASE AND CAUSES	PATHOPHYSIOLOGY	SIGNS AND SYMPTOMS

Renal tubular acidosis
▶ *Distal (type I):* familial with another genetic disease or an isolated autosomal dominant disease
▶ *Proximal (type II):* accompanies several inherited diseases, multiple myeloma, vitamin D deficiency, chronic hypocalcemia, after renal transplantation, and following treatment with certain drugs

In type I, the distal tubule is unable to secrete hydrogen ions across the tubular membrane, causing decreased excretion of titratable acids and ammonium and increased loss of potassium and bicarbonate. Prolonged acidosis leads to hypercalciuria and renal calculi.
In type II, defective reabsorption of bicarbonate in proximal tubule causes bicarbonate to flood the distal tubule, leading to impaired formation of titratable acids and ammonium for excretion, resulting in metabolic acidosis.

In infants:
▶ Vomiting, fever, constipation, anorexia, weakness, polyuria, growth retardation, nephrocalcinosis, and rickets
In children and adults:
▶ Growth problems, urinary tract infections, and rickets

Retinal detachment
▶ Caused by trauma, after cataract surgery, severe uveitis, and primary or metastatic choroidal tumors

The neural retina separates from the underlying retinal pigment epithelium.

▶ Floaters, flashing lights, scotoma in peripheral visual field (painless) and, eventually, a curtain or veil occurs in the field of vision

Retinitis pigmentosa
▶ Autosomal recessive disorder in 80% of affected children
▶ Less commonly transmitted as an X-linked trait

Slow, degenerative changes in the rods cause the retina and pigment epithelium to atrophy. Irregular black deposits of clumped pigment are in equatorial region of retina and eventually in the macular and peripheral areas.

▶ Progressive night blindness, visual field constriction with ring scotoma, and loss of acuity progressing to blindness

Reye's syndrome
▶ Cause unknown
▶ Viral agents and drugs (especially salicylates) have been implicated

Mitochondrial dysfunction and fatty vacuolization of the liver and renal tubules lead to hepatic injury and CNS damage.

▶ Vomiting
▶ Change in mental status progressing from lethargy to disorientation to coma

Rocky Mountain spotted fever
▶ Infection due to *Rickettsia rickettsii* carried by several tick species

R. rickettsii multiplies within endothelial cells and spreads via the bloodstream. Focal areas of infiltration lead to thrombosis and leakage of RBCs into surrounding tissue.

▶ Fever, headache, mental confusion, and myalgia
▶ Rash develops as small macules progress to maculopapules and petechiae. Initially, rash starts on wrists and ankles and spreads to trunk. Especially diagnostic is rash noted on palms and soles.

DISEASE AND CAUSES	PATHOPHYSIOLOGY	SIGNS AND SYMPTOMS
Rosacea ▶ Cause unknown	Small blood vessels of the face, usually the nose and cheeks, become flushed and dilated; also associated with papules and pustules.	▶ Pronounced flushing of nose and cheeks ▶ Papules, pustules, telangiectases can be superimposed
Sarcoidosis ▶ Cause unknown ▶ Evidence suggests that disease is result of exaggerated cellular immune response to limited class of antigens	Organ dysfunction results from an accumulation of T lymphocytes, mononuclear phagocytes, and nonsecreting epithelial granulomas, which distort normal tissue architecture.	▶ Mainly generalized, most commonly involving the lung with resulting respiratory symptoms ▶ Fever, fatigue, and malaise
Scabies ▶ Human itch mite (*Sarcoptes scabiei* var. *hominis*)	Mite burrows superficially beneath stratum corneum depositing eggs that hatch, mature, and reinvade the skin.	Occur from sensitization reaction against excreta that mites deposit ▶ Intense itching, worsens at night; threadlike lesions on wrists, between fingers, and on elbows, axillae, belt line, buttocks, and male genitalia
Schistosomiasis ▶ Blood flukes of the class Trematoda; *Schistosoma mansoni* and *S. japonicum* infect intestinal tract whereas *S. haematobium* infects the urinary tract	Infection follows contact by bathing with free swimming cercariae of the parasite, which penetrate the skin, migrate to intrahepatic portal circulation, and mature. Adult worms then lodge in venules of bladder or intestines.	Depends on infection site and stage of disease. *In initial stage:* ▶ Pruritic papular dermatitis at penetration site; fever; cough *In later stage:* ▶ Hepatosplenomegaly and lymphadenopathy ▶ May cause seizures and skin abscesses
Shigellosis ▶ Acute infectious inflammatory colitis due to *Shigella* organisms	*Shigella* is orally ingested or transmitted by fecal-oral route. Invasion of colonic epithelial cells and cell-to-cell spread of infection results in characteristic mucosal ulcerations.	*In children:* ▶ Fever, watery diarrhea, nausea, vomiting, irritability, and abdominal pain and distention *In adults:* ▶ Intermittent severe abdominal pain, tenesmus and, in severe cases, headache and prostration; fever is rare ▶ Stools may contain pus, blood, or mucous

DISEASE AND CAUSES	PATHOPHYSIOLOGY	SIGNS AND SYMPTOMS
Silicosis ▶ Exposure to high concentrations of respirable silica dust	Alveolar macrophages engulf respirable particles of free silica, which causes cytotoxic enzymes to be released. This attracts other macrophages into the area and produces fibrous tissue in the lung parenchyma. *Note:* Silicosis is associated with a high incidence of active tuberculosis.	*In simple nodular silicosis:* ▶ Cough and raise sputum, usually no symptoms *In conglomerate silicosis:* ▶ Severe shortness of breath, cough, and sputum; may lead to pulmonary hypertension and cor pulmonale
Sjogren's syndrome ▶ Autoimmune rheumatic disorder with unknown cause; genetic and environmental factors may be involved	Lymphocytic infiltration of exocrine glands causes tissue damage that results in xerostomia and dry eyes.	*In xerostomia:* ▶ Dry mouth, difficulty swallowing and speaking, ulcers of tongue, buccal mucosa and lips, and severe dental caries *In ocular involvement:* ▶ Dry eyes; gritty, sandy feeling; decreased tearing; burning, itching, redness, and photosensitivity *Extraglandular:* ▶ Arthralgias, Raynaud's phenomenon, lymphadenopathy, and lung involvement
Sleep apnea ▶ Caused by occlusion of airway (obstructive), absence of respiratory effort (central) or both	In obstructive sleep apnea, airflow ceases due to upper airway narrowing and glottal obstruction as a result of obesity or congenital abnormalities of the upper airway. When primary brain stem medullary failure occurs, patient may breathe insufficiently or not at all while asleep.	*In obstructive sleep apnea:* ▶ Snoring, excessive daytime sleepiness, intellectual impairment, memory loss, and cardiorespiratory symptoms *In central sleep apnea:* ▶ Sleeping poorly, morning headache, and daytime fatigue
Spasmodic torticollis ▶ Cause unknown, believed to be a form of focal dystonia	Rhythmic muscle spasms of sternocleidomastoid neck muscles possibly due to irritation of the nerve root.	▶ Unilateral, intermittent, or continuous painful spasms of the neck muscles

DISEASE AND CAUSES	PATHOPHYSIOLOGY	SIGNS AND SYMPTOMS

Spinal ischemia/infarction

▶ Caused by direct vascular compression (tumors and acute disc compression) or by remote occlusion (aortic surgery and dissecting aneurysm)

Major arterial branches that supply the spinal cord can become compressed or occluded, decreasing blood flow to the spinal cord, causing ischemia of cord and resulting in motor and sensory deficiencies.

▶ Sudden back pain and pain in distribution of affected segment followed by bilateral flaccid weakness and dissociated sensory loss below level of infarct

Sporotrichosis

▶ Fungal infection due to *Sporothrix schenckii,* which occurs in soil, wood, sphagnum moss, and decaying vegetation

Inflammatory response includes both the clustering of neutrophils and a marked granulomatous response, with epithelioid cells and giant cells producing nodular erythematous primary lesions and secondary lesions along lymphatic channels in cutaneous lymphatic type. In pulmonary sporotrichosis, inflammatory response produces pulmonary lesions and nodules. In disseminated sporotrichosis, multifocal lesions spread from skin or lungs.

In cutaneous or lymphatic sporotrichosis:
▶ Subcutaneous, movable, painless nodule on hands or fingers that grows progressively larger, discolors, and eventually ulcerates; additional lesions form on the adjacent lymph node chain
In pulmonary sporotrichosis:
▶ Productive cough, lung cavities and nodules, pleural effusion, fibrosis, and formation of fungus ball
In disseminated sporotrichosis:
▶ Weight loss, anorexia, synovial or bony lesions and, possibly, arthritis or osteomyelitis

Spurious polycythemia

▶ Due to conditions that promote dehydration (persistent vomiting or diarrhea, burns, and renal disease), hemoconcentration due to stress, or low normal plasma volume with high normal RBC mass

Concentration of RBCs in the circulating blood increases due to loss of blood plasma.

No specific symptoms; following may present:
▶ Headache, dizziness, and fatigue
▶ Diaphoresis, dyspnea, and claudication
▶ Ruddy complexion and short neck
▶ Slight hypertension

Stomatitis

▶ Acute herpetic due to herpes simplex virus
▶ Aphthous cause is unknown; predisposing factors include stress, fatigue, fever, trauma, and solar overexposure

Inflammation of the cells of the oral mucosa, buccal mucosa, lips and palate with resulting ulcers.

▶ Papulovesicular ulcers in mouth and throat, mouth pain, malaise, anorexia, and swelling of mucous membranes

DISEASE AND CAUSES	PATHOPHYSIOLOGY	SIGNS AND SYMPTOMS
Strabismus ▶ Eye malalignment that is frequently inherited; controversy exists whether amblyopia is caused by or results from strabismus	In paralytic (nonconcomitant) strabismus, paralysis of one or more ocular muscles may be due to oculomotor nerve lesion. In nonparalytic (concomitant) strabismus, unequal ocular muscle tone is due to supranuclear abnormality within the CNS.	▶ Noticeable eye malalignment by external eye examination, ophthalmoscopic observation of the corneal light reflex in center of pupils, diplopia, and other visual disturbances
Thrombocythemia ▶ *Primary:* cause unknown ▶ *Secondary:* due to chronic inflammatory disorders, iron deficiency, acute infection, neoplasm, hemorrhage, or postsplenectomy	A clonal abnormality of a multipotent hematopoietic stem cell results in increased platelet production, although platelet survival is usually normal. If combined with degenerative vascular disease, may lead to serious bleeding or thrombosis.	▶ Weakness, hemorrhage, nonspecific headache, paresthesia, dizziness, and easy bruising
Thrombophlebitis ▶ Caused by endothelial damage, accelerated blood clotting, and reduced blood flow	Alteration in epithelial lining causes platelet aggregation and fibrin entrapment of RBCs, WBCs, and additional platelets; the thrombus initiates a chemical inflammatory process in the vessel epithelium that leads to fibrosis, which may occlude the vessel lumen or may embolize.	▶ Varies with site and length of affected vein ▶ Affected area usually extremely tender, swollen, and red
Tinea versicolor ▶ Caused by *Pityrosporum orbiculare* (*Melassezia furfur*), which occurs normally in human skin ▶ Unclear whether disorder is due to infectious cause or a proliferation of normal skin fungi	Nondermatophyte dimorphic fungus converts to the hyphal form and causes characteristic lesions. Invasion of the stratum corneum by the yeast produces C9and C11 dicarboxylic acids that inhibit tyrosinase in vitro.	▶ Asymptomatic, well-delineated, hyperpigmented or hypopigmented macules occur on upper trunk and arms

DISEASE AND CAUSES	PATHOPHYSIOLOGY	SIGNS AND SYMPTOMS

Torticollis

▶ *Congenital:* include malposition of head in utero, prenatal injury, fibroma, and interruption of blood supply
▶ *Acquired/acute:* inflammatory diseases and cervical spinal lesions that produce scar tissue
▶ *Hysterical:* psychogenic inability to control neck muscles
▶ *Spasmodic:* organic CNS disorder

Contraction of the sternocleidomastoid neck muscles produces twisting of the neck and unnatural position of the head.

Congenital:
▶ Firm, nontender, palpable enlargement of the sternocleidomastoid muscle visible at birth
Acquired:
▶ Recurring unilateral stiffness of neck muscles
▶ Drawing sensation that pulls head to affected side
▶ Severe neuralgic pain of head and neck

Tourette's syndrome

▶ Autosomal dominant multipletic disorder

Obscure pathology; dopaminergic excess has been suggested because tics may respond to treatment with dopamine-blocking drugs.

▶ Single or multiple motor tics that commonly affect the face, and phonic tics

Trachoma

▶ Infection due to *Chlamydia trachomatis*

Chronic conjunctivitis due to *C. trachomatis* that leads to inflammatory leukocytic infiltration and superficial vascularization of the cornea, conjunctival scarring, and distortion of the eyelids. This cause lashes to abrade the cornea, which in turn progresses to corneal ulceration, scarring, and blindness.

▶ Mild infection resembling bacterial conjunctivitis; red and edematous eyelids, pain, photophobia, tearing, and exudation

Trichomoniasis

▶ Infection of the genitourinary tract due to *Trichomonas vaginalis*

T. vaginalis infects the vagina, urethra and, possibly, the endocervix, bladder, or Bartholin's or Skene's glands. In males, it infects the lower urethra and possibly the prostate gland, seminal vesicles, and epididymis.

In females:
▶ Malodorous, greenish-yellow vaginal discharge; irritation of vulva, perineum, and thighs; dyspareunia; and dysuria
In males:
▶ Generally asymptomatic; some transient frothy or purulent urethral discharge with dysuria and frequency

DISEASE AND CAUSES	PATHOPHYSIOLOGY	SIGNS AND SYMPTOMS
Trigeminal neuralgia ▶ Cause unknown, possibly a compression neuropathy ▶ At surgery or autopsy, the intracranial arterial and venous loops are found to compress the trigeminal nerve root at the brain stem	Painful disorder along the distribution of one or more of the trigeminal nerve's sensory divisions, most often the maxillary.	▶ Searing or burning pain lasting seconds to 2 minutes at the trigeminal nerve distribution ▶ Touching a trigger point often elicits pain
Trisomy 13 syndrome ▶ Chromosomal disorder due to an extra chromosome 13 that is usually maternally derived; risk increases with advanced maternal age	Genes and chromatin carried on chromosome 13 occur in triple dose rather that the normal double dose, resulting in many developmental abnormalities.	▶ Microcephaly with holoprosencephaly ▶ Microphthalmia and orbital defects ▶ Cleft lip and palate ▶ Congenital heart defects ▶ Flat, broad nose, low set ears, and inner ear abnormalities ▶ Polydactyly of hands and feet; club feet ▶ Cystic kidneys, hydronephrosis, cystic hygroma, omphaloceles, and genital abnormalities
Uterine leiomyomas ▶ Benign tumors with unknown cause ▶ Linked to steroid hormones and growth factors as regulators of growth	Smooth muscle tumors of the uterus (submucosal, intramural, subserosal) that are also found in the broad ligaments and, rarely, in the cervix.	▶ Usually asymptomatic ▶ Abnormal bleeding (common) ▶ Pain or pressure, urinary or bowel complaints
Uveitis ▶ Cause unknown but associated with many autoimmune and infectious diseases	An inflammation of any part of the uveal tract. Inflammatory cells floating in aqueous humor or deposited on corneal endothelium affect the uveal tract.	*Anterior:* ▶ Pain, redness, photophobia, and decreased vision *Intermediate:* ▶ Painless, presents with floaters and decreased vision *Posterior:* ▶ Diverse symptoms, most commonly floaters and decreased vision
Vaginal cancer ▶ Cause unknown; tumor development has been linked to intrauterine exposure to diethylstilbestrol and to human papilloma virus	Presents mainly as squamous cell carcinoma (sometimes as melanoma, sarcoma, and adenocarcinoma) and progresses from an intraepithelial tumor to an invasive cancer.	▶ Abnormal bleeding and discharge ▶ Firm, ulcerated lesion in the vagina

DISEASE AND CAUSES	PATHOPHYSIOLOGY	SIGNS AND SYMPTOMS
Vaginismus ▶ Cause related to physical (hymenal abnormalities, genital herpes, obstetric trauma, or atrophic vaginitis) or psychological (conditioned response to traumatic sexual experience) factors	An involuntary spastic constriction of the lower vaginal muscles.	▶ Muscle spasm with pain when an object is inserted into the vagina ▶ Lack of sexual interest or desire
Variola ▶ Infection due to *Poxvirus variola*	Virus is transmitted by respiratory droplets or direct contact. In the body, the virus replicates and causes viremia to develop.	▶ Fever, vomiting, sore throat, CNS symptoms (such as headache, malaise, stupor, and coma), macular rash progressing to vesicular, and pustular lesions
Velopharyngeal insufficiency ▶ Caused by an inherited palate abnormality or acquired from pharyngeal surgery or palatal paresis	Impaired closure of the velopharyngeal sphincter between the oropharynx and the nasopharynx results in abnormal speech and nasal emission of air.	▶ Unintelligible speech, hypernasality, nasal emission, poor consonant definition, and weak voice ▶ Dysphagia and, if severe, regurgitation through the nose
Vitiligo ▶ Cause unknown; usually acquired but may be familial (autosomal dominant) ▶ Possible immunologic and neurochemical basis suggested	Destruction of melanocytes (humoral or cellular) and circulating antibodies against melanocytes results in hypopigmented areas.	▶ Progressive, symmetric areas of complete pigment loss with sharp borders, generally appearing in periorifical areas, flexor wrists, and extensor distal extremities
Vulvovaginitis ▶ Caused by bacterial or viral infection, vaginal atrophy, or various traumas or irritations	Infectious diseases and other conditions cause an inflammatory reaction of the vaginal mucosa and vulva.	▶ Vaginal discharge (most common) ▶ Appearance of discharge (consistency, odor, and color) varies with causative agent ▶ May be accompanied by vulvar irritation, pain, or pruritus
Wegener's granulomatosis ▶ Cause unknown; resembles infectious process but no causative agent found ▶ Histologic changes suggest hypersensitivity as basis of disorder	Localized granulomatous inflammation of upper or lower respiratory tract mucosa may progress to generalized necrotizing granulomatous vasculitis and glomerulonephritis.	▶ Fever, malaise, anorexia, and weight loss ▶ Upper respiratory tract complaints including nosebleeds, sinusitis, nasal ulcerations, cough, hemoptysis, and pleuritis ▶ Necrotizing granulomatous skin lesions ▶ Pulmonary infiltrates with cavitation ▶ Glomerulonephritis with hypertension and uremia

DISEASE AND CAUSES	PATHOPHYSIOLOGY	SIGNS AND SYMPTOMS

Wilson's disease
▶ Inherited copper toxicosis

Defective mobilization of copper from hepatocellular lysosomes for excretion via the bile allows excessive copper retention in the liver, brain, kidneys, and corneas, leading to tissue necrosis and subsequent hepatic and neurologic disorders.

Kayser-Fleischer ring:
▶ Rusty brown ring of pigment at periphery of corneas
▶ Signs of hepatitis leading to cirrhosis
▶ Tremors, unsteady gait, muscular rigidity, inappropriate behavior, and psychosis
▶ Hematuria, proteinuria, and uricosuria

Wiskott-Aldrich syndrome
▶ X-linked recessive immunodeficiency disorder
▶ Defective B-cell and T-cell functions

Deficiency in both B-cell and T-cell function allows for susceptibility to infection. Metabolic defect in platelet synthesis causes production of small, short-lived platelets resulting in thrombocytopenia.

In newborn:
▶ Hemorrhagic symptoms such as bloody stools, bleeding from circumcision site, petechiae, and purpura
In older children:
▶ Recurrent systemic infections and eczema

X-linked infantile hypogammaglobulinemia
▶ Congenital disorder

Deficiency or absence of B cells leads to defective immune response and depressed production of all five immunoglobulin types.

Recurrent infections starts at about age 6 months:
▶ Otitis media, pneumonia, dermatitis, bronchitis, meningitis, conjunctivitis, abnormal dental caries, and polyarthritis

Selected references

Alcoser, P., and Burchett, S. "Bone Marrow Transplantation: Immune System Suppression and Reconstitution," *AJN* 99(6):26-32, 1999.

American Cancer Society. *Guidelines on Diet, Nutrition, and Cancer Prevention.* www2.cancer.org/prevention/index.cfm Updated 5/99. Accessed 10/4/99.

American Cancer Society. *Recommendations for the Early Detection of Cancer. Cancer Facts and Figures 1999.* www.cancer.org/statistics/cff99/data/data_recommendDetect.html. Updated 5/99. Accessed 10/4/99.

American Heart Association. *Cardiomyopathy.* Dallas: American Heart Association, 1999.

American Heart Association. *Rheumatic Heart Disease Statistics.* Dallas: American Heart Association, 1999.

Assessment Made Incredibly Easy. Springhouse, Pa.: Springhouse Corp., 1998.

Beattie, S. "Management of Chronic Stable Angina," *Nurse Practitioner* 24(5):44-53, 1999.

Beers, M., and Berkow, R. *The Merck Manual,* 17th ed. Whitehouse Station, N.J.: Merck and Co., Inc., 1999.

Bertolet, B.D., and Brown, C.S. "Cardiac Troponin: See Ya Later, CK!" *Chest* 111(1):2, January 1997.

Bickley, L.S., and Hoekelman, R.A. *Bates' Guide to Physical Examination and History Taking,* 7th ed. Philadelphia: Lippincott, 1999.

Bone, R. *Pulmonary and Critical Care Medicine Core Updates.* St. Louis: Mosby-Year Book, Inc., 1998.

Cairns, J., et al. "Coronary Thrombolysis," *Chest* 114(5): 634-657, 1998.

Chiramannil, A. "Clinical Snapshot: Lung Cancer," *AJN* 98(4):46-47, 1998.

Coats, U. "Management of Venous Ulcers," *Critical Care Nursing Quarterly* 21(2):14, August 1998.

Coudrey, L. "The Troponins," *Archives of Internal Medicine* 158(11):1173-1180, June 1998.

Diseases, 3rd ed. Springhouse, Pa.: Springhouse Corp., 2000.

Dugan, K.J. "Caring for Patients with Pericarditis," *Nursing98* 28(3):50-51, 1998.

Fauci, A.S., et al., eds. *Harrison's Principles of Internal Medicine,* 14th ed. New York: McGraw-Hill Book Co., 1998.

Fluids and Electrolytes Made Incredibly Easy. Springhouse, Pa.: Springhouse Corp., 1997.

Fox, S.I. *Laboratory Guide Human Physiology: Concepts and Clinical Applications.* Dubuque, Iowa: Brown, William, 1999.

Halper, J., and Holland, N. "Meeting the Challenge of Multiple Sclerosis-Part 1," *AJN* 98(10): 26-32, 1998.

Halper, J., and Holland, N. "Meeting the Challenge of Multiple Sclerosis-Part 2," *AJN* 98(11): 39-45, 1998.

Halperin, M. *Fluid, Electrolyte, and Acid-Base Physiology: A Problem-Based Approach.* Philadelphia: W.B. Saunders Co., 1998.

Handbook of Geriatric Nursing Care. Springhouse, Pa.: Springhouse Corp., 1999.

Handbook of Medical-Surgical Nursing. Springhouse, Pa.: Springhouse Corp., 1998.

Hanson, M. *Pathophysiology: Foundations of Disease and Clinical Interven-*

tion. Philadelphia: W.B. Saunders Co., 1998.

Healthy People 2000 Progress Review: Cancer. Department of Health and Human Services, Public Health Service, April 7, 1998. www.odphp.osophs.dhhs.gov/pubs/hp2000. Accessed 8/16/00.

Huether, S.E., and McCance, K.L. *Understanding Pathophysiology.* St. Louis: Mosby-Year Book, 1996.

Huston, C.J. "Emergency! Cervical Spine Injury," *AJN* 98(6): 33, 1998.

Ignatavicius, D.D., et al. *Medical-Surgical Nursing: Nursing Process Approach,* 2nd ed. Philadelphia: W.B. Saunders Co., 1995.

Jastremski, C.A. "Trauma! Head Injuries," *RN* 61(12): 40-46, 1998.

Lewandowski, D.M. "Myocarditis," *AJN* 99(8):44-45, 1999.

Lewis, A.M. "Cardiovascular Emergency!" *Nursing99* 29(6):49, 1999.

Mastering Geriatric Care. Springhouse, Pa.: Springhouse Corp., 1997.

McKinney, B.C. "Solving the Puzzle of Heart Failure," *Nursing99* 29(5):33-39, 1999.

Murray, S. *Critical Care Assessment Handbook.* Philadelphia: W.B. Saunders Co., 1999.

Pathophysiology Made Incredibly Easy. Springhouse, Pa.: Springhouse Corp., 1998.

Porth, C.M. *Pathophysiology: Concepts of Altered Health States,* 5th ed. Philadelphia: Lippincott-Raven Pubs., 1998.

Price, S.A., and Wilson, L.M. *Pathophysiology: Clinical Concepts and Disease Processes,* 5th ed. St. Louis: Mosby-Year Book, Inc., 1997.

Professional Guide to Diseases, 6th ed. Springhouse, Pa.: Springhouse Corp., 1998.

Professional Guide to Signs and Symptoms, 2nd ed. Springhouse, Pa.: Springhouse Corp., 1997.

Safety and Infection Control, Springhouse, Pa., Springhouse Corp., 1998.

Wakeling, K.S. "The Latest Weapon in the War Against Cancer," *RN* 62(7):58-60, July 1999.

Sparacino, P.S.A. "Cardiac Infections: Medical and Surgical Therapies," *Journal of Cardiovascular Nursing* 13(2):49, January 1999.

Sussman, C., and Bates-Jensen, B.M. *Wound Care, A Collaborative Practical Manual for Physical Therapists and Nurses,* Gaithersburg, Md.: Aspen Pubs., Inc., 1998.

Taylor, R., et al. *Family Medicine Principles and Practice,* 5th ed. New York: Springer Publishing Co., 1998.

Woods, A.D. "Managing Hypertension," *Nursing99* 29(3):41-46, March 1999.

Index

A

ABO incompatibility as cause of erythroblastosis fetalis, 384
Absorption as cell function, 6
Absorption atelectasis, 204
Accommodation, 487-488
Achalasia, 304
Achilles tendon rupture, 370
Acid-base balance, 110
  arterial blood gas values and, 124-125t
  buffer systems and, 123-124
  compensation by kidneys and, 124
  compensation by lungs and, 124
  conditions related to problems with, 123-124
  disorders of, 125-132
Acidemia, 123
Acidosis, 123
  metabolic, 124-125t, 129-131
  respiratory, 124-125t, 125-127
Acne, 506, 510-511
Acquired immunodeficiency syndrome, 403-406
  conditions associated with, 407
  opportunistic infections in, 406t
Acromegaly, 439-440
Acute idiopathic polyneuritis. *See* Guillain-Barré syndrome.
Acute leukemia, 43t
Acute pyelonephritis, 75t
Acute renal failure, 460-464
  phases of, 462-463
  type of, 461-462
Addisonian crisis, 425. *See also* Adrenal hypofunction.
Addison's disease, 424-427

Adrenal crisis, 425. *See also* Adrenal hypofunction.
Adrenal hypofunction, 424-427
Adult chorea, 281-282
Adult respiratory distress syndrome, 206-211
  progression of, 209i
Aging
  biological theories of, 12-13t
  of cells, 11
Agnosia, 246
AIDS. *See* Acquired immunodeficiency syndrome.
Air pollution as cancer risk factor, 18
Akathisia, 250-251t
Akinesia, 249
Alcohol
  as cancer risk factor, 18
  as teratogen, 90t
Alkalemia, 123
Alkalosis, 123
  metabolic, 124-125t, 131-132
  respiratory, 124-125t, 127-129
Allergic dermatitis, 511-512
Alloimmune reactions, 403
Alpha-fetoprotein as tumor cell marker, 32t
Alzheimer's disease, 257, 260-261
Amblyopia, 487
Amenorrhea, 545-547
Amyotrophic lateral sclerosis, 261-263
Anaphylactic shock, 187. *See also* Shock.
Anaphylaxis, 406-412. *See also* Hypersensitivity reactions.
  progression of, 409-411i
Anaplasia, cancer cells and, 22, 22i
Anemia as cancer sign, 28

i refers to an illustration; t refers to a table; C refers to a color plate page

---

i refers to an illustration; t refers to a table; C refers to a color plate page

---

i refers to an illustration; t refers to a table; C refers to a color plate page

---

i refers to an illustration; t refers to a table; C refers to a color plate page

---

i refers to an illustration; t refers to a table; C refers to a color plate page

i refers to an illustration; t refers to a table; C refers to a color plate page

---

i refers to an illustration; t refers to a table; C refers to a color plate page

i refers to an illustration; t refers to a table; C refers to a color plate page

i refers to an illustration; t refers to a table; C refers to a color plate page

W

X

Z

---

i refers to an illustration; t refers to a table; C refers to a color plate page

NOTES

NOTES

NOTES

NOTES